PREDICTIVE MODELLING FOR ENERGY MANAGEMENT AND POWER SYSTEMS ENGINEERING

PREDICTIVE MODELLING FOR ENERGY MANAGEMENT AND POWER SYSTEMS ENGINEERING

Edited by

Ravinesh Deo
School of Sciences, University of Southern Queensland, QLD, Australia

Pijush Samui
Department of Civil Engineering, NIT Patna, Patna, Bihar, India

Sanjiban Sekhar Roy
School of Computer Science and Engineering, Vellore Institute of Technology, Vellore, Tamil Nadu, India

ELSEVIER

Elsevier
Radarweg 29, PO Box 211, 1000 AE Amsterdam, Netherlands
The Boulevard, Langford Lane, Kidlington, Oxford OX5 1GB, United Kingdom
50 Hampshire Street, 5th Floor, Cambridge, MA 02139, United States

Notices
Knowledge and best practice in this field are constantly changing. As new research and experience broaden our understanding, changes in research methods, professional practices, or medical treatment may become necessary.

Practitioners and researchers must always rely on their own experience and knowledge in evaluating and using any information, methods, compounds, or experiments described herein. In using such information or methods they should be mindful of their own safety and the safety of others, including parties for whom they have a professional responsibility.

To the fullest extent of the law, neither the Publisher nor the authors, contributors, or editors, assume any liability for any injury and/or damage to persons or property as a matter of products liability, negligence or otherwise, or from any use or operation of any methods, products, instructions, or ideas contained in the material herein.

British Library Cataloguing-in-Publication Data
A catalogue record for this book is available from the British Library

Library of Congress Cataloging-in-Publication Data
A catalog record for this book is available from the Library of Congress

ISBN: 978-0-12-817772-3

Publisher: Candice Janco
Acquisitions Editor: Graham Nisbet
Editorial Project Manager: Leticia M. Lima
Production Project Manager: Kamesh Ramajogi
Cover Designer: Greg Harris

Typeset by MPS Limited, Chennai, India

Contents

12. Support vector machine model for multistep wind speed forecasting

13. MARS model for prediction of short- and long-term global solar radiation

14. Wind speed forecasting in Nepal using self-organizing map-based online sequential extreme learning machine

15. Potential growth in small-scale distributed generation systems in Brazilian capitals

16. Trend of energy consumption in developing nations in the last two decades: a case study from a statistical perspective

List of Contributors

Tilda Akiki Holy Spirit University of Kaslik, USEK, Jounieh, Lebanon

Mumtaz Ali Deakin-SWU Joint Research Center on Big Data, School of Information Technology, Deakin University, VIC, Australia

Youssouf Amrane Laboratory of Electrical and Industrial Systems, University of Sciences and Technology Houari Boumediene, Algiers, Algeria

Viorel Badescu Candida Oancea Institute, Polytechnic University of Bucharest, Bucharest, Romania

Dilki T. Balalla School of Sciences, University of Southern Queensland, Springfield, QLD, Australia

Kwok-wing Chau Department of Civil and Environmental Engineering, Hong Kong Polytechnic University, Kowloon, Hong Kong

Ravinesh Deo School of Sciences, University of Southern Queensland, Springfield, QLD, Australia

Alireza Ghaemi Department of Civil Engineering, Graduate University of Advanced Technology, Kerman, Iran

Harshna Gounder School of Sciences, University of Southern Queensland, Springfield, QLD, Australia

Fatemeh Hosseini-Fashami Department of Agricultural Machinery Engineering, Faculty of Agricultural Engineering and Technology, University of Tehran, Karaj, Iran

Sagthitharan Karalasingham School of Sciences, University of Southern Queensland, Springfield, QLD, Australia

Sungwon Kim Department of Railroad Construction and Safety Engineering, Dongyang University, Yeongju, South Korea

Anshuman Dey Kirty Vellore Institute of Technology, Vellore, India

Nour EL Yakine Kouba Laboratory of Electrical and Industrial Systems, University of Sciences and Technology Houari Boumediene, Algiers, Algeria

Timothy Littler Department of Energy, Power and Intelligent Control (EPIC), IEEE and EEECS Research Society, Queen's Belfast University, Belfast, Northern Ireland

Alejandro López-González Institute of Industrial and Control Engineering, Universitat Politècnica de Catalunya—BarcelonaTech, Barcelona, Spain; Department of Electrical Engineering—Campus Terrassa (ESEIAAT)—BarcelonaTech, Tarrassa, Spain; Socioeconomic Centre of Petroleum and Alternative Energies, Universidad del Zulia, Maracaibo, Venezuela

Nacer K M'Sirdi Aix Marseille Université, Marseille, France

Aziz Naamane Aix Marseille Université, Marseille, France

Ashkan Nabavi-Pelesaraei Department of Agricultural Machinery Engineering, Faculty of Agricultural Engineering and Technology, University of Tehran, Karaj, Iran; Management of Fruit and Vegetables Organizations, Tehran Municipality, Tehran, Iran

Bechara Nehme Holy Spirit University of Kaslik, USEK, Jounieh, Lebanon

Ananta Neupane School of Sciences, University of Southern Queensland, Toowoomba, QLD, Australia

Thong Nguyen-Huy School of Sciences, University of Southern Queensland, Springfield, QLD, Australia; Centre for Applied Climate Sciences, University of Southern Queensland, Toowoomba, QLD, Australia; Vietnam National Space Center, Vietnam Academy of Science and Technology, Hanoi, Vietnam

Eugenia Paulescu Faculty of Physics, West University of Timisoara, Timisoara, Romania

Marius Paulescu Faculty of Physics, West University of Timisoara, Timisoara, Romania

Ramendra Prasad Department of Science, School of Science and Technology, The University of Fiji, Saweni, Lautoka, Fiji

Shobna Mohini Mala Prasad School of Sciences, University of Southern Queensland, Springfield, QLD, Australia

Shahin Rafiee Department of Agricultural Machinery Engineering, Faculty of Agricultural Engineering and Technology, University of Tehran, Karaj, Iran

Nawin Raj School of Sciences, University of Southern Queensland, Springfield, QLD, Australia

Mohammad Rezaie-Balf Department of Civil Engineering, Graduate University of Advanced Technology, Kerman, Iran

Paula D. Rigo Post-Graduate Program in Production Engineering, Federal University of Santa Maria (UFSM), Santa Maria, Brazil

Carmen B. Rosa Post-Graduate Program in Production Engineering, Federal University of Santa Maria (UFSM), Santa Maria, Brazil

Neelesh Sharma University of Southern Queensland, Springfield, Springfield, QLD, Australia

Julio Cezar M. Siluk Post-Graduate Program in Production Engineering, Federal University of Santa Maria (UFSM), Santa Maria, Brazil

Qingyuan Tan Department of Mechanical, Automotive & Materials Engineering, University of Windsor, Windsor, ON, Canada

Jean Ubertalli IEEE PES member, Queen's Belfast University (QUB), Belfast, Northern Ireland

Linyan Wang Department of Mechanical, Automotive & Materials Engineering, University of Windsor, Windsor, ON, Canada

Meiping Wang Department of Mechanical, Automotive & Materials Engineering, University of Windsor, Windsor, ON, Canada

Zaher Mundher Yaseen Sustainable Developments in Civil Engineering Research Group, Faculty of Civil Engineering, Ton Duc Thang University, Ho Chi Minh City, Vietnam

Xiao Yu Department of Mechanical, Automotive & Materials Engineering, University of Windsor, Windsor, ON, Canada

Barbar Zeghondy Holy Spirit University of Kaslik, USEK, Jounieh, Lebanon

Ming Zheng Department of Mechanical, Automotive & Materials Engineering, University of Windsor, Windsor, ON, Canada

About the editors

Professor Ravinesh Deo is an Associate Professor at the University of Southern Queensland, Australia, the Program Director for the Postgraduate Science Program, and Research Leader in Artificial Intelligence. He also serves as an Associate Editor for two international journals: *Stochastic Environmental Research and Risk Assessment* and the *ASCE Journal Hydrologic Engineering* journal (USA). As an Applied Data Scientist with proven leadership in artificial intelligence, his research develops decision systems with machine learning, heuristic, and metaheuristic algorithms to improve real-life predictive systems especially using deep learning explainable AI, convolutional neural networks, and long short-term memory networks. He was awarded internationally competitive fellowships including the Queensland Government US Smithsonian Fellowship, Australia–India Strategic Fellowship, Australia–China Young Scientist Exchange Award, Japan Society for Promotion of Science Fellowship, Chinese Academy of Science Presidential International Fellowship and Endeavour Fellowship. He is a member of scientific bodies, and has won Publication Excellence Awards, Head of Department Research Award, Dean's Commendation for Postgraduate Supervision, BSc Gold Medal for Academic Excellence, and he was the Dux of Fiji in Year 13 examinations. Professor Deo has held visiting positions at the United Stations Tropical Research Institute, Chinese Academy of Science, Peking University, Northwest Normal University, University of Tokyo, Kyoto and Kyushu University, University of Alcala Spain, McGill University, and National University of Singapore. He has undertaken knowledge exchange programs in Singapore, Japan, Europe, China, the United States, and Canada and secured international standing by researching innovative problems with global researchers. He has published books with Springer Nature, Elsevier, and IGI and over 190 publications of which over 140 are Q1 including refereed conferences, edited books, and book chapters. Professor Deo's papers have been cited over 4000 times with a Google Scholar H-Index of 36 and a Field Weighted Citation Index exceeding 3.5.

Professor Pijush Samui is currently an Associate Professor at the National Institute of Technology, Patna, India. He is an established researcher in the application of Artificial Intelligence (AI) for solving different problems in engineering. He has developed a new method for prediction of response of soil during an earthquake. He has produced charts for the prediction of the response of soil during an earthquake and has developed equations for the prediction of lateral spreading of soil due to an earthquake. He has developed equations for the determination of the seismic liquefaction potential of soil based on strain energy and prediction of seismic attenuation. He has developed efficient models for the prediction of the magnitude of reservoir-induced earthquakes. He has developed models for the determination of medical waste generation in hospitals with equations used for practical purpose. The developed models can be used for the Clean India project. He has determined frequency effects on liquefaction by using the Shake Table. He has applied AI techniques for the determination of bearing capacity and settlement of foundation and equations for the determination of bearing capacity and settlement of shallow foundations. He also developed equations for the determination of compression index and angle of shearing resistance of soil. He has developed equations for the prediction of uplift capacity of suction caisson. He also developed equations for the determination of fracture parameters of concrete. His active research activity is evident from his extensive citation of publications in Google Scholar (total frequency of 1280) with an H-Index of 22. Dr. Samui has published journal papers, books/book chapters, and peer reviewed conference papers with coauthors from Australia, India, Korea, and several other nations. He also holds the position of Visiting Professor at the Far East Federal University (Russia).

Sanjiban Sekhar Roy is an Associate Professor in the School of Computer Science and Engineering, Vellore Institute of Technology. He joined VIT in the year 2009 as an Asst. Professor. His research interests include deep learning and advanced machine learning. He has published around 50 articles in reputed international journals (with SCI impact factors) and conferences. He also is an editorial board member for a handful of international journals and a reviewer for many highly reputed journals such as *Neural Processing Letters* (Springer), *IEEE Access: The Multidisciplinary Open Access Journal*, *Computers & Security* (Elsevier), *International Journal of Advanced Intelligence Paradigms* (Inderscience International Publishers), *International Journal of Artificial Intelligence and Soft Computing* (Inderscience International Publishers), *Ad Hoc Networks* (Elsevier), *Evolutionary Intelligence* (Springer), *Journal of Ambient Intelligence and Humanized Computing* (Springer), *Iranian Journal of Science and Technology, Transactions of Electrical Engineering* (Springer). He uses deep learning and machine learning techniques to solve many complex engineering problems, especially those related to imagery. He is specialized in deep convolutional neural networks and generative adversarial networks. Dr. Roy also has edited many books with reputed international publishers such as Elsevier, Springer, and IGI Global. Very recently, the Ministry of National Education, Romania in collaboration with "Aurel Vlaicu" University Arad Faculty of Engineers, Romania has awarded Dr. Roy with a "Diploma of Excellence" as a sign of appreciation for the special achievements obtained in scientific research in 2019.

Foreword

The demand for electrical power has been rising globally, both at national and community scales, and there is a need to find more sustainable and newer forms of electrical power generation resources. The world is concurrently faced with the challenge of mitigating climate change, a large portion of which is due to the emission of greenhouse gases arising from the use of fossil fuels.

Renewable energy is in the unique position of addressing both these issues simultaneously. The inclusion of renewable energy technologies (RETs) such as hydropower, wind, solar, and biomass to the generation mix of power grid supplies is routine practice. Such technologies currently supply some 26% of the global electrical power generation. As they displace almost the same fraction of fossil fuel power from the generation mix, these RETs reduce global greenhouse gas emissions by a comparable proportion.

Renewable energy generation consists of dispatchable (synchronous) power such as hydropower and biomass, and variable (or asynchronous) generation such as wind and solar. While synchronous generation may be added seamlessly to the generation mix, the inclusion of asynchronous generation requires more care. The variable nature of such renewable sources makes the total output of the grid supply unpredictable, and their integration into the system leads to system instabilities. These two issues necessitate, amongst other things, the predictive modeling of variable renewable energy resources as well the use of new methodologies for enhancing system strength.

This Edited Book considers the development of computational tools for prediction and optimization of energy production for power systems using computer-aided algorithms and energy management methodologies. The choice of the chapter contributions has been meticulously executed by the Editors. They consist of a wide range of topics specific to energy optimization and forecasting, and include the forecasting and nowcasting of wind and solar energy resources, enhancing the system viability and strength via digital inertia in the form of battery storage and providing algorithms for the management of community-scale rural energy systems.

Amongst the expected ultimate outcomes of this publication is the improvement of power grid system efficiency and its performance. This will have immediate consequences on efficiency of energy distribution at the national and community levels, and make a positive impact on countries' emissions reduction programs.

The publication of this book comes in the wake of the launch of Sustainable Development Goals (SDGs) and the Paris Agreement in 2015. It synergizes well with Goal 7 of SDG, which is to ensure access to affordable, reliable, sustainable, and modern energy for all. The substantive agreement reached in Paris with regard to climate change mitigation was the undertaking by all Parties to Nationally Determined Contributions (NDCs) to greenhouse gas reductions. Following the release of the IPCC Special Report on Global Warming of 1.5°C in October 2018 and the subsequent Climate Action Summit of

September 23, 2019, there has been vigorous debate regarding the adequacy of the NDCs in achieving the agreed goal of net zero emissions by 2050. The outcome of these deliberations was the realization that much greater reductions in emissions were necessary than those proposed originally in the Paris Agreement. The present book will go a long way toward facilitating the design and implementation of power grid systems that improve national and community-scale distribution as well as reducing overall national GHG emissions.

This book is of great relevance to two of the major ongoing discussions on global issues, and is an immensely valuable and timely addition to the scientific literature on energy modeling and management.

Professor Anirudh Singh
Lautoka, Fiji Islands
22 March 2020

Preface

Today climate advocates are advising energy industries to embed renewable energies into power grids; the role of artificial intelligence in demand side management remains paramount but the success of this vision is at the heart of latest modeling or optimization techniques. National electricity markets, energy management experts, electronic, electrical, and mechatronic engineers should be familiar with advanced optimization techniques that can be used to improve existing energy demand systems, and also to integrate renewable energy into real power grids.

This book provides ideas on optimization techniques as an interdisciplinary concept. The book acts as a common platform required by practitioners to become familiar with the latest developments of energy optimization techniques based on artificial intelligence. It is written to provide a "one-spot" collective resource for practitioners to learn about predictive models in the energy sector, their practical applications, and case studies.

The purpose is to provide the modeling theory in an easy-to-read format verified with onsite models (i.e., case studies) for specific geographic regions and scenarios. There is a need for this sort of text because we currently have several models in isolated contexts. Putting the theory of energy simulation models and applying those optimization techniques in a single bound book will help novice readers to grasp the concepts more easily than highly technical publications.

What problem does this book solve?

Currently, technical papers and books present materials in a way such that both a beginner reader and energy experts find it too hard to grasp the ideas. A requirement for postgraduate, early and mid-career researchers is to read and understand energy modeling in a way that they can quickly relate the theory and practical applications. This book will provide such a platform whereby readers will appreciate both the theory and practical applications, and also see the comparison of different energy management and optimizations in different chapters.

Why would readers choose this book?

Readers will choose this book because it contains both theory and practice related to energy demand management in a single document, has several optimization models in this area, provides easy-to-understand chapters, and supports people new to the field. For experts, the book will be appealing as it gives first-hand experience about artificial intelligence models—an area that is growing in the current phase.

The book is written as a practical guide focused for postgraduate teaching (case studies, modeling, and simulations), early and mid-career research and teaching scholars, academics, renewable energy practitioners, electrical and electronic engineers, climate energy scientists, and future energy policy makers. It will serve as a highly summarized text on latest developments in energy, consumer energy simulations, and energy demand side management.

The book provides the latest approaches for energy exploration, advanced predictive models, and case studies in geographically diverse locations, modern techniques, and demonstrations to apply artificial intelligence in decision-making for the renewable and conventional energy sector. It will therefore be a useful resource for the energy industry—particularly for engineering and energy management experts.

Rigor

This book is compiled carefully with highly focused chapters that will present to the readers the modern-day optimization techniques in energy exploration (particularly a balanced account of theory and case studies) applied in the energy demand side and real-life power management system. It will make a significant contribution to the development of mathematical tools and data simulation models, and their relevance to different geographic power distributions and case studies that will support modern-day energy engineering applications.

The text will be a useful resource for power systems engineering and the design of energy management platforms in complex consumer markets, for scientific application of real-time energy prediction and management systems, and for integrating artificial intelligence tools for real-time adaptive systems incorporated in energy predictions and management environments. The book will assist modern-day engineers and scientists to become familiar with advanced optimization techniques for better power systems designs, optimization techniques, and different algorithms for consumer power management.

It is our hope that all readers will benefit significantly in learning about the state-of-the-art machine learning models and decision support systems, including energy management science and energy policy perspectives.

Happy reading and learning!

Ravinesh Deo
University of Southern Queensland, Towoomba, QLD, Australia
June 24, 2020, Email: ravinesh.deo@usq.edu.au

CHAPTER

1

A Multiobjective optimal VAR dispatch using FACTS devices considering voltage stability and contingency analysis

Youssouf Amrane and Nour EL Yakine Kouba

Laboratory of Electrical and Industrial Systems, University of Sciences and Technology Houari Boumediene, Algiers, Algeria

1.1 Introduction

Numerous complex power system planning and operations optimization problems have to be solved by the power system engineers and researchers. Optimal reactive power planning (ORPP) is one example of an optimization problem which is concerned with the security and economy of a power system operation. The ORPP is one of the most complex problems of power systems since it requires the simultaneous minimization of two objective functions. The first one deals with the minimization of operation cost by reducing real power loss and improving the voltage profile. The second objective is to minimize the allocation cost of additional reactive power sources (capacitive or inductive banks, FACTS devices, etc.). Also, the ORPP problem must satisfy a number of physical and operational limitations constraints. The latter include the load flow equations, real and reactive power generator, lower and upper limits of the tap ratios of transformers, shunt capacitor or reactor outputs, and generator voltages (Amrane et al., 2014).

The ORPP is modeled as a large-scale nonlinear programming problem (NLP). To solve the ORPP problem many conventional and intelligent optimization algorithms have been proposed, such as quadratic and sequential quadratic programming (QP/SQP) (Grudinin, 1998), interior point method (IPM) (Amrane et al., 2014; Oliveira et al., 2015), particle swarm optimization (PSO) (Amrane and Boudour, 2015; Pourjafari and Mojallali, 2011), differential evolution algorithm (DEA) (Amrane et al., 2015), and bacterial foraging algorithm (BFA) (BelwinEdward et al., 2013).

Predictive Modelling for Energy Management and Power Systems Engineering
DOI: https://doi.org/10.1016/B978-0-12-817772-3.00001-X

In this paper, a hybrid PSO and gravitational search algorithm (PSO-GSA) (Mirjalili and Hashim, 2010) is proposed to solve the ORPP problem. The PSO-GSA has been found to be robust and flexible in solving the complex optimization problem. To validate the robustness of this method 23 benchmark functions have been used to validate the performance of the PSO-GSA algorithm and compare it with standard PSO and GSA (Mirjalili and Hashim, 2010). The obtained results show that the number of functions which the PSO-GSA performs well is nearly twice that of PSO and GSA, which shows it is robust and effective. Lenin et al. (2014) applied the PSO-GSA method to solve the optimal reactive power dispatch (ORPD) problem for real power loss and the minimization of bus voltage deviations. Mangaiyarkarasi and Raja (2014) showed another use of PSO-GSA, which cartels the exploiting and exploring features of the PSO and GSA to achieve the objective of determining the optimal location and optimal size of the static volt ampere reactive compensator (SVC), and thereby minimize the voltage deviation from the nominal value. The hybrid PSO-GSA algorithm also has been applied on the economic load dispatch problem (ELD) problem considering transmission loss, prohibited zones and ramp rate limits (Ashouri and Hosseini, 2013; Hardiansyah, 2013; Jiang et al., 2014). A state-of-the-art of the proposed method in several electrical engineering domains is presented in the appendix section.

The voltage instability study is considered as one of the critical issues in the electric power system. In this chapter, the voltage instability study is based on two different stability indexes. Namely fast voltage stability index (FVSI) (Amrane et al., 2014; Musinin and Abdul Rahman, 2002) and line stability index (Lmn) (Moghavemmi and Omar, 1998) are studied and used to identify the weakest bus and the most critical line in the system. The proposed approach has been tested on ORPP problems using SVCs and TCSCs devices for the equivalent Algerian electric power system 114-bus. Two stability indexes, FVSI and Lmn, are used to identify the weakest buses and lines to install the SVC and TCSC devices.

1.2 Problem formulation

In this chapter, the global objective function of the ORPP problem is to minimize three objective functions that represent: (1) the investment cost of FACTS devices (SVC and TCSC) (BelwinEdward et al., 2013); (2) transmission real power losses; and (3) voltage stability, while satisfying several equality and inequality constraints.

1.2.1 Objectives functions

1.2.1.1 Minimizing the investment cost of SVC and TCSC devices

- The SVC costs in $\left(\frac{\$}{kVars}\right)$

$$f_{SVC}(X,U) = \sum_{I=1}^{N_{SVC}} \left(0.0003 \times Q_{SVC}^2 - 0.3051\left|Q_{SVC}\right| + 127.38\right)\left(\frac{\$}{kVars}\right) \tag{1.1}$$

- The TCSC costs in $\left(\frac{\$}{kVars}\right)$

$$f_{TCSC}(X,U)=\sum_{I=1}^{N_{TCSC}}\left(0.0015\times Q_{TCSC}{}^{2}-0.7130\left|Q_{TCSC}\right|+153.75\right)\left(\frac{\$}{kVars}\right) \tag{1.2}$$

1.2.1.2 Minimizing the transmission real power losses

$$f_{Ploss}=\sum_{i\in N_{Li}}G_k(i,j)(V_i^2+V_j^2)+2V_iV_j\cos(\theta_i-\theta_j) \tag{1.3}$$

1.2.1.3 Minimizing the voltage stability

$$F_{VSTA}=\sum\frac{4\times Q_{jr}\left(Z_{ij}^{line}\right)^2}{|V_{is}|X_{ij}^{line}} \tag{1.4}$$

where f_{FACTS} is the objective function of the FACTS devices investment cost; G_c and H_c represent equality and inequality constraints of the system; U is the vector of controls variables; and X is the vector of state variables. f_{SVC} is the cost function of SVC; f_{TCSC} is the cost function of $TCSC$; Q_{SVC} is the SVC reactive power; Q_{TCSC} is the TCSC reactive power; f_{Ploss} is the objective function of real power losses problem; V_i,V_j are the voltage magnitudes; G_k, is the conductance of branch k; θ_i,θ_j are the voltage angel at buses i and j; N_{Li} is the number of transmission lines; f_{STA} is the objective function of voltage stability; $Z_{ij}{}^{line}$ is the line impedance; and $X_{ij}{}^{line}$ is the line reactance connecting bus i and bus j, while Q_{jr} is the reactive power at the receiving end and V_{is} is the sending end voltage.

1.2.2 Equality and inequality constraints

1.2.2.1 Equality constraints (the load flow equations)

$$P_{Gi}-P_{Di}-V_i\sum_{j=1}^{Nbus}V_j(G_{ij}\cos\theta_{ij}+B_{ij}\sin\theta_{ij})=0\; i=1,2,\ldots,N_{bus} \tag{1.5}$$

$$Q_{Gi}-Q_{Di}-V_i\sum_{i=1}^{Nbus}V_j(G_{ij}\sin\theta_{ij}-B_{ij}\cos\theta_{ij})=0\; j=1,2,\ldots,N_{bus} \tag{1.6}$$

1.2.2.2 Inequality constraints (technical limitations)

1. Generator constraints:

$$V_{Gi}^{\min}\le V_{Gi}\le V_{Gi}^{\max} \quad i=1,2,\ldots,N_G \tag{1.7}$$

$$Q_{Gi}^{\min}\le Q_{gi}\le Q_{Gi}^{\max} \tag{1.8}$$

2. FACTS device constraints:

$$Q_{SVCi}^{\min} \le Q_{SVCi} \le Q_{SVCi}^{\max} \quad i = 1, 2, \ldots, N_{SVC} \tag{1.9}$$

$$X_{TCSCi}^{\min} \le X_{TCSCi} \le X_{TCSCi}^{\max} \quad i = 1, 2, \ldots, N_{TCSC} \tag{1.10}$$

3. Transformer constraints:

$$T_i^{\min} \le T_i \le T_i^{\max} \quad i = 1, 2, \ldots, N_T \tag{1.11}$$

4. Security constraints:

$$V_{Li}^{\min} \le V_{Li} \le V_{Li}^{\max} \quad i = 1, 2, \ldots, N_{Load} \tag{1.12}$$

$$\left\{ S_t^{from}, S_t^{to} \right\} \le S_{\max}^i \quad i = 1, 2, \ldots, N_{Li} \tag{1.13}$$

with

$$S_i^t = \sqrt{(P_i^{\,t})^2 + (Q_i^{\,t})^2} \tag{1.14}$$

$$\begin{aligned} P_{ij}^{\,t} &= V_i^2 G_{ii} + V_i V_j (G_{ij}\cos\theta_{ij} + B_{ij}\sin\theta_{ij}) \\ Q_{ij}^t &= -V_i^2 B_{ii} + V_i V_j (G_{ij}\sin\theta_{ij} - B_{ij}\cos\theta_{ij}) \; i = 1, 2\ldots, N_{bus} \\ P_{ji}^t &= V_j^2 G_{jj} + V_i V_j (G_{ji}\cos\theta_{ji} + B_{ij}\sin\theta_{ji}) \; j = 1, 2\ldots, N_{bus} \\ Q_{ji}^t &= -V_j^2 B_{jj} + V_i V_j (G_{ji}\cos\theta_{ji} - B_{ji}\sin\theta_{ji}) \end{aligned} \tag{1.15}$$

where P_{Di}, Q_{Di} are real and reactive power at bus i; P_{Gi}, Q_{Gi} are real and reactive powers of the ith generator; V_i is the voltage magnitude at bus i; N_{Bus} is the number of buses; G_{ij} and B_{ij} are the conductance and susceptance between i and j; θ_{ij} is the phase angle difference between the voltages at i and j;θ_{ji} is the phase angle difference between the voltages at j and i; N_{bus} is the number of buses; V_G is the generator voltages; Q_G is the reactive power outputs; N_G is the number of generators;Q_{SVCi} is the SVC reactive power and X_{TCSCi} is the TCSC susceptance; N_{SVC} and N_{TCSC} are respectively the number of SVC and TCSC devices; T_i is the transformer tap settings; N_T is the number of transformers; V_L is the voltage at load bus and $\{S_t^{from}, S_t^{to}\}$ is the transmission line loading; N_{load} is the number of load buses and N_{Li} the number of transmission lines.

The equality constraints given by Eqs. (1.5) and (1.6) are satisfied by running the power flow Newton-–Raphson algorithm. The control variables presented in (1.7), (1.9), (1.10), and (1.11) are self-controlled, and the dependent variables are added in the quadratic penalty terms to the objective function in order to keep their final value close to their operating limits.

The objectives functions are standardized in a comparative manner with the base case, but by considering the fact that the objective of active power losses and the voltage stability problem are more important than the FACTS devices costs. For this reason, we use different coefficients for each objective (Amrane et al., 2013).

$$F(U,X) = \alpha \times \frac{F_{FACTS}}{\sum \Delta Q_{FACTS_{base}}} + \beta \times \frac{F_{PER}}{\sum \Delta_{PER_{Base}}} + \lambda \times \frac{F_{Vsta}}{\sum \Delta_{Vsta_{Base}}} \\ + \sigma_{v_i} \sum_{i \epsilon N_{PQ}} (V_i - V_i^{\lim})^2 + \sigma_{Q_{Gi}} \sum_{i \epsilon N_G} (Q_{Gi} - Q_{Gi}^{\lim})^2 + \sigma_{S_i^t} \sum_{i \epsilon N_{2Li}} (S_i^t - S_t^{\lim})^2 \tag{1.16}$$

With $\alpha = 20/100$, $\beta = 40/100$ and $\lambda = 40/100$.

In the above objective function $V_i^{\lim}$, $Q_{Gi}^{\lim}$, and $S_i^{\lim}$ are defined in the following equations.

$$V_i^{\lim} = \left\{ \begin{array}{c} V_i^{\min}, if\ V_i < V_i^{\min} \\ V_i^{\max}, if\ V_i > V_i^{\max} \\ V_i, if\ V_i^{\min} < V_i < V_i^{\max} \end{array} \right\} \tag{1.17}$$

$$Q_{Gi}^{\lim} = \left\{ \begin{array}{c} Q_{Gi}^{\min}, if\ Q_{Gi} < Q_{Gi}^{\min} \\ Q_{Gi}^{\max}, if\ Q_{Gi} > Q_{Gi}^{\max} \\ Q_{Gi}, if\ Q_{Gi}^{\min} < Q_{Gi} < Q_{Gi}^{\max} \end{array} \right\} \tag{1.18}$$

$$S_i^{\lim} = \left\{ \begin{array}{c} S_i^{\min}, if\ S_i^t < S_i^{\min} \\ S_i^{\max}, if\ S_i^t > S_i^{\max} \\ S_i^t, if\ S_i^{\min} < S_i^t < S_i^{\max} \end{array} \right\} \tag{1.19}$$

where $\sum \Delta Q_{FACTS_{base}}$ is the base case of the cost of FACTS devices; $\sum \Delta_{PER_{Base}}$ represents the base case transmission active power losses in the network; $\sum \Delta_{Vsta_{Base}}$ represents the total base case voltage stability; λ_{v_i}, λ_{QGi}, and λ_s are the penalty factors which can be increased in the optimization procedure; $V_i^{\lim}$, $Q_{Gi}^{\lim}$, and $S_t^{\lim}$ are defined in the following equations. Fig. 1.1 shows the global problem formulation.

1.3 A proposed hybrid particle swarm optimization and gravitational search algorithm

1.3.1 Particle swarm optimization

PSO is a swarm intelligence method inspired by the social behavior of bird flocking or fish schooling, and developed for global optimization algorithm by J. Kennedy and R. Eberhart in 1995 (Kennedy, 1995). It has become one of the most popular techniques applied in various optimization problems, due to its ease and capability to find near-optimal solutions. The PSO uses a number of particles that constitute a swarm. Each particle traverses the search space looking for the global optima (minimum or maximum). The particles that constitute the PSO system fly around in a multidimensional search space; during this flight each particle adjusts its position according to its own experience, and the experience of the neighboring particles, making use of the best position encountered by

$$\min\left\{\begin{array}{l} F(U,X) = 0.20\times\dfrac{F_{FACTS}}{\sum\Delta Q_{FACTS_{Base}}} + 0.40\times\dfrac{F_{PER}}{\sum\Delta_{PER_{Base}}} + 0.40\times\dfrac{F_{Vsta}}{\sum\Delta_{Vsta_{Base}}} + \\ \sigma_{V_i}\sum_{i\epsilon N_{PQ}}(V_i - V_i^{\lim})^2 + \sigma_{Q_{Gi}}\sum_{i\epsilon N_G}(Q_{Gi} - Q_{Gi}^{\lim})^2 + \sigma_{S_i^t}\sum_{i\epsilon N_{2Li}}(S_i^t - S_t^{\lim})^2 \end{array}\right\}$$

$$s.t\left\{\begin{array}{l} P_{Gj} - P_{Di} - V_i\sum_{j=1}^{NB} V_j(G_{ij}\cos\theta_{ij} + B_{ij}\sin\theta_{ij}) = 0 \\ Q_{Gj} - Q_{Di} - V_i\sum_{j=1}^{NB} V_j(G_{ij}\cos\theta_{ij} - B_{ij}\sin\theta_{ij}) = 0 \\ T_i^{\min} \le T_i \le T_i^{\max} \\ Q_{Gi}^{\min} \le Q_{Gi} \le Q_{Gi}^{\max} \\ V_{Gi}^{\min} \le V_{Gi} \le V_{Gi}^{\max} \\ b_i^{\min} \le b_i \le b_i^{\max} \\ X_i^{\min} \le X_i \le X_i^{\max} \\ V_{Li}^{\min} \le V_{Li} \le V_{Li}^{\max} \\ S_i \le S_i^{\max} \\ \textbf{with} \\ S_i = \sqrt{(P_i)^2 + (Q_i)^2} \\ P_{ij} = V_i^2 G_{ii} + V_iV_j(G_{ij}\cos\theta_{ij} + B_{ij}\sin\theta_{ij}) \\ P_{ji} = V_j^2 G_{jj} + V_iV_j(G_{ji}\cos\theta_{ji} + B_{ji}\sin\theta_{ji}) \\ Q_{ij} = -V_i^2 B_{ii} + V_iV_j(G_{ij}\sin\theta_{ij} - B_{ij}\cos\theta_{ij}) \\ Q_{ji} = -V_j^2 B_{jj} + V_iV_j(G_{ji}\cos\theta_{ij} - B_{ij}\sin\theta_{ij}) \end{array}\right.$$

FIGURE 1.1 Global formulation.

itself and its neighbors. The swarm direction of a particle is defined by the set of neighboring particles and its history experience (Soliman and Mantawy, 2012).

The flowchart of PSO is shown in Fig. 1.2.

1.3.2 Gravitational search algorithm

The GSA is a novel metaheuristic searching algorithm which was proposed by E. Rashedi et al. in 2009 (Rashedi et al., 2009). The basic physical theory, from which this algorithm is inspired, is based on Newton's theory (law of gravity and of motion). In

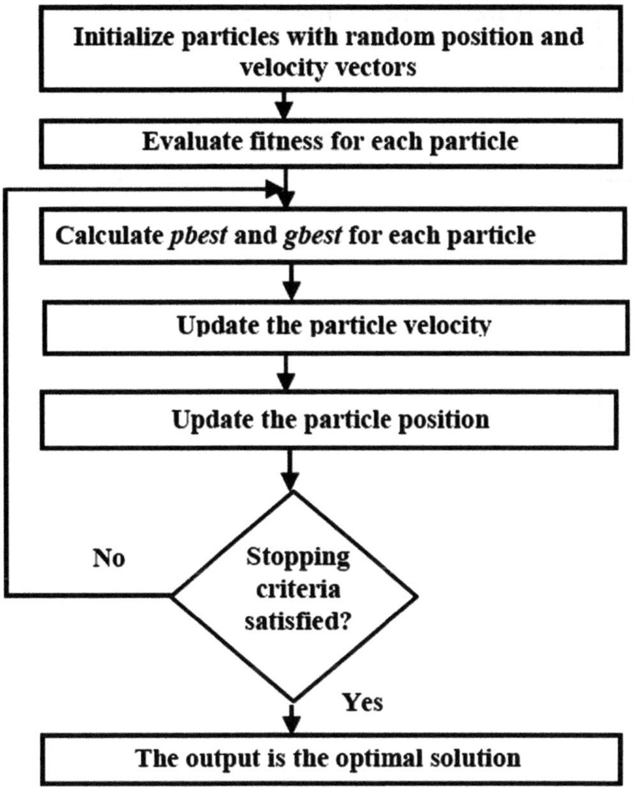

FIGURE 1.2 Flowchart of the PSO algorithm.

Newton's theory every particle in the universe attracts every other particle with a force that is directly proportional to the product of masses and inversely proportional to the square of the distance between them (Lenin et al., 2014; Rashedi et al., 2009).

The flowchart of PSO is shown in Fig. 1.3.

1.3.3 A hybrid particle swarm optimization gravitational search algorithm

The hybrid algorithm proposed in this study is a combination of a PSO algorithm and a GS algorithm. The PSO-GSA is a hybrid method which was proposed by S. Mirjalili et al. in 2010 (Lenin et al., 2014). Its basic idea is to combine the ability of social thinking (*gbest*) in PSO (Soliman and Mantawy, 2012) with the local search capability of GSA (Rashedi et al., 2009). The flowchart of PSO-GSA is shown in Fig. 1.4.

The details of the PSO-GSA-based optimization algorithm are as follows:

Step 1: A set of initial populations are created randomly within the minimum and maximum limits of the control variables and chosen as a parent population.
Step 2: The objective function for each particle in the initial population is evaluated.
Step 3: Calculation of the gravitational force, gravitational constant, and resultant forces among particles using (1.20), (1.21), and (1.22) respectively:

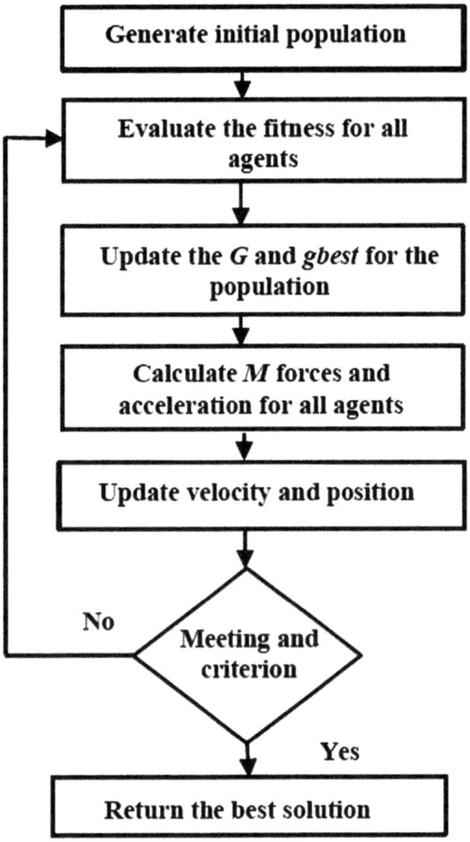

FIGURE 1.3 Flowchart of the GSA algorithm.

Gravitational force

$$F_{ij}^d(t) = G(t)\frac{M_{pi}(t) \times M_{aj}(t)}{R_{ij}(t) + \varepsilon}\left(x_j^d(t) - x_i^d(t)\right) \tag{1.20}$$

Gravitational constant

$$G(t) = G_0 \times \exp(-\alpha \times iter/\text{max}iter) \tag{1.21}$$

Resultant forces

$$F_i^d(t) = \sum_{\substack{j=1 \\ j \neq i}}^{N} rand_j F_{ij}^d(t) \tag{1.22}$$

Step 4: The calculation of M acceleration for all particles of particles as defined in (1.23).

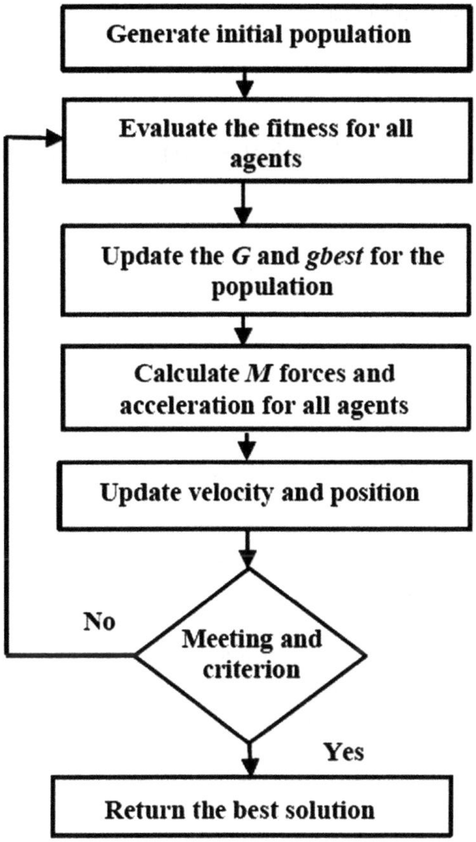

FIGURE 1.4 Flowchart of Hybrid PSO-GSA algorithm.

M acceleration

$$M_i(t) = \frac{m_i(t)}{\sum_{j=1}^{N} m_j(t)} \tag{1.23}$$

with

$$m_i = \frac{Fit(t) - worst(t)}{best(t) - worst(t)} where \begin{cases} best(t) = \begin{cases} \min(fit_j(t)) \\ j \in \{1, \ldots, m\} \end{cases} \\ worst(t) = \begin{cases} \max(fit_j(t)) \\ j \in \{1, \ldots, m\} \end{cases} \end{cases} \tag{1.24}$$

Step 5: The calculation of velocities of all particles using (1.26).

Velocities of all particles

$$V_i(t+1) = w \times V_i(t) + c_1{}' \times rand \times ac_i(t) + c_2{}' \times rand \times (gbest + X_i(t)) \quad (1.25)$$

with

$$ac_i^d(t) = \frac{F_i^d(t)}{M_{ii}(t)'} \quad (1.26)$$

Step 6: Update position of each particle according to (1.27).

Updating position

$$X_i(t+1) = X_i(t) + V_i(t+1) \quad (1.27)$$

Step 7: The objective function of the new searching points and the evaluation values are calculated. The process of updating velocities and positions will be stopped by meeting the end criterion.

Step 8: If the stopping criterion is met (maximum number of generations is reached or the optimal point is achieved), then print the results. Otherwise, go to Step 2.

Step 9: Return the best solution.

Where F_{ij}^d is the gravitational forces from particle j on particle i at a specific time t; G is the gravitational constant; F_i^d is the total force acting on particle i in a dimension d; t is a specific time and M_i is the mass of object I; ac_i^d is the acceleration of all particles; *pbest* is the valuation of fitness function and *gbest* is the best particle among all particles. X is the particle coordinates; V is the velocity; W is the inertia weight factor; *worst* and *best* are respectively the worst and best fitness.

1.4 Stability index

In this section, the FVSI and Lmn are reviewed and used to estimate the maximum loadability and to identify the critical lines and buses to install the FACTS controllers.

1.4.1 Fast voltage stability index

The FVSI was proposed by I. Murisin et al. (Musinin and Abdul Rahman, 2002), and is based on the concept of power flow through a single line.

Taking the symbol i as the sending bus and j as the receiving bus, the FVSI can be defined by:

$$\text{FVSI}_{ij} = \frac{4(Z_{ij}^{\text{line}})^2 Q_{jr}}{{V_{is}}^2 X_{ij}^{\text{line}}} \quad (1.28)$$

1.4.2 Lmn

The Lmn index was proposed by Moghavemmi and Omar (1998), and is formulated on the base of a power transmission concept in a single line.

The Lmn can be reproduced as:

$$\text{Lmn}_{ij} = \frac{4Q_{jr}X_{ij}^{line}}{[|V_{is}|\sin(\Theta-\delta)]^2} \tag{1.29}$$

It is important to note that the value of FVSI and Lmn must be kept lower than 1.00 in aim to preserve a stable system.

The steps implemented for identifying the critical buses and lines (Amrane et al., 2014) are described below.

1.5 Flexible alternating current transmission systems modeling

In this chapter, two typical FACTS devices have been used: SVC and TCSC.

1.5.1 Thyristor controlled series compensator model

The basic idea behind the power flow control with the TCSC is to decrease or increase the overall lines' effective series transmission impedance, by adding a capacitance or inductance correspondingly (Amrane et al., 2014).

This model represents the TCSC by a variable reactance X_{TCSC}. The active and reactive power injected to the nodes are represented by the following equations:

$$P_i = V_iV_jB_{ij}\sin(\theta_i - \theta_j). \tag{1.30}$$

$$Q_i = -V_i^2B_{ii} - V_iV_jB_{ij}\sin(\theta_i - \theta_j) \tag{1.31}$$

with

$$\begin{aligned}
B_{ii} = B_{jj} &= -\frac{1}{X_{xtcsc}} \quad \text{For inductive operation.}\\
B_{ij} = B_{ji} &= \frac{1}{X_{xtcsc}}\\
B_{ii} = B_{jj} &= \frac{1}{X_{xtcsc}} \quad \text{For capacitive operation.}\\
B_{ij} = B_{ji} &= -\frac{1}{X_{xtcsc}}
\end{aligned} \tag{1.32}$$

To avoid overcompensation, the working range of the TCSC should be limited between (Amrane et al., 2014):

$$-0.8X_{ij}^{line} \le X_{ij}^{TCSC} \le 0.2X_{ij}^{line} \tag{1.33}$$

TABLE 1.1 Minimization parameter settings.

Parameters	PSO-GSA
C1	1.5
C2	0.5
W	[0-1]
GO	100
A	10
Iteration and population	Algerian network 114-nodes
No. iteration	300
Population	60

TABLE 1.2 Optimal penalty factor.

λ	λv_i	λQ_{Gi}	λs
Algerian network 114-nodes	1000	400	400

TABLE 1.3 Stability index results.

Rank	Buses	$Q_{MaxFYSI}$ (p.u.)	Q_{MaxLMN} (p.u.)	Lines	Lines From-to	FVSI (p.u.)	LMN (p.u.)
1	67	0.2691	0.2591	27	17–27	0.9999	0.9999
2	43	0.36076	0.37076	10	15–16	0.9999	0.9999
3	66	0.38422	0.36422	72	24–25	0.9997	0.9999
4	93	0.41	0.42	43	42–48	0.9995	0.9999
5	77	0.41346	0.44346	33	21–60	0.9995	0.9998
6	41	0.47024	0.48024	25	17–21	0.9995	0.9997
7	50	0.48564	0.48564	146	92–93	0.9999	0.9995
8	55	0.49038	0.50038	61	35–29	0.9995	0.9994
9	51	0.49782	0.51782	28	17–31	0.9995	0.9994
10	89	0.51294	0.51294	112	49–41	0.9993	0.9994
11	56	0.5164	0.5164	26	17–72	0.9998	0.9994
12	69	0.5182	0.5182	117	85–87	0.9992	0.9991
13	68	0.5291	0.5491	142	99–102	0.9995	0.9991
14	12	0.5373	0.5173	153	110–112	0.9990	0.9990
15	54	0.5582	0.5682	14	8–4	0.9990	0.9990

1.5.2 Static volt ampere reactive compensator model

In this chapter, the SVC is modeled as a variable shunt reactive susceptance jb_{svc} installed at the node i (Fig. 1.6). In this case, only one term of the nodal admittances matrix is modified, corresponding to the node where the SVC is connected (Amrane et al., 2014).

The current generated or absorbed by SVC is represented based on the total susceptance by the following equation:

$$I_{svc} = jb_{svc}V_k \tag{1.34}$$

Reactive power injected by the SVC is presented as follows:

$$Q_{svc} = Q_i = -b_{svc}V_k^2 \tag{1.35}$$

1.6 Simulation results

In order to verify the effectiveness of the proposed approach, the hybrid PSO and gravitational search algorithm (PSO-SGA) has been tested for the equivalent Algerian electric power system 114-bus (220/60 kV). For comparison purposes, two other algorithms are also implemented for solving the problem, namely PSO and GSA. Table 1.1 shows the parameters, number of iterations, and population size of these algorithms. The penalty factors in (1.15) are listed in Table 1.2. The programs have been written in MATLAB-7 language and executed on a 2.91 GHz CPU dual-core with 4 GO RAM.

TABLE 1.4 FACTS devices placements.

Buses				
SVC	67	43	66	93
	77	41	50	55
	51	89	56	
Lines				
TCSC	17–27	15–16	24–25	42–48
	21–60	17–21	92–93	35–29
	17–31	49–41	17–72	

TABLE 1.5 Setting of control variables.

Var	Min	Max
T (p.u.)	0.9	1.1
V_G (p.u.)	0.9	1.1
Q_{svc} (p.u.)	0	0.5
X_{tcsc} (p.u.)	−0.8	0.2

TABLE 1.6 Test system description.

Variables	Values
P_g (MW)	3146.2
Q_g (MVar)	1799.4
P_{PER} (MW)	67.4567
Q_{PER} (Mvar)	265.840

TABLE 1.7 Optimal setting of control variables of Case 1 with SVC in p.u.

Var	Base	PSO	GSA	PSO-GSA	Var	Base	PSO	GSA	PSO-GSA
V_4	1.0700	1.1000	1.0931	1.1000	$T_{60\text{-}59}$	1.0300	1.0187	1.0119	0.9842
V_5	1.0500	1.0944	1.0859	1.0944	$T_{64\text{-}63}$	1.0300	0.9856	1.0069	0.9732
V_{11}	1.0500	1.0704	1.0591	1.0663	$T_{72\text{-}71}$	1.0300	0.9714	1.0035	0.9716
V_{15}	1.0400	1.0987	1.0807	1.1000	$T_{17\text{-}18}$	1.0300	0.9927	0.9954	0.9734
V_{17}	1.0800	1.0881	1.0344	1.1000	$T_{21\text{-}20}$	1.0300	1.0174	1.0072	0.9846
V_{19}	1.0300	1.0514	0.9745	1.0962	$T_{27\text{-}26}$	1.0300	1.0313	1.0033	0.9638
V_{52}	1.0400	1.0654	1.0066	1.1000	$T_{28\text{-}26}$	1.0300	1.0010	1.0258	0.9945
V_{22}	1.0500	1.0633	0.9935	1.1000	$T_{31\text{-}30}$	1.0300	1.0060	1.0119	0.9875
V_{80}	1.0800	1.0384	0.9813	1.0808	$T_{48\text{-}47}$	1.0300	0.9653	0.9887	0.9619
V_{83}	1.0500	1.0519	0.9993	1.1000	$T_{74\text{-}76}$	1.0300	0.9964	0.9949	0.9867
V_{98}	1.0500	1.0658	1.0114	1.0959	Q_{c67}	0	0.1355	0.2666	0.2082
V_{100}	1.0800	1.0671	1.0114	1.1000	Q_{c43}	0	0.2098	0.1921	0.4500
V_{101}	1.0800	1.0732	1.0203	1.1000	Q_{c66}	0	0.2350	0.2305	0.0120
V_{109}	1.0500	1.0999	1.0383	1.1000	Q_{c93}	0	0.2817	0.1879	0.2197
V_{111}	1.0200	1.0375	0.9896	1.1000	Q_{c77}	0	0.0378	0.1433	0.1238
$T_{80\text{-}88}$	0.9800	0.9442	0.9975	0.9703	Q_{c41}	0	0.3093	0.2519	0.3503
$T_{81\text{-}90}$	0.9500	1.0058	0.9931	1.1000	Q_{c50}	0	0.1008	0.1707	0.0155
$T_{86\text{-}93}$	1.0300	1.0060	0.9976	1.0687	Q_{c55}	0	0.1093	0.2018	0.1082
$T_{42\text{-}41}$	1.0300	0.9105	0.9048	0.9267	Q_{c51}	0	0.0208	0.1681	0.0259
$T_{58\text{-}57}$	1.0300	0.9902	1.0275	0.9812	Q_{c89}	0	0.2828	0.2161	0.2125
$T_{44\text{-}43}$	1.0300	0.9845	0.9969	1.0143	Q_{c56}	0	0.2800	0.3183	0.2340

To validate the effectiveness of the proposed approach, three case studies are considered:

- Case 1: Base case (nominal point).
- Case 2: Heavy case (20% of the base case).
- Case 3: Increased reactive power at critical nodes (10% of Q_{Max} FVSI).

1.6.1 Description of the test system and simulation results

To prove the robustness of the proposed technique in solving larger power systems, the equivalent Algerian electric power system is considered (Amrane et al., 2015). The equivalent network consists of 175 transmission lines, 15 generator-buses, 99 load buses, and 17

TABLE 1.8 Optimal setting of control variables of Case 1 with TCSC in p.u.

Var	Base	PSO	PSO-GSA	Var	Base	PSO	PSO-GSA
V_4	1.1000	1.0776	1.0776	$T_{60\text{-}59}$	1.0029	0.9883	0.9883
V_5	1.0944	1.0706	1.0706	$T_{64\text{-}63}$	0.9659	0.9559	0.9559
V_{11}	1.0810	1.0259	1.0259	$T_{72\text{-}71}$	0.9745	1.0100	1.0100
V_{15}	1.0982	1.0667	1.0667	$T_{17\text{-}18}$	0.9931	0.9799	0.9799
V_{17}	1.0856	1.0303	1.0303	$T_{21\text{-}20}$	1.0230	1.0014	1.0014
V_{19}	1.0382	1.0075	1.0075	$T_{27\text{-}26}$	1.0304	0.9822	0.9822
V_{52}	1.0625	1.0190	1.0190	$T_{28\text{-}26}$	1.0176	1.0032	1.0032
V_{22}	1.0452	1.0081	1.0081	$T_{31\text{-}30}$	1.0039	0.9888	0.9888
V_{80}	1.0203	0.9868	0.9868	$T_{48\text{-}47}$	0.9674	1.0224	1.0224
V_{83}	1.0414	1.0095	1.0095	$T_{74\text{-}76}$	0.9929	1.0060	1.0060
V_{98}	1.0564	1.0126	1.0126	$\text{TCSC}_{17\text{-}27}$	−0.0067	−0.0121	−0.0121
V_{100}	1.0591	1.0209	1.0209	$\text{TCSC}_{15\text{-}16}$	−0.0108	−0.0248	−0.0048
V_{101}	1.0691	1.0227	1.0227	$\text{TCSC}_{24\text{-}25}$	−0.0383	−0.0249	−0.0249
V_{109}	1.0817	1.0223	1.0223	$\text{TCSC}_{42\text{-}48}$	−0.0220	−0.0187	−0.0187
V_{111}	1.0216	0.9911	0.9911	$\text{TCSC}_{21\text{-}60}$	−0.0079	−0.0110	−0.0110
$T_{80\text{-}88}$	0.9031	0.9483	0.9483	$\text{TCSC}_{17\text{-}21}$	−0.0225	−0.0126	−0.0126
$T_{81\text{-}90}$	0.9681	0.9781	0.9781	$\text{TCSC}_{92\text{-}93}$	−0.0076	−0.0140	−0.0140
$T_{86\text{-}93}$	0.9723	0.9864	0.9864	$\text{TCSC}_{35\text{-}29}$	−0.1589	−0.1422	−0.1422
$T_{42\text{-}41}$	0.9186	0.9183	0.9183	$\text{TCSC}_{17\text{-}31}$	−0.1151	−0.0968	−0.0968
$T_{58\text{-}57}$	0.9620	0.9723	0.9723	$\text{TCSC}_{49\text{-}41}$	−0.0097	−0.0061	−0.0061
$T_{44\text{-}43}$	0.9768	0.9787	0.9787	$\text{TCSC}_{17\text{-}72}$	−0.2939	−0.2578	−0.2578

tap changer transformers. The total system's real and reactive power demands are 3146.2 MW and 1799.4 MVar. The top 15 weakest buses and lines for the equivalent Algerian electric power system 114-bus are listed in Table 1.3. The chosen buses and lines which will receive the FACTS devices (SVCs and TCSCs devices) are listed in Table 1.4. The control variable limits and the description of the test systems are listed respectively in Tables 1.5 and 1.6. The lower and upper limits of the load-bus voltages were set to 0.90 and 1.1 p.u., respectively.

Tables 1.7–1.11 show the optimal settings of control variables obtained with the proposed method, PSO and GSA, using two types of FACTS devices (SVC and TCSC) for the three proposed case studies. From these tables we can see that all the control variables obtained by the proposed method and the two compared methods are within the safe

TABLE 1.9 Optimal setting of control variables of Case 2 with SVC in p.u.

Var	Base	PSO	PSO-GSA	Var	Base	PSO	PSO-GSA
V_4	1.1000	1.0999	1.1000	$T_{60\text{-}59}$	1.0019	0.9826	0.9814
V_5	1.0877	1.0883	1.0888	$T_{64\text{-}63}$	1.0117	0.9820	0.9839
V_{11}	1.0637	1.0643	1.0648	$T_{72\text{-}71}$	0.9560	0.9539	0.9526
V_{15}	1.0899	1.0945	1.1000	$T_{17\text{-}18}$	1.0184	0.9661	0.9632
V_{17}	1.0936	1.0990	1.1000	$T_{21\text{-}20}$	1.0620	0.9855	0.9851
V_{19}	1.0070	1.0878	1.0924	$T_{27\text{-}26}$	1.0366	0.9436	0.9395
V_{52}	1.0446	1.0976	1.1000	$T_{28\text{-}26}$	1.0807	1.0074	1.0041
V_{22}	1.0249	1.0963	1.1000	$T_{31\text{-}30}$	1.0299	0.9874	0.9863
V_{80}	1.0498	1.0735	1.0632	$T_{48\text{-}47}$	0.9785	0.9366	0.9351
V_{83}	1.0702	1.0995	1.0880	$T_{74\text{-}76}$	1.0429	0.9841	1.1000
V_{98}	1.0811	1.0898	1.0900	Q_{c67}	0.1812	0.1762	0.1232
V_{100}	1.0884	1.0986	1.1000	Q_{c43}	0.1089	0.4907	0.4999
V_{101}	1.0995	1.0988	1.1000	Q_{c66}	0.4873	0.4168	0.3012
V_{109}	1.0842	1.0963	1.1000	Q_{c93}	0.4181	0.4636	0.4596
V_{111}	1.0648	1.0692	1.0720	Q_{c77}	0.3082	0.1538	0.0183
$T_{80\text{-}88}$	0.9000	0.9124	0.9642	Q_{c41}	0.4996	0.4851	0.5000
$T_{81\text{-}90}$	0.9454	0.9544	1.1000	Q_{c50}	0.3174	0.2426	0.2265
$T_{86\text{-}93}$	0.9577	0.9734	1.1000	Q_{c55}	0.2245	0.2791	0.2807
$T_{42\text{-}41}$	0.9133	0.9029	0.9000	Q_{c51}	0.2927	0.2204	0.2298
$T_{58\text{-}57}$	1.0405	0.9928	0.9834	Q_{c89}	0.2548	0.2394	0.3410
$T_{44\text{-}43}$	0.9524	0.9590	0.9578	Q_{c56}	0.4947	0.4728	0.5000

TABLE 1.10 Optimal setting of control variables of Case 2 with TCSC in p.u.

Var	Base	PSO	PSO-GSA	Var	Base	PSO	PSO-GSA
V_4	1.1000	1.0995	1.1000	$T_{60\text{-}59}$	0.9845	0.9828	0.9506
V_5	1.0887	1.0884	1.0887	$T_{64\text{-}63}$	0.9484	0.9530	0.9197
V_{11}	1.0645	1.0622	1.0631	$T_{72\text{-}71}$	1.0022	0.9597	0.9276
V_{15}	1.0970	1.0934	1.0913	$T_{17\text{-}18}$	1.0159	0.9601	0.9578
V_{17}	1.0785	1.0694	1.0713	$T_{21\text{-}20}$	1.0226	0.9947	0.9960
V_{19}	1.0320	1.0638	1.0999	$T_{27\text{-}26}$	1.0647	0.9681	0.9601
V_{52}	1.0606	1.0756	1.1000	$T_{28\text{-}26}$	1.0233	1.0169	0.9000
V_{22}	1.0470	1.0779	1.0991	$T_{31\text{-}30}$	1.0258	0.9920	0.9561
V_{80}	1.0461	1.0637	1.0589	$T_{48\text{-}47}$	0.9166	0.9414	0.9000
V_{83}	1.0785	1.0998	1.0897	$T_{74\text{-}76}$	1.0255	1.0210	1.0097
V_{98}	1.0670	1.0780	1.0805	$TCSC_{17\text{-}27}$	−0.0032	−0.0088	−0.0130
V_{100}	1.0842	1.0972	1.1000	$TCSC_{15\text{-}16}$	−0.0019	−0.0058	−0.0108
V_{101}	1.0888	1.0954	1.1000	$TCSC_{24\text{-}25}$	−0.0214	−0.0321	−0.0486
V_{109}	1.1000	1.0882	1.1000	$TCSC_{42\text{-}48}$	−0.0263	−0.0262	−0.0276
V_{111}	1.0090	1.0645	1.0843	$TCSC_{21\text{-}60}$	−0.0129	−0.0128	−0.0157
$T_{80\text{-}88}$	0.9630	0.9285	0.9000	$TCSC_{17\text{-}21}$	−0.0148	−0.0152	−0.0181
$T_{81\text{-}90}$	0.9587	0.9611	0.9105	$TCSC_{92\text{-}93}$	−0.1378	−0.1629	−0.1175
$T_{86\text{-}93}$	1.0125	0.9616	0.9106	$TCSC_{35\text{-}29}$	−0.0745	−0.1138	−0.1165
$T_{42\text{-}41}$	0.9000	0.9065	0.9000	$TCSC_{17\text{-}31}$	−0.0057	−0.0129	−0.0019
$T_{58\text{-}57}$	0.9888	0.9576	0.9152	$TCSC_{49\text{-}41}$	−0.2997	−0.2972	−0.3000
$T_{44\text{-}43}$	0.9355	0.9403	0.9000	$TCSC_{17\text{-}72}$	−0.0122	−0.0139	−0.0057

range. Thus the load voltages obtained by the three algorithms using SVC and TCSC are given in Figs. 1.5–1.9. Also the load voltages obtained are within the permissible limits (0.9 and 1.1 p.u.). These results are encouraging and show the effectiveness of the proposed approach (Tables 1.12–1.15).

The minimum active power losses, voltage stability, and investment cost of FACTS devices obtained by the application of the proposed method, PSO, and GSA for the three case studies using the SVC and TCSC are listed in Tables 1.12 and 1.16. The results show that in all case studies, the minimum active losses found by the installation of SVC device are greater than those achieved by the installation of TCSC.

In Case 1, the investments on SVCs devices of 1.96, 2.34, and 2.00 p.u. are respectively obtained by the application of PSO-GSA methods, PSO, and GSA, where the active power

TABLE 1.11 Optimal setting of control variables of Case 3 with SVC in p.u.

Var	Base	PSO	PSO-GSA	Var	Base	PSO	PSO-GSA
V_4	1.1000	1.1000	1.1000	$T_{60\text{-}59}$	1.0002	0.9851	0.9846
V_5	1.0930	1.0943	1.0945	$T_{64\text{-}63}$	1.0370	0.9713	0.9717
V_{11}	1.0707	1.0681	1.0681	$T_{72\text{-}71}$	1.0245	0.9695	0.9689
V_{15}	1.0919	1.0985	1.1000	$T_{17\text{-}18}$	1.0337	0.9723	0.9717
V_{17}	1.0903	1.0998	1.1000	$T_{21\text{-}20}$	1.0237	0.9834	0.9820
V_{19}	1.0237	1.0976	1.1000	$T_{27\text{-}26}$	0.9789	0.9587	0.9557
V_{52}	1.0419	1.0996	1.1000	$T_{28\text{-}26}$	1.0998	0.9944	0.9905
V_{22}	1.0299	1.0997	1.1000	$T_{31\text{-}30}$	1.0055	0.9884	0.9880
V_{80}	1.0615	1.0841	1.0846	$T_{48\text{-}47}$	1.0257	0.9546	0.9535
V_{83}	1.0701	1.0997	1.1000	$T_{74\text{-}76}$	0.9724	0.9848	0.9855
V_{98}	1.0828	1.0985	1.1000	Q_{c67}	0.4999	0.4968	0.5000
V_{100}	1.0885	1.0999	1.1000	Q_{c43}	0.3200	0.4816	0.5000
V_{101}	1.0917	1.0997	1.1000	Q_{c66}	0.4969	0.4978	0.5000
V_{109}	1.0980	1.0993	1.1000	Q_{c93}	0.4961	0.4996	0.5000
V_{111}	1.0782	1.0910	1.0934	Q_{c77}	0.0346	0.0417	0.0419
$T_{80\text{-}88}$	0.9699	0.9011	0.9000	Q_{c41}	0.4566	0.4867	0.4999
$T_{81\text{-}90}$	0.9214	0.9262	0.9262	Q_{c50}	0.0994	0.0352	0.0317
$T_{86\text{-}93}$	0.9003	0.9001	0.9000	Q_{c55}	0.1607	0.1135	0.1131
$T_{42\text{-}41}$	0.9006	0.9003	0.9000	Q_{c51}	0.0635	0.0332	0.0315
$T_{58\text{-}57}$	1.0027	0.9845	0.9861	Q_{c89}	0.4669	0.3215	0.3254
$T_{44\text{-}43}$	1.0271	0.9838	0.9857	Q_{c56}	0.0959	0.2414	0.2432

losses are reduced respectively by 15.80%, 7.616%, and 15.92%. The power losses are reduced by 3.03%, 5.77%, and 11.51% respectively by installing 0.69, 0.62, and 0.60 p.u. of TCSC.

Also we can show from the obtained results that the minimum voltage stability obtained by the application of the TCSC is greater than that reached by SVC. In Case 2, investments of 0.67, 0.70, and 0.65 p.u. of TCSC, respectively, are obtained by the application of PSO-GSA, PSO, and GSA. The minimum voltage stability is reduced to 0.40, 0.41, and 0.37 p.u., and by using SVC devices it is reduced to 0.69, 0.70, and 0.65 p.u.

The results obtained by increasing the reactive power at critical buses (Case 3), resulted in an increase of the installed reactive power. By comparing the results of Case 3 with the Case 1, we notice a net increase of reactive power. In Case 1, the total reactive energy

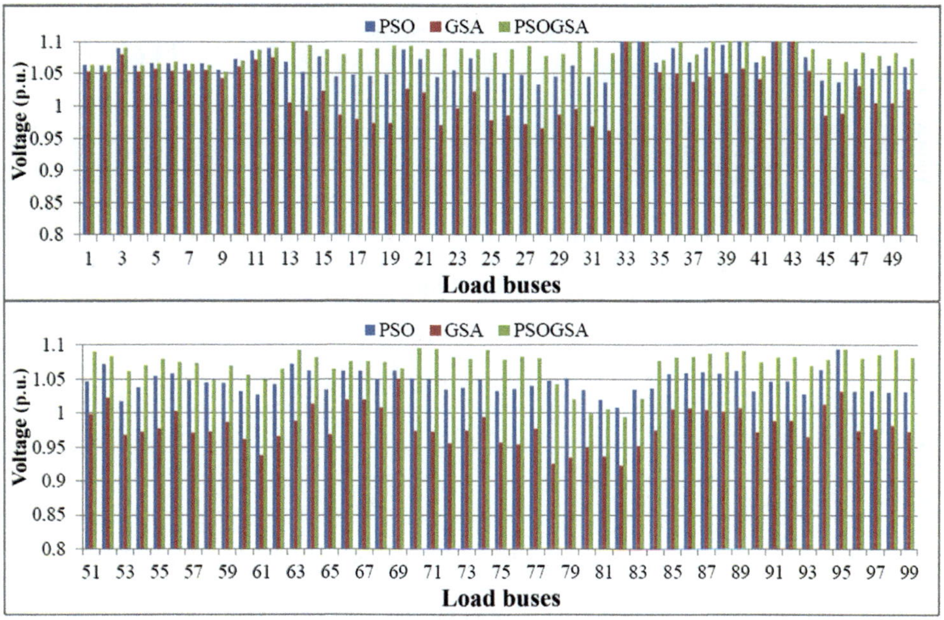

FIGURE 1.5 Load voltage of Case 1 with SVC.

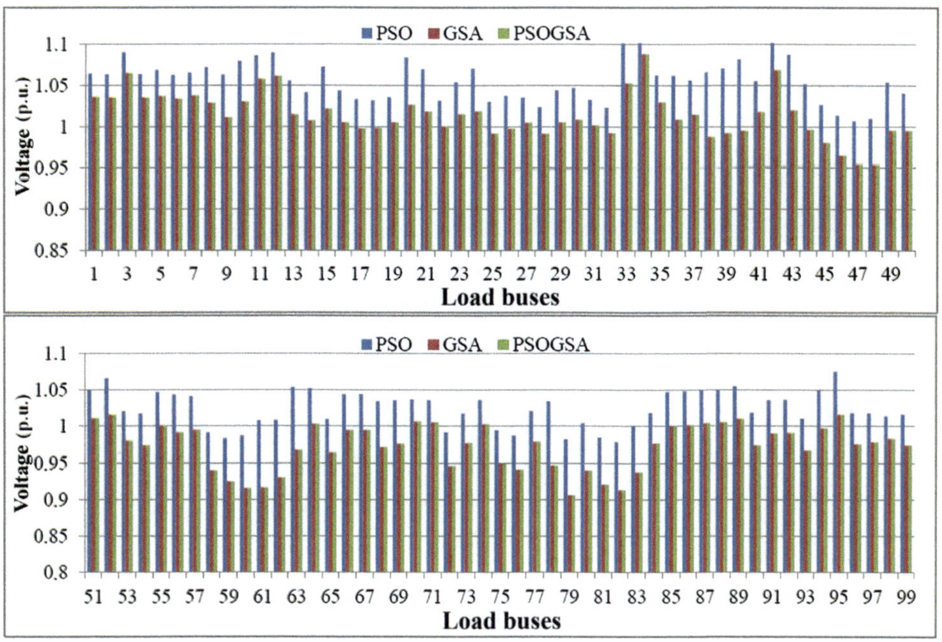

FIGURE 1.6 Load voltage of Case 1 with TCSC.

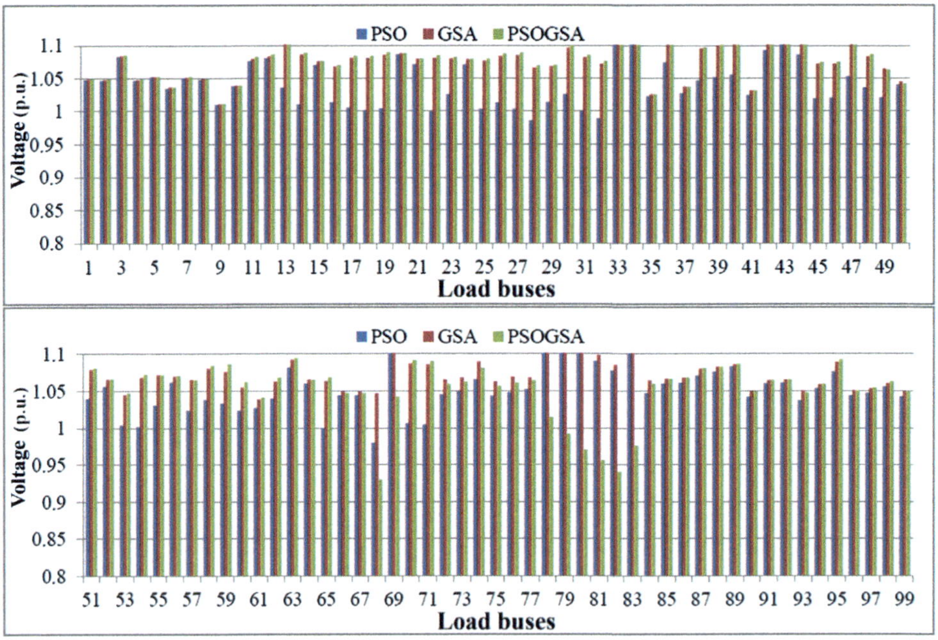

FIGURE 1.7 Load voltage of Case 2 with SVC.

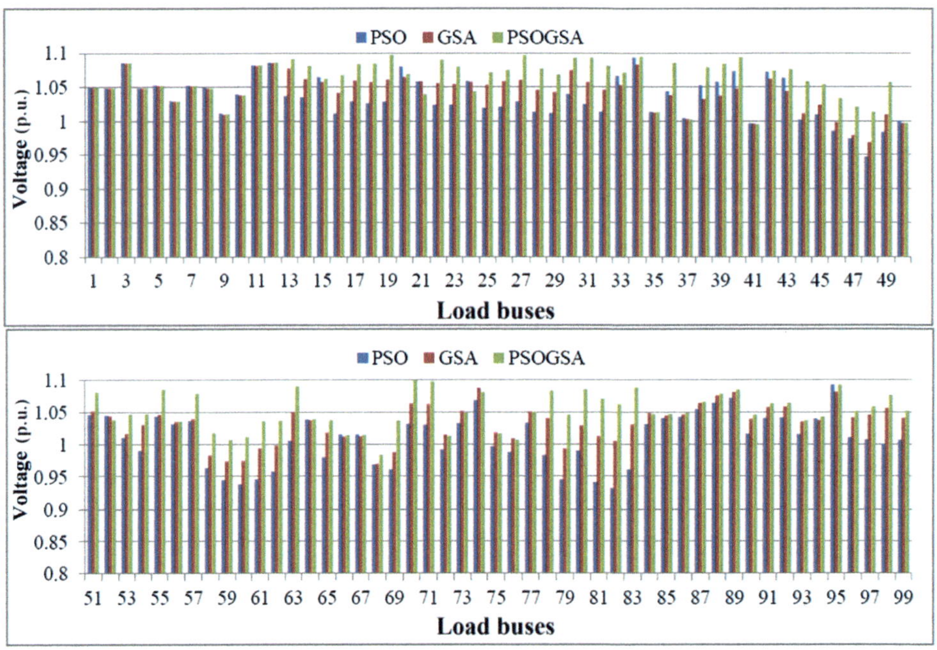

FIGURE 1.8 Load voltage of Case 2 with TCSC.

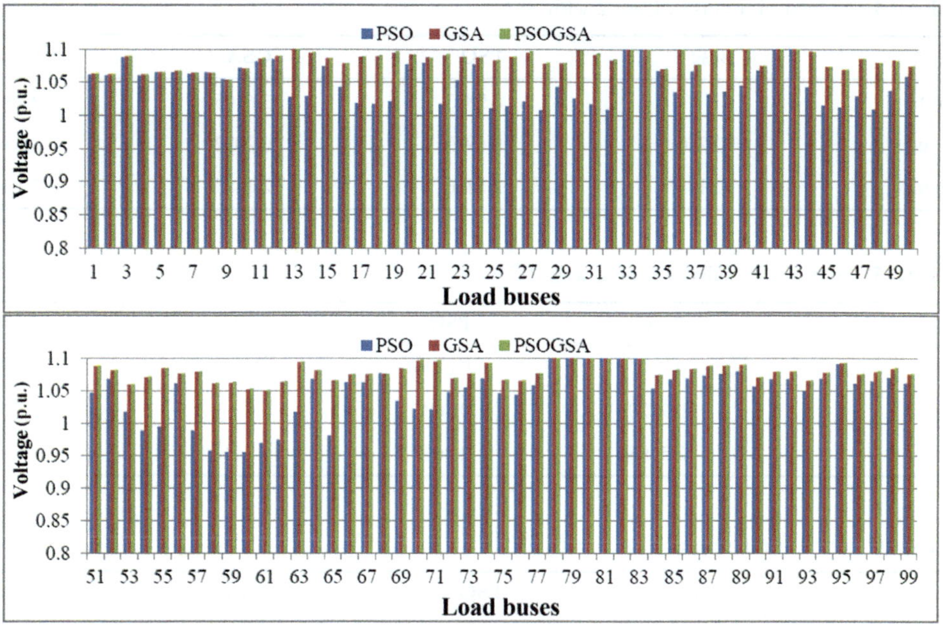

FIGURE 1.9 Load voltage of Case 3 with SVC.

TABLE 1.12 Statistical results of Case 1 with SVC.

	Base	PSO	GSA	PSO-GSA
Q_C (p.u.)	0	2.0028	2.3473	1.9601
F_{PER} (p.u.)	0.7159	0.5398	0.5703	0.5285
F_{Vsta} (p.u.)	0.6788	0.5717	0.6271	0.5707

TABLE 1.13 Statistical results of Case 1 with TCSC.

	Base	PSO	GSA	PSO-GSA
Abs(X_{TCSC}) (p.u.)	0	0.6931	0.6210	0.6010
F_{PER} (p.u.)	0.7159	0.3475	0.3450	0.3293
F_{vsta} (p.u.)	0.6788	0.6582	0.6396	0.5990

invested by the proposed method is 1.96 p.u. and reactive power installed in Case 3 is 3.15 p.u. which shows the effectiveness and robustness of the proposed algorithm. The convergence characteristics of the proposed method, PSO, and GSA by the SVC and TCSC are respectively presented in Figs. 1.10–1.14.

TABLE 1.14 Statistical results of Case 2 with SVC.

	Base	PSO	GSA	PSO-GSA
Q_C (p.u.)	0	3.5873	3.6405	3.4803
F_{PER} (p.u.)	0.9720	0.6995	0.7032	0.6565
F_{Vsta} (p.u.)	2.5195	2.1496	2.1683	2.1080

TABLE 1.15 Statistical results of Case 2 with TCSC.

	Base	PSO	GSA	PSO-GSA
Abs(X_{TCSC}) (p.u.)	0	0.6745	0.7070	0.6557
F_{PER} (p.u.)	0.9720	0.4087	0.4162	0.3747
F_{Vsta} (p.u.)	2.5195	2.2687	2.2801	2.2506

TABLE 1.16 Statistical results of Case 3 with SVC.

	Base	PSO	GSA	PSO-GSA
Q_C (p.u.)	0	3.1907	3.2482	3.1547
F_{PER} (p.u.)	0.7106	1.2000	1.1505	1.1459
F_{Vsta} (p.u.)	0.5784	0.6114	0.6196	0.5735

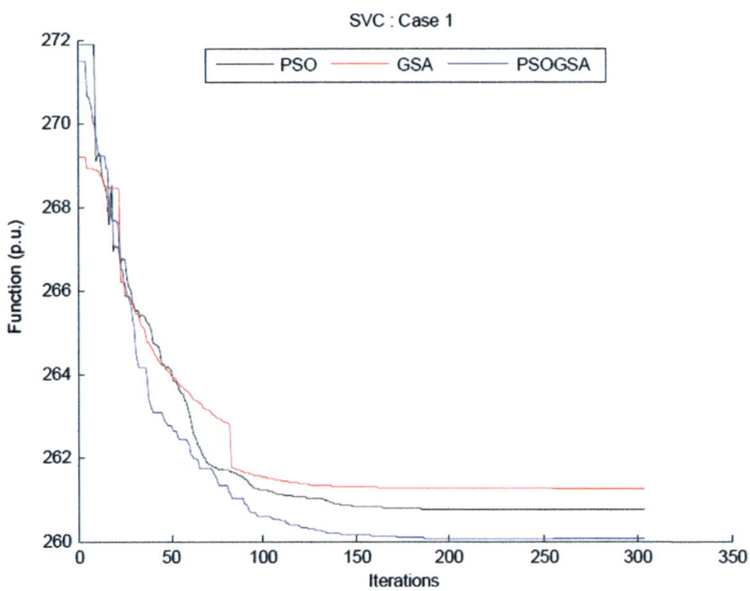

FIGURE 1.10 Objective function convergence characteristic for Case 1 using *SVC*.

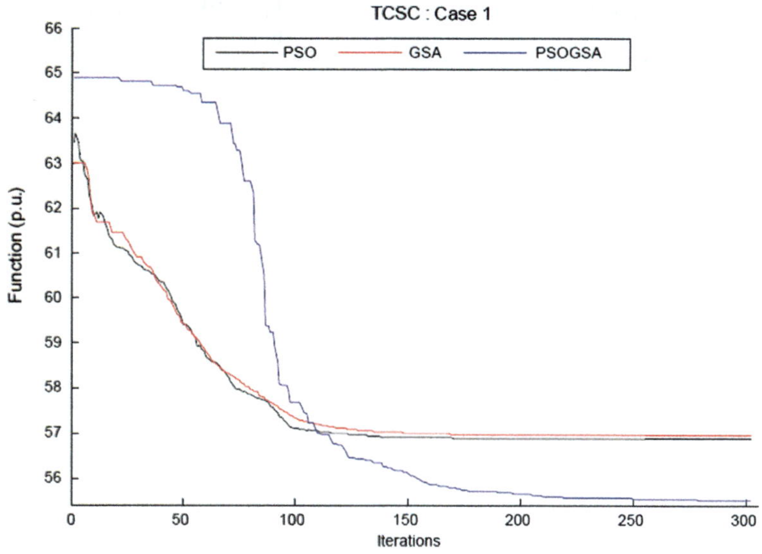

FIGURE 1.11 Objective function convergence characteristic for Case 1 using *TCSC*.

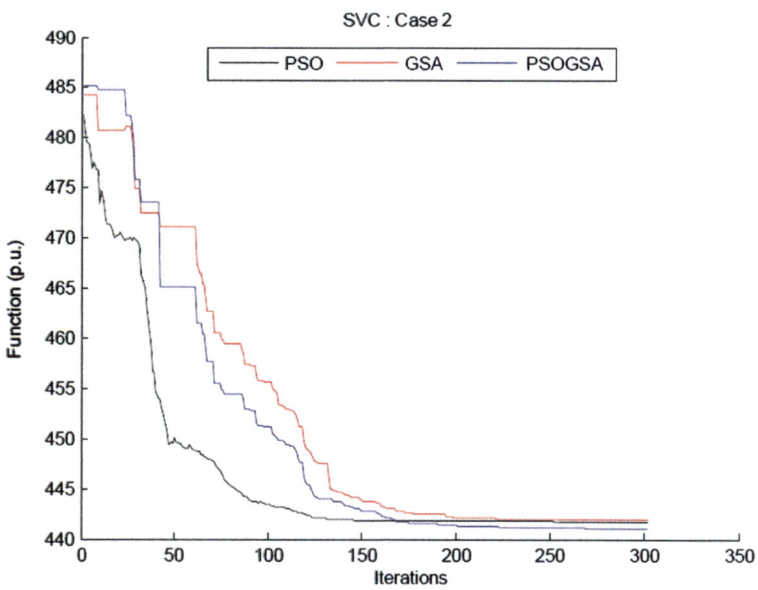

FIGURE 1.12 Objective function convergence characteristic for Case 2 using *SVC*.

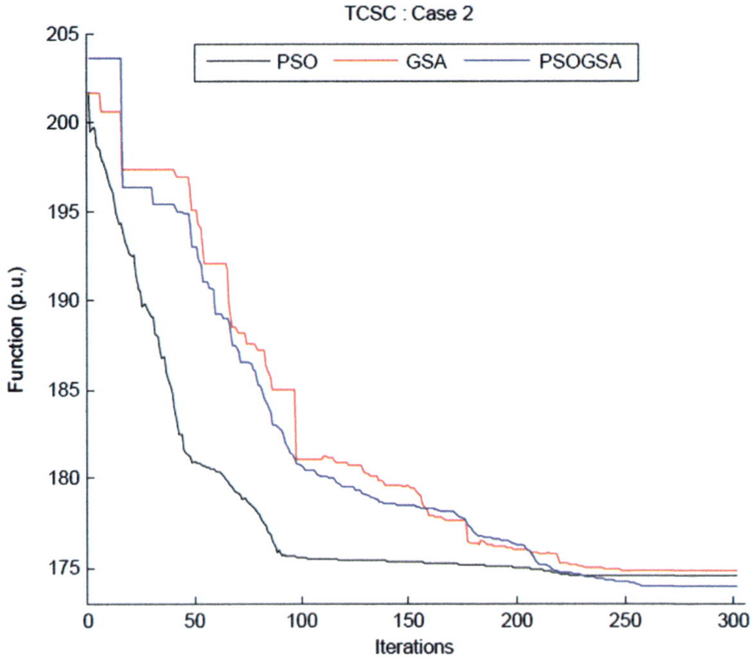

FIGURE 1.13 Objective function convergence characteristic for Case 2 using *TCSC*.

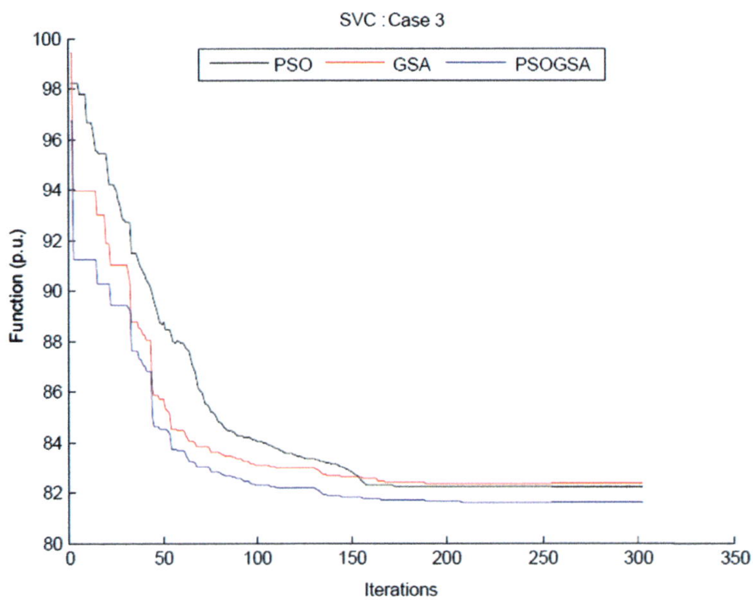

FIGURE 1.14 Objective function convergence characteristic for Case 3 using *SVC*.

1.7 Conclusion

The hybrid PSO-GSA has been used for solving the ORPP problem using two kinds of the FACTS (SVC and TCSC) devices. Two stability indexes are used to determine the optimal placement of FACTS devices.

The obtained results show that the installation of SVC devices reduces the system active power losses more than by the installation of TCSC devices. On the other hand, to obtain good minimum voltage stability, TCSC is better than SVC devices.

The simulation results show the performance of the hybrid PSO-SGA algorithm which minimizes the FACTS devices amount, transmission power losses, and the system voltage stability, even in the most critical cases.

References

Amrane, Y., Boudour, M., 2015. Particle swarm optimization based reactive power planning for line stability improvement. In: 3rd International Conference on Control, Engineering & Information Technology, May, pp. 1–7.

Amrane, Y., Boudour, M. Belazzoug, M., 2013. Optimal reactive power planning based on particle swarm applied to the Algerian electrical power system, systems and control (ICSC). In: 2013 3rd International Conference on, pp. 804–809.

Amrane, Y., Boudour, M., Belazzoug, M., 2014. A New Hybrid Technique for Power Systems Multi-Facts Optimization Design. International Transactions on Electrical Energy Systems.

Amrane, Y., Boudour, M., Ladjici, A.A., Elmaouhab, A., 2015. Optimal VAR control for real power loss minimization using differential evolution algorithm. Electr. Power Energy Syst. 66, 262–271.

Ashouri, M., Hosseini, S.M., 2013. Application of new hybrid particle swarm optimization and gravitational search algorithm for non convex economic load dispatch problem. J. Adv. Comput. Res. 4 (2), 41–51.

BelwinEdward, J., Rajasekar, N., Sathiyasekar, K., Senthilnathan, N., Sarjila, R., 2013. An enhanced bacterial foraging algorithm approach for optimal power flow problem including FACTS devices considering system loadability. ISA Trans. 52, 622–628.

Grudinin, N., 1998. Reactive power optimization using successive quadratic programming method. Power systems. IEEE Trans. on 13 (4), 1219–1225.

Hardiansyah, 2013. A novel hybrid PSO-GSA method for non-convex economic dispatch problems. I.J. Inf. Eng. Electron. Bus. 5, 1–9.

Jiang, S., Ji, Z., Shen, Y., 2014. A novel hybrid particle swarm optimization and gravitational search algorithm for solving economic emission load dispatch problems with various practical constraints. Int. J. Electr. Power Energy Syst. 55, 628–644.

Kennedy, J., 1995. Particle swarm optimization. In: Proceedings of the IEEE International Conference on Neural Networks, vol. IV, pp. 1942–1948.

Lenin, K., Ravindranath Reddy, B., Surya Kalavathi, M., 2014. A new hybrid PSOGSA algorithm for solving optimal reactive power dispatch problem. Int. J. Mechatron. Electr. Comput. Technol. 4 (10), 111–125. ISSN: 2305-0543.

Mallick, S., Ghoshal, S.P., Acharjee, P., Thakur, S.S., 2013. Optimal static state estimation using improved particle swarm optimization and gravitational search algorithm. Int. J. Electr. Power Energy Syst. 52, 254–265.

Mangaiyarkarasi, S.P., Raja, T.S.R., 2014. Optimal location and sizing of multiple static VAR compensators for voltage risk assessment using hybrid PSO-GSA algorithm. Arab. J. Sci. Eng. 39 (11), 7967–7980.

Mirjalili, S., Hashim, S.Z.M., 2010, A new hybrid PSOGSA algorithm for function optimization. In: IEEE International Conference on Computer and Information Application, ICCIA 2010, China, pp. 374–377.

Moghavemmi, M., Omar, F.M., 1998. Technique for contingency monitoring and voltage collapse prediction. EE Proc. Gener. Transm. Distrib. 145 (6), 634–640.

Musinin, I., Abdul Rahman T.K., 2002. Novel fast voltage stability index (FVSI) for voltage stability analysis in power transmission system. In: IEEE Student Conference Research and Development, Shah Alam, Malaysia, pp. 7803–7565.

Oliveira, E.J., et al., 2015. An optimal power flow based on safety barrier interior point method. Electr. Power Energy Syst. 64, 977–985.

Pourjafari, E., Mojallali, H., 2011. Predictive control for voltage collapse avoidance using a modified discrete multi-valued PSO algorithm. ISA Trans. 50, 195–200.

Precup, R.E., David, R.C., Stinean, A.I., Radac, M.B., Petriu, E.M., 2014. Adaptive hybrid particle swarm optimization-gravitational search algorithm for fuzzy controller tuning. In: Innovations in Intelligent Systems and Applications (INISTA) Proceedings, 2014 IEEE International Symposium on. IEEE, pp. 14–20.

Purcaru, C., Precup, R.-E., Iercan, D., et al., 2013. Hybrid PSO-GSA robot path planning algorithm in static environments with danger zones. In: System Theory, Control and Computing (ICSTCC), 2013 17th International Conference. IEEE, pp. 434–439.

Rashedi, E., Nezamabadi-pour, H., Saryazdi, S., 2009. GSA: a gravitational search algorithm. Inf. Sci. 179, 2232–2248.

Soliman, S.A.-H., Mantawy, A.-A.H., 2012. Modern Optimization Techniques With Applications in Electric Power Systems. Springer, Energy Systems.

Appendix

This section contains a state-of-the-art of the hybrid PSO-GSA technique that has arisen in the recent state-of-the-art literature (Table 1.A1).

TABLE 1.A1 A state-of-the-art of the hybrid PSO-GSA technique.

References		Paper title	Year
Mangaiyarkarasi and Raja (2014)	Optimal location and optimal size of the SVC	A Novel Algorithm for Optimal Location of FACTS Devices in Power System Planning.	2008
Mangaiyarkarasi and Raja (2014)		Optimal Location and Sizing of Multiple Static VAr Compensators for Voltage Risk Assessment Using Hybrid PSO-GSA Algorithm.	2014
Precup et al. (2014)	Optimal timing of Takagi–Sugeno–Kang Pi-fuzzy controllers (T-S-K PI-FCs).	Adaptive Hybrid Particle Swarm Optimization-Gravitational Search Algorithm for Fuzzy Controller Tuning.	2014
Mallick et al. (2013)	Static State Estimation (SE) problem.	Optimal static state estimation using improved particle swarm optimization and gravitational search algorithm.	2013
Lenin et al. (2014)	Optimal reactive power dispatch (ORPD) problem for real power loss and the bus voltage deviations minimization.	A New Hybrid PSO-GSA Algorithm for Solving Optimal Reactive Power Dispatch Problem.	2014
Purcaru et al. (2013)	Optimal path planning algorithm for mobile robots.	Hybrid PSO-GSA Robot Path Planning Algorithm in Static Environments with Danger Zones.	2013
Ashouri and Hosseini (2013)	Economic Load Dispatch Problem (ELD) problem.	Application of New Hybrid Particle Swarm Optimization and Gravitational Search Algorithm for Non Convex Economic Load Dispatch Problem.	2013
Hardiansyah (2013)		A Novel Hybrid PSO-GSA Method for Non-convex Economic Dispatch Problems.	2013
Jiang et al. (2014)	Economic emission load dispatch (EELD) problems	A novel hybrid particle swam optimization and gravitational search algorithm for solving economic emission load dispatch problems with various practical constraints.	2014

CHAPTER

2

Photovoltaic panels life span increase by control

Bechara Nehme[1], Nacer K M'Sirdi[2], Tilda Akiki[1], Aziz Naamane[2] and Barbar Zeghondy[1]

[1]Holy Spirit University of Kaslik, USEK, Jounieh, Lebanon [2]Aix Marseille Université, Marseille, France

Acronyms

AMU	Aix Marseille Université
AUF	Agence Universitaire de la Francophonie
a-Si:H	hydrogenated amorphous silicon
COP	Conference Of the Parties
DAQ	data acquisition system
DEE	differential equation editor
DH	damp heat test
DYI	delta yellowness index
EVA	ethylene vinyl acetate
HCR	higher center for research
LCOE	levelized cost of energy
LID	light-induced degradation
LSIS	Laboratoire des Science de l'Information et des Systèmes
MID	moisture-induced degradation
MLP	maximum life span point
MOSFET	metal oxide–semiconductor field-effect transistor
MPP	maximum power point
MPPT	maximum power point tracking
NOCT	nominal operating cell temperature
OC	open circuit
PID	potential-induced degradation
PO	perturb and observe
PV	photovoltaic
PWM	pulse width modulation
RES	renewable energy sources

Predictive Modelling for Energy Management and Power Systems Engineering
DOI: https://doi.org/10.1016/B978-0-12-817772-3.00002-1

RH	relative humidity
SC	short circuit
STC	standard test conditions
SWE	Staebler–Wronski effect
TCO	transparent conductive oxide
TEPSI	thermal electric solar panel integration
USEK	Holy Spirit University of Kaslik
UV	ultraviolet
UVD	ultraviolet light degradation
WVTR	water vapor transmission rate
YI	yellowness index

Chapter outcome

After completing this chapter, the reader will be able to:

- Understand the degradation process of photovoltaic panels and their causes (Section 2.2)
- Carry out the integrated modeling of degradation rates (Sections 2.3–2.5)
- Understand the thermal behavior of photovoltaic panels (Section 2.4)
- Identify new control strategies to apply to photovoltaic panels to increase their life span (Section 2.5)

2.1 Introduction

As humanity is in continuous need of energy, the leaders have three main challenges to accomplish in the coming years: energy security, social equity, and environmental impact mitigation. Photovoltaic (PV) panels present a serious solution to these challenges. We can list here a number of advantages of using PV technology:

- Solar energy is the most abundant and equitable source on earth.
- PV panels constitute a green energy source that does not pollute or contribute to climate change.
- PV panels can support electricity in rural places.
- We must note that one third of the Earth's population does not have access to the electric grid.
- Electricity is the most convertible form of power.
- PV panels are safe and reliable.
- PV panels can be produced from scrap materials of the electric industry.
- With automated and intelligent systems, no operators are needed.
- Once installed, no operating cost is required.
- No movable parts exist in an installed PV panel.
- PV panels can be integrated in any new or existing building.

Conversion efficiency of solar cells is now on average reaching 15%. Despite these acceptable yields, it remains important to raise a technological limitation: the improvement of the life span of PV modules. Degradation effects are observed during the use of these components in intermittent weather conditions: rain, snow, molds, dust ultraviolet (UV) rays, shock, corrosion, rapid losses of optoelectronic properties depending on the usage conditions. The objective of this chapter is to try to improve the life span of solar panels.

Our approach begins by taking into account the degradation modes of PV modules which are related to temperature, moisture, UV light, cracks, etc. Different modes of degradation have been analyzed to develop simulation models that take into account external environmental conditions.

Control algorithms have been developed for best utilization, avoiding defects and allowing PV modules to operate in optimal conditions for the mitigation of degradation processes.

Several reasons and arguments exist to justify the utility of this research project. The economic aim always being flagrant does not dominate the environmental causes which may cause threats to human life. Our work was also added to the 20-20-20 European project.

First, economically speaking, the cause that encourages a person to invest in a project is to obtain an increase on the amount of money originally paid. The investor also calculates the payback period and the levelized cost of energy (LCOE). For the extended life span of PV panels, investors will be more encouraged to adopt solar PV solutions. Currently PV panels are not gainful in some countries. However, in a few years with the depletion of fossil resources, the production of PV electricity will be essential to humanity.

Second, ecologically speaking, the increase in the life span of any product is generally favorable. In fact, factories are a major source of pollution and require a substantial amount of electrical energy. The increase in production also causes overexploitation of raw materials and the added costs/pollution of the transportation of PV panels and their accessories from the factory to the consumer increases with a lower life span.

Third, in 2025 the PV panels already installed will reach their end of life. They will become electrical waste, generating toxic substances to the environment and to humans. Recycling of PV panels is required by regulations (Study on photovoltaic panels supplementing the impact assessment for a recast of the WEEE directive, 2011). However, according to experts, recycling is not economically viable due to the low volumes of waste generated.

Fourth, the main cause that requires an increase in the life span of PV panels is related to the depletion of raw materials. Some components of PV panels are aluminum, lead, silver, indium, gallium, and germanium.

Fifth, another cause that supports the achievement of this work is the danger of the leaching of lead. Lead is a toxic and mutagenic element. It was classified as potentially carcinogenic in 1980 and as probably carcinogenic to humans and animals in 2004. Another cause that involves this research in preserving nature and wildlife is the leaching of cadmium. Cadmium is a white metal that is soft and malleable. When cadmium concentrations in the soil are high, they can influence the microorganisms' soil processes and threaten the entire ecosystem of the soil. Animals eating or drinking cadmium can sometimes have high blood pressure, liver disease, and problems with the central nervous system.

In this chapter we present first a review of the degradation modes of PV panels along with their modeling. Then, a model is presented that estimates the equivalent circuit of the panels as a function of external and operational conditions. Temperature is a key common parameter that affects the degradation modes of PV panels. For that, a new operating point for PV panels is proposed. It reduces the operating temperature of the panels. It is called the maximum life span point (MLP). Finally, a control algorithm is presented that tracks the MLP.

2.2 Degradation modes of photovoltaic panels

2.2.1 Introduction

PV panels convert the sun's energy into electrical energy. Even though the primary energy (solar irradiation) is free, the conversion efficiency of PV panels plays an important role in their development, market penetration, and energy share. Many faults and failures lead to a decrease in the conversion efficiency of PV panels. Efficiency loss is mainly noticed during the first 2 years of operation.

In this section, we survey known degradation modes that affect PV panels. We distinguish between degradation modes that continually reduce their efficiency over time and faults that suddenly occur and sharply reduce power production (Nehme et al., 2016).

We focus on the mechanisms that decrease the efficiency of the PV panels with time. Potential-induced degradation (PID), the light-induced degradation (LID), UV light degradation (UVD), moisture-induced degradation (MID), and cell cracks are the degradation modes that affect PV panels. Degradation generally affects PV panels during their life span, which is predicted to be between 25 and 30 years. Time compression is needed to study material behavior during the entire life span. Researchers apply accelerated life testing by increasing the stresses that PV panels are subject to, for example, they increase the level of irradiance, temperature, or humidity (Lall, 2004).

We will develop a PV panel physical model that takes into consideration degradations and we produce a degradations mathematical model. After that, the real-time simulations problem will be considered with the model's validation. We will start by developing a model of an ideal PV panel. Then we model degradations and their effect on the equivalent circuit.

2.2.2 Potential-induced degradation

PID increases the leakage current of a PV panel. By definition it is the current that leaks from the base to the emitter, it does not traverse the load. The leakage current is the sum of four currents (Luqueand and Hegedus, 2003): (1) A current passing through the soda-lime glass and then through water present on the surface of the panel (I1). (2) A current passing through ions or electrons present at the front surface of the active cell (I2). (3) A current passing through the ethylene vinyl acetate (EVA) encapsulation layer (I3). (4) Finally, a current closing the circuit by passing in the back contact (I4) (Fig. 2.1) (Nehme et al., 2014a,b).

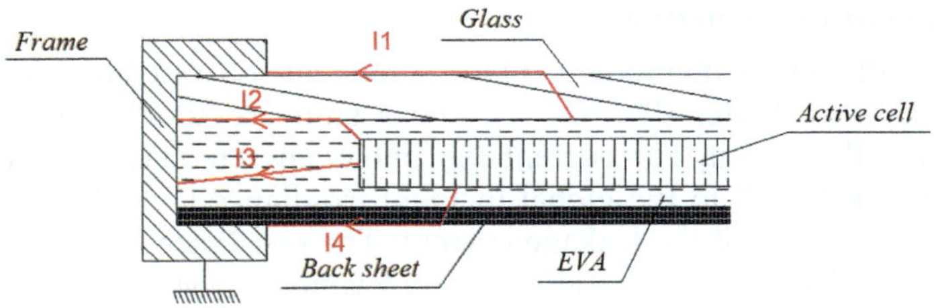

FIGURE 2.1 Leakage current of a panel.

2.2.2.1 Potential-induced degradation causes

Many causes lead to PID, they are divided into cell factors, module factors, system factors, and environment factors.

The cell factors are: the ARC, the emitter depth, and the base doping. Sodium ions present in the glass and in the ARC diffuse due to applied voltage into the front surface of the emitter. The sodium ions induce an electric field and cancel the passivation effect. This leads to an increased surface recombination current. Sodium may also diffuse into the n-doped emitter and acts as an electron acceptor and neutralizes the n-doping. Tests showed that cells without ARC (SiN_x Layer) are not subject to PID (Koch et al., 2011). Other tests prove that high resistivity (low doping rate) of the base increases immunity to PID (Berghold et al., 2010).

The module factors are related to the resistivity and isolation of the encapsulation. Degradation of EVA and of the soda-lime glass results in a decrease of the module encapsulation resistivity.

The system factor that affects PID is called surface polarization effect. All frames are grounded and the active cell is polarized. An array is made of modules in series and in parallel. The potential difference between the cell and the frame increases with the module position in the string. The voltage can reach 600 V in US standards or 1500 V in European standards. This high voltage will lead to the accumulation of electrons on the front surface of the cell, leading to an increased surface recombination current and leakage current.

The environment factors are mainly the humidity and the temperature. In fact, water molecules at the surface of the module increase the conductivity between the cell front surface and the frame. The temperature also increases the conductivity of the EVA and the encapsulation.

To mitigate or to reverse the PID effect, we can invert the polarization of the module in correspondence to the ground. In this case, the high electric voltage between the cell and the frame will be inverted, thus reversing mobile charges (ions and electrons) diffusion. The cell surface will be clean thus decreasing the leakage current.

The installation of the panels can avoid PID. In a string of PV panels, the frames are grounded and one side of the active panels is grounded. If the positive side is grounded, the cells operate at negative voltage with respect to the frame. If the negative side is grounded, the cells operate at positive voltage with respect to the frame. In order to avoid PID, n-type front surface panels must be grounded from the positive side and p-type front surface panels must be grounded from the negative side. However, with new transformless inverters, the active panels must not be grounded to avoid ground fault current. In fact, the DC and the AC compartments are not isolated. In this case, the system operates at floating potential.

2.2.2.2 Potential-induced degradation modeling

A PV cell is modeled as a current source. Its model contains a current source [short circuit current (I_{SC})] in parallel with two diodes (D_1 and D_2) and a shunt resistor (R_{sh}). A resistor (R_s) is mounted is series to the circuit. The current I_{SC} (photogenerated current) is proportional to the temperature T in K and to the global incident irradiance G in W/m^2. The shunt resistance (R_{sh}) models the path of the leakage current. The resistance R_s represents the resistance of the fingers and interconnects (Hersch and Zweibel, 1984). The current I of a cell is given by (2.1) (Luqueand and Hegedus, 2003):

$$I = I_{SC} - I_{01}\left(\exp\left(\frac{q(V + R_s \times I)}{K_B \times T_{pv}}\right) - 1\right) - I_{02}\left(\exp\left(\frac{q(V + R_s \times I)}{2 \times K_B \times T_{pv}}\right) - 1\right) - \frac{V + R_s \times I}{R_{sh}} \quad (2.1)$$

where I_{01} is the saturation current of diode D1 in A; I_{02} is the saturation current of the diode D2 in A; q is the electron charge 1.6×10^{-19} C; V is the output voltage of the cell in V; K_B is the Boltzmann constant 1.3806×10^{-23} J/K; T is the temperature in K.

The effect of temperature and Irradiance on the short circuit current, I_{SC} is given by (2.2):

$$I_{SC} = I_{SC,STC} \times \frac{G}{G_{STC}} \times (1 + K_0 \times (T - T_{1,\text{ref}})) \quad (2.2)$$

where $I_{SC,STC}$ is the short circuit current in standard test conditions (STC) in A; G is the irradiance in W/m^2; G_{STC} is the irradiance in STC 1000 W/m^2; K_0 is the temperature coefficient of I_{SC} in A/K; T is the temperature in K; $T_{1,\text{ref}}$ is the reference temperature in STC in K.

The effect of temperature on the saturation current of the first diode, I_{01} is given by (2.3):

$$I_{01} = I_{01,STC} \times \left(\frac{T}{T_{1,\text{ref}}}\right)^3 \times \exp\left(\frac{q \times EG_{T_{1,\text{ref}}} \times \left(\frac{1}{T_{1,\text{ref}}} - \frac{1}{T}\right)}{K_B}\right) \quad (2.3)$$

where $I_{01,STC}$ is the diode saturation current of the first diode in STC in A; q is the electron charge 1.6×10^{-19} C; $EG_{T1,\text{ref}}$ is the band gap energy in eV.

The effect of temperature on the saturation current of the second diode, I_{02} is given by (2.4):

$$I_{02} = I_{02,STC} \times \left(\frac{T}{T_{1,\text{ref}}}\right)^{3/2} \times \exp\left(\frac{q \times EG_{T_{1,\text{ref}}} \times \left(\frac{1}{T_{1,\text{ref}}} - \frac{1}{T}\right)}{K_B}\right) \quad (2.4)$$

where $I_{02,STC}$ is the diode saturation current of the second diode in STC in A.

The effect of temperature on the series resistor, R_s is given by (2.5):

$$R_s = R_{s,STC} \times \left(\frac{T}{T_{1,\text{ref}}}\right)^{TRs} \quad (2.5)$$

where $R_{s,STC}$ is the series resistance in STC in Ω; TRs is the temperature coefficient of R_s.

The effect of temperature on the shunt resistor, R_{sh} is given by (2.6):

$$R_{sh} = R_{sh,STC} \times \left(\frac{T}{T_{1,\mathrm{ref}}}\right)^{TRsh} \tag{2.6}$$

where $R_{sh,STC}$ is the shunt resistance in STC in Ω; $TRsh$ is the temperature coefficient of R_{sh}.

In order to calculate the output current given by Eq. (2.1) we need the elements of the equivalent circuit (Eqs. 2.2–2.6), the operating temperature, and the operating voltage. The Newton–Raphson method is used to calculate the output current.

A PV panel is generally composed of cells mounted in series. When working on the maximum power point (MPP) the panel's generated current is limited by the cell generating the lowest current. Current mismatch may be caused by no uniform temperature, no uniform irradiance, or no uniform cell degradation (Nehme et al., 2014a,b). We consider all cells operating at the same voltage. N_s being the number of cells in series in the panel the voltage across each cell is approximated to (Tina and Scrofani, 2008):

$$V = \frac{V_{pv}}{N_s} \tag{2.7}$$

where V_{pv} is the PV panel voltage.

In the above we presented the modeling of an ideal PV panel. We start now to express how the elements of the PV cell change as a function of external and operational conditions. But first, let us give a brief reminder of accelerated testing and the Arrhenius equation.

In the literature, researchers apply accelerated tests to evaluate the effect of environmental and operational conditions on the degradation rate of PV panels. Accelerated testing is used to get an idea about a product's lifetime. It decreases testing time and testing cost. In this section, we will develop an analytical model for each of the elements of the equivalent circuit as function of time, environmental, and operational conditions. Using the equivalent circuit instead of developing a general efficiency equation of the panel is justified as follows: each element comprising the panel ages at different rates, elements are composed of different materials, and may encounter different stresses according to their position in the panel (Caruso and Dasgupta, 1998).

A reaction rate is exponentially dependent to temperature. It is expressed by the Arrhenius equation, which was presented by Svante Arrhenius in 1889. The equation relates the reaction rate constant to the temperature as follows:

$$k = A \times \exp\left(\frac{-E_a}{R \times T}\right) \tag{2.8}$$

where k is the constant rate; A is the preexponent factor; E_a is the activation energy in eV; R is the gas constant 8.314 J/(mol K); T is the temperature in K.

Many experiments were done to understand the PID. Swanson et al. explained the PID of PV panels. They showed how the operating voltage of the system affects PID. They started by undertaking field observation. They noticed how the PID increases with the position of the panel in the string. Then they applied a set of laboratory tests to show the effect of module to ground voltage on PID (Swanson et al., 2005). Hacke et al. showed the effect of temperature on PID. They also started by field observation and conducted experiments in test chambers to assess the effect of temperature on the PID speed (Hacke et al., 2011). Berghold et al. showed the effect of relative humidity on PID. They also showed the effect of resistance and cell factors on the PID. They showed how to recover from PID by inverting the cell to frame potential (Berghold et al., 2010).

The effect of PID on PV cells is the increase of their leakage current. This corresponds to a decrease in the shunt resistance of the equivalent circuit. Studying the experiments done by researchers in literature we conclude that:

- The leakage current is proportional to the square of the panel voltage (panel to ground voltage) (Hacke et al., 2011).
- The leakage current is proportional to the square of the panel lifetime (Berghold et al., 2010).
- The leakage current is proportional to the square of the relative humidity (Berghold et al., 2010).
- The leakage current follows an Arrhenius equation with an activation energy of 0.94 eV (Hacke et al., 2011).

The variation of the shunt resistance is then given by the following equation:

$$R_{sh1,\deg} = \frac{1}{7 \times 10^{-6} \times V_{PG}^2 \times RH^2 \times \exp\left(\frac{-90{,}700}{R \times T}\right) \times t^2} \tag{2.9}$$

where $R_{sh1,\deg}$ is the value of the degraded shunt resistance in Ω; V_{PG} is the panel to ground voltage in V; RH is the relative humidity; t is the degradation time in seconds.

The 7×10^{-6} factor depends on the cell factors, especially the resistivity of the base, and it is used to tune the curve to the experiments.

2.2.3 Light-induced degradation

LID augments the recombination current of the base. LID affects c-Si cells by 3% and affects a-Si cells by 30%.

2.2.3.1 Light-induced degradation in c-Si cells

In n-type emitter crystalline silicon cells, the base is doped with an acceptor: boron. During the Czochralski manufacturing process, oxygen atoms diffuse into silicon. When exposed to light, boron atoms lose the hole and then attract the oxygen atom. A B-O complex is formed; it is a trap to electrons and holes, which increases the recombination current. LID is proportional to the concentration of oxygen and boron in the base. Oxygen is present due to the quartz crucible melting during hot silicon processing at 1412°C. The concentration of oxygen is about 5×10^{17} to 1×10^{18} cm^{-3}. The concentration of boron affects the base resistivity and the total efficiency of the cell. That is why a tradeoff must be adopted and a base resistivity of 3–6 Ω cm is used (Herguth et al., 2006). The degradation takes about 72 hours of illumination. Several actions may be undertaken to recover from LID. First we may apply a forward bias current; this phenomenon is called "current-induced regeneration." Or we may anneal the bulk at a temperature of 200°C (Sopori et al., 2012). Or we can wait for the B-O complex to regenerate with time under light (Herguth et al., 2006). To understand this mechanism, we define three states of the B-O complex. State 1 of the complex represents no activity toward electron hole recombination, this is the annealed state. State 2 of the complex represents high activity toward electron hole recombination, this is the degraded state. State 3 of the complex represents low activity toward electron hole recombination; this is the regenerated state (Herguth et al., 2006). The B-O complex can degrade under light from state 1 to state 2. The degradation follows an Arrhenius equation with an activation energy of 0.45 eV. The B-O complex can anneal from state 2 to state 1 under 200°C for 30 minutes (Sopori et al., 2012). The B-O complex can regenerate from state 2 to state 3 under light soaking. The regeneration follows an Arrhenius equation with an activation energy of 1.4 eV. It is proved by experiments that the regenerated state (state 3) is stable at solar cell operating conditions; however, the annealed state (state 1) is unstable (Fig. 2.2).

2.2.3.2 *Light-induced degradation in a-Si:H cells*

LID that occurs to hydrogenated amorphous silicon (a-Si:H) cells is called the Staebler–Wronski effect (SWE). It was discovered by Staebler and Wronski in 1977. The excessive light soaking increases the dangling bonds which increases the recombination rate. This effect is noted in multijunction micromorph solar cells. These cells are built by a μc-Si cell at the bottom and an a-Si:H cell at the top. This structure is used because the a-Si:H cell absorbs high-energy photons.

Two possible ways exist to recover from the SWE. The first is by annealing the bulk at 160°C for a few minutes (Swanson et al., 2005). And the other is by applying a bias voltage under illumination for 30 minutes. The abovementioned information proves the increase in efficiency of a-Si:H cells with high temperature. The bias voltage and the photogenerated electron will allow the reformation of the Si-H bond, which will recover the initial status of the cell.

2.2.3.3 *Light-induced degradation modeling*

Bhushan Sopori et al. explained LID in c-Si cells. They showed how LID affects the I–V characteristic and the equivalent circuit of the cell. They also showed how the system recovers by annealing (Sopori et al., 2012). Herguth et al. showed the effect of temperature on LID, they also defined new states of the B-O complex in c-Si cells. They showed how the form factor (FF), open circuit (OC) voltage V_{OC}, and I_{01}, I_{02} vary with time under LID (Herguth et al., 2006).

Studying the experiments done by researchers in literature we conclude that:

- The recombination current change is proportional to the irradiance (Herguth et al., 2006).
- The recombination current change follows an Arrhenius equation with an activation energy of 0.45 eV (Herguth et al., 2006).
- The recombination current change is proportional to the illumination time (Herguth et al., 2006).

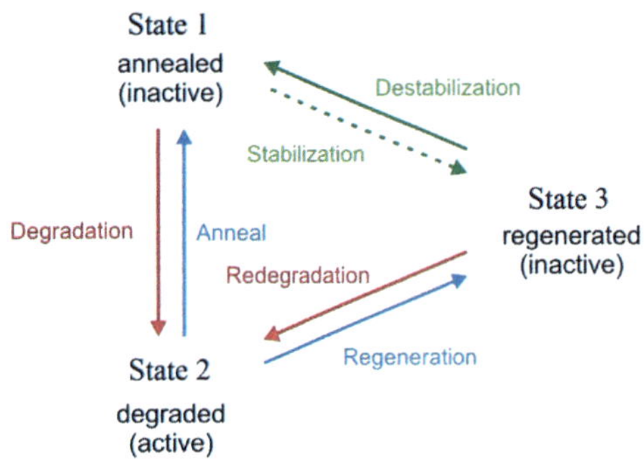

FIGURE 2.2 States of the B-O complex.

The saturation currents I_{01} and I_{02} in the equivalent circuit of the PV cell model are the recombination current. Their variation is fitted to the following equation:

$$I_{01,\text{deg}} = I_{01,STC} + 1.11 \times 10^{-10} \times \frac{G}{1000} \exp\left(\frac{-43,268}{R \times T}\right) \times t \quad (2.10)$$

$$I_{02,\text{deg}} = I_{02,STC} + 1.11 \times 10^{-10} \times \frac{G}{1000} \exp\left(\frac{-43,268}{R \times T}\right) \times t \quad (2.11)$$

where $I_{01,STC}$ is the saturation current of D1 before degradation in STC in A; $I_{02,STC}$ is the saturation current of D2 before degradation in STC in A; G is the irradiance in W/m^2.

The 1.11×10^{-10} factor depends on the dimensions of the cell and the boron concentration, it is used to tune the curve to the experiments.

2.2.4 Ultraviolet light degradation

UV light is electromagnetic radiation with a low wavelength (10–400 nm). It carries high energy and is partially present in the terrestrial solar spectrum. Si represents a low spectral response (SR) toward UV light, the latter is harmful to PV panels. In fact, PV panels are subject to yellowing after some years of operating under sunlight. Precisely, the EVA encapsulation degrades under UV illumination and becomes light yellow, then yellow–brown, and finally dark brown. This is called EVA discoloration.

2.2.4.1 Ultraviolet light degradation causes

The encapsulation is used for optical coupling, electrical isolation, physical isolation protection, mechanical support, and thermal conduction for the PV panel (Hidalgo and Medlege, 2014). The EVA is made up of 70% gel content and 0.3 wt.% (weight percentage) of Cyasorb UV 531. Cyasorb UV 531 absorbs UV light between 240 and 340 nm. Its melting point is 49°C. EVA degradation is calculated based upon gel and Cyasorb UV 531 content. In fact, after UV light illumination, the gel content increases and the Cyasorb UV 531 content decreases. The latter is no longer found in dark brown EVA. Acetic acid and volatile organic components are also produced by UV light exposure (Pern et al., 1991).

The effect of EVA discoloration on PV modules resides in the decrease of the EVA transmittance. In fact, with time the transmittance of the EVA encapsulation decreases for high-energy photons; which reduces the PV panel efficiency.

2.2.4.2 Ultraviolet light degradation modeling

F.J. Pern explained UVD and showed how the encapsulant discoloration occurs over time. He also showed how encapsulant transmittance varies as it degrades under UV light. In fact, as it degrades, the encapsulation will become more transparent to UV radiation and less transparent to useful radiation (Pern, 1994). Jae-Seong Jeong et al. showed the effect of temperature on UVD. They also studied the Fourier-transform infrared spectroscopy of EVA before and after exposure to UV light. The Fourier-transform infrared spectroscopy analysis showed a transformation in the ethylene and vinyl acetate bonding (Jeong and Park, 2013). Pern et al. showed the effect of UVD on the equivalent circuit of a PV cell; with encapsulant discoloration, the series resistor increases and the shunt resistor decreases. They also measured the quantum efficiency for different encapsulant discoloration levels; the quantum efficiency noticeably decreases with EVA discoloration (Pern et al., 1991).

We use the yellowness index (YI) to describe the encapsulation discoloration. We define the delta YI (DYI) as a variable that changes as a function of time to evaluate mathematically the degradation of EVA. Studying the experiments done by researchers in literature we conclude that:

- The DYI follows an Arrhenius equation with an activation energy of 0.93 eV (Jeong and Park, 2013).
- The DYI is proportional to the irradiance intensity (Pern, 1994).
- The DYI is proportional to the logarithm of the illumination time (Pern, 1994).

The variation of the DYI is fitted to the following equation:

$$\text{DYI} = 5.68 \times 10^{11} \exp\left(\frac{-90,000}{R \times T}\right) \times G \times \log\left(\frac{t}{3600}\right) \tag{2.12}$$

As EVA discoloration is a very long-term process, the average temperature and irradiance are used. The 5.68×10^{11} factor is used to tune the curve to the experiments.

The variation of R_s and R_{sh} are fitted to the following equations (Pern et al., 1991):

$$R_{s,\deg} = R_{s,STC} + 9.9 \times 10^{-3} \times \text{DYI} \tag{2.13}$$

$$R_{sh2,\deg} = R_{sh,STC} - 193 \times \text{DYI} \tag{2.14}$$

where $R_{s,STC}$ is the series resistance before degradation in STC in Ω; $R_{sh,STC}$ is the shunt resistance before degradation in STC in Ω.

2.2.5 Moisture-induced degradation

2.2.5.1 Moisture-induced degradation causes

PV panels are subject to moisture degradation. Water particles can enter the panel and cause degradation. This degradation mainly affects transparent conductive oxide (TCO)-

based PV panels. Humidity causes delamination and mainly increases the series resistance of the equivalent circuit. In fact, it degrades the TCO layer. The TCO layer is used in thin-film technologies because of its conductivity, transparency, and low cost. The TCO degrades under the effect of humidity, whereby its resistance increases. This is why PV panels undergo a damp heat test (DH). In the DH test, the panels or cells are exposed to an environment specified by an 85% RH (relative humidity) and 85°C. The test lasts for 1000 hours. According to IEC61646 standard, the efficiency of the panels must not decrease more than 5% (Westin et al., 2006).

2.2.5.2 *Moisture-induced degradation modeling*

Yaklin et al. showed the effect of humidity and temperature on TCO. They used a test chamber where they can change the humidity and temperature. Artificial illumination is also used (Yaklin et al., 2010). Kempe showed the effect of temperature on moisture ingress into PV panels. He noted that different materials have different diffusivity toward water that varies by orders of magnitudes. Using materials with low diffusivity means using high-cost materials (Kempe, 2005). Jane Kapur et al. showed the geometrical repartition of water fraction in the PV panel after water ingress. They showed that water concentration in the panel is not uniform; water is more present at the edges of the panel. They also showed the effect of temperature on the diffusion of water for different types of materials (Kapur et al., 2013).

Studying the experiments done by researchers in literature we conclude that:

- TCO degradation follows an Arrhenius equation with an activation energy of 0.91 eV (Yaklin et al., 2010).
- TCO degradation is proportional to the relative humidity (Yaklin et al., 2010).
- Water vapor transmission rate (WVTR) follows an Arrhenius equation with an activation energy of 0.38 eV (Kapur et al., 2013).
- Moisture repartition in the panel follows a sine function having a lower value at the center of the panel and a higher value near the edges (Kapur et al., 2013).

The variation of R_s is fitted to the following equation:

$$R_{s,\deg} = R_{s,STC} + 1.11 \times 10^7 \times \exp\left(\frac{-87,500}{\mathrm{R} \times T}\right) \times \sin(Dist) \times \exp\left(\frac{-33,678}{\mathrm{R} \times T}\right) \times RH \times t \quad (2.15)$$

where *Dist* refers to the normalized distance from the center of the panel.

As WVTR is a very long-term process, the average temperature is used. The factor 1.11×10^7 is used to tune the curve to the experiments.

2.3 Real-time simulation model

2.3.1 Development of the model

In the previous section the external conditions (temperature, irradiance, and relative humidity) were taken as average values. In this section, we present a more precise model; the model takes into consideration the variation of temperature, irradiance, and relative humidity across the day and the year. Thus the degradation equations are derived with

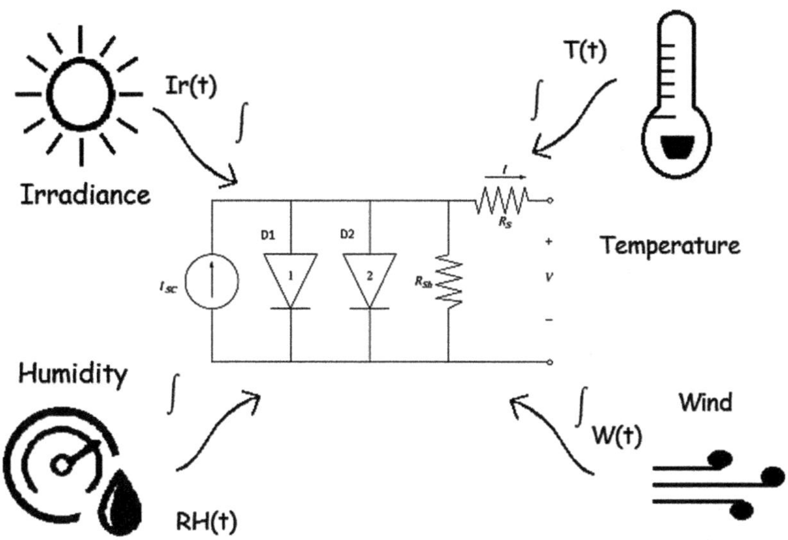

FIGURE 2.3 Summary representation of the real-time simulation model.

respect to time. The proposed model will generate the elements of the equivalent circuit of the PV cell, which vary nonlinearly. The model is time varying (Fig. 2.3).

We choose to apply the model for monocrystalline silicon cells. In fact, crystalline-based panels are the most used in the market. They present the highest efficiency in commercial PV panels. In addition, tests and research are most applied to crystalline-based PV panels. PID, LID, UVD, and cell cracks are the degradation modes that affect monocrystalline PV panels.

Eqs. (2.9) and (2.14) model two parallel resistances that represent PID and UVD. The equivalent shunt resistance is given by:

$$R_{sh,\deg} = \frac{R_{sh,STC} - 193 \times \text{DYI}}{\left(7 \times 10^{-6} \times V_{PG}^2 \times RH^2 \times \exp\left(\frac{-90{,}700}{R \times T}\right) \times t^2\right) \times \left(R_{sh,STC} - 193 \times \text{DYI}\right) + 1} \tag{2.16}$$

The system takes as input the irradiance G, the cell temperature T, the relative humidity RH, and the operating voltage V_{PG}. The values of Eqs. (2.10–2.13, 2.15, 2.16) are calculated. The initial values ($I_{SC,STC}$, $R_{sh,STC}$, $I_{01,STC}$, $I_{02,STC}$, and $R_{s,STC}$) and the constants (G_{STC} and R) are internal values of the system. The outputs of Eqs. (2.10–2.13, 2.15, 2.16) are integrated to generate the elements of the equivalent circuit of the PV cell (I_{SC}, R_{sh}, I_{01}, I_{02}, and R_s).

The inputs of the system are set as follows. The irradiance varies following a sine function during the day (M'Sirdi et al., 2014). In regions situated in the northern hemisphere, during June–September period, the irradiance increases. In regions situated in the southern hemisphere, during June–September period, the irradiance decreases. The temperature varies alongside the irradiance during the day. The relative humidity varies inversely to temperature (Kempe and Wohlgemuth, 2013) (Fig. 2.4).

The above equations were implemented in a differential equation editor (DEE) block in Matlab/simulink software. The outputs of the DEE block are the elements of the equivalent circuit of a PV cell. The elements are then updated according to the conditions of

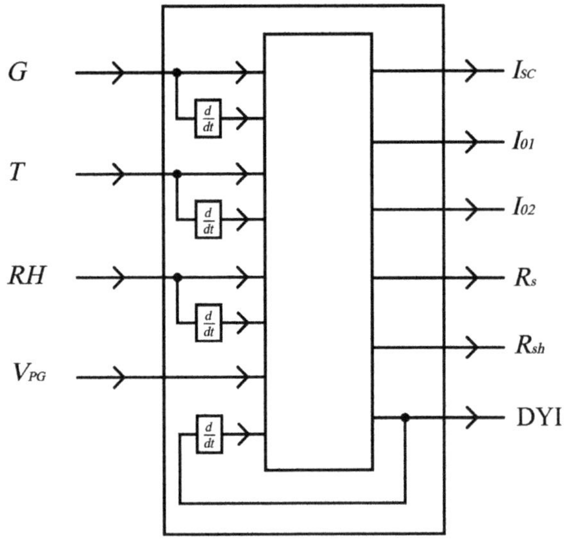

FIGURE 2.4 Degradation model of crystalline silicone cells.

temperature and irradiance. The current generated is calculated using the Newton–Raphson method. The state space modeling gives us the facility to express the equivalent circuit of the PV cell as states. In fact, the system is nonlinear and time varying and we have expressed these states as function of inputs (irradiance, temperature, relative humidity, etc.) and time. The degradation state at a certain time t results from the addition of degradations that took place before t.

2.3.2 Simulation results

In this section we present the simulation results using the developed model. The key performance index chosen is the normalized efficiency NE(t). The normalized efficiency at an instant t is the efficiency at that instant divided by the efficiency at time t_0, where t_0 corresponds to the time before degradation starts.

$$NE(t) = \frac{\text{Eff}(t)}{\text{Eff}(t_0)} \tag{2.17}$$

where Eff(t) is the efficiency of the panel at time t; Eff(t_0) is the efficiency of the panel at time t_0.

In the first simulation, we present (Fig. 2.5) the normalized efficiency of a PV panel for 40 years. The location of the site has an average irradiance of 709 W/m^2, an average RH of 50%, and the average panel temperature is 45°C. We can see a fast decrease in efficiency for the first 4–5 years, then a slow decrease that is marked by a logarithmic shape. We can also see that the curve has a wave-like shape.

In fact, the rapid decrease in efficiency during the first years of operation is caused by the PID and LID. The leakage current caused by the PID will saturate and the LID

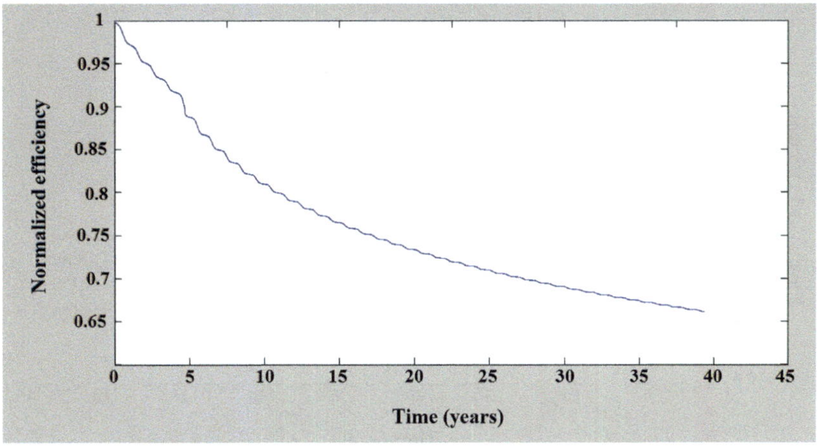

FIGURE 2.5 Normalized efficiency for 40 years ($G = 709\ W/m^2$, $RH = 50\%$, $T = 45°C$).

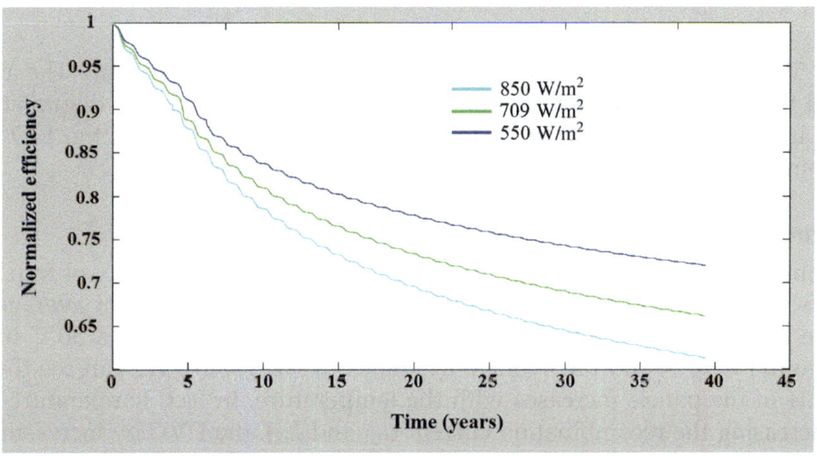

FIGURE 2.6 Normalized efficiency over 40 years ($G = 850\ W/m^2$ in *cyan*, $G = 709\ W/m^2$ in *green*, $G = 550\ W/m^2$ in *blue*).

degradation is limited to the first days of operation. The UVD, which depends algorithmically as a function of time, continues to decrease the panel's efficiency for all 40 years of operation. The wave-like shape is explained by the fact that during summer, high temperature and high irradiance are noted, which will cause a severe decrease in efficiency during this period. During winter, lower temperature and irradiance are noted which will cause a lower degradation rate.

2.3.2.1 *Irradiance variation*

In the second simulation, we present (Fig. 2.6) the normalized efficiency of three PV panels for 40 years. The location of the sites of the three panels have different average

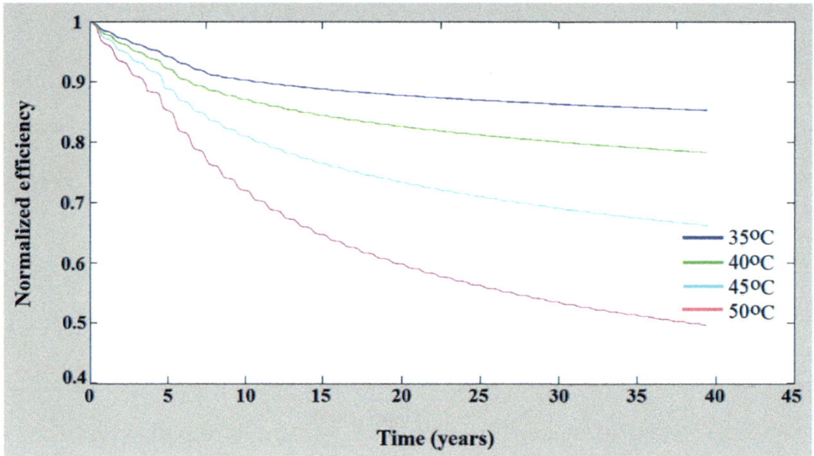

FIGURE 2.7 Normalized efficiency over 40 years (50°C in *magenta*, 45°C in *cyan*, 40°C in *green*, and 35°C in *blue*).

irradiance: 550 W/m^2 in blue, 709 W/m^2 in green, and 850 W/m^2 in cyan. The average RH is 50%, and the average panel temperature is 45°C. We can see that the degradation rate of the panels increases with the irradiance. In fact, irradiance influences the LID increasing the recombination current (I_{01} and I_{02}) and the UVD by increasing the DYI.

2.3.2.2 *Temperature variation*

In the third simulation, we present (Fig. 2.7) the normalized efficiency of four PV panels for 40 years. The location of the sites of the four panels imply different average temperatures of the four panels: 35°C in blue, 40°C in green, 45°C in cyan, and 50°C in magenta. The average RH is 50%, and the average irradiance is 709 W/m^2. We can see that the degradation rate of the panels increases with the temperature. In fact, temperature influences the LID, increasing the recombination current (I_{01} and I_{02}), the UVD by increasing the DYI, and the PID by decreasing the shunt resistance.

2.3.3 Validation of the model

In this paragraph we try to validate our developed real-time simulation model. It is literally impossible to use a panel for 40 years and measure its efficiency. We try to compare our results with gathered data from researchers who recorded the degradation rate of PV panels. Recorded degradation rates are nearly absent in literature since the related topic is new (Osterwald et al., 2006). The rule-of-thumb of the PV industry averages the degradation rate of PV panels to be -1%/year (Osterwald et al., 2006).

Researchers measured power generated from PV panels, irradiance, and temperature. When these data are available, they can calculate the efficiency of PV panels (Jordan et al., 2015). Sometimes they may require analytical methods to determinate the efficiency of PV panels (Jordan and Kurtz, 2010).

TABLE 2.1 Summary of recorded degradation rates.

Location	Years of operation (years)	Degradation rate (%/year)
Utah, USA	20	− 0.5
Riyadh, Saudi Arabia	20	− 2.6
Tokyo, Japan	5	− 0.95
Hamamatsu, Japan	10	− 0.062

The degradation rates recorded over time showed the same result of our real-time simulation model. In some references we see the average degradation rate to be close to −0.8%/year, which is close to our simulated result (Jordan et al., 2015; Jordan and Kurtz, 2013). The degradation rate started with high levels −0.8%/year for c-Si PV panels, then the degradation rate decreased logarithmically with time, which aligns with our simulated results (Jordan and Kurtz, 2010). In other reported results, c-Si modules present degradation rates between −0.5%/year and −1%/year, unfortunately no weather conditions were reported (Osterwald et al., 2006). The degradation rates reported for the first years of installation are high compared to the average degradation over a 10-year operation period (Osterwald et al., 2006).

In Table 2.1 we can see a summary of recorded degradation rates of PV panels with their corresponding location. We can see a varying degradation rate between regions, where hotter regions witness higher degradation rates, which aligns with our developed model. In addition, a high degradation rate is noted for the first operational years which aligns with our developed model.

2.4 Thermal model of a photovoltaic panel

As stated previously, temperature plays an important factor in degrading and causing faults to PV panels. Besides, temperature increases the degradation rate of PV panels exponentially, following an Arrhenius equation. This emphasizes the need to develop a thermal model of a PV panel in order to understand its thermal behavior. In this section, we will present the heat fluxes that affect the temperature of the panel. Then we will build the model under COMSOL Multiphysics software. At the end, a validation of the thermal behavior is achieved with the experimentation apparatus.

2.4.1 Thermal model development

In this section we present a detailed thermal model of PV panels where we take into consideration external heat sources, operating conditions, internal heat sources, and heat dissipation. The external heat source is represented by the sun. The operating conditions of the PV panel (operating voltage) directly affect the output power generated by the PV panel. The internal heat sources are due to the internal series and shunt resistors that generate heat due to the Joule effect. Heat is being dissipated to the outside of the panel via convection and radiation (Nehme et al., 2017).

A PV panel resembles a flat plate that receives solar irradiation and generates electrical power. In literature, many researchers developed thermal models of PV panels. Jones et al. developed a thermal model of PV panels based on energy balance. They considered the panel to be a uniform flat plate that receives irradiation, generates electrical power, and dissipates heat to the exterior (Jones and Underwood, 2001). Tina et al. added the effect of the operating voltage or the extracted power on the thermal energy balance of a PV panel. They divided the panel into three layers consisting of glass, cells, and backsheet. They also developed an electrical model of the PV panel (Tina and Scrofani, 2008). Ruhi Bharti et al. conducted experiences that showed the effect of the electrical load on the nominal operating cell temperature (NOCT). They noted the effect of internal heat generation of PV cells (Bharti et al., 2009). Rosa-Clot et al. studied a hybrid thermal–PV panel: the thermal electric solar panel integration (TEPSI). They showed how solar energy can be transformed into thermal and electrical energy (Rosa-Clot et al., 2011). Armstrong et al. developed the equivalent RC thermal circuit of a PV model. They also investigated the time constant of the thermal behavior of PV panels (Armstrong and Hurley, 2010).

The incident irradiation is the main heat source that contributes to heating the panel. This source is reduced by a factor of 5% due to reflection.

$$\dot{Q}_{sun} = 0.95 \times A \times G \tag{2.18}$$

where A is the total area of the panel in m^2.

The electrical output power generated by the PV panel depends on its operating voltage. In fact, the P–V characteristic is nonlinear and presents a mountain shape. The electric output power represents a sink to the thermal system model and is given by:

$$P_{ele} = V_{pv} \times I \tag{2.19}$$

The conversion efficiency of conventional PV panels is limited to 18%. The remaining input power will generate heat to the panel. This heat will be dissipated to the exterior via convection and radiation (White, 1991). The conduction is neglected because the panel nearly presents no contact to solids. The convection occurs on the front, back, and edge surfaces. The thermal convection flux to ambient air is expressed by the following formula:

$$\dot{Q}_{conv} = h \times A \times (T_{amb} - T_{pv}) \tag{2.20}$$

where h is the heat transfer coefficient in $W/(m^2\,K)$.

The heat transfer coefficient depends on many parameters, such as the wind velocity, the air density, the tilt angle of the panel, and the dimensions of the panel. Using some approximations we can correlate the heat transfer coefficient to the wind velocity by the following formula (Mattei et al., 2006):

$$h = 5.67 + 3.86 \times Wv \tag{2.21}$$

where Wv is the wind velocity in m/s.

The radiation occurs from the front and the back surfaces. The thermal radiation flux is expressed by the following formula:

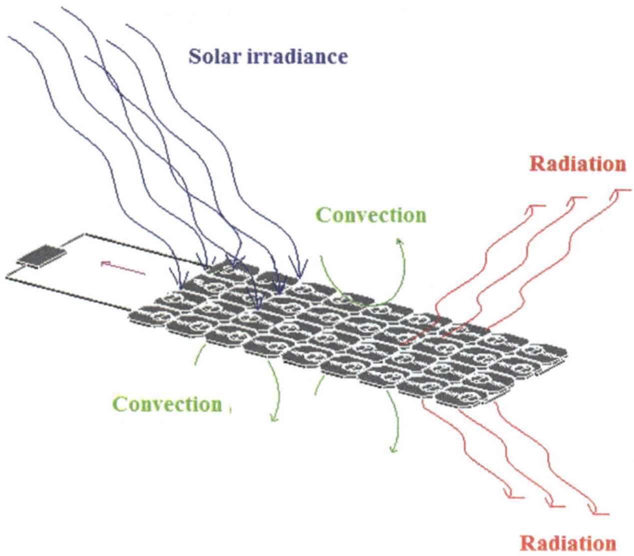

FIGURE 2.8 Thermodynamics of a PV panel.

$$\dot{Q}_{rad} = \varepsilon \times \sigma \times A \times \left(T_{amb}^4 - T_{pvs}^4\right) \tag{2.22}$$

where ε is the emissivity; σ is the Stephan–Boltzmann constant $5.67 \times 10^{-8}\,\mathrm{W/(m^2\,K^4)}$; T_{pvs} is the temperature of PV panel surface in K.

The front surface emissivity is taken to be 0.9 (Jones and Underwood, 2001; Armstrong and Hurley, 2010), and the back surface emissivity is taken to be 0.84 (Armstrong and Hurley, 2010).

The internal heat sources of the panel are due to the Joule effect of the series and shunt resistors. In fact, current passing through the printed Ag fingers and through the Cu busbars will lead to a localized heat generation modeled by the series resistor R_s. Besides, current passing through a path around the cell will lead to a localized heat generation modeled by the series resistor R_{sh}. The Joule heat generated is given by the following formula:

$$P_j = n \times R_s \times I^2 + n \times \frac{(V + R_s \times I)^2}{R_{sh}} \tag{2.23}$$

where n is the number of cells; R_s is the series resistance of the cell in Ω; R_{sh} is the shunt resistance of the cell in Ω.

In summary, a PV panel is a flat plate that heats due to the difference between incident irradiance and generated electrical power. The panel also heats because of internal heat generation due to the Joule effect. Because of the temperature difference between the panel and ambient air, heat dissipates from the panel to the exterior via convection and radiation (Fig. 2.8).

2.4.2 Model exploration

In order to explore the developed model under different operational and environmental conditions, we develop it under COMSOL Multiphysics. And then we simulate its behavior.

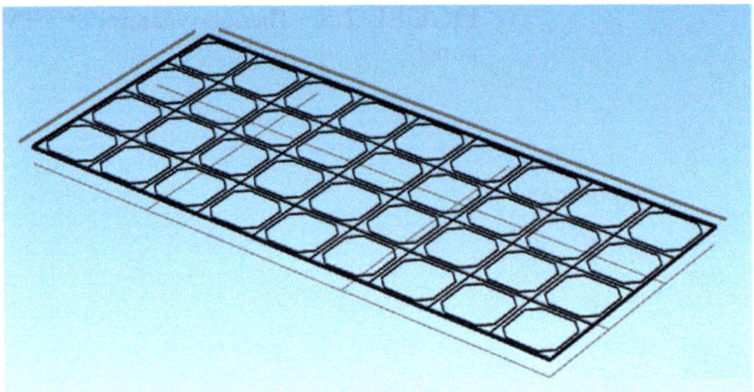

FIGURE 2.9 Modeling of the panel under study.

2.4.2.1 *Model development*

The finite element analysis software COMSOL Multiphysics is used. We build a panel consisting of 36(4 × 9) cells. Each cell has dimensions of 12.5 cm × 12.5 cm. A distance of 2 cm separates two adjacent cells (Fig. 2.9). The panel is constituted of three layers: the glass 0.3 cm, the PV cell (cell, ARC, and EVA) 0.2 cm, and the backsheet 0.2 cm. Under STC the panel can deliver 5.5 A.

The heat transfer (The Heat Transfer with Surface-to-Surface Radiation) physics module is used. A heat flux on the front surface models the incident irradiance. Convective heat flux is added to the front, back, and edge surfaces. Surface-to-Ambient Radiation is added to the front and back surfaces with their corresponding emissivity. A negative heat source is added, localized in each cell and modeling the generated electrical energy. Another heat source is added, localized in each cell and modeling the internal generated heat (Table 2.2).

2.4.2.2 *Simulation and result extraction*

In the first simulation we consider an irradiance of 1000 W/m^2, an external temperature of 20°C and a wind speed of 1 m/s. The panel is operating at a voltage of 18 V. We show the temperature at the surface of the panel in Fig. 2.10. We can see a slight difference in color between the core of the cells and the space between the cells. The cells are hotter than the space between them. This is explained by the fact that cells are a thermal source due to the Joule effect of the series resistance.

In the second simulation we consider variant environmental conditions. The irradiance is swept from 200 to 1200 W/m^2 and the ambient temperature is swept from 10°C to 42°C. The wind speed is fixed to 1 m/s with a pressure of 1 atm. The panel is operating at a voltage of 18.2 V. The cell temperature at the center of the panel is shown in figure fig:Tamb G variation (Fig. 2.11). It is clear that the cell panel temperature increases with the intensity of irradiance and with the ambient temperature. In fact, the irradiance affects the primary or external heat source. And with the increase of ambient temperature, heat dissipation or heat transfer from the panel to the external medium will decrease. The convection heat flux decrease is proportional to the ambient temperature and the radiation heat flux decrease is proportional the ambient temperature at the fourth power (Fig. 2.11).

TABLE 2.2 Model parameters.

Parameter	Description	Value
D	Dimension of the cell	0.125 m
Dx	Thickness of the cell	0.00007351 m
R_{SSTC}	Series resistance in STC	0.075 Ω
R_{ShSTC}	Shunt resistance in STC	50,000 Ω
d_{glass}	Thickness of the glass	0.003 m
d_{back}	Thickness of the backsheet	0.0001 m
k_{PV}	Thermal conductivity of the cell	93 W/(m K)
ρ_{PV}	Density of the cell	2000 kg/m^3
Cp_{PV}	Heat capacity at constant pressure of the cell	840 J/(kg K)
k_{glass}	Thermal conductivity of the glass	1.4 W/(m K)
P_{glass}	Density of the glass	3000 kg/m^3
Cp_{glass}	Heat capacity at constant pressure of the glass	500 J/(Kh K)
$K_{backsheet}$	Thermal conductivity of the backsheet	0.2 W/(m K)
$\rho_{backsheet}$	Density of the backsheet	1200 kg/m^3
$Cp_{backsheet}$	Heat capacity at constant pressure of the backsheet	1250 J/(kg K)
k_{copper}	Thermal conductivity of copper	400 W/(m K)
ρ_{copper}	Density of copper	8700 kg/m^3
Cp_{copper}	Heat capacity at constant pressure of copper	385 J/(kg K)
$\varepsilon_{front\ surface}$	Emissivity of the front surface	0.9
$E_{back\ surface}$	Emissivity of the back surface	0.84

In the third simulation, the ambient temperature is fixed to 20°C with a pressure of 1 atm. The wind speed is fixed to 1 m/s (Armstrong and Hurley, 2010). The operating voltage is swept from 3 to 32 V. The irradiance is fixed to 1000 W/m^2. The temperature is shown for the different points (front surface, cell, and back surface) in Fig. 2.12. The back temperature is plotted in green, the cell temperature is plotted in blue, and the front temperature is plotted in red. The violet curve represents the electric power generated by the panel.

We can see a difference of about 5°C between the front and the back surfaces. The temperature of the cell is about 1°C lower than the temperature of the front surface. We can see clearly that the temperature has a global minimum around 33 V which is close to the MPP. In reality the minimum temperature occurs when the power driven from panel and the Joule effect together reach a minimum.

In the fourth simulation, we repeat the third simulation with a variant irradiance from 100 to 1100 W/m^2. The temperature of the cell is shown in Fig. 2.13.

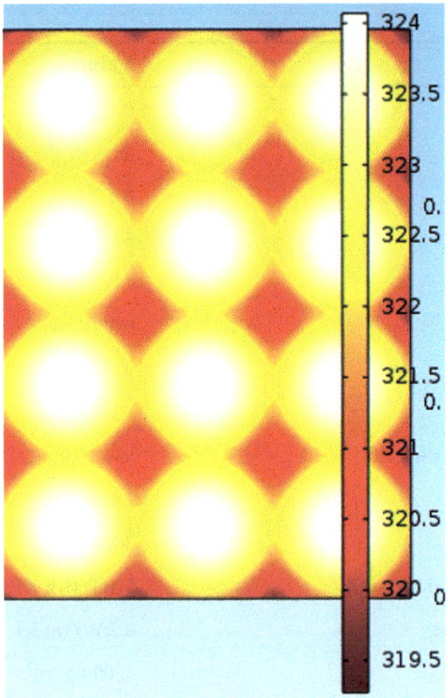

FIGURE 2.10 Surface temperature of the right side of the PV panel.

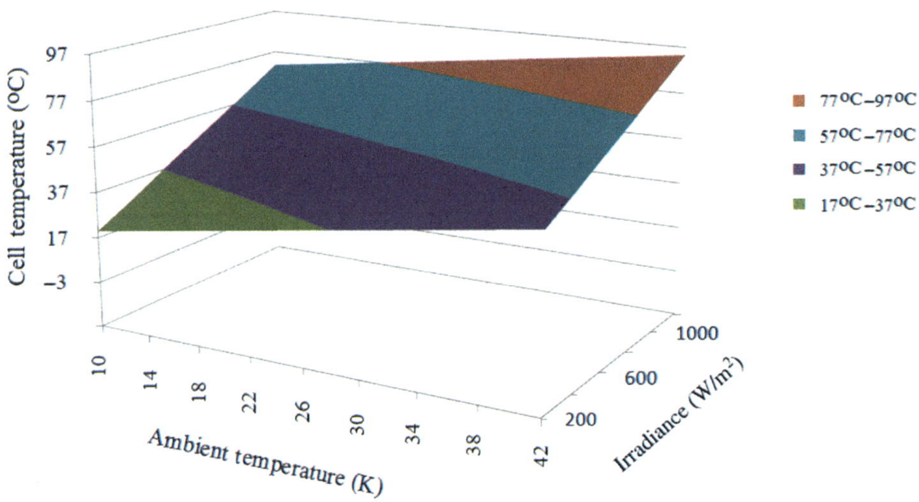

FIGURE 2.11 Cell temperature with irradiance and ambient temperature variations.

2.4.3 Experimental validation

2.4.3.1 Experimental apparatus

In this section we propose to validate the above developed model with real hardware experimentation. We build a PV system composed of a 100 W PV panel (monocrystalline,

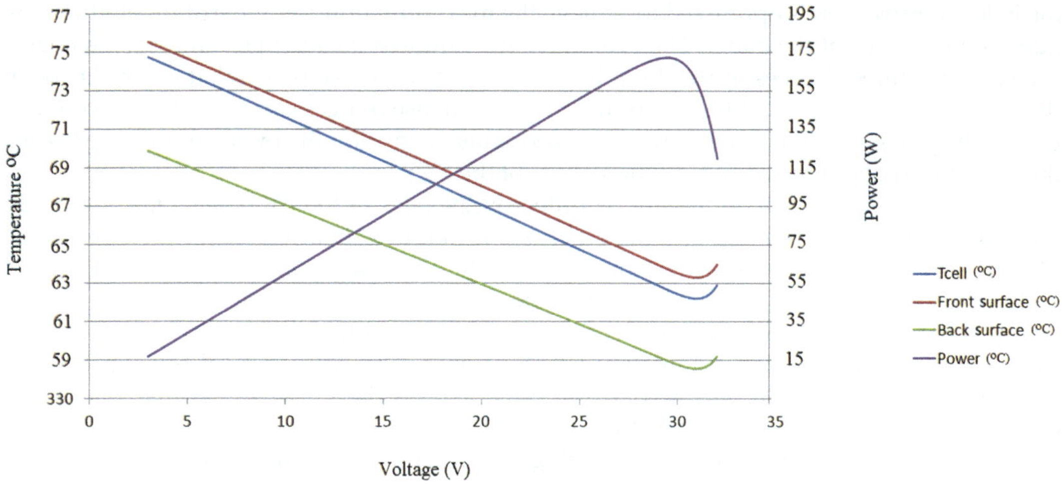

FIGURE 2.12 Panel temperature for different points with output power as a function of operating voltage.

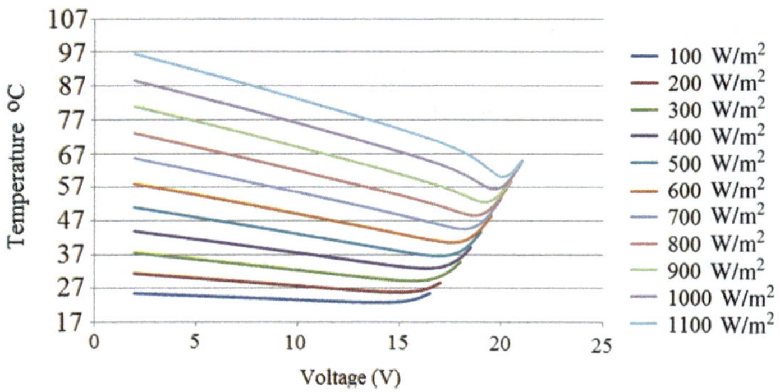

FIGURE 2.13 Temperature of the cell as a function of operating voltage for different irradiance levels.

Solar Innova SI-ESF-M-M90–100 W), halogen projectors, resistive load, DC/DC converter, back surface mounted thermocouples. Two thermocouples are used to measure the temperature of the panel. We use a pyranometer (Hukseflux-LP02) to measure the illumination. The PV panel is mounted on a two-axis rotating table. The table can rotate 180 degrees for each axis. The halogen projectors act as artificial lighting for the PV panel. One limitation in the laboratory testing is the limited illumination of artificial lighting. The DC/DC converter was developed in order to meet the requirements of the experimentation.

The panel is constituted of a string of 36 cells. Two bypass diodes are implemented. Each one bypasses 18 cells in cases of partial shading. The panel can deliver at STC and at MPP 18.4 V and 5.43 A. The OC voltage is 22.72 V and the short circuit current is 5.64 A. The dimensions of the panel are 119.5 cm × 54.1 cm × 3.5 cm. The panels' weight is 8 kg. The external frame is composed of aluminum (AL6063-T5). Each cell size is 12.5 cm × 12.5 cm. The panels' ingress protection rating is IP-65.

The halogen incandescent lamps (3300K) act as an emulator of the sun. Even though they present low illumination (529 W/m^2 with four lamps) they present good conditions to test the operation of the panel regarding its electrical and thermal properties. The repartition of the halogen projectors is optimized to achieve the maximum uniform repartition of illumination on the surface of the panel. The projectors are mounted on a support and each halogen projector has three degrees of freedom. It can rotate over the z-axis (yaw), it can rotate over the x-axis (roll), and its height can vary in order to increase or decrease the distance from the panel. The purpose of the three degrees of freedom is to position the projector in a way to get the best uniform light repartition on the surface of the PV panel.

The two K-type thermocouples are mounted on the back of the PV panel. One of them is mounted at the center of a cell and the other one is mounted in a region between two cells. The precision of the temperature sensor and the Data Acquisition (DAQ) system is 0.1°C.

The DC/DC converter is constituted of a buck–boost converter (Fig. 2.14). The control circuit is constituted of two NE555. The first one operates in the astable mode generating a clock reference pulse width modulation (PWM) signal. The duty cycle of the signal must be high; it is fixed to 0.87. The second NE555 generates the PWM signal that is used to control the switch of the buck–boost converter. The reference voltage of the duty cycle is adjusted via a potentiometer or via an analog computer signal using a NI USB-6008 DAQ. A two-way switch is used to choose between the two operations. The DAQ is also used to measure the output current and output voltage of the PV panel. A shunt resistor (0.1 Ω) is mounted in series with the panel. The voltage across the resistor represents an image of the current. A differential amplifier is used to measure the voltage across the panel. The switching devise is the IRFZ44N metal oxide–semiconductor field-effect transistor (MOSFET). We present the values of the elements of the buck–boost converter in Table 2.3 (Nehme and Akiki, 2016).

2.4.3.2 Experimental result

The experimentation was conducted in a laboratory room, where the wind speed was null, the ambient temperature was 21°C, the pressure was 1 atm, and the relative humidity was 92%. We illuminate the panel with the halogen projectors. The short circuit current is

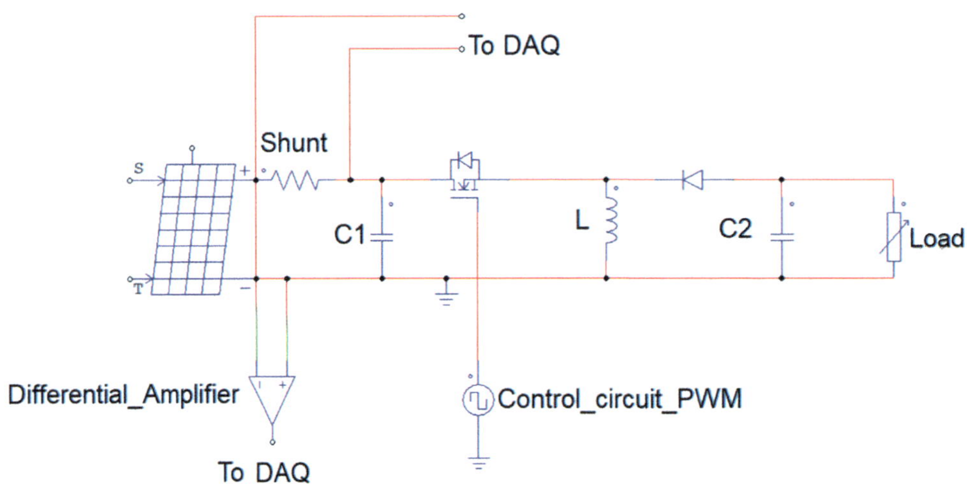

FIGURE 2.14 Block diagram of the buck–boost converter.

TABLE 2.3 Elements of the buck–boost converter.

Element	Value
Shunt	0.1 Ω
C1	470 μF
C2	470 μF
L	400 μH
Switching frequency	5 kHz

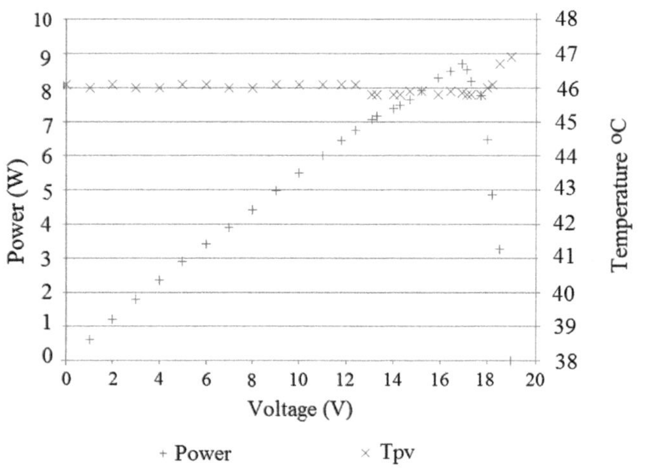

FIGURE 2.15 Thermal behavior of the PV panel as a function of operating voltage.

$I_{SC} = 1.6$ A. We start varying the operating voltage (from short circuit to OC) from 0 to 19.37 V and we note the temperature. The results are shown in Fig. 2.15. They align well with the simulation results. We can see that the minimum temperature occurs at a voltage slightly higher than the MPP.

A second experimentation was performed with variable wind speed. Two fans were used. The ambient temperature was 27°C, and the relative humidity was 92%. The results show the same thermal behavior. Figs. 2.16–2.18 present the simulation results with low, medium, and high wind speeds respectively. We can see that the temperature of the back of the cell (points in red) is higher than the temperature of the back of the spaces between the cells (points in green). The latter aligns with the simulation results shown in Fig. 2.10.

2.5 Mitigation of degradation via control

In this section we present our preventive actions that must be undertaken in order to increase the life span of PV panels via control. As previously stated, the degradation modes of PV panels follow Arrhenius equations where the constant rate of degradation is

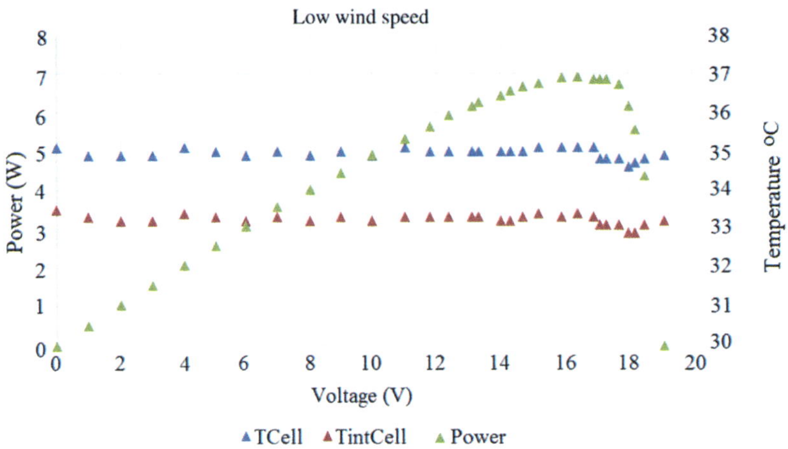

FIGURE 2.16 Thermal behavior of the PV panel as a function of operating voltage with low wind speed.

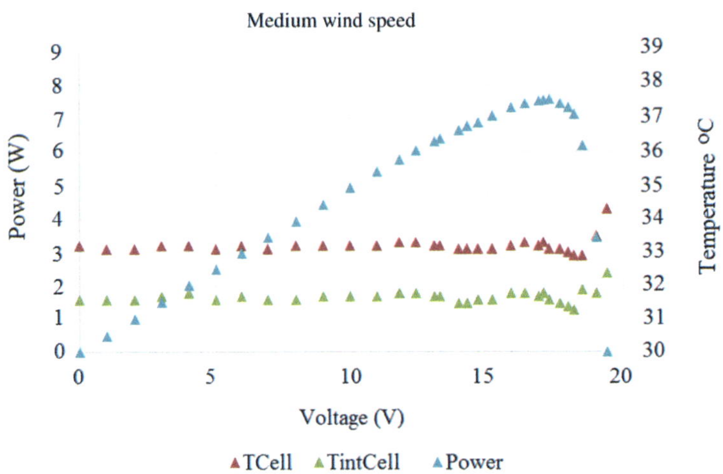

FIGURE 2.17 Thermal behavior of the PV panel as a function of operating voltage with medium wind speed.

proportional to the exponential of the temperature. The latter encourages us to operate the panel at a lower temperature.

2.5.1 Real-time simulation model with thermal behavior

In this part, we include the thermal model of PV panels in the previously developed real-time model Section 2.3. The general dynamic equation of heat transfer is given by the following equation:

$$\sum \dot{Q} = m \times C \times \Delta T \tag{2.24}$$

where $\sum \dot{Q}$ is the sum of all heat fluxes in W; m is the mass in kg; C is the thermal capacity in J/(kg K); T is the temperature in K.

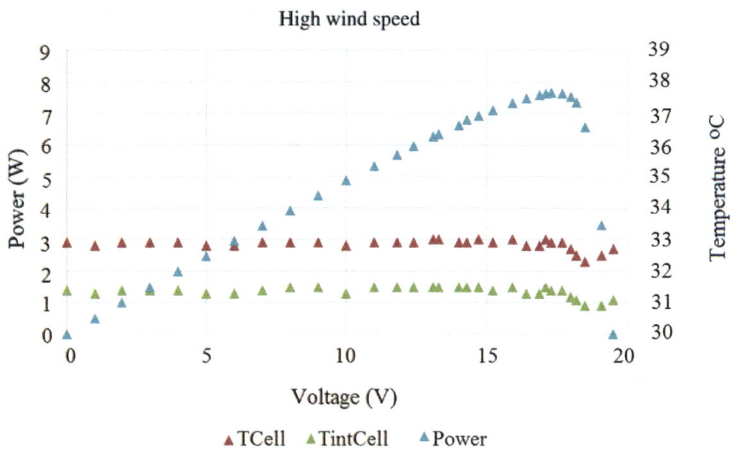

FIGURE 2.18 Thermal behavior of the PV panel as a function of operating voltage with high wind speed.

$$\sum \dot{Q} = \dot{Q}_{sun} + P_j - P_{ele} - \dot{Q}_{conv} - \dot{Q}_{rad}$$

$$\sum \dot{Q} = G \times A \times 0.95 + n \times R_s \times I^2 + n \times \frac{\left(\frac{V}{n} + R_s \times I\right)^2}{R_{sh}} - V \times I - h \times A \times \left(T_{amb} - T_{pv}\right)$$
$$- \varepsilon_f \times \sigma \times A \times \left(T_{amb}^4 - T_{pvs}^4\right) - \varepsilon_b \times \sigma \times A \times \left(T_{amb}^4 - T_{pvs}^4\right) \quad (2.25)$$

where ε_f is the emissivity of the front surface; ε_b is the emissivity of the back surface.

The real-time simulation model can now represent a real PV panel whose degradation rate can be changed by a change in operational conditions. Moreover, a change in operational conditions will lead to a change in temperature.

2.5.2 Maximum life span point

We define the MLP as the point where the PV panel operates at its minimum temperature. Recalling the thermal model previously developed, the MLP can be found at the right side of the MPP. It corresponds to a higher voltage. Fig. 2.19 shows the power curve with its corresponding MPP and the temperature curve with its corresponding MLP.

2.5.3 Tracking the maximum life span point

We start first by understanding the thermodynamic physics behind the MLP. The dynamic equation of heat transfer given by Eq. (2.25) can be divided into the following:

$$\dot{Q}_1 = \dot{Q}_{sun} = G \times A \times 0.95 \quad (2.26)$$

is the incident irradiation that cannot be controlled.

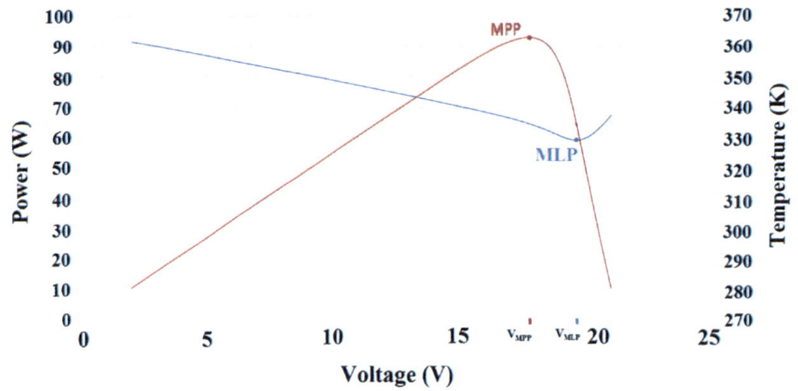

FIGURE 2.19 Power and temperature curves with MPP and MLP.

$$\dot{Q}_2 = -\dot{Q}_{conv} - \dot{Q}_{rad}$$
$$= -h \times A \times (T_{amb} - T_{pv}) - \varepsilon_f \times \sigma \times A \times \left(T_{amb}^4 - T_{pvs}^4\right) - \varepsilon_b \times \sigma \times A \times \left(T_{amb}^4 - T_{pvs}^4\right) \quad (2.27)$$

is the convection and radiation heat fluxes that depend directly on the temperature of the panel. These fluxes depend indirectly on the operation point of the panel but present complexity and low effect.

$$\dot{Q}_3 = P_j - P_{ele} = n \times R_s \times I^2 + n \times \frac{\left(\frac{V}{n} + R_s \times I\right)^2}{R_{sh}} - V \times I \quad (2.28)$$

is the internal heat sources and the converted electrical power. These fluxes depend directly from the operation point, as they depend on V and I. Our approach is to operate at the point where $\dot{Q}_3$; is at its minimum value every time.

We start by plotting the flux $\dot{Q}_3$ as a function of the operating voltage. In the same graph we plot the generated power. We can see in Fig. 2.20 the MPP and the MLP. The latter corresponds to the minimum value of the heat flux $\dot{Q}_3$. Our goal is to operate the panel at the MLP in order to decrease its temperature. $\dot{Q}_3$ presents a nonlinear behavior. In Fig. 2.21 we present the breakdown of $\dot{Q}_3$ flux. We can see separately the heat generated by R_s and R_{sh} and the extracted electrical power. The heat caused by R_{sh} presents low variation and negligible value. The heat caused by R_s presents high values for low voltages and decreases with voltage. The electrical power generated (that acts as a heat sink to the thermal model) presents a minimum point that corresponds to the MPP. When adding the three curves, we obtain the nonlinear blue curve of Fig. 2.20.

2.5.3.1 *Applying maximum power point tracking techniques to track the maximum life span point*

In this section, we would like to operate our panel on the MLP. The task is similar to operating the panel at the MPP. We would like to fix our operating point at the point

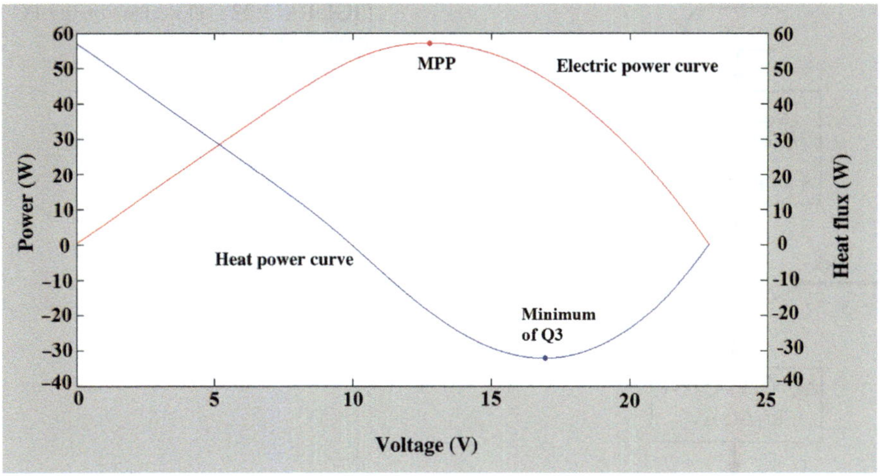

FIGURE 2.20 Electric power and $\dot{Q}_3$ heat flux.

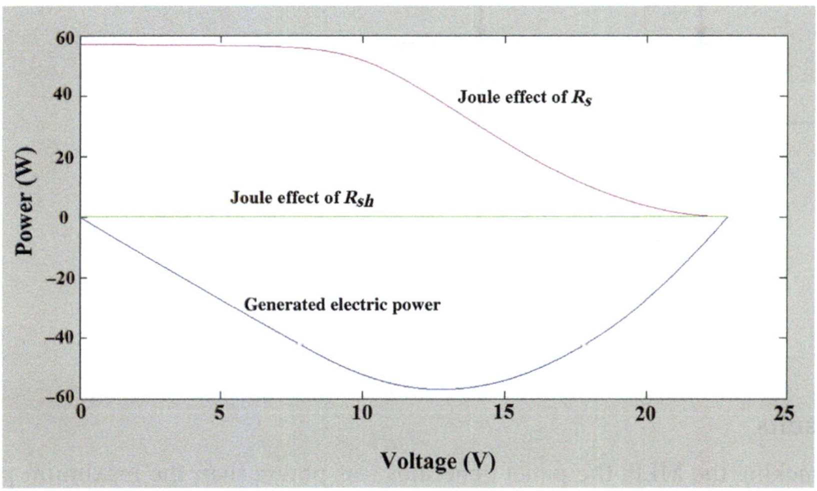

FIGURE 2.21 $\dot{Q}_3$ heat flux breakdown.

where the temperature becomes minimal, that is, in other words where $\dot{Q}_3$ becomes minimal. $\dot{Q}_3$ has one global minimum (MLP) that is defined by:

$$\frac{d\dot{Q}_{3MLP}}{dV} = 0 \tag{2.29}$$

At the left of the MLP $d\dot{Q}_{3MLP}/dV < 0$ at the right of the MLP $d\dot{Q}_{3MLP}/dV > 0$. A perturb and observe (PO) algorithm can be used. If the value of the $\dot{Q}_3$ slope has a negative value, we must increase the operating voltage. If the value of the $\dot{Q}_3$ slope has a positive value, we must decrease the operating voltage. Fig. 2.22 shows the MLPT algorithm.

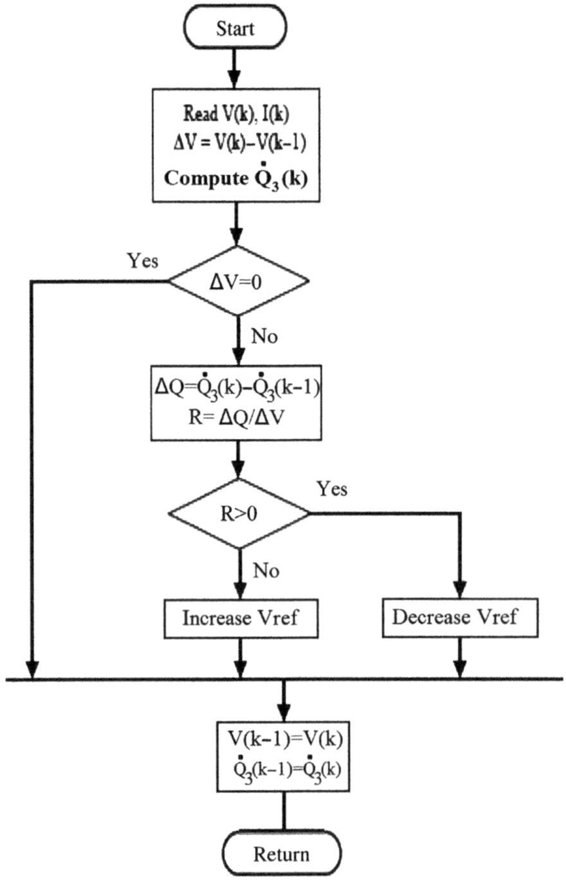

FIGURE 2.22 Flowchart of the PO algorithm applied for MLPT.

2.5.4 Results

When tracking the MLP, the panel generates less power than the maximum power. The latter represents a major drawback. For that, the MPP tracking (MPPT) algorithm is applied when the panel temperature reaches 60°C. Switching back to the MPPT requires that the temperature falls lower than 59.5°C.

We applied our MLP algorithm for the previous modeled monocrystalline 100 W PV panel. The panel model takes into account all the degradation modes with all the environmental and operational parameters affecting their magnitude. We choose an average irradiance of 709 W/m^2, an average ambient temperature of 28°C, and an average relative humidity of 50%. We start by showing the temperature of the panel T_{pv} for the 10th day of operation in Fig. 2.23. Then we show the temperature of the panel T_{pv} for the 1000th day of operation in Fig. 2.24. Next, we show the power produced for the 1000th day of operation in Fig. 2.25. Then we show the normalized efficiency for all 40 years of operation of the panel in Fig. 2.26. At the end, we show the energy production for the 40 years of

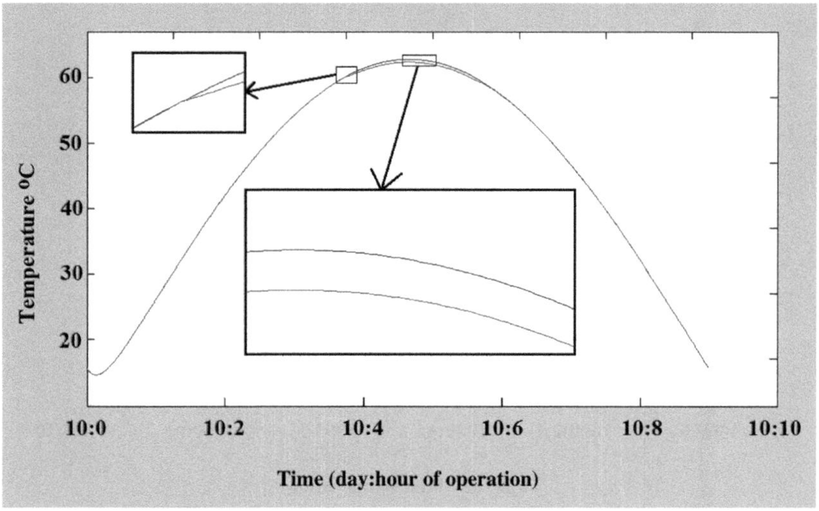

FIGURE 2.23 Temperature of the panel for the 10th day.

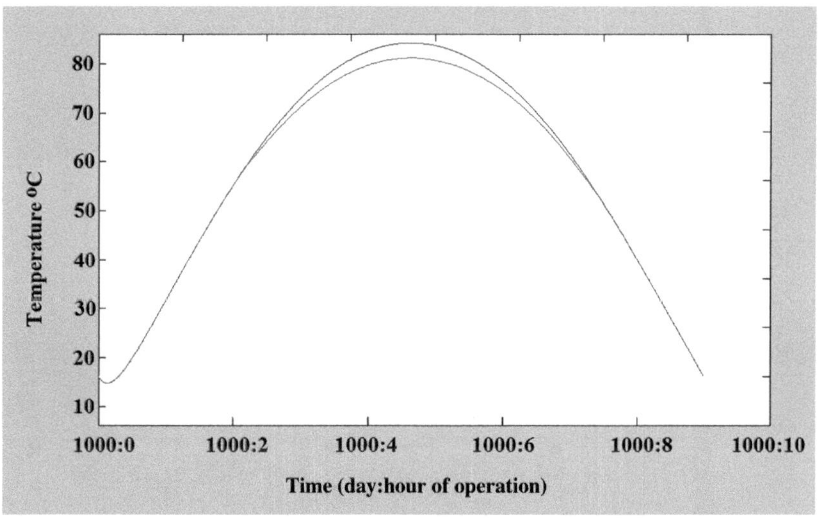

FIGURE 2.24 Temperature of the panel for the 1000th day.

operation in Fig. 2.27. For all the simulation results, the curves in blue represent the simulation using the MPPT algorithm and the curves in magenta represent the simulation using the MLPT algorithm.

In Fig. 2.23 the temperature of the panel is shown for the 10th day of operation. The temperature rises and falls down, it lags the ambient temperature due to the thermal capacitance of the panel. We can see in the left magnification that a small decrease of temperature is achieved when applying the MLPT algorithm. We can see also in the right

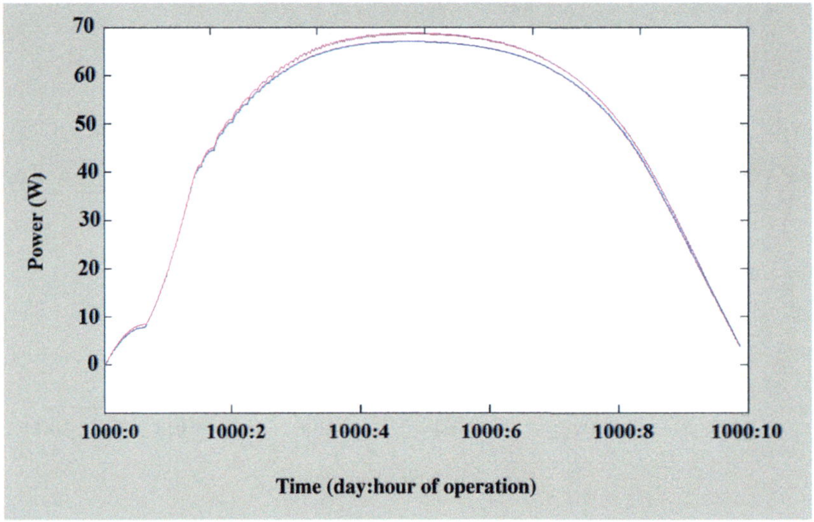

FIGURE 2.25 Power of the panel for the 1000th day.

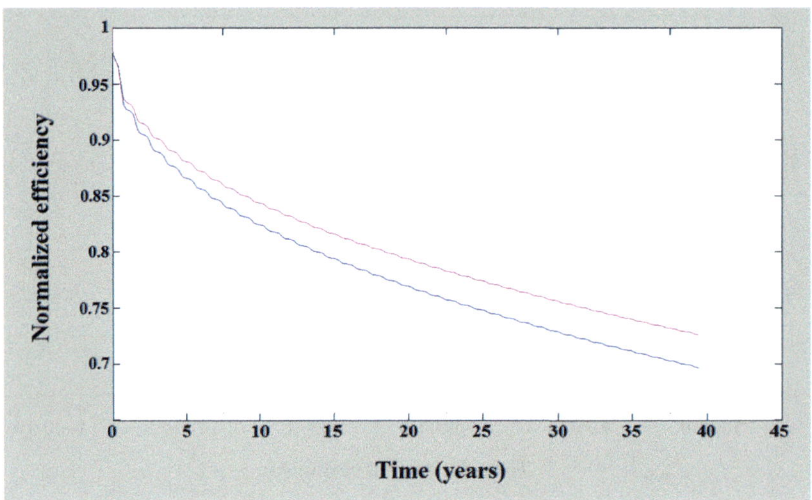

FIGURE 2.26 Normalized efficiency over 40 years.

magnification that the temperature starts differing between the blue and the magenta curves when the temperature of the panel reaches 60°C.

In Fig. 2.24, the temperature of the panel is shown for the 1000th day of operation. We can see the high temperature decreases when we apply the MLPT algorithm. This is due to the fact that the degraded panel (in blue) will heat up more than the less degraded panel (in magenta).

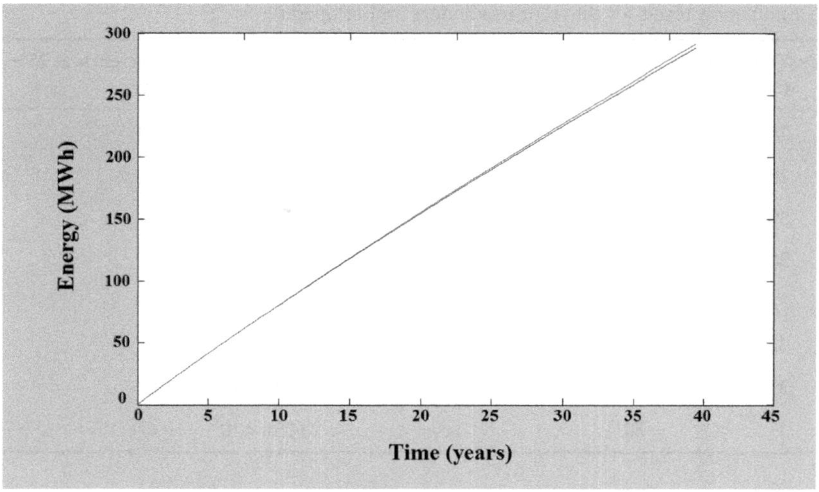

FIGURE 2.27 Energy produced for 40 years of operation.

During the operation for 40 years, a maximum temperature of 85°C has been recorded for the degraded panel. And a maximum temperature of 82°C has been recorded for the less degraded panel.

In Fig. 2.25, we can see the power produced during the 1000th day. The power produced by the degraded panel (in blue) is 4 W lower than the power produced by the less degraded panel (in magenta). Two reasons are behind this fact. First is that the degraded panel has a lower efficiency (in STC) than the less degraded panel; it has been working on higher temperatures for the past 999 days. Second, the actual operating temperature at of the degraded panel is higher than the actual operating temperature of the less degraded panel and temperature decreases the efficiency of the panel.

In Fig. 2.26, we can see the normalized efficiency of the two panels. The degraded panel (in blue) and the less degraded panel (in magenta). We can clearly observe that when applying our control strategy, the efficiency of the panel is maintained higher. We also see that the efficiency of the less degraded panel at 40 years is almost the same as that of the degraded panel at around 25 years. The latter proves that we achieved an increase in the life span of PV panels up to 40 years.

In Fig. 2.27 the energy produced by the two panels is shown for 40 years of operation. The degraded panel (in blue) produced 287 MWh in 40 years. The less degraded panel (in magenta) produced 291 MWh in 40 years. We can see an increase in energy production due to our control MLPT algorithm. By looking at the energy curve, we see a wavelet shape, especially in the first 4 years of operation. This shape is due to the fact that summer seasons produce more power than winter seasons. We can also see that the slope of the first years is higher than the slope of the later years. This is due to the degradation process of PV panels that decreases power production with time.

We now apply our approach for different irradiances and different ambient temperatures. The simulation results are shown in Table 2.4. When the control approach is Yes it

TABLE 2.4 Simulation result for different irradiances and temperatures.

Irradiance (W/m^2)	Temperature (°C)	Relative humidity (%)	Control approach	Energy at 40 years	Efficiency at 25 + years	Efficiency at 40 years
709	28	50	No	2.8798×10^8	0.73	0.69
709	28	50	Yes	2.9124×10^8	0.76	0.72
709	30	50	No	2.8002×10^8	0.71	0.67
709	30	50	Yes	2.8400×10^8	0.74	0.7
709	32	50	No	2.7126×10^8	0.68	0.64
709	32	50	Yes	2.7626×10^8	0.71	0.67
800	35	50	No	2.5821×10^8	0.63	0.6
800	35	50	Yes	2.6381×10^8	0.67	0.64

means that we are applying our algorithm to decrease the degradation rate of PV panels. When the control approach is No it means that the old MPPT algorithm is used all the time. We can see that for the same weather conditions the efficiency of the system is increased when applying our control approach. The efficiency at 40 years of the panels with our control approach is nearly equal to the efficiency at 25 + years of the panels without our control approach. The latter proves that we succeeded in reaching our objective. In addition, we can see the difference in energy production between the panels with the control approach and the panels without the control approach. This difference in energy production increases with the increase of ambient temperature.

2.5.5 Discussion

By applying our control algorithm, we succeeded in generating more energy from the same panel. Although the MLPT algorithm allows the panel to operate at a point generating less power than the MPP the degradation process of the panel has been reduced. As the panel operates at a lower temperature, the degradation speed is reduced. We also noticed that the more the panel degrades the more it heats up. This is explained by the fact that the series resistance R_s increases, generating more heat with the current passing through. Besides, the shunt resistance R_{sh} decreases, increasing the shunt current and the heat generated.

2.6 Conclusion

In this chapter, we presented our contribution to the modeling of degradation modes of PV panels. A complete model has been developed that takes into consideration the degradation process of PV panels, the effect of the environmental and operational conditions, and it also integrates the thermal model of the panel.

In order to mitigate the degradation of PV panels, we defined a new operating point of PV panels; the MLP. This point is characterized by operating the panel at a minimum temperature. As temperature is the major parameter affecting the degradation speed and as temperature affects the degradation rate following an Arrhenius equation, choosing to reduce the operating temperature is a key solution for increasing the PV life span. By operating the panel at the MLP during temperature higher than 60°C we ended by increasing the power production of the panel over a 40-year operation duration.

In a perspective to our works, the proposed MLP algorithm should be integrated in energy management systems. A decrease in the operational temperature of PV panels can lead to a better efficiency, higher life span, fewer faults, better reliability, and less hazards.

Acknowledgments

The work presented in this chapter would not be possible without the financial contribution of the Agence Univeritaire de la Francophonie (AUF) and the Higher Center of Research (HCR) of the USEK university. Moreover, the authors would like to thank the direction of the Laboratoire des Sciences de l'Information et des Systèmes (LSIS) of the Aix Marseille Université (AMU). The authors would also like to thank the direction of the Faculty of Engineering at the Holy Spirit University of Kaslik (USEK).

References

Armstrong, S., Hurley, W.G., 2010. A thermal model for photovoltaic panels under varying atmospheric conditions. Appl.Therm.Eng. 30 (11), 1488–1495.

Berghold, J., Frank, O., Hoehne, H., Pingel, S., Richardson, B., Winkler, M., 2010. Potential induced degradation of solar cells and panels. In: 25th EUPVSEC, pp. 3753–3759.

Bharti, R., Kuitche, J., TamizhMani, M.G., 2009. Nominal Operating Cell Temperature (NOCT): Effects of Module Size, Loading and Solar Spectrum. In: 34th IEEE Photovoltaic Specialists Conference (PVSC). IEEE, pp. 001657-001662.

Caruso, H., Dasgupta, A., 1998. A fundamental overview of accelerated testing analytical models. J. IEST, 41(1), 16-20.

Hacke, P., Terwilliger, K., Smith, R., Glick, S., Pankow, J., Kempe, M., et al., 2011. System voltage potential-induced degradation mechanisms in PV modules and methods for test. In: 37th IEEE Photovoltaic Specialists Conference. IEEE, pp. 000814-000820.

Herguth, A., Schubert, G., Kaes, M., Hahn, G., 2006. Avoiding boron-oxygen related degradation in highly boron doped Cz silicon. In: 21st European Photovoltaic Solar Energy Conference: 21th EC PVSEC, pp. 530-537.

Hidalgo, M., Medlege, F., 2014. Procede de caracterisation d'un copolymere d'ethylene et d'acetate de vinyle. <https://www.google.com.lb/patents/EP2715327A1?cl = fr>.

Hersch, P., Zweibel, K., 1982. Basic photovoltaic principles and methods (No. SERI/SP[HYPHEN]290[HYPHEN] 1448). Golden, CO: Solar Energy Research Inst.

Jeong, J., Park, N., 2013. Field discoloration analysis and UV/temperature accelerated degradation test of EVA for PV. In: IEEE 39th Photovoltaic Specialists Conference (PVSC). IEEE, pp. 3010-3013.

Jones, A.D., Underwood, C.P., 2001. A thermal model for photovoltaic systems. Sol. Energy 70 (4), 349–359.

Jordan, D.C., Kurtz, S.R., 2010. Analytical improvements in PV degradation rate determination. In: 35th IEEE Photovoltaic Specialists Conference. IEEE, pp. 002688-002693.

Jordan, D.C., Kurtz, S.R., 2013. Photovoltaic degradation rates – an analytical review. Prog. Photovolt. Res. Appl. 21 (1), 12–29.

Jordan, D.C., Sekulic, B., Marion, B., Kurtz, S.R., 2015. Performance and aging of a 20-year-old silicon PV system. IEEE J. Photovolt. 5 (3), 744–751.

Kapur, J., Norwood, J.L., Cwalina, C.D., 2013. Determination of moisture ingress rate through photovoltaic encapsulants. In: IEEE 39th Photovoltaic Specialists Conference (PVSC). IEEE, pp. 3020-3023.

Kempe, M.D., 2005. Control of moisture ingress into photovoltaic modules. In: Conference Record of the Thirty-first IEEE Photovoltaic Specialists Conference, 2005. IEEE, pp. 503-506.
Kempe, M.D., Wohlgemuth, J.H., 2013. Evaluation of temperature and humidity on PV module component degradation. In: IEEE 39th Photovoltaic Specialists Conference (PVSC). IEEE, pp. 0120-0125.
Koch, S., Seidel, C., Grunow, P., Krauter, S., Schoppa, M., 2011. Polarization effects and tests for crystalline silicon cells. In: 26th European Photovoltaic Solar Energy Conference and Exhibition, pp. 1726-1731.
Lall, P., 2004. Challenges in accelerated life testing. In: The Ninth Intersociety Conference on Thermal and Thermomechanical Phenomena In Electronic Systems (IEEE Cat. No. 04CH37543) (Vol. 2, pp. 727-Vol). IEEE.
Luqueand, A., Hegedus, S., (Eds.), 2003. Handbook of Photovoltaic Science and Engineering. John Wiley & Sons.
Mattei, M., Notton, G., Cristofari, C., Muselli, M., Poggi, P., 2006. Calculation of the polycrystalline PV module temperature using a simple method of energy balance. Renew. Energy 31 (4), 553. Available from: http://www.sciencedirect.com/science/article/pii/S096014810500073X.
M'Sirdi, N.K., Nehme, B., Abarkan, M., Rabbi, A., 2014. The best mppt algorithms by vsas approach for renewable energy sources (res). In: 3rd International Symposium on Environmental Friendly Energies and Applications (EFEA). IEEE, pp. 1-7.
Nehme, B., Akiki, T., 2016. Implementing a didactic photovoltaic energy laboratory for developing countries. In: 2016 3rd International Conference on Renewable Energies for Developing Countries (REDEC), September 29, pp. 1–4.
Nehme, B.F., Akiki, T.K., Naamane, A., M'Sirdi, N.K., 2017. Real-time thermoelectrical model of PV panels for degradation assessment. IEEE J. Photovolt. 7 (5), 1362–1375.
Nehme, B., M'Sirdi, N.K., Akiki, T., Naamane, A., 2014a. Contribution to the modeling of ageing effects in PV cells and modules. Energy Procedia 62, 565–575.
Nehme, B., M'Sirdi, N.K., Akiki, T., 2014b. A geometric approach for PV modules degradation. In: International Conference on Renewable Energies for Developing Countries. IEEE, pp. 49-53.
Nehme, B., Msirdi, N., Naamane, A., Akiki, T., 2016. Analysis and characterization of faults in PV panels. Energy Procedia 111, 1020–1029.
Osterwald, C.R., Adelstein, J., Del Cueto, J.A., Kroposki, B., Trudell, D., Moriarty, T., 2006. Comparison of degradation rates of individual modules held at maximum power. In: IEEE 4th World Conference on Photovoltaic Energy Conference (Vol. 2, pp. 2085-2088). IEEE.
Pern, F.J., 1994. Factors that affect the EVA encapsulant discoloration rate upon accelerated exposure. In: Proceedings of 1994 IEEE 1st World Conference on Photovoltaic Energy Conversion-WCPEC (A Joint Conference of PVSC, PVSEC and PSEC) (Vol. 1, pp. 897-900). IEEE.
Pern, F.J., Czanderna, A.W., Emery, K.A., Dhere, R.G., 1991. Weathering degradation of EVA encapsulant and the effect of its yellowing on solar cell efficiency. In The Conference Record of the Twenty-Second IEEE Photovoltaic Specialists Conference-1991 (Vol. 1, pp. 557-561). IEEE.
Rosa-Clot, M., Rosa-Clot, P., Tina, G.M., 2011. TESPI: thermal electric solar panel integration. Sol. Energy 85 (10), 2433–2442.
Sopori, B., Basnyat, P., Devayajanam, S., Shet, S., Mehta, V., Binns, J., et al., 2012. Understanding light-induced degradation of c-Si solar cells. In: 38th IEEE Photovoltaic Specialists Conference. IEEE, pp. 001115-001120.
Study on Photovoltaic Panels Supplementing the Impact Assessment for a Recast of the WEEE Directive. 2011.
Swanson, R., Cudzinovic, M., DeCeuster, D., Desai, V., Jurgens, J., Kaminar, N., et al., 2005. The surface polarization effect in high-efficiency silicon solar cells. In: 15th PVSEC. Shanghai, China.
Tina, G.M., Scrofani, S., 2008. Electrical and thermal model for PV module temperature evaluation. In: MELECON 2008-The 14th IEEE Mediterranean Electrotechnical Conference (pp. 585-590). IEEE.
Westin, P., Neretnieks, P., Edoff, M., 2006. Damp heat degradation of CIGS-based PV modules. In: 21st European Photovoltaic Solar Energy Conference, pp. 2470-2473.
White, F.M., 1991. Heat and Mass Transfer. Reading, MA: Addison-Wesley.
Yaklin, M.A., Schneider, D.A., Norman, K., Granata, J.E., Staiger, C.L., 2010. Impacts of Humidity and Temperature on the Performance of Transparent Conducting Zinc Oxide, p. 2493.

CHAPTER

3

Community-scale rural energy systems: General planning algorithms and methods for developing countries

Alejandro López-González[1,2,3]

[1]Institute of Industrial and Control Engineering, Universitat Politècnica de Catalunya—BarcelonaTech, Barcelona, Spain [2]Department of Electrical Engineering—Campus Terrassa (ESEIAAT)—BarcelonaTech, Tarrassa, Spain [3]Socioeconomic Centre of Petroleum and Alternative Energies, Universidad del Zulia, Maracaibo, Venezuela

List of Acronyms

AARE	Autonomous Authorities in Rural Electrification
AEP	Annual Energy Production
AT	Available Terrains
CGCC	Centralized Generation Cost Curve
COE	Cost of Energy
GCC	Generation Cost Curve
GHI	Global Horizontal Irradiance
HOMER	Hybrid Optimization of Multiple Energy Resources
HWT	Home Wind Turbines
IGCC	Isolated Generation Cost Curve
iHOGA	Improved Hybrid Optimization by Genetic Algorithms
LV	Low Voltage
MASS	Mesoscale Atmospheric Simulation System
MCP	Measure-Correlate-Predict
MILP	Mixed-integer Linear Programming
MV	Medium Voltage
NPV	Net Present Value
NWP	Numerical Weather Prediction
O&M	Operation and Maintenance
PV	Photovoltaic
RET	Renewable Energy Technologies

Predictive Modelling for Energy Management and Power Systems Engineering
DOI: https://doi.org/10.1016/B978-0-12-817772-3.00003-3

RHMG	Rural Hybrid Microgrids
SDGs	Sustainable Development Goals
SEPLAN	Sustainable Energy Planning
SHS	Solar Home Systems
SMARTS	Simple Model of the Atmospheric Radiative Transfer of Sunshine
SOC	State of charge
SWT	Small Wind Turbine
TY	Typical Meteorological Year
ViPOR	Village Power Optimization Model for Renewables
VMM	Virtual Met Masts
WAsP	Wind Atlas Analysis and Application Program
WRG	Wind Resources Grid

3.1 Introduction

One in every five people in the world lives without access to electricity, this is 1.06 billion people (SE4All, 2017). Most of them live in developing countries so this energy access gap, between developing and developed countries, will profoundly compromise global agenda goals on poverty and inequality reduction, education, public health, and climate change, among others. Sustainable Development Goals (SDGs) established 2030 as the target year for reaching universal energy access (United Nations, 2015). However, considering the current electrification rate, 674 million people will still be without access to electricity in 2030 due to not enough measures being taken by some governments and the international cooperation funds (SE4All, 2017). Considering that most unelectrified rural houses are in extremely scattered communities, the conventional fuel-based electricity supply options within a centralized power generation and transmission networks must be reconsidered. Some other technologies and grid configurations, including renewable energy technologies (RET) and off-grid, centralized, or individual solutions, and configurations for rural electrification must be prioritized. One of these alternatives is rural hybrid microgrids (RHMG), which can combine different RETs, backup and storage technologies to centrally feed a set of local loads within a small rural settlement or community, using a low- or medium-voltage network (Cader et al., 2016). On the other hand, extremely scattered and isolated rural houses can be fed using RET-based individual off-grid configurations. These could be wind or solar based, through home wind turbines systems (HWT) or solar home systems (SHS), respectively. The optimal configuration for a particular rural location is usually composed of one or more RHMG and some other HWT and/or SHS, within one or more electrification projects (Domenech et al., 2014).

Several numerical methods have been discussed in the literature for achieving the optimal configurations in a rural electrification project using more than one RHMG, SHS, and/or HWT. Most of them consider the minimum total net present cost of the system subjected to technical constraints, such as minimum state of charge (SOC) of the battery bank and capacity shortage (Ferrer-Martí et al., 2013). However, many issues considered in some methods are excluded in others. For example, some studies consider in detail the RETs and grid equipment cost throughout the life span of the project, but not the important relation to energy demand and energy storage in batteries during the design stage (Ashok, 2007; Bala and Siddique, 2009). For optimal coordination among multiple types of RETs, storage and backup components, while satisfying the underlying capacity and technical constraints, some mixed-integer linear programming (MILP) models have been developed (Dai and Mesbahi, 2013). The MILP approach is generally applicable to both continuous and integer variables in small and medium size optimization problems. In rural electrification MILP is formulated for capacity and operating constraints for generating, storage and load units, and technical constraints, aiming to find the optimum continuous and integer variables that define a numerical solution (Dai and Mesbahi, 2013). Ferrer-Martí et al. (2011) developed a MILP model to optimize a community system, considering load points demand in terms of both power and energy, including also characteristics and constraints of complementary equipment (batteries, inverters, controllers, and meters). Recently, Falcón-Roque et al. (2017) presented an energy planning model based on multicriteria optimization techniques, named Sustainable Energy Planning (SEPLAN). The SEPLAN model allows the incorporation of objectives that could be in conflict, for example, economic, environmental, or social objectives, as well as universal access to energy. Finally, due to the increasing interest in rural electrification, universal energy access, and RETs for off-grid applications, several commercial software have been developed and continuously improved for evaluating different aspects of these projects (Singh et al., 2016), for example, HOMER, ViPOR, RETScreen, and iHOGA (Falcón-Roque et al., 2017). In addition, some authors have pointed out that nowadays several commercial programs provide regional wind resource assessment: WAsP, Openwind, Wind PRO, Wind Farm, Farm visualization, etc. Such computing models help in predicting the wind characteristics of the locations for which measurements are not available (Murthy and Rahi, 2017). Regarding solar energy, some precise data are now freely available on the web (IRENA, 2018).

According to the revised literature, a rural electrification project could be divided into three basic stages (Ashok, 2007; Huang et al., 2015), which are (1) terrain recognition, including load points consumption estimations and resources assessment (Bala and Siddique, 2009); (2) optimum RET technologies configuration for each load point or set of load points (Baños et al., 2011); and (3) optimum technologies allocation, in individual and centralized systems, considering distribution network layout and voltage level (Domenech et al., 2015). We realized all the previously described stages, and their different issues, could be accomplished using one or more together of the previously mentioned software which are worldwide available.

Regarding the first stage, OpenWind is chosen due to its performance, which is similar or better than many others available (mentioned before). In this sense, González-Longatt et al. (2014) demonstrated that OpenWind is useful to model the horizontal and vertical extrapolation of wind data, considering the elevation and roughness length. This way, they made an estimation of the wind resource at each virtual anemometer mast or load point in their study to create regional wind resources grid (WRG) maps in Venezuela. According to Casella (2015), Openwind's technique is efficient in particular conditions, concerning sites with complex topography like Andean mountains or coastal regions, where the OpenWind measure–correlate–predict (MCP) analyses tended to provide higher accuracy than physical modeling methods like Wind Atlas Analysis and Application Program (WAsP). Regarding solar energy assessment, topography and roughness length are not as important for wind when community-scale projects are developed. Therefore solar resources data, which is freely available on the web, can be perfectly useful for the planning and design of solar-based rural electrification projects (IRENA, 2018).

Regarding the second stage, Hybrid Optimization Model for Electric Renewables (HOMER), developed by National Renewable Energy Laboratory (NREL, United States), is the preferred tool for this research due to its demonstrated efficiency obtaining the optimum configuration from a large set of technologies (PV, wind, diesel, batteries, etc.), loads (AC/DC), and with respect to the particular characteristics of a consumption point or a set of them (Bhattacharyya, 2012). HOMER has been validated as a more precise method than MILP or any others. In addition, HOMER provides easy implementation of sensitivity analysis and generates a ranked list of alternative configurations which may be of interest to decision-makers (Zachar and Daoutidis, 2015). This sensitivity analysis over a changing load, resources, and/or RETs provides an optimum cost configuration that can be obtained for each load level, under some wind and solar resources' availability. Conceptually, the graph composed by the lowest cost configuration (*Y* axis) for each load level (*X* axis) is called a Generation Cost Curve (GCC). This GCC output from HOMER is the most important part of our proposed method.

Finally, in the third stage, the optimum technologies allocation can be accomplished using another NREL software tool which is the Village Power Optimization for Renewables (ViPOR), due to its widely extended and free availability all over the world. This program objective is the design of community-level microgrids. ViPOR optimally designs the distribution grid to the village level (Mendes et al., 2011). ViPOR uses the GCC output from HOMER to design a distribution system combining microgrids and individual systems, according to an optimal allocation of RETs (Ranaboldo et al., 2014). Thus HOMER designs the generation system, with great detail regarding equipment and including many technologies, while ViPOR plans the distribution scheme (Domenech et al., 2015). This way, three widely available software, working by stages, could be used to reach precise results un rural electrification projects' planning and this is particularly important for the knowhow extension in Autonomous Authorities in Rural Electrification (AARE) from developing countries.

This chapter proposes a modular methodology based on three software (OpenWind, HOMER, and ViPOR) synergies for rural electrification planning at a community scale, considering the developing countries' perspectives. From a general perspective, three basic stages are considered, which are not usually explicit in most of the community-scale rural electrification projects methodologies. This modular scope is proposed to find the GCC of the RETs in the conditions of the community and its terrain and optimize them along the load points in the terrain where the community is allocated. Some other software can be used for these three methodologies' proposed stages if the corresponding outputs are useful to build GCC and optimize the RETs allocation in the community accordingly.

3.1.1 Theoretical framework

Our proposed methodology, within this chapter, is divided into three basic stages, which are included in most of the rural electrification methodologies but not in an explicit way. The first stage is terrain recognition, including load points, consumption estimations, and resources assessment. The second stage is the optimum RET technologies configuration for each load point or set of load points. The third stage is the optimum technologies allocation, in individual and centralized systems, considering the distribution network layout and voltage level (these are described in Fig. 3.1). We realized all the previously described stages could be accomplished using together OpenWind (and/or WAsP), HOMER, and ViPOR in the first, second, and third stages, respectively. Fig. 3.1 shows the planning stages of the method based on the optimization of the GCC configuration in centralized and individual systems. In the next sections, a general description of our methodology is presented and the particular description of the stages' implementation.

Every electrification plan of a rural community, within a well-defined territory inside a country or region, must be framed in a master plan or energy policy assumed by the authority in rural electrification that executes the project. In this sense, it is not possible to elaborate an electrification plan of a community without starting from the premises that are given by the general master plan or the ruling rural electrification policies. It is not convenient to change the premises for each community electrification project that takes place in a specific country. The premises should include:

- Energy access threshold, that is the average daily minimum consumption to which a home is considered to have an effective access to electricity with favorable consequences for its sustainable development.
- Technologies that are available or that can be accessed for the electrification of the community. This is PV-panels, SWTs, batteries, gensets, and inverters, considering their available sizes, electrical capacities, and characteristics.

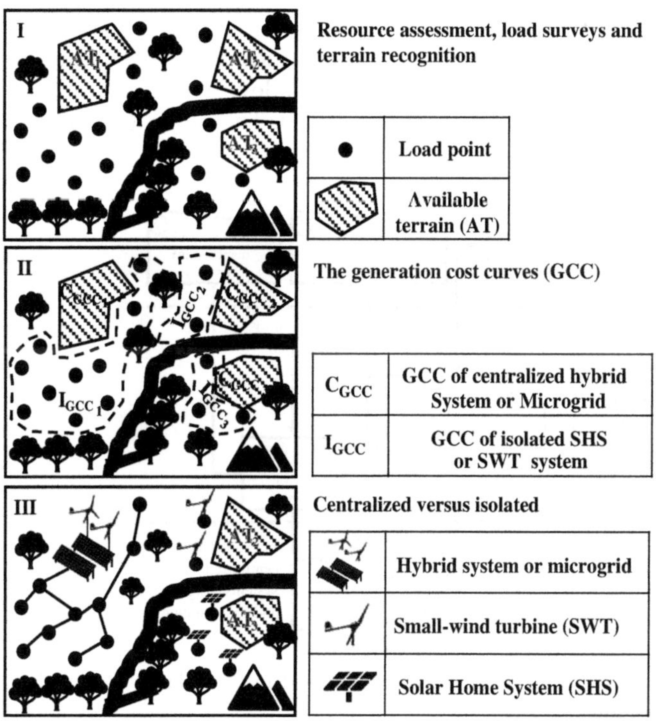

FIGURE 3.1 Planning stages for the generation cost curve optimization method.

- Maximum level of emissions per kWh generated, in cases where diesel gensets are used.
- Available terrains (AT) in the community for the installation of the generation systems of one or several possible hybrid microgrids. Maximum dimensions of the generation centers, in the case of hybrid rural microgrids.

Considering the premises, the first step is *resource assessment, load surveys, and terrain recognition* (I). Directly (measurement campaign) or indirectly, by means of software based on solar incidence algorithms and/or satellite data in the case of solar energy and based on the MCP method, the estimation of available resources is made. It is possible to obtain them through an intermediate solution, in which MCP uses values from a short-term wind measurement campaign (3 months). In this stage, a visit to the community is carried out and the previously recognized AT are measured (AT), in this way their accessibility and physical conditions are determined in order to be able to serve as the location of the center of generation of a hybrid microgrid. The geographic coordinates of the houses and all the load points are lifted and possible difficulties and facilities for the distribution system of a possible microgrid are detected. Through surveys the number of people per house and its dimensions are determined, in order to have an indirect estimate of the energy needs, based on previously defined access thresholds. In Fig. 3.2, numbered as 1, this first stage is shown as a manual input to the methodology implementation. The purpose is to obtain the solar and wind resources assessment. This way, stage I is completed.

The second stage consists in obtaining the GCC for the hybrid and centralized source types or generation systems, in each AT (centralized GCC; CGCC) and for each load point, in isolated source types (isolated GCC; IGCC) for each of the houses. The procedure consists of specifying a GCC for each CGCC you define. Conceptually, a GCC provides the information of the configuration composed by RETs (wind, solar, and/or hydro), batteries, inverters, and backup diesel of lower cost for each electric load level, in a determined point with a known availability of renewable energy sources (RES). Therefore it requires a

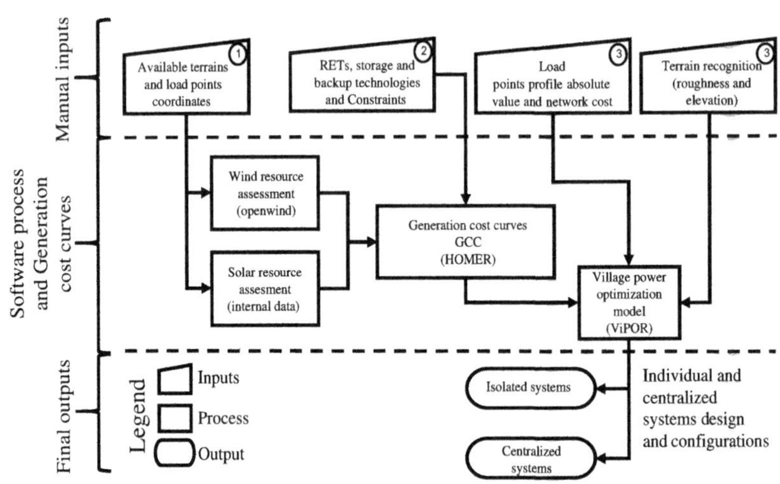

FIGURE 3.2 Community-scale rural electrification generation cost curve-based method algorithm. Numbers show the corresponding stage in the manual inputs.

highly computational power, so this cost information can be created using HOMER. The procedure for creating GCC is slightly different for isolated and centralized source types, because it is not usual to use backup systems in individual systems. It is possible that for a set of load points, very close to each other, the I_{GCC} can be virtually the same, thus simplifying the necessary calculation requirements. In Fig. 3.2, numbered as 2, the second stage needs the available RETs technical data as manual input to the GCC assessment. The purpose is to obtain the GCC from solar and wind resources assessment considering the allocated RET technoeconomic performance. This way, stage II is completed.

Finally, in the third stage, the different IGCC of each load point and of each CGCC centralized point are optimized, through a linear optimization procedure, to obtain the lowest cost configuration, considering isolated and centralized configurations, which satisfy the requirements of loading all the points. This is achieved by accounting for the previously defined design premises in the energy policies framework of the rural electrification master plan, within which this project is framed. However, different data from terrain are now needed. In Fig. 3.2, numbered as 3, roughness and elevation data from the terrain are needed as manual inputs to this final analysis. In this way, the *simulation results: centralized grid and isolated* are obtained.

3.1.2 Methodology

In this section, the methodological procedure proposed for the evaluation of solar and wind energy resources is described. Furthermore, the terrain recognition objectives and the load surveys procedure are also explained. The descriptions in this section refer to manual inputs 1 and 3 in Fig. 3.2.

3.1.2.1 Resource assessment

Regarding solar energy, satellite data collection, such as Global Solar Atlas by World Bank Group (Pillot et al., 2013) and the Global Solar Dataset by IRENA (2018), provides average annual global horizontal irradiance (GHI) considering it constant at a 7 and 3 km spatial resolution, respectively. GHI space resolution would be usually available over approximately 3–7 km depending on the latitude, which could be enhanced by downscaling to a nominal resolution of approximately 1 km, using numerical methods. Therefore, for community electrification projects within an area below 7 km, the same GHI could be considered using satellite data. Detailed information about clear-sky spectral irradiances and how atmospheric changes affect the distribution of photon energy for each wavelength of light could be precisely estimated using the Simple Model of the Atmospheric Radiative Transfer of Sunshine, or SMARTS, developed by Gueymard (NREL, 2013). Therefore the solar resources assessment can be based on satellite open access data provided that the area extension of the community is below 7 km. More specific adjustment of the solar panels to the radiative spectrum of the place can be achieved using programs like SMARTS.

Regarding wind energy, onsite measurement campaigns on long timescales are often impractical and prohibitively expensive, considering community microgrid projects have comparatively lower investment costs than larger wind farms that widely justify a long and expensive measurement campaign. In these cases, indirect methods are needed to obtain a reliable, long-term, time data series to perform an adequately precise wind resource assessment (Weekes and Tomlin, 2014). In our case, a MCP analyses are considered, using the software OpenWind developed by AWS Truepower. The MCP strategy involves a previous correlation assessment between the target site and concurrent measurements at a local long-term reference site, such as an airport or meteorological station, and the prediction of the long-term wind resource at the target site using long-term historical data from the reference site (Weekes and Tomlin, 2014). This is achieved through a Mesoscale Atmospheric Simulation System (MASS), which is a mesoscale numerical weather prediction model (NWP) that can simulate and capture a broad range of meteorological phenomena from microscale phenomena. Therefore a high spatial resolution is achieved employing substantial computing power and linear "microscale" wind flow models. We achieved a 10-m grid resolution for average wind speed and direction distribution in "La Macolla" for the small wind turbines (SWT) annual energy production (AEP). This is provided in the form of a wind resource grid (WRG), a table of regularly spaced parameters, through OpenWind.

Additional wind resource information was obtained in tabular files (TAB) representing measurements at one or more "virtual met masts" (VMM), corresponding to each of the SWT locations (coordinates). TAB files, based on VMM, are made from a typical meteorological year (TY) containing a sample of 365 days' hourly data. Therefore TAB files, and its TY-VMM data series, corresponding to our case study location yield much more accurate results for energy production than Weibull modeled distributions. The TY-VMM time series data is derived from global reanalysis data from three public sources, MERRA, ERAI, and CFSR, after the NWP method application. These public data sources cover anywhere in the world spanning 1979 to the present. All of the previously explained procedures could be similarly carried out using WAsP (DTU Wind Energy, 2018). The previously described process is completely included in the manual input block, numbered as 1 in Fig. 3.2.

3.1.2.2 Load surveys

Through surveys, the number of people per house and its dimensions are quantified, to have an indirect estimate of the energy needs, based on previously defined access thresholds or those included in national energy policies or rural electrification master plans. Regarding energy access thresholds, the International Energy Agency (IEA) considers electricity access includes a household having an electricity supply connection with a minimum level of consumption of 250 kilowatt-hours (kWh) per year for a rural household and 500 kWh for an urban household, which increases over time to reach the national average (IEA, 2017). For example, in Venezuela, national government policies consider 730 kWh per year for 4–6 persons in a rural household (Fundelec, 2012). Surveys will adjust the estimation considering the number of persons per house below or above the estimated standard range.

The following case study has considered the estimations from energy policies made by the Venezuelan government. The base estimate must take into account an interannual increase in demand equal to or greater than the estimated annual growth of the population for that region or for the rural areas of the country, over a 25-year lifetime period (IEA,

2017). The previously described process is completely included in the manual input block, numbered as 3 in Fig. 3.2.

3.1.2.3 *Terrain recognition*

On the ground, the geographic coordinates of the houses and all the loads must be pointed out, as well as detecting the possible difficulties and facilities for the electrical distribution system (dummy points) of a possible microgrid. Regarding resources, considering GHI as constant all over the terrain, only obstacles that will affect it by creating shadows and interferences should be noted during terrain recognition. On the other hand, it is known that the power curve of a wind turbine depends upon a large number of meteorological and topographic parameters, and among them wind shear is one of the most important parameters that influence the uncertainty of the power curve measurement (Wagner et al., 2009). Digital topographic maps need to be checked during terrain recognition. Other useful tools are open access and official digital maps; all of them must be checked to complete the terrain recognition, avoiding negative impacts on energy production. The effect of turbulence intensity on the SWT performance is even more complicated and only partly understood. It depends on the aerodynamics and the inertia of the rotor and SWT are more sensitive to them than large wind turbines. To avoid turbulence, no obstacles must be near the SWT and locations next to forest and/or mountains could be susceptible to turbulence. Therefore when no other alternative energy source is available in a load point where many obstacles prevail, it could be better to make a short measurement campaign (3–6 months). Terrain recognition and load surveys are carried out simultaneously. The previously described process is completely included in number 1 and 3 manual input blocks in Fig. 3.2.

3.1.3 The generation cost curves

Different procedures for obtaining GCC are carried out for each generation system type, that is, isolated (IGCC) or centralized (CGCC). Defining expected load shape is the first step in GCC calculation. However, the load profile shape specified for the individual load points is different than that for centralized system types. The first reason for doing that is that customers supplied with a centralized grid are usually expected to use more energy because the supply is less limited. Second, the aggregation of many individual loads tends to considerably smooth the load profile. Finally, some community loads together with the residential ones makes change noticeable. GCC, both for individual and centralized, must be obtained over a sensitivity load range into the search space.

Component cost and performance data information (constraints) are uploaded in HOMER V2.67 and it is slightly different for isolated and centralized generation system types. Centralized systems are much larger than the isolated systems, therefore components are likely to be different. For example, battery strings and capacity configuration considered for centralized systems are likely to be larger than those for the isolated ones, and no more than one SWT is usually installed in isolated systems. Finally, search space must be appropriate for the entire range of sizes you are considering for the loads and the available components (RETs, storage, and backup technologies) should be able to supply all the load sizes considered over the range and its costs.

Fig. 3.3 shows the general structure of a GCC both for individual or centralized generation system and the search space results. Search space involves several PV panel sizes, a few battery bank sizes, and a few inverter sizes. The search space may also contain two or more types of wind turbines and diesel sizes. Each point of the GCC is related to the solution or configuration of elements of generation, storage, and optimal backup for a specific load (x axis), individual, or centralized for a set of community load points. The primary y axis corresponds to cost of energy (COE, \$/kWh) for the optimum solution for each load point and the secondary axis is related to their net present value (NPV).

Therefore optimum results for each load level were previously obtained in HOMER (optimization results) and allocated for its corresponding load point in the GCC. Search space may be broken into two or more intervals as the number of aggregated loads increases.

3.1.3.1 *Centralized generation cost curves*

For centralized GCC (CGCC), hybrid generation (Wind–PV), storage using batteries strings, and backup generators using diesel generators are considered in all the possible configurations. Defining expected load shape starts by making the load profile shape by considering all load types in the village and all the load points (sensitivity maximum load in Fig. 3.2). Then, the sensitive range must be defined by considering the lowest average valued load type as the minimum scale range. Therefore the sensitivity analysis range is performed between the sensitivity minimum load point unit to the entire village load taken together. Community-scale solar data values are considered for all the centralized systems $CGCC_{1-n}$ in $AT_{1\text{-}n}$ to be the same as those specified for all the centralized or isolated source type. The wind data may be different if the proposed site of the centralized system has a different wind regime than the load points themselves, due to topographic and elevation considerations. According to the NREL, it is a good idea to specify one small value (1%–5% of the largest value) to account for the fixed cost of the centralized power system.

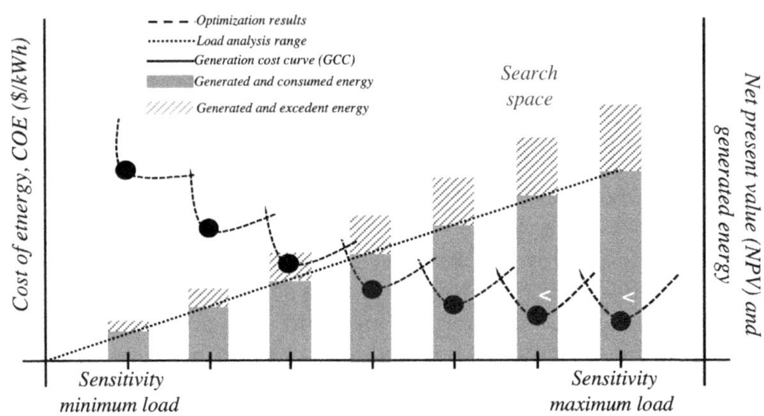

FIGURE 3.3 Description of the generation cost curve and the search space.

3.1.3.2 Isolated generation cost curves

For isolated GCC (IGCC), hybrid generation could be considered, but only PV alone or SWT alone are usually preferable with battery strings storage systems. No backup generators using diesel generators are considered for isolated or individual systems. Defining expected load shape starts by making the load profile shape by considering the minimum individual value for load points (sensitivity by minimum load in Fig. 3.2).

Then, the sensitive range must be defined by considering the highest average valued load type as the maximum scale range. Therefore the sensitivity analysis range is performed between the sensitivity minimum individual load point unit and the maximum load allocated for an individual load point in the community. Solar data values are considered for all the individual systems $I_{GCC,1\text{-}n}$ in load points to be the same as those specified for all the centralized or isolated source type. The wind data may be different if the proposed site has a different wind regime due to topographic and elevation considerations. Elsewhere, wind turbines would presumably be placed very near the load points, so wind resources cannot be optimally allocated as their location is always next to the load point. This is not the case with CGCC, in which SWTs may be placed some distance from the load points to capitalize on a better wind resource.

3.1.3.3 Simulation results: centralized grid and isolated systems

In this final stage, the optimal configuration of the distribution network is considered in centralized systems. Lines of both medium and low voltage are evaluated, considering the maximum and/or minimum extensions according to the design criteria of each project and/or program. In the first place, it is defined where the individual and where centralized systems will be used and then the configuration of the distribution network in both medium and low voltage is added to the analysis. At the end, points are obtained with individual systems, points where to locate centralized systems and a distribution network for both medium and low voltage. There are no fixed design criteria for all cases, these are defined according to the promoter of the program and/or project. In this final stage we use the ViPOR to get the results based on a linear optimization that makes this program. The procedure is described in detail in the following paragraphs.

A village power optimization model is used to choose the optimum configuration within the search space containing all the GCC for centralized and isolated systems in the target community (one, two, or more). In this case, ViPOR software, developed by NREL, is used to decide which houses should be powered by isolated systems (SWT or SHS) and which should be included in the RHMG (Bopp and Lippkau, 2008). Therefore some other data is needed, particularly with reference to the electrical distribution systems' operational and maintenance costs, considering medium- and low-voltage power lines, distribution transformers, power electronic devices, connectors, and all the initial investment needs, and consideration of local terrain and some data about the local geography is also needed. ViPOR requires a geographical description of the local terrain including the load points such as houses, schools, stores, and health posts.

Therefore terrain is considered as a distribution network cost multiplying factor, so a completely flat terrain, without a slope will have a multiplication factor of one. This is a manual input numbered as 3 in Fig. 3.2. As far as the terrain becomes more irregular, with taller vegetation and/or irregular and inclined planes, the multiplication factor of the distribution costs will be greater. According to Bopp and Lippkau (2008), ViPOR is the only readily available tool for evaluating trade-offs between SHS or centralized PV minigrids. It can be very useful for designing village electrification systems, but it is necessary to be skilled in village electrification, especially for setup costs, and good information about local conditions is needed for good results.

3.1.4 Case Study

A case study corresponding to one community in Venezuela is described. La Macolla is a small community of 12 houses scattered in an estimated total area of 1 km^2 (95 ha), on the northeastern coast of the Paraguaná peninsula (Falcón State). Then distance between the first and the last house of the community is 1.5 km. The community is located at sea level, on the Caribbean coast (0 m.a.s.l). The annual average temperature is 26.72°C and the average temperature of the warmest month is 27.65°C in September, and the minimum is 25.66°C in December. According to the Köppen classification (Rubel and Kottek, 2010), the climate in La Macolla is category *Bwh* (Hot desert), where precipitation is too low so vegetation is at most a very scanty shrub. In the case study, houses have been electrified since 2012. This rural electrification project was executed by the government of Venezuela, using RET-based generation systems. These systems were installed based on an standardized method which stablished four RHMG configurations, corresponding to villages composed of 10, 20, 30, and 40 houses (López-González et al., 2017a,b). Therefore La Macolla's rural electrification project had to be designed in 2012 accordingly and adapted to the previously designed corresponding RHGM and sized HWT (1.5 kW for HWT) and PV modules (160 Wp per module).

Fig. 3.4 shows the actual electrification system in La Macolla which is composed of one RHMG and a low-voltage grid feeding eight houses and one school (Fig. 3.4A). The low-voltage microgrid is 898 m long with no transformers. Therefore the low-voltage distribution network investment cost was only 3% of the total initial investment. The remaining

FIGURE 3.4 Community of "La Macolla" (Falcón, Venezuela). (A) Rural hybrid wind–PV–diesel–batteries microgrid and its distribution network. (B) Small wind turbine at the end of the microgrid distribution network. (C) Solar PV modules in the community fishing center of "La Macolla."

TABLE 3.1 Centralized generation microgrids configuration.

	Solar PV generation		Small wind generation		Battery storage		Diesel backup
La Macolla	Quantity	Total capacity	Quantity	Total capacity	Quantity	Total capacity	Total capacity
Hybrid microgrid (1)	36 × 160 W	5.76 kW	1 × 6 kW	6 kW	48 × 800 Ah	38,400 Ah	14 kW
Home wind turbine (4)	–	–	4 × 1.5 kW	6 kW	12 × 1080 Ah	12,960 Ah	–
Hybrid individual (1)	24 × 160 W	3.84 kW	1 × 3 kW	3 kW	24 × 1080 Ah	25,920 Ah	–

four houses are individually electrified using 1.5 kW HWT (Fig. 3.4B). An individual hybrid system was built to support a community fishing center using solar PV modules and a 3-kW wind turbine (Fig. 3.4C). In total, four HWT of 1.5 kW were installed, along with a RHMG (wind–solar–diesel–batteries), and an individual hybrid system for the community fishing center. COE ($/kWh) in La Macolla is 1.832 $/kWh due to oversized RETs in microgrid and HWT individual systems (40% of total initial investment cost). Table 3.1 shows the installed systems.

During visits and surveys in La Macolla, it became evident that low-voltage distribution network (LV lines) failures are costly and prolonged, due to the rural and scattered location of the community. These failures are usually sectorized, only affecting pasrt of the community. Failure consequences can be further prolonged because not all members of the community are willing to contribute financially to cover the costs of repair, only the affected houses support the partially damaged grid. Another important issue is when meteorological conditions imply a longer use of the diesel backup system. In these cases, the difficulty to move to the nearest city to buy gasoil and transport it to their community implies prolongation of the service interruptions.

The purpose of the case study is to apply our methodology and compare the results with those observed on site, drawing conclusions about the differences in results. Our design is based on the same RET standardized sizes, stablished by Fundelec, as described in the previous paragraph and referenced literature (López-González et al., 2017a,b). In the following three sections, the proposed stages for the project planning methodology are explained and applied to the case study in La Macolla. First, the resource assessment, load surveys, and terrain recognition were implemented during visits to this community in Falcon State in northwestern Venezuela. Secondly, the GCC were generated and through ViPOR the simulation results were achieved. These steps are detailed in each section.

3.1.5 Results and discussion

3.1.5.1 Resource assessment, load surveys, and terrain recognition

First, a load and topographic survey of the location of each house in the community is carried out, using a GPS device. During the load survey, the average daily consumption per household is estimated (kWh/d) and compared with the previous estimations made by the Foundation for the Electrical Development of Venezuela (Fundelec). Fundelec estimated daily consumption per household to be about 2 kWh/d (Fundelec, 2012). However, surveys show the daily average consumption of five homes in La Macolla is 1.39 kWh/d.

In addition, there is a school and a community fishing center which have an average daily consumption of 4 and 6 kWh/d, respectively (MPPEE, 2013). These are named as Load Type 1 and Load Type 2 icons, respectively, which are shown in Fig. 3.5. Additionally, it shows where the productive activities are concentrated, which in this case refer to goat breeding and fishing. This is an important consideration after obtaining the mathematical results of the model since modifications can always be made to prioritize or expand the generation capacity according to the perspectives of productive growth in a sector.

Secondly, through the terrain recognition, three AT are identified as possible locations for centralized generation systems through one or more RHMG. Terrains are selected according to their proximity to the optimum point to serve the loads considering the availability restrictions of them and their topographic conditions. The three AT identified during the visit are shown in Fig. 3.5: AT_0, AT_1, and AT_2. Using satellite images, a 25×25-m square terrain type allocation is carried out. Four terrain types are usually enough: water (blue) is extremely expensive; the area along the beach or swamp (dark brown) is moderately expensive; forest (green) is less expensive; and, finally, grass or dry land (light brown) is the least expensive at all. Different terrain types than these could be found in different projects, however, four cost multipliers factors are enough. In this case, water is 5, swamp is 3, forest 2, and grass or dry land is 1.

Finally, a renewable resources assessment regarding wind and solar energy was made. Regarding wind energy resources, OpenWind software outputs were used. As for the solar resource, considering that the territory is small enough to have the same weather conditions during the year, the same solar resources are considered for the three ATs. In this

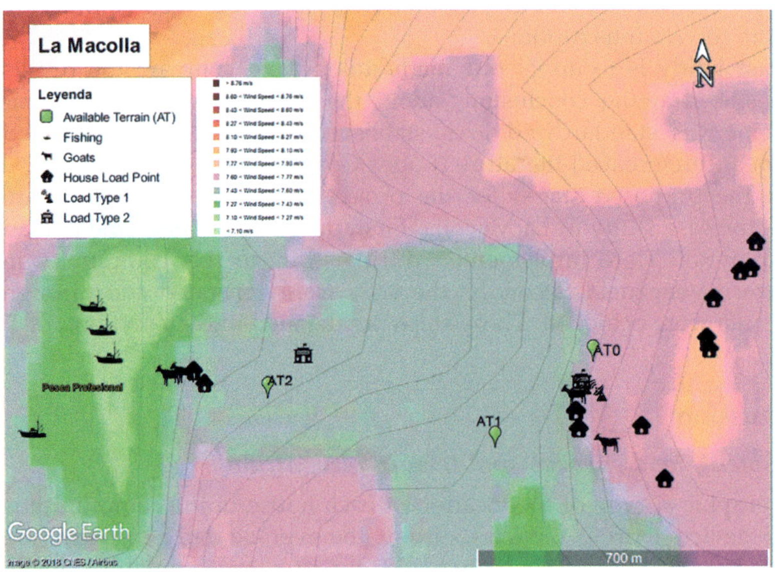

FIGURE 3.5 Wind resource grid, elevation isolines and small wind turbines allocation in "La Macolla" (Paraguaná, Falcón). Figure composed by OpenWind outputs and Google Earth satellite images.

case, the radiation, according to IRENA data bases is 6.06 kWh/m^2/d and the clarity index is 0.667 (IRENA, 2017). As for the wind potential, it varies very slightly between the three points from 7.43 to 7.77 m/s. This is because there is not much difference in the elevation or in the roughness factors between these three AT points, as can be seen in Fig. 3.5.

3.1.5.2 The generation cost curves

Considering both wind and solar generation, HOMER v2.67 outputs provide the GCC corresponding to the centralized generation for ATs and possible isolated domestic systems per house. In the case of centralized generation, battery storage and diesel backup are considered, while for individual or domestic systems only battery storage is considered. In Fig. 3.6, the GCC is shown for all ATs, taking into account that wind and solar resources are similar in all the ATs. COE ($/kWh) performs a potential function that decreases when increasing the average daily demand for electricity and increases when increasing the initial costs.

From total daily average demands of 30–60 kWh/d (approximately 30 houses), centralized systems are cheaper using wind turbines; for higher consumption (>60 kWh) a hybrid system is better (wind–PV–diesel–batteries). In all cases, a diesel backup system is required to supply energy gaps between demand and available energy resources due to the daily load profile. In all the wind turbines design cases, considering the constraints regarding capacity shortages, there are surpluses of electrical generation from this RET, which are dissipated by electrical resistance.

In Fig. 3.7 the GCC for the individual systems is shown, taking into account that the wind and solar resources in each of the points are similar. COE ($/kWh) decreases when the average daily demand for electricity increases and COE increases when the initial costs increase. Up to 4.5 kWh/d, the appropriate solution for individual domestic electrification

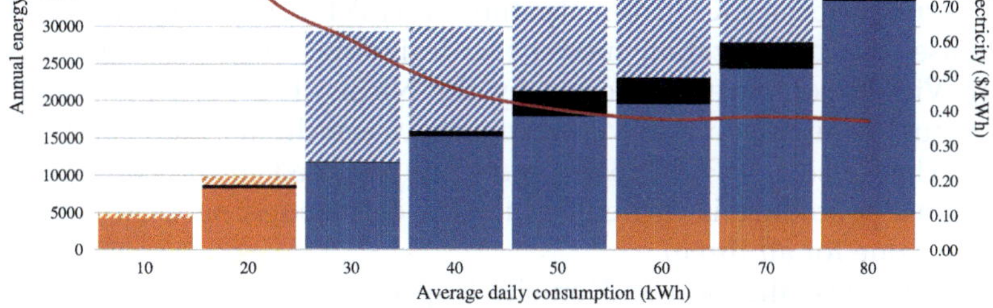

FIGURE 3.6 Generation cost curve based on the cost of electricity ($/kWh) for all the three available terrains in "La Macolla" and the total amount of electricity generation for each of the average daily demands between 10 and 80 kWh/d (5–40 houses).

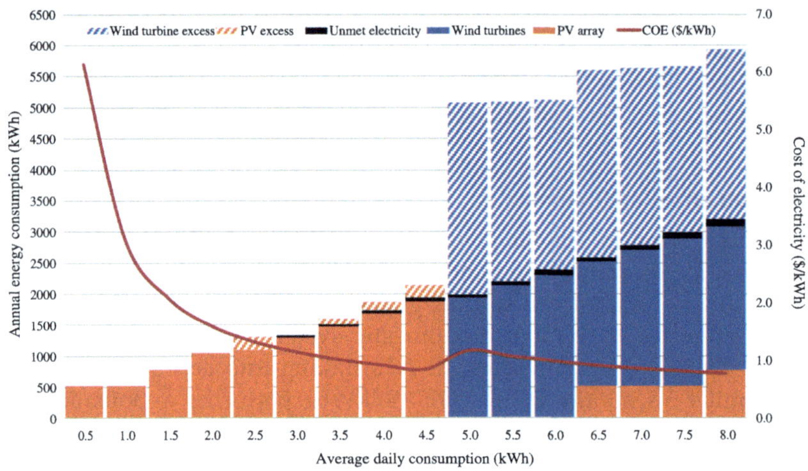

FIGURE 3.7 Generation cost curve based on the cost of electricity ($/kWh) for all of the houses load points and the total amount of electricity generation, considering and average daily demand between 0.5 and 8 kWh/d per house (25% to 400% or 1/4 to 4 times the 2 kWh typically estimated).

is solar panels, however, between average daily demands of 5 and 6.5 kWh/d, HWT are the lowest cost solution. For larger demands (>6.5 kWh/d), a hybrid system (wind–PV–batteries) is better. In all cases, storage in batteries is required and the lags between demand and available energy resources due to the load daily profile characteristics cause interruptions in the electricity supply that never exceed 5% of the total AEP. In all cases where wind turbines are applied, when considering the electrical capacity shortages constraints, there are surpluses of electrical generation that are dissipated using electrical resistance.

In all cases, the lowest battery SOC (%) occurs between 6:00 p.m. and 10:00 p.m., when the centralized generation system could eventually require the backup genset starts up. The months of lower availability of wind and solar resources are October and November. Therefore during these 2 months the probability of the use of backup gensets increases and average values of SOC (%) are lower than the annual average values.

3.1.5.3 Simulation results: centralized grid and isolated

Using the corresponding GCC from the three AT: AT_0, AT_1, AT_2, and GCC from four individual houses, an optimization model for designing village electrification system is applied using ViPOR software developed by the NREL. The geographic coordinates are used for the visual comparison and verification between digital maps and real placement of the houses. Then, using ViPOR, load points are assigned to the corresponding houses location. The same procedure is carried out for the corresponding GCC curve for each AT (in this case the same for all three).

Based on satellite visualization and on-site inspection, a color is assigned to each type of terrain in 25 × 25-m squares, as shown in the methodology. In Fig. 3.7 La Macolla community land is shown as it was loaded in ViPOR. Then, considering technical and economic restrictions in the construction of medium- and low-voltage lines, as well as distribution transformers, according to the costs considered, the configuration shown in Fig. 3.7 is obtained. This configuration is composed of five pole-mounted distribution transformers, medium- and low-voltage lines. There are 987 m of medium-voltage power distribution lines (MV) and 448 m of low-voltage power distribution lines (LV). (Fig. 3.8).

The results shown in Table 3.2 differ from what has been implemented in the community. Instead of a hybrid microgrid that feeds eight houses and four individual systems with HWT for four houses, the optimal solution, according to our methodology, would have been a wind–diesel microgrid for 10 houses and the school and another for the fishing center and two homes. This implies a reduction of 64.7% in RETs initial investment costs and an increase in 276% in power distribution network initial investment cost. Globally, our proposed methodology provides a configuration which requires a 54.7% lower investment cost than Fundelec's previously applied solution.

3.1.5.4 General discussion

The comparison among the previously made project and the GCC method proposed is fairly made considering the same stock or standardized technologies for HOMER simulations and the GCC results. Therefore the differences found cannot be assigned to RETs

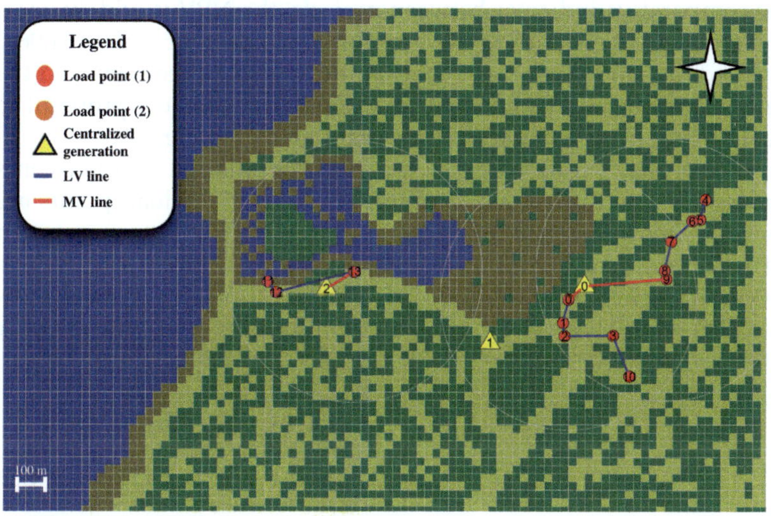

FIGURE 3.8 Wind resource grid, elevation isolines, and small wind turbines allocation in "La Macolla" (Paraguaná, Falcón). Figure composed by OpenWind outputs and Google Earth satellite images.

TABLE 3.2 Selected terrains are AT_0 and AT_2.

Type of system	Placement	Load type	Consumption (kWh/d)	Solar PV	Microwind	Diesel	Batteries
Wind–diesel–batteries	AT_0	10 houses; 1 load type 1	24 kWh/d	–	1 × 6 kW	1 × 9 kW	24 × 800 A-h
PV–diesel–batteries	AT_2	2 houses; 1 load type 2	10 kWh/d	3 kW	–	1 × 9 kW	24 × 800 A-h

choices nor even to terrain types, because it has been also considered according to a real in situ inspection. Differences found are related to cost optimization under common technical constraints for both projects. The main differences come due to the grouping of load points in microgrids being privileged before the individual systems and this is basically due to the costs. It is important to realize that the previously made project was designed under a national electrification program using standardized configurations according to the number of houses to be electrified (López-González et al., 2017a,b). Thus no cost optimization was possible at a local level and that is why NPV is 277% higher than the proposed GCC method project. However, these economic aspects have technical implications, thus the technical and economic results are explained and discussed in this section.

Regarding the *economic* analysis, the GCC-based method favors the grouping of load points, because of the low comparative costs of the low- and medium-voltage distribution network (LV and MV, respectively), in comparison to electrical generation RETs costs. Therefore the GCC method results increase the distribution network costs by 276% in comparison to the 2012 project. However, the proposed total initial investment for the GCC-based method is 45.3% lower. This is due to the distribution network costs representing barely 3% in the project implemented in 2012, but 24.5% in the proposal of the GCC method. The largest extension of the electrical distribution network is the determining factor in the O&M annual cost increase, which rises by 23.86% with respect to the 2012 project. As a global economic comparative indicator between projects, the NPV is widely used. In this case, even though the O&M costs are higher, the NPV of the proposal is 64.06% less than the design applied in 2012 and the COE parameter is reduced by 70.14%. Therefore, from a merely economic point of view, the GCC-based method proposal is more efficient. The presentation of these comparative results is shown in Fig. 3.9.

Regarding the technical analysis, the topology of the electrical distribution networks of both projects (previously made and proposed) and the locations of the load points are shown in Fig. 3.10. Thus both centralized and individual systems of electricity generation are shown. When considering the real topography and elevation of terrain, the results are quite like the previously implemented project, in terms of the topology of the electrical distribution network. Therefore the GCC-based method results favor the quicker

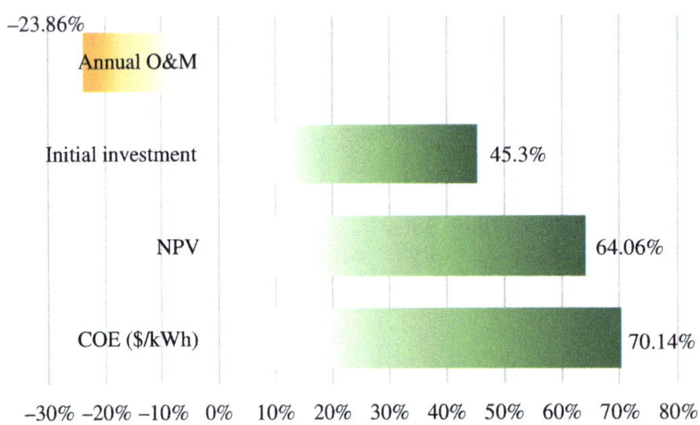

FIGURE 3.9 Savings in initial investment costs, O&M, net present value, and Cost Of Energy, COE ($/kWh) of the generation cost curve-based model, with respect to the actual implementation of the project in "La Macolla."

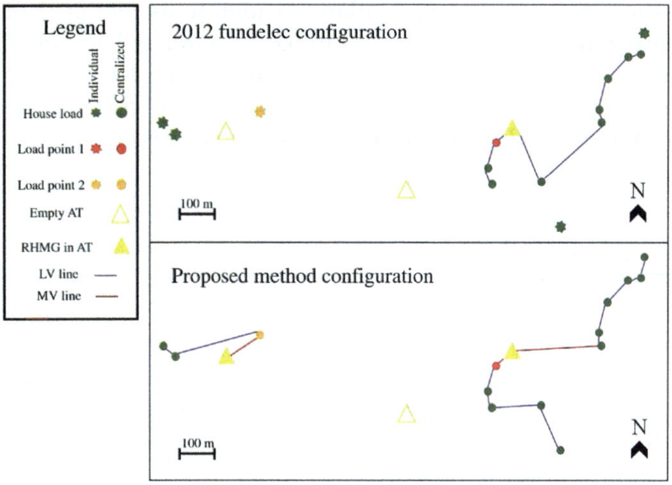

FIGURE 3.10 Wind resource grid, elevation isolines, and small wind turbines allocation in "La Macolla" (Paraguaná, Falcón).

obtaining of a much more realistic result than other computer-based methods. Due to the previously explained reasons, the GCC-based method privileges the centralized systems before the individual ones, sacrificing savings in the electrical distribution network length.

Based on the obtained GCC, for both individual and centralized generation (Figs. 3.6 and 3.7), a common trend can be commonly observed. As for centralized generation, for groups of more than 10 houses and/or total consumption of more than 20 kWh, the use of wind turbines (with storage in batteries and diesel backup) or hybrid systems (wind–solar) is much more effective than photovoltaic solar systems only with battery storage and diesel backup (Fig. 3.6). As for generation for individual load points, this happens from a daily average consumption higher than 5 kWh/d (Fig. 3.7). This does not have to do with the resources availability, since in this territory there is a huge availability of both wind and solar resources (IRENA, 2017). Similar results have been obtained by Hosseinalizadeh et al. (2017).

By raising the voltage level, up to medium voltage (MV) values, in the first sections of the distribution line it is possible to reach the extreme load points that were previously fed by HWT individual systems in 2012, as shown in Fig. 3.10. Therefore the proposed electricity network has 1438 m while the current one is 898 m, that is, there is an increase of 59.8% in the extension of the distribution network. This has very little relevance from an economic point of view, however, its importance increases in terms of the quality of the electric service. In this type of project, there are usually delays in preventive maintenance and many delays in corrective actions. Therefore longer lines will increase failures, this is due to faults in the electrical distribution lines being proportional to length (IEEE, 2012). This way, individual systems would be more reliable than centralized ones, when no electrical network maintenance is made by the community.

Finally, the GCC-based method shows wind–diesel and PV–diesel as economically better choices than hybrid ones. However, other considerations, such as the difficult access to diesel, are not quantifiable issues in an objective way to be included in this model.

Therefore these and some other issues are related to particular considerations that are not technical or directly economic. This way, there is a nonquantified cost related to social adaptation of the project that, in this case, could be as much as 64,6% of the NPV of the project. This means that it is possible to conclude that the social adaptation cost for the project could be higher than 50% of the total NPV of a rural electrification project.

3.2 Conclusion

The main achievement of this work has been to show the basic structure of the rural electrification optimization algorithm, through an easily understandable method in three sharply defined stages. A simplified algorithm for RET-based rural electrification has been proposed and applied, employing widely available and known software. Beginning from some easily found data, this GCC-based method algorithm provides a technically and economically optimized solution for community-scale projects.

The GCC can be obtained from many methods other than using HOMER and, later, the optimal GCC allocation according to load points is a simpler exercise. This GCC allocation could be carried out using linear optimization software where the distribution network layout is constrained by terrain types, previously known during in situ inspection. Other software than ViPOR could be used for linear optimization. Therefore this is a method where real data and software simulations are used together to reach a terrain-adapted solution. This method is easy to share and apply in developed and developing countries, by just understanding the simplified algorithm and applying the clearly shaped procedure stages.

The GCC has been proven to be an effective function to achieve a global optimum at the community scale, using RET configurations for the local optimum for the neighborhood. The GCC-based method considers the local optimum corresponding to each of the RETs configurations for the individual or clustered load points and its optimum allocation in a community-scale optimum (global optimum). Considering the GCC-based method, many other issues could be included taking account of social constraints. However, all these social or other qualitative issues are usually so locally centered that no general model reaches universally accepted results. Therefore whatever the users are looking for, the GCC-based method provides the basic technoeconomic optimum results to start from a realistic base. Modifications on the different stages could be used to reach better results according to circumstances.

Acknowledgments

This chapter was funded by the Spanish Ministry of Science and Innovation under the project title: "Optimization of Micro networks with Renewable Energies under Uncertainty and Future Network Integration," RTI2018–097962-B-I00. This research was cofinanced by the Centre for Cooperation Development of the Universitat Politècnica de Catalunya-Barcelona TECH under the project title "Development of tools for the evaluation of energy projects", 2019-B014. Some wind data was kindly provided by UL-AWS TruePower. This work has been possible thanks to the kind collaboration of the energy resources evaluation engineers of the Department of Alternative Energy Sources of Corpoelec and the State Directorate of the Venezuelan Ministry of Electricity in Maracaibo, Zulia State.

References

Ashok, S., 2007. Optimised model for community-based hybrid energy system. Renew. Energy 32 (7), 1155–1164.

Bala, B.K.K., Siddique, S.A., 2009. Optimal design of a PV-diesel hybrid system for electrification of an isolated island-Sandwip in Bangladesh using genetic algorithm. Energy Sust. Dev. 13 (3), 137–142. Available from: https://doi.org/10.1016/j.esd.2009.07.002.

Baños, R., et al., 2011. "Optimization methods applied to renewable and sustainable energy: a review.". Renew. Sust. Energy Rev. 15 (4), 1753–1766.

Bhattacharyya, S.C., 2012. Review of alternative methodologies for analysing off-grid electricity supply. Renew. Sust. Energy Rev. 16 (1), 677–694.

Bopp, G., Lippkau, A. 2008. World-wide overview about design and simulation tools for hybrid PV systems. IEA PVPS Task 11. Report IEA-PVPS T11-01:2011.

Cader, C., Blechinger, P., Bertheau, P., 2016. Electrification planning with focus on hybrid mini-grids—a comprehensive modelling approach for the global south. Energy Procedia 99 (March), 269–276. Available from: https://doi.org/10.1016/j.egypro.2016.10.116.

Casella, L., 2015. "Performance analysis of the first method for long-term turbulence intensity estimation at potential wind energy sites.". Renew. Energy 74, 106–115. Available from: https://doi.org/10.1016/j.renene.2014.07.031.

Dai, R., Mesbahi, M., 2013. Optimal power generation and load management for off-grid hybrid power systems with renewable sources via mixed-integer programming. Energy Convers. Manag. 73, 234–244. Available from: https://doi.org/10.1016/j.enconman.2013.04.039.

Domenech, B., et al., 2014. A community electrification project: combination of microgrids and household systems fed by wind, PV or micro-hydro energies according to micro-scale resource evaluation and social constraints. Energy Sust. Dev. 23, 275–285. Available from: https://doi.org/10.1016/j.esd.2014.09.007.

Domenech, B., Ferrer-Martí, L., Pastor, R., 2015. Hierarchical methodology to optimize the design of stand-alone electrification systems for rural communities considering technical and social criteria. Renew. Sust. Energy Rev. 51, 182–196. Available from: https://doi.org/10.1016/j.rser.2015.06.017.

DTU Wind Energy, 2018. Wind Energy Industry-Standard Software – WAsP. <http://www.wasp.dk/> (accessed 16.04.18.).

Falcón-Roque, E.J., et al., 2017. Energy planning model with renewable energy using optimization multicriteria techniques for isolated rural communities: Cajamarca Province, Peru. J. Renew. Sust. Energy 9, (6).

Ferrer-Martí, L., Pastor, R., Capó, G.M., Velo, E., 2011. Optimizing microwind rural electrification projects. a case study in Peru. J. Glob. Optim. 50 (1), 127–143.

Ferrer-Martí, L., Domenech, B., García-Villoria, A., Pastor, R., 2013. A MILP model to design hybrid wind-photovoltaic isolated rural electrification projects in developing countries. Eur. J. Oper. Res. 226 (2), 293–300.

Fundelec, 2012. Sembrando Luz: Promoviendo las Energías Renovables en Zonas Aisladas, Indígenas y Fronterizas del país (No. 1), Energía que transforma tu vida. Caracas. Ministerio del Poder Popular para la Energía Eléctrica (MPPEE).

González-Longatt, F., González, J.S., Payán, M.B., Santos, J.M.R., 2014. Wind-resource atlas of Venezuela based on on-site anemometry observation. Renew. Sust. Energy Rev. 39, 898–911. Available from: https://doi.org/10.1016/j.rser.2014.07.172.

Hosseinalizadeh, R., Rafiei, E. sadat, Alavijeh, A.S., Ghaderi, S.F., 2017. Economic analysis of small wind turbines in residential energy sector in Iran. Sust. Energy Technol. Assess. 20, 58–71. Available from: https://doi.org/10.1016/j.seta.2017.02.018.

Huang, Z., Yu, H., Peng, Z., Zhao, M., 2015. Methods and tools for community energy planning: a review. Renew. Sust. Energy Rev. 42 (4800), 1335–1348. Available from: https://doi.org/10.1016/j.rser.2014.11.042.

IEA, 2017. Energy access outlook 2017: from poverty to prosperity, Paris. <http://www.iea.org/publications/freepublications/publication/WEO2017SpecialReport_EnergyAccessOutlook.pdf>.

IEEE, 2012. In: Society IEEE Power & Energy (Ed.), 1997 IEEE Guide for Electric Power Distribution Reliability Indices (IEEE Std 1366TM-2012). The Institute of Electrical and Electronics Engineers.

IRENA, 2017. Global atlas gallery 3.0. <https://irena.masdar.ac.ae/gallery/#gallery> (accessed 19.06. 17.).

IRENA, 2018. VAISALA global solar dataset provides average annual GHI at a 3km spatial resolution. global atlas gallery 3.0. <https://irena.masdar.ac.ae/gallery/#map/2651> (accessed 13.04. 18.).

López-González, A., Domenech, B., Gómez-Hernández, D., Ferrer-Martí, L., 2017a. Renewable microgrid projects for autonomous small-scale electrification in Andean countries. Renew. Sust. Energy Rev. 79, 1255–1265. Available from: http://linkinghub.elsevier.com/retrieve/pii/S1364032117308407.

López-González, A., Domenech, B., Ferrer-Martí, L., 2017b. Rural electrification with renewable energy and sustainable development in isolated, indigenous and frontier communities of Venezuela. In: Energy for Society (Ed.), 1st International Conference on Energy Research & Social Science. Sitges: Energy Research and Social Science.

Mendes, G., Ioakimidis, C., Ferrão, P., 2011. On the planning and analysis of integrated community energy systems: a review and survey of available tools. Renew. Sust. Energy Rev. 15 (9), 4836–4854.

MPPEE, 2013. Plan de Desarrollo Del Sistema Electrico Nacional 2013–2019. Caracas.

Murthy, K.S.R., Rahi, O.P., 2017. A comprehensive review of wind resource assessment. Renew. Sustain. Energy Rev. 72 (May 2015), 1320–1342. Available from: http://linkinghub.elsevier.com/retrieve/pii/S1364032116306918.

NREL, 2013. NREL: Renewable Resource Data Center—SMARTS. Renewable Resource Data Center. <https://www.nrel.gov/rredc/smarts/> (accessed 16.04.18.).

Pillot, B., et al., 2013. Solar energy potential atlas for planning energy system off-grid electrification in the Republic of Djibouti. Energy Convers. Manag. 69, 131–147. Available from: https://doi.org/10.1016/j.enconman.2013.01.035.

Ranaboldo, M., et al., 2014. Renewable energy projects to electrify rural communities in Cape Verde. Appl. Energy 118, 280–291. Available from: https://doi.org/10.1016/j.apenergy.2013.12.043.

Rubel, F., Kottek, M., 2010. Observed and projected climate shifts 1901-2100 depicted by world maps of the Köppen-Geiger climate classification. Meteorol. Z. 19 (2), 135–141. Available from: http://www.schweizerbart.de/papers/metz/detail/19/74932/Observed_and_projected_climate_shifts_1901_2100_de?af = crossref (accessed 30.05. 17.).

SE4All, 2017. Seizing the energy access dividend. Vienna.

Singh, S., Singh, M., Kaushik, S.C., 2016. Feasibility study of an islanded microgrid in rural area consisting of PV, wind, biomass and battery energy storage system. Energy Convers. Manag. 128, 178–190. Available from: https://doi.org/10.1016/j.enconman.2016.09.046.

United Nations, 2015. A/RES/70/1 transforming our world: the 2030 agenda for sustainable development. <https://sustainabledevelopment.un.org/content/documents/21252030Agendafor Sustainable Development web.pdf>.

Wagner, R., et al., 2009. The influence of the wind speed profile on wind turbine performance measurements. Wind Energy 12 (4), 348–362.

Weekes, S.M., Tomlin, A.S., 2014. Data efficient measure-correlate-predict approaches to wind resource assessment for small-scale wind energy. Renew. Energ. 63, 162–171. Available from: https://doi.org/10.1016/j.renene.2013.08.033.

Zachar, M., Daoutidis, P., 2015. Understanding and predicting the impact of location and load on microgrid design. Energy 90, 1005–1023. Available from: https://doi.org/10.1016/j.energy.2015.08.010.

CHAPTER

4

Proven energy storage system applications for power systems stability and transition issues

Jean Ubertalli[1] *and Timothy Littler*[2]

[1]IEEE PES member, Queen's Belfast University (QUB), Belfast, Northern Ireland [2]Department of Energy, Power and Intelligent Control (EPIC), IEEE and EEECS Research Society, Queen's Belfast University, Belfast, Northern Ireland

4.1 Introduction

Energy storage systems (ESSs) are evolving as one of a number of key enabling technologies delivering grid and network solutions in support of electric power system modernization. Accommodating growth in demand and diversity in generation requires active system adaptation. ESSs are essential in delivering a range of ancillary services to maintain system stability, lessen intermittent variability from renewable sources, including wind and solar generation, and improve renewable energy dispatchability while reducing curtailment. The benefits of energy storage exceed simple generation augmentation and their acceptance is increasingly based on ESS units offering value stacking preserving traditional infeed while delivering optional services, including rapid responses (<500 ms) to events which can affect system frequency stability. In particular, events which occur in response to sudden changes in generation or demand can directly affect system frequency. Thus ESSs present potential solutions for grid and network applications on the transmission and distribution systems, respectively (EPRI, 2005). Moreover, ESSs can also help to integrate different generation technologies, including gas boilers, CHP, EHP, and TES units.

The increasing contribution of nonsynchronous generation has resulted in a rising number of issues potentially threatening and affecting power grids. Transmission congestion, energy balancing needs, frequency regulation, voltage limit violation, and overloading of network operating resources represent some of the challenges resulting from greater penetration of renewable generation. In addition, Electric vehicles (EVs) pose problems on the

Predictive Modelling for Energy Management and Power Systems Engineering
DOI: https://doi.org/10.1016/B978-0-12-817772-3.00004-5

distribution demand side due to their high need for access to large power capacities for recharging.

Large power systems often source significant amounts of renewable energy from isolated regions, these regions can then experience particularly low inertia issues which can result in systems-wide events, such as the September 29, 2016 blackout in South Australia (Bloom et al., 2017; Brogan et al., 2018). The expansion of SNSP on large systems will eventually cause system-wide low inertia issues (Ulbig et al., 2014), similar to those addressed on smaller networks unless these are arrested by countermeasures. Although Alaska, for example, has a number of isolated power systems that rely on diesel generators, wind turbines, or hybridized wind–diesel systems, it is particularly difficult for conventional diesel generators to keep good power quality in a closed grid since the generator output cannot follow the demand change rapidly, which renders a slow response to grid events (ACEP, 2012).

Ireland has set itself the challenge of sourcing 40% of its electrical energy from renewable sources by 2020 (Flynn et al., 2016). To achieve this goal it is anticipated that the Irish power system will need to operate at a system nonsynchronous penetration (SNSP) of 75% for some of the time (Eirgrid/Soni). At present, the Irish power system regularly operates at an SNSP of 60% and many valuable lessons have been learned in recent years in accommodation of greater distributed and low-inertia energy penetration (Flynn et al., 2016). In particular shortfall in inertia has been identified as a significant challenge for increasing SNSP (Brogan et al., 2018; Flynn et al., 2016). Therefore as the SNSP continues to increase, and lessen large thermal synchronous units, the level of centralized spinning inertia will decrease, which potentially exposes system frequency to volatile and real-time changes, hence affecting stability. The nature of major power systems challenges determines to what degree ESSs are required alongside the necessity to deliver efficient and low-carbon solutions to sustain power systems and aging power systems infrastructure (Fig. 4.1).

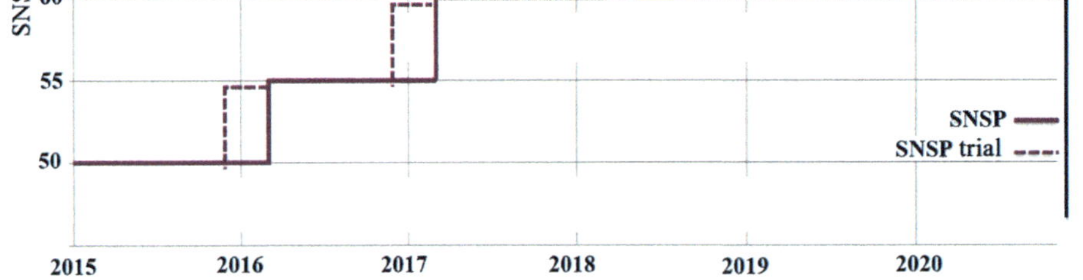

FIGURE 4.1 SNSP increase from 50% to 75% (2015–2020 +).

Thus ESSs will play a key role in regulating the frequency in electric power systems (Greenwood et al., 2017). Significant enhancements in the features of ESS units have been recently achieved, with improvements in efficiency, power output, energy density, longer life cycles, and reduced cost (Rastler, 2010; Barai et al., 2018). Despite progression in storage technologies, however, the applications are still limited by the lack of versatility and flexibility among ESS unit services (Barai et al., 2018). In Ireland, for example, the Northern Irish power system uses a 10 MW BESS at Kilroot Power Station designed only to provide frequency response services, including fast frequency response (FFR), POR, and TOR, and this plant is intended to be increased to 140 MW in future (Brogan et al., 2017). However, diversity in ancillary services exists, particularly among BESS suppliers with more comprehensive units delivering arbitrage, capacity, and rapid responses to maintain frequency and system stability: multiple services are generally considered as value stacking options, to enhance storage applications. The hybridization of ESS units with different technical interconnection can compensate for single service drawbacks and guarantee power systems energy balancing and spinning reserve (hybrid-ESS units are maintained at a level of charge ready to respond to a generation or transmission outage, for example, grid short circuit or significant overcurrent diversion). Most of the recent published work focuses on the integration of renewable sources in smaller grids (Wang et al., 2014a; Chong et al., 2016) or microgrids (Ariyaratna et al., 2018; Jing et al., 2018) and distributed ESS (Arvind Parwal et al., 2018).

The material in this chapter includes a generic transmission system model of the Danish TSO Energinet.dk, which is implemented in the simulation tool DigSilent Power Factory. A small test model of the transmission system with a large offshore wind farm is used to investigate specific applications of proven ESSs, which can be a benefit to the power grid under transient events which cause grid frequency excursion and transition issues from a large penetration of distributed energy sources. This chapter also demonstrates the use of a hybrid energy storage approach, integrating storage technologies with supplementary operating characteristics, which can be beneficial for applications in decentralized grid support services.

4.2 Proven energy storage for increased service provision

ESSs are principally used in the power systems in three different operating modes: charge, storage, and discharge. In each mode, a balance between power and energy in the power system must be maintained so that energy storage retains the appropriate rated power to have the appropriate rated power and energy capacity. Different energy storage technologies coexist due to a range of technical operating offerings, which make them attractive for various applications to help mitigate frequency-related problems and preserve good power quality. Units can manage immediate or evolving transient events, particularly grid frequency excursion, and limit the rate of change of frequency (RoCoF) events, as well as maintaining voltage deviations.

Table 4.1 lists proven storage services to provide services at the transmission grid level. Proven energy storages turnkey solutions are commercially available from reputable companies, contract vendors, and grid agencies.

TABLE 4.1 Proven energy storage for grid-scale services.

Technology class	Services applicable	Reliability	Nonsynchronous	Synchronous
Batteries (e.g., Li-Ion, NaS, Metal-air, VRB, etc.)	FFR, PR, SR, TR, & reserve	Proven	Digital inertia	–
Ultrabattery	FFR & PR	Marketed	Digital inertia	–
Supercapacitors	FFR & PR	Proven	Digital Inertia	–
Flywheels	FFR, PR, & SR	Proven	Digital inertia	–
SMES	FFR, PR, & SR	Marketed & proven	Digital inertia	–
CAES	PR, SR, & TR	Proven	–	Inertial
PHS	PR, SR, & TR	Proven	–	Inertial
Rotating stabilizers	FFR, PR	Unproven at grid level		Inertial
Synchronous compensators	PR	Proven	–	Inertial

The increasing level of distributed energy penetration in power systems, such as the Irish power grid and South Australia systems, demands a critical imperative for large-scale and reliable storage. At grid level, high-power batteries (yypically Lithium and Lithium-ion), supercapacitors, and rotating flywheels stabilizers are all applicable technologies. The synchronous machine can be designed with a high number of pole pairs and possess significant mass to enable it to contribute synchronous inertial response (IR) to a system to limit frequency deviations. In addition, synchronous compensators can provide IRs to guarantee better dynamic voltage recovery after severe system faults and compensators can be considered as energy storage devices due to their ability to provide "inertia" response for a short period of time to maintain grid stability.

Emerging technologies include flow batteries and cryogenic storage, pumped hydro storage (PHS) and several newer and unproven technologies, including rotating stabilizers and demand-side control for dynamic loading shifting and participatory options. New storage emerging technologies promise improvements in cost, lifecycle, energy density, safety, and market services.

4.3 Grid functions for energy storage system

The increasing share of SNSP poses a major challenge to frequency dynamic stability due to intermittent connection, availability, and variability which creates load-generation imbalance: presently, SNSP does not contribute directly to frequency control. SNSP is usually operated in such a way to produce as much power as possible (Egido et al., 2015). Therefore SNSP does not provide standing reserves. Hence ESSs are of high value to be used to deliver frequency support to maintain grid stability.

Furthermore, ESS can reinforce existing services and create new products to bridge the gap in power generation over longer durations. ESSs can further enhance the reliability of power systems (Eyer et al., 2004) and make more efficient use of the network (Saboori et al., 2015).

The adoption of ESSs delivers value applications for transmission grids (Oskilly et al., 2015), principally improving available energy capacity. However, ESSs can contribute significantly to distribution networks on the demand-side applications and provide ancillary energy services (Fig. 4.2).

ESSs are designed to meet performance criteria, bridging timescales from milliseconds to hours and even days with potential MW contribution (Editorial, 2015). At different levels of the power system, supporting applications demand different services from ESS units. At generation level, for example, ESSs can improve grid absorption of excess renewable capacity (Olabi, 2017), thus avoiding the curtailment of wind farm output and maintaining demand profiles. ESSs can provide active power to counter the instability as a consequence of wind farm variability, helping shift generation during periods of low demand to peak loading (Zhao et al., 2015). Storage complements existing on-site generation by delivering fast-acting frequency responses within transient (<200 ms), primary (>1 seconds), secondary (>2 minutes), and tertiary (>15 minutes) services. This allows commercial enhancement of energy efficiency and optimization of other generation (https://www.powerstar.com/virtue/virtue-energy-storage/), hence reducing operating costs.

At the transmission level, ESS units can reduce congestion on the grid infrastructure by increasing transmission network capacity (Saboori et al., 2015), by increasing the proportion of renewable generation. This can help provide stabilizing control influences and reduce variability and intermittency and deliver ancillary services (frequency response

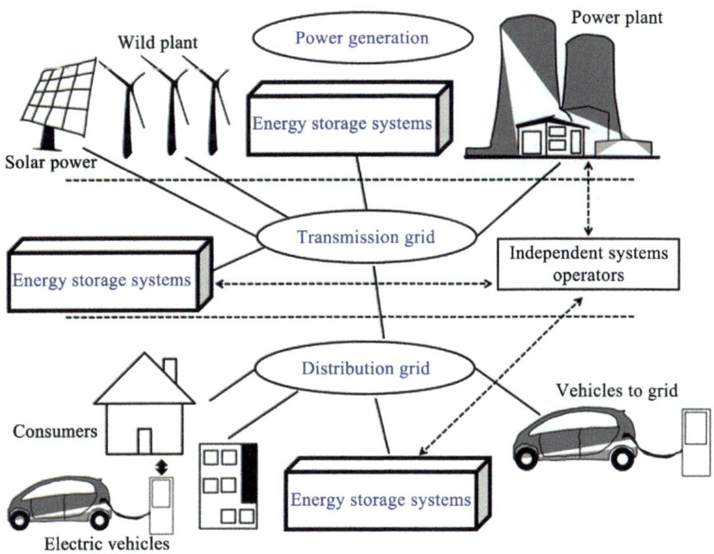

FIGURE 4.2 Application of ESS in power systems.

and standing reserve) to be connected to the grid, and increase the reliability of RES. Therefore it helps to stabilize the integration of RES by removing the variability and intermittency associated with wind farm plants. It further can be designed to provide ancillary services (frequency response/standing reserves) for the grid (Greenwood et al., 2017; Egido et al., 2015), potentially lessening the need for spinning reserve and hot/cold and banked generation.

At the distribution level, ESSs contributes directly to the deployment of distributed generation to maximize generation and limit unpredictability in preserving dynamic stability. Storage units can also minimize distribution costs, reduce outages and energy-related failures, which can exceed 17% of annual revenue (https://www.powerstar.com/virtue/virtue-energy-storage/). ESSs typically installed from substation level downward provide essential to demand-side services (Nguyen and Flueck, 2012), as in Fig. 4.3.

In Fig. 4.3 the yellow boxes represent new support services in which ESSs can maintain dynamic stability. ESSs are often classified into six umbrella groups (Lyons et al., 2015), but with an emerging trend of delivering fast-acting services. ESS options and services can be summarized as follows:

- Voltage control: supporting heavily loaded feeders, power factor correction, reducing generator curtailment, minimizing on-load tap changer (OLTC) operations, mitigating flicker, sags, and swells (Mitsuki Sagara et al., 2016; Wade et al., 2010; Wang et al., 2014b; Yi et al., 2012);
- Power flow management: deferring network reinforcement, reducing reverse power flows generator curtailment and losses (Wade et al., 2010);

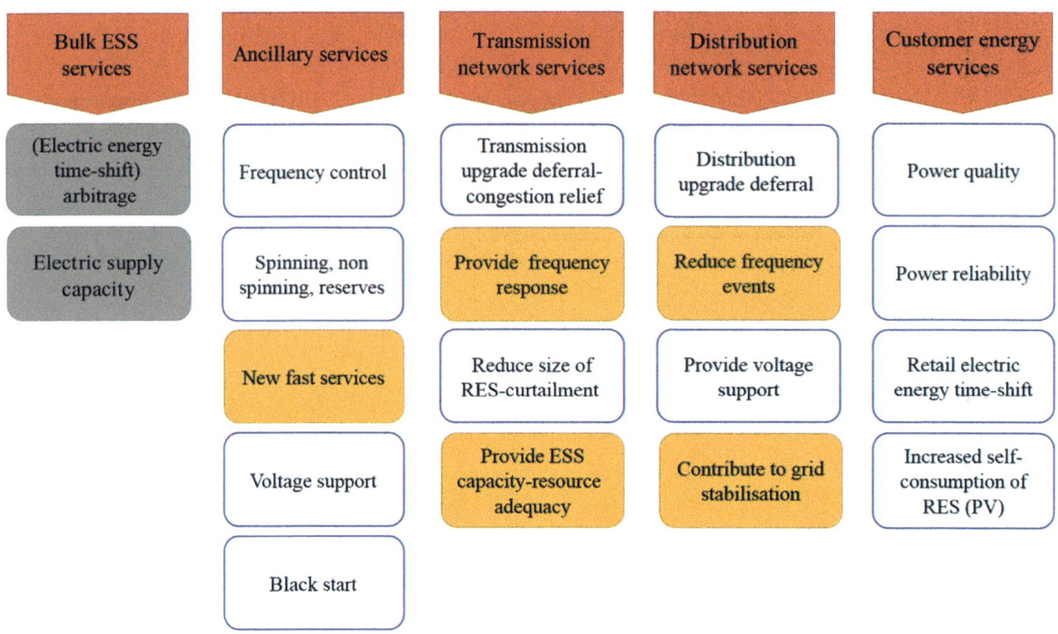

FIGURE 4.3 Key applications of ESS.

- Energy/ancillary markets: energy arbitrage, balancing market participation, reducing intermittent generation variability, increasing intermittent generation yield from nontemporary connections, and providing ancillary services (frequency response/ operating reserves) (Greenwood et al., 2017; Koller et al., 2015; Knap et al., 2014b; Brogan et al., 2018);
- New fast ancillary services: provision of inertia (SIR) from synchronous generators, condensers, rotating stabilizers (Australian Energy Market Operator, 2017), FFR or enhanced frequency response (Greenwood et al., 2017; DS3 System Services, 2014; Australian Energy Market Operator, 2017), fast reserve and postfault active power recovery (FPF-APR) mitigating the impact of voltage disturbances on system frequency (DS3 System Services, 2014), dynamic reactive response (DRR), the ability of a unit when connected to deliver reactive current for voltage dips in excess of 30% (DS3 System Services, 2014);
- Commercial/regulatory: assisting in compliance with energy security standard (ER P2/ 6) (Energy Networks Association, 2006), reducing customer minutes lost(CML), and reduction in generator curtailment (Blake et al., 2013);
- System restoration: voltage control and power flow management in a postfault network (Blake et al., 2013);
- Network management: facilitating islanded networks, supporting black starts, and switching ESS between alternative feeders (at a normally open point) (Wade et al., 2010).

Other than electrical storage coupled through inversion, technologies such as PHS and compressed air storage (CAES) can be connected synchronously alongside synchronous generators, but are heavily dependent on topographical resources. However, supercapacitors, batteries, superconducting magnetic storage, and flywheels are typically connected nonsynchronously with power converter interfaces, such as inverters. These technologies can ramp up to full response within 500 ms and can sustain a response for >10 minutes. ESSs present features which must be taken into consideration for different applications, including capital cost, power and energy rating, power and energy density, ramp rate, efficiency, response time, self-discharge losses, and life and cycle (Wang et al., 2012).

4.4 Energy storage characterization for digital inertia

Although the concept of *analogue inertia* is considered as traditionally derived from large-mass, spinning units with control and governance for synchronization and rotational speed, *digital inertia* sourced typically from battery storage resources can take different forms for grid availability:

- Frequency triggered digital inertia with droop control
- Frequency responses, which provide enhanced governor responses
- Rate of change of frequency responses, which emulate real IR
- Step response, which is effectively a combination of frequency and RoCoF

Although batteries and supercapacitors are able to deliver each of these attributes, batteries offer systems operators with a high level of flexibility with systems subject to transient events. Batteries-based units can step, droop, or provide emulated IR through grid supporting power inverters, which are typically equipped with controllers to regulate P and Q to maintain the grid stability and voltage. Digital inertia is established through the power inverter controls, which interface the power grid and energy storage. Moreover, power electronics are used to release and absorb energy from the grid or the connected storage machines. BESS can potentially operate a system with up to 75% instantaneous penetration of nonsynchronous generation (SNSP) and maintain curtailment at low levels.

FFR or enhanced frequency response energy storage devices have the potential to provide power responses to help mitigate high RoCoF events. Analysis in Table 4.2 indicates the suitability of energy storage devices to maintain frequency within grid-code limits and reduce RoCoF issues. This is highly dependent on the device response characteristics and digital inertia storage is currently one of the most proven methods to manage the frequency excursions, reduce nadir and prevent high RoCoF during grid faults, having been proven in places such as Ireland, South Australia, and Great Britain.

The deployment of proven storage can help reduce the cost of spinning inertia. Energy storage using battery technologies can require significant up-front investment, but value stacking of services offers enhanced options, including FFR.

ESS applications can also deliver technical features which help limit the environmental impact. Table 4.2 displays technical response characteristics of storage systems, which contribute to projected services in future grids and networks. Battery energy storage in particular currently has the greatest potential due to its relatively small footprint compared to other storage methods.

4.4.1 Size analysis of energy storage

Generally, for consumer-oriented mass markets, domestic sizing of renewable sources and the geographical and regional location will dictate the proportional sizing for energy storage devices. At present, however, large grid-scale systems have no such equivalent scaling and are generally based on required capacity, location, and invariability cost.

In a more general context, and at grid level, ESS sizing is governed by the level of contributory inertia (IR) and required primary frequency response (PFR): for example, Delille et al. (2012), used an ESS for the IR that was sized to deliver arbitrarily chosen rated power for at least 15 seconds, whereas Knap et al. (2014a) performed simulation studies for various ESSs for sizing in IR applications, with the unit specified to fulfil target limits of RoCoF and minimum system frequency. A probabilistic approach was adopted by Yue and Wang (2015) to size an ESS for IR required for frequency variations caused by high solar penetration, based on a series of simulations. Optimization of the ESS size by Oudalov et al. (2007) was determined to maximize profit income through an ancillary service market. Moreover, in Knap et al.'s study (2016) ESS sizing was based on system parameters, the inertia constant, and power/frequency characteristics, which were used in the estimation of ESS power and energy ratings.

TABLE 4.2 Storage technical response characterization.

ESS technology	Available MW	Maturity	Efficiency	Response time	Charge time	Runtime	Lifetime, years (cycle)
Ultracaps	0–0.05	Commercialized	90–95	5–20 ms	s–h	10 s^{-1} min	20 + (> 100,000)
Supercaps	1 +	Commercialized	90–95	10–20 ms	s–h	ms–60 mins	20 + (> 100,000)
SMES	10	Developing	95–98	1–100 ms	min–h	ms–8 s	20 + (> 100,000)
PHS	100–5000	Very mature	75–85	3 mins	h–months	1–24 h +	40–60 (> 13,000)
CAES (Large-scale)	5–1000	Mature	70–89	1–15 mins	h–months	1–24 h +	20–40 (> 13,000)
FES	20	Commercialized	93–95	< 4 ms–s	s–min	ms–15 mins	15 + (> 100,000)
Cryogenic (Large-	10–200	Developing	60–75	< mins	–	1–24 h +	30
scale)	50–200	Marketed	–	s	2–10 mins	s–h	3–15 (2000)
Rotating stabilizers	(MVA)	Mature	70–90	5–10 ms	min–days	s–h	3–15 (3000)
Lead-acid	0–40	Developing	–	5 ms	min–days	s–h	10–15 (2500–4500)
Ultra-battery	0–36	Commercialized	80–90	3–5 ms	s–h	min–h	5–15 (1000–20,000)
Nas	0.05–34	Mature	85–90	20–100 ms	min–days	s–h	10–20 (2000–3500)
Lithium-Ion	0–200	Commercialized	60–65	ms	min–days	s–24 h +	–(100–300)
NiCd	0–40	Developing	~50	ms	h–months	s–10 h	5–10 (12,000 +)
Metal-air	0–0.01	Commercialized	~85	< 80 ms^{-1} min	h–months	s–10 h	5–10 (2000 +)
VRB	250	Marketed	~75	< 80 ms^{-1} min	h–months	0–24 h +	– (–)
ZnBr	0.05–10	Developing	~20–30, planned eff.	–	h–months	s–24 h +	5–20 +
Solar fuel	0–10	Research/	>54	<1 s	h–months	1–8 h	(1000–20,000 +)
H_2 fuel cell	0–58.8	developing	25–58	–	min–days	1–8 h	20–40 (> 13,000)
CES	0.1–300	Developing	40–50	–	min–days	1–24 h +	10–20 (–)
AL–TES	0–5, 1103	Developing	50–90	–	min–days		5–15 (> 13,000)
HT–TES	0–60	Developing	30–60				

The studies above are based on different approaches for various problems, but in reality, ESS sizing must be based on key performance indicators. The first step of every sizing process is the determination of technical and technology requirements for the ESS. The outcome of this process sets requirements in terms of power and capacity. Different physical ESS principles and actual properties of the final ESS can vary widely from these requirements.

Technological characteristics for battery-based ESSs include ramp rate, number of load cycles, state of charge (SoC), and the depth of discharge (DoD)—all determined within a local environmental context and based on the assessment of ambient temperature and humidity. A critical issue with ESS integration is how to accommodate parameter uncertainties, including power delivery and capacity, which are not entirely predictable due to a dependence on variability in renewable and distributed generation sources.

Power, energy capacity, and discharge time should be carefully considered for sizing. Lower than optimal values can result in the inability to meet application-required performance and may accelerate aging; higher than optimal parameters may be economically unsustainable. In addition to battery assessment, the power converter is the essential link for ESS integration. To appropriately size the ESS for target applications, the parameters in Table 4.3 are considered:

Factors in Table 4.3 determine ESS performance. Moreover, the size of energy storage depends on the required specific grid functions.

TABLE 4.3 Parameters essential for sizing ESS.

Description	Units
General information	
Expected power range of operation	MG or kW
Expected energy storage range	MWh or kWh
Environmental conditions,predominantly shall operate temperature range	°C or Kelvin
Available space	m^3
The marginal cost of power converters (DC-DC & DC-AC)	Currency unit/kW
Definition of EoL	Various
Battery-specific information	
Calendar life	Years
Cyclic life	Number of full cycles
Flywheel-specific information	
Operational life span (MTBF)	Hours
Standby power loss	kW
LIC-specific information	
Calendar life	Years
Cyclic life	Number of full cycles

4.4.2 Hybridized energy storage systems

This section provides an analytical demonstration of the hybridization of ESS for its application benefits in the power systems. A hybridized energy storage scheme is able to deliver a broad range of grid services and, in certain instances, serve as a virtual source of synchronous response, for example, in releasing energy to synthesis inertia. Also providing power and energy-related grid services offers versatility to arrest transient instability. A hybridized storage typically comprises complementary technologies:

- Battery: battery (e.g.-ion-flow battery)
- Supercapacity: battery
- Flywheel (FES): battery
- Battery: CAES
- Battery: PHS
- Flywheel: synchronous compensator
- Battery: rotating stabilizers

Hybridized storage can comprise two different technologies to deliver the range of services at grid (transmission) and network-level (distribution). Batteries, specifically Lithium-ion, play a key role in various hybridization schemes of ESSs. Flow batteries (in particular vanadium units, VFBs) are emerging as effective technologies due to their extended lifetime, low-carbon footprint, and increasing access to high capacity energy, with commercial units available >200 MW for grid applications and <30 kW for demand-side and network services.

Ultracapacitors are used in combination with a lead-acid battery to form an *ultrabattery*, a recent advance in lead-acid battery technology developed by the Australian Commonwealth and Industrial Research Organization. It is a hybridized energy storage that incorporates a supercapacitor with a lead-acid battery in a unit cell. Hybridized storage can resolve the increasingly complex operational requirement of the electric power delivery system and alternate between energy absorbing and release to maintain voltage and frequency within required ranges (Carnegie et al., 2013), operating within an entire energy management system.

4.4.3 Increased service provision to transmission systems operator

ESSs offer service products, which deliver operational flexibility to transmission systems operators (TSOs). For example, in Ireland ESS units have provided rapid MW-level responses faster than existing PFRs or primary operating reserve times to limit frequency excursion transients. Provision of this and related services can include IRs, thus TSOs can maintain or increase outputs with time frames set by grid codes.

An increasing share of variable generation connected at distribution level (at low and medium voltage levels) requires distribution system operators (DSOs) to guarantee system stability security. Deployment of storage into transmission grids is crucial in maintaining system stability while accommodating distributed generation during and after extreme grid faults.

TABLE 4.4 Theoretical BESS output with varying performance parameters.

Delay time [ms]	Ramp [S^{-1}]	360 MW BESS [MW.s]
0	∞	2160
250	5	2034
250	0.5	1710
500	5	1944
500	0.5	1620
1000	5	1764
1000	0.5	1440
2000	5	1404
2000	0.5	1080

As an example, the quantity of power delivered by a 360 MW BESS, acting with different delay times and ramp rates, is presented in Table 4.4, based on a study by the Irish Eirgrid and System Operator Northern Ireland (SONI) which concluded that this level of capacity is appropriate to maintain system stability with 75% SNSP (EirGrid/SONI, 2016).

It can be mentioned that from Table 4.4 a 360 MW BESS operating with no delay and an infinite ramp would, in principle, provide an additional 2.16 GW.s of energy to the system over 6 seconds. If the BESS target is to provide 75% of the 2.16 GW.s response, then this would require 1.65 GW.s. ESS connected to the transmission network offers more flexibility and active response during frequency disturbances.

4.5 Test model of the transmission system

A test model of a transmission system with a large offshore wind farm was implemented in the DigSilent Power Factory Simulation and contains 17 buses with voltage range from 0.7 to 400 kV, four central power plants and control devices, a static VAR compensator (SVC), several consumption centers, and Type-1 fixed-speed asynchronous generators (local wind farm) and WECC Type-3 full converter wind turbine generators (WTG, 2nd generation)—equivalent to a large offshore wind farm.

This test model in Fig. 4.4 resembles the Eastern Danish Transmission system and was constructed for the investigation of the storage applications. The central power plants have four large synchronous generators G1, G2, G3, and G4 equipped with AVR_IEEET1, GOV_IEEEG1, and PSS _CONV. Local wind farm contribution was 60%, indicating an active power supply of 500 MW. Buses 115–117 are located offshore.

The large offshore wind farm was connected to bus 117, with rating of 165 MW. Normally, the large offshore wind farm is commissioned with fixed-speed, active-stall controlled wind turbines that are similar to the Danish offshore wind farm commissioned at Nysted (Rodsand). However, in this study, the fixed-speed supplied by the local embedded

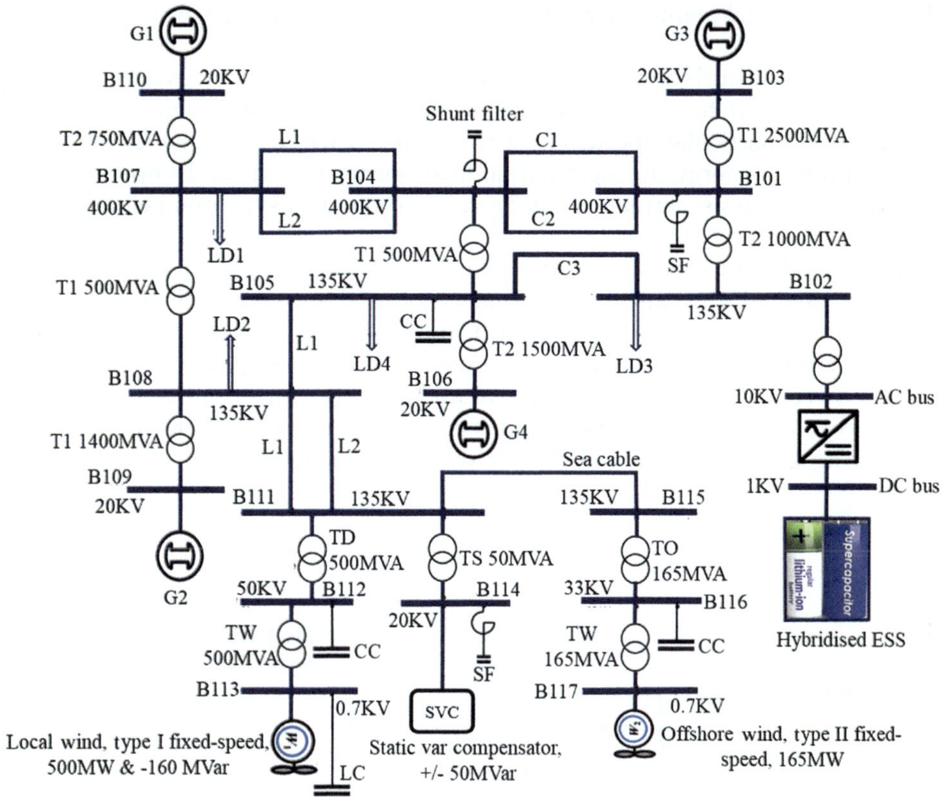

FIGURE 4.4 Small test model of the transmission system with a large offshore wind farm (165 MW rating).

generation is 60 % of the installed power capacity, being 500 MW. Buses 115-117 are located offshore. The fixed-speed, active stall controlled wind turbines were replaced with Type-3, WECC full converter WTGs. The SVC unit has a dynamic range of + /-50 MVar, applied to improve the voltage profile in the on-land connection point of the large offshore wind farm-bus 111. SVC enables independent control of reactive power import and export, which is used to manage voltage and power flow in the transmission network.

4.5.1 Embedded generation

Embedded generation resources used in the test model study are described as follows.

Type-1, a wind turbine is driven by asynchronous generators within H_{WTR} 5 seconds (inertia constant of wind turbine rotor); this parameter does not provide responses to the overall system following a disturbance and eventually H_{WTR} has no influence on the frequency response. These asynchronous generators deliver power at a constant frequency and voltage. The rate of the mechanical power reduction is limited to -0.5 MWs/MVA, which makes it possible to attain the required 20% of the rated mechanical power in less than 2 seconds from any arbitrary operational point. The reactive power is balanced using a switched capacitor, as shown in Fig. 4.5.

Full converter wind turbine generation achieves a better efficiency over a wide range of wind speeds compared to a Type-1 unit, fixed-speed wind farm, which can only reach peak efficiency at a particular speed. This variable speed wind farm enables the maximization of the capture of energy during a partial load operation and can be operated with different generators, such as permanent magnet synchronous, asynchronous, and induction generators.

The Type-3 model consists of four components, as shown in Fig. 4.6 (Clark et al., 2010). This model is used to analyze the performance of offshore wind and the full converter WTGs interaction with the transmission grid. In this generic Type-3 generator/converter model, flux dynamics are eliminated to preserve the rapid response of the power converter. The power converter dictates the real and reactive power delivered to the power grid. These two wind farms (Type-1 and Type 3) do not contribute to the frequency response during grid faults but can deliver real and reactive power during normal grid operation.

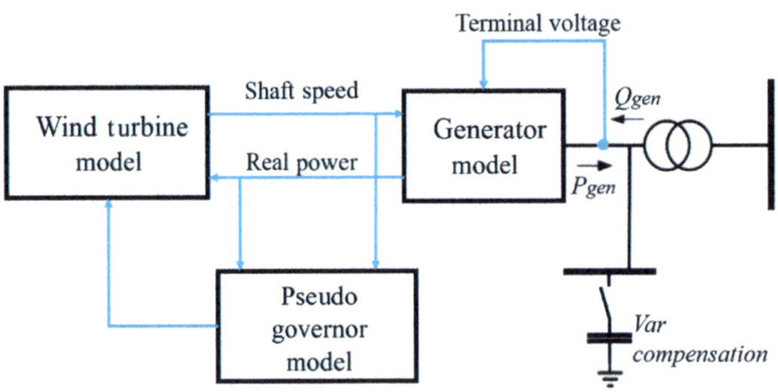

FIGURE 4.5 Type-1 fixed-speed, asynchronous generators.

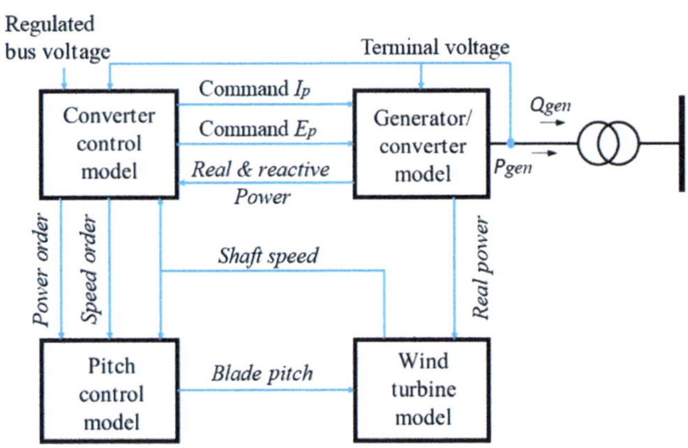

FIGURE 4.6 Type-3 WTG, full converter.

4.5.2 BESS operation

The structure of BESS units consists of a battery bank, AC/DC power converter, and grid-connected transformer. This battery model has four individual control sections: frequency controller, charge controller, active power control, and voltage control.

The BESS imports and exports power and the BESS can engage in two types of market-incentive services: service provision (e.g., FFR, primary response) and energy trading or energy arbitrage. One technical aspect for the battery is the SoC, which preserves the battery, for example, Li-ion batteries can be damaged when their SoC drops below certain levels (in reality 80%–20% standard limit). Here, the BESS is modeled as a SoC-dependent voltage source with internal resistance and can be computed as shown by Ubertalli and Littler (2018). However, it is not directly considered in this investigation.

4.5.3 Droop response and deadband for frequency quality

The BESS releases power when *dpref* is positive and absorbs power from the grid when *dpref* is negative. The BESS frequency response takes action when the grid frequency differs from the nominal 50 Hz as expressed in (4.1)

$$dpref = fmeas - fgrid \tag{4.1}$$

The droop response sets the limits of the battery full active power response to be triggered following grid faults causing frequency excursion. It can be expressed as in (4.2).

$$Presponse = \frac{d_{pref}}{droop(R)} \tag{4.2}$$

where *fmeas* is the measured frequency, *fgrid* is the grid frequency before disturbance, and *Presponse* is the BESS power response to contribute to the frequency response during grid events and to deliver digital inertia (Fig. 4.7).

FFR is needed when inertia is lower (due to high SNSP 50%–75% and distributed energy sources) and to enhance the frequency response during grid faults. A MW-level BESS can provide services to manage the frequency stability during and after faults. The control system interacts with both the BESS inverter interface and the storage unit on the power grid (Fig. 4.8).

A deadband is implemented to prevent excessive droop responses and acts only if the response starts to move outside of the range.

4.5.4 PV control

In power and voltage control, the output power from the BESS is compared against the active power reference from the frequency control. The frequency control defines the active power reference (*dpref*) signal corresponding to a frequency excursion measured at the grid connection point (of BESS). The signal (Δid) is the difference between the charge controller input and the output current on the d-axis.

The reference voltage value (*Vmeas*) and the voltage at BESS (*Vgrid*) point of connection is used as input to another PI control to produce the reactive power current (*Iq, ref*) on the q-axis. The antiwind PI control scheme is necessary to reduce performance degradation,

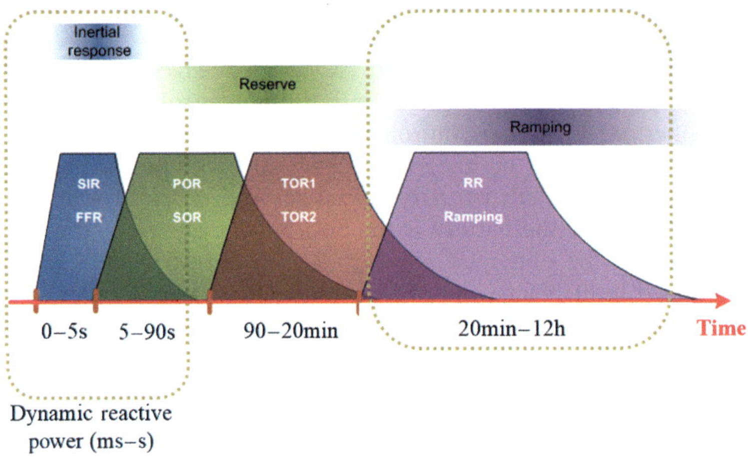

FIGURE 4.7 Frequency response services on the Irish power grid (DS3).

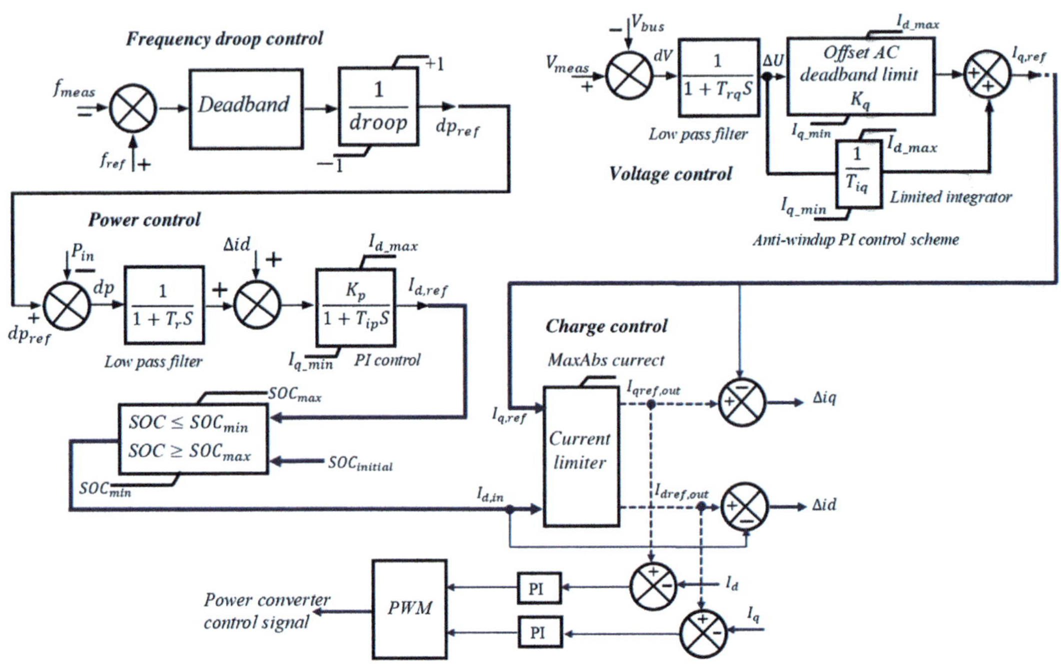

FIGURE 4.8 BESS control system.

reduce integral errors, and prevent divergence of the integral error when the control cannot keep up with the reference. However, the main target of this study is to demonstrate the application of ESS to provide power response to the power network following transient events.

4.5.5 Charge control

The BESS delivers active power if battery SoC satisfies the operating conditions range. A current limiter restricts the total amount of current flow through *d* and *q*-axis within the design limits to avoid overloading the BESS converter. Reactive power is not dependent on the battery, so the reactive current is not considered in the charge control as it is generated by the power converter PI control scheme. When the frequency is within the grid-defined deadband region the battery can be charged. The battery is charged if SoC is less or equal to SoCmax and discharged if SoC is lower than SoCmin. The current limiter regulates the active and reactive power output reference with the total BESS converter capability.

4.5.6 Ultracapacitor storage

Ultracapacitor storage is characterized by a much higher specific power rating but lower specific energy compared to batteries. Its specific energy is in the range of a few watt-hours per kilogram. However, the specific power can reach up to 10 kW/kg, much higher than any type of battery. The low specific energy density and dependence of the terminal voltage on SoC make it difficult for the ultracapacitor alone to serve as the primary energy storage, but there are benefits to using the ultracapacitor storage as a supplementary source. The ultracapacitor equivalent circuit is shown in Fig. 4.9.

The performance of an ultracapacitor can be represented by the terminal voltage during discharge and charge with different current rates. The capacitor has three parameters: the capacitance (its electric potential *VC*), series resistance *Rs*, and dielectric leakage resistance *RL*, as shown in Fig. 4.9. The terminal voltage of the ultracapacitor during discharge can be expressed as

$$Vt = Vc - \mathrm{IRs} \tag{4.3}$$

Some ultracapacitor ESSs (UESSs) have been developed with high power ratings for a variety of grid applications, such as peak power reserve, frequency control, digital inertia, voltage regulation, power outages, and to improve generator response. The ultracapacitor storage device can deliver much needed fast response during grid faults and capture

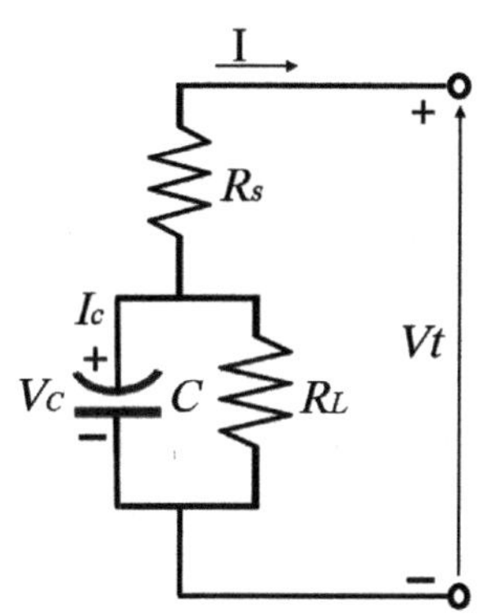

FIGURE 4.9 Ultracapacitor equivalent circuit model.

excess power when the grid is operating normally. Furthermore, it efficiently complements a primary energy source in emerging applications as it releases and captures rapidly, as displayed in Table 4.2 (storage technical response characterization). In this work, the use of the ultracapacitor is simplified as an ideal capacitor with small resistance.

4.5.7 Case studies

The deployment of large-scale ESS connected through a power inverter to the grid permits fast power responses during and after grid contingencies to maintain the dynamic stability of a system. ESS units connected at a transmission network increase reliability, stability, and ensure the power response when the frequency quality becomes more volatile due to the lack of inertia and governor response absence of synchronous MW, variability, unpredictability, and curtailment of distributed energy resources (DERs) (wind power and PV). BESS units are extremely important to avoid potential blackout, harvest excess DERs, and enhance applications at local levels. A secondary response is delivered by both the BESS (SoC 80%–20%) and the remainder of synchronous units.

- *Case study 1*: Participation of the BESS for secondary frequency control. In this case, the modified active power P is calculated as:

$$P = Pset + \left(Kp \times dpsco - Kpf \times dFdev\right) \times signPQRorient \tag{4.4}$$

where $Pset$ is the active power set point, Kpf is the primary frequency bias, $dFdev$ is the frequency deviation ($dFdev = fmeas - fgrid$), Kp is the PWM converter participation factor for secondary frequency control, as defined in the power-frequency controller, $dpsco$ is the power unbalance, as calculated by the power-frequency controller, and $signPQRorient$ specifies the sign of the active and reactive power flow at the controlled remote cubicle (at the connection point). In this case, the load disturbance was set in DigSilent Power Factory as the underfrequency transient.

Fig. 4.10 presents a frequency response with a 40% reduced inertia on the largest synchronous machine SG4. The integration of the BESS reduces the frequency nadir dip before and after 40% inertia rate on SG4 is reduced. The power response is activated when the frequency deviation ($dFdev$) is more than 15 mHz in a 50 Hz grid. Thus the deadband is small and droop is 200 mHz or 0.004 pu. The full active power from the BESS PWM converter is within 200 mHz. With the fast BESS, the frequency dip is stopped and settled at 49.762 Hz. The synchronous generator SG4 output active power during and after the disturbance is shown in Fig. 4.11; active power released by BESS improves the performance of the synchronous generator and provides a secondary frequency response control after disturbance. The BESS can control active power output much faster than a power plant SG4 because there are no mechanical inertia or delays.

BESS effectively reduces the active power generated (solid green line) from the synchronous generator due to the effect of the BESS response. There is a slight decrease in power response (solid green line SG4) at time 5.791 seconds, which delivers 1253.67 MW from the fast response of the BESS (compared to the grid without BESS connection).

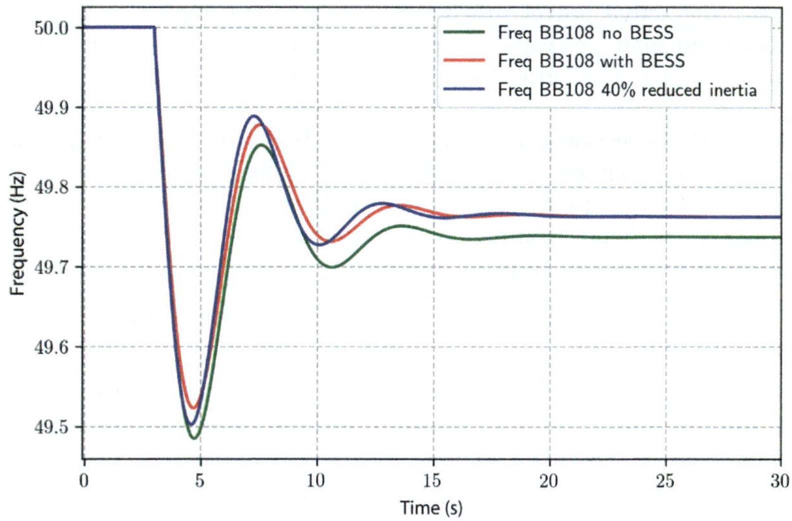

FIGURE 4.10 Frequency response with BESS following an underfrequency event.

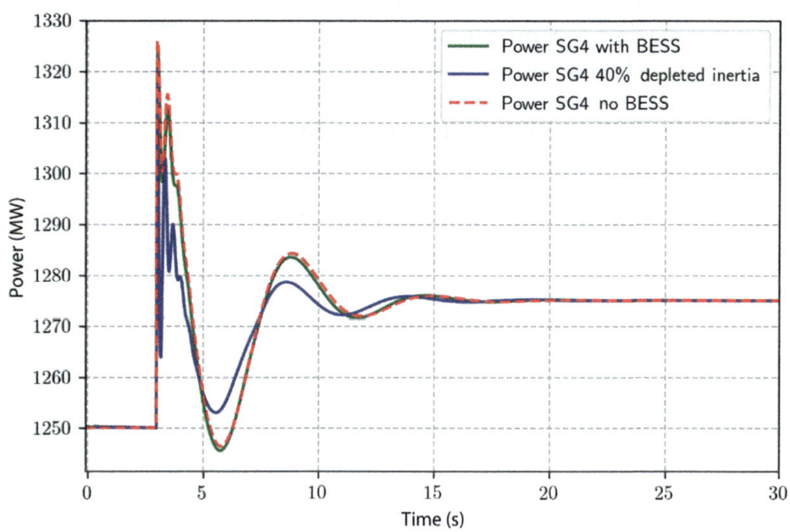

FIGURE 4.11 Synchronous generators SG4 power response.

Moreover, when there is a depletion of inertia in the system the BESS unit delivers power as shown in Fig. 4.11, which enables recovery and the power settles close to the same just after 15 seconds.

In Fig. 4.12 the connection of the BESS into the grid improves the behavior of the SG2 and SG3 and contributes actively to the secondary frequency response with less power response being drawn from SG2 (solid green line) and SG3 (dotted purple line)—7.714 and 13.768 MW less power, respectively, than in the case without BESS. This means the

BESS can control the active power much faster than a power plant SG2 and SG3. A fast response power plant manages rapid frequency volatility during major disturbances. The Soc is 80% or 0.8/unit but it varies slightly as the BESS releases the power response as shown in Fig. 4.13.

In Fig. 4.13 the BESS response remains close for both 40% reduced inertia (solid blue line) on SG4 and power delivered (dashed red line), as the BESS capacity is set up to participate in the secondary frequency response during and after disturbance. The BESS can deliver a secondary frequency control service, which depends completely on settings of the power converter to achieve the particular target.

- *Case study 2*: Participation of ultracapacitor storage to the frequency response following underfrequency disturbance.

The ultracapacitor contributes to a reduction in the frequency nadir (dashed green line) at the instant of fault, compared with the frequency response from the BESS, and provides high power and a fast response just after 3 seconds when the disturbance occurred (Fig. 4.14).

The ultracapacitor provides less power at 6 seconds than its counterparts BESS due to its fast runtime (estimated 10 seconds–1 minute). BESS and ultracapacitor perform the task of improving the frequency response as shown in Fig. 4.14. The BESS, in terms of speed of response and power (MW) deployment, permitted the frequency to settle at 49.762 Hz, compared with the ultracapacitor scheme where frequency settled 49.757 Hz (Fig. 4.15).

The RoCoF of 0.0321 Hz/second was achieved just after the contingency, compared to the case without storage, 0.034 Hz/second. The energy storage schemes provide transient response when the disturbance occurs and arrest the high RoCoF, with thhe frequency nadir reduced at 49.5231 Hz (with BESS) and 49.532 Hz (with ultracapacitor storage).

In Fig. 4.16 the power delivered by the ultracapacitor scheme (dashed red line) in the system compares well with the BESS response (solid blue line). The synchronous response (SG4 green line) ramped from 3 seconds until it settled 10 seconds after the disturbance.

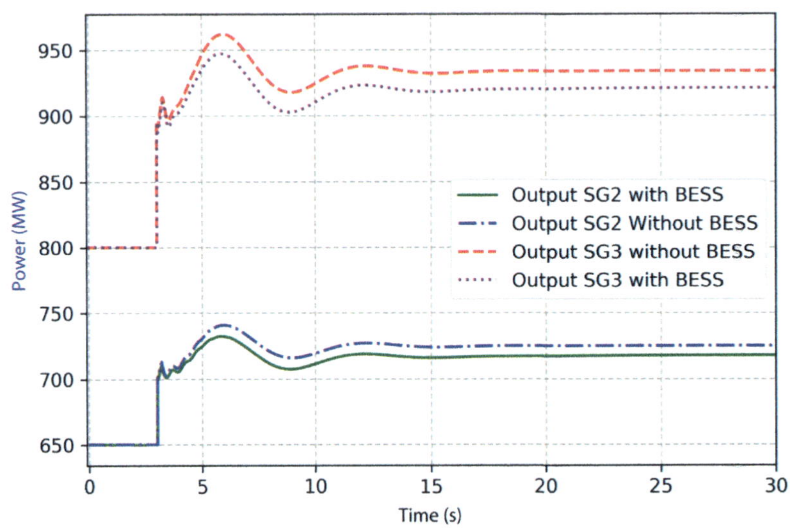

FIGURE 4.12 Synchronous generators power response with BESS.

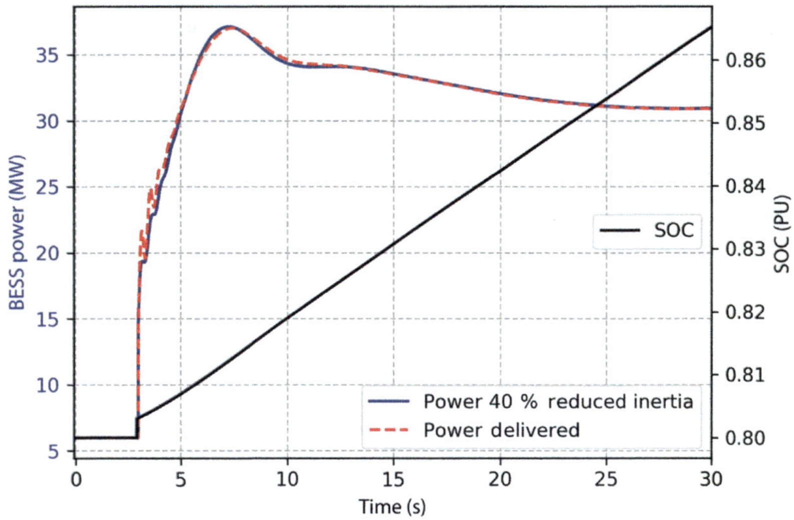

FIGURE 4.13 BESS power and SOC measured during a disturbance.

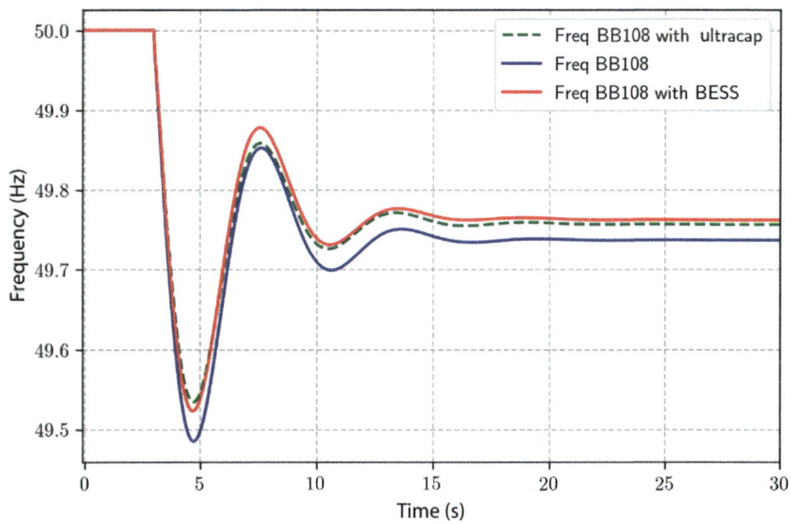

FIGURE 4.14 Frequency response with BESS and ultracapacitor.

There is a wide range of devices designed for reactive power provision and voltage support. The grid test model used one SVC for voltage support. The grid storage inverter coupled to the transmission allows sufficient voltage support. It can also deliver voltage capabilities during and after disturbance, as in Fig. 4.17, in addition to frequency response and reserve services. ESSs enable independent control of real and reactive power import and export.

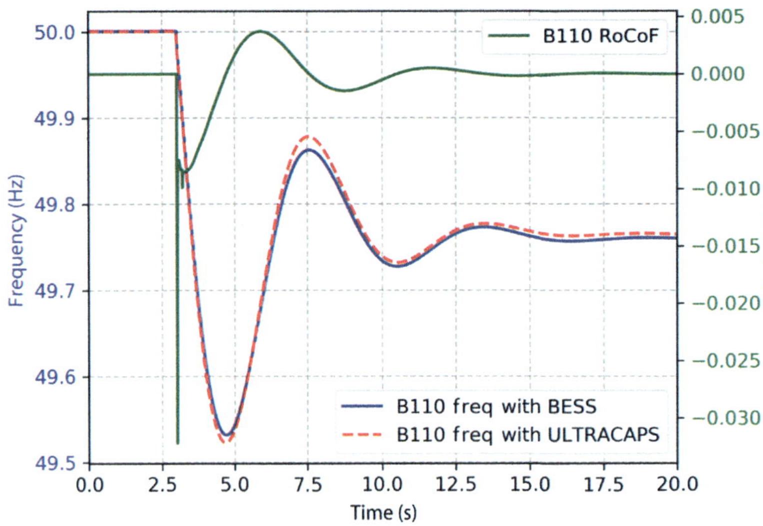

FIGURE 4.15 RoCoF and frequency response with energy storage measured at B110.

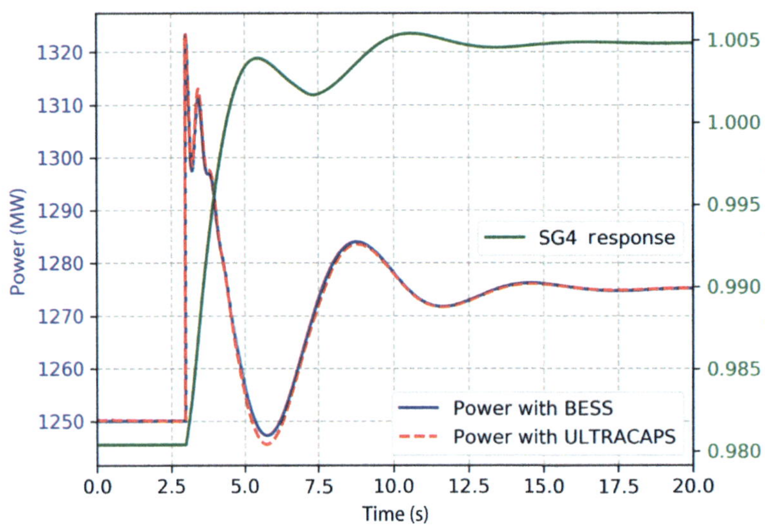

FIGURE 4.16 BESS, ultracapacitor, and synchronous response to underfrequency event.

- *Case study 3*: Local wind farm connected to the bus 112, rating 250 MW, tripped and the use of underfrequency load shedding (UFLS) relay on load D1 connected to the bus 107.

In this case study, a loss of 250 MW from wind farm generation depleted the power delivered under normal operating conditions. In Fig. 4.18 the frequency quality has significantly deteriorated (dashed blue) with the nadir reaching 49.465 Hz at 3.021 seconds and the rate of change of frequency 0.04133 Hz/second. A large disturbance increases the

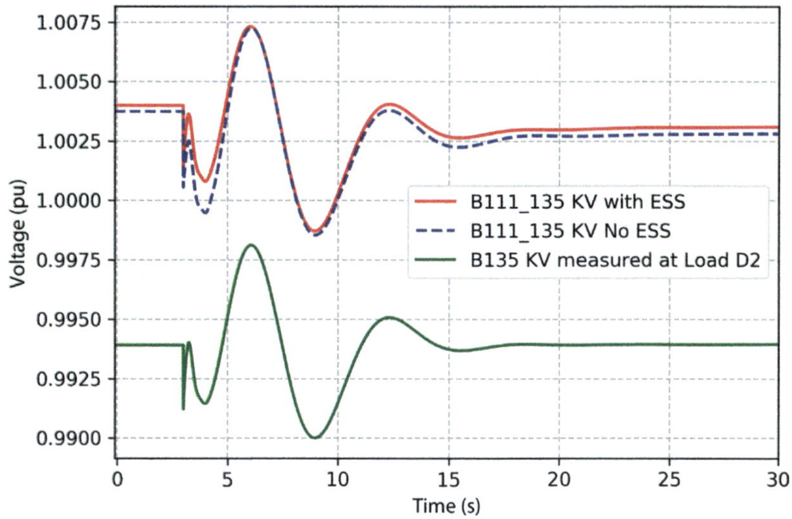

FIGURE 4.17 Voltage profile (135 kV infeed), with ESS (bus 111 and load D2) during a disturbance.

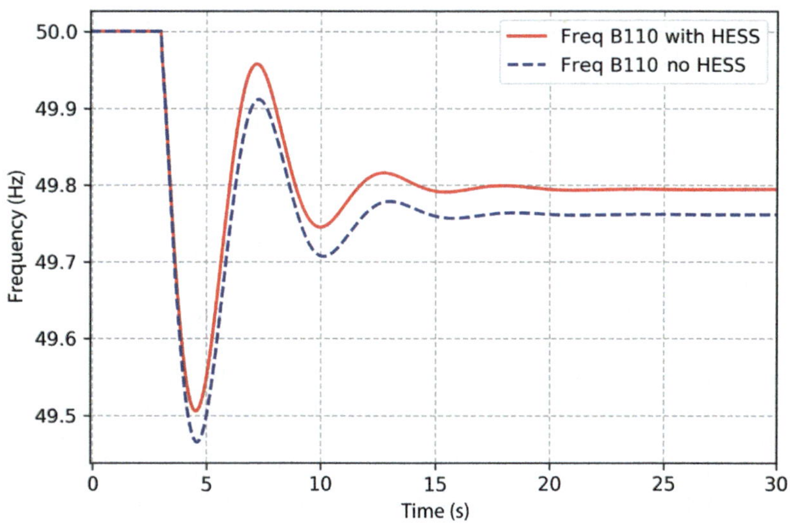

FIGURE 4.18 Frequency response measured at B110 when 250 MW local wind farm tripped.

RoCoF but a hybridized scheme (red solid line) can participate in the frequency control by providing a proportional response, which limits the nadir at 49.506 Hz.

A simple UFLS relay is deployed in the transmission network for the load D1 at bus 110, which represents a disconnection of 100 MW in response to the frequency decline. In this case, the load is set to trip if the frequency falls below 49.80 Hz for 4 seconds or more.

As in Fig. 4.19, when there is no UFLS and HESS (dash-dotted blue line) is not in service the nadir dips to 49.466 Hz, while the long-term response displays that the frequency settled at 49.762 Hz, but still below 50 Hz after 30 seconds. With the hybridized storage

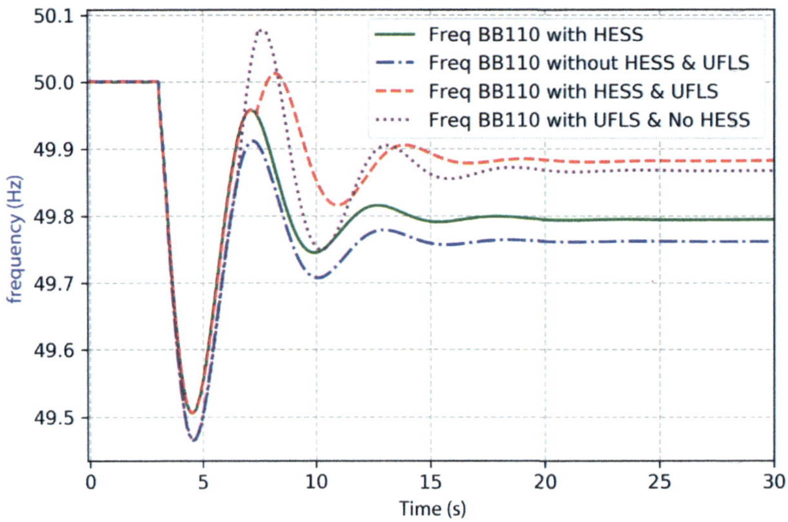

FIGURE 4.19 Frequency response and UFLS in the system.

scheme connected (green solid line), the nadir is improved at 49.505 Hz and the frequency response settled at 49.795 Hz. When 100 MW is disconnected by the UFLS (dotted purple line) the nadir dips close to 49.465 Hz and frequency settled at 49.867 Hz. The frequency can recover back to 50 Hz if a large load is shed by the UFLS or large energy storage capacity is deployed in the system to deliver the transient response, such as VRB, CAES, and HPS. As seen in Fig. 4.19 (dashed red line), the connection of a hybridized scheme can limit the tripping of the UFLS relay and improve the frequency response.

Table 4.5 demonstrates improvement when ESSs are connected and parametized to respond to transient events, participating in primary then secondary control through the power inverter in control modes P–Vac. Table 4.6 illustrates the load shedding relay settings.

In this case study, a 230 MW load D3 at bus 108 is shed by the UFLS: as the frequency fell below 49.80 Hz for 4 seconds, the load tripped to arrest a large frequency dip. When the 230 MW load was switched off, the frequency recovery was close to 50 Hz with the hybrid ESS and UFLS online (dotted purple line), as shown in Fig. 4.20. When the UFLS was individually online (dashed red line) the frequency settled at 49.979 Hz. There is also a slight reduction in frequency nadir in both cases.

The results provide evidence of the combination of energy storage in providing a response to a significant system grid. The HESS scheme can respond rapidly and maintain grid frequency stability during a major contingency, hence the scheme demonstrates flexibility in arresting transient events, while critically ensuring postfault system resilience.

High penetration of variable and distributed energy sources can lead to a considerable increase of the frequency deviation due to reduced inertia. Table 4.7 presents frequency and standard deviations for each case.

It can be seen in this case study that when a significant quantity, 230 MW, of generation is removed from the system, the frequency deviation is lower compared with the case without the HESS connected.

TABLE 4.5 Shows results of the frequency response with UFLS and HESS connected.

	UFLS stage activated	Freq. nadir [Hz]	Settling freq. [Hz]	Load shed [MW]
No HESS & UFLS	4	49.466	49.761	100
HESS	4	49.504	49.794	100
UFLS	4	49.465	49.867	100
HESS & UFLS	4	49.506	49.883	100
UFLS	4	49.465	49.979	230
HESS & UFLS	4	49.506	49.997	230

TABLE 4.6 UFLS relay main settings.

UFLS stage activated	Frequency [Hz]	Time [s]
1	49.80	4
2	49.50	2.30
3	49.20	1
4	48.70	0.00

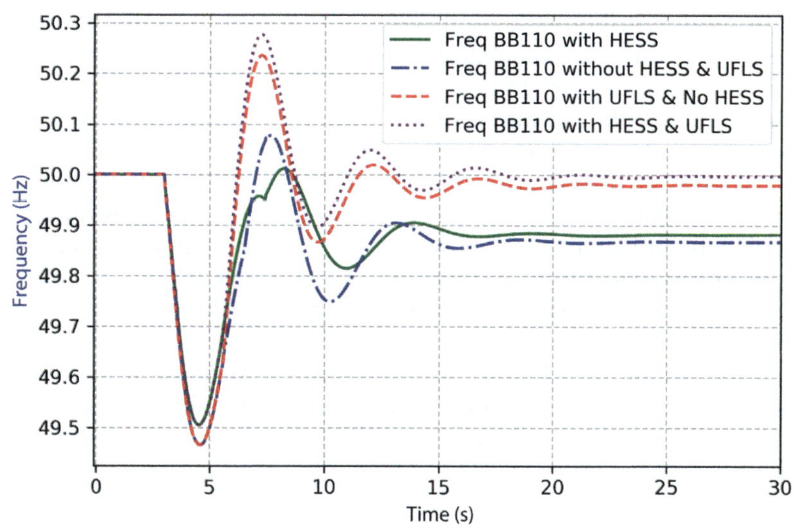

FIGURE 4.20 Frequency response following the large underfrequency transient event.

As demonstrated in all cases studies presented, an ESS can help solve multiple problems, but chiefly those encountered from greater levels of variable, renewable, and distributed energy sources across a system. Furthermore, ESSs can participate in improving the performance of frequency control under a lower influence of governor responses.

TABLE 4.7 Summary statistics of frequency deviation.

	10% load event		230 MW loss generation	
	Freq ESS online 1	Freq No ESS 1	Freq ESS online 2	Freq no ESS 2
Average	49.8768 Hz	49.8600 Hz	49.9745 Hz	49.9534 Hz
Std dev.	0.0972	0.1123	0.1288	0.1318
20%	49.8800 Hz	49.8592 Hz	49.9894 Hz	49.9720 Hz
50%	49.8820 Hz	49.8674 Hz	49.9970 Hz	49.9791 Hz
75%	49.9007 Hz	49.8842 Hz	50.0000 Hz	49.9917 Hz

4.6 Future implications of hybridized scheme to transition issues

The dynamic stability of the power system with a high penetration of distributed energy source is a work in progress. How it will shape, reshape, and operate will be determined by the role of key system stakeholders and the business models developed in response to regulations and policy impacting energy and infrastructure transition. It is essential to exploit the valuable and increasing level of sustainable resources connected at the distribution level, particularly wind, solar, and ESSs to deliver appropriate system-wide benefits.

The connection of DERs on both sides of the customer meter, improves demand-side participation in local distribution-level markets and ensures optimal tariffs and energy trading to preserve and develop reliability and ancillary services. New societal needs have become important, including recognition of large-scale EV connection and greater awareness of the impact of greenhouse gas emissions.

Current developments and future power grids and networks must be required to accommodate wide-scale distributed generation sources, smart appliances, EVs, and energy storage devices to transform passive consumers to active prosumers with greater influence than before and, as financial stakeholders, an expectation of good power quality and backup power services at local and national levels (Fig. 4.21).

By making full use of the flexibility of emerging DERs, including demand-side response (DSR) technology and storage, flexible services will help guarantee dynamic stability with cost savings delivered through active consumer load management, and real-time tariff pricing signals used to microassist demand profiting and improvement of system resilience using plug and participate devices (in addition to smart meters) to ensure better security in systems during and after grid faults.

4.6.1 Dynamic system stability

Distribution faults do not necessarily remain at this level and can propagate to consumer loads, if nearby, but more critically to transmission level. High volume penetration of a distributed generation source can bring a number of simultaneous effects, which will directly affect the quality of supply, particularly voltage dips, but which may also threaten

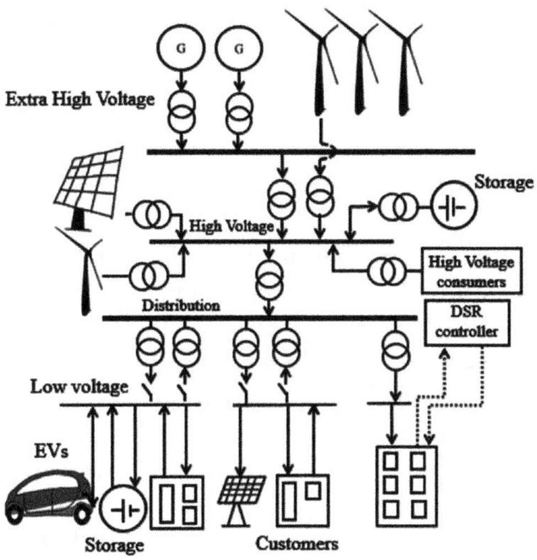

FIGURE 4.21 Energy to and from power systems consumers.

system dynamic stability. The depletion of system inertia is usually one of the major issues affecting energy transition, as well as decreasing the system-wide inverse droop due to greater displacement of centralized and thermal generation and more low-inertia generating units being connected. System inertia is related to aggregate droop and there is evident correlation between depleted inertia and increase in frequency fluctuation, as demonstrated in the case studies presented.

4.6.2 Impact of HESS responses

Power systems are challenged by problems which affect dynamic stability, among these active power oscillations, and frequency and voltage control. As shown in this work, the use of ESSs for frequency problems in particular can present viable solutions, since energy storage can potentially deliver fast responses. Hybridized ESS units can deliver solutions during major contingencies, and following disturbances in postfault recovery.

Hybridized storage is able to deliver ramp-up power responses to preserve frequency stability. Ultracapacitors can supply rapid demand ramp and fast responses at the instant of transient events during large disturbances. Large BESS capacities (MW and MWh) can help to contribute to frequency control solutions and in many instances supply the equivalent system services delivered by traditional synchronous generation more efficiently and without some of the drawbacks, including reduced postevent recovery oscillations.

Battery and ultracapacitor storage can also be hybridized with mechanical storage (PHS and CAES) to deliver equivalent analogue inertia. By hybridizing storage technologies, digital inertia services can be delivered reliably and with reduced maintenance impacts. Ultracapacitors, in particular, provide a form of kinetic energy which, though batteries offer potential energy, can improve system damping to limit active power oscillations and strengthen the grid to better resist major events which undermine system frequency and threaten event cascading.

4.7 Chapter summary

The test model developed in DigSilent Power factory in this chapter provided a number of case studies to examine storage services. The case studies investigated the impact of battery and ultracapacitor or supercapacitor storage in the system to improve the transient responses and render primary and secondary responses to strengthen the grid after large under-frequency events. Deployment of ESSs lessens the need for contributions from synchronous generation, which has fringe benefits in limiting impacts on asset ageing. UFLS relays tackle frequency transients caused by load-generation imbalance, which can mean high or low-frequency events. Load shedding is proportional to the frequency deviation before it reaches its minimum nadir during an under-frequency event. The work has investigated energy storage systems for the provision of grid service, and it can be used to maintain the system stability following under frequency transient (e.g. generation loss). The application of energy storage devices contributes to project service into the grid and storage devices are characterised by their power response, which can take different forms for inverter - connected energy storage such batteries, super capacitor and flywheel. The synthetic inertia is delivered through the inverter controls mechanism, which is the interface between the storage and the power grid.Many studies have sized grid ESS on different approaches to resolve various problems. However, this work presents the sizing of grid ESS based on key parameters indicators, which can allow to meet the application required performance. Hybridised storage unit can play a significant role in systems frequency stability as it can have two or more different storage technologies to provide multiple service and benefit at transmission and distribution level.Increasing investment in smart energy solutions at distribution level using deployment of intelligent systems is transforming the traditional model of energy use and pricing from heavily centralised to decentralised and demand-side networks, where passive consumers are becoming active prosumers. This transition is only possible with emerging technologies, among which storage schemes are one of the keys enabling assets which are helping bridge the gap between 1) the option of installing and operating greater levels of generation to meet rising demand; or 2) the more favoured option of educating consumers (domestic and industrial) to use energy optimally and efficiently and participate in available time-scheduled demand-reduction by connecting load when capacity is inexpensive and plentiful.

References

ACEP, 2012. Advanced battery storage systems testing at ACEP, the prudent energy VRB-ESS. University of Alaska, March 2012.

Ariyaratna, P., Muttaqi, K.M., Sutanto, D., 2018. A novel control strategy to mitigate slow and fast fluctuations of the voltage profile at common coupling point of rooftop solar PV unit with an integrated hybrid energy storage system. J. Energy Storage 20, 409–417.

Arvind Parwal, M., Fregelius, I., Temiz, M., Goteman, J.G., de Oliveira, C., Bostrom, et al., 2018. Energy management for a grid-connected wave energy park through a hybrid energy storage system. Appl. Energy 231, 399–411.

Australian Energy Market Operator, 2017. Fast Frequency Response in the NEM/Future Power System Security Program.

Barai, G.R., Venkatesh, B., Awadallah, M.A., 2018. Optimisation of hybrid energy storage systems for power curve smoothening at grid-scale. Can. J. Electr. Comp. Eng. 41 (2), Spring.

Blake, S., Jialiang, Y., Taylor, P., Miller, D., Lloyd, I., 2013. Using electrical energy storage to support customers under faulted network conditions. In: 22nd International Conference and Exhibition in 2013. Electricity Distribution (CIRED 2013), pp. 1–4.

Bloom, A., Helman, U., Holttinen, H., Summers, K., Bakke, J., Brinkman, G., et al., 2017. "It's indisputable: five facts about planning and operating modern power systems. IEEE Power Energy Mag. 15 (6), 22–30.

Brogan, P.V., Alikhanzadeh, A.H., Best, R.J., Morrow, D.J., Kubik, M.L., 2017. Fast frequency response requirements for replacements of observed generator response during under frequency transients. CIGRE Symposium, Dublin, vol.26.

Brogan, P.V., Member, R.B., Member, J.M., McKinley, K., Kubik, M.L., 2018. Effect of BESS response on frequency and RoCoF during under frequency transients. IEEE Trans. Power Syst. 34, 575-583.

Carnegie, R., Gothan, D., Nderitu, D., Paul, V.P., 2013. Utility-scale energy storage systems: benefits, applications, and technologies, state utility forecasting group, June 2013.

Chong, L.W., Wong, Y.W., Rajkumar, R.K., Rajkumar, R.K., Isa, D., 2016. Hybrid energy storage systems and control strategies for stand-alone renewable energy power systems. Renew. Sustain. Energy Rev. 66, 174–189.

Clark, K., Miller, N.W., Sanchez-Gasca, J.J., 2010. Modeling of GE Wind Turbine-Generators for Grid Studies. Version 4.5, April 2010, General Electric International, Inc.

Delille, G., et al., 2012. Dynamic frequency control support by energy storage to reduce the impact of wind and solar generation on isolated power system's inertia. IEEE Trans. Sustain. Energy 3 (4), 931–939.

DS3 System Services, 2014. Portfolio capability analysis/EirGrid-Soni. November 2014.

Editorial, 2015. Energy storage systems for a low carbon future—in need of an integrated approach. Appl. Energy 137, 463–466.

Egido, I., Sigrist, L., Lobato, E., Rouco, L., Barrado, A., 2015. An Ultra-capacitor for frequency stability enhancement in small-isolated power systems: models, simulation and fields tests. App. Energy 137, 670–676.

Eirgrid/Soni. RoCoF alternative & complementary solutions project phase 2 study report. <http://www.eirgridgroup.com/sitefiles/library/EirGrid/RoCoF-Alternative-Solutions-Project-Phase-2-Report-Final.pdf>.

EirGrid/SONI, 2016. RoCof alternative & complementary solutions project phase 2 study report, March 31, 2016. <www.eirgridgroup.com/sitefiles/library/EirGrid/RoCoF-Alternative-Solutions-Project-Phase-2-Report-Final.pdf>.

Energy Networks Association, 2006. Engineering Recommendation P2/6—Security of Supply. Energy Networks Association.

EPRI, 2005. VRB Energy Storage for Voltage Stabilisation, Testing and Evaluation of the PacifiCorp Vanadium Redox Battery Energy Storage System at Castle Valley. Utah.

Eyer, J.M., Iannucci, J.J., Corey, G.P., 2004. Energy Storage Benefits and Market Analysis Handbook, A Study for the Doe Energy Storage Systems Program. Sandia National Laboratories.

Flynn, D., Power, M., O'Malley, M., 2016. Renewables Integration, Flexibility Measures and Operational Tools for the Ireland and N. Ireland Power System. Société de l'électricité, de l'électronique et des technologies de l' Information et de la communication.

Greenwood, D.M., Lim, K.Y., Patsios, C., Lyons, P.F., Lim, Y.S., Taylor, P.C., 2017. Frequency response services designed for energy storage. Appl. Energy 203, 115–127.

https://www.powerstar.com/virtue/virtue-energy-storage/ (accessed 15.11.18).

Jing, W., Lai, C.H., Wong, W.S.H., Wong, M.L.D., 2018. A comprehensive study of battery-supercapacitor hybrid energy storage system for standalone PV power system in rural electrification. Appl. Energy 224, 340–356.

Knap, V., et al., 2014a. Grid inertial response with lithium-ion battery energy storage systems. In: IEEE Internaional Symposium on Industrial Electronics (ISIE), June 2014a, pp. 1813–1818.

Knap, V., Sinha, R., Swierczynski, M., Stroe, D.-I., Chaudhary, S., 2014b. Grid inertial response with lithium-ion battery energy storage systems. In: 23rd IEEE International Symposium on Industrial Electronics, ISIE.

Knap, V., Chaudhary, Sanjay, K., Stroe, D.-I., Swierczynski, M.J., Craciun, B.-I., et al., 2016. Sizing of an energy storage system for grid inertial response and primary frequency reserve. IEEE Trans. Power Syst. 31 (5).

Koller, M., Borsche, T., Ulbig, A., Anderson, G., 2015. Review of grid applications with the Zurich 1 the. Electr. Power Syst. 120, 128–135.

Lyons, P.F., wade, N.S., Jiang, T., Taylor, P.C., Hashiesh, F., Michel, M., et al., 2015. Design and analysis of electrical energy storage demonstration projects on UK distribution networks. Appl. Energy 137, 677–691.

Mitsuki Sagara, M., Danish, M.S.S., Funabashi, T., 2016. Voltage stability improvement to power systems with energy storage systems. In: 17th International Conference on Harmonics and Quality of Power (ICHQP), October 16–19, 2016.

Nguyen, C.P., Flueck, A.J., 2012. Agent based restoration with distributed energy storage support in smart grids. IEEE Trans. Smart Grid 3 (2), 1029–1038.

Olabi, A.G., 2017. Renewable energy and energy storage systems. Energy 136, 1–6. Available from: https://doi.org/10.1016/j.energy.2017.07.054.

Oskilly, A.P., Taylor, P.C., Yan, J., 2015. Energy storage systems for a low carbon future in need of an integrated approach. Appl. Energy 137, 463–466.

Oudalov, A., et al., 2007. Optimizing a battery Eeergy storage system for primary frequency control. IEEE Trans. Power Syst. 22 (3), 1259–1266.

Rastler, D., 2010. Electricity Energy Storage Technology Options: A White Paper Primer on Applications, Costs and Benefits. Palo Alto, CA: Electric Power Research Institute.

Saboori, H., Hemmati, R., Jirdehi, M.A., 2015. Reliability improvement in radial electrical distribution network by optimal planning of energy storage systems. Energy 93, 2299–2312.

Ubertalli, J., Littler, T.B., 2018. Protecting frequency stability using local inertial mapping. IET J. Eng. DPSP, 2018.

Ulbig, A., Borsche, T.S., Anderson, G., 2014. Impact of low rotational inertia on power system stability and operation. IFAC Proc. Vols. 47 (3), 7290–7297.

Wade, N., Taylor, P., Lang, P., Jones, P., 2010. Evaluating the benefits of an electrical energy storage system in a future smart grid. Energy Policy .

Wang, D., Ren, C., Sivasubramaniam, A., 2012. Energy storage in datacentres: what, where, and how much? ACM Sigmetrics. 40 (1), 187–198.

Wang, P., Liang, D.H., Yi, J., Lyons, P.F., Davison, P.J., Taylor, P.C., 2014b. Integrating electrical energy storage into coordinated voltage control schemes for distribution networks. IEEE Trans. Smart Grid 5, 1018–1032.

Wang, G., Ciobotaru, M., Agelidis, V.G., 2014a. Power smoothing of large solar PV plant using hybrid energy storage. IEEE Trans. Sustain. Energy 162, 971–974.

Yi, J., Wang, P., Davison, P.J., Lyons, P.F., Liang, D., Taylor, P.C., 2012. Distribution network voltage control using energy storage and demand-side response. In: IEEE PES Innovative Smart Grid Technologies (ISGT) Europe Conference, Berlin, Germany.

Yue, M., Wang, X., July 2015. Grid inertial response-based probabilistic determination of energy storage system capacity under high solar penetration. IEEE Trans. Sustain. Energy 6 (3), 1039–1049.

Zhao, H., Wu, Q., Hu, S., Xu, H., Rasmussen, C.N., 2015. Review of energy storage system for Wind power integration support. Appl. Energy 137, 545–553.

CHAPTER

5

Design and performance of two decomposition paradigms in forecasting daily solar radiation with evolutionary polynomial regression: wavelet transform versus ensemble empirical mode decomposition

Mohammad Rezaie-Balf[1], *Sungwon Kim*[2], *Alireza Ghaemi*[1] *and Ravinesh Deo*[3]

[1]Department of Civil Engineering, Graduate University of Advanced Technology, Kerman, Iran [2]Department of Railroad Construction and Safety Engineering, Dongyang University, Yeongju, South Korea [3]School of Sciences, University of Southern Queenland, Springfield, QLD, Australia

5.1 Introduction

Owing to the remarkable influence of destructive fossil fuels on the environment and the rapid growth in the global energy demands, resources of renewable energy are receiving further attention of scholars, governments, and industries. There are various forms of energy resources (e.g., nuclear, geothermal, hydropower, wind, and solar), of which solar energy, as a sustainable, reliable, and renewable energy resource, has attracted researchers' attention.

Solar energy is considered to be a promising renewable energy resource due to its unique feature which is widely accessible around the world. Although, the sun releases this crucial energy, only some of it reaches the ground, which can be analytically

Predictive Modelling for Energy Management and Power Systems Engineering
DOI: https://doi.org/10.1016/B978-0-12-817772-3.00005-7

investigated using the geometric relationship between the sun and Earth. Hence, for a useful implementation and management of solar energy investment, accurate forecasting of the long-term amount of solar radiation (SR) which reaches the Earth is as the prominent issue (Mohammadi et al., 2016; Bayrakçı et al., 2018; Baser and Demirhan, 2017; Salcedo-Sanz et al., 2018). Solar energy is recruited either directly or indirectly in a wide range of applications such as air and water heating/cooling and electricity production. It is a fact that because of the developments in technology, solar energy could play a crucial role in providing large portions of the current and future global energy demands for both private and public sectors. Currently, the harnessing of solar energy is getting cheaper and more efficient because of technological changes, market competition, and enhancements in the capture, conversion, and distribution processes. It would be significant for mitigating global warming and offer sustainability. The knowledge of global SR is one of the important and fundamental requirements in various technological and scientific applications of solar energy such as the evaluation of established solar technologies as well as predictions of the installations' feasibilities in the future (Mohammadi et al., 2016; Wu and Wang, 2016; Despotovic et al., 2016).

Generally, the SR information related to every specific location should be acquired from measurement stations using the highly accurate instruments. However, the reliable measured global SR data are scarce in a number of locations around the world due to the paucity of instruments and fiscal issues (Shamshirband et al., 2015; Gairaa et al., 2016). In this regard this has necessitated the development of appropriate models for predicting the global SR with acceptable accuracy by using a significant number of input elements (Mohammadi et al., 2016; Gueymard, 2014). These parameters contain geographical and meteorological variables, such as ambient temperatures, SD, water vapor, sea-level pressures (SLP), relative humidity (RH), cloud covering, longitude, latitude, altitude, as well as extraterrestrial radiation.

During the last decades, a vast number of artificial intelligence (AI) and computational intelligence approaches have been considered by scholars as highly efficient techniques for the prediction of global SR in various locations across the globe. A couple of reviews of machine learning (ML) methods used for estimating the solar energy problems are described. Yadav and Chandel (2014) reported an exhaustive review on SR estimation using artificial neural networks (ANN). They described the different case studies that utilized atmospheric input parameters. Mellit and Kalogirou (2008) recruited various AI methods [e.g., ANN, genetic algorithm (GA), fuzzy logic, and expert systems] to evaluate their performance in the sizing of photovoltaic power systems, power simulations, photovoltaic systems control, and the prediction of photovoltaic-based power by meteorological and atmospheric datasets. Moreover, the research performed by Behrang et al. (2010) included the various combinations of input variables into the radial basis function neural network (RBF-NN) and multilayer perceptron neural network (MLP-NN) models. Estimation of daily SR (DSR) was investigated using a hybrid model that was presented by Kim et al. (2016). They employed the wavelet transform (WT) to decompose the SR dataset into subseries data, and then used these new datasets to develop support vector machines (SVM). Bou-Rabee (2017) assessed the performance of ANN in Kuwait, whereas Alsina et al. (2016) assessed the ANN performance to forecast the monthly DSR at 45 measuring stations in Italy. Lou et al. (2016) designed a data-intelligent model with boosted regression tree (RT) to estimate diffuse SR in Denver and Hong Kong. Mohammadi et al. (2015) applied an integration of WT and SVM to assess the model accuracy for the prediction of horizontal global SR. The research carried out by Olatomiwa et al. (2015) combined support vector regression (SVR) with firefly algorithm (FFA) to validate its suitability for SR forecasting in Nigeria.

Belaid and Mellit (2016) tackled a problem of SR estimation using SVR, considering different forecasting horizons, while Monteiro et al. (2017) compared the results of ANN and SVR techniques to generate the photovoltaic power. In the study of Yacef et al. (2012), a Bayesian network (BN) in comparison with the ANN models was applied, whereas a new approach based on temporal Gaussian processes (GP) was presented by Salcedo-Sanz et al. (2014a,b). In the latter, the GP model was seen to improve alternative methods (e.g., ANN, SVR, and RTs). Another approach for SR estimation used "all-sky" conditions and satellite images. For example, Fu and Cheng (2013) estimated SR with features extracted from all-sky images, cloud pixels, frame differences, gradient magnitude, intensity levels, accumulated intensity along the vertical line of the Sun, and the number of corners in a cloud image. The solar ultraviolet index was estimated by Deo and Şahin (2017) using various models, namely extreme learning machine (ELM), multivariate adaptive regression spline (MARS), and M5 model tree (M5Tree). It is interesting to say that they used a very short-term reactive (e.g., 10 minutes lead-time) ultraviolet index to develop their proposed models. Recently, Salcedo-Sanz et al. (2018) applied a hybrid approach called the neuroevolutionary wrapper-based model to predict the daily global SR in the solar-rich sunshine state of Queensland, Australia. They employed the optimization of Coral Reefs for the feature selection processes guided using an ELM algorithm as the model's fitness function to screen an optimal set of predictor variables. In early 2018 Ghimire et al. (2018) presented a new model to forecast the monthly mean DSR by constructing ELM integrated with the moderate resolution imaging spectroradiometer (MODIS)-based satellite. They introduced self-adaptive differential evolutionary ELM as a benchmark, which was compared with several optimization algorithm-based models.

Nevertheless, during the last decades, a vast number of researches have been carried out to prognosticate the SR diffusion using conventional and empirical techniques, but there is still a major challenge regarding the development of suitable methods with acceptable adaptability and reliability for increasing the accuracy of predictions. Recently, integrating various techniques to create a hybrid model has received remarkable attention in the SR area. It is generally possible to take the merits of the specific nature of each technique for improving the accuracy. In fact, the significant features of each technique are the abilities to capture various patterns in the data series. Regarding both empirical and theoretical findings, it has been proved that integration of techniques would be a particularly successful method for estimating SR with the permissible precision and reliability (Deo and Şahin, 2017; Ji and Chee, 2011; Mostafavi et al., 2013; Salcedo-Sanz et al., 2014a,b).

Appling the hybrid approaches for SR forecasting has reached enormous popularity since it takes advantage of different methods. Consequently, in the present study, a new model, which was created by the hybridization of evolutionary polynomial regression (EPR) and two preprocessing procedures (e.g., EEMD and WT), is recruited to forecast the daily global SR. Basically, EPR is taken into account as a soft computing technique which has lately obtained importance in the different hydrological applications. To verify the capability of the implemented hybrid W-H-EPR, W-D-EPR, W-C-EPR, and EEMD-EPR models, long-term measured databases containing the daily global SR and different meteorological variables for Busan and Seoul stations, South Korea are utilized. The performances of W-H-EPR, W-D-EPR, W-C-EPR, and EEMD-EPR models are assessed by the diverse statistical measures together and the diagnostic plots, and this is benchmarked against the four primary standalone models. Additionally, to evaluate the predictive capacity of the proposed models, uncertainty analysis as a fair indication is applied. The organization of this study is as follows:

Section 5.2 includes the description of the datasets utilized in this study, and employs statistical criteria for evaluating the model performances. Section 5.3, which offers the utilized methodology, is divided into three parts: WT decomposition is described (Section 5.3.1); EEMD is defined (Section 5.3.2); and EPR as a soft computing technique is presented (Section 5.3.3). After describing the models' implementation in Section 5.4, the results and discussion are explained in Section 5.5. Ultimately, Section 5.6 presents the conclusions.

5.2 Study area and evaluation criterion

In this study, meteorological data were gathered from different weather sites, namely Busan (longitude, 129°03′ E; latitude, 35°10′ N; altitude, 69.2 m) and Seoul (longitude, 126°96′ E; latitude, 37°57′ N; altitude, 85.5 m) in South Korea. These stations are operated by the Korea Meteorological Administration (KMA) (https://data.kma.go.kr/), and the regions used to test the performance of the proposed models are illustrated in Fig. 5.1. The weather data consisted of 16 years (from 2000 to 2016) of daily records of air temperature (TA), RH, vapor pressure (VP), SLP, pan evaporation (PE), SD, and SR.

Generally, the availability of long-term measured data is of particular significance to provide accurate estimation of SR. Therefore the lengths of utilized datasets for this study can be considered sufficiently long for modeling the global SR properly. Furthermore, the accuracy of the models developed for SR estimation is mainly affected by the quality of the raw SR data (Mohammadi et al., 2016). In the used measured global SR data, there were some inconsistencies and abnormalities in the values, typically due to instruments' malfunction. After the data sorting, the missing values for the global sequences were replaced with interpolated values using time series analysis techniques through automated programs. More generally, this technique provides that if there are certain days that have either missing value or data is erroneous from sunrise to sunset, in this case, the value will be replaced with the average of the two nearest nonmissing values (e.g., nearest previous and following measured data) (Bailek et al., 2018).

In the present study, the datasets were divided into parameter estimation and testing datasets. Hence, the first dataset covers the period from 2000 to 2012, while the second dataset covers the period from 2013 to 2016. The first dataset is used for the DSR model construction and statistical performance analysis, while the second dataset is used to validate the accuracies of the selected models. Table 5.1 represents the statistical parameters of

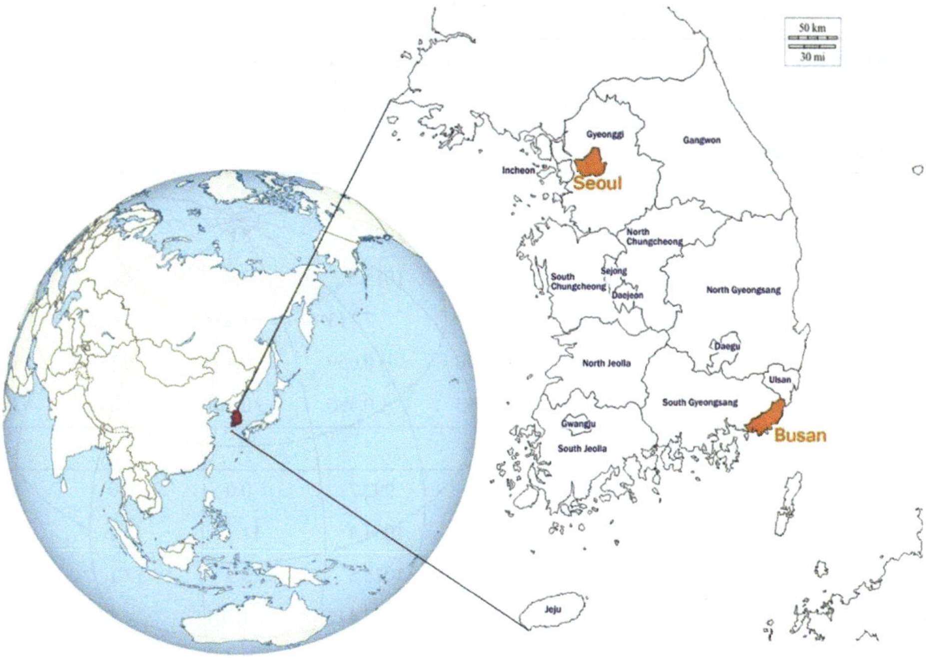

FIGURE 5.1 A map of the study area.

the used climatic variables (X) in the studied stations. Moreover, Fig. 5.2 illustrates the distribution of climatic variables.

5.3 Methodology

In the present study, hybrid models (e.g., W-H-EPR, W-D-EPR, W-C-EPR, and EEMD-EPR) were implemented by coupling the standalone EPR model with WT and EEMD for the DSR forecasting. Additionally, the methodology of the proposed approaches is illustrated briefly in this section.

5.3.1 Wavelet transform

WT analysis is one of the preprocessing techniques for extracting the prominent characteristics of the original dataset, such as discontinuities, trends, and breakdown points. However, these characteristics may be missed in some signal analysis techniques (Rajaee and Shahabi, 2016).

One of the advantages of the WT method is decomposing a time series signal to separate subsignals. WT is divided into continuous WT (CWT) and discrete WT (DWT). Generally, DWT is more beneficial than CWT for solving the practical issues in the hydrology field (Shafaei and Kisi, 2017; Rezaie-balf et al., 2017).

TABLE 5.1 Statistic parameters of used climatic variables (2000–2016).

Statistical parameters	Meteorological variables with unit						
	TA (°C)	RH (%)	VP (hPa)	SLP (hPa)	PE (mm)	SD (h)	SR (MJ/m^2)
Busan (calibration dataset)							
Minimum	− 7.2	11.3	0.8	992.2	0.2	0.0	0.0
Maximum	30.1	99.0	37.2	1036.2	8.8	13.1	31.3
Mean	14.8	62.2	12.9	1015.5	3.1	6.1	14.1
S_x	8.144	18.826	8.392	7.143	1.483	3.863	7.033
C_{sx}	− 0.249	− 0.238	0.496	− 0.059	0.455	− 0.343	0.101
K_x	− 0.926	− 0.889	− 0.865	− 0.496	− 0.401	− 1.248	− 0.824
Busan station (validation dataset)							
Minimum	− 7.6	17.6	1.1	992.7	0.0	0.0	0.2
Maximum	31.7	99.9	33.1	1034.4	11.5	13.1	28.7
Mean	15.4	64.1	13.4	1015.8	3.4	7.1	14.0
S_x	8.125	17.854	8.282	7.442	1.701	4.123	6.952
C_{sx}	− 0.254	− 0.217	0.389	− 0.057	0.460	− 0.552	0.088
K_x	− 0.979	− 0.742	− 1.031	− 0.642	− 0.263	− 1.064	− 0.891
Seoul station (calibration dataset)							
Minimum	− 15.5	19.9	0.8	990.7	0.0	0.1	0.0
Maximum	31.8	96.5	30.6	1039.2	13.5	13.0	31.1
Mean	12.7	61.3	11.3	1016.1	5.2	3.1	12.1
S_x	10.397	14.885	7.941	8.126	3.756	2.003	6.594
C_{sx}	− 0.298	0.009	0.610	− 0.001	0.005	0.798	0.331
K_x	− 1.067	− 0.621	− 0.862	− 0.673	− 1.278	0.158	− 0.659
Seoul station (validation dataset)							
Minimum	− 14.4	23.8	0.9	990.8	0.0	0.0	0.5
Maximum	31.4	99.8	32.2	1037.6	13.5	15.0	28.6
Mean	13.3	60.5	11.5	1016.2	6.8	3.0	12.3
S_x	10.677	14.523	8.041	8.363	3.921	1.8404	6.027
C_{sx}	− 0.287	0.073	0.618	− 0.022	− 0.427	0.639	0.193
K_x	− 1.136	− 0.422	− 0.817	− 0.752	− 1.036	0.609	− 0.750

Note: X_{min}, X_{max}, X_{mean}, S_x, C_{sx}, and K_x denote the minimum, maximum, mean, standard deviation, skewness, and kurtosis, respectively.

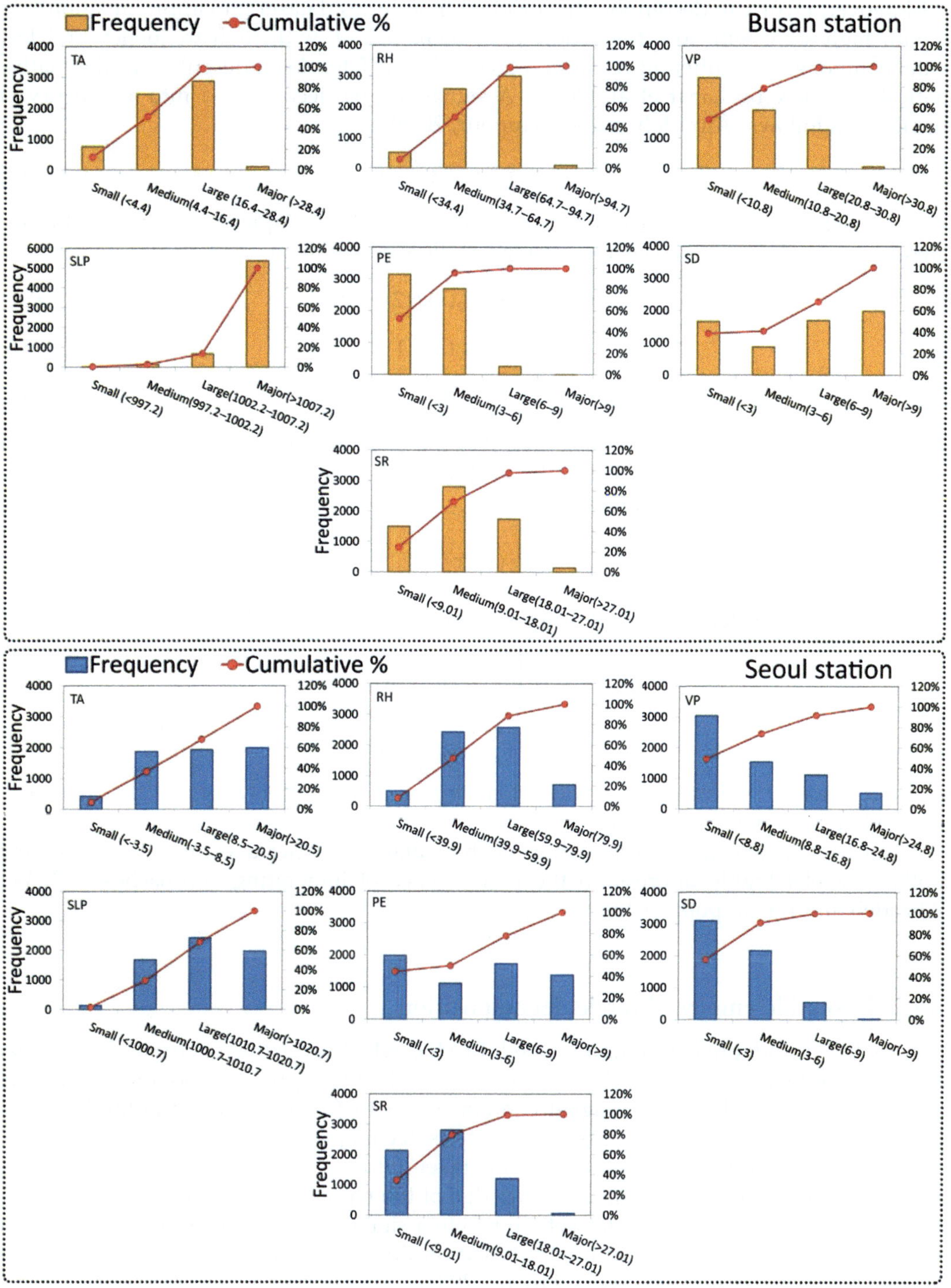

FIGURE 5.2 Histograms of the parameters.

Moreover, DWT can decompose the original time series dataset into approximation (i.e., a_i with high scale with low frequency) and detail (i.e., d_i with high frequency and low scale) components at different resolution levels. The WT has been described in the mathematical overview by Labat et al. (2000). A signal can be decomposed into a category of functions as follows (Alizadeh and Kavianpour, 2015):

$$\psi_{i,j}(x) = 2^{\frac{i}{2}}\psi_{i,j}\left(2^i x - j\right) \tag{5.1}$$

where $\psi_{i,j}(x)$ has been generated using the mother wavelet ψ implemented by i and translated by j. ψ is also provided by the scaling function $\varphi(x)$ which is written as,

$$\varphi(x) = \sqrt{2}\sum h_0(n)\varphi(2x - n) \tag{5.2}$$

$$h_1(n) = nh_0(n - 1) \tag{5.3}$$

$$\psi(x) = \sqrt{2}\sum h_1(n)\varphi(2x - n) \tag{5.4}$$

in which $h_0(n)$ plays a prominent role in DWT and is computed corresponding to wavelet bases with different characteristics. In addition, ψ must determine the condition of Eq. (5.5) (Gupta and Gupta, 2007).

$$\int \psi(x)dx = 0 \tag{5.5}$$

The discrete wavelet function of a signal $f(x)$ is evaluated using Eqs. (5.6) and (5.7) as,

$$C_{i,j} = \int_{-\infty}^{+\infty} f(x)\psi^*_{i,j}(x)dx \tag{5.6}$$

$$f(x) = \sum C_{i,j}\psi_{i,j}(x) \tag{5.7}$$

where ψ^* and $C_{i,j}$ indicate the convoluted conjugate of mother wavelet and the coefficient of signal approximate, respectively.

It can be said that the determination of the suitable wavelet decomposition level and mother wavelet highly impress on the performance of integrating approaches with WT (Alizadeh and Kavianpour, 2015).

5.3.2 Ensemble empirical mode decomposition

Recently, an adaptive technique called EEMD, which is extracted from the popular empirical mode decomposition (EMD), has been applied to represent a nonlinear and nonstationary signal as the sum of signal components with frequency and amplitude-modulated parameters with a noise-assisted analysis technique (Huang et al., 1998; Wu & Huang, 2009; Seo & Kim, 2016). The EMD is taken into account as a self-adaptive decomposition method that without initial knowledge of the nature and intrinsic mode functions (IMFs), numbers embedded in datasets illustrate a signal as a sum of zero-mean behavior of both slow and fast fluctuation modes based on IMFs that defined using two conditions as follows (Huang et al., 1998; Wu & Huang, 2009):

1. Throughout length, the number of extrema zero-crossings should be equal or greater than one.
2. At any point, the average value of the signal and the envelope determined using the local maxima and minima, respectively, are zero. According to these requirements, several meaningful IMFs may be determined. Generally, each IMF indicates a simple oscillatory mode in comparison with the function of simple harmonic. Regarding the definition, each data series $y(t)$ $(t = 1, 2, \ldots, n)$ can be decomposed by a shifting trend that can be described briefly as follows (Huang et al., 1998):
   1. Recognize all local minima and maxima points for a provided time series $y(t)$ and select them as lower envelope $e_{min}(t)$ and upper envelope $e_{max}(t)$, respectively. Hence, the average between upper and lower envelopes $a(t)$ and its difference from $y(t)$ are computed as follows:

$$a(t) = \left(e_{max}(t) + e_{min}(t)\right)/2 \tag{5.8}$$

$$h(t) = y(t) - a(t) \tag{5.9}$$

   2. Regarding the benchmarks, if $h(t)$ meets the two IMF conditions, $h(t)$ is determined as the first IMF, which is written as $c1(t)$ and 1 is indicated its index; otherwise, the position of $y(t)$ and $h(t)$ is replaced and steps 1–4 are iterated until $h(t)$ meets the two conditions of IMF.
3. After that, the remainder $r_1(t) = y(t) - c_1(t)$ is behaved as novel data subjected to the same shifting trends as the description of the next IMF from $r_1(t)$. Eventually, the shifting process can be ceased, when the remainder $r(t)$ becomes a monotonic function or at most has one local extreme point from which no more IMF can be extracted (Huang et al.; 2003).

As a result of this shifting process, the original time series $y(t)$ can be computed using the sum of IMFs and a final residue as follows:

$$y(t) = \sum_{i=1}^{m} c_i(t) + r_m(t) \tag{5.10}$$

where $r_m(t)$ and m indicate the ultimate residue and IMF numbers, respectively, and $c_i(t)$ is nearly orthogonal to the former, and roughly the average of them is continuously equal to zero. For more information about EMD technique and stopping criteria, refer to Huang et al. (1998, 2003) literatures.

However, recent research showed that EMD suffers from the mode mixing problem (Wu & Huang, 2009). Mode mixing is defined not only as a single IMF including disparate scale components but also as a component of a similar scale residing in various IMFs (Lei et al., 2009). EEMD is a developed technique for overcoming the mode mixing suffered in EMD, illustrated by Wu & Huang (2009). In EEMD model, the added white noise would fill in the whole time-frequency space uniformly, which leads to the facilitation of a natural separation of the frequency scales and reduces the occurrence of mode mixing. Regarding the EMD technique features, the stages of EEMD can be explained as follows (Wu & Huang, 2009):

1. Adding a series of white noise to the target datasets.
2. Decomposing the datasets including white noise to IMFs.

3. Repeating stages (1) and (2) over and over again with various white noise series every time.
4. Obtaining the (ensemble) means relevant to decompositions IMFs as the final finding.

5.3.3 Evolutionary polynomial regression

EPR as a nonlinear stepwise regression technique of AI methods is the combination of numerical regression with a GA to provide a mathematical expression based on evolutionary computing. Generally, the form of the EPR mathematical expression is presented as (Laucelli and Giustolisi, 2011; Balf et al., 2018),

$$y = \sum_{j=1}^{m} F(X, C_j, f(X)) + C \tag{5.11}$$

where X and C_j indicate independent and constant values, respectively; y is the forecasted value; f is a function defined by user; F indicates a function built into the process; and m is the sample number.

Through EPR modeling processes, a model is constructed by a training dataset and is evaluated by testing. A variety of factors, such as function type (e.g., tangent hyperbolic and exponential) and number of input variables, have remarkabl effects on the EPR performance (Rezaie-Balf and Kisi, 2018). Actually, one of the advantages for the EPR modeling process is the application of the GA and the least squares (LS) techniques. The GA is applied to find the best models with appropriate accuracy according to the input variables. Furthermore, the LS method is recruited to determine the constant values using the minimization of the sum of squared errors (SSE) approach which is written as follows:

$$\mathrm{SSE} = \frac{\sum_{i=1}^{M} (O_i - P_i)^2}{M} \tag{5.12}$$

in which P and O are the forecasted and observed values related to the training dataset, respectively, and M is the samples number (Rezania et al., 2008; Ahangar-Asr et al., 2011). The ultimate structure of the EPR model can be expressed as Eqs. (5.13–5.16).

$$\hat{y} = a_0 + \sum_{j=1}^{m} a_j (X_1)^{\mathrm{ES}(j,1)} \ldots (X_K)^{\mathrm{ES}(j,K)} f\left((X_1)^{\mathrm{ES}(j,K+1)}\right) \ldots f\left((X_K)^{\mathrm{ES}(j,2K)}\right) \tag{5.13}$$

$$\hat{y} = a_0 + \sum_{j=1}^{m} a_j f\left((X_1)^{\mathrm{ES}(j,1)} \ldots (X_K)^{\mathrm{ES}(j,K)}\right) \tag{5.14}$$

$$\hat{y} = a_0 + \sum_{j=1}^{m} a_j (X_1)^{\mathrm{ES}(j,1)} \ldots (X_K)^{\mathrm{ES}(j,K)} f\left((X_1)^{\mathrm{ES}(j,K+1)}\right) \ldots (X_K)^{\mathrm{ES}(j,2K)} \tag{5.15}$$

$$\hat{y} = f\left(a_0 + \sum_{j=1}^{m} a_j (X_1)^{\mathrm{ES}(j,1)} \ldots (X_K)^{\mathrm{ES}(j,K)}\right) \tag{5.16}$$

where $\hat{y}$ is the forecasted value, f is user-determined function, $ES(j, K)$ is defined as an exponent of this function corresponding to the Kth input of the jth term with its range is selected by user, and m and a_0 are the number of samples and constant value, respectively. A typical flow diagram for the EPR procedure is presented in Fig. 5.3. To know detailed information of the EPR model, the prior research has been provided by Giustolisi and Savic (2009) and Laucelli et al. (2016).

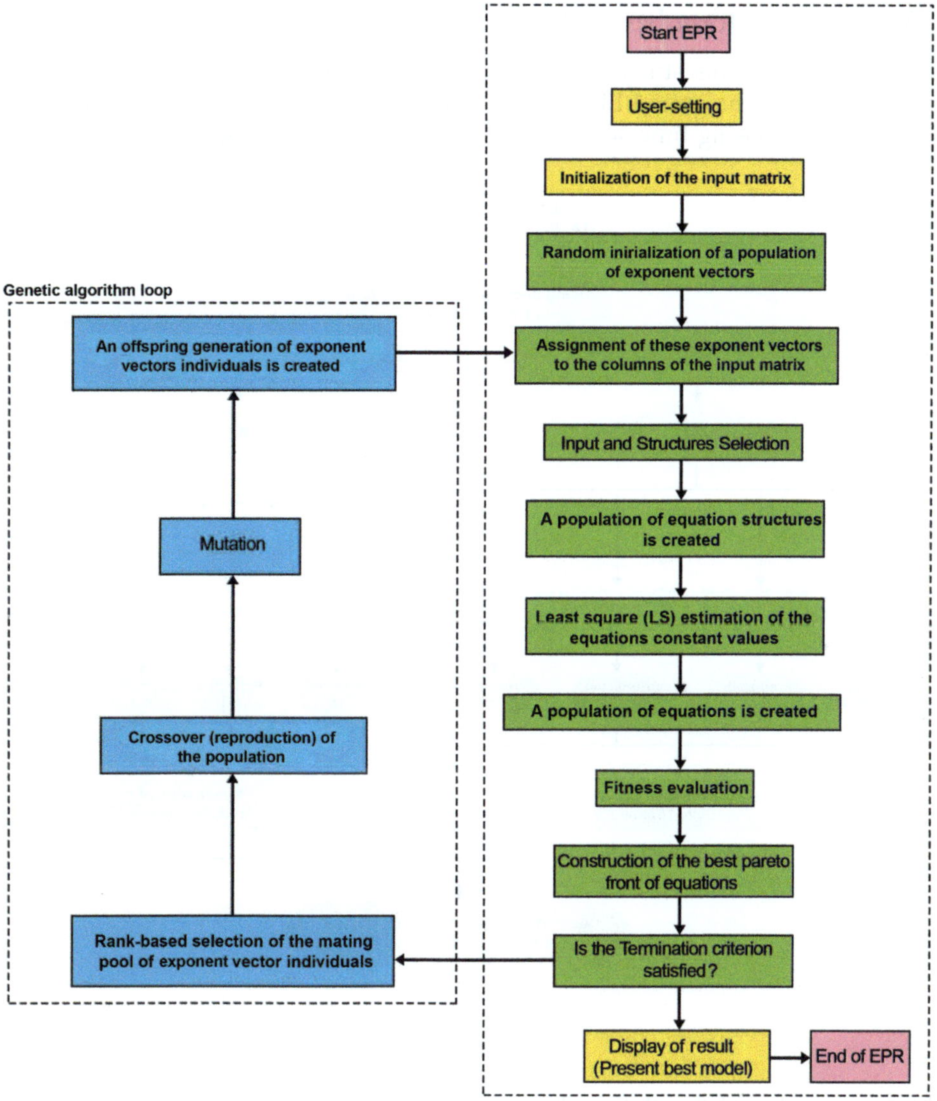

FIGURE 5.3 Flow diagram for the evolutionary polynomial regression procedure.

5.4 Models implementation and application

In developing the EPR model with WT and EEMD techniques, DSR was selected as the target variable in the relationship of the climatic variables including TA, RH, VP, SLP, PE, SD, and SR. To improve the proposed model's performance, the decomposition of input variables into IMF and wavelet components for the original datasets was performed (Fig. 5.4).

5.4.1 Wavelet transform-based DSR forecasting

The main target of this study is expressed as the performance of forecasting the DSR time series using the integrated EPR and WT. The selections of appropriate mother wavelet types for decomposing time series signal for approximation and detail components are

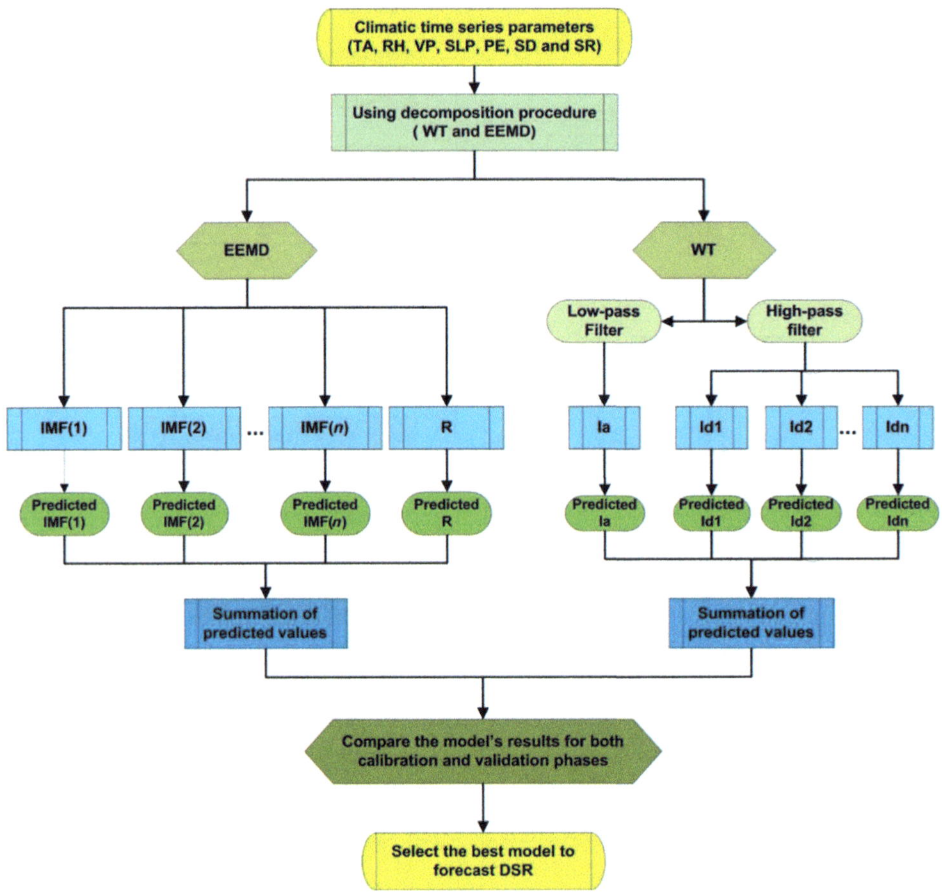

FIGURE 5.4 Work flow of present study. PE, *pan evaporation;* RH, *relative humidity;* SD, *sunshine duration;* SLP, *sea-level pressure;* SR, *solar radiation;* TA, *air temperature;* VP, *vapor pressure.*

absolutely essential for the wavelet implementation of hybrid methods (Shoaib et al., 2015). To evaluate the influence of wavelet type on the models' precision, three types of mother wavelets listing Coiflets (Coif), Haar, and Discrete Meyer (Dmay) were employed to forecast the DSR. Therefore the hybrid models combining WT and EPR can be categorized as W-H-EPR (EPR using Haar mother wavelet), W-D-EPR (EPR using Dmay mother wavelet), and W-C-EPR (EPR using Coiflets mother wavelet), respectively. To achieve this aim, first of all, the DSR time series datasets were decomposed to approximate subseries (A_i) and detail subseries (D_i), where i is the decomposition level. After that, approximation and detail subseries were employed as input matrices for W-H-EPR, W-D-EPR, and W-C-EPR models.

The determinations of decomposition level and mother wavelet types play crucial roles for model performance. In general, the higher decomposition may lead to a slow training process and decrease the model accuracy. To determine the level of decomposition using mathematical approaches, Eq. (5.17) was applied (Wang and Ding, 2003; Seo et al., 2015).

$$L = Int[\log(N)] \tag{5.17}$$

where L and N are level of decomposition and the number of time series datasets. In this study, proposed models were developed by applying MATLAB and Statistics Toolbox Releas (2014) software (Misiti et al., 1996).

5.4.2 Ensemble empirical mode decomposition-based DSR forecasting

The main goal of the EEMD-EPR model was its use for DSR forecasting at two different regions in South Korea. Based on Fig. 5.4, the three main steps of the proposed EEMD-EPR model for the forecasting paradigm can be summarized as follows:

1. At first, the original DSR time series dataset $y(t)$ ($t = 1, 2, \ldots, n$) is decomposed into m number of IMF components $c_i(t)$, $i = 1, 2, \ldots, m$, and a residual component $r_m(t)$ by using EEMD technique.
2. Secondly, the EPR model as one of the forecasting tools is employed to predict each IMF extracted and the residual components, separately, corresponding to each component.
3. Finally, the obtained results using the EPR model are combined to create a unit output which can be applied as the final DSR forecasting results relevant to the original ones.

To summarize, the EEMD-EPR model emphasizes the "decomposition and ensemble" idea. The decomposition is to facilitate the forecasting process, whereas the ensemble is to formulate a consensus estimation on the original datasets. In this work, for verifying and making the pattern of extracted IMFs and residual components with regard to reflecting the forecasting technique and improving the forecasting process, two DSR time series for two stations in South Korea (e.g., Busan and Seoul) were selected.

Furthermore, Table 5.2 shows the EPR parameters (i.e., expression structure, inner function, bias presence, as well as the number of model terms and constants) which were applied for developing the proposed models. As seen in Table 5.2, among the various EPR structures corresponding to each technique (e.g., W-H-EPR, W-D-EPR, W-C-EPR, and

TABLE 5.2 Parameters of evolutionary polynomial regression used for the development of the proposed models.

Models	Expression structure	Inner function	Number of model constants	Presence of bias	Number of terms
Busan station					
EPR	$Y = \text{sum}(ai \times X1 \times X2 \times f(X1 \times X2)) + a0$	Log	2	No	8
W-H-EPR	$Y = \text{sum}(ai \times f(X1 \times X2)) + a0$	Log	1	Yes	7
W-D-EPR	$Y = \text{sum}(ai \times X1 \times X2 \times f(X1) \times f(X2)) + a0$	Exp	1	Yes	9
W-C-EPR	$Y = \text{sum}(ai \times X1 \times X2 \times f(X1 \times X2)) + a0$	Log	2	Yes	8
EEMD-EPR	$Y = \text{sum}(ai \times X1 \times X2 \times f(X1) \times f(X2)) + a0$	Exp	2	Yes	8
Seoul station					
EPR	$Y = \text{sum}(ai \times f(X1 \times X2)) + a0$	Exp	1	Yes	7
W-H-EPR	$Y = \text{sum}(ai \times X1 \times X2 \times f(X1) \times f(X2)) + a0$	Log	2	No	8
W-D-EPR	$Y = \text{sum}(ai \times f(X1 \times X2)) + a0$	Exp	1	Yes	8
W-C-EPR	$Y = \text{sum}(ai \times X1 \times X2 \times f(X1) \times f(X2)) + a0$	Exp	1	Yes	9
EEMD-EPR	$Y = \text{sum}(ai \times X1 \times X2 \times f(X1 \times X2)) + a0$	Exp	1	No	8

EEMD-EPR), the one which had the highest accuracy was selected. Finally, these models were evaluated to determine the best one.

5.5 Results and discussions

The proposed models were employed to forecast the DSR, and the corresponding results from each of the proposed models for both mentioned stations (e.g., Busan and Seoul) are discussed in the following section. In this way, the measured daily weather data (e.g., TA, RH, VP, SLP, PE, and SD) during 16 years (from 2000 to 2016) have been selected for the DSR forecasting. Among all the datasets, 4749 datasets were used for the calibration stage and the remainder of them (1461 datasets) were applied to validate the proposed model. In order to forecast the DSR using the proposed predictive models, the input and output variables were separately decomposed for each calibration and validation (Du et al., 2017; Quilty and Adamowski, 2018).

In terms of wavelet preprocessing, Haar, Dmey, and Coif mother wavelets were used to decompose the input variables. Fig. 5.5 shows the decomposed data series using WT and EEMD methods.

Thereafter, the performances of W-H-EPR, W-D-EPR, W-C-EPR, and EEMD-EPR models for the DSR forecasting were investigated through some criteria including Nash–Sutcliffe efficiency (NSE), root mean square error (RMSE), mean absolute error (MAE), Willmott's index (WI), and Legates–McCabe's index (LMI) for two stages (calibration and validation). The results of proposed models were reported in Table 5.3 (Busan

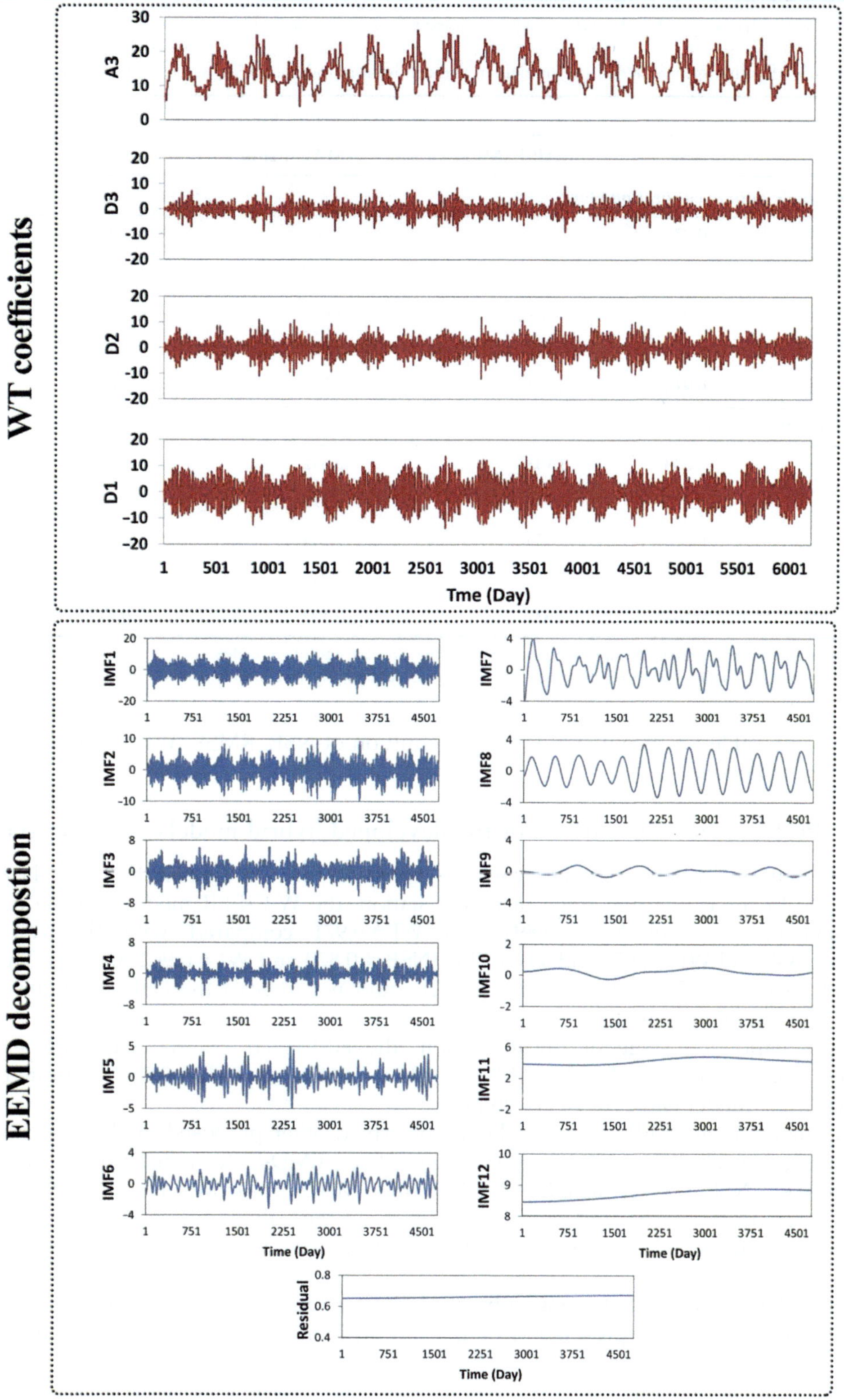

FIGURE 5.5 Time series data for the DSR decomposition using (A) WT and (B) EEMD methods.

TABLE 5.3 Evaluation metrics of the models in the calibration and validation stages at Busan station.

Models	Statistical error indices				
	NSE	RMSE (MJ/m^2)	MAE (MJ/m^2)	WI	LMI
Total available data in calibration stage					
EPR	0.852	2.707	2.103	0.952	0.646
W-H-EPR	0.870	2.399	1.917	0.962	0.672
W-D-EPR	0.887	2.321	1.843	0.970	0.684
W-C-EPR	0.891	2.282	1.823	0.971	0.689
EEMD-EPR	**0.916**	**2.253**	**1.748**	**0.982**	**0.721**
Total available data in validation stage					
EPR	0.803	3.183	2.542	0.934	0.575
W-H-EPR	0.813	2.918	2.376	0.956	0.592
W-D-EPR	0.810	3.035	2.394	0.952	0.584
W-C-EPR	0.848	2.677	2.210	0.968	0.621
EEMD-EPR	**0.864**	**2.637**	**2.127**	**0.973**	**0.679**

Note: The bold numbers represent the values of performance criteria for the best-fitted models.
LMI, Legates–McCabe's index; *NSE*, Nash–Sutcliffe efficiency; *RMSE*, root mean square error; *WI*, Willmott's index of agreement.

station) and Table 5.4 (Seoul station). The definition of NSE, RMSE, WI, and LMI are presented in Appendix.

5.5.1 Performance comparison of the developed hybrid models at Busan station

The results of different mother wavelets types for the DSR forecasting are shown at Table 5.3. In the calibration stage of Busan station, the W-C-EPR model had the highest precision in terms of NSE = 0.891 and WI = 0.971 compared with the W-D-EPR (NSE = 0.887 and WI = 0.97) and W-H-EPR (NSE = 0.870 and WI = 0.962) models. The single EPR model (NSE = 0.852 and WI = 0.952) could not forecast the DSR accurately compared with the hybrid models.

In the validation stage of Busan station, the addressed criteria (e.g., NSE = 0.848, RMSE = 2.677 MJ/m^2, MAE = 2.210 MJ/m^2, WI = 0.968, and LMI = 0.621) indicated that the W-C-EPR model had better performance.

The combination of EPR and EEMD outperformed other proposed models for the DSR forecasting in calibration and validation stages. The EEMD-EPR model provided the best results (i.e., NSE = 0.916, RMSE = 2.253 MJ/m^2, MAE = 1.748 MJ/m^2, WI = 0.982, and LMI = 0.721) in the calibration stage. In addition, in the validation stage, the EEMD-EPR model was suggested as the first rank for the DSR forecasting (NSE = 0.864 and MAE = 2.127 MJ/m^2).

Scatterplots of forecasted versus observed DSR values at Busan station in calibration and validation stages are presented in Fig. 5.6. Considering Fig. 5.6, the DSR values using

TABLE 5.4 Evaluation metrics of the models in the calibration and validation stages at Seoul station.

Models	Statistical error indices				
	NSE	RMSE (MJ/m^2)	MAE (MJ/m^2)	WI	LMI
Total available data in calibration stage					
EPR	0.877	2.303	1.785	0.966	0.673
W-H-EPR	0.902	2.125	1.634	0.971	0.704
W-D-EPR	0.880	2.277	1.819	0.966	0.667
W-C-EPR	0.886	2.220	1.769	0.969	0.676
EEMD-EPR	**0.913**	**2.021**	**1.577**	**0.978**	**0.774**
Total available data in validation stage					
EPR	0.827	2.606	1.944	0.959	0.613
W-H-EPR	0.906	1.838	1.565	0.976	0.698
W-D-EPR	0.864	2.219	1.715	0.967	0.658
W-C-EPR	0.866	2.204	1.704	0.968	0.661
EEMD-EPR	**0.921**	**1.723**	**1.468**	**0.982**	**0.718**

Note: The bold numbers represent the values of performance criteria for the best-fitted models.

those proposed by the EEMD-EPR model are approximately closest to the perfect line. For instance, in this stage, the forecasted DSR value by the EEMD-EPR model corresponding to a roughly measured value of two was under 10, whereas this value was forecasted as above 13 using the W-EPR model. It can be also seen that the forecasted DSR values using all proposed models in the validation stage were overestimated. Time series plots of observed and forecasted DSR in the validation stage can be seen in Fig. 5.7.

5.5.2 Performance comparison of the developed hybrid models at Seoul station

Similar to the methodology used for previous section, five proposed models (i.e., EPR, W-H-EPR, W-D-EPR, W-C-EPR, and EEMD-EPR) were provided for the purpose of DSR forecasting at Seoul station. The performances of EPR, W-H-EPR, W-D-EPR, W-C-EPR, and EEMD-EPR models for forecasting the DSR are presented in Table 5.4. In the case of Seoul station, it can be found that the W-H-EPR model (i.e., NSE = 0.902, RMSE = 2.125 MJ/m^2, MAE = 1.634 MJ/m^2, WI = 0.971, and LMI = 0.704) was more robust compared with the W-C-EPR and W-D-EPR models for forecasting the DSR in the calibration stage within the hybrid models. In the validation stage, the forecasted DSR values using the W-H-EPR and W-C-EPR models provided 17.16% and 0.67% lower values of RMSE, respectively, than the W-D-EPR model. It indicated that the W-H-EPR model possessed a more powerful ability than the W-D-EPR and W-C-EPR models for forecasting the DSR. The results of the statistical error indices were worst for the single EPR model in the calibration and validation stages.

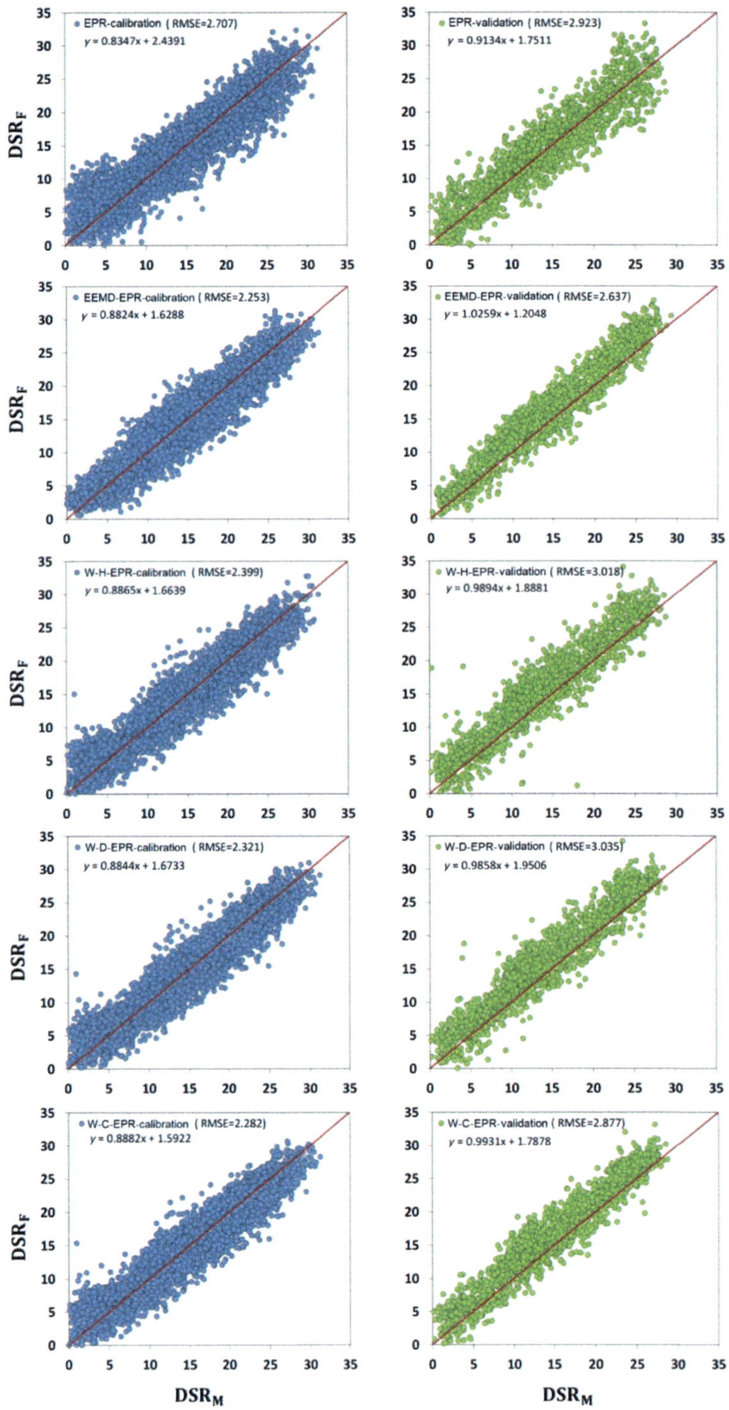

FIGURE 5.6 Scatterplots of observed and forecasted daily solar radiation for proposed models at Busan station in the calibration and validation stages.

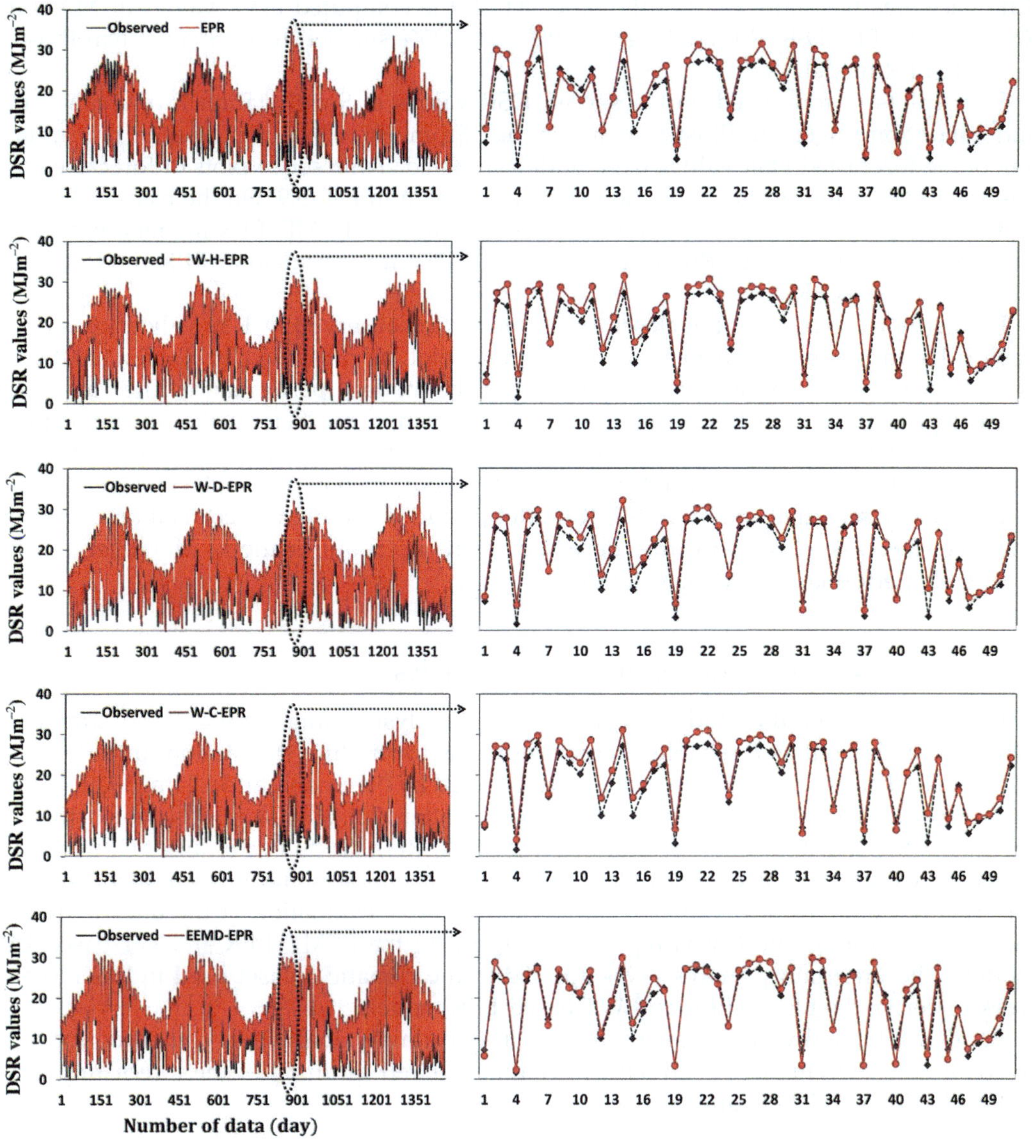

FIGURE 5.7 Time series plots of observed and forecasted DSR for proposed models at Busan station in the validation stage.

In addition, it was clear that the EEMD-EPR model accomplished better performance compared with the ETR, W-H-EPR, W-D-EPR and W-C-EPR models in the calibration stage.

Similar to the performance of the EEMD-EPR model in the calibration stage, it roughly had more precision (LMI = 0.718 and MAE = 1.468 MJ/m^2) in comparison with the W-H-EPR model (LMI = 0.698 and MAE = 1.565 MJ/m^2) in the validation stage. However, this difference was not sigificant. On the other hand, the computed error values of MAE and RMSE by the EEMD-EPR model were 6.19% and 6.25% smaller, respectively, than those of the W-H-EPR model.

Scatterplots of forecasted and observed DSR values at Seoul station in the calibration and validation stages are illustrated in Fig. 5.8. From Fig. 5.8, it can be seen that quite a few of the forecasted DSR values were underestimated. Time series plots of forecasted and observed DSR values in the validation stage are shown in Fig. 5.9. On the basis of Figs. 5.8 and 5.9, the forecasted and observed DSR values using the EEMD-EPR model reveal better agreement than the W-H-EPR, W-D-EPR, W-C-EPR, and standalone EPR models, especially for the peak SR.

Furthermore, to confirm the closet technique for the observed DSR in validation stage, a Taylor diagram is provided in Fig. 5.10. A polar plot, as presented by Taylor (2001), has been drawn to obtain a visual understanding of the models' performance. The Taylor diagram depicts three statistics within itself: (1) correlation coefficient (the azimuth angle); (2) normalized standard deviation (radial distance from the origin); and (3) RMSE (distance from the reference observed point) (Taylor, 2001; Heo et al., 2014). From Fig. 5.10 it can be clearly seen that the EEMD-EPR model provided the highest correlation coefficient and the forecast closest to the observed DSR at Seoul station. Moreover, it has acceptable performance at Busan station.

5.5.3 Monte Carlo simulation-uncertainty analysis

In this section of the study, the uncertainty analysis using Monte Carlo simulation (MCS) is carried out as a reliable alternative means for finding the random uncertainty of the models. The MCS technique was first developed by Neumann and Ulam (1949) to simulate probabilistic events for military purposes (Sattar, 2014). It should be noted that the reported DSR and its characteristics, including TA, RH, VP, SLP, PE, and SD, may have uncertainty owing to their values. Poor forecasting techniques and measurements are two uncertainty examples of E values. Hence, a quantitative evaluation of the uncertainties in the DSR forecasting rate E was reported by applying the EEMD-EPR and WT-based EPR models. The uncertainty analysis was applied to the climatic dataset used in this research work to derive the EEMD-EPR and WT-based EPR models.

Eq. (5.18), the individual estimation error in uncertainty analysis, is computed for all of the dataset to determine the standard deviation (S_e) and mean ($\overline{e}$) of the forecasting error, as Eqs. (5.19) and (5.20) (Sattar, 2014),

$$e_i = \log_{10}(\mathrm{DSR}^{\mathrm{pre}}{}_i) - \log_{10}(\mathrm{DSR}^{\mathrm{obs}}{}_i) \tag{5.18}$$

$$\overline{e} = \sum_{i=1}^{n} e_i \tag{5.19}$$

$$S_e = \sqrt{\sum_{i=1}^{n} \left(\frac{(e_i - \overline{e})^2}{n-1} \right)} \tag{5.20}$$

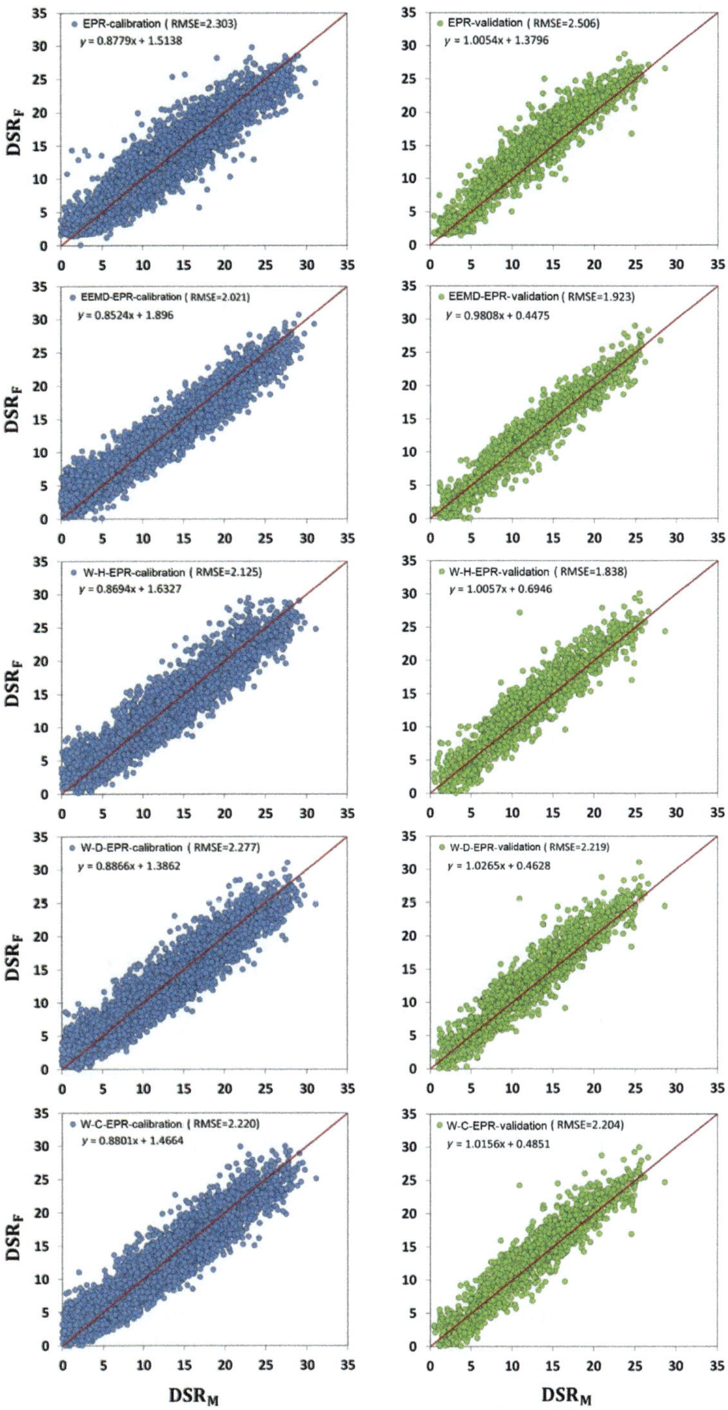

FIGURE 5.8 Scatterplots of observed and forecasted DSR for proposed models at Seoul station in the calibration and validation stages.

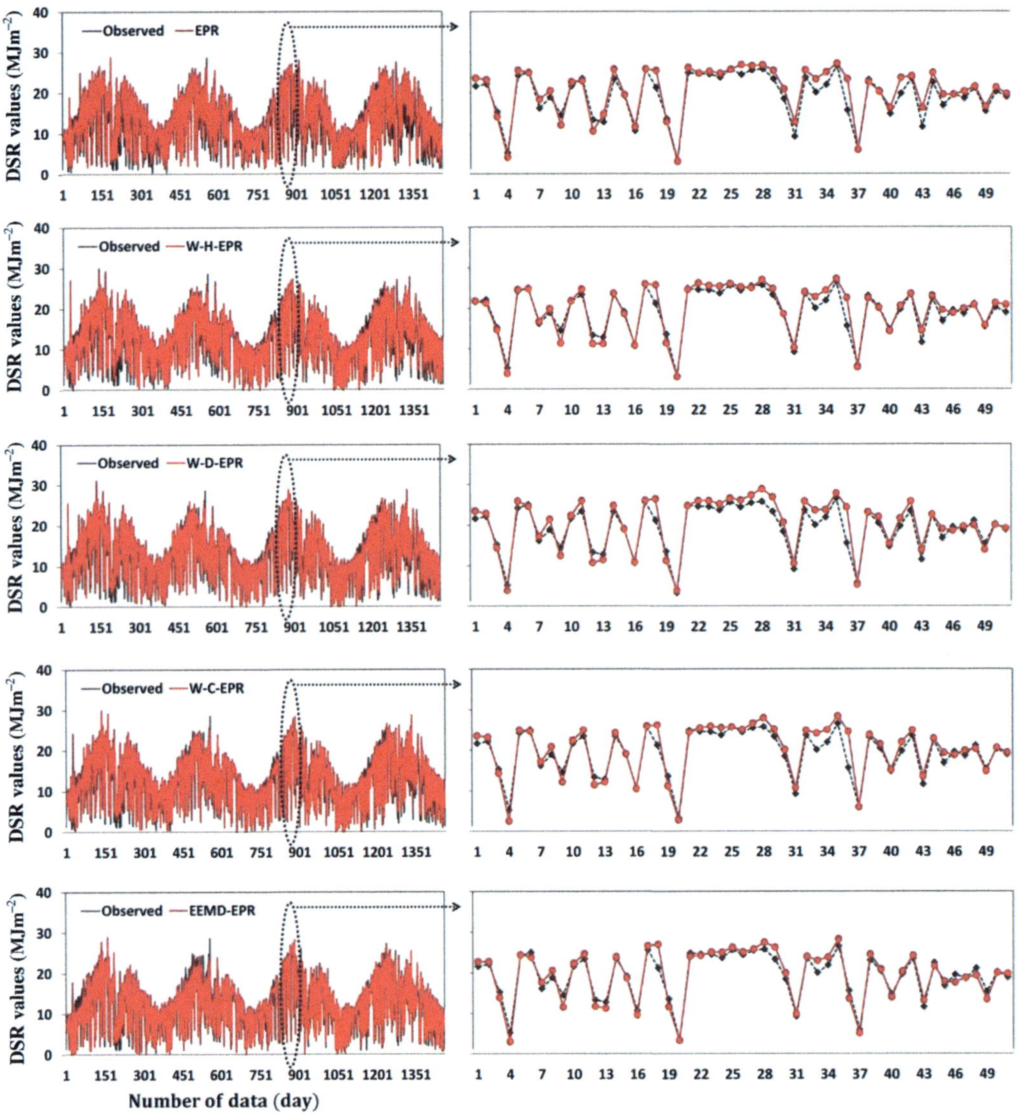

FIGURE 5.9 Time series plots of observed and forecasted DSR for proposed models at Seoul station in the validation stage.

where n is the number of dataset and DSR^{pre} and DSR^{obs} are predicted and observed values of DSR, respectively. It can be said that positive and negative values of the mean prediction error indicate that the forecasted DSR values suggest overestimation and underestimation, respectively, compared with observed values. Thus a confidence band is determined around the estimated error values by help of the Wilson score technique (Julious, 1998). Moreover,

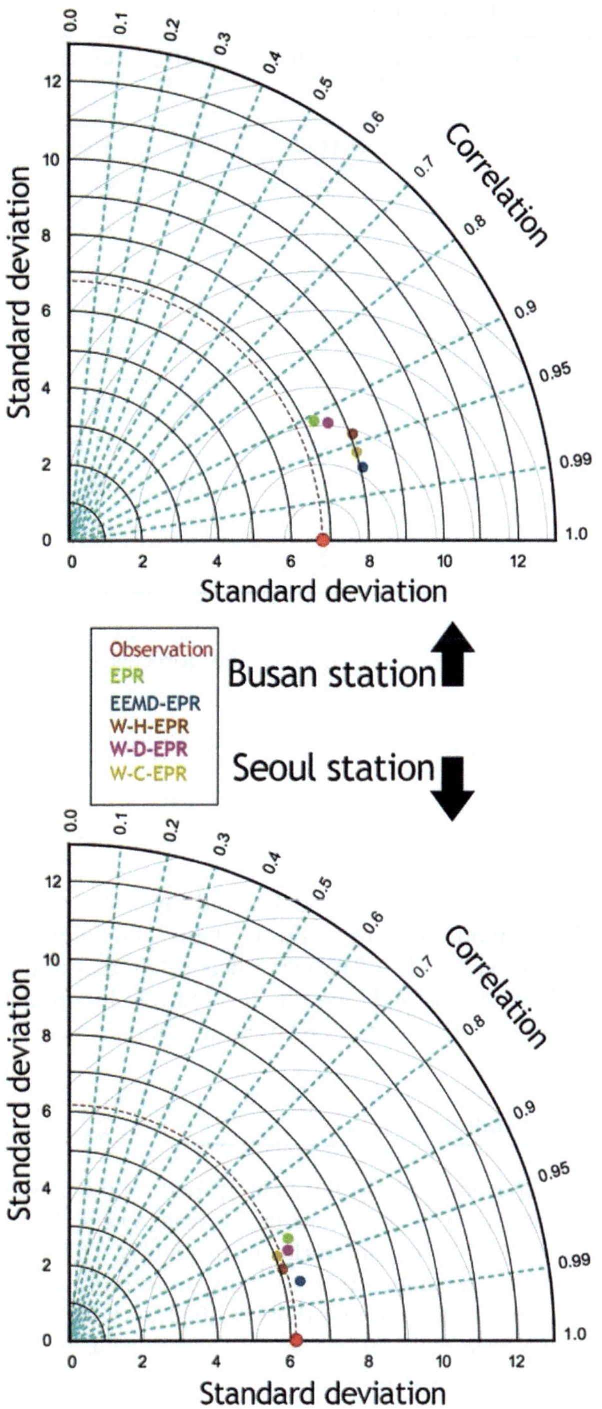

FIGURE 5.10 Taylor diagram showing the correlation coefficient between the forecasted and observed DSR values, and standard deviation for each model.

TABLE 5.5 Uncertainty analysis using Monte Carlo simulation for DSR forecasting models.

Models	Mean prediction error	Width of uncertainty band	Median	MAD	Uncertainty
Busan station					
EPR	0.53	±2.87	14.12	5.93	42.02
W-H-EPR	1.87	±2.36	15.33	6.09	39.74
W-D-EPR	1.73	±2.49	15.53	6.03	38.83
W-C-EPR	1.74	±2.28	15.38	5.72	37.23
EEMD-EPR	1.28	±2.06	15.41	5.62	36.61
Seoul station					
EPR	1.44	±2.25	13.11	5.37	40.96
W-H-EPR	0.20	±1.82	12.81	5.06	39.48
W-D-EPR	3.51	±6.65	14.95	5.97	39.93
W-C-EPR	0.78	±2.06	12.91	5.39	41.73
EEMD-EPR	0.67	±2.05	13.89	5.33	38.35

using $\pm 1.96S_e$ provides a roughly 95% confidence band around the forecasted P_i in Eq. (5.21) (Sattar and Gharabaghi, 2015),

$$\left\{P_i \times 10^{-\bar{e}-1.96S_e}, P_i \times 10^{-\bar{e}+1.96S_e}\right\}. \tag{5.21}$$

The result of the uncertainty analysis, such as mean forecasting errors of the EPR, W-H-EPR, W-D-EPR, W-C-EPR, and EEMD-FPR models, the range of uncertainty band, and mean absolute deviation (MAD), are shown in Table 5.5. It can be found from Table 5.5 that the positive values of mean estimation error indicated that all proposed models could forecast the DSR values compared with the observed ones. In addition, in terms of Busan station, the width of the uncertainty band related to the EEMD-EPR model with ±2.06 was smaller than that of the W-C-EPR model with ±2.28 and single EPR with ±2.87. However, the EEMD-EPR model with ±2.05 has a higher range than the W-H-EPR model with ±1.82 at Seoul station.

5.6 Conclusions

Due to highly destructive effects of on the environment and the steep growth in the global energy demands, renewable energy resources have been the focus of researchers. Solar energy, a renewable energy resource, is available all over the world. In this study, DSR has been forecast using an AI approach called EPR. To achieve this goal, six daily inputs (i.e., TA, RH, VP, SLP, PE, and SD) and one output (DSR), measured from 2000 to 2016, have been decomposed into new variables using two preprocess processes, WT and EEMD.

The results of EPR, WT-based EPR, and EEMD-based EPR models have been compared using comparative statistics containing NSE, RMSE, MAE, WI, and Legates–LMI. A holistic evaluation via statistical assessment and diagnostic plots indicates that the EEMD-EPR model generates superior forecasting compared with the standalone EPR model and WT-based EPR models. The comparison reveals that the EEMD-EPR model provides the best performance at Seoul and Busan (calibration and validation stages) stations. The performances of single EPR and hybrid EPR models are evaluated based on the error size and the uncertainty analysis of model forecasting. The forecasting errors and uncertainties associated with the proposed EEMD-EPR model are smaller than those associated with the W-H-EPR, W-D-EPR, and W-C-EPR models.

Appendix

Nash–Sutcliffe efficiency: This criterion is taken into account to evaluate the ability of hydrological models. The highest value of unity demonstrates a perfect fit between observed and predicted DSR, where a negative values means that the model performs worse than the arithmetic mean of the time series (Nash and Sutcliffe, 1970).

$$\mathrm{NSE} = 1 - \frac{\sum_{i=1}^{N} (\mathrm{DSR}_{\mathrm{for}} - \mathrm{DSR}_{\mathrm{obs}})^2}{\sum_{i=1}^{N} \left(\mathrm{DSR}_{\mathrm{obs}} - \overline{\mathrm{DSR}_{\mathrm{obs}}}\right)^2}$$

Root mean square error: RMSE shows the difference between the forecasted and observed values. RMSE varies from zero for perfect estimates to large positive values for poor estimates.

$$\mathrm{RMSE} = \sqrt{\frac{1}{N}\sum_{i=1}^{N} (\mathrm{DSR}_{\mathrm{for}} - \mathrm{DSR}_{\mathrm{obs}})^2}$$

Willmott's index of agreement: WI is a standardized measure for model prediction error and ranges from zero to one. The values close to zero show poor accuracy while the values close to one reveal good performance (Willmott et al., 2012).

$$\mathrm{WI} = 1 - \frac{\sum_{i=1}^{N} (\mathrm{DSR}_{\mathrm{obs}} - \mathrm{DSR}_{\mathrm{for}})^2}{\sum_{i=1}^{N} \left(\left|\mathrm{DSR}_{\mathrm{for}} - \overline{\mathrm{DSR}_{\mathrm{obs}}}\right| - \left|\mathrm{DSR}_{\mathrm{obs}} - \overline{\mathrm{DSR}_{\mathrm{obs}}}\right|\right)^2}$$

Legates–McCabe's index: This criterion considers absolute values for computation and gives errors and differences the appropriate weights (Legates and McCabe, 1999). Therefore LMI is not inflated by the squared values and is insensitive to outliers, making it simple and easy to interpret.

$$\mathrm{LMI} = 1 - \left[\frac{\sum_{i=1}^{N} |\mathrm{DSR}_{\mathrm{for}} - \mathrm{DSR}_{\mathrm{for}}|}{\sum_{i=1}^{N} \left|\mathrm{DSR}_{\mathrm{obs}} - \overline{\mathrm{DSR}_{\mathrm{obs}}}\right|}\right]$$

where $\mathrm{DSR}_{\mathrm{obs}}$ and $\mathrm{DSR}_{\mathrm{for}}$ are the observed and forecasted DSR values, respectively; $\overline{\mathrm{DSR}_{\mathrm{obs}}}$ and $\overline{\mathrm{DSR}_{\mathrm{for}}}$ represent the mean of the observed and forecasted DSR values, respectively; and N is the number of time series data.

References

Ahangar-Asr, A., Faramarzi, A., Javadi, A.A., Giustolisi, O., 2011. Modelling mechanical behaviour of rubber concrete using evolutionary polynomial regression. Eng. Comput. 28 (4).

Alizadeh, M.J., Kavianpour, M.R., 2015. Development of wavelet-ANN models to predict water quality parameters in Hilo Bay, Pacific Ocean. Mar. Pollut. Bull. 98 (1-2), 171–178.

Alsina, E.F., Bortolini, M., Gamberi, M., Regattieri, A., 2016. Artificial neural network optimisation for monthly average daily global solar radiation prediction. Energy Convers. Manag. 120, 320–329.

Bailek, N., Bouchouicha, K., Al-Mostafa, Z., El-Shimy, M., Aoun, N., Slimani, A., et al., 2018. A new empirical model for forecasting the diffuse solar radiation over Sahara in the Algerian Big South. Renew. Energy 117, 530–537.

Balf, M.R., Noori, R., Berndtsson, R., Ghaemi, A., Ghiasi, B., 2018. Evolutionary polynomial regression approach to predict longitudinal dispersion coefficient in rivers. J. Water Supply: Res. Technol. Aqua 67 (5), 447–457.

Baser, F., Demirhan, H., 2017. A fuzzy regression with support vector machine approach to the estimation of horizontal global solar radiation. Energy 123, 229–240.

Bayrakçı, H.C., Demircan, C., Keçebaş, A., 2018. The development of empirical models for estimating global solar radiation on horizontal surface: a case study. Renew. Sustain. Energy Rev. 81, 2771–2782.

Behrang, M.A., Assareh, E., Ghanbarzadeh, A., Noghrehabadi, A.R., 2010. The potential of different artificial neural network (ANN) techniques in daily global solar radiation modeling based on meteorological data. Sol. Energy 84 (8), 1468–1480.

Belaid, S., Mellit, A., 2016. Prediction of daily and mean monthly global solar radiation using support vector machine in an arid climate. Energy Convers. Manag. 118, 105–118.

Bou-Rabee, M., Sulaiman, S.A., Saleh, M.S., Marafi, S., 2017. Using artificial neural networks to estimate solar radiation in Kuwait. Renew. Sust. Energ. Rev 72, 434–438.

Deo, R.C., Şahin, M., 2017. Forecasting long-term global solar radiation with an ANN algorithm coupled with satellite-derived (MODIS) land surface temperature (LST) for regional locations in Queensland. Renew. Sustain. Energy Rev. 72, 828–848.

Despotovic, M., Nedic, V., Despotovic, D., Cvetanovic, S., 2016. Evaluation of empirical models for predicting monthly mean horizontal diffuse solar radiation. Renew. Sustain. Energy Rev. 56, 246–260.

Du, K., Zhao, Y., Lei, J., 2017. The incorrect usage of singular spectral analysis and discrete wavelet transform in hybrid models to predict hydrological time series. J. Hydrol. 552, 44–51.

Fu, C.L., Cheng, H.Y., 2013. Predicting solar irradiance with all-sky image features via regression. Sol. Energy 97, 537–550.

Gairaa, K., Khellaf, A., Messlem, Y., Chellali, F., 2016. Estimation of the daily global solar radiation based on Box–Jenkins and ANN models: a combined approach". Renew. Sustain. Energy Rev. 57, 238–249.

Ghimire, S., Deo, R.C., Downs, N.J., Raj, N., 2018. Self-adaptive differential evolutionary extreme learning machines for long-term solar radiation prediction with remotely-sensed MODIS satellite and Reanalysis atmospheric products in solar-rich cities. Remote. Sens. Environ. 212, 176–198.

Giustolisi, O., Savic, D.A., 2009. Advances in data-driven analyses and modelling using EPR-MOGA. J. Hydroinform. 11 (3-4), 225–236.

Gueymard, C.A., 2014. A review of validation methodologies and statistical performance indicators for modeled solar radiation data: towards a better bankability of solar projects. Renew. Sustain. Energy Rev. 39, 1024–1034.

Gupta, K.K., Gupta, R., 2007. Despeckle and geographical feature extraction in SAR images by wavelet transform. ISPRS J. Photogramm. Remote. Sens. 62 (6), 473–484.

Heo, K.Y., Ha, K.J., Yun, K.S., Lee, S.S., Kim, H.J., Wang, B., 2014. Methods for uncertainty assessment of climate models and model predictions over East Asia. Int. J. Climatol. 34 (2), 377–390.

Huang, N.E., Shen, Z., Long, S.R., Wu, M.C., Shih, H.H., Zheng, Q., et al., 1971. The empirical mode decomposition and the Hilbert spectrum for nonlinear and non-stationary time series analysis. Proc. R. Soc. Math. Phys. Eng. Sci. 1998 454, 903–995.

Huang, N.E., Wu, M.L., Long, S.R., Shen, S.S., Qu, W., Gloersen, P., et al., 2003. A confidence limit for the empirical mode decomposition and Hilbert spectral analysis. Proc. R. Soc. Math. Phys. Eng. Sci. 459 (2037), 2317–2345.

Ji, W., Chee, K.C., 2011. Prediction of hourly solar radiation using a novel hybrid model of ARMA and TDNN. Sol. Energy 85 (5), 808–817.

Julious, S.A., 1998. Two-sided confidence intervals for the single proportion: comparison of seven methods by Robert G. Newcombe. Stat. Med. 17, 857–872. Statistics in medicine 2005; 24(21): 3383-4.

Kim, S., Seo, Y., Singh, V.P., 2016. Computation of daily solar radiation using wavelet and support vector machines: a case study. Harmony Search Algorithm. Springer, Berlin, Heidelberg, pp. 279–284.

Labat, D., Ababou, R., Mangin, A., 2000. Rainfall–runoff relations for karstic springs. Part II: continuous wavelet and discrete orthogonal multiresolution analyses. J. Hydrol. 238 (3-4), 149–178.

Laucelli, D., Giustolisi, O., 2011. Scour depth modelling by a multi-objective evolutionary paradigm. Environ. Model. Softw. 26 (4), 498–509.

Laucelli, D., Romano, M., Savić, D., Giustolisi, O., 2016. Detecting anomalies in water distribution networks using EPR modelling paradigm. J. Hydroinform. 18 (3), 409–427.

Legates, D.R., McCabe Jr, G.J., 1999. Evaluating the use of "goodness-of-fit" measures in hydrologic and hydroclimatic model validation. Water Resour. Res. 35 (1), 233–241.

Lei, Y., He, Z., Zi, Y., 2009. Application of the EEMD method to rotor fault diagnosis of rotating machinery. Mech. Syst. Signal. Process. 23 (4), 1327–1338.

Lou, S., Li, D.H., Lam, J.C., Chan, W.W., 2016. Prediction of diffuse solar irradiance using machine learning and multivariable regression. Appl. Energy 181, 367–374.

MATLAB and Statistics Toolbox Release 2014a, The MathWorks, Inc., Natick, Massachusetts, United States.

Mellit, A., Kalogirou, S.A., 2008. Artificial intelligence techniques for photovoltaic applications: a review". Prog. Energy Combust. Sci. 34 (5), 574–632.

Misiti, M., Misiti, Y., Oppenheim, G., Poggi, J.M., 1996. Wavelet Toolbox. The MathWorks Inc., Natick, MA.

Mohammadi, K., Shamshirband, S., Kamsin, A., Lai, P.C., Mansor, Z., 2016. Identifying the most significant input parameters for predicting global solar radiation using an ANFIS selection procedure. Renew. Sustain. Energy Rev. 63, 423–434.

Mohammadi, K., Shamshirband, S., Tong, C.W., Arif, M., Petković, D., Ch, S., 2015. A new hybrid support vector machine–wavelet transform approach for estimation of horizontal global solar radiation. Energy Convers. Manag. 92, 162–171.

Monteiro, R.V., Guimarães, G.C., Moura, F.A., Albertini, M.R., Albertini, M.K., 2017. Estimating photovoltaic power generation: performance analysis of artificial neural networks, support vector machine and Kalman filter. Electr. Power Syst. Res. 143, 643–656.

Mostafavi, E.S., Ramiyani, S.S., Sarvar, R., Moud, H.I., Mousavi, S.M., 2013. A hybrid computational approach to estimate solar global radiation: an empirical evidence from Iran. Energy 49, 204–210.

Nash, J.E., Sutcliffe, J.V., 1970. River flow forecasting through conceptual models part I—A discussion of principles. J. Hydrol. 10 (3), 282–290.

Neumann, J.V., Ulam, S., 1949. The monte Carlo method. J. Am. Stat. Asso. 44 (247), 335–341.

Olatomiwa, L., Mekhilef, S., Shamshirband, S., Mohammadi, K., Petković, D., Sudheer, C., 2015. A support vector machine–firefly algorithm-based model for global solar radiation prediction. Sol. Energy 115, 632–644.

Quilty, J., Adamowski, J., 2018. Addressing the incorrect usage of wavelet-based hydrological and water resources forecasting models for real-world applications with best practices and a new forecasting framework. J. Hydrol. 563, 336–353.

Rajaee, T., Shahabi, A., 2016. Evaluation of wavelet-GEP and wavelet-ANN hybrid models for prediction of total nitrogen concentration in coastal marine waters. Arab. J. Geosci. 9 (3), 176.

Rezaie-Balf, M., Kisi, O., 2018. New formulation for forecasting streamflow: evolutionary polynomial regression vs. extreme learning machine. Hydrol. Res. 49 (3), 939–953.

Rezaie-Balf, M., Naganna, S.R., Ghaemi, A., Deka, P.C., 2017. Wavelet coupled MARS and M5 model tree approaches for groundwater level forecasting. J. Hydrol. 553, 356–373.

Rezania, M., Javadi, A.A., Giustolisi, O., 2008. An evolutionary-based data mining technique for assessment of civil engineering systems. Eng. Comput. 25 (6).

Salcedo-Sanz, S., Casanova-Mateo, C., Muñoz-Marí, J., Camps-Valls, G., 2014a. Prediction of daily global solar irradiation using temporal Gaussian processes. IEEE Geosci. Remote. Sens. Lett. 11 (11), 1936–1940.

Salcedo-Sanz, S., Casanova-Mateo, C., Pastor-Sánchez, A., Sánchez-Girón, M., 2014b. Daily global solar radiation prediction based on a hybrid coral reefs optimization–extreme learning machine approach. Sol. Energy 105, 91–98.

Salcedo-Sanz, S., Deo, R.C., Cornejo-Bueno, L., Camacho-Gómez, C., Ghimire, S., 2018. An efficient neuro-evolutionary hybrid modelling mechanism for the estimation of daily global solar radiation in the Sunshine State of Australia. Appl. Energy 209, 79–94.

Sattar, A.M., 2014. Gene expression models for the prediction of longitudinal dispersion coefficients in transitional and turbulent pipe flow. J. Pipeline Syst. Eng. Pract. 5 (1), 04013011.

Sattar, A.M., Gharabaghi, B., 2015. Gene expression models for prediction of longitudinal dispersion coefficient in streams. J. Hydrol. 524, 587–596.

Seo, Y., Kim, S., 2016. Hydrological forecasting using hybrid data-driven approach. Am. J. Appl. Sci. 1 (3), 8.

Seo, Y., Kim, S., Kisi, O., Singh, V.P., 2015. Daily water level forecasting using wavelet decomposition and artificial intelligence techniques. J. Hydrol. 520, 224–243.

Shafaei, M., Kisi, O., 2017. Predicting river daily flow using wavelet-artificial neural networks based on regression analyses in comparison with artificial neural networks and support vector machine models. Neural Comput. Appl. 28 (1), 15–28.

Shamshirband, S., Mohammadi, K., Chen, H.L., Samy, G.N., Petković, D., Ma, C., 2015. Daily global solar radiation prediction from air temperatures using kernel extreme learning machine: a case study for Iran. J. Atmos. Sol. Terr. Phys. 134, 109–117.

Shoaib, M., Shamseldin, A.Y., Melville, B.W., Khan, M.M., 2015. Runoff forecasting using hybrid wavelet gene expression programming (WGEP) approach. J. Hydrol. 527, 326–344.

Taylor, K.E., 2001. Summarizing multiple aspects of model performance in a single diagram. J. Geophys. Res. Atmos. 106 (D7), 7183–7192.

Wang, W., Ding, J., 2003. Wavelet network model and its application to the prediction of hydrology. Nat. Sci. 1 (1), 67–71.

Willmott, C.J., Robeson, S.M., Matsuura, K., 2012. A refined index of model performance. Int. j. Climatol 32 (13), 2088–2094.

Wu, Z., Huang, N.E., 2009. Ensemble empirical mode decomposition: a noise-assisted data analysis method. Adv. Adapt. Data Anal. 1 (01), 1–41.

Wu, Y., Wang, J., 2016. A novel hybrid model based on artificial neural networks for solar radiation prediction. Renew. Energy 89, 268–284.

Yacef, R., Benghanem, M., Mellit, A., 2012. Prediction of daily global solar irradiation data using Bayesian neural network: a comparative study. Renew. Energy 48, 146–154.

Yadav, A.K., Chandel, S.S., 2014. Solar radiation prediction using artificial neural network techniques: a review. Renew. Sustain. Energy Rev." 33, 772–781.

CHAPTER

6

Development of data-driven models for wind speed forecasting in Australia

Ananta Neupane[1], Nawin Raj[2], Ravinesh Deo[2] and Mumtaz Ali[3]

[1]School of Sciences, University of Southern Queensland, Toowoomba, QLD, Australia [2]School of Sciences, University of Southern Queensland, Springfield, QLD, Australia [3]Deakin-SWU Joint Research Center on Big Data, School of Information Technology, Deakin University, VIC, Australia

6.1 Introduction

Emissions of greenhouse gas could be double by 2050 (Marchal et al., 2011). Thus climate-change researchers, environmental policy makers, and energy utilities have attempted to develop relatively clean and environment-friendly energy resources. The current energy market mainly consists of the renewable energy sources, like solar, wind, hydro, geothermal, and marine. Due to their availability and potentiality, solar and wind are the most widespread sources of renewable energy in the world. Around 20% of electricity is generated from renewable sources but only 1.4% of electricity is generated from wind (GWEC, 2017) (Table 6.1 Fig. 6.1).

In Australia about 15% of energy is generated from renewable sources where wind energy shares 12.1% (4.1 GW) of the total renewable energy (AE, 2017). In 2016, 54 GW power was added to the global energy grid (GWEC, 2017). Although Australia has huge wind energy potential, its contribution in 2017 was insignificant (0.14 GW) (GWEC, 2017).

The westerly wind belt of Australia (also known as the roaring 40s) has among the world's best wind energy potential. The southern part of Australia lies in this belt (Geoscience Australia, n.d.). Moreover, the statistical data show that coastal areas (both onshore and offshore) of western, southeastern, southwestern, and southern Australia

Predictive Modelling for Energy Management and Power Systems Engineering
DOI: https://doi.org/10.1016/B978-0-12-817772-3.00006-9

TABLE 6.1 Data sites of the research with wind farm name, location, capacity, and topography.

Site	Name of the wind farm	States	Capacity (MW)	Location	Topography (m)
Silverton (NSW)	Silverton wind farm	New South Wales (NSW)	1000 (proposed)	31.97°S, 141.46°E	305
Macarthur (VIC)	Macarthur wind farm	Victoria	420	38.05°S, 142.19°E	137
Kemmiss Hill (SA)	Kemmiss Hill Wind farm (1,2,3)	South Australia (SA)	30	33.5°S, 138.1°E	71
Walkway (WA)	walkway wind farm	Western Australia (WA)	89	28.92°S, 114.93°E	191
Woolnorth (TAS)	Woolnorth wind farm	Tasmania (TAS)	140	40.69°S, 144.72°E	7

Note that these sites are referred to by the respective state (e.g., VIC, NSW etc.) in the discussions presented in the body of the thesis.

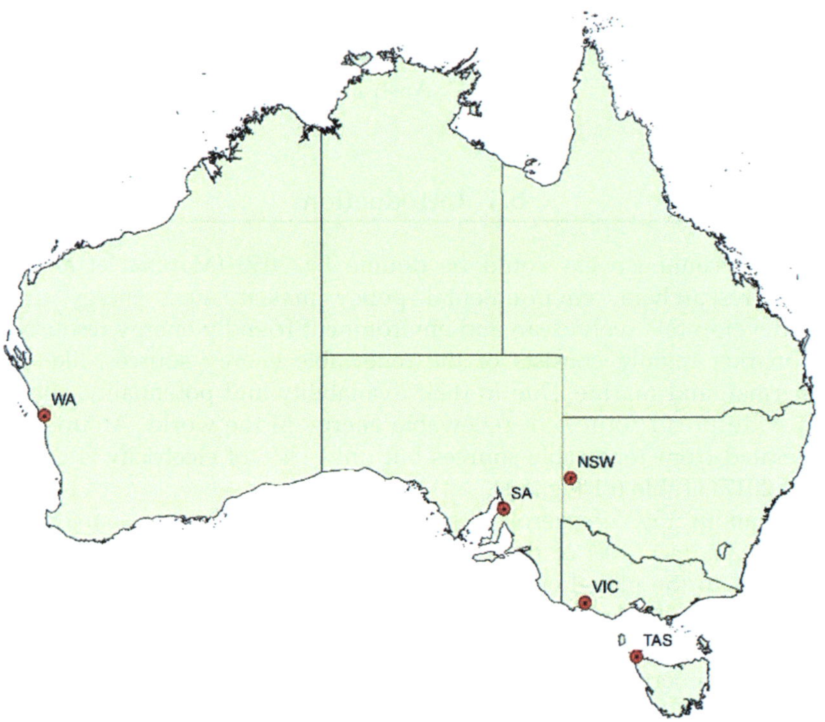

FIGURE 6.1 Locations of wind speed data (*Red* dotted points).

have an average wind speed more than 7.5 m/s (80 m high from the ground) (GENI, 2016). Compared to other parts of the world, this speed is considered to be very high for wind energy generation (Pham et al., 2016).

Although Australia has the potential to develop and use wind power, statistical data shows that it is not progressing fast enough in its development of new wind (renewable) sources for several reasons ("Australian Energy Resources Assessment," n.d.). One of the main reason is, limited research has been conducted in the wind energy sector, including wind speed forecasting technology.

Wind speed forecasting is difficult because of the intermittent and random character of the wind. In addition, data-driven models are not available in the Australian context. There are several data-driven models providing wind speed predictions on a global scale but these models are not generalized to apply in other locations because every location has specific geographic conditions. This chapter aims to develop and evaluate different data-driven forecasting algorithms that are most accurate and replicable in different wind potential areas in Australia ranging from short-term (i.e., 6 hourly) to medium-term horizons.

There are several types of models for wind speed forecasting but the physical model (where equations related to wind and its covariates are used) and the computer-based model (data-driven or artificial intelligence-based model) are the two main types. Wind speed of present time is the effect of different environmental parameters such as temperature, air pressure, airflow obstructions, etc. Thus physical models can be very complex to formulate these parameters, and may have limitations and cannot be generalized in different locations even if the environmental parameters are the same (Yang and Wang, 2018). On the other hand, computer-based models are very convenient, working in an efficient way to achieve better results in terms of prediction accuracy with broader adaptability (Xydas et al., 2016). Data-driven models, such as the ANN, MLR, RF models, and the M5 tree models are widely used for environmental system forecasting. Therefore the focus of this chapter is to develop and to compare these models to predict wind speed.

6.1.1 ANN model

The nature of environmental variables means that they are in a nonlinear form, in relation to different time scales (e.g., hourly and daily). The unpredictable nature of the wind is one of the fundamental difficulties for wind prediction (Lei et al., 2009). The measurement of these variables requires efficient machine learning models, such as the ANN, which can provide more accurate predictions (Lek and Guegan, 1999). In particular, the implementation of the ANN in several other areas, such as water quality forecasting (Palani et al., 2008), energy consumption forecasting (Ahmad et al., 2014), and pollution index forecasting (Jiang et al., 2004), has been very successful. Thus the potential of developing these models in the wind forecasting area has become a desirable task for data-driven modelers.

The development of the ANN model was the outcome of the research performed by Warren McCulloch and Pitts Walter in 1943 from their work *"A Logical Calculation of Ideas Immanent in Nervous Activity"* (McCulloch and Pitts, 1943). The basic functions of the ANN model are pattern classification, pattern recognition, and prediction (Bandyopadhyay and Chattopadhyay, 2007). The ANN is used as a way to measure and understand the process in the biological brain and application of artificial intelligence. The ANN model is based on the connection of units called artificial neurons in the processing unit; neurons transmit the signal from one unit to other. The ANN model has been widely applied for environmental forecasting. For example, it is used for (1) flood forecasting in the Ajay River Basin in Jharkhand (Mukerji et al., 2009); (2) short-term wind speed forecasting in La Venta, Oaxaca, Mexico (Cadenas and Rivera, 2009); (3) streamflow forecasting in the drought-prone Murray–Darling basin region in Australia (Prasad et al., 2017); and (4) standardizing the prediction and evapotranspiration index using hydrometeorological parameters and climate indices in eastern Australia (Deo and Şahin, 2015). Scholars have claimed that the performance of the ANN is better than MLR and ARIMA models in solar radiation forecasting because it can learn and identify complex data patterns (Deo and Şahin, 2017).

The ANN is a flexible computing framework and a universal approximation method that can be applied to a wide range of forecasting problems with a high degree of accuracy (Khashei and Bijari, 2011). It is one of the most important types of nonparametric nonlinear time series models which have been proposed and examined for time series forecasting. Another advantage, as Khashei and Bijari (2011) have highlighted, is that the ANN can be used in the data generation process and is less susceptible to model misspecification problems.

6.1.2 MLR model

The MLR model can be used to predict the dependent variable on the bases of other integrated independent variables (Bottenberg and Ward, 1963). This model is able to explain the "causal and effect" relationships between one influential variable and two or more causal variables in the form of mathematical regression coefficients (Deo and Şahin, 2017). The number of causal variables can be used to determine the number of regression coefficients in an MLR model. Generally, the independent variables might be continuous or categorical variables but must be numeric in nature.

The forecasted result from this model is good when the independent variables are in a linear form (Baroni et al., 1992). It has been recently used for long-term global solar radiation forecasting in Australia (Deo and Şahin, 2017) and electricity consumption forecasting in Italy (Bianco et al., 2009).

6.1.3 RF model

The RF model was first introduced by Leo Breiman in his paper "*Random Forest*" in 2001 (Breiman, 2001). For the RF model, three parameters have been used: (1) the number of predictor variables which have been randomly selected from the data (*fboot*); (2) number of decision trees (*ntree*); and (3) optimum terminal nodes (*leaf*). There are five steps to generate the results from RF model: (1) generate training sample subsets; (2) feature vector selection randomly; (3) prediction results of testing data; (4) generalized error estimation; and (5) results of combined forecasting (Shi et al., 2018). Therefore the RF model creates decision trees from the random selection of a subset of huge data, makes predictions in each subset, then centralizes the prediction (Prasad et al., 2018).

Generally, the RF model is designed with an algorithm that can estimate the unknown mapping between its inputs and outputs from the observed data (Shi et al., 2018). Thus the RF is a powerful statistical classifier and is a well-established algorithm that is useful for environmental variables. It has shown an excellent performance where the number of variables is much larger than the number of observations. Further, it can cope with complex interaction structures as well as highly correlated variables and can return measures of variable importance (Boulesteix et al., 2012). According to Vincenzi et al. (2011), the application of the fitted RF model to continuous maps of the entire environmental variable is comparably appropriate.

In the context of the present research, the RF model can be one of the most competitive forecasting tools used for wind forecasting due to its usage elsewhere, such as stock index forecasting (Kumar and Thenmozhi, 2006), short-term energy load forecasting (Cheng et al., 2012), and urban water demand forecasting (Herrera et al., 2010).

6.1.4 M5 tree model

Quinlan first introduced the M5 tree model in 1992 in his contribution *"Learning with continuous classes"* (Quinlan, 1992). According to Quinlan, model trees are the extension of regression trees with multivariate linear regression models (Quinlan, 1992). They can deal with a continuous class problem with clearly and easily structured decision trees. The M5 tree model divides the training set of data into subsets to calculate the preceding learning values (Etemad-Shahidi and Mahjoobi, 2009).

In the M5 tree model, the division of the training set of data is used to minimize the intrasubset variation of the output variables values through the branch of the node (Prasad et al., 2018). This model uses standard deviation to measure the variability of the values where the node passes from the root to the branch (Etemad-Shahidi and Mahjoobi, 2009). The M5 tree model has been widely applied for: (1) measuring the water level–discharge relationship (Bhattacharya and Solomatine, 2005); (2) generating soil moisture forecasts (Prasad et al., 2018); and (3) predicting evapotranspiration (Pal and Deswal, 2009), and the results are very competitive.

In this chapter, the M5 tree can be one of the most competitive forecasting tools used for wind forecasting due to its usage elsewhere such as generating soil moisture forecasts (Prasad et al., 2018) and evapotranspiration forecasting (Pal and Deswal, 2009).

6.1.5 ARIMA model

In the early 1970s, Box and Jenkins popularized the AutoRegressive Integrated Moving-Average (ARIMA) model for time series forecasting and analysis (Box and Jenkins, 1970) (Cadenas et al., 2016). Since then, the ARIMA model has been continuously used worldwide for forecasting in different sectors. The ARIMA has very highly competitive statistical properties and applications in the model building process (Zhang, 2003). It is a linear and quite flexible model for the time series representation that has been used for the last four decades

(Zhang, 2003). According to its preassumed linearity, it can be separated into complete autoregressive (AR), complete moving average (MA), and their combination (ARMA).

In the ARIMA model, the forecasting value of the variable is a chain combination of the past values and their errors (Pai and Lin, 2005). The ARIMA model has been applied for (1) wind speed prediction in México (Cadenas et al., 2016); (2) stock exchange forecasts in the United States (Ariyo et al., 2014); and (3) short-term traffic flow prediction in India as one of the most precise methods (Kumar and Vanajakshi, 2015).

Thus the ARIMA model is one of the most well-known statistical tools for the time series prediction used in electricity price forecasting (Conejo et al., 2005) and stock price forecasting (Pai and Lin, 2005), so it has been used for wind speed forecasting in this chapter.

A review of the literature shows that the ANN, MLR, RF, M5 tree, and ARIMA models used in this chapter have been widely used in data processing algorithms for different aspects of forecasting in different countries and various environmental modeling contexts. However, these models are still not widely used for wind forecasting mainly for different time horizons (e.g., short-term or daily), especially in the context of wind speed prediction in Australia, and therefore, this chapter primarily focuses on addressing this gap.

A brief description of data sources, materials, and methodologies are presented in Section 6.2, results of different time horizons from different models are described in Sections 6.3 and 6.4 and the conclusion is presented in the last section.

6.2 Materials and methods

6.2.1 Data sources

The data have been acquired from the European Centre for Medium-Range Weather Forecasts (ECMWF) reanalysis (REN) repository where daily wind data had already been converted to 6-hourly using atmospheric models. The reanalysis data have been taken from the zonal wind (*U* wind) which moves from west to east and the meridional wind (*V* wind) which moves from south to north 10 m from the ground. Net wind (*N* wind) has been calculated from *U* wind and *V* wind using Pythagoras theorem.

This chapter used the dataset ranging from January 1, 1988 to December 31, 2017, which provides a lengthy record to develop reliable data-driven models. The *U*, *V*, and *N* wind speeds, which correspond to zonal, meridional, and net wind speed components, respectively, have been extracted for 6-hourly intervals for developing short-term wind speed prediction models. Similarly, the *U*, *V*, and *N* wind speeds have extracted for daily (i.e., medium-term) periods.

The study area is located in five different regions in the southern part of Australia where the wind speed magnitude is potentiality higher than the other parts of the continent. These study locations are follows

1. Silverton, New South Wales (NSW) (31.97°S, 141.46°E). Silverton wind farm is positioned 25 km northwest of Broken Hill and about 5 km north from the town of Silverton in the state of NSW. Its total wind energy generation capacity was proposed to be 1000 MW but 200 MW is expected to be completed in 2018 (AGL, 2018b). This

wind farm will be built by AGL Energy Limited and its topography is about 305 m above sea level (AGL, 2018b).

2. Kemmiss Hill, South Australia (SA) (33.5°S, 138.1°E). The proposed Kemmiss Hill wind farm site lies on Kemmiss Hill south of Adelaide in the state of South Australia. It was proposed by Origin Energy Australia to be 30 MW, but Yankalilla council rejected it in 2004 due to the impact in the community (ABCNews, 2004). Its topography is about 71 m above sea level.
3. Woolnorth, Tasmania (TAS) (40.69°S, 144.72°E). Woolnorth wind farm lies around 7 m above sea level in the northwest of TAS having a nameplate capacity of 180 MW, owned by Hydro Tasmania and Guohua Energy Investment Co. Ltd. Woolnorth wind farm has been completed in three phases: 2002, 2004, and 2007 (AussieRenewables, 2018).
4. Macarthur, Victoria (VIC) (38.05°S, 142.19°E). Macarthur wind farm is located in Macarthur in the state of Victoria, having a total nameplate capacity of 420 MW, and is the biggest wind farm in the southern hemisphere, operated by AGL Energy and completed in 2013, having a topography around 137 m above sea level (AGL, 2018a).
5. Walkway in Western Australia (WA) (28.92°S, 114.93°E). Walkway wind farm lies south of Geraldton in the state of Western Australia with a total nameplate capacity of 89 MW, operated by Infigen Energy, completed in 2004. The topography of this wind farm is around 191 m above sea level (InfigenEnergy, 2018).

6.2.1.1 Characteristics of short-term wind data

The mean of short-term *U* wind of Silverton (NSW) and Walkway (WA) were negative so most of the wind moved from east to west in these locations. For the rest of the data points, the average wind speed was positive, that is, most of the time in Kemmiss Hill (SA), Woolnorth (TAS), and Macarthur (VIC) *U* wind movement was from west to east. The average value for short-term *U* wind was very low because wind velocity was measured as negative when it moved from east to west.

The standard deviation of short-term *U* wind varied from 2.7 m/s in Macarthur (VIC) and 4.8 m/s in Kemmiss Hill (SA). For *U* wind, degree of symmetry in Silverton (NSW) was very high (0.85 m/s) but in Woolnorth (TAS), it was very low (−0.08 m/s) where most of the sites (four) were positively skewed. Outlier or the degree of peak wind speed in the frequency distribution in Silverton (NSW) was very high (0.7 m/s) but in Macarthur (VIC) it was very low (−0.16 m/s) for *U* wind as depicted in the Table 6.2.

TABLE 6.2 *U* wind characteristics for the data with short-term time horizon.

	NSW	SA	TAS	VIC	WA
Mean (m/s)	− 0.25	0.79	1.63	0.87	− 0.53
SD (m/s)	3.23	4.75	3.71	2.69	3.68
Skewness (m/s)	0.85	0.58	− 0.08	0.41	0.12
Kurtosis (m/s)	0.71	− 0.33	− 0.30	− 0.16	0 − 0.20

Note that the accurate site names for these states where models were tested were shown in Table 6.1 and plotted in Fig. 6.1.

TABLE 6.3 *V* wind characteristics of the short-term time horizon.

	NSW	SA	TAS	VIC	WA
Mean (m/s)	1.38	1.10	0.13	0.09	2.04
SD (m/s)	4.01	4.84	3.52	2.92	3.33
Skewness (m/s)	− 0.40	− 0.45	− 0.20	− 0.36	− 0.44
Kurtosis (m/s)	− 0.34	− 0.40	− 0.49	− 0.56	− 0.16

Note that the accurate site names for these states where models tested were shown in Table 6.1 and plotted in Fig. 6.1.

TABLE 6.4 *N* wind characteristics of the data for short-term time horizon (all variables are in m/s).

	NSW	SA	TAS	VIC	WA
Mean (m/s)	4.88	6.31	4.87	3.73	5.02
SD (m/s)	2.17	2.82	2.25	1.63	1.99
Skewness (m/s)	0.60	0.53	0.44	0.83	0.31
Kurtosis (m/s)	0.45	0.20	− 0.16	0.95	− 0.11

Note that the accurate site names for these states where models were tested were shown in Table 6.1 and plotted in Fig. 6.1.

Similarly, for *V* wind, the mean of all data sites was positive, that is, in all the data sites, *V* wind moved from south to north most of the time. The highest mean value was in Walkway (WA 2.04 m/s) and the lowest average was in Macarthur (VIC −0.09 m/s). The scatter of short-term data for *V* wind from the central value was very high in Kemmiss Hill (SA) (4.8 m/s) whereas in Macarthur (VIC), it was very low (2.9 m/s). Skewness of *V* wind was negative in all sites, that is, most of the data lay to the left of the central value. Likewise, the kurtosis value of all the data sites were also negative, that is, short-term data distribution of *V* wind was tailed on the left side of the central value, whereas Walkway (WA) had a small tail and Macarthur (VIC) had a bigger tail in these data locations (Table 6.3).

For short-term *N* wind, average wind speed varied from 6.03 to 3.7 m/s. In this case, Kemmiss Hill (SA) had the highest wind speed, whereas Macarthur (VIC) had the lowest value. The average scatter of *N* wind was also the highest in Kemmiss Hill (SA) and the lowest in Macarthur (VIC), that is, in Macarthur (VIC) *N* wind was more stable but it was less stable in Kemmiss Hill (SA). All the skewness values were positive in *N* wind, indicating that the short-term *N* wind data was elongated to the right side of the central value. Similarly, the degree of "peak wind speed" of the short-term *N* wind was positive in three sites and negative in two sites. Kurtosis varied from 0.9 to −0.1 m/s in Macarthur (VIC) and Walkway (WA), respectively (Table 6.4).

6.2.1.2 Characteristics of daily wind data

Average zonal wind speeds of NSW and WA were negative, that is, the wind moved from west to east more than the wind moved from east to west in these data locations.

The arithmetic means of *U* wind in SA, TAS, and VIC were positive, that is, wind moved from east to west in these data locations most of the time.

For daily time horizons as well, the average wind speed was very low due to the measurement of wind from both directions. The highest average was (1.63 m/s) in TAS and the lowest was (−0.61 m/s) in WA. Standard deviation (SD) of daily *U* wind varied from 2.3 to 4.3 m/s in VIC and SA, respectively, that is, in the site in SA data were more scattered from the central value compared to the site in VIC. Similarly, the degree of symmetry was the highest in NSW (0.73 m/s) and the lowest in TAS (−0.12 m/s).

Overall, all the data sites except in TAS were positively skewed for daily *U* wind. IN a comparison of the degree of "peak wind speed" in the frequency distribution of site, NSW was the highest (0.32 m/s) and the lowest site was in VIC (−0.16 m/s) (Table 6.5).

For daily meridional wind (*V* wind), the mean of all the data sites was positive, that is, in all study sites *V* wind moved from south to north more than north to south. The highest mean wind was 2.02 m/s in the site in WA and the lowest was 0.1 m/s in TAS. Standard deviation measured the distance of the data points from the central value; here the highest scatter was in the site in SA (4.20 m/s) and the lowest was in the site in VIC (2.54 m/s).

Similarly, like *U* wind, *V* wind also had negative skewness in all data sites. that is, most of the data were elongated to the left from the central value. Likewise, the measurement of the kurtosis was also negative in all study points, that is, daily *V* wind was tailed in the left side (Table 6.6).

TABLE 6.5 Zonal wind (*U* wind) characteristics for daily wind of all the data sites.

	NSW	SA	TAS	VIC	WA
Mean (m/s)	− 0.39	0.45	1.63	0.83	− 0.61
SD (m/s)	2.76	4.30	3.43	2.30	3.17
Skewness (m/s)	0.73	0.61	− 0.12	0.30	0.04
Kurtosis (m/s)	0.32	− 0.34	− 0.35	− 0.60	− 0.22

Note that the accurate site names for these states where models were tested were shown in Table 6.1 and plotted in Fig. 6.1.

TABLE 6.6 Meridional (*V*) wind characteristics for daily time series of all data sites.

	NSW	SA	TAS	VIC	WA
Mean (m/s)	1.39	1.06	0.10	0.13	2.02
SD (m/s)	3.48	4.20	3.09	2.54	2.87
Skewness (m/s)	− 0.35	− 0.41	− 0.09	− 0.33	− 0.34
Kurtosis (m/s)	− 0.51	− 0.42	− 0.58	− 0.79	− 0.22

Note that the accurate site names for these states where models were tested were shown in Table 6.1 and plotted in Fig. 6.1.

TABLE 6.7 Net wind (*N* wind) characteristics for medium-term (daily) time scale.

	NSW	SA	TAS	VIC	WA
Mean (m/s)	4.82	6.21	4.92	3.69	5.00
SD (m/s)	1.74	2.36	1.94	1.18	1.55
Skewness (m/s)	0.50	0.66	0.44	0.85	0.32
Kurtosis (m/s)	0.11	0.32	− 0.22	1.05	− 0.22

Note that the accurate site names for these states where models were tested were shown in Table 6.1 and plotted in Fig. 6.1.

For daily measurements of *N* wind, the mean wind varied from 3.69 to 6.21 m/s and average wind speed was the highest in SA followed by WA, TAS, NSW, and VIC, respectively. In the site in VIC, standard deviation was less, compared to the other study sites, that is, *N* wind was more stable in VIC and less stable in SA.

The value of skewness of all the data sites was positive and varied from 0.32 to 0.85 m/s, that is, in all the data sites, data points were elongated to the right (Table 6.7) for *N* wind. Kurtosis values in the sites in WA and TAS were negative and the other data sites were positive, that is, flatness (or peak wind speed) data character of the frequency distribution lay between −0.22 to 1.05 m/s in WA/TAS and VIC, respectively (Table 6.7).

6.2.2 ANN model

The ANN model has been widely used after 1985 when the training of the neural network algorithm was introduced (Arce-Medina and Paz-Paredes, 2009). However, it should be noted that the ANN model does not identify automatically the most suitable training algorithm without an iterative model identification process; that is, it works as a complete black-box process-based model (Deo and Şahin, 2017).

The mathematical equation of the ANN model can be written as (Deo and Şahin, 2017).

$$f(w) = g\sum\nolimits_{n=1}^{m} t_n(s) \times u_n(s) + a \tag{6.1}$$

where $u_n(s)$ is the inputs lag wind; $t_n(s)$ is the weight of the nth neuron, a is the bias of the neurons; $f(w)$ is the forecasted wind; s is the time horizon and g is the hidden transfer function, $n = 1...m$ is hidden neurons.

This research study has attempted to develop an ANN model and apply several MATLAB R2016a-based algorithms to generate the best performance of the data-driven model. The training performance function is generally based on minimum value of the mean square error (*MSE*) after which the best model is selected. In this chapter, the net train parameters = 100, maximum epoch = 500, and goal = 0.0001 (i.e., threshold error) have been used in the MATLAB R2016a software and create a feed-forward backpropagation network.

For the model development process, the logarithmic sigmoid function (logsig) has been used for the input function, positive linear transfer function (purelin) has been used for

TABLE 6.8 ANN data division and function selection. Red is the best combination.

Data division			Function selection		
Training (%)	**Validation (%)**	**Testing (%)**	**Input function**	**Hidden layer**	**Output function**
90	5	5	purelin	logsig	trainlm
80	10	10	logsig	purelin	trainlm
70	15	15	purelin	logsig	trainbfg
60	20	20	logsig	purelin	trainbfg
40	30	30	purelin	logsig	traingdx
			logsig	purelin	traingdx
			purelin	logsig	trainscg
			logsig	purelin	trainscg
			purelin	logsig	traincgp
			logsig	purelin	traincgp

Note: Logarithmic Sigmoid function (logsig), Positive Linear transfer function (purelin), training Levenberg–Marquardt function (trainlm), training BFGS Quasi-Newton function (trainbfg), training Variable Learning Rate Gradient Descent function (traingdx), training Scaled Conjugate Gradient function (trainscg) and training Fletcher-Powell Conjugate Gradient function (traincgp). The data partition and function selection for ANN model shown in red is the most optimal one.

the hidden layer function, and training Levenberg–Marquardt function (trainlm) has been used for output function. Following this, different data partitions have been used as a trial and error method to get the optimum ANN model. The training portion started from 90% and decreased by 10% in every trial. At the same time, the validation and testing portion increased by 5% in every step until it provided the best result. Table 6.8 shows the full data partition combination.

After generating the most appropriate data partition (i.e., 40% for training, 30% for validation, and 30% for testing) again a trial and error method has been established for the selection of the best function. The research study has therefore used 10 combinations of the input function, the hidden layer function, and the output function (Table 6.8) until it optimized the best results (i.e., logsig for input function, purelin for hidden layer function, and trainlm for the output function was the best combination). This process was applied for a short-term time horizon (6-hourly) data.

This most suitable data partition (i.e., 40%, 30%, 30%) was allocated for the entire chapter and the function selection was used for all the ANN models (daily time horizon as well).

6.2.3 MLR model

The multiple linear regression (MLR) model is based on the cause and effect relationship between inputs and their output (Deo and Şahin, 2017). It creates the linear relationship between cause variables and their effect variables. Mathematically it can be defined as follows (Deo and Şahin, 2017).

$$f(x) = a_0 + a_1 f(x_1) + a_2 f(x_2) + a_3 f(x_3) + a_4 f(x_4) + a_5 f(x_5) \ldots + a_n f(x_n) \quad (6.2)$$

where a_0 is the y-intercept (constant); a_i is coefficients of the input variables $\forall i = 1, 2, 3 \ldots n$; (fx_1) is the correlated inputs lag wind; $f(x)$ is the forecasted wind.

Here, the MLR model has been developed for the training period where it can generate coefficients a_i and the y-intercept a_0, then these values are applied in the testing period by using MATLAB R2016a software. In this research, the first 40% of data has been used for the training period and the last 30% of data has been used for the testing period for MLR model development.

6.2.4 RF model

The random forest model is a family of ensemble-based data mining tools employed to generate accurate predictions without overfitting (Breiman, 2001). Bootstrapping procedures can help modelers to increase their model performances since it performs random selection of training samples, and therefore can reduce the scatterings of data from the central value without increasing the bias, that is, fewer trees are highly influenced by the "noise" in the training set but many trees are less affected by "noise." Therefore training on many trees on a single training set might be highly correlated (Ho, 1995).

Bootstrapping (or bagging) agreement, a machine learning algorithm which reduces variance and helps to avoid overfitting, has also been applied for training algorithms for the RF model.

Mathematically, it can be defined as (Biau, 2012):

Here, $A = a_1, a_2, a_3 \ldots a_i$ is a training set with responses $B = b_1, b_2, b_3 \ldots b_i$, bagging repeatedly selects a random sample with replacement for the training with fitting trees for the sample $m = 1 \ldots M$.

Here i training samples from A and B are written as A_m, B_m respectively.

The prediction of unseen sample a^i is defined by

$$g = \frac{1}{M} \sum_{m=1}^{M} x^m(a^i) \quad (6.3)$$

where x^m the is regression tree onA_m, B_m, etc. and a^i is generated by averaging the prediction from the entire individual regression tree on a^i.

There are three parameters required for the RF model to generate the forecast results. They are (1) the maximum number of terminal nodes (leaf), (2) number of trees (ntrees), and (3) randomly assigned predator variables (fboot) (Prasad et al., 2018). To select the best leaf, ntrees, and fboot, this research used a trial and error method.

In this study, *leaf* = 1, 3, 5, 20; *ntrees* = 50, 100, 200, 400; and *fboot* = 0.4, 0.8,1 were assumed and the model was run in MATLAB R2016a software for every amalgamation to generate the best combination (Table 6.9) in the short-term wind speed. The best combination of parameters to generate optimum RF model is *leaf* = 1, *ntrees* = 400, and *fboot* = 1. For all these runs, the best data partition generated from the ANN model (i.e., training 40%, validation 30%, and testing 30%) is used.

TABLE 6.9 Best function selection in RF model where blue is the best combination.

leaf	*ntrees*	*fboot*
1	50	0.4
3	100	0.8
5	200	1
20	400	

6.2.5 M5 tree model

The M5 tree model is based on binary decision trees where cause and effect variables are mapped from linear regression functions in the terminal node (Rahimikhoob et al., 2013). The M5 tree divides the training dataset into subsets to construct the model tree. The splitting is contingent on the minimization of the intrasubset variation in the output variable's values down each branch (Prasad et al., 2018). Intrasubset variation depends on the standard deviation of the subset value that reaches a node as a measure of error value in a particular node. Then the M5 tree calculates standard deviation reduction (SDR) as the expected reduction, which is mathematically defined as:

$$SDR = \mathrm{sd}(M) - \sum \frac{M_i}{M} \mathrm{sd(M)} \tag{6.4}$$

where sd is the standard deviation, M is the set of examples that reach the node, and M_i is the subsets of ith input of the potential set. In the division process, subtree (child node) has a smaller sd than the parent's node (Alipour et al., 2014).

In the MATLAB R2016a software, the minimum number of cases has been chosen as 5; smoothing = 15 and split threshold = 0.05 was chosen for the development of the M5 tree model. Data partition was taken as 70% for training and 30% for testing.

6.2.6 ARIMA model

Forecasting from the ARIMA model is the combination of the past value and their errors (Pai and Lin, 2005). Mathematically, it can be expressed by the following:

$$\boldsymbol{x_s} = b_0 + \boldsymbol{a_1} x_{s-1} + \boldsymbol{a_2} x_{s-2} + \boldsymbol{a_3} x_{s-3} + \ldots \boldsymbol{a_q} x_{s-q} + \varepsilon_s - b_1 \varepsilon_{s-1} - b_2 \varepsilon_{s-2} - b_3 \varepsilon_{s-3} - \ldots b_{s-r} \varepsilon_{s-r} \tag{6.5}$$

where $\boldsymbol{x_s}$ is the given value, ε_s is the random error at time s, a_i and b_j are the coefficients, and q and r are polynomials of the ARIMA model (Pai and Lin, 2005).

For example, ARIMA (1, 0, 1) can be defined as

$$\boldsymbol{x_s} = b_0 + \boldsymbol{a_1} x_{s-1} + \varepsilon_s - b_1 \varepsilon_{s-1} \tag{6.6}$$

In this research, the ARIMA (q d s) was calculated; autoregressive degree q, differencing degree d, and moving-average degree s were selected according to minimum Akaike information criteria (AIC) and Bayesian information criteria (BIC) values for every time series data.

To choose the minimum AIC and BIC and maximum likelihood, the numerical value of q, d, and s were considered less than or equal to 30 if the process proceeded for longer. The ARIMA model was compared with other computer-based models (for example, ANN, RF, M5 tree, MLR) and in some data sites. R software was used to generate the forecast result for the ARIMA model.

6.2.7 Performance evaluation

The model performance of wind forecasting was ascertained by using the evaluation metrics. The following statistical criteria were considered.

6.2.7.1 Correlation coefficient

Correlation coefficient gives the strength of the linear relationship between two variables. It also describes whether the linearity was strong enough to use the model for the data. The value of R lies between -1 and 1 whereas if $R=0$, then the variables have no relation and if $R=1$, then they have perfect positive relation, that is, the change in one variable indicates the change in the other variable in the same direction with the same ratio.

Correlation coefficient can be formulated by the following (Chai and Draxler, 2014):

$$R^2 = \left[\frac{\sum_{i=1}^{n}(DO_i - < DO >)(DS_i - < DS >)}{\sqrt{\sum_{i=1}^{n}(DO_i - < DO >)^2}\sqrt{\sum_{i=1}^{n}(DS_i - < DS >)^2}} \right]^2 \tag{6.7}$$

where DO_i is the data observed; DS_i is the data simulated; <DO> is the mean of observed data; <DS> is the mean of simulated data.

6.2.7.2 Nash–Sutcliffe coefficient

Nash–Sutcliffe coefficient (ENS) ranges from zero to one; if $ENS=1$ then it perfectly matches the modeled simulation with the observed data. If $ENS=0$ then it indicates that the model predictions are as accurate as the mean of the observed data. ENS can be used to quantitatively describe the accuracy of model outputs (Prasad et al., 2018). ENS is sensitive to the extreme value and can be defined mathematically by the following (Nash and Sutcliffe, 1970):

$$E_{NS} = 1 - \left[\frac{\sum_{i=1}^{n}(DO_i - DS_i)^2}{\sum_{i=1}^{n}(DO_i - < DO >)^2} \right], \quad -\infty \le E_{NS} \le 1 \tag{6.8}$$

6.2.7.3 *Willmott's index of agreement*

Willmott's Index (d) of agreement is the standardized measure of the degree of model prediction error and varies between 0 and 1. $d=1$ represents the perfect agreement between the simulated data and the observed data and decreases the value of d from 1 to 0, which tends to decrease their agreement as well. The index of agreement can detect additive and proportional difference in the observed and simulated means and variance (Bennett et al., 2013). However, it is overly sensitive to the extreme values due to the square difference. Mathematically, it can be defined as (Willmott et al., 2012):

$$d = 1 - \left[\frac{\sum_{i=1}^{n} (DO_i - DS_i)^2}{\sum_{i=1}^{n} (|DS_i - <DO>| + |DO_i - <DS>|)^2} \right] \tag{6.9}$$

$$0 \leq d \leq 1$$

6.2.7.4 Root mean square error

Root mean square error ($RMSE$) represents the sample standard deviation of the difference between predicted values and observed wind speed values. These individual differences are called residuals. $RMSE$ is a good measure of accuracy but only to compare forecasting errors of different models for particular variables as it is scale dependent. It measures overall performance across the entire range of the dataset and provides a good measure of the model. For a perfect model $RMSE=0$.

It can be defined by the following (Chai and Draxler, 2014):

$$RMSE = \sqrt{\left(\frac{1}{n}\right) \sum_{i=1}^{n} (DS_i - DO_i)^2} \tag{6.10}$$

6.2.7.5 Mean absolute error

Mean absolute error (MAE) is used to measure how close forecast values are to observed wind speed outcomes but it is not weighted toward higher and lower magnitude events. It is the average of the absolute errors. Thus MAE always 0, where $MAE=0$ represent a perfect fit model.

The formula for MAE is (Chai and Draxler, 2014)

$$MAE = \frac{1}{n} \sum_{i=1}^{n} |(DS_i - DO_i)| \tag{6.11}$$

6.2.7.6 Relative root mean square error

$RRMSE$ is calculated by dividing $RMSE$ by the mean of the observed data. If $RRMSE < 10\%$ it is considered excellent, $10\%-20\%$ is considered good, $20\%-30\%$ is fair, and more than 30% is poor (Despotovic et al., 2016).

$$RRMSE = \frac{\sqrt{\left(\frac{1}{n}\right)\sum_{i=1}^{n}(DS_i - DO_i)^2}}{\frac{1}{n}\sum_{i=1}^{n} DO_i} \times 100 \tag{6.12}$$

6.2.7.7 Relative mean absolute error

RMAE is also called mean absolute percentage error (*MAPE*) according to Despotovic et al. (2016), and it calculates the absolute value of the average between the simulated and observed wind data. It ranges $(0-\infty)$ and is defined by (Chai and Draxler, 2014),

$$RMAE = \frac{1}{n}\sum_{i=1}^{n} \left| \frac{(DS_i - DO_i)}{DO_i} \right| \times 100 \tag{6.13}$$

6.2.7.8 Legates and McCabes index

The value of L lies between $(-\infty, 1)$ and has an ideal value of 1. L considers the absolute values for calculated and given errors and differences for the appropriate weights (Legates and McCabe, 1999).

$$L = 1 - \left[\frac{\sum_{i=1}^{n} |(DS_i - DO_i)|}{\sum_{i=1}^{n} |DO_i - \langle DS \rangle|} \right] \tag{6.14}$$

6.3 Results of short-term wind speed forecasting

Partial autocorrelation function (PACF) is a correlation between observed values in a time series with its observed value of preceding same time steps. PACF was used to generate significant lag correlation from observed wind for the short-term and daily wind speed prediction.

Before the training and testing process, data were normalized into [0, 1] (Deo and Sahin 2016) by using the following formula

$$A_{Norm} = \frac{A - A_{\text{Min}}}{A_{\text{Max}} - A_{\text{Min}}}$$

where A, A_{Min}, and A_{Max} are variable, minimum of A, and maximum of A, respectively.

6.3.1 Selection of net winds

In this section, the results for 6-hourly forecasting models are presented where the ANN, RF, MLR, M5 tree, and ARIMA models were designed and tested for the first time at specific data locations in Australia in the wind speed prediction problem over the short-term horizon. The conclusions were drawn from the comparison of error matrices (Eq. 6.3.7–6.3.14).

In this chapter, *N* wind was forecasted for a short-term time scale in two ways:

1. *N* wind was calculated from raw *U* and *V* wind using Pythagoras' theorem ($N = \sqrt{U^2 + V^2}$) and then forecasted through models, as described above (*N* wind).
2. *U* and *V* (using different models like ANN, MLR, RF, and M5 tree) were forecasted and then Pythagoras' theorem was used to establish the value of the forecasted wind (*NF* wind).

Table 6.10 shows the comparison between two types of net winds (*N* and *NF*) with different forecasting accuracy indicators. For example, in the ANN model in Silverton (NSW), indicators of *N* wind forecasting were $R = 0.71$, $d = 0.62$, $ENS = 0.51$, $RRMSE = 31\%$ $RMAE = 38\%$, $L = 0.33$, $RMSE = 1.51$ m/s, and $MAE = 1.14$ m/s, whereas *NF* wind had $R = 0.71$, $d = 0.71$, $ENS = 0.41$, $RRMSE = 34\%$, $RMAE = 34\%$, $L = 0.27$, $RMSE = 1.65$ m/s, and $MAE = 1.25$ m/s.

Indicators of *N* wind forecasting of *R, ENS, RMSE, MAE, RRMSE,* and *L* were better compared to *NF* wind forecasting indicators but *d* and *RMAE* were slightly better in *NF* wind forecasting. However, overall, most of the index indicators were more accurate for *N* wind forecasting for all four models (ANN, RF, MLR, and M5 tree) of all the data sites.

Similarly, the correlation coefficient (*R*) was also better for *N* wind for most of the models in most of the sites (Table 6.10). On the other hand, in most of the data sites and models, the indicator of *RMAE* and *d* was better in *NF* wind compared to *N* wind. Overall, at least five indicators out of eight were better in *N* wind compared to *NF* wind in all the data sites and all four models. Thus, in this research, *N* wind had been forecasted in all the data sites and all models for short-term and daily wind forecasting.

6.3.2 Model design for short-term prediction

For the short-term prediction of wind speed data, the study period covered January 1, 1988 to December 31, 2017 with the number of datum points set at 43,832 for all models.

The number of significant lags for short-term wind (for *U, V,* and *N*) was five (i.e., number of data points $= t - 5 = 43,832 - 5 = 43,827$) but to develop the model, only highly correlated lags were used for the input variables (Table 6.11).

According to lag correlation, input was selected. For example, in relation to *V* wind in SA, lag-length was five but significant lag correlation was four (i.e., input variables of *V* wind in SA were four) which was represented by 5L4 (Table 6.12).

For every site, *N* wind was highly correlated to the succeeding 6-hour *N* wind [i.e., $(t-1)$] but other lags varied in different sites. For example, in WA, there were five significant lags [$(t-1), (t-2), (t-3), (t-4)$, and $(t-5)$] but in SA, there were only three significant lags [$(t-1), (t-2)$, and $(t-5)$] (Table 6.12 and Fig. 6.2).

6.3.3 Model performance

6.3.3.1 ANN model performance for short-term time series

The ANN model, developed by trialing a maximum number of hidden neurons (hn) was set to 30, with an incremental step of 1, and used the best data partition with the best

TABLE 6.10 Short-term net wind comparison; *NF* (Net wind was forecasted from forecasted *U* and forecasted *V*) *N* (net wind from raw *U* and raw *V*).

Site	Model	Type	*R*	*d*	*ENS*	*RMSE* (m/s)	*MAE* (m/s)	*RRMSE* (%)	*RMAE* (%)	*L*
NSW	ANN	NF	0.71	0.71	0.41	1.65	1.25	34	34	0.27
		N	0.71	0.62	0.51	1.51	1.14	31	38	0.33
	RF	NF	0.69	0.70	0.37	1.70	1.29	36	35	0.24
		N	0.70	0.62	0.49	1.54	1.17	32	39	0.32
	M5	NF	0.70	0.70	0.40	1.67	1.26	35	34	0.26
		N	0.70	0.62	0.49	1.54	1.17	32	38	0.32
	MLR	NF	0.71	0.71	0.38	1.70	1.31	35	35	0.23
		N	0.71	0.62	0.50	1.52	1.15	32	38	0.32
SA	ANN	NF	0.80	0.78	0.57	1.79	1.40	29	31	0.35
		N	0.80	0.73	0.64	1.63	1.25	27	33	0.42
	RF	NF	0.78	0.76	0.54	1.85	1.44	30	32	0.33
		N	0.79	0.72	0.62	1.67	1.29	27	34	0.40
	M5	NF	0.79	0.77	0.56	1.80	1.39	30	31	0.35
		N	0.79	0.72	0.63	1.65	1.27	27	33	0.41
	MLR	NF	0.79	0.77	0.53	1.85	1.47	30	32	0.31
		N	0.79	0.72	0.62	1.67	1.28	27	33	0.40
TAS	ANN	NF	0.76	0.76	0.51	1.64	1.26	33	34	0.34
		N	0.77	0.72	0.59	1.50	1.15	30	38	0.39
	RF	NF	0.74	0.74	0.47	1.70	1.31	34	36	0.31
		N	0.76	0.71	0.58	1.52	1.17	31	39	0.38
	M5	NF	0.75	0.75	0.50	1.66	1.28	33	36	0.33
		N	0.75	0.71	0.56	1.55	1.19	31	39	0.38
	MLR	NF	0.76	0.71	0.57	1.53	1.17	31	39	0.38
		N	0.76	0.71	0.57	1.53	1.17	31	39	0.38
VIC	ANN	NF	0.59	0.60	0.18	1.46	1.14	40	37	0.08
		N	0.62	0.46	0.39	1.26	0.96	34	41	0.22
	RF	NF	0.56	0.59	0.13	1.50	1.17	41	39	0.06
		N	0.62	0.46	0.38	1.26	0.97	35	41	0.22
	M5	NF	0.57	0.59	0.16	1.47	1.14	40	38	0.08
		N	0.61	0.46	0.37	1.27	0.98	35	41	0.21

(*Continued*)

TABLE 6.10 (Continued)

Site	Model	Type	*R*	*d*	*ENS*	*RMSE* (m/s)	*MAE* (m/s)	*RRMSE* (%)	*RMAE* (%)	*L*
	MLR	NF	0.58	0.60	0.08	1.54	1.21	42	39	0.02
		N	0.60	0.44	0.36	1.28	0.99	35	42	0.20
	ANN	NF	0.74	0.75	0.45	1.46	1.11	30	29	0.31
		N	0.69	0.61	0.48	1.43	1.09	29	33	0.33
	RF	NF	0.73	0.75	0.44	1.48	1.13	30	28	0.30
WA		N	0.68	0.60	0.47	1.44	1.10	29	34	0.32
	M5	NF	0.72	0.74	0.43	1.49	1.13	30	29	0.30
		N	0.66	0.59	0.44	1.48	1.13	30	34	0.30
	MLR	NF	0.70	0.72	0.37	1.57	1.22	32	31	0.25
		N	0.66	0.57	0.43	1.49	1.14	30	34	0.29

Note that the accurate site names for these states where models tested were shown in Table 6.1 and plotted in Fig. 6.1, where *d*, *ENS*, and *L* measured [0, 1].

TABLE 6.11 Data characteristics of short-term wind.

	Study period	Number of datum point	Time series	Wind type	Data source	Data partition (%)	No of lags
ANN	01-01-1988 To 31-12-2017	43,832	6 h interval	*U*, *V*, *N*	Reanalysis datasets (ERA-interim)	40–30–30	5
RF						40–30–30	
M5						70–30	
MLR						40–30–30	
ARIMA						–	–

TABLE 6.12 Total number of statistically significant lags and the number of input variables based on these lagged data used for short-term wind speed prediction, where first prefix of L represented total number lags used and suffix of L represented highly correlated lag inputs.

Wind type	NSW	SA	TAS	VIC	WA
U	5L4	5L3	5L3	5L3	5L5
V	5L4	5L4	5L4	5L4	5L5
N	5L4	5L3	5L5	5L5	5L5

Note that accurate site names for these states where models tested were shown in Table 6.1 and plotted in Fig. 6.1.

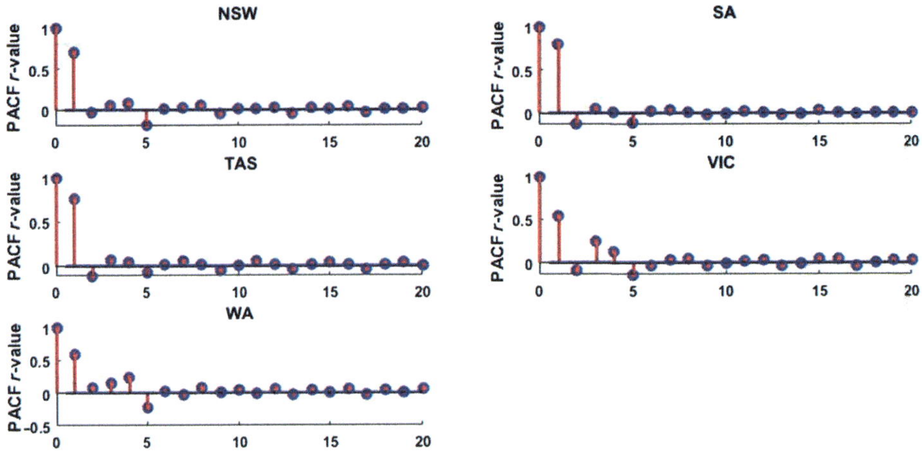

FIGURE 6.2 Partial autocorrelation function (PACF) of 6-hourly *N* wind of all study sites.

function combination (logsig for input function, purelin for hidden function, and trainlm for output function). The lowest mean square error for an optimal model was sought and used several evaluation matrices, as discussed in Section 6.2.3, to verify the results. Numerically, different data sites had different numerical quantifications of model evaluation matrices for different winds (i.e., *U*, *V*, and *N*).

For *U* wind model evaluation matrices comparisons, WA was the best site where $R = 0.89$, $d = 0.89$, $ENS = 0.80$, $RMSE = 1.65$, $MAE = 1.21$, and $L = 0.60$. Similarly, TAS provided the least performance for *U* wind where $R = 0.77$, $d = 0.72$, $ENS = 0.59$, $RMSE = 1.50$ m/s, $MAE = 1.15$ m/s, and $L = 0.40$.

Likewise, for *V* wind, there was 0.86 correlation and Willmott's index of agreement was calculated with $L = 0.54$ in WA, which was the highest in all five sites. However, in VIC, the lowest correlation and Willmott's Index (around 0.80) and *L* index of 0.47 was calculated.

Likewise, for *N* wind, highest *R*, *d*, *ENS*, and *L* and the lowest *RRMSE* and *RMAE* were calculated in SA but the lowest *R*, *d*, *ENS*, and *L* and the highest *RRMSE* and *RMAE* were calculated in VIC (Table 6.13).

Thus for the ANN model, as a site to site comparison, WA was the best for *U* and *V* winds but SA was the best for *N* wind.

6.3.3.2 RF model performance for short-term time series

The RF model was developed to find the best leaf, ntrees, and fboot by using a trial and error method and it determined that ntree = 400, fboot = 1, and leaf = 1 was the best combination function by using evaluation matrices for the RF model (Section 6.2.5).

For *U* wind in the RF model, forecasting indicators *R*, *d*, *ENS*, and *L* were all the highest in Walkway (WA) but the lowest in Macarthur (VIC), that is, compared to the performance of the RF model for all five study sites, the best performance presented in Walkway (WA) and the least in Macarthur (VIC) for *U* wind. Similarly, for *V* wind, the RF model provided

TABLE 6.13 Short-term the ANN model performance in the testing period.

Site	Wind	Data division	Inputs	hn (1:30)	*R*	*d*	*ENS*	*RMSE* (m/s)	*MAE* (m/s)	*RRMSE* (%)	*RMAE* (%)	*L*
	U	40–30–30	4	24	0.82	0.76	0.68	1.80	1.31	−448	420	0.47
NSW	*V*	40–30–30	4	13	0.84	0.85	0.71	2.15	1.52	162	263	0.53
	N	40–30–30	4	19	0.72	0.64	0.52	1.50	1.14	31	37	0.33
	U	40–30–30	3	18	0.80	0.76	0.64	1.58	1.20	194	276	0.44
VIC	*V*	40–30–30	4	30	0.80	0.81	0.65	1.72	1.29	1442	321	0.47
	N	40–30–30	5	26	0.63	0.47	0.39	1.25	0.96	34	41	0.23
	U	40–30–30	3	19	0.88	0.84	0.78	2.18	1.60	666	205	0.57
SA	*V*	40–30–30	4	17	0.84	0.85	0.71	2.53	1.83	230	363	0.52
	N	40–30–30	3	16	0.80	0.73	0.64	1.62	1.25	27	33	0.42
	U	40–30–30	5	28	0.89	0.89	0.80	1.65	1.21	−215	718	0.60
WA	*V*	40–30–30	5	28	0.86	0.86	0.74	1.65	1.22	84	195	0.54
	N	40–30–30	5	26	0.69	0.61	0.48	1.43	1.09	29	33	0.33
	U	40–30–30	3	18	0.77	0.72	0.59	1.50	1.15	30	38	0.40
TAS	*V*	40–30–30	4	25	0.82	0.81	0.68	2.04	1.47	3029	261	0.51
	N	40–30–30	5	18	0.77	0.72	0.59	1.50	1.15	30	38	0.40

the best performance indices in Walkway (WA) and the worst in Macarthur (VIC) (Table 6.14). On the other hand, for *N* wind, Kemmiss Hill (SA) provided best performance by the RF model where $R = 0.79$, $d = 0.72$, $ENS = 0.62$, $RRMSE = 27\%$, $RMAE = 34\%$, and $L = 0.40$. However, the performance indicator showed that forecasting from the RF model was the least accurate in Macarthur (VIC).

6.3.3.3 M5 tree model performance for short-term time series

For the M5 tree model, the *U* wind, $R = 0.87$, $ENS = 0.76$, and $L = 0.56$ were calculated in Kemmiss Hill (SA) station, which was the highest of all the five sites but all of these performance indices were the lowest in Macarthur (VIC), that is, through the M5 tree model, forecasting of *U* wind in Kemmiss Hill (SA) was the best and in Macarthur (VIC) was the worst (Table 6.15).

Similarly, for *V* wind, $R = 0.84$, $d = 0.84$, and $ENS = 0.71$ were the highest in Walkway (WA) but *L* was slightly lower in Walkway (WA) compared to Silverton (NSW). In general, the performance of the M5 tree for the Walkway (WA) site was the most accurate. Compared to all the five data sites, the M5 tree provided less accurate forecasting in Macarthur (VIC). Likewise, for *N* wind in Kemmiss Hill (SA), M5 tree provided the best forecasting model ($R = 0.79$, $d = 0.72$, $ENS = 0.62$, $L = 0.40$, $RRMSE = 27\%$, and

TABLE 6.14 Short-term the RF model performance in the testing period.

Sites	Wind	Inputs	*R*	*d*	*ENS*	*RMSE* (m/s)	*MAE* (m/s)	*RRMSE* (%)	*RMAE* (%)	*L*
	U	4	0.82	0.75	0.66	1.83	1.35	−458	413	0.46
NSW	*V*	4	0.83	0.84	0.69	2.21	1.58	167	231	0.52
	N	4	0.70	0.62	0.49	1.54	1.17	32	39	0.32
	U	3	0.78	0.75	0.61	1.64	1.25	202	366	0.42
VIC	*V*	4	0.80	0.80	0.63	1.76	1.32	1468	341	0.46
	N	5	0.62	0.46	0.38	1.26	0.97	35	41	0.22
	U	3	0.87	0.83	0.76	2.25	1.67	690	215	0.56
SA	*V*	4	0.83	0.84	0.70	2.59	1.88	236	360	0.51
	N	3	0.79	0.72	0.62	1.67	1.29	27	34	0.40
	U	5	0.89	0.88	0.79	1.67	1.23	−218	781	0.59
WA	*V*	5	0.85	0.85	0.73	1.69	1.25	86	204	0.52
	N	5	0.68	0.60	0.47	1.44	1.10	29	34	0.32
	U	3	0.85	0.85	0.73	1.99	1.46	122	225	0.53
TAS	*V*	4	0.81	0.80	0.66	2.10	1.53	3132	271	0.49
	N	5	0.76	0.71	0.58	1.52	1.17	31	39	0.38

$RMAE = 33\%$) and for the Macarthur (VIC) site, it was the least accurate forecasting model ($R = 0.60$, $d = 0.44$, $ENS = 0.36$, $L = 0.20$, $RRMSE = 35\%$, and $RMAE = 42\%$).

6.3.3.4 MLR model performance for short-term time series

For the MLR model, correlation coefficient (*R*) varied from 0.79 to 0.87, Willmott index (*d*) varied from 0.74 to 0.86, and Legates and McCabes index (*L*) varied from 0.43 to 0.56 for *U* wind. The MLR model provided the best forecasting in the Kemmiss Hill (SA) data site where forecasting indicators *R*, *d*, *ENS*, and *L* were higher but in Macarthur (VIC), all these indicators were lower compared to other sites so forecasting by the MLR model for *U* wind was less accurate in Macarthur (VIC)(Table 6.16).

Similarly, for *V* wind, the fifth data site [Walkway (WA)] had the best prediction using the MLR model because *R*, *d*, *ENS* were the highest with the lowest errors in *RMSE* and *MAE*, but *L* in Silverton (NSW) was slightly higher than in Walkway (WA). Likewise, for *N* wind, Kemmiss Hill (SA) had the best wind prediction through the MLR model with 0.79, 0.72, 0.62, and 0.40 of *R*, *d*, *ENS*, and *L*, respectively. In Kemmiss Hill (SA), the percentage errors *RRMSE* and *RMAE* were also less compared to other sites, followed by Walkway (WA), Woolnorth (TAS), Silverton (NSW), and Macarthur (VIC). Other forecasting indices were also poor in Macarthur (VIC) compared to other sites. Thus using the MLR model for *N* wind, the best forecasting result was in Kemmiss Hill (SA) and the worst was in Macarthur (VIC).

TABLE 6.15 Short-term the M5 tree model performance in the testing period.

Site	Wind	Inputs	*R*	*d*	*ENS*	*RMSE* (m/s)	*MAE* (m/s)	*RRMSE* (%)	*RMAE* (%)	*L*
	U	4	0.81	0.75	0.66	1.84	1.36	−459	366	0.46
NSW	*V*	4	0.83	0.83	0.69	2.22	1.57	168	261	0.52
	N	4	0.71	0.62	0.50	1.52	1.15	32	38	0.32
	U	3	0.77	0.74	0.62	1.62	1.23	199	357	0.43
VIC	*V*	5	0.80	0.80	0.63	1.76	1.33	1470	313	0.46
	N	5	0.60	0.44	0.36	1.28	0.99	35	42	0.20
	U	3	0.87	0.83	0.76	2.23	1.66	684	197	0.56
SA	*V*	3	0.82	0.84	0.67	2.70	2.01	245	376	0.48
	N	3	0.79	0.72	0.62	1.67	1.28	27	33	0.40
	U	5	0.86	0.85	0.75	1.85	1.39	−242	723	0.54
WA	*V*	5	0.84	0.84	0.71	1.75	1.31	89	232	0.5
	N	5	0.66	0.57	0.43	1.49	1.14	30	34	0.29
	U	3	0.86	0.86	0.74	1.95	1.41	119	211	0.54
TAS	*V*	4	0.81	0.82	0.66	2.1	1.50	3126	246	0.50
	N	5	0.76	0.71	0.57	1.53	1.17	31	39	0.38

6.3.4 Model comparison for short-term wind speed prediction

The ANN, RF, MLR, M5 tree, and ARIMA models were tested for the first time in these data locations in Australia for short-term wind speed prediction for *U*, *V*, and *N* winds. The model efficiency was tested according to the comparison of error matrices, for example, *R*, *d*, *ENS*, *RMSE*, *MAE*, *RRMSE*, *RMAE*, and *L*. For short-term wind speed prediction, there were variable performances for different artificial intelligence-based models in different data sites.

6.3.4.1 Comparison of different models for U wind

Model performance of artificial intelligent-based models in different sites fluctuated differently for *U* wind. For example, correlation coefficient of ANN, MLR, RF, and M5 tree varied from 0.78–0.80 in Macarthur (VIC), 0.81–0.82 in Silverton (NSW), 0.87–0.88 in Kemmiss Hill (SA), 0.77–0.86 in Woolnorth (TAS), and 0.86–0.89 in Walkway (WA), respectively.

Similarly, *L* index also fluctuated from 0.46–0.47 in Silverton (NSW), 0.56–0.57 in Kemmiss Hill (SA), 0.40–0.54 in Woolnorth (TAS) 0.42–0.44 in Macarthur (VIC), and 0.54–0.60 in Walkway (WA). *d* also fluctuated from 0.72–0.89 in different locations and *ENS* also lay between 0.61 and 0.80.

Error indicators *RMSE* and *MAE* also lay between 1 and 2 m/s. However, compared to all the data-driven models through different testing matrices, the ANN gave the best result

TABLE 6.16 Short-term the MLR Model performance in the testing period.

Site	Wind	Inputs	*R*	*d*	*ENS*	*RMSE* (m/s)	*MAE* (m/s)	*RRMSE* (%)	*RMAE* (%)	*L*
	U	4	0.81	0.75	0.66	1.84	1.36	−459	366	0.46
NSW	V	4	0.83	0.83	0.69	2.22	1.57	168	261	0.52
	N	4	0.71	0.62	0.50	1.52	1.15	32	38	0.32
	U	3	0.79	0.74	0.62	1.62	1.23	199	357	0.43
VIC	V	5	0.80	0.80	0.63	1.76	1.33	1470	313	0.46
	N	5	0.60	0.44	0.36	1.28	0.99	35	42	0.20
	U	3	0.87	0.83	0.76	2.23	1.66	684	197	0.56
SA	V	3	0.82	0.83	0.67	2.70	2.01	245	376	0.48
	N	3	0.79	0.72	0.62	1.67	1.28	27	33	0.40
	U	5	0.86	0.85	0.75	1.85	1.39	−242	723	0.54
WA	V	5	0.84	0.84	0.71	1.75	1.31	89	232	0.50
	N	5	0.66	0.57	0.43	1.49	1.14	30	34	0.29
	U	3	0.86	0.86	0.74	1.95	1.41	119	211	0.54
TAS	V	4	0.81	0.80	0.66	2.10	1.50	3126	246	0.50
	N	5	0.76	0.71	0.57	1.53	1.17	31	39	0.38

in Silverton (NSW), Kemmiss Hill (SA), Macarthur (VIC), and Walkway (WA) data sites for *U* wind. However, in Woolnorth (TAS), indicators like *R*, *d*, *ENS*, and *L* were higher for the MLR model but error indicators like *RMSE*, *MAE*, *RRMSE*, and *RMAE* were lower for the ANN model.

Overall, according to the result, the ANN model provided the best short-term *U* wind forecasting. Other data-driven models like MLR, RF, and M5 tree were also highly competitive in terms of forecasting accuracy.

However, compared to the artificial intelligent models, the statistical model, that is, the ARIMA model, had very poor performance. For example, in Macarthur (VIC), the value of correlation coefficient was more than 0.78 for all artificial intelligent models but the ARIMA model was just 0.18 (Table 6.17).

Similarly, for the Willmott index indicator, the ARIMA had 0.25 whereas other data-driven models had more than 0.74. All the forecasting indicators of data-driven models had far better performance compared to the ARIMA model for *U* wind for short-term forecasting and the ANN was the most accurate.

6.3.4.2 Comparison of different models for V wind

Performance of *V* wind forecasting also varied based on different models and sites. For example, the correlation coefficient (*R*) of *V* wind was 0.80–0.86 among all the data-driven models. Similarly, the values of *L*, *d*, and *ENS* varied from 0.44–0.54, 0.76–0.86, and

TABLE 6.17 U wind comparison for all the data sites through all artificial intelligent models and statistical model.

Models	*R*	*d*	*ENS*	*RMSE* (m/s)	*MAE* (m/s)	*RRMSE* (%)	*RMAE* (%)	*L*
				VIC				
ANN	0.80	0.76	0.64	1.58	1.20	194	276	0.44
MLR	0.79	0.74	0.62	1.62	1.23	199	357	0.43
RF	0.78	0.75	0.61	1.64	1.25	202	366	0.42
M5	0.79	0.76	0.62	1.61	1.23	199	309	0.43
ARIMA	0.18	0.25	0.03	2.57	2.12	317	560	0.01
				NSW				
ANN	0.82	0.76	0.68	1.80	1.31	−448	420	0.47
MLR	0.81	0.75	0.66	1.84	1.36	−459	366	0.46
RF	0.82	0.75	0.66	1.83	1.35	−458	413	0.46
M5	0.82	0.76	0.66	1.83	1.35	−458	375	0.46
				SA				
ANN	0.88	0.84	0.78	2.18	1.60	666	205	0.57
MLR	0.87	0.83	0.76	2.23	1.66	684	197	0.56
RF	0.87	0.83	0.76	2.25	1.67	690	215	0.56
M5	0.88	0.84	0.77	2.22	1.64	682	210	0.56
				TAS				
ANN	0.77	0.72	0.59	1.50	1.15	30	38	0.40
MLR	0.86	0.86	0.74	1.95	1.41	119	211	0.54
RF	0.85	0.85	0.73	1.99	1.46	122	225	0.53
M5	0.86	0.86	0.74	1.97	1.43	120	226	0.54
				WA				
ANN	0.89	0.89	0.80	1.65	1.21	−215	718	0.60
MLR	0.86	0.85	0.75	1.85	1.39	−242	723	0.54
RF	0.89	0.88	0.79	1.67	1.23	−218	781	0.59
M5	0.88	0.88	0.78	1.72	1.25	−224	880	0.58

0.63–0.74, respectively, in different locations. However, error indicators *RMSE* and *MAE* ranged from 1.58–2.7 m/s and 1.2–2 m/s respectively.

The ANN forecasting provided the best results at Silverton (NSW), Kemmiss Hill (SA), Woolnorth (TAS), and Walkway (WA) data sites. Whereas in the study site at Macarthur (VIC), all error indicators, that is, *RMSE MAE*, *RRMSE*, and *RMAE* were at a minimum but *L* was higher in the M5 tree model. Nevertheless, for the *V* wind, the ANN provided the best forecasting model for all the data sites for short-term time horizons and other data-driven models (RF, MLR, and M5 tree) had performance close to the ANN model.

Accuracy of the ARIMA model was very low in *V* wind forecasting. For example, in the Macarthur (VIC) data site, the correlation coefficient was less than 0.1, whereas other data-driven models' performance for correlation coefficient was around 0.80. Similarly, *L* index of the ARIMA model was less than 0.01 but other artificial intelligence-based models had achieved accuracy of more than 0.44 on the same site. Similarly, other forecasting indicators, e.g., *d*, *ENS*, *RMSE*, *MAE*, *RRMSE*, and *RMAE* also performed at a very low level (Table 6.18) by using the ARIMA model.

Thus for short-term forecasting of *V* wind, the ANN model was better and the ARIMA model was the least accurate for short-term wind speed forecasting.

6.3.4.3 Comparison of different models for N wind

Performance of *N* wind also varied in different locations using different models. For example, artificial intelligence-based techniques for wind speed forecasting (e.g., ANN, MLR, RF, and M5 tree) provided better forecasting indicators, like *R*, *d*, *ENS*, *RMSE*, *MAE*, *RRMSE*, *RMAE*, and *L* compared with a statistical model like the ARIMA. By using data-driven models, the value of *R*, *d*, *ENS*, and *L* lay within 0.61–0.80, 0.44–0.73, 0.36–0.64, and 0.21–0.42, respectively, in different locations. Similarly, error parameters like *RMSE*, *MAE*, *RRMSE*, and *RMAE* lay within 1.25–1.67 m/s, 0.96–1.29 m/s, 27%–35%, and 33%–41%, respectively, overall.

However, while comparing the short-term *N* wind forecasting through artificial intelligence-based technique in all the sites, error matrices were not vastly different to each other for data-driven models. For example, in site one [Silverton (NSW)], *R* was 0.70–0.72, *d* was 0.62–0.64, *ENS* was 0.49–0.52, and *L* was 0.32–0.33. A similar trend was evident for *RMSE*, *MAE RRMSE*, and *RMAE* model indicators in the Silverton (NSW) wind site as well.

Model preciseness was assigned using scatter plots of observed *N* wind and forecasted *N* wind for short-term wind speed prediction with the least squares fit of the fifth wind site [Walkway (WA)] (Fig. 6.3).

While the correlation coefficient of the ANN model was 0.69, the correlation coefficient of other models was 0.68 (the RF model) and 0.66 for the rest of the models.

Similarly, the box plot in Fig. 6.4 illustrates the spread of the wind data with respect to its quartiles, of observed and forecasted *N* wind speed in the third wind site [Woolnorth (TAS)]. The character of the dataset of observed and forecasted *N* wind shows in the figure that the predicted dataset had fewer outliers compared with the observed dataset, that is, the first quartile of forecasted *N* wind was higher than the first quartile of observed *N* wind. Similarly, the forecasted third quartile was smaller than the observed third quartile.

The results showed that performance was vastly different between data-driven models and the statistical models. For example, in VIC for *N* wind, the value of *R* for the lowest data-driven model (MLR) was 0.60 but for the statistical model it was just 0.02 where the *L* index was negative performed by the ARIMA model.

TABLE 6.18 *V* wind comparison for all the data sites through all artificial intelligent models and statistical model.

Models	*R*	*d*	*ENS*	*RMSE* (m/s)	*MAE* (m/s)	*RRMSE* (%)	*RMAE* (%)	*L*
				NSW				
ANN	0.84	0.85	0.71	2.15	1.52	162.00	263.00	0.53
MLR	0.83	0.83	0.69	2.22	1.57	168.00	261.00	0.52
RF	0.83	0.84	0.69	2.21	1.58	167.00	231.00	0.52
M5	0.83	0.84	0.69	2.21	1.57	167.00	263.00	0.52
				SA				
ANN	0.84	0.85	0.71	2.53	1.83	230.00	363.00	0.52
MLR	0.82	0.83	0.67	2.70	2.01	245.00	376.00	0.48
RF	0.83	0.84	0.70	2.59	1.88	236.00	360.00	0.51
M5	0.82	0.83	0.67	2.71	2.00	246.00	374.00	0.48
				TAS				
ANN	0.82	0.81	0.68	2.04	1.47	3029.00	261.00	0.51
MLR	0.81	0.80	0.66	2.10	1.50	3126.00	246.00	0.50
RF	0.81	0.80	0.66	2.10	1.53	3132.00	271.00	0.49
M5	0.81	0.80	0.66	2.10	1.51	3135.00	265.00	0.49
				VIC				
ANN	0.80	0.76	0.64	1.58	1.20	194.00	276.00	0.44
MLR	0.80	0.80	0.63	1.76	1.33	1470.00	313.00	0.46
RF	0.80	0.80	0.63	1.76	1.32	1468.00	341.00	0.46
M5	0.80	0.80	0.63	1.76	1.32	1474.00	346.00	0.46
ARIMA	0.09	0.32	− 0.08	3.02	2.43	2525.00	256.00	0.00
				WA				
ANN	0.86	0.86	0.74	1.65	1.22	84.00	195.00	0.54
MLR	0.84	0.84	0.71	1.75	1.31	89.00	232.00	0.50
RF	0.85	0.85	0.73	1.69	1.25	86.00	204.00	0.52
M5	0.85	0.85	0.72	1.72	1.28	88.00	210.00	0.51

The following time series plot (Fig. 6.5) of *N* wind on the last 30 data points could also easily demonstrate the difference between the artificial intelligence-based technique of forecasting and the statistical method of forecasting. The ANN, MLR, RF, and M5 tree models followed the wind speed pattern but the ARIMA model did not follow the wind

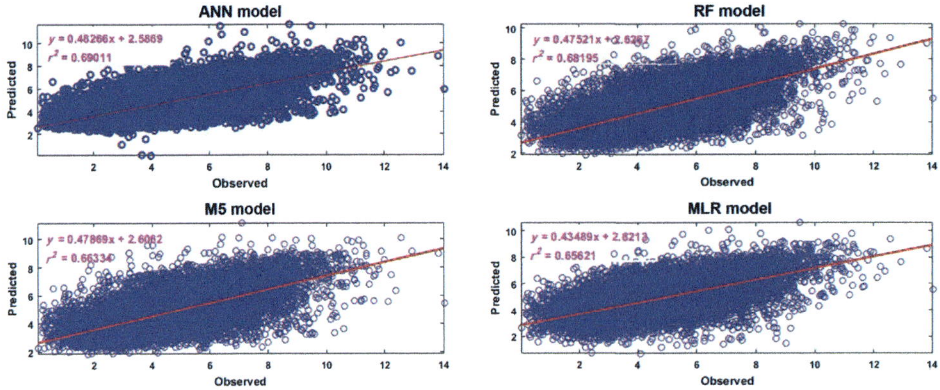

FIGURE 6.3 Scatter plots of observed *N* wind and forecasted *N* wind for station Walkway (WA) of all artificial intelligent technique (ANN, RF, M5 tree, and MLR models). (Note: *red* line represented the least-square fit line to the respective scatter plots).

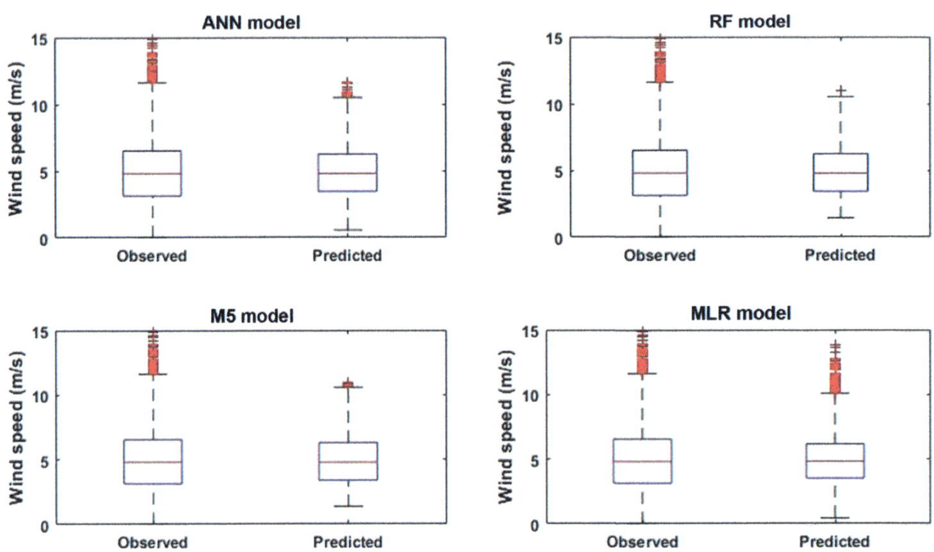

FIGURE 6.4 Box plot of observed *N* wind and forecasted *N* wind in the testing period of data-driven models respectively the ANN, RF, M5 tree and MLR models of Woolnorth (TAS).

pattern. Compared with the other data-driven models, forecasted wind from the ANN model exhibited the observed wind pattern more accurately.

Overall, model performance indicators of the ANN model performed relatively better (Table 6.19, Figs. 6.2–6.5) in all wind sites compared to other artificial intelligence techniques for short-term *N* wind prediction.

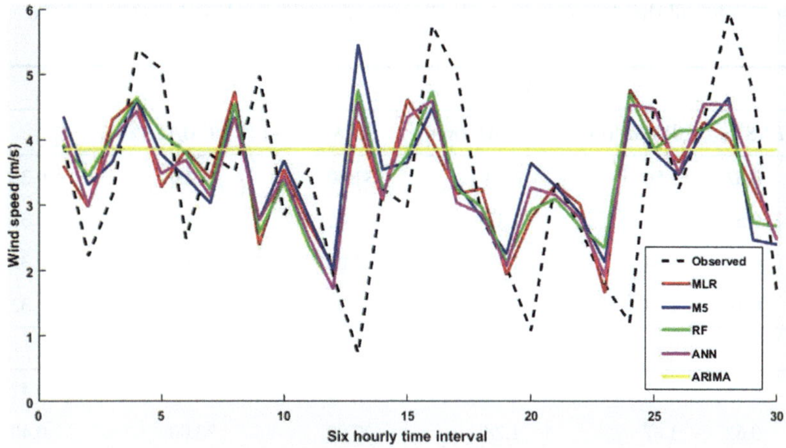

FIGURE 6.5 Time series plot of observed and forecasted N wind in the last 30 data points of testing period of Macarthur (VIC) wind site.

6.4 Results of daily wind speed forecasting

6.4.1 Model design for daily wind speed prediction

For the daily wind speed prediction, study period (January 1, 1988 to December 31, 2017) and the study sites (five study sites) were the same as for the short-term wind speed prediction. Daily datum was calculated by using an arithmetic mean of 6-hourly data taken from the Reanalysis dataset (ERA-Interim).

The total number of datum points of daily wind was 10,958 for each study site. To generate the input variables, PACF was used for all kinds of winds (i.e., *U*, *V*, and *N*). Most of the sites and winds had three significant lags except *V* wind of SA and VIC. Thus the number of data points $= t - 3 = 10,958 - 3 = 10,955$ except *V* wind of SA and *V* wind of VIC (Table 6.21).

Here, for *V* wind of SA, the number of data points $= t - 17 = 10,958 - 17 = 10,941$ and for *V* wind of VIC, the number of data points $= t - 12 = 10,958 - 12 = 10,546$. However, to develop the model, only highly correlated lags were used for the input of each model (Table 6.20).

For each kind of wind (*U*, *V*, and *N* wind) for daily, every $t - 1$ lag was highly correlated with observed wind but other lags were varied (Table 6.21). For example, the following figure (Fig. 6.6) showed that PACF of *N* wind of all five data sites $t - 1$ lag (second line from the left of each figure) was highly correlated compared with other lags.

6.4.2 Model performance

6.4.2.1 ANN model performance for daily time series

To develop the ANN model for daily wind speed prediction, 30 hidden neurons were taken with an increment step of one, the same as the short-term wind speed prediction. Similarly, the best data partition generated by short-term wind speed prediction were used for "daily" wind speed prediction.

TABLE 6.19 *N* wind comparison for all the data sites through all artificial intelligent-based models and the statistical models.

Models	*R*	*d*	*ENS*	*RMSE* (m/s)	*MAE* (m/s)	*RRMSE* (%)	*RMAE* (%)	*L*
				NSW				
ANN	0.72	0.64	0.52	1.50	1.14	31.00	37.00	0.33
MLR	0.71	0.62	0.50	1.52	1.15	32.00	38.00	0.32
RF	0.70	0.62	0.49	1.54	1.17	32.00	39.00	0.32
M5	0.70	0.62	0.49	1.54	1.17	32.00	38.00	0.32
				SA				
ANN	0.80	0.73	0.64	1.62	1.25	27.00	33.00	0.42
MLR	0.79	0.72	0.62	1.67	1.28	27.00	33.00	0.40
RF	0.79	0.72	0.62	1.67	1.29	27.00	34.00	0.40
M5	0.79	0.72	0.63	1.65	1.27	27.00	33.00	0.41
				TAS				
ANN	0.77	0.72	0.59	1.50	1.15	30.00	38.00	0.40
MLR	0.76	0.71	0.57	1.53	1.17	31.00	39.00	0.38
RF	0.76	0.71	0.58	1.52	1.17	31.00	39.00	0.38
M5	0.75	0.71	0.56	1.55	1.19	31.00	39.00	0.38
				VIC				
ANN	0.63	0.47	0.39	1.25	0.96	34.00	41.00	0.23
MLR	0.60	0.44	0.36	1.28	0.99	35.00	42.00	0.20
RF	0.62	0.46	0.38	1.26	0.97	35.00	41.00	0.22
M5	0.61	0.46	0.37	1.27	0.98	35.00	41.00	0.21
ARIMA	0.02	0.16	−0.02	1.62	1.28	44.00	57.00	−0.03
				WA				
ANN	0.69	0.61	0.48	1.43	1.09	29.00	33.00	0.33
MLR	0.66	0.57	0.43	1.49	1.14	30.00	34.00	0.29
RF	0.68	0.60	0.47	1.44	1.10	29.00	34.00	0.32
M5	0.66	0.59	0.44	1.48	1.13	30.00	34.00	0.30

Function selection for daily prediction was also used for the best function combination generated from short-term wind speed prediction (logsig for input function, purelin for hidden layer function, and trainlm for output function) by the ANN model.

For comparison of each data site, the ANN model performance varied for different winds (i.e., *U*, *V*, and *N*) and different sites. For example, The ANN model provided the

TABLE 6.20 Data characteristics (study period, lags, datum point, time series, wind type, and data source) of daily wind.

	Study period	Number of datum point	Time series	Wind type	Data source	Data partition (%)	No of lags
ANN	January 1, 1988–December 31, 2017	10958	Daily	*U*, *V*, *N*	Reanalysis datasets (ERA-interim)	40–30–30	3–17
RF						40–30–30	
M5						70–30	
MLR						40–30–30	
ARIMA						–	–

TABLE 6.21 Total number of lags and number of input variables for daily wind speed prediction where first prefix of *L* represented total number lags used and suffix of *L* represented highly correlated lags input.

Wind type	NSW	SA	TAS	VIC	WA
U	3L3	3L3	3L3	3L3	3L3
V	3L3	17L17	3L3	12L12	3L3
N	3L3	3L3	3L3	3L3	3L3

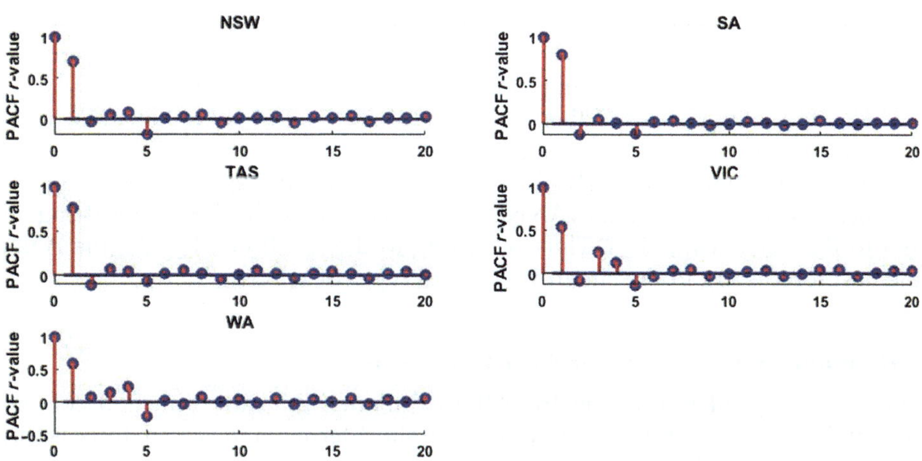

FIGURE 6.6 Partial autocorrelation function (PACF) of *N* wind of all study sites for daily wind speed prediction.

best forecasting result for WA for *U* wind where the value of *R*, *d*, *ENS*, and *L* were 0.70, 0.67, 0.49, and 0.30 respectively. However, error indicators like *RMSE* and *MAE* were less in VIC for *U* wind. Similarly, the wind site in VIC had the lowest *R*, *d*, *ENS*, and *L* and the least *ENS* and *RMSE*.

TABLE 6.22 Daily ANN model forecasting performance matrices for testing period (*green* for *U* wind, *red* for *V* wind, and *black* for *N* wind).

Site	Wind type	*hn*	*R*	*d*	*ENS*	*RMSE* (m/s)	*MAE* (m/s)	*RRMSE* (%)	*RMAE* (%)	*L*
	U	21	0.64	0.54	0.41	2.13	1.67	−585	336	0.25
NSW	*V*	8	0.58	0.50	0.33	2.83	2.23	216	296	0.22
	N	28	0.47	0.31	0.22	1.55	1.22	32	31	0.14
	U	27	0.66	0.58	0.44	3.20	2.54	537	341	0.28
SA	*V*	27	0.65	0.65	0.41	3.28	2.63	304	425	0.25
	N	3	0.50	0.28	0.25	2.02	1.59	33	32	0.14
	U	1	0.64	0.57	0.40	2.59	2.03	158	457	0.25
TAS	*V*	2	0.46	0.33	0.21	2.76	2.25	4195	270	0.13
	N	4	0.48	0.35	0.23	1.73	1.41	35	35	0.12
	U	13	0.61	0.52	0.37	1.84	1.48	221	316	0.24
VIC	*V*	15	0.58	0.53	0.34	2.05	1.66	1794	258	0.22
	N	9	0.45	0.23	0.20	1.05	0.82	29	25	0.10
	U	19	0.70	0.67	0.49	2.28	1.79	−440	389	0.30
WA	*V*	29	0.68	0.64	0.47	2.16	1.66	108	688	0.31
	N	8	0.59	0.48	0.35	1.23	0.96	25	23	0.23

Likewise, WA had better forecasting indicators for *V* wind. Here for *V* wind, *R*, *d*, *ENS*, *RMSE*, *MAE*, and *L* were, respectively, 0.68, 0.64, 0.47, 2.16 m/s, 1.66 m/s, and 0.31. However, for the sites in TAS the evaluation indicators of *R*, *d*, *ENS*, *RMSE*, *MAE*, and *L* were the lowest for *V* wind. Similarly, for *N* wind of daily wind speed forecasting, all forecasting indicators showed that the ANN model performed the best forecasting in WA compared to the other wind sites. Thus according to the result (Table 6.22) the ANN provided the best forecasting in the site in WA for all types of winds.

6.4.2.2 RF model performance for daily time series

For daily wind speed forecasting from the RF model, the same function combination was used which was generated and tested for short-term wind speed forecasting (i.e., $ntrees = 400, leaf = 1,$ and $fboot = 1$). For *U* wind forecasting in the site in WA, most of the indicators (*R*, *d*, *ENS*, and *L*) were relatively better than the other sites. However, error matrices *RMSE* and *MAE* were slightly higher in WA compared to the other sites for *U* wind.

In relation to the performance of the RF model for *U* wind for daily forecasting, the majority of the indicators performed the lowest for the site in VIC. Similarly, for *V* wind forecasting, most of the indicators (five out of eight) provided the better performance in WA and worse performance in TAS (five out of eight worst performance indicators).

TABLE 6.23 Daily RF model forecasting performance matrices for testing period (*green* for *U* wind, *red* for *V* wind, and *black* for *N* wind).

Sites	Wind type	*R*	*d*	*ENS* (%)	*RMSE* (m/s)	*MAE* (m/s)	*RRMSE* (%)	*RMAE* (%)	*L*
	U	0.58	0.48	0.33	2.27	1.76	−565	370	0.21
NSW	*V*	0.54	0.48	0.28	2.94	2.33	223	455	0.18
	N	0.42	0.29	0.16	1.61	1.26	34	32	0.11
	U	0.59	0.48	0.34	3.40	2.71	1053	440	0.21
SA	*V*	0.60	0.53	0.35	3.28	2.56	300	275	0.23
	N	0.45	0.28	0.19	2.10	1.66	35	33	0.11
	U	0.59	0.53	0.35	2.81	2.21	172	340	0.22
TAS	*V*	0.39	0.30	0.14	2.88	2.34	4382	285	0.09
	N	0.43	0.32	0.17	1.79	1.45	36	36	0.10
	U	0.55	0.47	0.30	1.91	1.55	236	402	0.19
VIC	*V*	0.57	0.49	0.32	2.07	1.68	1809	231	0.21
	N	0.33	0.19	0.06	1.14	0.89	31	27	0.02
	U	0.66	0.63	0.43	2.37	1.88	−311	427	0.26
WA	*V*	0.64	0.60	0.40	2.13	1.67	109	353	0.25
	N	0.52	0.43	0.26	1.31	1.03	27	24	0.18

Likewise, for *N* wind performance for daily wind speed forecasting by the RF model, the wind site in WA was the best, where $R = 0.52$, $d = 0.43$, $ENS = 0.26$, $MAE = 1.03$ m/s, $RRMSE = 27\%$, $RMAE = 24\%$, and $L = 0.18$. *L* index and *ENS* were very low in VIC (both less than 0.1). However, error indicator *MAE* was better in this site (0.89 m/s) (Table 6.23). Overall, daily wind speed forecasting from the RF model was better in the site in WA for all kinds of winds.

6.4.2.3 M5 tree model performance for daily time series

Here the correlation coefficient varied from 0.68–0.76 for *U* wind by the M5 tree model forecasting and the Willmott index lay between 0.60 and 0.73. For *U* wind, forecasting from the M5 tree model was comparatively better in WA and worse in VIC (*R*, *d*, *ENS*, and *L* are lowest). However, the performance of error indicators like *RMSE* and *MAE* were better in VIC compared to other data sites.

Similarly, daily *V* wind forecasting from the M5 tree model was better in SA where, $R = 0.76$, $d = 0.71$, $ENS = 0.56$, and $L = 0.38$ although *RMSE* and *MAE* were not the lowest in this site. However, in the data site in TAS, six out of eight indicators showed the least performance for daily *V* wind forecasting. Likewise, for *N* wind, six indicators out of eight demonstrated better forecasting in WA through the M5 tree model for daily wind speed forecasting, where *RMSE* and *MAE* were slightly higher in WA compared with VIC (Table 6.24).

TABLE 6.24 Daily M5 tree model forecasting performance matrices for testing period (*green* for *U* wind, *red* for *V* wind, and *black* for *N* wind).

Sites	Wind type	*R*	*d*	*ENS*	*RMSE* (m/s)	*MAE* (m/s)	*RRMSE* (%)	*RMAE* (%)	*L*
	U	0.70	0.61	0.49	1.97	1.52	−541	305	0.31
NSW	*V*	0.67	0.60	0.44	2.65	2.08	190	200	0.29
	N	0.60	0.44	0.35	1.42	1.12	29	28	0.20
	U	0.73	0.64	0.53	2.94	2.31	493	327	0.35
SA	*V*	0.76	0.71	0.56	2.82	2.16	262	239	0.38
	N	0.62	0.45	0.37	1.90	1.49	31	29	0.22
	U	0.70	0.64	0.49	2.41	1.87	147	430	0.31
TAS	*V*	0.59	0.47	0.34	2.50	2.00	2426	1071	0.22
	N	0.60	0.46	0.35	1.54	1.23	32	32	0.21
	U	0.68	0.60	0.46	1.70	1.36	204	293	0.30
VIC	*V*	0.73	0.67	0.52	1.76	1.39	1352	374	0.35
	N	0.54	0.35	0.28	1.00	0.77	27	23	0.16
	U	0.76	0.73	0.57	2.10	1.64	−405	316	0.36
WA	*V*	0.74	0.70	0.55	1.99	1.52	100	773	0.37
	N	0.65	0.55	0.42	1.20	0.92	24	21	0.28

6.4.2.4 MLR model performance for daily time series

For the daily time horizon by the MLR model forecasting, all three types of winds (*U*, *V*, and *N*) were more accurate in WA. For *N* wind, six indicators out of eight showed the better performance in WA for daily time horizon forecasting.

For *V* wind, seven indicators were better in WA whereas for *U* wind four out of eight indicators were the best for WA. Through using the MLR model, forecasting of *V* wind involved less accurate daily forecasting in TAS (the lowest *R*, *d*, *ENS*, and *L*). Similarly, in VIC, *U* and *N* winds forecasting from the MLR model was more complex because it had the lowest *R*, *ENS*, and *L*, however, it did have the lowest *MAE* and *RMSE* (Table 6.25).

6.4.3 Model comparison for daily wind speed prediction

6.4.3.1 Comparison of different models for U wind forecasting for daily time scale

Different data-driven models had different performance for *U* wind speed forecasting on the same data in the same data site for the daily time horizon. For example, the value of *L* index by the ANN, MLR, RF and the M5 tree varied from 0.29−0.36 in WA, 0.21−0.31 in NSW, 0.21−0.35 in SA, 0.22−0.31 in TAS, and 0.19−0.30 in VIC.

The correlation coefficient (*R*) varied from 0.58−0.70 in NSW, 0.59−0.73 in SA, 0.59−0.70 in TAS, 0.55−0.68 in VIC, and 0.66−0.76 in WA. The value of *d* also ranged

TABLE 6.25 Daily MLR model forecasting performance matrices for testing period (*green* for *U* wind, *red* for *V* wind, and *black* for *N* wind).

Sites	Wind type	*R*	*d*	*ENS*	*RMSE* (m/s)	*MAE* (m/s)	*RRMSE* (%)	*RMAE* (%)	*L*
	U	0.60	0.47	0.36	2.22	1.74	−554	442	0.21
NSW	*V*	0.56	0.47	0.31	2.88	2.29	219	304	0.20
	N	0.45	0.27	0.20	1.57	1.25	33	32	0.12
	U	0.63	0.49	0.39	3.28	2.63	1014	353	0.23
SA	*V*	0.60	0.55	0.36	3.26	2.56	298	273	0.23
	N	0.49	0.29	0.24	2.04	1.60	33	32	0.13
	U	0.64	0.57	0.40	2.69	2.11	165	312	0.25
TAS	*V*	0.45	0.32	0.20	2.77	2.27	4211	252	0.12
	N	0.48	0.34	0.23	1.73	1.41	35	35	0.12
	U	0.59	0.48	0.34	1.84	1.51	228	339	0.21
VIC	*V*	0.58	0.51	0.33	2.06	1.66	1796	258	0.22
	N	0.45	0.21	0.20	1.05	0.82	29	25	0.10
	U	0.68	0.62	0.46	2.31	1.83	−303	427	0.29
WA	*V*	0.67	0.62	0.45	2.05	1.61	105	611	0.28
	N	0.58	0.45	0.33	1.25	0.98	25	23	0.22

from 0.47–0.61 in NSW, 0.48–0.64 in SA, 0.53–0.64 in TAS, 0.47–0.60 in VIC, and 0.62–0.73 in WA.

Error indicators like *MAE* and *RMSE* also varied differently in different locations. In the location NSW, the M5 tree provided the best forecasting results compared with ANN, MLR, and RF models, where all the parameters (R, *d*, *ENS*, *RMSE*, *MAE*, *RRMSE*, *RMAE*, and *L*) were superior to the other models, followed by the ANN model. However, the MLR and the RF models had mixed indicators (R of RF < R of MLR, d of RF > d of MLR, *ENS* of MLR > *ENS* of RF, *RMSE* and *MAE* of MLR < *RMSE* and *MAE* of RF). In the data site in SA, performance of the M5 tree was the best followed by ANN model for daily *U* wind forecasting.

Overall, according to the results from Table 6.26, the M5 tree provided the best results for *U* wind forecasting for the daily time horizon followed by the ANN model but the MLR and the RF models had mixed results for the third and fourth performance position.

6.4.3.2 Comparison of different models for V wind forecasting for daily time scale

For *V* wind forecasting, the performance of different models was different in different data sites. For example, the data site in SA, *L* index, fluctuated from 0.23 to 0.38 among all the data-driven models. The performance of the M5 tree model was the best, whereas the MLR and the RF models were at the bottom on the same data site. On the same site, the

TABLE 6.26 *U* wind comparison for all the data sites for daily forecasting through artificial intelligence-based model (*red* color is the best performance).

Models	*R*	*d*	*ENS*	*RMSE* (m/s)	*MAE* (m/s)	*RRMSE* (%)	*RMAE* (%)	*L*
				NSW				
ANN	0.64	0.54	0.41	2.13	1.67	−585	336	0.25
MLR	0.60	0.47	0.36	2.22	1.74	−554	442	0.21
RF	0.58	0.48	0.33	2.27	1.76	−565	370	0.21
M5	0.70	0.61	0.49	1.97	1.52	−541	305	0.31
				SA				
ANN	0.66	0.58	0.44	3.20	2.54	537	341	0.28
MLR	0.63	0.49	0.39	3.28	2.63	1014	353	0.23
RF	0.59	0.48	0.34	3.40	2.71	1053	440	0.21
M5	0.73	0.64	0.53	2.94	2.31	493	327	0.35
				TAS				
ANN	0.64	0.57	0.40	2.59	2.03	158	457	0.25
MLR	0.64	0.57	0.40	2.69	2.11	165	312	0.25
RF	0.59	0.53	0.35	2.81	2.21	172	340	0.22
M5	0.70	0.64	0.49	2.41	1.87	147	430	0.31
				VIC				
ANN	0.61	0.52	0.37	1.84	1.48	221	316	0.24
MLR	0.59	0.48	0.34	1.84	1.51	228	339	0.21
RF	0.55	0.47	0.30	1.91	1.55	236	402	0.19
M5	0.68	0.60	0.46	1.70	1.36	204	293	0.30
				WA				
ANN	0.70	0.67	0.49	2.28	1.79	−440	389	0.30
MLR	0.68	0.62	0.46	2.31	1.83	−303	427	0.29
RF	0.66	0.63	0.43	2.37	1.88	−311	427	0.26
M5	0.76	0.73	0.57	2.10	1.64	−405	316	0.36

correlation coefficient also differed by 0.16 among the models (performance of RF/MLR was 0.60 and M5 tree was 0.76). *d* and *ENS* were also comparatively higher from the M5 tree and lower from the RF model. Error indicators showed that *MAE* and *RMSE* were also the lowest from the M5 tree model and highest in the ANN model. Likewise, in other data sites (NSW, TAS, VIC, and WA), performance of the M5 tree model was the best (highest *L*, *R*, *d*, and *ENS* in all data sites) followed by the ANN model for daily *V* wind forecasting for the testing period.

Similarly, all indicators showed that the RF model's performance was the least (lowest *L* in all the data sites) compared to the other data-driven models for daily *V* wind prediction (Table 6.27).

TABLE 6.27 *V* wind comparison for all the data sites for daily forecasting through artificial intelligence-based models (*red* color is the best indicator).

Models	*R*	*d*	*ENS*	*RMSE* (m/s)	*MAE* (m/s)	*RRMSE* (%)	*RMAE* (%)	*L*
NSW								
ANN	0.58	0.50	0.33	2.83	2.23	216	296	0.22
MLR	0.56	0.47	0.31	2.88	2.29	219	304	0.20
RF	0.54	0.48	0.28	2.94	2.33	223	455	0.18
M5	0.67	0.60	0.44	2.65	2.08	190	200	0.29
SA								
ANN	0.65	0.65	0.41	3.28	2.63	304	425	0.25
MLR	0.60	0.55	0.36	3.26	2.56	298	273	0.23
RF	0.60	0.53	0.35	3.28	2.56	300	275	0.23
M5	0.76	0.71	0.56	2.82	2.16	262	239	0.38
TAS								
ANN	0.46	0.33	0.21	2.76	2.25	4195	270	0.13
MLR	0.45	0.32	0.20	2.77	2.27	4211	252	0.12
RF	0.39	0.30	0.14	2.88	2.34	4382	285	0.09
M5	0.59	0.47	0.34	2.50	2.00	2426	1071	0.22
VIC								
ANN	0.58	0.53	0.34	2.05	1.66	1794	258	0.22
MLR	0.58	0.51	0.33	2.06	1.66	1796	258	0.22
RF	0.57	0.49	0.32	2.07	1.68	1809	231	0.21
M5	0.73	0.67	0.52	1.76	1.39	1352	374	0.35
WA								
ANN	0.68	0.64	0.47	2.16	1.66	108	688	0.31
MLR	0.67	0.62	0.45	2.05	1.61	105	611	0.28
RF	0.64	0.60	0.40	2.13	1.67	109	353	0.25
M5	0.74	0.70	0.55	1.99	1.52	100	773	0.37

6.4.3.3 Comparison of different models for N wind forecasting for daily time scale

Performance of *N* wind also varied in different locations in the models for daily forecasting. For data-driven models, *L* index varied from 0.02–0.28 in different data locations and different models. Similarly, *R*, *d*, and *ENS* also fluctuated from 0.33–0.65, 0.19–0.55, and 0.06–0.42, respectively, in different data locations and different artificial intelligence-based models.

Compared to data-driven models, the model performance of statistical models such as the ARIMA was very poor. For example, in the data site in NSW, *L* index of the lowest data-driven model (RF model in this site) was 0.11 but *L* from the ARIMA model was 0. Similarly, other indicators like *R*, *d*, and *ENS* were also below 0.03 in the same site in NSW from the ARIMA model (Table 6.28).

However, the performance of error indicators by the ARIMA model like *RMSE*, *MAE*, *RRMSE*, and *RMAE* were slightly better ($RMSE = 1.76$ m/s, $MAE = 1.42$ m/s, $RRMSE = 37\%$, $RMAE = 37\%$) but still higher compared to artificial intelligence-based models for daily *N* wind forecasting. Likewise, in the study site in TAS, performance indicators of the ARIMA models were $R = 0.01, d = 0.05, ENS = 0,$ and $L = 0$, but in the same data site performances of the M5 tree model for the same indicators were 0.60, 0.46, 0.35, and 0.21, respectively.

In the data site in TAS, error indicators ($RMSE = 1.97$ m/s, $MAE = 1.6$ m/s, $RRMSE = 40\%$, $RMAE = 41\%$) were slightly better but still less compared to other data-driven models.

On the other hand, *N* wind forecasting model indicators were quite different for the training period compared to the testing period. Table 6.29 shows the model performance indicators of both types of models (data-driven models and the ARIMA model) in the study site in WA.

Performance of the RF model was the best where $R = 0.97$, $d = 0.93$, $ENS = 0.90$, and $L = 0.70$. Similarly, error indicators were also very good generated by RF model ($RMSE = 0.5$ m/s, $MAE = 0.39$ m/s, $RRMSE = 10\%$, $RMAE = 9\%$). Performance of the ANN model was the least for the training period ($R = 0.58$, $d = 0.49$, $ENS = 0.33$, $L = 0.22$, $RMSE = 1.29$ m/s, $MAE = 1.01$ m/s, $RRMSE = 26\%$, $RMAE = 23\%$).

For the training period, the performance of the ARIMA model was also better than the ANN model but worse than the other artificial intelligence-based models for daily *N* wind forecasting in the training period (Table 6.29).

Finally, in these all data sites, compared to data-driven models (ANN, MLR, RF and M5 tree models), the model performance of the M5 tree was the best for the testing period because it had the highest values of *L*, *R*, *d*, and *ENS* and the lowest error indicators (*RMSE*, *MAE*, and *RRMSE*), followed by the ANN, MLR, and RF models, respectively. However, the statistical model ARIMA had the least accurate performance for *N* wind prediction compared to data-driven models for the testing period.

In Fig. 6.7 model preciseness was assigned using a scatterplot of observed *N* wind and forecasted *N* wind for the testing period of the data site in NSW for daily wind speed prediction from the data-driven model with the least squares fit.

In this case, the y-intercept of the model lay at 3.7–3.9 whereas the correlation coefficient and the gradient ranges were from 0.42–0.46 and 0.20–0.22, respectively. However, on the same data site (NSW) the statistical model (Fig. 6.8) had a negative correlation and gradient and most of the forecasted data generated from the

TABLE 6.28 *N* wind comparison for all the data sites for daily through artificial intelligence-based model for testing period (*red* color is the best and *green* is the worst).

Models	*R*	*d*	*ENS*	*RMSE* (m/s)	*MAE* (m/s)	*RRMSE* (%)	*RMAE* (%)	*L*
NSW								
ANN	0.47	0.31	0.22	1.55	1.22	32	31	0.14
MLR	0.45	0.27	0.20	1.57	1.25	33	32	0.12
RF	0.42	0.29	0.16	1.61	1.26	34	32	0.11
M5	0.60	0.44	0.35	1.42	1.12	29	28	0.20
ARIMA	− 0.03	0.03	0.00	1.76	1.42	37	37	0.00
SA								
ANN	0.50	0.28	0.25	2.02	1.59	33	32	0.14
MLR	0.49	0.29	0.24	2.04	1.60	33	32	0.13
RF	0.45	0.28	0.19	2.10	1.66	35	33	0.11
M5	0.62	0.45	0.37	1.90	1.49	31	29	0.22
TAS								
ANN	0.48	0.35	0.23	1.73	1.41	35	35	0.12
MLR	0.48	0.34	0.23	1.73	1.41	35	35	0.12
RF	0.43	0.32	0.17	1.79	1.45	36	36	0.10
M5	0.60	0.46	0.35	1.54	1.23	32	32	0.21
ARIMA	0.01	0.05	0.00	1.97	1.60	40	41	0.00
VIC								
ANN	0.45	0.23	0.20	1.05	0.82	29	25	0.10
MLR	0.45	0.21	0.20	1.05	0.82	29	25	0.10
RF	0.33	0.19	0.06	1.14	0.89	31	27	0.02
M5	0.54	0.35	0.28	1.00	0.77	27	23	0.16
WA								
ANN	0.59	0.48	0.35	1.23	0.96	25	23	0.23
MLR	0.58	0.45	0.33	1.25	0.98	25	23	0.22
RF	0.52	0.43	0.26	1.31	1.03	27	24	0.18
M5	0.65	0.55	0.42	1.20	0.92	24	21	0.28

ARIMA model lay at around 4.83 m/s, which was not a good indicator for wind forecasting.

Furthermore, the model strength had also been presented by using a box plot (Fig. 6.9) to illustrate the distribution of the data with respect to their quartiles for observed and

TABLE 6.29 Comparison of all study models in the data site in WA for daily *N* wind prediction in training period (*green* color was the worst and *red* color was the best).

Models	*R*	*d*	*ENS*	*RMSE* (m/s)	*MAE* (m/s)	*RRMSE* (%)	*RMAE* (%)	*L*
ANN	0.58	0.49	0.33	1.29	1.01	26	23	0.22
RF	0.97	0.93	0.90	0.50	0.39	10	9	0.70
M5	0.65	0.55	0.42	1.20	0.92	24	21	0.28
ARIMA	0.57	0.71	0.32	1.28	0.99	25	22	0.22

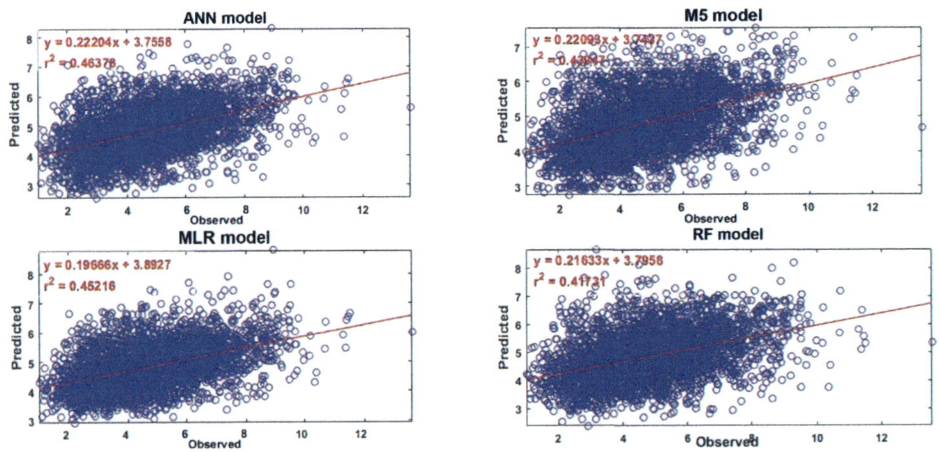

FIGURE 6.7 Scatterplot of observed *N* wind and forecasted *N* wind of station NSW of all data-driven models (ANN, M5 tree, MLR, and RF models) for daily wind forecasting (Note: *red* line represented the least-square fit line to the representative scatterplot).

forecasted data from the data-driven models. This box plot had clearly indicated that the distribution of the forecasted dataset was squeezed compared to their observed dataset, that is, the distance of first quartile and the third quartile was shorter from the second quartile in the forecasted dataset compared to their observed dataset for the testing period of *N* wind of the SA data site.

Finally, data-driven models were compared with the statistical model (ARIMA) in the site in NSW for daily wind speed prediction through a time series plot for the last month of the study period (December 2017).

Fig. 6.10 clearly shows that the prediction of wind speed curve from the artificial intelligence-based models followed the observed wind speed curve but the prediction of wind speed curve from the statistical model (ARIMA) was unable to follow the observed wind speed curve.

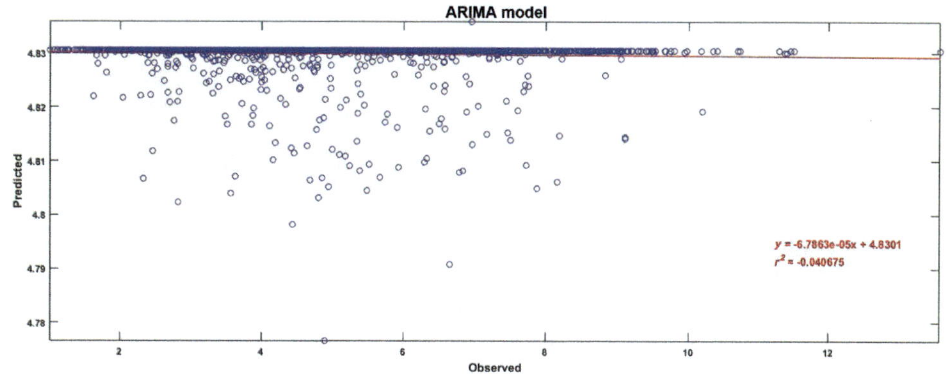

FIGURE 6.8 Scatterplot of observed *N* wind and forecasted *N* wind of station NSW of statistical model (ARIMA model) for daily wind forecasting (Note: *red* line represented the least-square fit line to the representative scatterplot).

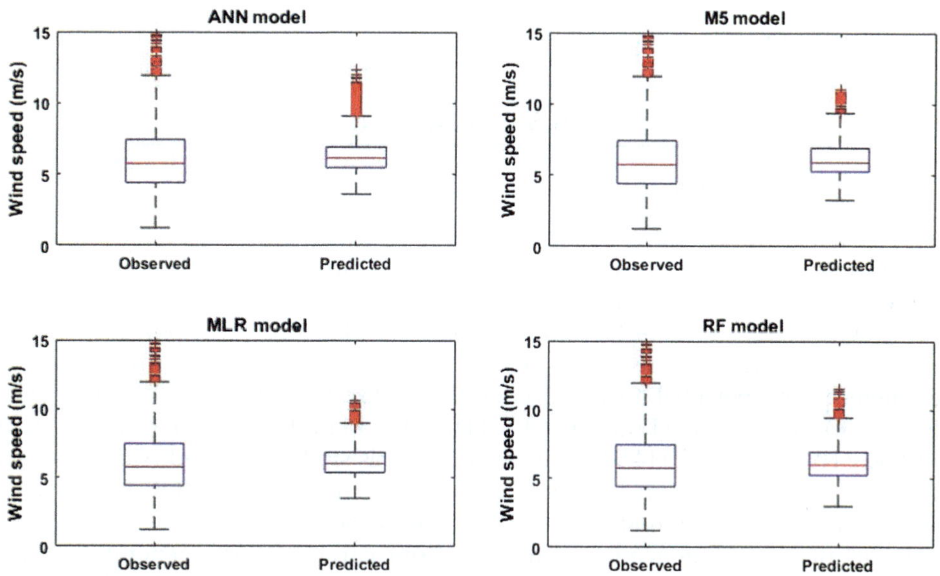

FIGURE 6.9 Box plot of observed *N* wind and forecasted *N* wind of station SA of data-driven model for daily wind speed forecasting for the testing period (ANN, M5 tree, MLR and RF respectively).

The scatterplot, box plot, time series plot, and Table 6.28, which compare each and every data site with different forecasting accuracy measurements (e.g., *R*, *d*, *ENS*, *RMSE*, *MAE*, *RRMSE*, *RMAE*, and *L*) show that the M5 tree model was the most accurate followed by the ANN model for daily wind speed prediction in the testing period.

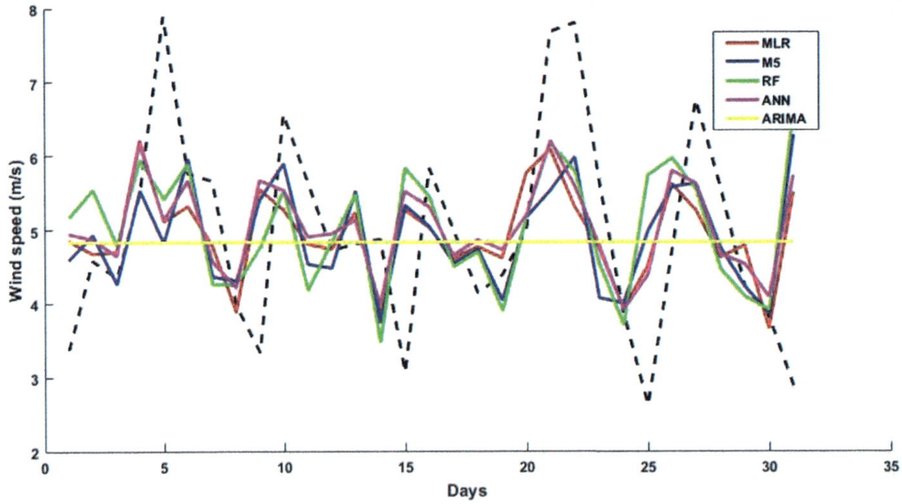

FIGURE 6.10 Time series plot of observed and forecasted *N* wind in the last month (December 2017) of study period in the first data site (NSW).

6.5 Summary

6.5.1 Concluding remark

The prediction of wind in different time scales is a challenging task due to the unpredictability of its movement but it is of great importance to wind farms for wind energy generation. In addition, wind prediction is also crucial for many other sectors like meteorological forecasting, sustainable agriculture, and infrastructure development. This study was primarily focused on applied artificial intelligence-based wind prediction models (the ANN, M5 tree, RF, and MLR) in five different wind potential areas [Silverton (NSW) (31.97°S, 141.46°E), Kemmiss Hill (SA) (33.5°S, 138.1°E), Woolnorth (TAS) (40.69°S, 144.72°E), Macarthur (VIC) (38.05°S, 142.19°E), and Walkway (WA) (28.92°S, 114.93°E)] in Australia. This study compared ARIMA—a statistical model with four data-driven models (the ANN, M5 tree, RF and MLR) in different time horizons, that is, short-term (6-hourly) and daily in the same data sites.

For the artificial intelligence-based model, input variables were generated according to significant lag correlation of wind. Wind data (zonal and meridional winds) was taken from reanalysis datasets of ECMWF from January 1, 1988 to December 31, 2017. For the ANN model development, different data partitioning was used to generate the best data partition for training, validation, and testing (40%, 30%, and 30% for training, validation, and testing partitions, respectively) and used for daily forecasting as well. Similarly, the best function selection was calculated in the short-term time horizon by using a trial and error method (logsig for input function, purelin for hidden layer function, and trainlm for the output function was the best combination) and used for daily time scales.

The performance of models were assessed by using *R*, *d*, *ENS*, *RMSE*, *MAE*, *RRMSE*, *RMAE*, and *L* indices (discussed in Section 6.2.8) with a run of 30 hidden neurons step by step and the best performer was chosen. Correspondingly, for the RF model, the best function selection was also calculated in different run combinations of leaf, fboot, and ntrees by a trial and error method in a short-term time horizon and the best function combination was used ($leaf = 1 fboot = 1, \text{and} ntrees = 400$ was the best combination) for the daily time scale.

For the ARIMA model, autoregressive degree *p*, differencing degree *d*, and moving-average degree *q* were calculated according to minimum AIC and BIC values for every time series data, that is, short-term and daily forecasting.

6.5.2 Summary of the findings

This study is unique in the sense that the ANN, RF, MLR, M5 tree, and the ARIMA models were tested for the first time in these data locations (Table 6.1) in Australia for wind speed prediction.

6.5.2.1 Summary of findings of short-term (6-hourly) wind speed prediction

1. Forecasting of *U* wind for short-term time horizons from artificial intelligence-based models in the data site in WA was the best, where the value of *R* ranged from 0.86–0.89 and *L* ranged from 0.54–0.60. Similarly, forecasting of *U* wind was less accurate in the data site in VIC by the data-driven model, where the value of *R* ranged from 0.78–0.80 and the value of *L* ranged from 0.42–0.44. *U* wind forecasting from the ARIMA model was less accurate compared to data-driven models. For example, in the data site in VIC, the value of *R* was 0.18 and *L* was 0.01, generated from the ARIMA model for short-term wind speed forecasting.

   Likewise, for short-term *U* wind forecasting, the ANN model was the best in four data sites and the MLR was the best in the data site in TAS. In the data site in VIC, the best performer, the ANN model had $RMSE = 1.58$ m/s and the worst performer of the data-driven model was the RF model ($RMSE = 1.64$ m/s) for short-term *U* wind forecasting. However, *RMSE* from the ARIMA model at the same site was 2.57 m/s, which was relatively higher compared to the artificial intelligence-based models.
2. For short-term *V* wind forecasting, the data site in NSW was the best because in this site, the value of *R* and *L* ranged from $0.83 - 0.84$ and $0.52 - 0.53$, respectively. At the same time, the data site in VIC had the least accurate forecasting for short-term *V* wind prediction where value of *L* ranged from $0.44 - 0.46$ and the value of *R* lay around 0.80. Likewise, the ANN model was the best for *V* wind forecasting in the short-term time horizon because it had the best performance in all five data locations in the testing period followed by the M5 tree, RF, and the MLR models. For example, in the data site in VIC, performance indicators of the ANN model like *RMSE* and *ENS* were 1.58 m/s and 0.64, respectively, whereas the poor data-driven indicator, the RF had $RMSE = 1.76\ \text{m/sand} ENS = 0.63$. However, the performance of the statistical model ARIMA had negative *ENS* and more than 3 m/s of *RMSE* value in the same data site (i.e., VIC) for short-term *V* wind forecasting.

3. For the short-term N wind forecasting, the data site in SA had the best prediction where indicators L and R varied between 0.40 and 0.42 and 0.79 and 0.80, respectively. However, the data site in VIC performed the least accurate forecasting for short-term N wind speed prediction where prediction indicators L and R varied $0.20 - 0.23$ and $0.60 - 0.63$, respectively. Likewise, the ANN model was the best performer in all five data sites followed by the RF model, M5 tree and the MLR model. The worst data-driven model performer varied with different artificial intelligence-based models in different locations. However, forecasting indicators showed that the statistical model, the ARIMA, had less accurate forecasting results. For example, short-term N wind forecasting by the ARIMA model in the data site in VIC showed that the value of $R = 0.02$ whereas all other data-driven models gave a performance of $R > 0.60$. Similarly the L index from the ARIMA model in the same data site was negative.

6.5.2.2 Summary of findings of daily wind speed prediction

1. For daily U wind forecasting from artificial intelligence-based models, the data site in WA had the best prediction where the value of L ranged from 0.26–0.36 and R ranged from 0.66–0.76. The data site in VIC had less accurate forecasting indicators where the value of L and R varied from 0.19–0.30 and 0.55–0.68, respectively. Likewise, the M5 tree model had the best performance indicators for all the data sites followed by ANN and MLR, whereas the RF had poor performance indicators in all the data sites in the testing period.
2. For V wind forecasting in the daily time scale, the data site in SA performed the best forecasting indicators where: $L = 0.23 - 0.38, R = 0.60 - 0.76, \text{and} RMSE = 2.82 - 3.28$ m/s. However, in the data site in TAS, performance indicators indicated that daily V wind forecasting was less where $L = 0.09 - 0.22, R = 0.39 - 0.59, \text{and} MAE = 2\text{–}2.34$ m/s. Similarly, V wind forecasting from the M5 tree model was the best in all data locations for daily time horizons followed by the ANN, MLR, and the RF models.
3. For daily N wind forecasting, the data site in WA had the best performance indicators where $L = 0.18\text{–}0.28$, $RMAE = 21\%\text{–}24\%$, $RRMSE = 24\%\text{–}27\%$. However, the data site in VIC had less accurate daily N wind forecasting indicators in the testing period where L, $RMAE$, and $RRMSE$ were $0.02\text{–}0.16$, $23\%\text{–}27\%$, and $27\%\text{–}31\%$, respectively. Comparison of data-driven models in the daily time scale for N wind forecasting, showed that the M5 tree model performed the best in all the data sites followed by the ANN, MLR, and the RF models. However, performance of the statistical model (the ARIMA model) was much less, for example, the forecasting indicator L was 0 in data site in TAS and NSW, at the same time, the correlation coefficient was 0.01 in the data site in TAS and negative in NSW.

6.5.3 Limitations and recommendations for future works

The following recommendations have been made for further research opportunities:

1. This study used meridional and zonal wind data and determined the significant lag correlation to generate the input variables for the prediction of U, V, and N wind. But

for the determination of wind speed prediction, other environmental variables, e.g., rainfall, humidity, and temperature, might have important roles to generate wind speed, which have not been covered by this chapter.

2. This research used 6-hourly time scale U and V wind but the changes of time scale may influence the results.
3. The data partitions used in this chapter were $90\% - 5\% - 5\%, 80\% - 10\% - 10\%, 70\% - 15\% - 15\%, 60\% - 20\% - 20\%, \text{and} 40\% - 30\% - 30\%$ to determine the best data partition but other division ratios might present better results for the models.
4. Similarly, for the ANN model, only 10 combinations of input functions, hidden layer functions, and output functions were used but other combinations may generate different results. In this study, hidden neurons of the ANN model were chosen from 1 to 30 but more than 30 neurons may change the model performance. For the RF model, leaf varied from 1 to 20, fboot from 0 to 1, and ntrees from 50 to 400 to determine the best combination which also could not cover all the possible combinations. For the ARIMA model, to determine the best combination of (p, d, q) (where p is the order (number of time lags), d is the degree of difference and q is the order of moving-average model) the least values of AIC and BIC, and the highest likelihood has been chosen but all of the possible combinations of p, d, q were not used to determine the least AIC and BIC and highest likelihood.
5. The average tip height of wind turbines in current commercial wind farms in Australia lies between 130 and 160 m from the ground but in this study, wind data had been taken at 10 m height from the ground. Thus these results cannot be used directly to calculate wind energy capacity of these study points. Similarly, wind speed has been found to change at different heights from the ground at the same study point. Hence, further studies may consider gathering data at different heights.
6. Recently introduced decomposition tools such as empirical wavelet transformation (EWT) for data preanalysis can be useful for the ANN, MLR, RF, and the M5 tree models to develop hybrid artificial intelligence-based models for wind speed forecasting. Other tools, e.g., multiverse optimizer (MVO) which provides the weight of the independent variables in the processing unit of models (the ANN, MLR, RF, M5 tree), may enhance the accuracy of the models. Therefore consideration of such hybrid tools, e.g., EWT + ANN models, and dual hybrid models, e.g., EWT + ANN + MVO, may improve the forecasting.

References

ABCNews. 2004. Council rejects Kemmiss Hill wind farm plan. Retrieved from <https://www.abc.net.au/news/2004-10-15/council-rejects-kemmiss-hill-wind-farm-plan/569212>.

AE. 2017. Australian energy update. Retrieved from <http://www.environment.gov.au/energy/publications/australian-energy-update-2017>.

AGL. 2018a. Macarthur wind farm. Retrieved from <https://www.agl.com.au/about-agl/how-we-source-energy/macarthur-wind-farm>.

AGL. 2018b. Silverton wind farm. Retrieved from <https://www.agl.com.au/about-agl/how-we-source-energy/silverton-wind-farm>.

Ahmad, A., Hassan, M., Abdullah, M., Rahman, H., Hussin, F., Abdullah, H., et al., 2014. A review on applications of ANN and SVM for building electrical energy consumption forecasting. Renew. Sust. Energ. Rev. 33 (1), 102–109.

Alipour, A., Yarahmadi, J., Mahdavi, M., 2014. Comparative study of M5 model tree and artificial neural network in estimating reference evapotranspiration using MODIS products. J. Climatol. 2014, 11. Available from: https://doi.org/10.1155/2014/839205.

Arce-Medina, E., Paz-Paredes, J.I., 2009. Artificial neural network modeling techniques applied to the hydrodesulfurization process. Mathe. Comput. Model. 49 (1-2), 207–214.

Ariyo, A.A., Adewumi, A.O., Ayo, C.K. 2014. Stock price prediction using the ARIMA model. Paper presented at the Computer Modelling and Simulation (UKSim), 2014 UKSim-AMSS 16th International Conference on.

AussieRenewables. 2018. Woolnorth Wind Farm. Retrieved from <https://www.aussierenewables.com.au/directory/woolnorth-wind-farm-103.html>.

Australian Energy Resources Assessment. n.d. Retrived Novemember 6, 2019, from <https://aera.ga.gov.au/#!/wind>.

Bandyopadhyay, G., Chattopadhyay, S., 2007. Single hidden layer artificial neural network models versus multiple linear regression model in forecasting the time series of total ozone. Int. J. Environ. Sci. Technol 4 (1), 141–149.

Baroni, M., Clementi, S., Cruciani, G., Costantino, G., Riganelli, D., Oberrauch, E., 1992. Predictive ability of regression models. Part II: selection of the best predictive PLS model. J. Chemom. 6 (6), 347–356.

Bennett, N.D., Croke, B.F., Guariso, G., Guillaume, J.H., Hamilton, S.H., Jakeman, A.J., et al., 2013. Characterising performance of environmental models. Environ. Model. Software 40, 1–20. Available from: https://doi.org/10.1016/j.envsoft.2012.09.011.

Bhattacharya, B., Solomatine, D.P., 2005. Neural networks and M5 model trees in modelling water level–discharge relationship. Neurocomputing 63, 381–396. Available from: https://doi.org/10.1016/j.neucom.2004.04.016.

Bianco, V., Manca, O., Nardini, S., 2009. Electricity consumption forecasting in Italy using linear regression models. Energy 34 (9), 1413–1421.

Biau, G., 2012. Analysis of a random forests model. J. Mach. Learn. Res. 13 (April), 1063–1095. Retrieved from http://www.jmlr.org/papers/volume13/biau12a/biau12a.pdf.

Bottenberg, R.A., & Ward, J.H. (1963). Applied multiple linear regression. Technical Documentary Report PRL-TDR-63–6. 6570th Personnel Research Laboratory, Aerospace Medical Division, Air Force Systems Command, Lackland Air Force Base.

Boulesteix, A.L., Janitza, S., Kruppa, J., König, I.R., 2012. Overview of random forest methodology and practical guidance with emphasis on computational biology and bioinformatics. Wires Data Min. Knowl. 2 (6), 493–507. Available from: https://doi.org/10.1002/widm.1072.

Box, G., Jenkins, G., 1970. Time Series Analysis: Forecasting and Control. Holden Day, San Francisco, CA, USA.

Breiman, L., 2001. Random forests. Mach. Learn. 45 (1), 5–32.

Cadenas, E., Rivera, W., 2009. Short term wind speed forecasting in La Venta, Oaxaca, México, using artificial neural networks. Renew. Energ 34 (1), 274–278. Available from: https://doi.org/10.1016/j.renene.2008.03.014.

Cadenas, E., Rivera, W., Campos-Amezcua, R., Heard, C., 2016. Wind speed prediction using a univariate ARIMA model and a multivariate NARX model. Energies 9 (2), 109. Available from: https://doi.org/10.3390/en9020109.

Chai, T., Draxler, R.R., 2014. Root mean square error (RMSE) or mean absolute error (MAE)?–Arguments against avoiding RMSE in the literature. Geosci. Model Dev. 7 (3), 1247–1250.

Cheng, Y.-Y., Chan, P.P., Qiu, Z.-W. 2012. Random forest based ensemble system for short term load forecasting. Paper presented at the Machine Learning and Cybernetics (ICMLC), 2012 International Conference on.

Conejo, A.J., Plazas, M.A., Espinola, R., Molina, A.B., 2005. Day-ahead electricity price forecasting using the wavelet transform and ARIMA models. IEEE Trans. Power Syst. 20 (2), 1035–1042.

Deo, R.C., Şahin, M., 2015. Application of the artificial neural network model for prediction of monthly standardized precipitation and evapotranspiration index using hydrometeorological parameters and climate indices in eastern Australia. Atmos. Res. 161, 65–81. Available from: https://doi.org/10.1016/j.atmosres.2015.03.018.

Deo, R.C., Şahin, M., 2016. An extreme learning machine model for the simulation of monthly mean streamflow water level in eastern Queensland. Environ. Monitoring Assess. 188 (2), 90.

Deo, R.C., Şahin, M., 2017. Forecasting long-term global solar radiation with an ANN algorithm coupled with satellite-derived (MODIS) land surface temperature (LST) for regional locations in Queensland. Renew. Sust. Energ. Rev. 72, 828–848. Available from: https://doi.org/10.1016/j.rser.2017.01.114.

Despotovic, M., Nedic, V., Despotovic, D., Cvetanovic, S., 2016. Evaluation of empirical models for predicting monthly mean horizontal diffuse solar radiation. Renew. Sust. Energ. Rev. 56, 246–260. Available from: https://doi.org/10.1016/j.rser.2015.11.058.

Etemad-Shahidi, A., Mahjoobi, J., 2009. Comparison between M5′ model tree and neural networks for prediction of significant wave height in Lake Superior. Ocean Eng. 36 (15-16), 1175–1181.

GENI. 2016. Wind energy potential in OCEANIA. Retrieved March 6, 2017, from <https://www.geni.org/globalenergy/library/renewable-energy-resources/world/oceania/wind-oceania/index.shtml>.

Geoscience Australia. n.d. Retrieved November 6, 2019, from <http://www.ga.gov.au/scientific-topics/energy/resources/other-renewable-energy-resources/wind-energy>.

GWEC. 2017. 2017 global wind energy council report. Retrieved from <http://gwec.net/global-figures/wind-energy-global-status/>.

Herrera, M., Torgo, L., Izquierdo, J., Pérez-García, R., 2010. Predictive models for forecasting hourly urban water demand. J. Hydrol. 387 (1-2), 141–150.

Ho, T.K. 1995. Random decision forests. Paper presented at the Document analysis and recognition, 1995, proceedings of the third international conference on.

InfigenEnergy. 2018. Annual report 2018. Retrieved from Sydney Australia <https://s3-ap-southeast-2.amazonaws.com/infigen/wp-content/uploads/2018/08/27091658/IFN-Annual-Report-2018-1.pdf>.

Jiang, D., Zhang, Y., Hu, X., Zeng, Y., Tan, J., Shao, D., 2004. Progress in developing an ANN model for air pollution index forecast. Atmos. Environ. 38 (40), 7055–7064.

Khashei, M., Bijari, M., 2011. A novel hybridization of artificial neural networks and ARIMA models for time series forecasting. Appl. Soft Comput. 11 (2), 2664–2675.

Kumar, M., Thenmozhi, M. 2006. Forecasting stock index movement: a comparison of support vector machines and random forest.

Kumar, S.V., Vanajakshi, L., 2015. Short-term traffic flow prediction using seasonal ARIMA model with limited input data. Eur. Trans. Res. Rev. 7 (3), 21. Available from: https://doi.org/10.1007/s12544-015-0170-8.

Legates, D.R., McCabe Jr, G.J., 1999. Evaluating the use of "goodness-of-fit" measures in hydrologic and hydroclimatic model validation. Water Resour. Res. 35 (1), 233–241.

Lei, M., Shiyan, L., Chuanwen, J., Hongling, L., Yan, Z., 2009. A review on the forecasting of wind speed and generated power. Renew. Sust. Energ. Rev. 13 (4), 915–920. Available from: https://doi.org/10.1016/j.rser.2008.02.002.

Lek, S., Guegan, J.F., 1999. Artificial neural networks as a tool in ecological modeling, an introduction. Ecol. Model. 120 (2–3), 65–73.

Marchal, V., Dellink, R., van Vuuren, D., Clapp, C., Château, J., Lanzi, E., et al. (2011). OECD environmental outlook to 2050: climate change chapter. Pre-release version.

McCulloch, W.S., Pitts, W., 1943. A logical calculus of the ideas immanent in nervous activity. The Bull. Mathematical Biophy. 5 (4), 115–133.

McCulloch, W.S., Pitts, W., 1943. A logical calculus of the ideas immanent in nervous activity. Bull. Mathe. Biophys. 5 (4), 115–133.

Mukerji, A., Chatterjee, C., Raghuwanshi, N.S., 2009. Flood forecasting using ANN, neuro-fuzzy, and neuro-GA models. J. Hydrol. Eng 14 (6), 647–652.

Nash, J.E., Sutcliffe, J.V., 1970. River flow forecasting through conceptual models part I—a discussion of principles. J. Hydrol. 10 (3), 282–290.

Pai, P.F., Lin, C.S., 2005. A hybrid ARIMA and support vector machines model in stock price forecasting. Omega-Int. J. Manage. Sci. 33 (6), 497–505. Available from: https://doi.org/10.1016/j.omega.2004.07.024.

Palani, S., Liong, S.-Y., Tkalich, P., 2008. An ANN application for water quality forecasting. Mar. Pollut. Bull. 56 (9), 1586–1597.

Pal, M., Deswal, S., 2009. M5 model tree based modelling of reference evapotranspiration. Hydrol. Process 23 (10), 1437–1443.

Pham, S.A.C.M.P., Ahn, I., Nowakowski, L.D.T.N.D. 2016. Australian energy update 2016. Director, Energy Statistics, Department of the Environment and Energy, GPO Box 787, CANBERRA, ACT, 2601.

Prasad, R., Deo, R.C., Li, Y., Maraseni, T., 2017. Input selection and performance optimization of ANN-based streamflow forecasts in the drought-prone Murray Darling Basin region using IIS and MODWT algorithm. Atmos. Res. 197, 42–63. Available from: https://doi.org/10.1016/j.atmosres.2017.06.014.

Prasad, R., Deo, R.C., Li, Y., Maraseni, T., 2018. Ensemble committee-based data intelligent approach for generating soil moisture forecasts with multivariate hydro-meteorological predictors. Soil Till. Res. 181, 63–81. Available from: https://doi.org/10.1016/j.still.2018.03.021.

Quinlan, J.R., 1992, November. Learning with continuous classes. In: 5th Australian Joint Conference on Artificial Intelligence, vol. 92, pp. 343–348.

Rahimikhoob, A., Asadi, M., Mashal, M., 2013. A comparison between conventional and M5 model tree methods for converting pan evaporation to reference evapotranspiration for semi-arid region. Water Resour. Manag. 27 (14), 4815–4826.

Shi, K., Qiao, Y., Zhao, W., Wang, Q., Liu, M., Lu, Z., 2018. An improved random forest model of short-term wind-power forecasting to enhance accuracy, efficiency, and robustness. Wind Energy 21 (12), 1383–1394. Available from: https://doi.org/10.1002/we.2261.

Vincenzi, S., Zucchetta, M., Franzoi, P., Pellizzato, M., Pranovi, F., De Leo, G.A., et al., 2011. Application of a Random Forest algorithm to predict spatial distribution of the potential yield of Ruditapes philippinarum in the Venice lagoon, Italy. Ecol. Modell. 222 (8), 1471–1478. Available from: https://doi.org/10.1016/j.ecolmodel.2011.02.007.

Willmott, C.J., Robeson, S.M., Matsuura, K., 2012. A refined index of model performance. Int. J. Climatol. 32 (13), 2088–2094.

Xydas, E., Marmaras, C., Cipcigan, L.M., Jenkins, N., Carroll, S., Barker, M., 2016. A data-driven approach for characterising the charging demand of electric vehicles: a UK case study. Appl. Energy 162, 763–771. Available from: https://doi.org/10.1016/j.apenergy.2015.10.151.

Yang, Z., Wang, J., 2018. A hybrid forecasting approach applied in wind speed forecasting based on a data processing strategy and an optimized artificial intelligence algorithm. Energy 160, 87–100.

Zhang, G.P., 2003. Time series forecasting using a hybrid ARIMA and neural network model. Neurocomputing 50, 159–175. Available from: https://doi.org/10.1016/S0925-2312(01)00702-0.

CHAPTER

7

Hybrid multilayer perceptron-firefly optimizer algorithm for modelling photosynthetic active solar radiation for biofuel energy exploration

Harshna Gounder[1], *Zaher Mundher Yaseen*[2] and *Ravinesh Deo*[1]

[1]School of Sciences, University of Southern Queensland, Springfield, QLD, Australia
[2]Sustainable Developments in Civil Engineering Research Group, Faculty of Civil Engineering, Ton Duc Thang University, Ho Chi Minh City, Vietnam

Acronyms

ANFIS	Adaptive Neuro Fuzzy Inference System
ANFIS-FFA	Adaptive Neuro Fuzzy Inference System–Firefly Algorithm
ANN	Artificial Neural Network
FFA	Firefly Algorithm
GES DISC	Goddard Earth Sciences Data and Information Services Center
GIOVANNI	Goddard Earth Sciences Data and Information Services Center's Interactive Online Visualization and Analysis Infrastructure—NASA's open source data repository
IRBF-FFA	Includes all of Imperialist Competition Algorithm (ICA), Radial Basis Function (RBF) and Firefly Algorithm (FFA) hybrid
MATLAB	Matrix Laboratory—a multiparadigm numerical computing environment and proprietary programming language developed by MathWorks
MFA	Modified Firefly Algorithm
MLP	Multilayer Perceptron
MLP-FFA	Multilayer Perceptron-Firefly Algorithm
MLP-LM	Multilayer Perceptron–Levenberg–Marquardt (LM) backpropagation learning algorithm
MLR	Multiple Linear Regression
NASA	The National Aeronautics and Space Administration
PACF	Partial Autocorrelation Function

Predictive Modelling for Energy Management and Power Systems Engineering
DOI: https://doi.org/10.1016/B978-0-12-817772-3.00007-0

PAR	Photosynthetic Active Radiation
RBF	Radial Basis Function
RF	Random Forest
SPEI	Standardized Precipitation Evapotranspiration Index
SVM	Support Vector Machines
SVR	Support Vector Regression
SVR-FA	Support Vector Regression-Firefly Algorithm
SVR-GA	Support Vector Regression–Genetic Algorithm
SVR-HBMO	Support Vector Regression–Honey Bee Mating Optimization
SVR-MFA	Support Vector Regression–Modified Firefly Algorithm
SVR-PSO	Support Vector Regression–Particle Swarm Optimization
TMY	Typical Meteorological Year Databank for Australia

7.1 Introduction

Prediction of solar energy is an essential research area for environmental scientists, engineers, energy experts, policy-makers, and climate advocates as it enables the use of carbon-free renewable energy sources with less negative environmental effects, providing a promising outlook for combating climate change (Bioenergy Australia, 2016b; Deo and Şahin, 2017; Solangi et al., 2011). Australia has particularly gained attention for solar resourcing due to ideal solar conditions including high insolation, limited rainfall, and low fraction of cloud cover causing less solar radiation scattering across large spatial areas (Beath, 2012; Deo and Şahin, 2017; Yusaf et al., 2011).

An important component of solar energy is its emitted photosynthetic active radiation (PAR), a key indicator of available usable sunlight for growth of photosynthetic microalgae (Sudhakar et al., 2013). Such algae have been used as a low pollution, renewable fuel source (biofuel) (Geoscience Australia, 2018; Pittman et al., 2011). PAR is the visible sunlight radiation (400–700 nm) as seen in Fig. 7.1, which is a vital component to the photosynthetic growth of plants and certain algae (Sudhakar et al., 2013).

Algal biofuel can produce much more biomass per unit area than other terrestrial plant sources (Fig. 7.2), hence it can provide an alternative that saves land space and grows very rapidly (Shurin et al., 2013). Additionally, algal bioenergy does not interfere with food production resources, as is the case for plants which are needed for human consumption

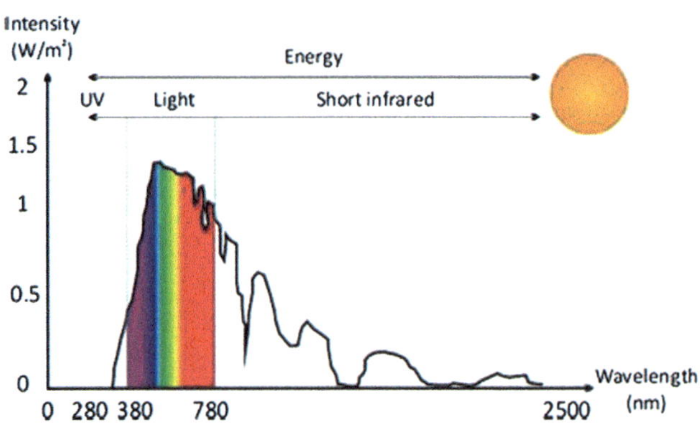

FIGURE 7.1 Solar radiation spectrum (Sudhakar et al., 2013).

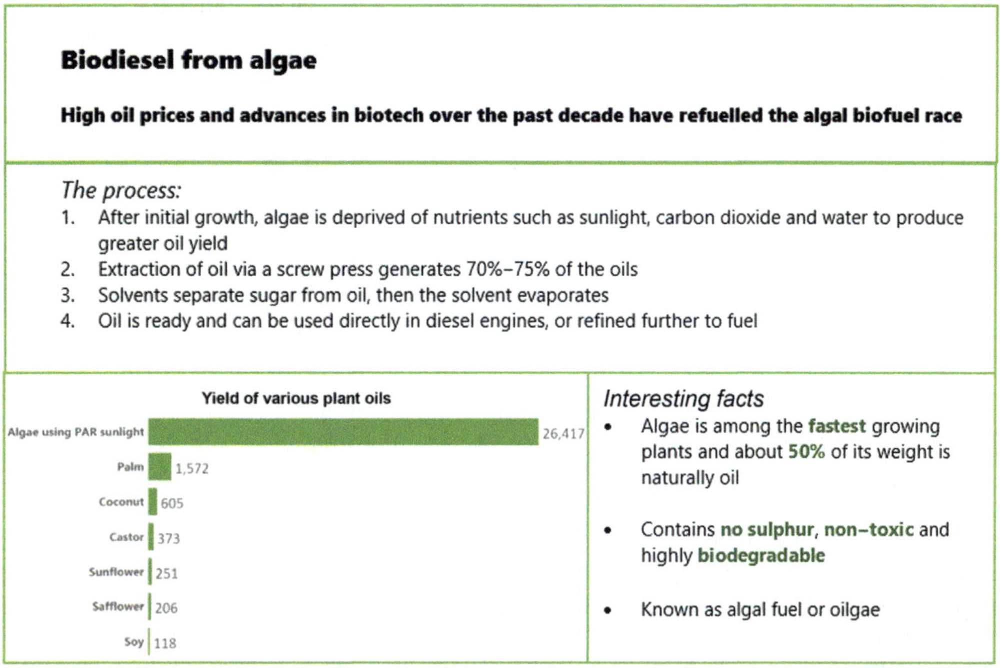

FIGURE 7.2 Infographic of how algae can produce biofuel. *From Harris, J., 2010. Science thursday. WRMA, Australia (2010). Available online at <http://hopeful-ink.blogspot.com.au/2010/09/science-thursday.html>.*

(Bioenergy Australia, 2016a). The process by which algal biofuel is produced is seen in Fig. 7.2.

In Australia and many other parts of the world, coal (for fuel production) is a vital, but depleting resource. This is seen in Fig. 7.3 where coal supply is decreasing, but demand is increasing in almost an opposite pattern in more recent years (Mason et al., 2013). There needs to be a clear shift from current nonrenewable fuel sources to sustain fuel usage.

Australia contains only 10% of worldwide coal resources—for which there is only around 100 years left of production at current rates (John, 2015). Currently the coal exports industry in Australia leads to more than $30 billion annually—with 87% of coal production being exported globally, without a renewable fuel source these values will decline proportionally to coal resources (John, 2015). Hence, Australia needs to maintain its fuel resources not only for its own stocks, but also for Australia's economy (John, 2015).

Even though such emphasis has been put on the use of biofuel (which is an emerging technology), as of 2017 Australia is still only a relatively small producer of biofuels, producing only 0.2% of worldwide bioethanol and 0.1% of biodiesel (Cochran, 2017). In thought of Australian coal levels depleting, Australia is a justified location for the growth of algal biofuel farms. Australia has the world's highest solar irradiation on average, with 35 $MJ/m^2/day$ or 9.7 $kWh/m^2/day$ (Geoscience Australia, 2018).

Australia's annual solar radiation is approximately equal to 58 million petajoules (PJ) which is equivalent to approximately 10,000 times Australia's annual energy consumption

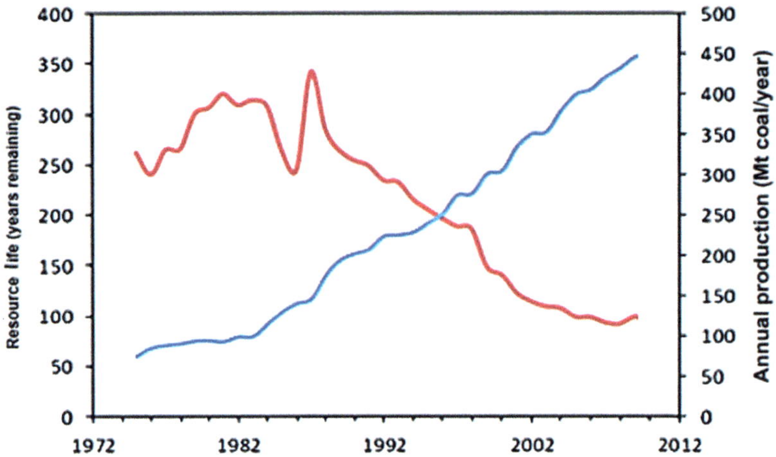

FIGURE 7.3 Australian coal supply and demand (Mason et al., 2013).

(Geoscience Australia, 2018). Hence, modeling PAR in Australia itself is a very reasonable purpose, that has not been yet performed using optimized algorithms. Additionally, in the "sunshine" state of Queensland, there are optimum levels of PAR sunlight reaching the ground—hence an ideal location for algal biofuel farms.

As a result, Australia has given much attention to this bioinitiative. The Australian Government's Geoscience Australia announced that as at November 2015, 812 MW of bioenergy generation capacity was installed, and emphasis was place on enhancing the commercialization and production of algae as a biofuel (Geoscience Australia, 2018). The Liquid Fuel Supply Act 1984 requires fuel sellers to have available for purchase sustainable biobased fuel in their stores, such as E10 fuel (Business Queensland, 2018). E10 fuel contains around 10% of ethanol which is made by distilling biomass (Bioenergy Australia, 2016c). As at July 1, 2018, the Biobased Petrol Mandate increased the amount of petrol sold by a store to be biobased to 4% from the previous 3% (Business Queensland, 2018). The Biobased Diesel Mandate as at the year 2015 requires 0.5% of all specifically diesel-type fuel to be biobased (known as biodiesel).

For utilization of freely available PAR to enhance the growth of photosynthetic algae as a biofuel, quantitative knowledge of PAR is paramount (Sudhakar et al., 2013). The amount of available PAR depends on location, time of year, season, and atmospheric conditions (Sudhakar et al., 2013). Understanding PAR and its availability is essential for modeling algal growth capabilities and yield (Sudhakar et al., 2013).

Methods applied for PAR prediction utilize two types of predictive models: the deterministic (mathematical) and artificial intelligence (AI) models, which are data-driven. Although deterministic models have been abundantly used, as described in Section 7.3, the use of AI models for a remote Queensland location (Toowoomba is such a location to be used in this chapter) are virtually nonexistent for PAR models. In particular, a hybrid-optimized AI model (to be used in this chapter) using advanced prediction techniques have not been used for PAR, but are gaining more attention these days for other environmental variables, as seen in Table 7.1 and discussed in Section 7.2.2.

TABLE 7.1 Summary of papers reviewed in this section.

Papers	Benchmark models	Better performing model
Deo et al. (2018)	MLP	MLP-FFA
Ghorbani et al. (2017b)	MLP-LM	MLP-FFA
Raheli et al. (2017)	MLP	MLP-FFA
Ghorbani et al. (2017a)	SVM, MLP	MLP-FFA
Yaseen et al. (2018)	ANFIS	ANFIS-FFA
Kavousi-Fard et al. (2014)	ARMA, ANN, SVR-GA, SVR-HBMO, SVR-PSO and SVR-FA	SVR-MFA (Modified FFA)
Soleymani et al. (2016)	MLP, SVM	RBF-FFA
Goudarzi et al. (2016)	MLP, SVM	IRBF-FFA
Khatibi et al. (2017)	MLP-LM	MLP-FFA
Mashaly and Alazba (2016)	MLR	MLP
Mouatadid et al. (2015)	MLR	ANN (MLP is a feed-forward ANN)
Choubin et al. (2016)	MLR, ANFIS	MLP

The advent of AI models which do not require complex mathematical equations, but still perform very well is gaining much attention, as seen in the Literature Review Section 7.2.2. AI models require little explanation on the mathematics behind the variability of data within the model (Deo and Şahin, 2017). Using AI models, the relationship between the inputs and target variable is related through a machine learning algorithm that uses pattern recognition methods (Santamouris et al., 1999). AI models are simple in nature and use (Deo and Şahin, 2017). They make no assumption of the underlying data distribution, although they do have a competitive performance over mathematical models (Sahin, et al., 2014; Deo et al., 2016; Şahin et al., 2014). In this study a multilayer perceptron (MLP) hybrid with a firefly algorithm (FFA) as an optimizer called MLP-FFA is used as a novel approach to PAR modeling in a regional city in Queensland, Australia called Toowoomba.

Even though mathematical models can be useful, they are limited in terms of being able to extract the features that are not incorporated in the original mathematical formulations to learn the data, even though they may predict better; whereas in AI models, features are provided and the best ones are chosen through training during the model (Deo and Şahin, 2017). The complexity of mathematical models generally also worsens with a larger number of predictors as models are not as capable, by manual changing and testing of features, instead of through training the model (Deo and Şahin, 2017). There may also be difficulties in relation to the validity of assumptions that are made for mathematical models, for example, the model parameters may be assumed to not vary with time, e.g., the same amount of duration of sunshine across several days may be assumed, but is not always the case (Deo and Şahin, 2017). Assumptions may also be needed on normality, linearity, distributions, and homoscedacity (variance constancy). AI models do not make such assumptions, as a data pattern is not required—the model simply learns from the data to make predictions (Deo and Şahin, 2017). Additionally, AI models can be very accurate, easy and quick to develop and run as opposed to mathematical models which take large amounts of time to build an accurate model, as they are created manually. AI models are more able to handle complex, noisy data as they predict through learning the data using several connections of data (Deo and Şahin, 2017). Mathematical models tend not to work as well for messy data (Deo and Şahin, 2017). AI models are universal (can be applied to other locations) as they do not require upper and lower limits in their formulas (which are location-specific limits) unlike mathematical models which are defined by these limits (Metzen, 2018). These limits can be harder to get a precise value for and make the mathematical models not universal. Hence, it can be noted that the use of AI models is following the current times, whereby these data-driven models are more advantageous than traditional mathematical models, especially in this new "data revolution" age. As a result, this study uses an AI model to predict PAR.

Considering all the aspects discussed here, the novelty of this chapter is to design the model and establish its effectiveness based on the MLP-FFA hybrid-optimized AI approach where historical lagged PAR is used as inputs to forecast observed PAR values in the sunshine state of Queensland where there is abundant solar energy resources (Geoscience Australia, 2018). This motivation is driven by state-wide policies and emphasis on using biofuels as a necessary renewable fuel resource to sustain declining fuel resources. The model used to predict PAR is necessary to map PAR availability in order to properly harness it in algal farms, which rely on PAR sunlight for productivity and can be employed as a strategy by decision-makers in agriculture.

The aims of this chapter are as follows: (1) develop an AI model for PAR forecasting in a regional subtropical Queensland location (Toowoomba), using historical lagged data as inputs; (2) develop the MLP model for PAR forecasting as a benchmark; (3) to improve the accuracy of this MLP model by applying the firefly nature-inspired optimizer (MLP-FFA hybrid model); (4) create the random forest (RF) and multiple linear regression (MLR) models to benchmark against MLP-FFA (objective model); and (5) assess and compare the performance of MLP-FFA, MLP, RF and MLR models using robust statistical metrics. These aims clearly will be able to overcome the research problem, whereby for a depleting coal fuel resource, no optimized AI models have been employed to predict growth conditions for an instead renewable biofuel source, in order to enhance its growth and sustainability. PAR is a growth necessity for photosynthetic algae as a biofuel, so definitely will aid in tackling this issue. A possible limitation of this study may be the availability of open source data, which literature has shown has been hard to find for PAR. Additionally, PAR data is not available at small time resolutions, such as daily or hourly, which would have been advantageous for shorter time predictions, instead of a monthly total (which was used).

7.2 Chapter background review

7.2.1 Statistical and mathematical models

Methods currently applied for PAR prediction utilize statistical linear models, mathematical relationship models or nonlinear AI (data-driven) models. Although linear models (especially of regression type) have been extensively used, as described next, the use of nonlinear AI models have been gaining more attention these days. AI models are more powerful, as (1) they are nonlinear, which is ideal for nonlinear highly seasonal PAR; (2) they create a model from training, unlike from a line of best fit in the case in regression models; (3) they are well-suited to complex data; (4) they are universal, unlike regression and physical models which are representative only of the data used to build them; and hence (5) they are very reliable and accurate (Deo et al., 2018). This section will discuss the current nonstatistical and mathematical PAR models used in the literature, which are clearly not as advantageous as AI models.

Current PAR models that are not of AI form are mainly of the regression type with some physical models, which are not very robust, especially as compared to AI models. Sudhakar et al. (2013) estimated daily (averaged from monthly) satellite PAR for different Indian latitudes ranging from 9 degrees to 34 degrees using daily (averaged from monthly) global radiation (Hg) as the input feature and a power regression model for PAR. The results for R^2 scores (R^2 score shows how similar the observed and simulated PAR are here) ranged from 0.97 to 0.768, and larger latitudes were not able to be estimated as well using this linear regression model. PAR is not linear, so a linear model actually doesn't make as much sense, when nonlinear AI models can be adopted instead. Vindel et al. (2018) also estimated PAR using a linear regression model with global horizontal irradiance (GHI) as input. They have instead used target PAR values simulated from empirical expressions, which they have actually identified as a limitation but stated that it was their best choice as observed PAR is hard to find online and takes a long time to self-measure. They found very high correlations greater than 0.99 for observed and simulated PAR, but it seems like this could be a case of overfitting due to performance based on calculated PAR (target) compared to estimated PAR that has also been simulated, and thus they would both be very similar using this unobserved target. This is definitely something to consider for future models. Alados et al. (2000) estimated PAR under varying cloud conditions. They instead created a mathematical (physical) model to predict PAR where total cloud transmittance was used as the variable, as well as flux of cloud sky conditions. This mathematical model would be location-specific and not universal, even though some good results were obtained (Deo et al., 2018). AI models have the added advantage that they are not location-specific and universal models (Deo et al., 2018). Qin et al. (2012) also built a physical transmittance model for PAR, based on how much PAR is transmitted through differing cloud cover throughout a day. Again, this would be a location-specific model (Deo et al., 2018). Rosati et al. (2004) used linear regression to predict PAR as an equation of daily photosynthesis of leaves. This is again a regression model, which is not as robust as nonlinear AI models, especially with PAR being nonlinear (Deo et al., 2018). This is an added advantage of AI models—that they are nonlinear, and do not require patterns in data to build good models (which regression requires), as training uses the data to learn and to create a model based on learning, without a predefined pattern required (Deo et al., 2018). Liang et al. (2006) estimated PAR based on calculated surface reflectance and TOA through the least trimmed squares regression technique to minimize the sum of the k smallest squared residuals as a slightly different technique. But, again, this is a linear regression technique, which is not as robust as an AI model (Deo et al., 2018).

As aforementioned, regression is not a very powerful model and in AI cases, is usually used as the first initial model only to find patterns for exploratory data analysis purposes (Wagner, 1959). Regression is not very powerful as it creates a line of best fit to the data, hence it does not account for variances that cannot be picked up as well and usually will only go through one iteration without a chance to learn through many iterations (as is the case in AI models) (Wagner, 1959). An AI model can be much more specific to the data as it can account for this variance, but still can be relatively fast (Deo et al., 2018). AI models also usually have much more powerful equations behind them than a simple regression of the form $y = mx + c$, where m is gradient, x is the independent variable, and c is the y-intercept (the AI equations are too detailed too discuss in this literature review, due to the word limits) (Wagner, 1959). Literature in Section 7.2.3 shows that has been outperformed by other AI models.

Overall, it can be seen that non AI models for PAR, tend to be somewhat similar, using some sort of statistical regression with related features, or another simple physical/mathematical equation. This is not as robust and versatile a method as using AI models, especially with optimizers as will be discussed further on (Deo et al., 2018). AI models do however have a disadvantage in that they are black-box, so the physical processes and equations are not known, but even though they are black-box they can outperform regression and are much more robust (Deo et al., 2018). Hence, this project will develop a novel optimized AI model for PAR, instead of using a regression or physical model as in the current literature. MLR regression type technique will be only used as a benchmark model to compare to the main AI model to be used (which will be defined through further sections).

7.2.2 Artificial intelligence models

This section provides the current literature that have used AI models for PAR, which are all utilizing AI models without any sort of optimization, even though this has the potential to improve performance as identified in the next section. Zempila et al. (2016) used a linear regression model as a comparison to their objective neural network (AI) model to estimate PAR using solar zenith angle, aersol optical depth, and water vapor as inputs. They found that the nonlinear neural network model outperforms the linear regression, which supports the above claims also in the previous section, as to regression not being as powerful. Lopez et al. (2001) instead used MLP (AI) to estimate PAR, using calculated PAR again (as the target variable) as they could not identify a good measured source. They created two MLP models: one with sunshine duration as inputs, and the second with several different meteorological and radiometric variables as inputs, which is a different approach. Yu and Guo (2016) estimated PAR using ANN and regression to compare with global solar radiation as the input, similar to the first literature described in this section. They also found that the AI model (ANN) predicted more accurately for both overcast and clear sky conditions. Zhu et al. (2013) estimated the fraction of PAR (fPAR) using again neural networks for modeling with land cover class, pixel-center latitude, pixel-center longitude, and normalized difference vegetation index. They instead identified that in their study, which created models for six locations, that the simulation of the seasonal cycle (i.e., patterns of seasonality in PAR) in the northern latitudes was not matching the satellite PAR (target) from NASA. They also confirmed that ground manual collection of PAR gave these same results. Perhaps, using an optimizer could have improved the model in terms of this. Wang et al. (2016) estimated PAR using improved ANN (AI) models, specifically MLP, Generalized regression neural network (GRNN), and radial basis neural network (RBNN) compared to an all-sky PAR model (ALSKY) using air temperature, relative humidity, dew point, water vapor pressure, and air pressure inputs. This is a more comparative study than others. They also found that the MLP model performed the best in their study.

Overall, whilst AI models have been used for PAR modeling, an optimizer method has not been used at all. Also, it proved a more difficult task to find papers using AI to model PAR than with regression models (which were of very high abundance), meaning that while regression models have been very often performed, AI models have not been as well studied, but are gaining attention in more current times. Nature-inspired optimizers can improve performance as seen in the next section, so should definitely be used, even though they haven't been used in the current literature for PAR.

In many works, PAR modeling has not been performed with the MLP-FFA algorithm or with any sort of optimizer applied, but definitely should be because, as seen below, they can noticeably improve performance. Papers that show that MLP-FFA outperformed the MLP/other models. Deo et al. (2018) predicted wind speed and they claim that MLP-FFA "leads to a significant improvement in the predictive accuracy, presumably due to the optimal weights attained in the hidden layer that allows a more robust feature extraction process" as compared to MLP. Ghorbani et al. (2017a) predicted pan evaporation over daily time horizons comparing MLP-FFA this time to both SVM and MLP separately. Results showed that MLP-FFA still outperforms even the high-performing SVM algorithm, and again MLP-FFA outperforms MLP.

MLP-FFA is usually compared to the standard MLP in papers, but FFA is used with other data-intelligent models also and has been compared to some other benchmark common data-intelligent models. Acronyms for models here are defined in the "Acronyms" section. Yaseen et al. (2018) forecasted monthly rainfall where the FFA algorithm with adaptive neuro fuzzy inference system (ANFIS), called ANFIS-FFA, outperformed the ANFIS model alone. Soleymani et al. (2016) predicted water level of rivers using a hybrid radial basis function (RBF) and FFA algorithm called RBF-FFA and found that RBF-FFA had more precise predictions compared to both MLP and SVM. Goudarzi et al. (2016) predicted vertical handover to minimize unnecessary handovers for mobile node during the vertical handover process. Performance evaluation found that the IRBF-FFA model performed better compared to MLP and SVM. Lastly, Khatibi et al. (2017) predicted stream flow using the hybrid MLP-LM and MLP-FFA algorithms. This paper showed that MLP-FFA still performed better than MLP-LM.

This section of the literature review clearly shows that using the advantageous AI MLP-FFA optimized algorithm to model PAR is novel, but should definitely be employed as the literature shows that MLP-FFA performs better than other standard models, as summarized in Table 7.1, and that FFA makes a big contribution to this. Also, MLP tends to perform better than other models, especially when with FFA. This project hence aims to develop the MLP-FFA model in a subtropical Toowoomba location to fill this gap.

7.2.3 Other par estimation methods

In terms of inputs, location, and data collection methods currently used for PAR models, some gaps have been identified, which this project aims to fill. Usually other features, such as global solar radiation (Sudhakar et al., 2013; Wang et al., 2015; Yu and Guo, 2016), GHI (Vindel et al., 2018; Zempila et al., 2016), air mass (Wang et al., 2015), cloud cover (Alados et al., 2000), solar zenith angle (Zempila et al., 2016), water vapor (Wang et al., 2016; Zempila et al., 2016), and sunshine duration (Lopez et al., 2001), are used to predict PAR and not historical lagged PAR (which is probably better than using features as it is actually using historical PAR to predict PAR, not features just related to PAR).

Historical lagged PAR inputs are possible as they are a stable (due to being seasonal) input, and allow for accurate predicting of PAR with its own lagged values, which has a great potential to be a very efficient learning during training. To further show that lags are an effective technique, literature will now be discussed that has effectively used lagged inputs for nonPAR models (as lags are novel for PAR models). Luk et al. (2000) used lagged inputs into their ANN model for rainfall forecasting, finding that the ANN with a lower number of well selected lags performed very well. Bowden et al. (2005) forecasted river salinity using a neural networks model also with lagged inputs, which they claimed led to lower error than in previous studies. Mittnik (1990) estimated GNP growth rates using a multivariate distributed lag model, which is also a specific and advanced technique.

Litterman (1986) performed economic forecasting using a Bayesian vector autoregression technique with lagged inputs, and claimed that the use of lags is an "inexpensive, reproducible statistical technique that is as accurate, on average, as those used by the best known commercial forecasting services." Clements and Galvão (2009) instead used a MIDAS (mixed-data sampling) model which also included a lag distribution for econometric regression predictions and found great performance also. These literature show that lagged inputs are very well used, as they are shown to be very effective, although astonishingly they have not been applied for PAR models which is a very possible technique as PAR is a stable input. Hence, this project will aim to fill this gap using lagged PAR inputs (lags are time minus 1 month, time minus 2 months, etc.). for the MLP-FFA model.

Papers predicting PAR in subtropical Queensland or even Australia are nonexistent, with papers being instead in locations all over the world, even though Australia has astounding levels of sunlight/PAR coverage (Beath, 2012; Deo and Şahin, 2017; Yusaf et al., 2011). PAR has been predicted instead in America (Alados et al., 2000; Yu and Guo, 2016), China (Wang et al., 2015; Wang et al., 2016), Greece (Zempila et al., 2016), India (Sudhakar et al., 2013), and Spain (Lopez et al., 2001; Vindel et al., 2018) as the main coutnries, but also in several others. As aforementioned, the Australian government is putting more focus on renewable energy, as Australia is the sunniest continent in the world, so the fact that PAR has not yet been modeled in Australia is quite astonishing. Subtropical locations in Australia (and also in general) have more sunlight due to less cloud cover and high insolation (Schmidt, 2019), although such a subtropical Australian region has not been used for PAR models. As a result, this project will build an MLP-FFA PAR model in subtropical Toowoomba, Queensland, Australia.

The papers reviewed have claimed that full online data was not available at the time of their studies for their required time scales, hence PAR values were instead calculated or averaged (averaged in the case where their time resolution required was not available), as manual collection is very time-, resource-, and effort-consuming for several years of data. This study uses total monthly PAR data as inputs and objective variables to overcome this.

7.2.4 Concluding remarks

Literature predicting PAR is quite limited in terms of many factors. The most important factor from literature is that no papers were found that predict PAR with nature-inspired optimizers applied at all. Papers seem to use physical models, such as mainly regression, which are definitely not as powerful as AI models. AI models with FFA optimization can clearly improve results, as seen in Section 7.2.2. Another major factor is that PAR has been modeled in various locations around the world, but not in Australia which is the sunniest country in the world and has subtropical regions and a government initiative has placed a focus on renewable bioenergy and biofuels. Lastly, it can be seen that whilst a lot of features have been used as inputs in PAR models, no lagged inputs are used—which can possibly provide very good results as it is learning from its own PAR data to predict, instead of features related to PAR.

These gaps are aimed to be filled by modeling PAR using an advanced well-performing AI MLP-FFA algorithm in subtropical Toowoomba, Australia, using satellite historical lagged PAR from NASA's Goddard Earth Sciences Data and Information Services Center's Interactive Online Visualization and Analysis Infrastructure—NASA's open source data repository (GIOVANNI) as inputs. This project has the potential to provide cutting-edge novel research, which can be applied to the biofuel industry as per recent Australian government initiatives (Geoscience Australia, 2018) with an advanced PAR model.

7.3 Materials and methodology

7.3.1 Study region and data

The study site is located in the "Sunshine State" of Queensland, Australia's second largest state, which is a subtropical region. In Queensland, there is an urgent need for properly mapping solar-related energy resources and considering the sustainability of this, for increased and enhanced use of this renewable resource (Martin and Rice, 2012; Troccoli, 2015). Toowoomba, in the southeast region of Queensland (Fig. 7.4), is ideal for the collection of PAR data due to its sunlight capacity. The rural Toowoomba is a subtropical region, meaning it has 10%–20% less cloud cover than tropics and temperate zones (Schmidt, 2019), hence a large amount of PAR sunlight would be reaching the surface. As a result, a Toowoomba location is ideal for algal biofuel production (Prasad et al., 2014), which is dependent on PAR. Despite this, commercial production of algal-derived materials is still very scarce in Queensland, and worldwide they are still not being produced in large quantities (Prasad et al., 2014). As stated in Section 7.1, PAR as an algal biofuel in Queensland has gained much attention in recent times (Al-lwayzy et al., 2014; Queensland Biofutures, 2016; Duong et al., 2015; Katter, 2015; McNamara, 2008; Prasad et al., 2014). Hence PAR collection and measurement in Toowoomba, Queensland is a justified research

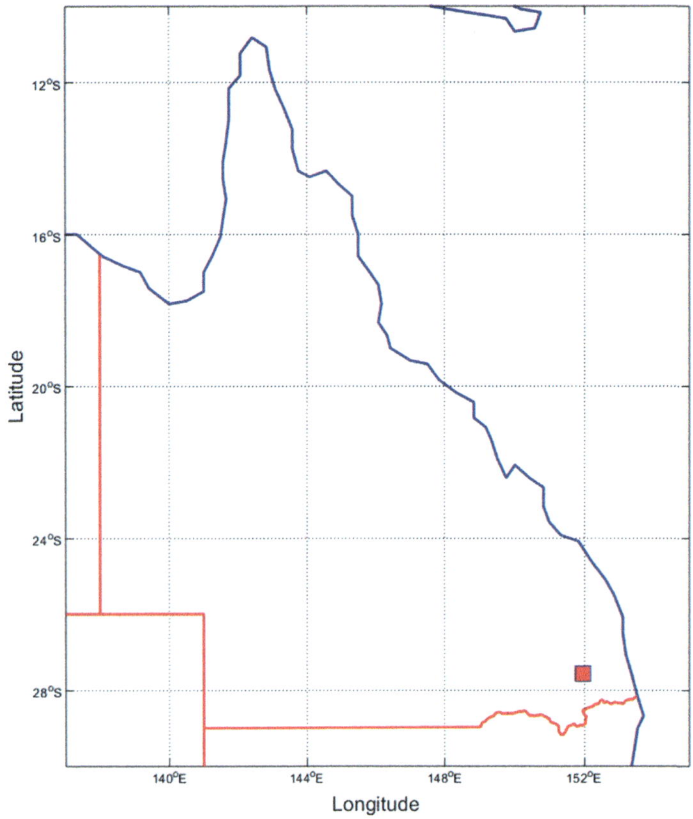

FIGURE 7.4 Toowoomba study region.

endeavor for the use of modeling PAR as an algal biofuel. Additionally, even though Toowoomba is a good location for PAR modeling, it was anticipated to use other locations also, such as the Great Barrier Reef, Red Sea, etc., but the data available did not provide fulfilling results with several missing values; a possible limitation that could be looked into for future work. PAR has been known to be hard to find data for online, as seen in Section 7.2.3. Another open source data other than that used (GIOVANNI) for PAR was not found. Also, no other time scale (apart from monthly, such as daily, hourly, etc.). was available with good data.

The collected Toowoomba PAR data is plotted in Fig. 7.5, and ranges across 184 months (approximately 15.5 years) from July 2002 to November 2017 (the latest data available). The data is in the form of satellite remote sensing from NASA's Goddard Earth Sciences Data and Information Services Center's (GES DISC) open source data repository called Interactive Online Visualization and Analysis Infrastructure (GIOVANNI) (Giovanni, 2018). The PAR data collected was total monthly data, unlike papers in Section 7.2.3 which have averaged or calculated data to make it fit to a time scale. Table 7.2 shows summary statistics for the data. PAR normally can range from 0–100 $\gamma/(\text{s m}^2)$. The minimum PAR

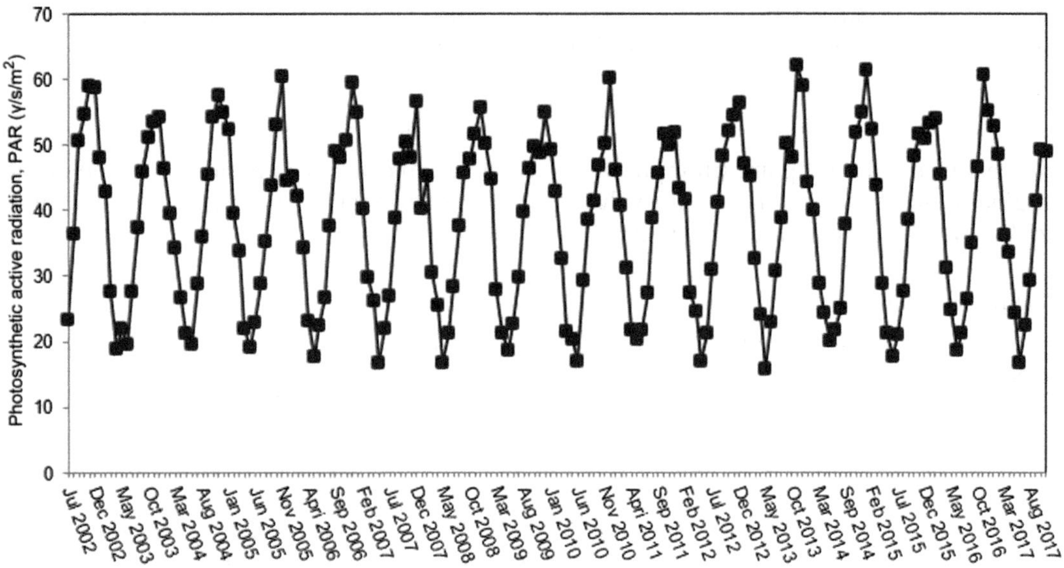

FIGURE 7.5 Monthly cycle of main predictor (photosynthetic active radiation, PAR).

TABLE 7.2 Descriptive statistics of monthly photosynthetic active radiation (PAR), where γ = photons.

Minimum	Maximum	Mean	Standard deviation	Skewness	Kurtosis
15.76 $\gamma/(\mathrm{s\ m}^2)$	61.95 $\gamma/(\mathrm{s\ m}^2)$	38.43 $\gamma/(\mathrm{s\ m}^2)$	13.103	− 0.099	− 1.310

tends to be during periods where less sunlight is prominent, during the winter months, and at higher levels during summer months, with a very noticeable cycle. as seen in Fig. 7.5. Hence, there are tendencies that the annual cycle is very consistent, so PAR can be easily modeled using its historical values, which are generally not stochastic. Standard deviation indicates that the values do differ from the mean, which indicates the changes across seasons, causing values not to be centralized around the mean. Skewedness is a low negative value, indicating a slightly longer left tail overall; and kurtosis is also low indicating overall lack of flatness (with prominent peaks due to seasonality).

7.3.2 Model description

7.3.2.1 Normalization, feature selection, and data partitions

Normalization

To develop a robust MLP-FFA forecasting model, or in fact any model, an essential activity is to optimize the architecture involved within the model to improve the cause-and-effect relationships between inputs and objective variable (Deo and Şahin, 2017).

Before testing any model parameters though, normalization of inputs and predictand must be performed. This scaling of the inputs and objective variables is necessary to avoid numeric issues caused by large fluctuations in data, by scaling data to be in a smaller range of zero and one (Deo and Şahin, 2017), using the formula:

$$x_{\text{normalized}} = \frac{x - x_{\min}}{x_{\max} - x_{\min}} \tag{7.1}$$

where x is any datum point (in input or objective variables), $x_{\min}$ is minimum value of the original dataset, $x_{\max}$ is maximum value of the original dataset and $x_{\text{normalized}}$ is normalized value of a particular datum point.

Feature selection

In terms of model development, firstly the inputs used in this objective model (and the benchmark models) were historical lagged PAR values, this means that for the entire time series of data, a shift was made upwards by a month in the entire data, then the remaining months after a cutoff (the level where the lag with the least months has values until) are removed for all lags. This is demonstrated in an exemplar in Table 7.3, where the cutoff is indicated by the red line. For data that is not stochastic, the historical lags can be used as a good predictor of the objective variable (as demonstrated in Section 7.2.3), which is in fact a novel approach to PAR forecasting. In the development of the MLP-FFA model, as well as all other models used, feature selection is an important technique. For this purpose, a partial autocorrelation function (PACF) was used. PACF uses cross-correlation of the lags with the original PAR data (Deo and Şahin, 2017) using MATLAB. For any discrete signal, the correlation between $x_i = (x_1, x_2, \ldots, x_{M-1})$ and $y = (y_1, y_2, \ldots, y_{N-1})$ is:

$$\varphi_{xy,k=} \sum_{j=\max(0,k)}^{\min(M-1+k,N-1)} x_{j-k}, k = -(M+1), \ldots, 0, \ldots, (N-1) \tag{7.2}$$

where cross-correlation, r_{cross} is:

$$r_{\text{cross}}(t) = \frac{\varphi_{xy}(t)}{\sqrt{\varphi_{xx}(o)\varphi_{yy}(0)}} \tag{7.3}$$

The cross-correlation is bounded by $[-1,1]$, where 1 indicates that both time series (of lag and original PAR) have the exact shape in which they both increase or decrease together (indicated by positive) at a similar rate, even though amplitude of values may differ based on the variable (Deo and Şahin, 2017). Whilst -1 indicates that both time series have the same shape but in opposite directions meaning that as one increases, the other decreases at a similar rate (Deo and Şahin, 2017). N and M are the respective sample length of the predictor (lags) and predictand (original) data.

Train, test, and validation data partitions

Another important factor for model development is train, test, and validation data splits that were trialed, at levels indicated in Table 7.4. When considering the splits, there should be thought put into overfitting and underfitting to the data points (Srivastava et al., 2014). Too much training can lead to overfitting and too little training can lead to underfitting (Srivastava et al., 2014), hence a balance should be reached for the final model.

TABLE 7.3 Exemplar of historical lagged inputs.

Original PAR values	$t-1$	$t-2$	$t-3$	$t-4$	$t-5$
1	2	3	4	5	6
2	3	4	5	6	7
3	4	5	6	7	8
4	5	6	7	8	9
5	6	7	8	9	10
6	7	8	9	10	11
7	8	9	10	11	12
8	9	10	11	12	13
9	10	11	12	13	14
10	11	12	13	14	15
11	12	13	14	15	16
12	13	14	15	16	17
13	14	15	16	17	18
14	15	16	17	18	19
15	16	17	18	19	20
16	17	18	19	20	
17	18	19	20		
18	19	20			
19	20				
20					

During overtraining, a high performance can be a result, although this does not necessarily mean prediction is good (Srivastava et al., 2014), but instead indicates that there was a high possibility that the model has fit to the noise in data, instead of to the true data pattern. Underfitting leads to not fitting enough to the data, hence also not a good result (Srivastava et al., 2014).

Hence, this is considered when evaluating performance to make the decision of the optimal model, where overly high and low training was not tested as a result and only the percentages shown in Table 7.4 were used. Also there needs to be enough data for validation and testing for predictions during these phases to predict on new data and come to the final model in testing, which the percentages in Table 7.4 allow, except probably for the 5% test and validation, which will not be used for the final model but was just run for comparison purposes. The 10% test and validation may be a bit low, but seem to be used in papers, so to see if it is valid its performance will be evaluated.

TABLE 7.4 Train, validation and test splits (months are a rounded estimate for percentages).

Number of months	Training	Validation	Testing
Train = 157 Validation = 9 Test = 9	90%	5%	5%
Train = 140 Validation = 17 Test = 18	80%	10%	10%
Train = 123 Validation = 26 Test = 26	70%	15%	15%
Train = 105 Validation = 35 Test = 35	60%	20%	20%
Train = 88 Validation = 43 Test = 44	50%	25%	25%
Train = 71 Validation = 52 Test = 51	40%	30%	30%
Train = 53 Validation = 61 Test = 61	30%	35%	35%

7.3.2.2 *Multilayer perceptron neural network*

This study applies a widely used artificial neural network (ANN)-based model called MLP (McCulloch and Pitts, 1943) with the aim to improve PAR prediction by coupling it with an optimizer algorithm. The standalone MLP model will also be used for forecasting for comparison purposes to the hybrid model. The MLP model is a type of ANN model in which parallel information processing systems are present, containing a set of neurons arranged into several hidden layers (McClelland and Rumelhart, 1989). The MLP model has been used in other studies previously for predictions of renewable energy variables, as seen in Table 7.1. In MLP, the multiple layer feed-forward perceptron backpropagation (FFBP) learning algorithm contains the input layer, hidden layer, and output layer and the backpropagation algorithm aims to minimize the global error of MLP (Deo and Şahin, 2017).

In each hidden layer, the neurons are connected by a weight to neurons in a subsequent layer in training (Deo and Şahin, 2017). The sigmoid and linear activation functions were used and trialled in the hidden and output layer (to come to the optimum combination). The FFBP is superior to the other category of ANN models, whereby in the FFBP the neuronal architecture has the purpose of successively validating the model parameters (i.e., weighted connections and neuron biases) until the empirical error reaches a set tolerance through each iteration (epoch) of the forward passing of the updated parameters (to hidden layer then outputs) and then backward propagation of errors to fine-tune/reduce them, then produce improved outputs (Abbot and Marohasy, 2012; Adamowski et al., 2012). Mathematically, the ANN algorithm is (Deo and Şahin, 2015; Kim and Valdés, 2003):

$$y(x) = F\left(\sum_{i=1}^{L} w_i(t) \times x_i(t) + b\right) \tag{7.4}$$

where $x_i(t)$ is the input variables (predictor lags) in discrete time space t, $y(x)$ is the forecasted PAR, L is the number of hidden neurons determined iteratively, $w_i(t)$ is the weight that connects the i-th neuron in the input layer, b is the neuronal bias, and $F()$ is the hidden transfer function. The MLP model applies this algorithm in the way described above using the FFBP.

Due to ANN being a black-box model, it does not explicitly identify the training (learning) algorithm without an iterative model identification process (Deo and Şahin, 2015). Hence, this study has applied several training algorithms whose performances were evaluated to select the best MLP model. MATLAB-based training algorithms are classified into three categories: the quasi-Newton (that uses trainlm and trainbfg functions) (Huang, 1970), the gradient descent (traingdx) (Fletcher and Reeves, 1964), and the conjugate gradient (trainscg, traincgf, traincgb and traincgp) (Hestenes and Stiefel, 1952; Ali and Smith, 2006). The quasi-Newton method is based on the Levenberg–Marquardt (LM) function that locates the minimum of the input data which is expressed as the sum of squares of nonlinear real-valued functions (HariKumar et al., 2009) and the Broyden–Fletcher–Goldfarb–Shanno (BFGS) (Dennis Jr and Schnabel, 1996; Marquardt, 1963) which minimizes the mean square error.

Additionally, resilient backpropagation (trainrp) (Riedmiller and Braun, 1993) is generally fast, requires a noticeable amount of memory, and does not store the updated values of weight/bias, while the gradient descent with momentum and adaptive learning rate (traingdx) (Riedmiller and Braun, 1993) combines adaptive learning with momentum training where momentum coefficient is included as a training parameter (Anusree and Binu, 2014). Training using the one-step secant (trainoss) as an alternative, faces a smaller computation overhead, without storing the Hessian matrix but rather assumes it at each iteration as a compromise between the quasi-Newton and gradient algorithm. Overall, several training algorithms have been tested within MLP to evaluate their performance against the FFA used in the objective model of this project, MLP-FFA. The performance of only the best performing training algorithms for MLP are relevant for comparison and have been shown further in Section 7.4.1.2.

Since the MLP model is black-box, a vital task is to determine the optimum transfer function, which is not known a priori. A series of functions that are available in MATLAB toolbox can be trialled (Vogl et al., 1988), although three popularly used main functions will be trialled in this project (Deo et al., 2016; Ghorbani et al., 2013; Raheli et al., 2017; Şahin et al., 2013):

$$\text{TangentSigmoid}(\text{tansig}): F(x) = \frac{2}{1+\exp(-2x)} - 1 \tag{7.5}$$

$$\text{LogSigmoid}(\text{logsig}): F(x) = \frac{1}{1 + \exp(-x)} \tag{7.6}$$

$$\text{PositiveLinear(purelin)}: F(x) = x, x \geq 0, \text{or0otherwise} \tag{7.7}$$

where x is the predictor dataset analyzed in accordance with the function $F(x)$ that is able to map the predictive features to produce a hidden layer weight for the suitable model (Deo and Şahin, 2017).

Note that all four models used in this project have their code included in the Appendix, created so that results are reproducible. A random seed was applied for all models to ensure results are reproducible and do not vary every time they are run, that is, random selections in the models start at the same seeded number and have the same sequence at each run.

7.3.2.3 *Hybrid multilayer perceptron-Firefly algorithm model*

To obtain improved and precise results for PAR prediction, and the MLP model (Section 7.3.2.1) was integrated with the FFA as an optimizer tool. MLP is a strong model, but as seen in Section 7.2.2, with an optimizer, can be improved. Fig. 7.6 illustrates the architecture of the hybrid MLP-FFA model. Basically, the FFA algorithm was created by Yang (2010a,b) and it is a category of swarm intelligence optimization techniques based on the movement of fireflies (Deo et al., 2018). The solution of the optimization can be assumed to be an agent, metaphorically like a firefly which glows in its proportion due to its quality (Deo et al., 2018). Consequently, each brighter firefly attracts its partners, disregarding their gender, which makes it more efficient to find the best partner (Łukasik and Żak, 2009). The fireflies are attracted toward light, so in terms of the algorithm, the entire swarm of fireflies moves toward the brightest firefly (optimal result) (Deo et al., 2018). Hence the attractiveness of the firefly is directly proportional to its brightness. Also, the brightness depends on the intensity of the agent (Kayarvizhy et al., 2014). For development of the FFA-based model, the creation of the optimal objective function and variation of light intensity must be considered (Deo et al., 2018). In regard to the MLP-FFA model directly, the FFA algorithm is used as a training algorithm to find optimal weights for the inputs in the model, that is, the brightest fireflies (Deo et al., 2018).

Following papers by Yang (Yang, 2010a,b), the light intensity $I(r)$, the attractiveness (b), and the Cartesian distance between any two fireflies i and j are equated by:

$$I(r) = I_0 \exp(-\gamma r^2) \tag{7.8}$$

$$\beta(r) = \beta_0 \exp(-\gamma r^2) \tag{7.9}$$

$$r_y = \|x_i + x_j\| = \sqrt{\sum_{k=1}^{d} (x_{i,k} - x_{j,k})} \tag{7.10}$$

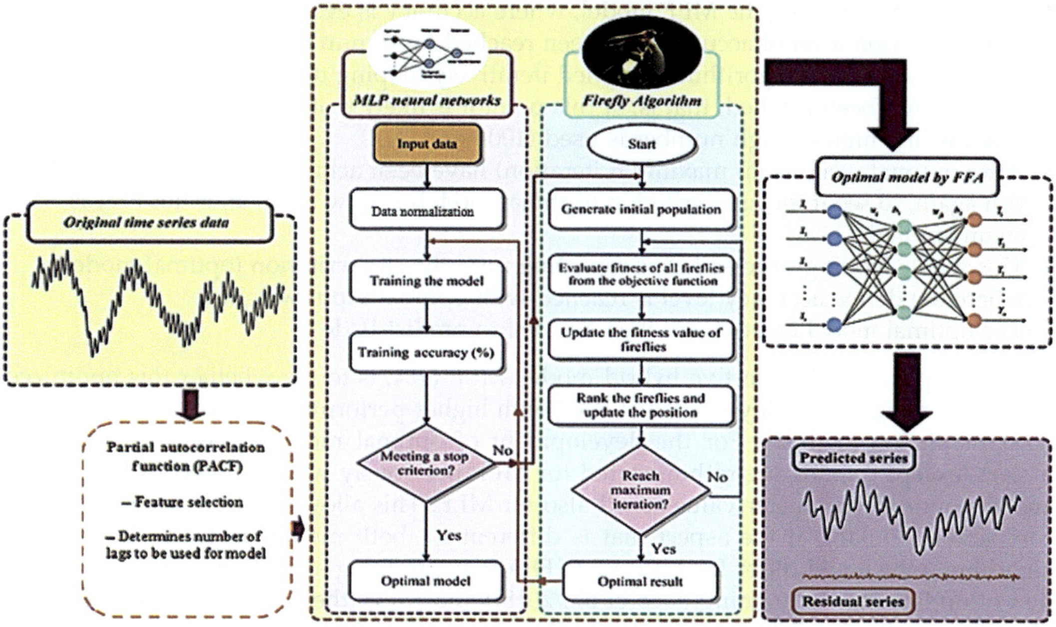

FIGURE 7.6 Flowchart of the construction of the hybrid multilayer perceptron-firefly optimizer (MLP-FFA) model structure (Deo et al., 2018).

where d is the dimensionality of the given problem, γ is the light absorption coefficient, $I(r)$ and I_0 are the light intensity at distance r and initial light intensity, respectively, from a "firefly" within the model, and $\beta(r)$andβ_0 are the attractiveness β at a distance r and $r=0$ (initial).

The next movement of the firefly i is equated by :

$$x_i^{i+1} = x_i + \Delta x_i \tag{7.11}$$

$$\Delta x_i = \beta_0 e^{-\gamma r^2}\left(x_j - x_i\right) + \alpha\varepsilon_i \tag{7.12}$$

where the first term in Eq. (7.12) is due to the attraction, and the second term represents the randomization ε_i where $\propto$ is the randomization coefficient with value in between zero and one and ε_i is the random number vector derived from a Gaussian distribution (Chai and Draxler, 2014). In this project, as mentioned in Section 7.3.2.1 the optimal values for the weights are selected using defined MLP training algorithms to optimize the performance of this classical model, in order to compare it to the performance of the more effectively known optimizer training algorithm, FFA.

As aforementioned, FFA is a nature-inspired optimizer that improves the selection of weights in the model. The steps of the MLP-FFA model to explain Fig. 7.6 briefly are (Deo et al., 2018):

1. Lags are trained using the MLP model, where accuracy is evaluated to decide whether a stop criterion level of accuracy has been reached (optimum model).
2. If not, then the FFA algorithm is applied iteratively looping until the defined maximum iteration has been reached, that is, at this point it is likely that there should be optimal weights, if a high enough number is used (100 here).
3. After optimal weights (at maximum iteration) have been achieved in FFA, then MLP is run again, to see if the stop criterion (optimal model) is now reached, if not FFA is again applied.
4. This whole looping process keeps repeating until the stop criterion (optimal model), where the define accuracy level is reached with optimal input weights.
5. The optimal model created will then be used to predict PAR.

The purpose of this objective hybrid model, MLP-FFA, is to see whether this optimized model has a high capability to predict PAR with higher performance than other more standard prediction methods. For the development of optimal model parameters, the parameters (except training algorithm) stated for MLP previously will be tested for MLP-FFA, and the optimal parameter values used also for MLP. This allows both these models to be comparable, looking at the aspect that is different for both models instead—the training algorithm, which will differ for MLP-FFA [FFA is the training algorithm that has the function of optimizing the training (Deo et al., 2018)] and MLP (the several described training algorithms in Section 7.3.2.2 will be trialled).

7.3.2.4 Random forest model

Another well-used standalone AI model is RF, that will be applied as a benchmark to MLP-FFA with the aim to show that the optimizer algorithm makes a more powerful model, than a widely used and high-performing model. Briefly to explain RF, it is a regression tree-based ensemble technique introduced by Breiman (2001) as an extension of the ensemble bagging method by Breiman (1996). Using a bootstrap aggregation (bagging) technique, the RF model is capable of resolving the overfitting issue of conventional solitary regression trees (Prasad et al., 2018). In the training phase, "*n*" bootstrap replicas are selected from the training dataset, using the random sampling with replacement technique. Utilizing the data that was not used during training, a single tree is created on every separate "*n-replicas*" with simultaneous computation of out-of-bag (OOB) errors of respective trees (Prasad et al., 2018), which is defined as:

$$\text{OOBerror} = \frac{1}{N}\sum_{t=1}^{N}(\text{PAR}_t^{\text{OBS}} - \text{PAR}_t^{\text{FOR}})^2 \tag{7.13}$$

where $\text{PAR}_t^{\text{OBS}}$ is the t-th observed PAR value and $\text{PAR}_t^{\text{FOR}}$ is the forecasted value.

To further explain the RF model, the single regression trees are then put together and the forecasted output from all trees in the ensemble are averaged, to come to a final output. At each decision tree split, a random subset of inputs are sampled for testing (Prasad et al., 2018). In RF, three model parameters need to be optimized: (1) the number of input variables randomly assigned at each node (fraction of in-bag observations), (2) the number of trees in the ensemble (forest), and (3) the tree size with respect to the maximum number of terminal nodes (leaves per tree) (Prasad et al., 2018). Additionally to allow RF to be comparable to MLP and MLP-FFA, the optimal training, test, and validation phase split from MLP-FFA was used in RF, and the RF model was run separately for each of these phases.

The objective model in this study is the MLP-FFA hybrid model. The purpose of using the MLP hybrid with the optimizer is as this MLP-FFA model has been used in papers before, as seen in Table 7.1 of Section 7.2.2 [where MLP-FFA performed better than standalone models in papers by Deo et al. (2018); Ghorbani et al. (2017b); Raheli et al. (2017); Ghorbani et al. (2017a); Khatibi et al. (2017)], whereas using RF with an optimizer has not been found as widely. Hence it seems a more sound choice to test whether the MLP-FFA model, which is known to perform well, can effectively forecast PAR as well, instead of a less used model.

7.3.2.5 Multiple linear regression model

To further compare the performance of the MLP-FFA optimized model, a standard statistical technique that examines the cause-and-effect relationship between the objective ($y = \text{PAR}$) and input variables (lags), called MLR, was employed. MLR is a regression-based technique in which multiple inputs can be used with the goal to create a model to be able to explain as much as possible the variations (due to noise) in the input dataset to determine their corresponding regression coefficients (Deo and Şahin, 2017). The MLR model has a regression equation of the form:

$$y = C + \beta_1 x_1 + \beta_2 x_2 + \ldots, + \beta_k x_k \tag{7.14}$$

where N is the number of observations, k is the number of inputs, $y(N \times 1)$ is a matrix of objective variable (PAR), $x(N \times k)$ is a vector of inputs, C is y-intercept, and β is multiple regression coefficient for each regressor (input) variable (Civelekoglu et al., 2007; Şahin et al., 2013). The magnitude for β is estimated through the least squares method (Özdamar, 2013). Briefly explaining further, for prediction purposes the MLR equation is fitted to a set of inputs and objective variables in the training phase (Deo and Şahin, 2017). Then, this fitted MLR model using its coefficients and y-intercept are used to forecast the PAR objective variable with an additional set of inputs in the testing period. For MLR, the model was run with the same number of train, test, and validation percentages that were optimal for MLP-FFFA (MLR run separately on each of these data partitions), to allow both these models to be comparable.

7.3.2.6 Justification of choice of models

There are some reasons why MLP-FFA was identified to model PAR. The data is supervised data meaning that the label (target variable) is provided. The PAR is numeric data and since it is not extremely large (175 rows of nine lag inputs), accuracy can be aimed for instead of speed (not required to be made speedier), hence accuracy is more important. Considering these conditions, the best models to use for the data in this investigation are RF, neural network, and gradient boosting tree (Liu, 2017).

MLP is a popular and for several applications a better type of neural network, because it can do everything that a single-layer perceptron can do but still provides the added

TABLE 7.5 Comparison of best modeling techniques for data (Liu, 2017).

Method	Advantages and disadvantages
MLP/ANN	Advantages – Good for complex noisy data without overfitting – Can create a vast amount of networks to classify which is accurate – Adaptive – Adaptive learning: can learn how to do tasks based on the data given for training – MLP/neural networks do not make assumptions in terms of the underlying probability density functions or other probabilistic information – The decision function is obtained directly from training – Is a universal approximate Disadvantages – Cannot handle well insufficient training data (not the case here)
Random forest/gradient boosting tree	Advantages – They are nonparametric so no specific data distribution necessary – Fast Disadvantages – Can overfit causing bias and false results – Are not so good with classifying numeric data which will cause several branches and nodes

capability of considering multiple hidden layers in a feed-forward neural network method (Deo et al., 2018). Table 7.5 shows that MLP is a good technique with several advantages, hence why it is so widely used (Liu, 2017). MLP is also commonly optimized with the FFA algorithm, compared to other techniques as seen in the Literature Review. FFA has several advantages also. FFA can deal with highly nonlinear, multimodal optimization problems efficiently (Deo et al., 2018). FFA is one of the optimization techniques that is well integrated to form a hybrid model. FFA does also not require a good initial solution to start its iteration process (Deo et al., 2018). FFA is an often-hybridized optimization technique, especially with MLP as seen in literature.

Hence, as the objective model MLP model is used with FFA. After a neural network, the next best model for the data would be RF or gradient boosting tree (Liu, 2017), although RF is much more widely used so was chosen as the benchmark AI model. MLR is a form of linear regression analysis for predictive modeling (Deo and Şahin, 2017) and is hence a standard model which is usually the first model tried and used as a benchmark against better models and hence MLR model was also used to model PAR.

7.3.3 Performance evaluation

To statistically evaluate the performance of MLP-FFA integrated with firefly optimizer against the three benchmark models—MLP, RF, and MLR—several statistical metrics were employed. These metrics are also used in model development to decide upon the optimal model parameters and input data. The performance metrics fall under two categories: (1) association metrics which measure the similarity between observed and forecasted PAR; and (2) error metrics which show the error in prediction. The mathematical representations of the association metrics are shown in Eqs. 7.15–7.18 and the error metrics are Eqs. 7.19–7.21.

1. Correlation coefficient (r):

$$r = \frac{\sum(O - \overline{O})(P - \overline{P})}{\sqrt{\sum(O-\overline{O})^2 \sum(P-\overline{P})^2}} \tag{7.15}$$

2. Willmott's index of agreement (WI):

$$WI = 1 - \left[\frac{\sum_{i=1}^{N}(O_i - P_i)^2}{\sum_{i=1}^{N}\left(\left|P_i - \overline{O}_i\right| + \left|O'_i - \overline{O}_i\right|\right)^2}\right] \tag{7.16}$$

3. Nash–Sutcliffe coefficient (E_{NS}):

$$E_{NS} = 1 - \left[\frac{\sum_{i=1}^{N}(O_i - P_i)^2}{\sum_{i=1}^{N}\left(O_i - \overline{O}_i\right)^2}\right] \tag{7.17}$$

4. Legate and McCabe's index (E_1):

$$E_1 = 1 - \left[\frac{\sum_{i=1}^{N} O_i - P_i}{O_i - \overline{O}_i}\right] \tag{7.18}$$

5. Absolute error:

$$\text{Error} = |O_i - P_i| \tag{7.19}$$

6. Root mean square error ($RMSE$):

$$RMSE = \sqrt{\frac{1}{N}\sum_{i=1}^{N}(P_i - O_i)^2} \tag{7.20}$$

7. Mean absolute error (MAE):

$$MAE = \frac{1}{N}\sum_{i=1}^{N}|P_i - O_i| \tag{7.21}$$

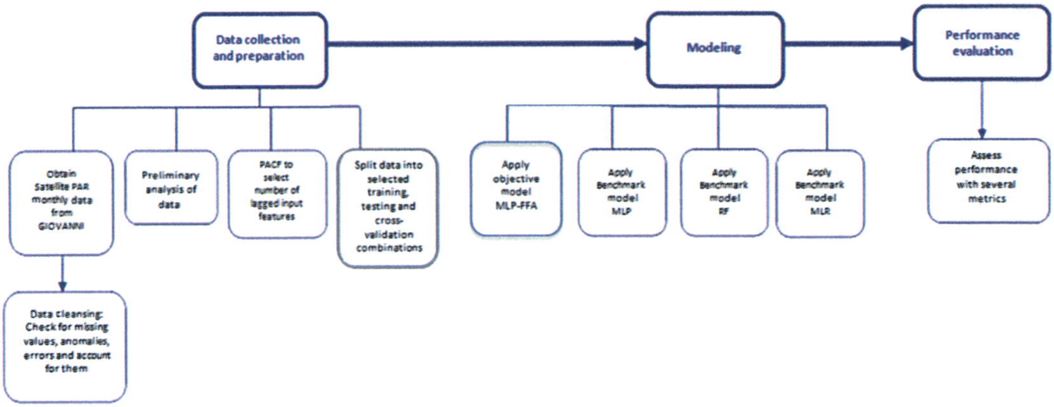

FIGURE 7.7 Schematic view of the entire methodology.

where P_i is predicted PAR values, O_i is observed PAR, $\overline{O}_i$ and $\overline{P}_i$ are the observed and forecasted mean PAR, and N is the number of datum points.

7.3.4 Methodology overview

Fig. 7.7. demonstrates the overall architecture of methods in this study. Firstly, satellite-based monthly data was collected and prepared into the right format (csv) to be utilized for forecasting with MATLAB software. This first section entails data cleaning in terms of checking for missing data, anomalies, and errors, and then accounting for them. Then a preliminary data analysis is performed to see data trends and decide the best methods. Feature selection was then performed using PACF, as aforementioned. Secondly, modeling of monthly PAR was carried out using the MLP-FFA objective model, and MLP, RF, and MLR benchmark models. Finally, performance evaluation was carried out to assess the predictive prowess of MLP-FFA and compare it to the benchmark models, with the aim of showing that it performs better. All modeling was carried out using MATLAB software.

7.4 Application results and analysis

7.4.1 Development of predictive models

In order to select the optimal model for the four used—MLP-FFA, MLP, RF, and MLR—several model parameters must be trialled, as well as choosing the right inputs and train, test, and validation data percentages, as mentioned in Section 7.3.2. The results of these are shown below.

7.4.1.1 Feature selection

PACF was used to select the number of historical lagged PAR inputs to be used for predictions. At $r_{\text{cross}}(t) = 0$, both the predictor and predictand are uncorrelated, but at $r_{\text{cross}}(t) \geq 0.7$, a good correlation between them can be concluded (Deo and Şahin, 2017).

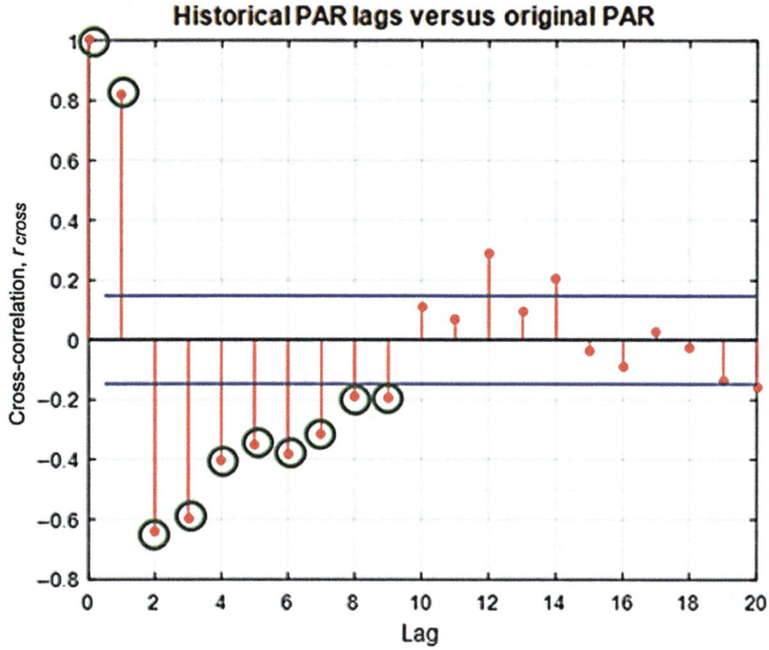

FIGURE 7.8 Cross-correlation coefficient (r_{cross}) between predictor variable (PAR) and lagged historical PAR (input variables) (going from time minus 1 month to time minus 20 months). 95% confidence interval for r is indicated in *blue*. Note: *green* circles indicate consecutive statistically significant lags.

As seen in Fig. 7.8 (which shows the results of PACF), above the 95% confidence interval (blue line), the first nine lags are statistically significant. Following this lags 12, 14, and 20 are also significant, although going to this extreme may cause too much removal of data (removing 20 months for lag of 20), and also a consecutive set of lags would allow the model to learn more based on pattern than separate nonconsecutive ones. Hence for all models used (namely, MLP-FFA, MLP, RF, and MLR) the first nine historical lags for monthly PAR were used with a lagged combination of PAR($t-1$), PAR($t-2$), PAR ($t-3$), PAR($t-4$), PAR($t-5$), PAR($t-6$), PAR($t-7$), PAR($t-8$), and PAR($t-9$). Historical lags are used to predict the oldest month that has been cutoff for a certain lag (i.e., a subsequent month from the inputs) (Ali et al., 2018). Although alone, just with good inputs an optimal model cannot be developed, and other parameters also need to be optimized.

7.4.1.2 Multilayer perceptron-firefly algorithm and multilayer perceptron

Train, test, and validation splits

The test period is where the final model is chosen and hence all performance metrics are considered to evaluate and choose the optimal models (Deo et al., 2018). Results from training and validation for all model parameters are shown in the Appendices, where for these two phases only r and *RMSE* are required as it is not yet the final model being chosen, it is just the initial phases. As seen in Table 7.6, during the testing period (when the model is finalized), the 60% training data, 20% testing, and 20% validation gave the best results in terms of highest association metrics (showing the level of association between

TABLE 7.6 Train, test and validation split performance for MLP-FFA during the testing period, based on E_{NS}, L, r, RMSE, and WI.

MLP-FFA train, test and validation split evaluation									
Train/test/ validation split	**Transfer function for hidden and output layer**	**Model architecture (input-hidden-output)**	r	*RMSE* $\gamma/(s\ m^2)$	*WI*	E_{NS}	L	*MAE* $\gamma/(s\ m^2)$
90%/5%/5%	logsig tansig	9-120-1	0.969	3.762	0.972	0.935	0.820	2.446
80%/10%/10%	logsig tansig	9-120-1	0.965	3.443	0.971	0.926	0.791	2.341
70%/15%/15%	logsig tansig	9-120-1	0.980	3.042	0.979	0.953	0.824	2.249
60%/20%/20%	logsig tansig	9-120-1	**0.983**	**2.795**	**0.981**	**0.960**	**0.832**	**2.141**
50%/25%/25%	logsig tansig	9-120-1	0.972	3.543	0.971	0.933	0.777	2.740
40%/30%/30%	logsig tansig	9-120-1	0.970	3.635	0.970	0.930	0.774	2.788

The best values are shown in boldface. E_{NS}, Nash–Sutcliffe efficiency, L = Legates–McCabe's index; r, Pearson's correlation coefficient; *RMSE*, root mean square error; *WI*, Willmott's index.

observed and simulated PAR): r, W, E_{NS}, and L and lowest error in *RMSE* and *MAE* for MLP-FFA. Hence, this was chosen for all MLP-FFA models run as the optimal percentages. During the testing phase, all performance metrics are assessed to ensure the optimal final model is decided upon. For MLP-FFA. a training of 60% is not too much data that it overfits, but also not too little data that satisfactory training cannot be performed upon it. Twenty percent of the data is also sufficient to assess further the model to see whether it is still optimal on new sets of data, as well as evaluating the final model during testing phase, and is a commonly used test percentage in papers. To allow MLP-FFA and MLP to be comparable all parameters are maintained as a control, except for the differing one which is the training algorithm (which is FFA compared against several for MLP).

Hidden layer size

Thirdly, a number of hidden layers were optimized, testing hidden layers in intervals of 10, from 10 until 150, as seen in Table 7.7. When trialling hidden layers, care should be taken to consider the time taken for building the model. Especially for MLP-FFA, the model is run for each maximum iteration (100 chosen here) times, for all "x" hidden layers. Hence, if too many hidden layers are chosen it is very time-consuming. Also, if the number of hidden layers is low than the model will not have enough capacity to predict it well. It can be seen in Table 7.7 that the hidden layer size of 120 gave the best results for MLP-FFA during the testing period for all metrics, and hence this value will be used for MLP-FFA and MLP to allow them to both be comparable. It was decided to show here only the intervals of 10 from 100 to 150, as this is a large enough hidden layer size for the model, that was still not overly time-consuming. The training and validation performance are also displayed in Table 7.7, but again the test is used to choose the final model.

TABLE 7.7 Hidden layer size performance for MLP-FFA, based on E_{NS}, L, r, RMSE, and WI.

MLP-FFA hidden layer size evaluation								
Train/test/ validation split	**Transfer function for hidden and output layer**	**Model architecture (input-hidden-output)**	***r***	***RMSE* γ/(s m²)**	***WI***	**E_{NS}**	***L***	***MAE* γ/(s m²)**
60%/20%/20%	logsig tansig	9-100-1	0.981	3.230	0.975	0.946	0.812	2.393
60%/20%/20%	logsig tansig	9-110-1	0.981	3.041	0.978	0.952	0.827	2.203
60%/20%/20%	logsig tansig	9-120-1	**0.983**	**2.795**	**0.981**	**0.960**	**0.832**	**2.141**
60%/20%/20%	logsig tansig	9-130-1	0.982	2.982	0.979	0.954	0.832	2.147
60%/20%/20%	logsig tansig	9-140-1	0.979	3.170	0.976	0.948	0.815	2.356
60%/20%/20%	logsig tansig	9-150-1	0.980	3.144	0.977	0.949	0.816	2.341

The best values are shown in boldface. E_{NS}, Nash–Sutcliffe efficiency, *L*, Legates–McCabe's index, *r*, Pearson's correlation coefficient, *RMSE*, root mean square error, *WI*, Willmott's index.

TABLE 7.8 Transfer function for output and hidden layer performance for MLP-FFA during the testing period, based on E_{NS}, L, r, RMSE, and WI.

MLP-FFA Transfer function evaluation								
Train/test/ validation split	**Transfer function for hidden and output layer**	**Model architecture (input-hidden-output)**	***r***	**RMSE γ/(s m²)**	**WI**	**ENS**	**L**	**MAE γ/(s m²)**
60%/20%/20%	tansig tansig	9-120-1	0.946	4.788	0.947	0.881	0.720	3.568
60%/20%/20%	tansig logsig	9-120-1	0.863	10.683	− 0.095	0.409	0.336	8.466
60%/20%/20%	tansig purelin	9-120-1	0.920	5.712	0.923	0.831	0.656	4.383
60%/20%/20%	logsig logsig	9-120-1	0.879	10.568	− 0.103	0.421	0.356	8.212
60%/20%/20%	logsig tansig	9-120-1	**0.983**	**2.795**	**0.981**	**0.960**	**0.832**	**2.141**
60%/20%/20%	logsig purelin	9-120-1	0.978	3.276	0.975	0.944	0.803	2.514
60%/20%/20%	purelin purelin	9-120-1	0.981	3.423	0.973	0.939	0.791	2.666
60%/20%/20%	purelin tansig	9-120-1	0.981	3.069	0.977	0.951	0.816	2.345
60%/20%/20%	purelin logsig	9-120-1	0.835	10.764	− 0.104	0.400	0.325	8.607

The best values are shown in boldface. E_{NS} = Nash–Sutcliffe efficiency, *L* = Legates–McCabe's index, *r* = Pearson's correlation coefficient, *RMSE* = root mean square error, *WI* = Willmott's Index.

Transfer function

The transfer function of input and output layers were chosen within the model, as explained in Section 7.3.2.1. Table 7.8 shows that for the testing period, the best metric values were for the hidden transfer function of logarithmic sigmoid (logsig) and output

TABLE 7.9 Results from MLP with several training algorithms (as in far left column), during the testing period, based on E_{NS}, L, r, RMSE, and WI.

MLP model parameters				Performance metric values					
Training (learning) algorithm	Train/test/ validation split	Transfer function for hidden and output layer	Model architecture (input-hidden-output)	r	*RMSE* $\gamma/(s\ m^2)$	*WI*	E_{NS}	L	*MAE* $\gamma/(s\ m^2)$
trainlm	60%/20%/20%	logsig tansig	9-120-1	0.960	5.126	0.939	0.864	0.707	3.734
trainbfg	60%/20%/20%	logsig tansig	9-120-1	0.976	3.416	0.972	0.940	0.801	2.531
traincgf	60%/20%/20%	logsig tansig	9-120-1	0.976	**3.211**	**0.974**	**0.947**	**0.810**	**2.424**
traincgb	60%/20%/20%	logsig tansig	9-120-1	0.977	3.250	0.974	0.945	0.817	2.339
trainrp	60%/20%/20%	logsig tansig	9-120-1	**0.978**	3.395	0.973	0.940	0.806	2.468
trainscg	60%/20%/20%	logsig tansig	9-120-1	0.972	3.386	0.972	0.941	0.794	2.632
trainoss	60%/20%/20%	logsig tansig	9-120-1	0.974	3.372	0.972	0.941	0.818	2.323
traincgp	60%/20%/20%	logsig tansig	9-120-1	0.978	3.543	0.971	0.935	0.772	2.909

The best values are shown in boldface. E_{NS}, Nash–Sutcliffe efficiency; L, Legates–McCabe's index; r, Pearson's correlation coefficient; *RMSE*, root mean square error, *WI*, Willmott's Index.

transfer function of tangent sigmoid (tansig) to be used for MLP-FFA. MLP will again use these same transfer functions to be comparable to MLP-FFA.

Training algorithm

The training algorithm within the model was optimized for MLP, as explained in Section 7.3.2.1. To compare it with the firefly training algorithm in MLP-FFA, several algorithms were tested in MLP, as seen in Table 7.9. It can be seen that during the testing phase trainrp had the highest r value and traincgf performed best for all other metrics. Hence a good choice would be to use traincgf for the final model and was used for all MLP models run. It will be seen in the results section that the same model architecture but except using the FFA as the training algorithm produced better results than the best MLP training algorithm (trainscg).

7.4.1.3 Comparative baseline models – random forest and multiple linear regression

RF and MLR also need to be developed with the optimization of parameters to select the optimal model, in order to compare to the optimal model for MLP-FFA. The lagged PAR inputs were also used to predict PAR. As described in Section 7.3.2.3, additionally to the train, test, and validation percentages, there are three parameters that need to be tested, as seen in Table 7.10. These are minimum number of observations per decision tree leaf (leaf), number of decision trees in the ensemble (ntrees), and fraction of in-bag observations (fboot). It can be seen in Table 7.10 that the optimal leaf observation is three during the testing phase. To avoid underfitting, leaf was trialled starting at 3 and ending at 20 to avoid overfitting. For the number of decision trees, all metrics were optimum at ntrees of

TABLE 7.10 RF model development, for comparison purposes using testing phase, based on E_{NS}, L, r, RMSE, and WI.

RF model parameters				Performance metric values					
Minimum number of observations per decision tree leaf	**Number of decision trees in the ensemble**	**Fraction of in-bag observations**	**Train/test/ validation split**	r	***RMSE*** **γ/(s m²)**	***WI***	E_{NS}	L	***MAE*** **γ/(s m²)**
3	800	1	60%/20%/20%	**0.9819**	**4.2249**	**0.9389**	**0.9675**	**0.7378**	**3.3426**
5	800	1	60%/20%/20%	0.9803	4.6615	0.9287	0.9633	0.7254	3.9701
10	800	1	60%/20%/20%	0.9807	4.2363	0.9385	0.9070	0.7362	3.3620
20	800	1	60%/20%/20%	0.0000	13.8917	0.0027	0.0000	0.0000	12.7469
3	50	1	60%/20%/20%	0.9763	4.2457	**0.9406**	0.9066	0.7317	3.4193
3	200	1	60%/20%/20%	0.9820	4.2426	0.9392	0.9067	0.7366	3.3576
3	800	1	60%/20%/20%	**0.9819**	**4.2249**	0.9389	**0.9675**	**0.7378**	**3.3426**
3	1600	1	60%/20%/20%	0.9807	4.2363	0.9385	0.9070	0.7362	3.3620
3	800	0.4	60%/20%/20%	NaN	13.8917	0.0048	0.0000	−0.0001	12.7474
3	800	0.8	60%/20%/20%	**0.9823**	5.1284	0.9033	0.8637	0.6649	4.2716
3	800	1	60%/20%/20%	0.9819	**4.2249**	**0.9389**	**0.9675**	**0.7378**	**3.3426**

The best values are shown in boldface. E_{NS}, Nash–Sutcliffe efficiency; L, Legates–McCabe's index; r, Pearson's correlation coefficient; *RMSE*, root mean square error index; *WI*, Willmott's index.

800, apart from *WI*, hence generally it can be concluded that this is the best value for the model. fboot was optimal for all metrics except r at a value of 1, so again can be considered the optimal value. The train split was kept at 60% train, 20% test, and 20% validation for RF, to allow it to be comparable with MLP-FFA running the models on the same datasets. As long as all other parameter values are constant (no matter what value they are), the same patterns should still be shown in terms of evaluation the parameter being tested.

For MLR, the train, test, and validation data partitions were the only factor considered during the model development stage, as there aren't parameters that require extensive trialling for this model. This is because all it does is use the input values to predict the objective PAR variable using a simple statistical regression equation; there is no training where parameters need to be chosen for the best training. For this study, the same partition will be used as MLP-FFA to allow it to be comparable, that is, 60% train, 20% test, and 20% validation. The MLR model will be run for the data in each of these partitions separately.

7.4.2 Model comparisons

In this chapter, the results obtained for evaluating an MLP-FFA model coupled with remote sensing data for forecasting of monthly PAR in a regional Queensland location are also presented. The MLP-FFA model is compared with respect to the performance of the MLP, RF, and MLR models to find whether the MLP-FFA model is able to accomplish the highest level of performance accuracy with the lowest error. In this section, the results of monthly PAR forecasting based on error, and the statistical performance metrics described in Eqs. 7.1–7.7 will be presented and their respective values assessed.

In order to compare directly the forecasted and observed monthly PAR, Fig. 7.9A–C plots the model error (the absolute value of $PAR_{forecasted}-PAR_{observed}$) for each tested month in testing (also known as cross-validation) period (March 2014 to February 2017) for MLP-FFA error compared to each other benchmark model, namely MLP, RF, and MLR. A scatterplot also comparing the models in a similar fashion, but for forecasted values comparing to observed values are also included. There is compelling evidence that MLP-FFA outperforms both the MLP and RF models for all months included in cross-validation. On a month-by-month basis, the forecasting error is of a noticeable larger magnitude for both MLP and RF. The MLP-FFA model has relatively lower error overall, with errors for either MLP or RF at times being double that for the same month for MLP-FFA. It can also be seen that generally, MLP seems to produce more errors in forecasting than RF. MLR has some higher errors, but also some lower errors than MLP-FFA, but not very distinct overall. MLR is still able to give very good results in some cases, as the data is very clean with a distinct pattern, hence it is easy to predict well with a good model. Hence, for future work it would be beneficial to use MLP-FFA on less patterned data, on which it is estimated that MLP-FFA will perform comparatively much better than MLR, as it is a more robust model, as discussed in Section 7.3.6. MLR, which is a deterministic linear statistical model, is not viable to be used in real application for PAR (which is nonlinear). The MLR model makes linear relations to nonlinear data, thus it is not reliable and has probably overfitted with the linear relationship. MLR was only used as a benchmark in this study. Despite this, the modeling in this project with the data acquired still is enough to show that MLP-FFA does better than MLP, RF, and MLR in most cases, hence overall overperforming them. This is still a valid conclusion and outcome of this research. A similar result is seen with the values, whereby MLP-FFA is more closely associated to the observed values than MLP and RF. The differentiation between MLP-FFA and MLR is again not as clear for the same reasons. It is unusual though for MLR to be performing better than such robust models such as MLP and RF, as well as MLP-FFA, so MLR could be a possible case of overfitting to the noise in the data, instead of more to the data pattern (Hawkins, 2004). Hence it may not be able to be generalized to more stochastic data (Hawkins, 2004), which is an important aim in model development (Hawkins, 2004).

A boxplot of model forecasting error for all four models is shown in Fig. 7.10 for a comparative view. In each boxplot, the outliers (indicated by +) represents the extreme values of forecasting error within the cross-validation set and the upper quartile, median, and lower quartile values are also indicated. The boxplot provides justification that the errors for the MLP-FFA model are over a much smaller range than for MLP and RF, as well as smaller overall for MLR. MLP-FFA correspondingly has a smaller magnitude of all quartile statistics than RF and MLP and has smaller maximum and minimum (disregarding outliers) than MLR. The trend in the monthly forecasting error values having larger magnitudes for MLP, RF, and MLR than for MLP-FFA are consistent with previous results. The MLP-FFA model is performing very well in terms of distribution toward smaller

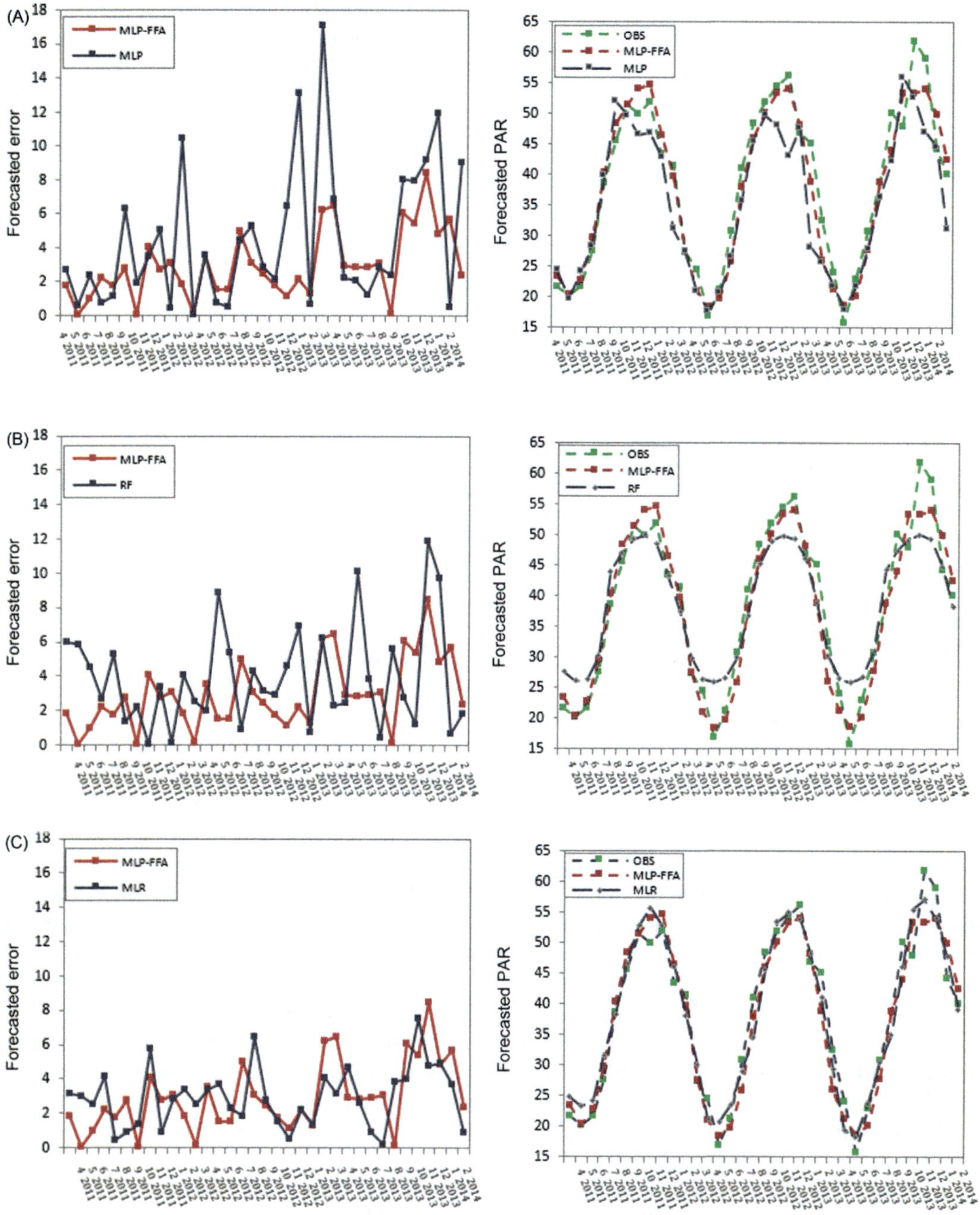

FIGURE 7.9 Times series of monthly PAR forecasting error (*left*) and plot of forecasted and observed PAR (*right*) during cross-validation (testing) phase, all in $\gamma/(\text{s m}^2)$. (A) MLP-FFA (objective model) and MLP (baseline model), (B) MLP-FFA and RF, (C) MLP-FFA and MLR.

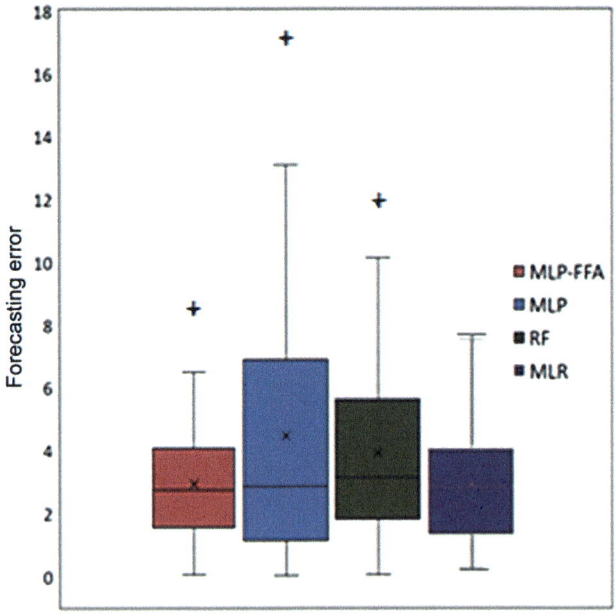

FIGURE 7.10 Boxplot of distribution of monthly forecasted error $\gamma/(s\,m^2)$ generated from MLP-FFA, MLP, RF, and MLR model during test phase. Mean is indicated by the x and median in the line inside the box.

magnitudes, and also the MLP model can be seen to be performing better than RF and also MLR in terms of upper error values (including maximum), in the boxplots.

In agreement with the monthly PAR forecasting previous results, Fig. 7.11A–D shows scatterplots of the forecasted values against observed PAR for all four models, including a goodness-of-fit regression line, $y = mx + C$ based on observed and forecasted data and the r^2 correlation coefficient. Note that r^2 and gradient (m) close to 1.00 and intercept (C) close to 0 are indicative of a perfect forecasting model. In this respect, MLP-FFA has performed better than the three benchmark models, including MLR, with the highest gradient and lowest intercept of $m = 0.9633$ and $C = 0.7808$ for MLP-FFA, compared to MLP having $m = 0.8219$ and $C = 3.8283$, RF having $m = 0.6901$ and $C = 11.949$, and MLR having $m = 0.9332$ and $C = 2.5695$, thus showing a better association between observed and forecasted values. Despite some degree of scatter between the observed and forecasted PAR data, a reasonably linear fit is evident for MLP-FFA, with differing levels of accuracy across the models. Compared to RF and MLP, MLP-FFA has a tighter fit of datum points to the goodness-of-fit regression line. MLP-FFA has a similar fit to the line of datum points to MLR, again due to nonstochastic data allowing for easy predictions, although gradient and intercept are still noticeable worse for MLR. Hence, the scatterplot shows that overall MLP-FFA has forecasted relatively well.

In Table 7.11 the forecasted models using normalized error and association metrics in terms of correlation coefficient, Willmott's index, Nash–Sutcliffe efficiency, Legate–McCabe's index, *RMSE*, and *MAE* are evaluated for all four models during the testing phase. For model development, the test is the final stage where the optimal model is finalized. All metrics are assessed in the testing phase as a result, to ensure the optimal

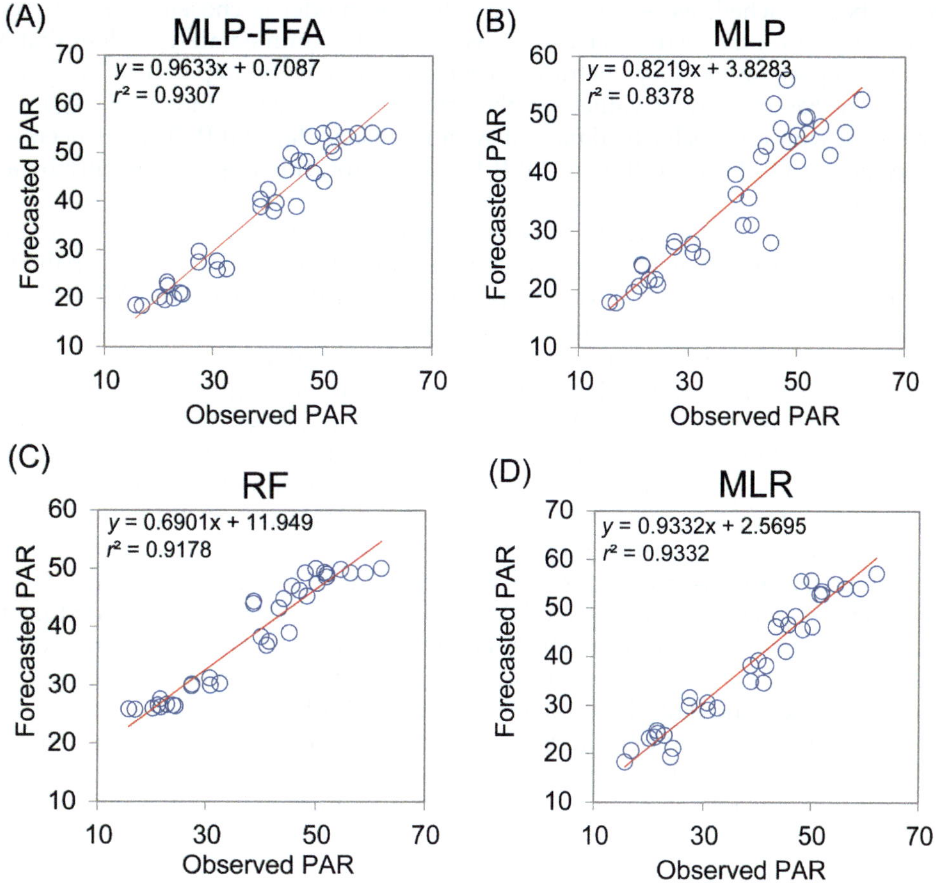

FIGURE 7.11 Scatterplot of MLP-FFA, MLP, RF, and MLR model observed against forecasted PAR $\gamma/(s\ m^2)$ during testing phase with a linear fit trendline showing its equation and r^2 value.

TABLE 7.11 Evaluation of monthly forecasting models, based on E_{NS}, L, MAE, r, RMSE, and WI.

Model	*r*	*WI*	*RMSE* $\gamma/(s\ m^2)$	*MAE* $\gamma/(s\ m^2)$
MLP-FFA	0.9825	0.9811	2.7950	2.1413
MLP	0.9759	0.9742	3.2111	2.424
RF	0.9819	0.9389	4.2249	3.3426
MLR	0.9675	0.9647	2.3872	1.9335

E_{NS}, Nash–Sutcliffe efficiency; *L*, Legates–McCabe's index; *MAE*, mean absolute error during testing phase; r = Pearson's correlation coefficient; *RMSE*, root mean square error; *WI*, Willmott's index.

model has been reached, as testing is where the best model is chosen. Hence to look at performance metrics, the testing set will be used here. It is immediately evident that MLP-FFA has the highest association metrics in terms of *r* and *WI*. Compared to RF and MLR, MLP-FFA also has the lowest values for the error metrics, *RMSE* and *MAE*. Again, MLR performs well with nonstochastic data, but with stochastic data MLP-FFA has been shown to be very powerful. But even though MLR has less error, this less error is only by a relatively small amount.

Fig. 7.12A displays scatter graphs of performance association metrics and Fig. 7.12B displays bar graphs of error metrics to evaluate the monthly PAR forecasting capabilities of the four models in a visual form. As compared with MLP, RF, and even MLR, there is a superior result with MLP-FFA having highest values for all four association metrics. MLP-FFA also has lowest values for the error metrics by a noticeable amount compared to MLP and RF, although MLR is slightly lower, but not by a very large amount. Hence, MLP-FFA is a more powerful model than MLP and RF, which agrees with papers from the Literature Review. Since the purpose here is comparing models, not finalizing the optimal model (like in the test phase), only the two association metrics with the highest values and the two error metrics are shown. MLP-FFA performs better than MLR for association metrics, which also agrees with literature.

Concluding this results section, it can be seen that MLP-FFA outperforms MLP and RF—two very popular and robust models used widely in papers for forecasting tasks. MLR has been seen to overperform MLP-FFA in a lot of cases, although there were some cases also where MLP-FFA overperformed. For MLR, data that is more difficult to forecast (such as more stochastic daily PAR data) may lead to MLP-FFA outperforming, as it is a more robust model (as mentioned in Section 7.3 and Section 7.2, showing that MLP-FFA, as well as the standalone MLP even outperforms MLR). MLR is not a strong model, so it is estimated that for more stochastic data it will not perform well. Unfortunately, daily or

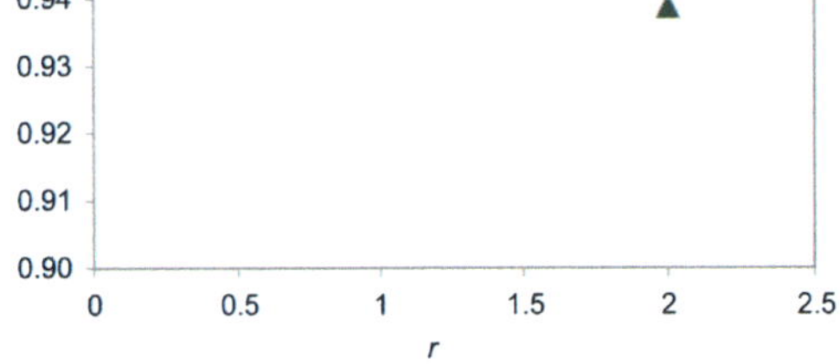

FIGURE 7.12 Performance evaluation metrics to compare MLP-FFA, MLP, RF, and MLR in testing phase using (A) association metrics: E_{NS}, Nash–Sutcliffe efficiency; *L*, Legates–McCabe's index and (B) error metrics (in $r/s/m^2$); *MAE*, mean absolute error; *r*, Pearson's correlation coefficient; *RMSE*, root mean square error; *WI*, Willmott's index.

hourly PAR data was not available without sparsity, so for future work manual PAR daily and hourly estimation is suggested, which was outside the scope of this project. MLR tends to overfit and fit to the random noise in the data. So generally, it can be considered that the MLP-FFA model is a very powerful model that has been shown to outperform even other robust models like MLP and RF.

7.5 Discussion

To help in the bid to battle the climate change, which will lead to environmental, health, and economic issues, more sustainable energy resources should be utilized, adhering to the nationwide attention to solar energy investments, consistent with global trends (Deo and Şahin, 2017). A desire to attain predictive models incorporated with remotely sensed data for forecasting global solar radiation is driving a lot of research momentum among scientists, who are exploring satellite data-coupled predictive models over deterministic models with ground-based (on-site) data (Deo and Şahin, 2017). The precise aim of this project was hence to model PAR using remote sensing data from a Toowoomba site in Queensland, Australia using moderate-resolution imaging spectroradiometer (MODIS), and a novel approach with data-intelligent (artificial intelligence) models for PAR forecasting. Incorporating data from merely 14.5 years (which was all that was available), still showed that a well-performing MLP-FFA model can be developed; and using only 35 months of data (for each of cross-validation and testing using a 60% train, 20% cross-validation, and 20% testing split) in the results section it is validated that MLP-FFA can outperform other robust models, including MLP and RF, with even this amount of historical PAR data as inputs. The fact that the MLP-FFA model is coupled with satellite-derived PAR data, is itself an advancement, as opposed to traditional deterministic models that fully rely on ground-based variables for forecasting. This study, which can be adapted to any location where a satellite footprint is available, is an advancement over Typical Meteorological Year (TMY) datasets that provide quantitative estimates of solar radiation mostly limited to certain ground-based measurement stations (Deo and Şahin, 2017).

In terms of the MLR model as compared to MLP-FFA, it was found that this weaker model seemed to outperform MLP-FFA in most cases, but only by a small amount. This is because the monthly PAR data used was not stochastic at all, and had a clear monthly pattern according to respective seasons. Even a poor model has the capability to predict non-stochastic data, which is very easy to predict, but to show the real robustness of a strong model like MLP-FFA, it is suggested that in future work more stochastic PAR data, such as daily or hourly, should be forecasted to show MLP-FFA's capabilities by giving it a hard dataset to predict, in which it is estimated MLR will perform worse. Unfortunately, PAR data at more specific temporal resolutions were not available from the website used (GIOVANNI). Another website, called ECMWF (which is an ERA-interim reanalysis data platform by the European Centre for Medium Range Weather Forecasts), was originally proposed to be utilized as they had more temporal and spatial resolutions and data from 1979 up to the current date, but for PAR data it gave several missing values, to the extent that the data was nonusable for the purpose. As of yet another open source satellite data of smaller temporal resolution PAR has not been found, and manual ground-based collections exceeded the time and cost limits of this study. But for future work hours should be investigated also as real-time forecasting requires this. Papers in Section 7.2 also seem to claim that satellite-derived data is hard to obtain for environmental variables. Additionally, another solution for future work would be to investigate if other subtropical Queensland locations have better PAR data available for smaller temporal resolutions. It was actually attempted to obtain PAR data also for other locations, but yet these gave very sparse data. Even though literature states that AI models are universal as opposed to physical model (Metzen, 2018), this could be further confirmed if more sites were modeled (Deo and Şahin, 2017). Another suggestion for future work would be to investigate how well the MLP-FFA model can forecast PAR for each season bucketed into a total PAR group, this would be appropriate to estimate the best seasons for growth of resources (such as algal biofuels) during the respective seasons where PAR is highest. Perhaps, even other imaging techniques (apart from MODIS) could be evaluated.

Despite these limitations in this research work, a very significant novel contribution has been produced, which shows that MLP-FFA has the capability to predict PAR (which has not been predicted using AI models), better than other very capable AI models (MLP and RF), using historical lagged PAR data as model inputs in a subtropical regional Queensland location, which has the potential to sustain algal biofuel farms. This affirms the research purpose which was to see if an optimized hybrid AI algorithm is effectively able to predict PAR better than the current methods. This result, which answers the original aims of this study, is able to overcome the research problem in which for a depleting coal fuel resource, no optimized AI models have been formed to predict growth conditions for an alternative renewable biofuel source, in order to enhance its growth and sustainability. PAR as a growth requirement for photosynthetic algae as a biofuel will definitely aid in tackling this issue if modeled using the best techniques that agree with the current data revolution era.

For the future project, our next steps will be to test MLP-FFA with RF-FFA and other models with FFA (yet to be decided which models). This was not the aim of this project which was to develop an MLP-FFA model and see if it (and an FFA optimizer hybrid with a very well-known algorithm) is able to well predict PAR first. Now that it is known that it can, the aim of the next project will be to test several models, to see which works best with FFA, now that it is known that FFA gives the ability to improve the performance as compared to standalone models.

7.6 Conclusion

This study established the preciseness of the robust MLP–firefly optimizer (MLP-FFA) artificial intelligence algorithm coupled with satellite-derived photosynthetic active radiation (PAR) data in order to forecast PAR itself using historical values for a regional Queensland location (Toowoomba). To optimize the MLP-FFA model, (1) hidden and output transfer functions; (2) number of hidden neurons; (3) training, cross-validation, and test percentage splits; and (4) number of historical lags were trialled, such that the logarithmic sigmoid hidden function and tangent sigmoid output function, with 120 hidden neurons, nine lags as inputs and a 60% training, 20% cross-validation, and 20% testing set, was finally adopted for thre forecasting of PAR. To meet the objective (Section 7.1), the MLP-FFA objective model was benchmarked with the MLP, RF, and MLR models for PAR forecasting. The findings of this study are that the performance of the MLP-FFA model, according to forecasting error directly, as well as association and error performance metrics, outperformed the MLP and RF models. MLR generally outperformed MLP-FFA, but this is most likely a result of nonstochastic monthly PAR data, and it is recommended for future work to investigate lower temporal resolution data (which in turn is more stochastic), which a more robust model like MLP-FFA is known to handle, but a poor model like MLR cannot. A lower resolution such as daily or hourly was not found at the time of this study, and manual data collection was not appropriate due to the time and cost limits of this research. Albeit, the MLP-FFA model has still been very effective in modeling PAR with very high accuracy and low error, leading to a significant contribution to research which is confirming something unknown—that the MLP-FFA model is in fact very effective at satellite-based PAR modeling with historical data as inputs for learning for a regional Queensland location. The results of this study are a significant research contribution, which can be used to forecast PAR conditions, a vital requirement for the growth of algal biofuel; these algae can be farmed in the ideal location of the sunny, subtropical Toowoomba region.

References

Abbot, J., Marohasy, J., 2012. Application of artificial neural networks to rainfall forecasting in Queensland, Australia. Adv. Atmos. Sci. 29, 717–730.

Adamowski, J., Fung Chan, H., Prasher, S.O., Ozga-Zielinski, B., Sliusarieva, A., 2012. Comparison of multiple linear and nonlinear regression, autoregressive integrated moving average, artificial neural network, and wavelet artificial neural network methods for urban water demand forecasting in Montreal, Canada. Water Resour. Res. 48.

Alados, I., Olmo, F., Foyo-Moreno, I., Alados-Arboledas, L., 2000. Estimation of photosynthetically active radiation under cloudy conditions. Agric. For. Meteorol. 102, 39–50.

Ali, M., Deo, R.C., Downs, N.J., Maraseni, T., 2018. Multi-stage committee based extreme learning machine model incorporating the influence of climate parameters and seasonality on drought forecasting. Comput. Electron. Agri. 152, 149–165.

Ali, S., Smith, K.A., 2006. On learning algorithm selection for classification. Appl. Soft Comput. 6, 119–138.

Al-lwayzy, S.H., Yusaf, T., Al-Juboori, R.A., 2014. Biofuels from the fresh water microalgae Chlorella vulgaris (FWM-CV) for diesel engines. Energies 7, 1829–1851.

Anusree, K., Binu, G., 2014. Analysis of training functions in a biometric system. Int. J. Recent. Innov. Trends Comput. Commun. 2, 150–154.

Beath, A.C., 2012. Industrial energy usage in Australia and the potential for implementation of solar thermal heat and power. Energy 43, 261–272.

Bioenergy Australia. (Ed.), 2016a. How Is Ethanol Made?.

Bioenergy Australia. (Ed.), 2016b. Mandate Remains Key to Biodiesel Industry.

Bioenergy Australia. (Ed.), 2016c. What Engines Can Use Ethanol?, Australia.

Bowden, G.J., Maier, H.R., Dandy, G.C., 2005. Input determination for neural network models in water resources applications. Part 2. Case study: forecasting salinity in a river. J. Hydrol. 301, 93–107.

Breiman, L., 1996. Bagging predictors. Mach. Learn. 24, 123–140.

Breiman, L., 2001. Random forests. Mach. Learn. 45, 5–32.

Business Queensland, 2018, Queensland Government, Australia (2018), Available online at <https://www.business.qld.gov.au/industries/manufacturing-retail/retail-wholesale/selling-fuel-qld/qld-biofuels-mandates>.

Chai, T., Draxler, R.R., 2014. Root mean square error (RMSE) or mean absolute error (MAE)?–Arguments against avoiding RMSE in the literature. Geosci. Model. Dev. 7, 1247–1250.

Choubin, B., Khalighi-Sigaroodi, S., Malekian, A., Kişi, Ö., 2016. Multiple linear regression, multi-layer perceptron network and adaptive neuro-fuzzy inference system for forecasting precipitation based on large-scale climate signals. Hydrolog. Sci. J. 61, 1001–1009.

Civelekoglu, G., Yigit, N., Diamadopoulos, E., Kitis, M., 2007. Prediction of bromate formation using multi-linear regression and artificial neural networks. Ozone: Sci. Eng. 29, 353–362.

Clements, M.P., Galvão, A.B., 2009. Forecasting US output growth using leading indicators: an appraisal using MIDAS models. J. Appl. Econom. 24, 1187–1206.

Cochran, M., 2017. Australian biofuels 2017 industry overview and developments. Asian Pacific Fuel Industry Forum. APAC Biofuel Consultants, Melbourne Convention Centre, Melbourne.

Dennis Jr, J.E., Schnabel, R.B., 1996. Numerical Methods for Unconstrained Optimization and Nonlinear Equations. Siam.

Deo, R.C., Şahin, M., 2015. Application of the artificial neural network model for prediction of monthly standardized precipitation and evapotranspiration index using hydrometeorological parameters and climate indices in eastern Australia. Atmos. Res. 161, 65–81.

Deo, R.C., Şahin, M., 2017. Forecasting long-term global solar radiation with an ANN algorithm coupled with satellite-derived (MODIS) land surface temperature (LST) for regional locations in Queensland. Renew. Sustain. Energy Rev. 72, 828–848.

Deo, R.C., Wen, X., Qi, F., 2016. A wavelet-coupled support vector machine model for forecasting global incident solar radiation using limited meteorological dataset. Appl. Energy 168, 568–593.

Deo, R.C., et al., 2018. Multi-layer perceptron hybrid model integrated with the firefly optimizer algorithm for windspeed prediction of target site using a limited set of neighboring reference station data. Renew. Energy 116, 309–323.

Duong, V.T., Thomas-Hall, S.R., Schenk, P.M., 2015. Growth and lipid accumulation of microalgae from fluctuating brackish and sea water locations in South East Queensland—Australia. Front. Plant. Sci. 6, 359.

Fletcher, R., Reeves, C.M., 1964. Function minimization by conjugate gradients. Comput. J. 7, 149–154.

Geoscience Australia, 2018. ABARE, Australian Energy Resource Assessment. Geoscience Australia and ABARE, Canberra, Australia.

Ghorbani, M., Khatibi, R., Hosseini, B., Bilgili, M., 2013. Relative importance of parameters affecting wind speed prediction using artificial neural networks. Theor. Appl. Climatol. 114, 107–114.

Ghorbani, M., Deo, R.C., Yaseen, Z.M., Kashani, M.H., Mohammadi, B., 2017a. Pan evaporation prediction using a hybrid multilayer perceptron-firefly algorithm (MLP-FFA) model: case study in North Iran. Theor. Appl. Climatol. 1–13.

Ghorbani, M.A., Deo, R.C., Karimi, V., Yaseen, Z.M., Terzi, O., 2017b. Implementation of a hybrid MLP-FFA model for water level prediction of Lake Egirdir, Turkey. Stoch. Environ. Res. Risk Assess. 1–15.

GIOVANNI, 2018. Earth data. NASA, United States. Available online at <https://giovanni.gsfc.nasa.gov/giovanni/>.

Goudarzi, S., et al., 2016. A novel RSSI prediction using imperialist competition algorithm (ICA), radial basis function (RBF) and firefly algorithm (FFA) in wireless networks. PLoS One 11, e0151355.

HariKumar, R., Vasanthi, N., Balasubramani, M., 2009. Performance analysis of artificial neural networks and statistical methods in classification of oral and breast cancer stages. Int. J. Soft Comput. Eng. (IJSCE) Volume 2.

Harris, J., 2010. Science thursday. WRMA, Australia (2010). Available online at <http://hopeful-ink.blogspot.com.au/2010/09/science-thursday.html>.

Hawkins, D.M., 2004. The problem of overfitting. J. Chem. Inf. Comput. Sci. 44, 1–12.

Hestenes, M.R., Stiefel, E., 1952. Methods of Conjugate Gradients for Solving Linear Systems. NBS, Washington, DC.

Huang, H.-Y., 1970. Unified approach to quadratically convergent algorithms for function minimization. J. Optim. Theory Appl. 5, 405–423.

John, A., 2018. Australian non-renewable energy resources. In: Parliament of Australia: Science, Technology, Environment and Resources, Australia, 2015. Available online at <https://www.aph.gov.au/About_Parliament/Parliamentary_Departments/Parliamentary_Library/pubs/BriefingBook44p/EnergyResources>.

Katter, H., 2015. Towards a clean energy economy: Achieving a biofuel mandate for Queensland. Google Scholar.

Kavousi-Fard, A., Niknam, T., Golmaryami, M., 2014. Short term load forecasting of distribution systems by a new hybrid modified FA-backpropagation method. J. Intell. Fuzzy Syst. 26, 517–522.

Kayarvizhy, N., Kanmani, S., Uthariaraj, R., 2014. ANN models optimized using swarm intelligence algorithms. WSEAS Trans. Comput. 13, 501–519.

Khatibi, R., Ghorbani, M., Pourhosseini, F.A., 2017. Stream flow predictions using nature-inspired Firefly Algorithms and a Multiple Model strategy–directions of innovation towards next generation practices. Adv. Eng. Inform. 34, 80–89.

Kim, T.-W., Valdés, J.B., 2003. Nonlinear model for drought forecasting based on a conjunction of wavelet transforms and neural networks. J. Hydrol. Eng. 8, 319–328.

Liang, S., et al., 2006. Estimation of incident photosynthetically active radiation from Moderate Resolution Imaging Spectrometer data. J. Geophys. Res. Atmos. 111.

Litterman, R.B., 1986. Forecasting with Bayesian vector autoregressions—five years of experience. J. Bus. Econ. Stat. 4, 25–38.

Liu, H., 2017. Which machine learning algorithm should I use? In: SAS (Ed.), Subconscious Musing: Advanced Analytics | Machine Learning. SAS, USA.

Lopez, G., Rubio, M., Martınez, M., Batlles, F., 2001. Estimation of hourly global photosynthetically active radiation using artificial neural network models. Agric. For. Meteorol. 107, 279–291.

Luk, K., Ball, J.E., Sharma, A., 2000. A study of optimal model lag and spatial inputs to artificial neural network for rainfall forecasting. J. Hydrol. 227, 56–65.

Łukasik, S., Żak, S., 2009. Firefly algorithm for continuous constrained optimization tasks. International Conference on Computational Collective Intelligence. Springer, pp. 97–106.

Marquardt, D.W., 1963. An algorithm for least-squares estimation of nonlinear parameters. J. Soc. Ind. Appl. Math. 11, 431–441.

Martin, N.J., Rice, J.L., 2012. Developing renewable energy supply in Queensland, Australia: A study of the barriers, targets, policies and actions, Renew. Energy, 44. pp. 119–127.

Mashaly, A.F., Alazba, A., 2016. MLP and MLR models for instantaneous thermal efficiency prediction of solar still under hyper-arid environment. Comput. Electron. Agri. 122, 146–155.

Mason, L., Mikhailovich, N., Sharpe, S., Mudd, G.M., Giurco, D., 2013. Advantage Australia: Resource Governance and Innovation for the Asian Century. Institute for Sustainable Futures at University of Technology, Sydney.

McCulloch, W.S., Pitts, W., 1943. A logical calculus of the ideas immanent in nervous activity. Bull. Math. Biophys. 5, 115–133.

McClelland, J.L., Rumelhart, D.E., 1989. Explorations in Parallel Distributed Processing: A Handbook of Models, Programs, and Exercises. MIT Press.

McNamara, A., 2008. In: Queensland, T.Ts (Ed.), Towards Oil Resilience Community Information Paper. Queensland Government.

Metzen, J.H., 2018. Universality, Robustness, and Detectability of Adversarial Perturbations under Adversarial Training.

Mittnik, S., 1990. Forecasting with balanced state space representations of multivariate distributed lag models. J. Forecast. 9, 207–218.

Mouatadid, S., Deo, R.C., Adamowski, J.F., 2015. Prediction of SPEI using MLR and ANN: A case study for Wilsons Promontory Station in Victoria. Machine Learning and Applications (ICMLA), 2015 IEEE14th International Conference on. IEEE, pp. 318–324.

Özdamar, K., 2013. Statistical Data Analysis with Package Software. April Publishing House, Ankara.

Pittman, J.K., Dean, A.P., Osundeko, O., 2011. The potential of sustainable algal biofuel production using wastewater resources. Bioresour. Technol. 102, 17–25.

Prasad, P., Pullar, D., Pratt, S., 2014. Facilitating access to the algal economy: mapping waste resources to identify suitable locations for algal farms in Queensland. Resour. Conserv. Recycl. 86, 47–52.

Prasad, R., Deo, R.C., Li, Y., Maraseni, T., 2018. Soil moisture forecasting by a hybrid machine learning technique: ELM integrated with ensemble empirical mode decomposition. Geoderma 330, 136–161.

Qin, J., Yang, K., Liang, S., Tang, W., 2012. Estimation of daily mean photosynthetically active radiation under all-sky conditions based on relative sunshine data. J. Appl. Meteorol. Climatol. 51, 150–160.

Queensland Biofutures, 2016. Queensland Biofutures 10-Year Roadmap and Action Plan Quuensland Government.
Raheli, B., Aalami, M.T., El-Shafie, A., Ghorbani, M.A., Deo, R.C., 2017. Uncertainty assessment of the multilayerperceptron (MLP) neural network model with implementation of the novel hybrid MLP-FFA method for prediction of biochemical oxygen demand and dissolved oxygen: a case study of LangatRiver. Environ. Earth Sci. 76, 503.
Riedmiller, M., Braun, H., 1993. A direct adaptive method for faster backpropagation learning: The RPROP algorithm. Neural Networks, 1993, IEEE International Conference on. IEEE, pp. 586–591.
Rosati, A., Metcalf, S., Lampinen, B., 2004. A simple method to estimate photosynthetic radiation use efficiency of canopies. Ann. Bot. 93, 567–574.
Şahin, M., Kaya, Y., Uyar, M., 2013. Comparison of ANN and MLR models for estimating solar radiation in Turkey using NOAA/AVHRR data. Adv. Space Res. 51, 891–904.
Şahin, M., Kaya, Y., Uyar, M., Yıldırım, S., 2014. Application of extreme learning machine for estimating solar radiation from satellite data. Int. J. Energy Res. 38, 205–212.
Santamouris, M., Mihalakakou, G., Psiloglou, B., Eftaxias, G., Asimakopoulos, D., 1999. Modeling the global solar radiation on the earth's surface using atmospheric deterministic and intelligent data-driven techniques. J. Clim. 12, 3105–3116.
Schmidt, G.A., 2019. Cloud climatology. In: Studies, N.A.a.S.A.G.I.f.S. (Ed)., National Aeronautics and Space Administration Goddard Institute for Space Studies, USA.
Shurin, J.B., et al., 2013. Industrial-strength ecology: trade-offs and opportunities in algal biofuel production. Ecol. Lett. 16, 1393–1404.
Solangi, K., Islam, M., Saidur, R., Rahim, N., Fayaz, H., 2011. A review on global solar energy policy. Renew. Sustain. Energy Rev. 15, 2149–2163.
Soleymani, S.A., et al., 2016. A novel method to water level prediction using RBF and FFA. Water Resour. Manag. 30, 3265–3283.
Srivastava, N., Hinton, G., Krizhevsky, A., Sutskever, I., Salakhutdinov, R., 2014. Dropout: a simple way to prevent neural networks from overfitting. J. Mach. Learn. Res. 15, 1929–1958.
Sudhakar, K., Srivastava, T., Satpathy, G., Premalatha, M., 2013. Modelling and estimation of photosynthetically active incident radiation based on global irradiance in Indian latitudes. Int. J. Energy Environ. Eng. 4, 21.
Troccoli, A., 2015. Integrated Solar Radiation Data Sources over Australia Final Report: Project Results and Lessons Learnt. *CSIRO*, p. 37.
Vindel, J., et al., 2018. Modeling photosynthetically active radiation from satellite-derived estimations over Mainland Spain. Remote. Sens. 10, 849.
Vogl, T.P., Mangis, J., Rigler, A., Zink, W., Alkon, D., 1988. Accelerating the convergence of the back-propagation method. Biol. Cybern. 59, 257–263.
Wagner, H.M., 1959. Linear programming techniques for regression analysis. J. Am. Stat. Assoc. 54, 206–212.
Wang, L., et al., 2015. Estimation of hourly and daily photosynthetically active radiation in Inner Mongolia, China, from 1990 to 2012. Int. J. Climatol. 35, 3120–3131.
Wang, L., Kisi, O., Zounemat-Kermani, M., Hu, B., Gong, W., 2016. Modeling and comparison of hourly photosynthetically active radiation in different ecosystems. Renew. Sustain. Energy Rev. 56, 436–453.
Yang, X.-S., 2010a. Firefly algorithm, stochastic test functions and design optimisation. arXiv preprint arXiv:1003.1409.
Yang, X.-S., 2010b. Nature-Inspired Metaheuristic Algorithms. Luniver Press.
Yaseen, Z.M., et al., 2018. Rainfall pattern forecasting using novel hybrid intelligent model based ANFIS-FFA. Water Resour. Manag. 32, 105–122.
Yu, X., Guo, X., 2016. Hourly photosynthetically active radiation estimation in Midwestern United States from artificial neural networks and conventional regressions models. Int. J. Biometeorol. 60, 1247–1259.
Yusaf, T., Goh, S., Borserio, J., 2011. Potential of renewable energy alternatives in Australia. Renew. Sustain. Energy Rev. 15, 2214–2221.
Zempila, M.-M., Taylor, M., Bais, A., Kazadzis, S., 2016. Modeling the relationship between photosynthetically active radiation and global horizontal irradiance using singular spectrum analysis. J. Quant. Spectrosc. Radiat. Transf. 182, 240–263.
Zhu, Z., et al., 2013. Global data sets of vegetation leaf area index (LAI) 3g and fraction of photosynthetically active radiation (FPAR) 3g derived from global inventory modeling and mapping studies (GIMMS) normalized difference vegetation index (NDVI3g) for the period 1981 to 2011. Remote. Sens. 5, 927–948.

Further reading

Bai, J., 2013. Photosynthetically active radiation loss in the atmosphere in North China. Atmos. Pollut. Res. 4, 411–419.

Bentzen, J., Engsted, T., 2001. A revival of the autoregressive distributed lag model in estimating energy demand relationships. Energy 26, 45–55.

Bernard, O., 2011. Hurdles and challenges for modelling and control of microalgae for CO2 mitigation and biofuel production. J. Process Control 21, 1378–1389.

Box, G.E., Jenkins, G.M., Reinsel, G.C., Ljung, G.M., 2015. Time Series Analysis: Forecasting and Control. John Wiley & Sons.

Clean Energy Australia, 2015. Clean Energy Australia Report 2014. Geoscience Australia, ABARE, Australian Energy Resource Assessment, Geoscience Australia and ABARE, Canberra, Australia, 2018.

Dawson, C.W., Abrahart, R.J., See, L.M., 2007. HydroTest: a web-based toolbox of evaluation metrics for the standardised assessment of hydrological forecasts. Environ. Modell. Software 22, 1034–1052.

ECMWF, 2017. ERA Interim, Daily. ECMWF, Europe, 2018. Available online at <http://apps.ecmwf.int/datasets/data/interim-full-daily/levtype = sfc/>.

Gordon, R., Seckbach, J., 2012. The Science of Algal Fuels: Phsycology, Geology, Biophotonics, Genomics and Nanotechnology. Springer Science & Business Media.

Holdren, J.P., 2006. The energy innovation imperative: addressing oil dependence, climate change, and other 21st century energy challenges. Innovat. Technol. Govern. Global. 1, 3–23.

Hu, B., Wang, Y., Liu, G., 2010. Long-term trends in photosynthetically active radiation in Beijing. Adv. Atmos. Sci. 27, 1380–1388.

Hughes, T.P., et al., 2003. Climate change, human impacts, and the resilience of coral reefs. Science 301, 929–933.

IPCC, 2007. Climate change 2007: Working group III: Mitigation of climate change. In: IPCC (Ed.), 4.3.3.3 Biomass and Bioenergy. Switzerland.

Jain, A., 2016. Fundamentals of deep learning – starting with artificial neural network. In: Vidhya, A. (Ed.), Analytics Vidhya. Analytics Vidhya, India.

Keskin, M.E., Terzi, Ö., 2006. Artificial neural network models of daily pan evaporation. J. Hydrol. Eng. 11, 65–70.

Legates, D.R., McCabe Jr, G.J., 1999. Evaluating the use of "goodness-of-fit" measures in hydrologic and hydroclimatic model validation. Water Resour. Res. 35, 233–241.

Markvart, T., 2000. Solar Electricity, 2nd ed. Wiley & Sons, Chichester; New York.

Meese, R.A., Rogoff, K., 1983. Empirical exchange rate models of the seventies: Do they fit out of sample? J. Int. Econom. 14, 3–24.

Mekanik, F., Imteaz, M., Gato-Trinidad, S., Elmahdi, A., 2013. Multiple regression and Artificial Neural network for long-term rainfall forecasting using large scale climate modes. J. Hydrol. 503, 11–21.

Menetrez, M.Y., 2012. An overview of algae biofuel production and potential environmental impact. Environ. Sci. Technol. 46, 7073–7085.

Nash, J.E., Sutcliffe, J.V., 1970. River flow forecasting through conceptual models part I—A discussion of principles. J. Hydrol. 10, 282–290.

Pankaew, P., Milton, E., Dash, J., 2014. Estimating hourly variation in photosynthetically active radiation across the UK using MSG SEVIRI data. IOP Conference Series: Earth and Environmental Science. IOP Publishing, p. 012069.

Queensland Bio-based Industrial Products Sector. 2010. Queensland Government, Queensland.

Queensland Biofutures. (Ed.), 2018. Queensland Biofuels Mandates. Queensland Government, Canberra.

Sedgwick, P., 2012. Pearson's correlation coefficient. BMJ Br. Med. J. 345.

Shachak, M., Steinberger, Y., 1980. An algae—desert snail food chain: energy flow and soil turnover. Oecologia 46, 402–411.

Shilwant, S., 2015a. Firefly algorithm. LinkedIn Learning: Engineering. Available online at <https://www.slideshare.net/supriyashilwant/firefly-algorithm-49723859>.

Shilwant, S., 2015b. Detail description on behavior of firefly optimization. Engineering, LinkedIn, India.

Stoddard, F.L., Mäkelä, P.S., Puhakainen, T., 2011. Adaptation of boreal field crop production to climate change. Climate Change-Research and Technology for Adaptation and Mitigation. InTech.

Tang, W., et al., 2013. Reconstruction of daily photosynthetically active radiation and its trends over China. J. Geophys. Res. Atmos. 118.

U.S. Department of Energy, 2017. Algal Biofuels. Department of Energy.

Wang, L., et al., 2013. Measurement and estimation of photosynthetically active radiation from 1961 to 2011 in Central China. Appl. Energy 111, 1010–1017.

Wei, W.W., 1978. The effect of temporal aggregation on parameter estimation in distributed lag model. J. Econom. 8, 237–246.

Willmott, C.J., 1981. On the validation of models. Phys. Geog. 2, 184–194.

Willmott, C.J., 1984. On the evaluation of model performance in physical geography. Spatial Statistics and Models. Springer, pp. 443–460.

Yu, X., Wu, Z., Jiang, W., Guo, X., 2015. Predicting daily photosynthetically active radiation from global solar radiation in the contiguous United States. Energy Convers. Manag. 89, 71–82.

Yu, H., et al., 2018. Comparative study of hybrid-wavelet artificial intelligence models for monthly groundwater depth forecasting in extreme arid regions, Northwest China. Water Resour. Manag. 32, 301–323.

Zhang, R., White, A.T., Pour Biazar, A., McNider, R.T., Cohan, D.S., 2018. Incorporating GOES satellite photosynthetically active radiation (PAR) retrievals to improve biogenic emission estimates in Texas. J. Geophys. Res. Atmos. 123, 1309–1324.

CHAPTER

8

Predictive modeling of oscillating plasma energy release for clean combustion engines

Xiao Yu, Qingyuan Tan, Linyan Wang, Meiping Wang and Ming Zheng

Department of Mechanical, Automotive & Materials Engineering, University of Windsor, Windsor, ON, Canada

8.1 Introduction

Future transportation will use versatile power sources from sustainable and renewable energy sources. Clean combustion engines can make major contribution to green house gas reduction in the transportation sector. Lean and diluted mixtures are effective to produce clean combustion efficiently (Takahashi et al., 2015; Turner et al., 2014; Gallon et al., 2013). Advanced combustion technology adopts lean and diluted mixtures to further improve engine efficiencies and reduce harmful emissions. Ignition difficulty and slow flame propagation speed become major challenges because of the much lower chemical reactivity of the air–fuel mixtures (Schumann et al. 2013; Huang et al., 2007).

Oscillating plasma discharge shows promise for the successful ignition of lean and diluted mixtures. As shown in Fig. 8.1, the streamers released from the plasma antenna can stretch beyond the size of the conventional spark gap and establish a much bigger ignition volume.

The high-frequency oscillating electric field generated at the tip of the antenna can force the distortion on the flame front. The enhancement of burning rate and acceleration of the flame kernel growth (Zheng and Yu, 2015) are contributed by the flame front wrinkles in the shadowgraph image. The shorter combustion duration contributes to further improvement of engine efficiency under lean/diluted conditions. The oscillating plasma system provides a quick response to the ignition command in microsecond order, compared with

Predictive Modelling for Energy Management and Power Systems Engineering
DOI: https://doi.org/10.1016/B978-0-12-817772-3.00008-2

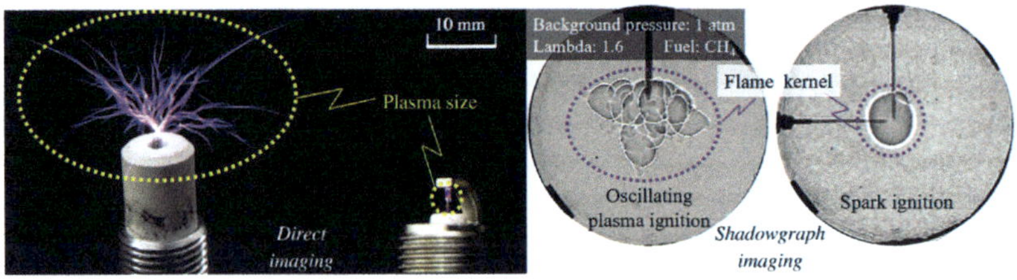

FIGURE 8.1 Plasma and traditional spark discharge and relevant flame kernel formation in shadowgraph images.

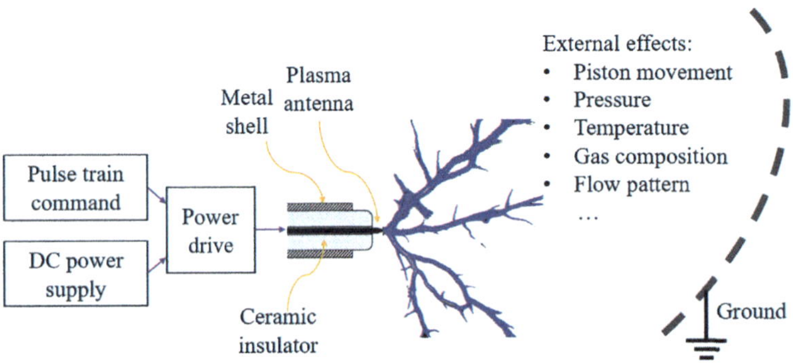

FIGURE 8.2 Basic structural scheme of a plasma ignition system.

millisecond order for a traditional transistor–coil–ignition system. The system can also continuously deliver ignition energy according to engine demands (Starikovskaia, 2006).

However, as the ground electrode is absent from the plasma antenna, the plasma ignition is susceptible to the surrounding conditions, because the dielectric medium around the plasma antenna becomes a part of the electrical circuit during energy release. An illustrative structural scheme of a basic plasma ignition system is shown in Fig. 8.2. Under engine conditions, the rapid change of in-cylinder pressure and temperature produces a complicated background condition, which challenges a stable plasma discharge (Fansler et al., 2015; Reinhart, 2016; National Highway Traffic Safety Administration, 2015). At the top dead center (TDC) of engine operation, the distance between the plasma antenna and the surface of the piston is greatly shortened, which can trigger unwanted arc discharge (Martin et al., 2016; Nakata et al., 2016). Contamination of the plasma antenna and its ceramic insulator can lead to surface discharge or direct arc to the metal shell of the plasma igniter (Steffen et al., 2014; Burrows et al., 2016; Wang, 2019). Therefore to ensure robust and reliable plasma discharge, understanding the discharge mechanism and operational boundaries of the plasma discharge process is crucial (Zheng and Yu, 2015).

A model that can describe the plasma discharge dynamics is important for understanding the plasma discharge process. The model will support the detection of abnormal (void and arc) discharge and thus provide possible strategies to overcome the discharge failure. For normal discharge events, the model is purposed to predict discharge energy profile, which is essential for combustion control and energy management.

In this work, the challenge of plasma discharge under engine-like conditions is discussed in detail. Subsequently, the plasma discharging process is characterized by electrical waveform measurements of the voltage and current. Plasma and its surrounding conditions are considered as a load of the oscillating circuit. By modeling the impedance change of the plasma discharge, the dynamics during the system operation are thus reviewed mathematically. Furthermore, the load impedance also provides an alternative parameter for the prediction of electrical energy that is supplied to the plasma discharge during system operation. Since the quantity of such electrical energy is closely related to the ignition and combustion performance, the prediction of the energy is critical for the control and optimization of the plasma ignition system.

8.2 Challenges of plasma discharge under engine conditions

As shown in Fig. 8.2, a typical plasma discharge system consists of a DC power supply, a power drive, and an antenna. A pulsed-train command with favorable oscillating frequency is applied to stimulate the plasma power drive, aiming to generate high voltage at the plasma antenna. Unlike traditional spark plugs utilizing a ground electrode to guide the direction of the plasma channel, the plasma ignition system releases the plasma streamers by ionizing the gaseous media around the antenna (Idicheria and Najt, 2016). The ionized surrounding media becomes a part of the discharging circuit, and plasma discharge is thus affected by the changes in background conditions (Singleton et al., 2010; Auzas et al., 2010). Fig. 8.3 lists potential factors that may affect the plasma discharge. These factors include the ignition system, background conditions, and external disturbances.

8.2.1 Ignition system impact

The oscillating frequency of the power drive can be adjusted by a pulse train command. However, the oscillating frequency needs to match the optimal working frequency of a system [resistor–inductor–capacitor (RLC) electrical resonance frequency] for plasma discharge. The optimal working frequency is determined by the hardware setup (Suess et al., 2012). Fig. 8.4 shows waveforms of secondary voltage and current with the impact of oscillating frequency on energy and secondary voltage. During the plasma oscillation, the amplitudes of secondary voltage and current waveforms increase initially until reaching the peak values. To sustain plasma discharging, the amplitudes decrease and stabilize subsequently. As shown in Fig. 8.4, the oscillating frequency needs to be tuned within a

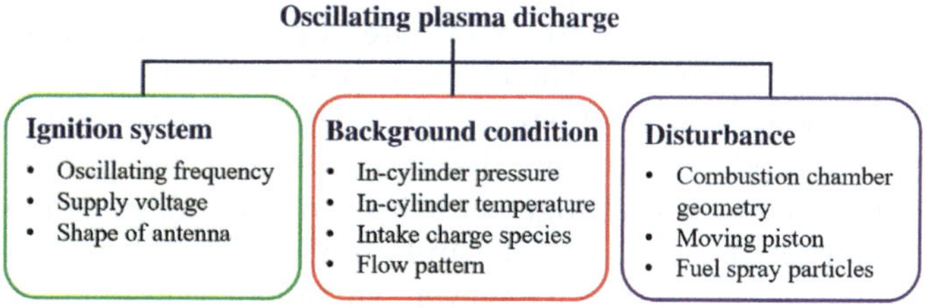

FIGURE 8.3 Factors affecting oscillating plasma discharge.

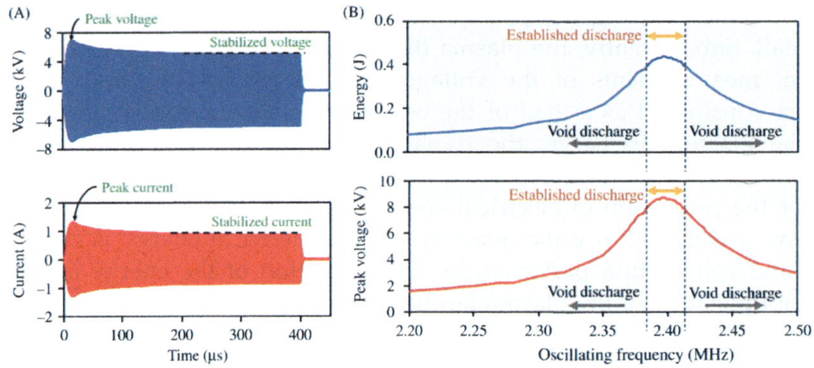

FIGURE 8.4 Secondary voltage and current measurements in time domain (A), discharge energy and peak voltage in frequency domain (B).

narrow bandwidth (~0.03 MHz) to generate sufficient secondary voltage to establish a plasma discharge. Therefore the secondary voltage is selected as the parameter to study the plasma status and modeling. Discharging bandwidth is affected by the background conditions and disturbances. Under engine conditions, elevated cylinder pressure can suppress the plasma discharge event, and the discharging bandwidth is narrowed consequently (Burrows and Kristapher, 2016; World Automotive Congress, 2013).

The peak secondary voltage is determined by the DC power supply voltage, which needs to be controlled within a strict range to establish plasma discharge. For instance, under the experimental setup in this work, the DC power supply voltage lower than 30 V is not sufficient to generate plasma discharge, however, a supply voltage higher than 80 V will trigger an arc discharge.

Since plasma discharge relies on initial ionization of the gas media at the vicinity of the plasma antenna, the electric field gradient is thus an important parameter affecting the plasma discharge event (Auzas et al., 2010; Bentaleb et al., 2015). Under similar discharge voltage, the electric field gradient is influenced by the shape of the antenna tip. Test results indicate that the plasma is easy to discharge from a sharp tip. The wear and erosion of the antenna will degrade the electric field gradient, leading to void discharge even under the same discharge voltage.

8.2.2 Background condition impact

Fig. 8.5 demonstrates the background pressure impact on the plasma discharge and flame kernel formation. It is noted that the plasma streamers are suppressed by background pressure both in length and branch numbers. The elevated pressure results in the volume shrinkage of the initial flame kernel area. A spark event with similar ignition energy and spark gap of 2 mm is compared with plasma discharge. Overall, the plasma discharge can generate a bigger initial flame kernel together with more wrinkles at the flame front. Under higher background pressure, both the size and number of the streamers decrease, resulting in a smaller initial flame kernel. However, small but dense flame

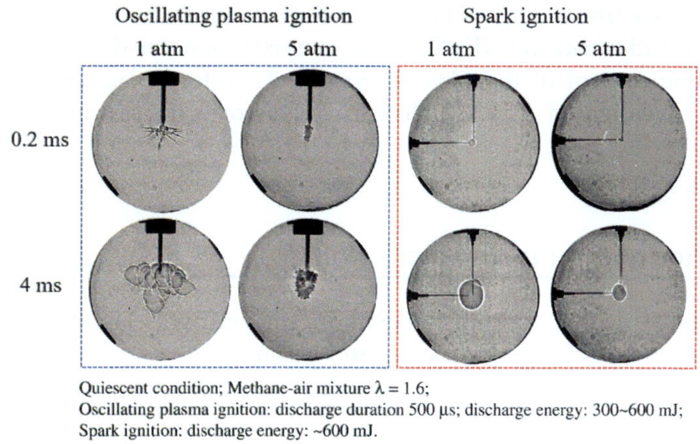

FIGURE 8.5 Shadowgraph imaging of flame kernel development initiated by oscillating plasma and spark ignition under pressurized mixture.

kernels are formed along the streamers, which further contribute to the dense flame front wrinkles and more intensified successful ignition sites.

The background gas composition and pressure often lead to different ionization coefficients, attachment coefficients, and the mobility status of the space charges (Schnyder et al., 2013; Howatson, 2013). Thus it is important to investigate the influence of background gas ionization and pressure on the plasma discharge. A higher background temperature enhances the kinetic energy of the air molecules, which benefits the plasma discharge. The minimum voltage required for plasma discharge decreases with the increase of the background temperature (Yan et al., 2016).

The impact of the flow pattern on plasma discharge is less understood, but preliminary research in the authors' work suggests marginal impact of flow speed on plasma discharge. Considering the plasma discharge with an oscillating frequency of 2 MHz, the building-up process is normally within microseconds. With cross flow speed of 30 m/s, which is considered as high in-cylinder flow motion, the moving distance within each oscillation cycle is around tens of micrometers. Because the size of normal plasma discharge is around 5–10 mm even under elevated background pressure conditions, the air movement is negligible.

8.2.3 External disturbance impact

Because of the absence of the ground electrode, the distribution of the electric field close to the antenna tip can be easily distorted by external disturbances. A touchdown of the plasma streamers to the ground is recognized as an arc discharge. The approach of a piston during spark event near engine compression TDC, or combustion chamber geometry can significantly increase the arcing tendency. Once the unwanted arc discharge takes place, the electric field around it changes rapidly and collapses, and a strong current surge can be noticed on the discharge waveform. The current surge during the arc discharge can generate heat, and hence accelerate the wear of the antenna tip (Wang et al., 2019).

Carbon deposits generated by incomplete combustion and unburned fuel droplets can attach to the surface of the ceramic insulator. The degraded insulation leads to surface discharge and arc discharge, which will compromise the ignition capability of the plasma streamers.

Therefore a stable plasma discharge under engine condition remains to be a challenge. A model that can predict normal and abnormal discharging events based on discharge waveforms will enable the dynamic control of the plasma system. Therefore plasma discharge characteristics need to be investigated first via electrical waveform measurements and analysis.

8.3 Experimental setup and methodology

An in-house designed oscillating plasma ignition system in the authors' laboratory (Zheng et al., 2019) provides adjustable discharge frequency and duration.

8.3.1 High-frequency plasma power drive

The plasma ignition system included an external source for generating a command pulse train, high-frequency plasma power drive, and plasma antenna. As determined by the coil parameters in the plasma power drive, which is presented in Fig. 8.6, the oscillating capability is from 1 to 3 MHz.

The high-frequency ignition drive for the oscillating plasma consists of a power MOSFET switch and the associated inductor–capacitor (LC) oscillation circuit. The inductor of the LC circuit is the primary coil of a step-up transformer with the inductance of L_1. The open and close of the MOSFET switch controlled by the command pulse train generates the charge and discharge of the DC-blocking capacitor (C_4) that is directly connected to L_1. The charging and discharge of C_4 induces a pulsed voltage across the inductor L_1. This primary voltage is amplified by the step-up transformer. A higher potential AC voltage is presented across the secondary coil. The secondary voltage usually lies within the range of 5–15 kV when the DC voltage supply varies from 12 to 60 V. In this work, the secondary voltage and current measurements are placed between the secondary coil and the plasma antenna. The parameters of electronic components in Fig. 8.8 are listed in Table 8.1.

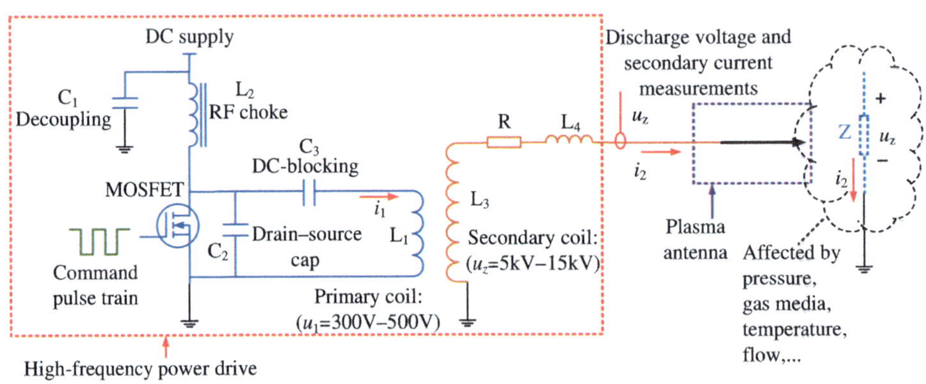

FIGURE 8.6 Experimental setup of oscillating plasma ignition system.

TABLE 8.1 Parameters of electronic components in high-frequency plasma power drive.

	Voltage		Current		Inductance		Capacitance		Resistance	
Primary	u_1	0.7–2 kV	i_1	10–15 A	L_1	4.4 μH	C_1	470 nF	–	–
					L_2	47 μH	C_2	330 pF	–	–
					–	–	C_3	47 nF	–	–
Secondary	u_z	5–15 kV	i_2	0.5–2 A	L_3	29 μH	–	–	R	17 Ω
					L_4	261 μH	–	–	–	–

8.3.2 Mathematical description and model assumption of the plasma ignition system

The plasma antenna together with the external disturbances is assumed to have an equivalent impedance of Z as shown in Fig. 8.6. u_1 represents the primary voltage. The voltage of the load Z is the measured secondary voltage that is assumed to be u_z. The currents in the primary and the secondary side of the transformer are represented by i_1 and i_2, respectively. The positive current directions of the red arrows are illustrated in the diagram. Based on the above assumption, if the coupling coefficient of the transformer is assumed to be k, the mutual inductance of the transformer as M can be defined as:

$$M = k \times \sqrt{L_1 \times L_2} \tag{8.1}$$

As a result, in the time domain, the voltage of the primary coil of the transformer can be derived as:

$$u_1 = L_1 \frac{di_1}{dt} - M \frac{di_2}{dt} \tag{8.2}$$

Similarly, for the secondary circuit, the secondary coil with the resistance R and the plasma antenna Z are connected in series with L_2. Therefore the net summation of voltage potential across all the components in this closed loop is zero. Hence:

$$M \frac{di_1}{dt} - L_2 \frac{di_2}{dt} - i_2 R - i_2 Z = 0 \tag{8.3}$$

Replacing $\frac{di_1}{dt}$ in Eq. (8.3) with the expression in Eq. (8.2) gives:

$$\left(\frac{M^2}{L_1} - L_2 \right) \frac{di_2}{dt} - i_2(R + Z) + \frac{M}{L_1} u_1 = 0 \tag{8.4}$$

In discrete time, the representation of i_2 at the nth sampling instance can be represented as:

$$i_2(n) = \frac{\left[\frac{M}{L_1} u_1(n) + \frac{\Gamma}{\Delta t} i_2(n-1) \right]}{(R + Z + \frac{\Gamma}{\Delta t})} \tag{8.5}$$

where $i_2(n)$ represents the secondary current at the nth sampling instance, and $u_1(n)$ represents the voltage of the primary coil obtained at the nth sampling instance. In Eq. (8.5) $\Gamma = L_2 - \frac{M^2}{L_1}$.

Since $u_1(n)$ is also the voltage supplied from the power supply, the value of the secondary current $i_2(n)$ can then be derived iteratively using Eq. (8.5). The expression for u_z at the nth sampling instance is:

$$u_Z(n) = Z \times i_2(n) \tag{8.6}$$

It is noted that Eqs. (8.5) and (8.6) are only valid for the circuit model illustrated in Fig. 8.6. During the plasma discharge, the intrinsic physical processes are much more complicated than the assumptions made above. The presence of the parasite elements and the interaction between the transformer and the driver circuit are unavoidable, but not considered yet in this simple model.

8.3.3 Experimental instruments

The secondary voltage was measured through a Tektronix P6015 high-voltage probe with 1000:1 attenuation attached to a socket that is plugged to the end of the plasma antenna. The secondary current is measured by a Pearson 411 current transformer on the wire between the secondary coil and the plasma antenna. The electrical waveforms were recorded by a PicoScope 4824 digital storage oscilloscope with a sampling frequency of 100 MHz. A Keysight E4990A impedance analyzer with 20 Hz to 20 MHz frequency range is applied to measure the capacitance and inductance of electronic components in the plasma ignition system.

8.4 Predictive modeling of oscillating plasma discharge

Fig. 8.7 shows a simplified structure of the modeling process. First-principle models are used to describe the working process of the electric circuit. The discharge process is sophisticated and involves factors from different domains. Empirical models become important to describe the operation boundary and working process through experiments. Electric waveforms of the discharge process together with optical observations can be used to establish and validate the models. Simulation results from the model are important to predict the energy release and system efficiency of the oscillating plasma discharge system.

8.4.1 Measurement of electrical waveforms

An example of a pulse train command and measured secondary voltage and current waveforms of a plasma discharge event is plotted in the logarithmic scale in Fig. 8.8.

The DC supply voltage is initially set to 36 V provided by a battery pack. The command pulse train is applied to control the MOSFET. During the experiments, the high-frequency power drive, the high-voltage probe, the current transformer, and the coaxial cables for

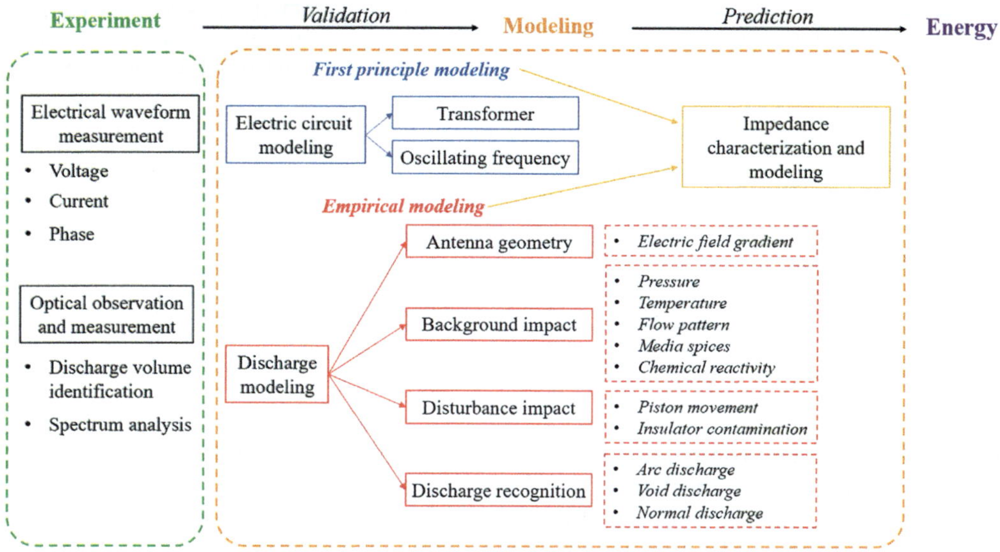

FIGURE 8.7 Predictive models of oscillating plasma energy release.

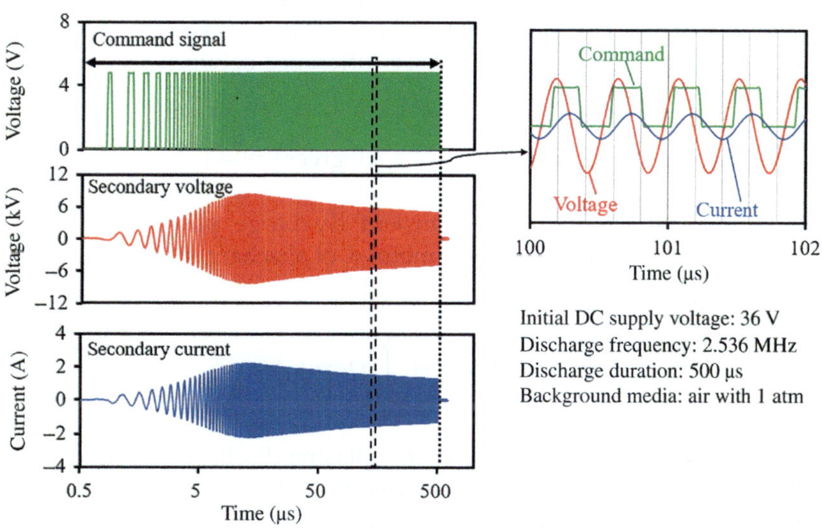

FIGURE 8.8 Secondary voltage and current waveforms in logarithmic scale during a plasma discharge event.

signal transmission may cause the signal propagation delay. The phase difference between the secondary voltage and current may change with different experimental setups. Consequently, the calculation of the corresponded power and discharge energy is affected. The accurate measurement of the phase difference remains challenging. However, with a fixed experimental setup, the comparison of phase difference in each case is still valid to

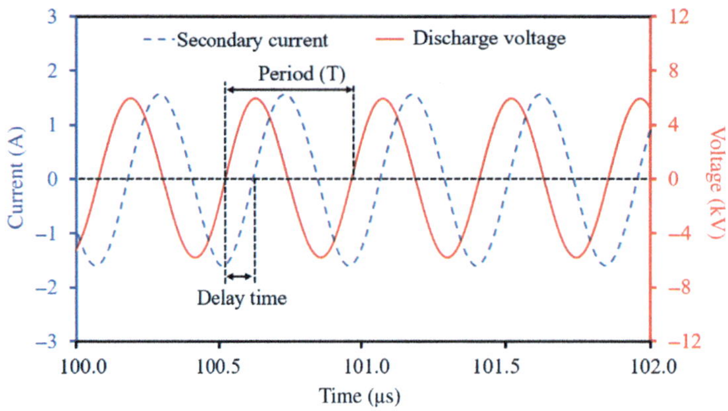

FIGURE 8.9 Demonstration of the phase difference between secondary voltage and current.

study the effects of environmental changes (e.g., pressure, temperature, and gas composition) on oscillating plasma discharge.

Fig. 8.9 demonstrates the phase difference between secondary voltage and current. Based on the experimental setup described in Section 8.3, a delay time exists between the secondary current and voltage.

The measurement of phase difference $\varphi = \varphi_V - \varphi_I$ is influenced by the experimental setups (e.g., length of the wires and impedance of the ignition driver). The experimental setup is fixed in this paper. The period (T) of the oscillating waveform is the length of one oscillating cycle in the time domain. The phase difference between secondary voltage and current can be calculated by

$$\text{Phase difference}(\varphi)(\text{degrees}) = \varphi_V - \varphi_I(\text{degrees}) = \frac{\text{Delay time}(\mu\text{s}) \times 360^\circ}{\text{Period}(T)(\mu\text{s})} \tag{8.7}$$

The phase difference influences the transient discharge power and subsequent discharge energy calculation. The calculation method of discharge power and energy of the oscillating plasma is presented as

$$E = \int_0^{t_{\text{end}}} P dt = \int_0^{t_{\text{end}}} U_z \times I_2 dt \tag{8.8}$$

where E is the discharge energy in an oscillating plasma discharge event, t_{end} is the ending time of the discharge duration, P is the discharge power of the oscillating plasma, and U_z and I_2 are the secondary voltage and current, respectively.

8.4.2 Oscillating frequency modulation

The oscillating plasma ignition system is configured as a series-resonant RLC circuit, that is, a resistor–inductor–capacitor circuit. The selection of operating frequency has significant effects on the performance of the system. Ideally, the oscillating frequency matches the resonance frequency of the system to generate a required voltage. In Fig. 8.10

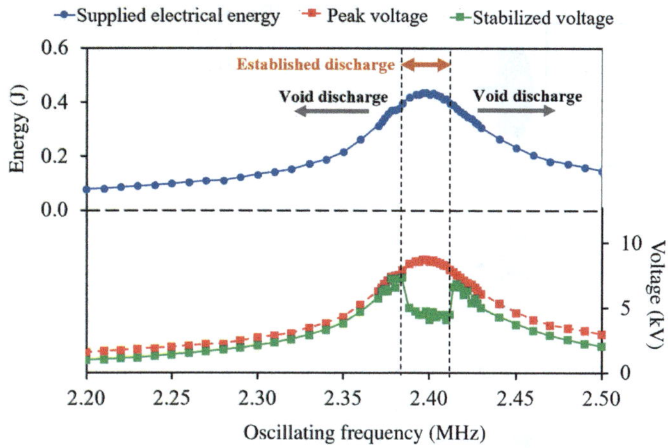

FIGURE 8.10 The frequency band and width of established oscillating plasma discharge.

the maximum value of the secondary voltage in each plasma discharge event varies with oscillating frequency. With the experimental setup described in Fig. 8.6, the discharge frequency in the range of 2.386–2.412 MHz produces the established plasma discharge. The peak voltage and energy under the oscillating frequency (2.4 MHz) indicates the largest plasma discharge volume.

8.4.3 Plasma discharge patterns and external effects

The plasma discharge is based on the ionization of the molecules and atoms in the gas media. When the plasma successfully establishes, higher voltage is required from the ignition system to maintain the ionized plasma channel. The secondary voltage waveforms of three typical oscillating plasma discharge events are plotted in Fig. 8.11, including the void discharge, established discharge, and arc discharge. The black dash lines shown in Fig. 8.12 indicate the amplitude shaping as the envelope of the waveforms. When the command pulse train is sent to the plasma power drive, an oscillation begins in the plasma ignition system. The secondary voltage keeps increasing until it reaches the peak value. When the plasma fails to release from the antenna, the amplitude of the measured voltage stays relatively stable throughout the commanded duration. However, with established plasma discharge, the waveform envelopes decline slightly and stabilize until the command is off. In an arc discharge event, when the plasma streamers touch down to a metal surface, the secondary voltage drops immediately from 6 to 2 kV.

8.4.4 Predictive modeling of oscillating plasma impedance

The secondary voltage is an important parameter that reflects the performance of the plasma ignition system. When the system is operating within the modulated oscillating frequency, a circuit model is proposed to describe the impedance change for the plasma discharge. Therefore the secondary voltage and current are the input and output parameters for the experimental model of this predictive modeling, respectively.

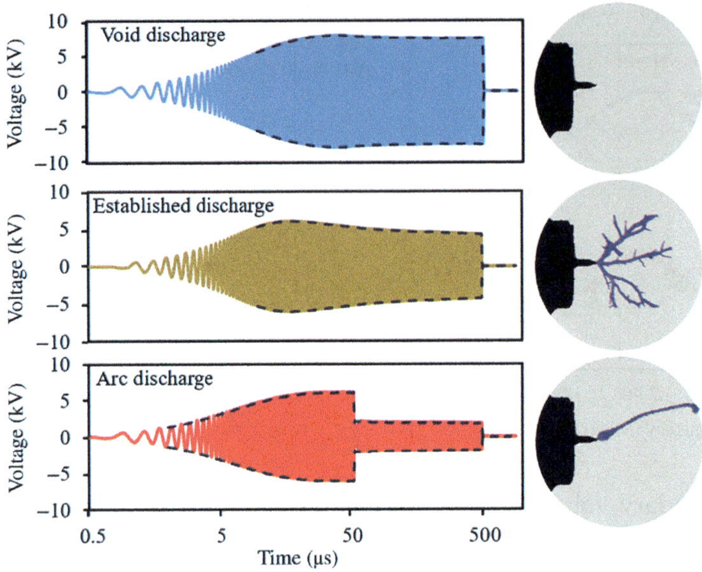

FIGURE 8.11 The comparison of void, established, and arc discharge events.

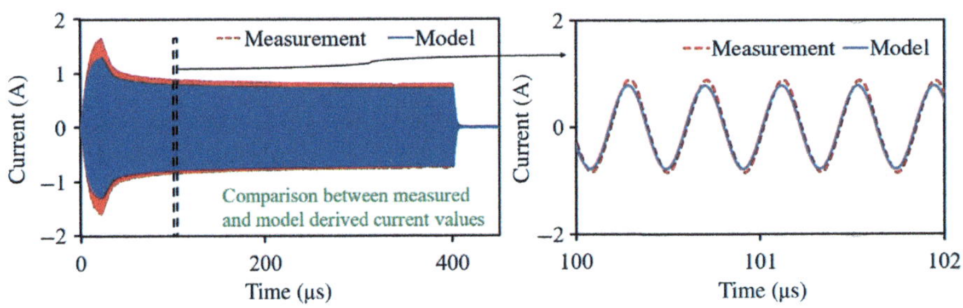

FIGURE 8.12 Comparison between the measured and calculated current.

The impedance for the plasma discharge is an important electrical parameter that can be used for the electrical energy estimation for the plasma discharge event and may also be used to reflect the interaction between the plasma ignition system and its surrounding. Using Fourier transformation, the amplitude, as well as the phase information of the voltage and current measurement, can be derived by:

$$Z = \frac{|A_V|}{|A_I|} e^{j(\varphi_V - \varphi_I)} \tag{8.9}$$

where $|A_V|$ and $|A_I|$ are the amplitude values of the secondary voltage U_z and current I_2. With the experimental setup explained in Section 8.3 under atmospheric condition, the measured phase difference between secondary voltage and current is approximate

60 degrees, in the polar coordinate (Remmert, 2013). Eq. (8.9) can also be presented in its Cartesian form as:

$$Z = \frac{|A_V|}{|A_I|}\cos(\varphi_V - \varphi_I) + j\frac{|A_V|}{|A_I|}\sin(\varphi_V - \varphi_I) \tag{8.10}$$

Since the impedance lies within the 1st quadrant of the Cartesian coordinate, the real part and the imaginary part of the impedance share the same sign. Therefore the impedance of the load is modeled by a resistor in series connection with an inductor. The resistance and inductance of the load can be derived by the following linear equations with binary variables R_Z and L_Z:

$$\begin{cases} {R_Z}^2 + (2\pi f \times L_Z)^2 = |Z_f|^2 \\ \dfrac{2\pi f \times L_Z}{R_Z} = \tan(\varphi_f) \end{cases} \tag{8.11}$$

where $|Z_f|$ and φ_f are the amplitude and the phase of the load Z obtained at the selected frequency f, and R_Z and L_Z represent the equivalent resistance and inductance of the load Z. In this work, only the impedance values obtained at the frequency 2.4 MHz are used to derive the values for both R_Z and L_Z, and the derived values are 2.65 kΩ and 36 mH respectively.

8.4.5 Predictive modeling of oscillating plasma discharge energy

The derived impedance of the load Z and the measured voltage U_z can also be used to calculate the current that flows through the plasma antenna:

$$\hat{I}_2 = \frac{U_z}{Z} \tag{8.12}$$

Since in the s-plane, $Z = R_Z + s \times L_Z$, thus in the discrete-time domain, Eq. (8.12) can be rewritten as:

$$\hat{i}_2(n) = \frac{u_z(n) + \hat{i}_2(n-1) \times L_Z/dt}{R_Z + L_Z/dt} \tag{8.13}$$

In Eq. (8.13), $\hat{i}_2(n)$ and $u_z(n)$ represent the calculated current and the measured voltage at the nth sampling instance. Thus for Eq. (8.12), using the measured voltage $u_z(n)$ and the current $\hat{i}_2(n-1)$ from the previous sampling instance, the current at the nth sampling time can be estimated. This equation describes the dynamics of secondary current $\hat{i}_2$, and the change of the current profile can be predicted. The comparison between the predicted current $\hat{I}_2$ and the measured current, I_2, $f = 2.4$ MHz is shown in Fig. 8.12. In the time domain, the predicted current has shown a good agreement with the measurement.

Furthermore, by using this impedance model, the energy release of the plasma antenna can be estimated using the voltage measurement with Eq. (8.8). The quantity of plasma energy is directly related to the success rate of the ignition and combustion. The estimation accuracy of the energy release is beneficial for the evaluation of the ignition system and is important for further combustion control of the plasma discharge.

8.5 Conclusions

The oscillating plasma discharge has shown promising advantages to ignite lean/diluted air–fuel mixtures. However, the discharging process is very sensitive to external conditions. For engine application, the change of in-cylinder pressure, temperature, flow pattern, and the approach of the piston affect the plasma discharge. The multidisciplinary nature of the discharging process and its sensitivity to the surroundings make the simulation work challenging. In this chapter, the challenges of stable plasma discharge under engine-like conditions are discussed. The electrical waveforms of the discharge events are measured and analyzed to characterize the plasma discharge process. An empirical model based on the impedance characterization is built and key parameters have been identified to provide discharge energy prediction for normal discharge events. Major conclusions are given below:

1. The oscillating plasma discharge establishes a large ignition volume to generate a bigger flame kernel leading to a faster flame propagation process.
2. The plasma antenna surrounding forms a part of the plasma discharging circuit. Under engine conditions, dynamic changes of the in-cylinder pressure, temperature, and piston positions provide major challenges for a successful plasma discharge event.
3. A first-principle model is provided to describe the working principle of the circuit, and an empirical model characterizing the discharge impedance is established to describe the dynamics of the plasma ignition system based on the impedance modeling.
4. The modeled and measured secondary current have shown a good agreement within the operating frequency bandwidth of the plasma ignition system. The energy release of the plasma antenna can be further predicted.

References

Auzas, F., Tardiveau, P., Puech, V., Makarov, M., Agneray, A., 2010. Heating effects of a non-equilibrium RF corona discharge in atmospheric air. J. Phys. D: Appl. Phys. 43 (49), 495204.

Bentaleb, S., Blin-Simiand, N., Jeanney, P., et al., 2015. Ignition of lean air/hydrocarbon mixtures at low temperature by a single corona discharge. J. AerospaceLab 10, 1–8.

Burrows, J., Kristapher, M., 2016. Analytical and experimental optimization of the advanced corona ignition system. International Conference on Ignition Systems for Gasoline Engines. Springer, Cham, pp. 267–292.

Burrows, J., Lykowski, J., Mixell, K., 2016. Corona ignition system for highly efficient gasoline engines. MTZ Worldwide 74 (6), 38–41.

Fansler, T.D., Reuss, D.L., Sick, V., et al., 2015. Combustion instability in spray-guided stratified charge engines: a review. Int. J. Engine Res. 16 (3), 260–305.

Gallon, E., Fontana, G., Palmaccio, R., 2013. Effects of exhaust gas recycle in a downsized gasoline engine. Appl. Energy 105, 99–107.

Howatson, A.M., 2013. An Introduction to Gas Discharges: Pergamon International Library of Science. Technology, Engineering and Social Studies. Elsevier.

Huang, C.C., Shy, S.S., Liu, C.C., et al., 2007. A transition on minimum ignition energy for lean turbulent methane combustion in flamelet and distributed regimes. Proc. Combustion Inst 31 (1), 1401–1409.

Idicheria, C.A., Najt, P.M., 2016. Potential of Advanced Corona Ignition System (ACIS) for Future Engine Applications. Ignition Systems for Gasoline Engines .

Martin, S., Schauer, F.X., Sauer, C., Weber, G., Hahn, J., Schwarz, C., 2016. Challenges to the ignition system of future gasoline engines–an application oriented systems comparison. International Conference on Ignition Systems for Gasoline Engines. Springer, Cham, pp. 3–25.

Nakata, K., Nogawa, S., Takahashi, D., et al., 2016. Engine technologies for achieving 45% thermal efficiency of SI engine. SAE Inter. J. Engines 9 (1), 179–192.

National Highway Traffic Safety Administration (NHTSA) and Environmental Protection Agency (EPA), 2015. NHTSA and EPA set standards to improve fuel economy and reduce greenhouse gases for passenger cars and light trucks for model years 2017 and beyond.

Reinhart, T.E., 2016. Commercial medium- and heavy-duty truck fuel efficiency technology study – Report #2 (Report No. DOT HS 812 194). National Highway Traffic Safety Administration, Washington, DC.

Remmert, R., 2013. Classical Topics in Complex Function Theory, 172. Springer Science & Business Media.

Schnyder, R., Howling, A.A., Bommottet, D., Hollenstein, C., 2013. Direct current breakdown in gases for complex geometries from high vacuum to atmospheric pressure. J. Phys. D: Appl. Phys. 46 (28), 285205.

Schumann, F., Sarikoc, F., Buri, S., Kubach, H., Spicher, U., 2013. Potential of spray-guided gasoline direct injection for reduction of fuel consumption and simultaneous compliance with stricter emissions regulations. Int. J. Engine Res. 14 (1), 80–91.

Singleton, D., Pendleton, S.J., Gundersen, M.A., 2010. The role of non-thermal transient plasma for enhanced flame ignition in C2H4–air. J. Phys. D: Appl. Phys. 44 (2), 022001.

Starikovskaia, S., 2006. Plasma assisted ignition and combustion. J. Phys. D: Appl. Phys 39, 65–99.

Steffen, B., Rixecker, G., Brichzin, V., Becker, M., 2014. High-frequency ignition system based on corona discharge. MTZ Worldwide 75 (1), 30–35.

Suess, M., Guenthner, M., Schenk, M., Rottengruber, H.S., 2012. Investigation of the potential of corona ignition to control gasoline homogeneous charge compression ignition combustion. Proc. Inst. Mech. Engineers, Part D: J. Automobile Eng. 226 (2), 275–286.

Takahashi, D., Nakata, K., Yoshhihara, Y., et al., 2015. Combustion development to achieve engine thermal efficiency of 40% for hybrid vehicles. SAE Technical Paper No. 2015-01-1254.

Turner, J., Popplewell, A., Patel, R., et al., 2014. Ultra-boost for economy: extending the limits of extreme engine downsizing. SAE Int. J. Engines 7 (1), 387–417.

Wang, L., 2019. Characterization of corona discharge for ignition improvement. Electronic Theses and Dissertations 7853, <https://scholar.uwindsor.ca/etd/7853>.

Wang, L., Tan, Q., Yu, S., Yu, X., Chen, X., Zheng, M., 2019. A framework for the active control of corona ignition systems. SAE Technical Paper Series .

World Automotive Congress, 2013. Lecture Notes in Electrical Engineering, vol. 189. Springer, Berlin, Heidelberg.

Yan, P., Zheng, C., Zhu, W., Xu, X., Gao, X., Luo, Z., et al., 2016. An experimental study on the effects of temperature and pressure on negative corona discharge in high-temperature ESPs. Appl. Energy 164, 28–35.

Zheng, M., Yu, S., 2015. Advanced ignition systems for future clean combustion engines: review. J. Autom. Saftey Energy 6 (4), 295–313.

Zheng, M., Yu, S., Wang, M., 2019. Active-control resonant ignition system. U.S. Patent 10,263,397.

CHAPTER

9

Nowcasting solar irradiance for effective solar power plants operation and smart grid management

Marius Paulescu[1], *Eugenia Paulescu*[1] *and Viorel Badescu*[2]

[1]Faculty of Physics, West University of Timisoara, Timisoara, Romania [2]Candida Oancea Institute, Polytechnic University of Bucharest, Bucharest, Romania

9.1 Introduction

Since the ancient times, humanity has used fossil fuels. This happened because fossil fuels are easy to exploit and increasing the volumetric density of energy can be achieved by relatively simple technologies (e.g., oil refinement). Presently, fossil fuels still supply over 80% of the world's energy consumption. However, the awareness of the increasing environment degradation together with events like the oil crisis in 1973 and the catastrophic nuclear accident in Chernobyl in 1986 have led to global sustainable development policies. The energy-saving strategies and the increasing use of renewable energy may synergistically lead to a slowdown in environment degradation. In any future scenario, it is generally accepted that solar energy will play a substantial role in the conservation of the energy resources concurrently with meeting the global energy demand. From a different perspective, closely related to our high-tech lifestyle, electricity has won a privileged place in our energy options.

In this context, no wonder that the weight of solar electricity in the energy mix has experienced an impressive increase and this trend is expected to continue. According to Solar Power Europe's Global Market Outlook 2019–23 (Schmela et al., 2019), the total installed solar power capacity in the world has multiplied in 10 years by 32 times, from 15.8 GWp in 2008 to 509.3 GWp by the end of 2018. Thus 2018 has become a milestone in photovoltaic (PV) history, as the year when the world cumulative PV capacity broke the threshold of 500 GWp. Furthermore, for the first time, in 2018 the annual installed capacity exceeded 100 GWp. Even in pessimistic scenarios, Solar Power Europe anticipates for 2023

Predictive Modelling for Energy Management and Power Systems Engineering
DOI: https://doi.org/10.1016/B978-0-12-817772-3.00009-4

a doubling of the world's solar power capacity to more than 1000 GWp. In an optimistic scenario, a tripling of the world's solar power capacity is expected.

A challenge standing against the growing share of PV electricity in the energy mix is the intrinsic nature of solar energy, which stochastically fluctuates in time, owing to irregular weather patterns (e.g, Perez et al., 2016). In order to reduce the cost of integrating solar plants into the existing power grid, forecasting their power output is a key issue. High-quality forecasts may enable the grid operators to plan the operation of other power plants, to compensate for the variations in the output power of PV plants. At the same time, the smart grid concept tends to become usual in power grid management. Forecasting the load (Sepasi et al., 2017) and the power generated from renewable energy sources such as wind (Yan et al., 2015) and solar (Pedro and Coimbra, 2015) are important tasks in providing smartness to the grid. A smart grid management targets a real-time balance of the variability in the output power of solar plants. Thus smart grid management requires intrahour forecasts of solar plant power output. The startup time of conventional power sources connected to the grid as balancing instruments represents the main compulsion for real-time management of the grid.

The forecasting procedure of PV plant power output involves two interrelated topics: (1) a realistic modeling of PV plant operation, and (2) forecasting the solar resource availability. Definitively, the quality of forecasting of PV plants' energy production closely follows the accuracy of forecasting of the solar resource. Some authors go even further and reduce the problem of PV plant output power forecasting to the problem of incoming solar energy forecasting. In the last years, a massive effort has been devoted to increasing the accuracy of the forecasting of solar resources. The work of Inman et al (2013) is an outstanding review on the theory behind the forecasting procedures, well illustrated with some successful applications. That review captures well the dimensions of this effort. From a different perspective, the study by Blaga et al. (2019) presents a comparative analysis of the performance in solar irradiance forecasting. The study overcomes the limitations in comparing different forecasting models caused by the large number of metrics used for reporting the performance of the models. More than 160 papers published between 2006 and 2017 reporting on the accuracy of forecasting of solar resources were analyzed. From those papers only the results presented in 40 papers were inserted into a purpose-built database. The selection was made on the basis of the following criterion: the accuracy of the results must be presented in terms of normalized root mean square error (*nRMSE*) and normalized mean bias error (*nMBE*). As defined (see Section 3.5), the two statistical indicators allow the comparison of samples of different sizes, including data of different magnitudes, thus ensuring a common basis for intercomparison analyzes. This selection criterion leads to an arbitrary sampling, and thus the analysis generates a reliable picture of the state-of-the-art in the accuracy of solar resources forecasting. Thus the study of Blaga et al. (2019) introduced a rigorous framework for comparing different classes of models. The study results are reported from four perspectives: (1) classes of forecasting models (persistence, classical statistics, artificial intelligence, cloud-motion tracking, numerical weather prediction (NWP) and hybrid models); (2) time horizon of the forecast (intrahour, intraday, and day ahead); (3) climate (tropical, arid, temperate, and snow); and (4) performance trend over time. The main conclusions of Blaga et al. (2019) are (1) unexpected at first glance, the forecasting models from the hybrid class (combining classical statistics with artificial intelligence and/or exogenous inputs) demonstrate the best performance; (2) the forecasting error (quantified by *nRMSE*) increases with the increase in the time horizon of the forecast. Intrahour forecasts reach on average smaller values of *nRMSE* while the day-ahead forecasts are less reliable; (3) the models from the artificial intelligence and hybrid classes are recommended for intrahour forecasts. The ARIMA models may also perform well. For intraday forecasts, the models from the artificial intelligence and classical statistics classes show low errors, over all climate types. The hybrid models perform well at sites located in tropical and snow climates; and (4) generally, during the last decades the accuracy of the forecasting models has increased over time. For instance, over the last 10 years the forecasting errors have decreased significantly, on average *nMBE* being reduced by two thirds and *nRMSE* being reduced by one third, as compared to the values at the beginning of the considered period.

This chapter is focused on the performance of intrahour forecasting of solar irradiance. Generally, the intrahour forecasts are elaborated based on two types of models: (1) statistical extrapolation of the solar irradiance time series measured in situ; and (2) all-sky imager-based algorithms. A brief survey of some results in intrahour forecasting of solar irradiance follows.

Dong et al. (2013) proposes an exponential smoothing model operating on high-resolution solar irradiance data. This model, together with other statistical models (autoregressive integrated moving average, linear exponential smoothing, simple exponential smoothing, and random walk) were tested against data measured at two stations: Singapore (humid tropical climate) and Colorado, US (temperate semiarid). For a time horizon less than 20 minutes and a data granularity less than 20 minutes, the reported *nRMSE* ranges between 20% and 30%. *nRMSE* increases to 50% for a time horizon of the forecast over 40 minutes and a data granularity of 20 minutes. Boland et al. (2016) reported results of testing three forecasting procedures: a purely statistical one, joining Fourier series and linear ARMA models; a second one, combining clear-sky index models and neural network models; and a third one, joining a clear-sky index model with an ARMA model. The tests were performed against data measured in four locations: two continental locations, one coastal, and one inland. A very interesting and rather surprising conclusion is that the three procedures gave similar forecasts. Reikard et al. (2017), after analyzing various time series, meteorological and time-varying parameter models, concluded that each forecasting model has peculiar strengths and weaknesses. Time-series models are able to predict more accurately at short-time horizons. In a direct comparison of time series and physical-based models, the ARIMA model is more accurate at short time horizons, while the meteorological models become more accurate as the time horizon increases. In the frame of artificial intelligence modeling, Pedro and Coimbra (2015) reported a solar irradiance forecasting procedure based on the optimization of the k-nearest neighbors and artificial neural networks algorithms for a lead time ranging between 15 minutes and 2 hours. The results show that the forecast error becomes more than double in locations with high solar irradiance variability compared with locations with a low variability. The authors concluded that the type of climate and the associated cloud patterns have a deep impact on the forecasting performance. The combination of statistical approaches and postprocessed data from all-sky images is another technique largely investigated for increasing the forecast accuracy. For example, Fu and Cheng (2013) report a solar irradiance forecasting procedure for 5–15 minutes ahead that utilizes features extracted from all-sky images. Data collected during a month on a costal site in Taiwan were used to validate the procedure. Results show that 5-minute ahead estimates can achieve a monthly averaged normalized mean absolute error of 21.9%.

This chapter begins with the definition of the physical quantities associated with solar radiation energy flux, that is, the solar irradiance components. Then, a brief introduction in radiometry is presented, followed by a description of the radiometric data series used in this study. The common postprocessing of solar irradiance data series (i.e., computing the clearness index, sunshine number, relative sunshine, sunshine stability number) is also discussed. Section 9.2 ends with a discussion on the solar irradiance variability. Section 9.3 begins with a presentation of the statistical models evaluated in this study (persistence, ARIMA forecasting of clearness index, and the two-state model). Various metrics commonly used in evaluating the performance of solar irradiance models are also defined. Section 9.4 is devoted to the analysis of the forecasts accuracy in the case of the three models. The analysis is made from three perspectives: (1) forecast accuracy at different time horizons; (2) forecast precision; and (3) the influence of the solar irradiance variability on the forecast accuracy. As a case study, this analysis carries on the study by Paulescu and Paulescu (2019). The results presented in Section 9.4 emphasize the complex influence of various factors (from the model structure to the variability in the state-of-the-sky) on the performance of a model in the intrahour forecasting of solar resources. The main conclusions are gathered in Section 9.5.

9.2 Solar irradiance

9.2.1 Solar irradiance—terms and definition

Due to the large distance between the Sun and the Earth, the beam of sunrays is nearly parallel at the upper limit of the Earth's atmosphere. This beam is referred to as extraterrestrial radiation (ETR). When passing through the atmosphere, ETR is separated into two components: (1) the beam component which is that part of ETR that reaches the Earth's surface without being scattered or absorbed through the atmosphere; and (2) the diffuse component which is that part of ETR scattered in the atmosphere. More precisely, in radiometry the following physical quantities associated with the energy of solar radiation are measured:

- *Direct–normal irradiance* G_n (W/m^2) the energy flux coming directly from the solid angle subtended by the Sun disc. It is measured at ground level on a unitary surface perpendicular on the direction Sun–receiver.
- *Direct horizontal irradiance* G_b (W/m^2) differs from G_n in that it is measured on a unitary surface in the horizontal plane.

Lambert's cosine law states that the energy flux density on a plane surface is directly proportional to the cosine of the incidence angle. Since the incidence angle of the solar beam striking the horizontal ground is equal to the Sun's zenith angle θ_z (Fig. 9.1), then:

$$G_b = G_n \cos\theta_z \tag{9.1}$$

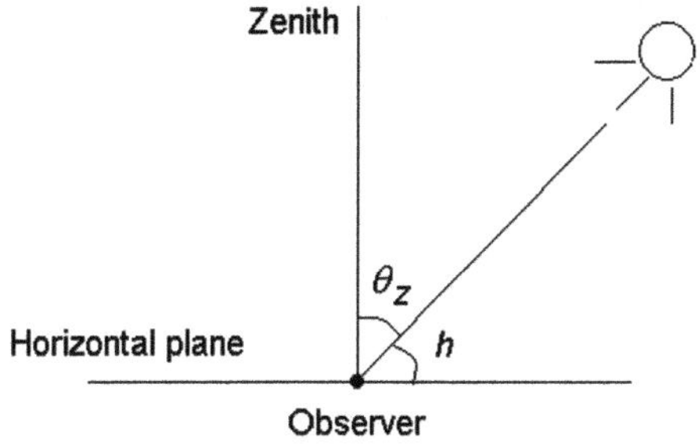

FIGURE 9.1 Solar geometry: θ_z denotes the zenith angle while h denotes the Sun's elevation angle.

- *Diffuse solar irradiance* G_d (W/m^2) represents the density of the solar energy flux coming from the entire sky dome, without the solid angle subtended by the Sun disc, that falls on a horizontal surface.
- *Global horizontal irradiance* G (W/m^2) is the sum of the direct horizontal and diffuse components:

$$G = G_b + G_d = G_n \cos\theta_z + G_d \tag{9.2}$$

The term "global" is associated with the fact that the solar radiation is received from the entire 2π solid angle of the sky vault.

For gaining further insight into solar geometry and the transfer of solar radiation through the atmosphere, we refer the reader to Chapter 5, Modeling Solar Radiation at the Earth Surface, of Paulescu et al. (2013).

9.2.2 Radiometric data used in this study

Radiometry is the science that studies the measurement of electromagnetic radiation. The instruments designed for measuring the energy flux of electromagnetic radiation are generically referred to as radiometers. Any of the quantities defined in Section 9.2.1 are measured with specific methods and instruments. The global solar irradiance is measured with pyranometers which are broadband spectral instruments, receiving the solar radiation incoming on a planar surface from the entire vault. A pyranometer can be also used to measure the diffuse solar irradiance by using a shadow-ring that prevents the direct–normal radiation reaching the sensor. Fig. 9.2 illustrates the two applications of a pyranometer. The present classification of the pyranometers with respect to their quality is defined by the International Standard ISO 9060/1990 that was also adopted by the World Meteorological Organization (WMO 2008). ISO 9060 standard defines three classes of pyranometers: the best quality is called (somewhat inadequately) *secondary standard*, the second best is called *first class*, while the third best is called *second class*. A very good introduction to the solar radiation measurements may be found in the book of Vignola et al. (2012) which details the strengths and weaknesses of instruments used to conduct solar and infrared radiation measurements. According to WMO (2008) the sunshine duration can be estimated on the basis

FIGURE 9.2 Photograph of the Kipp and Zonen SMP-10 pyranometers measuring the global and diffuse solar irradiances on the Solar Platform of the West University of Timisoara, Romania.

of high-frequency records of global and diffuse solar irradiances (the pyranometric method). WMO defines the sunshine duration during a given period Δt as the sum of the time intervals from Δt for which the direct–normal solar irradiance exceeds $120\,W/m^2$. This definition implies that the Sun is shining at time t if the direct–normal solar irradiance exceeds $120\,W/m^2$. The radiometric stations routinely measure global and diffuse solar irradiances. At such stations, for computing the sunshine duration, firstly Eq. (9.2) is employed to derive the direct–normal solar irradiance from global and diffuse solar irradiances, which is then compared with the WMO threshold.

In this study, the operation and the performance of three forecasting solar irradiance models (see Section 9.3) are illustrated on the basis of radiometric data recorded on the Solar Platform of the West University of Timisoara, Romania (Solar Platform, 2019). The town of Timisoara (latitude 45°46′N, longitude 21°25′E and 85 m asl) has a warm temperate climate, fully humid, with a warm summer, typical for the Pannonia Basin. The Köppen climate classification is *Cfb* (Kottek et al., 2006). Global G and diffuse G_d solar irradiances have been processed in this study. Secondary standard pyranometers Kipp and Zonen SMP-10 are employed. Fig. 9.2 shows the two pyranometers operating on the Solar Platform. The pyranometers are integrated into an acquisition data system based on the National Instruments PXI Platform. Measurements on the Solar Platform are performed all day long at equal time intervals of 15 seconds.

For illustrating this study, the data series were prepared using data recorded in June 2019. For this reason, Fig. 9.3 displays graphically the global solar irradiance for $h > 15$ degrees measured in every day of June 2019. Each curve was built from more than 2800 measurement points (12 hours × 240 measurements/h). On different days, solar irradiance followed very different path. Some days are mostly cloudy (June 01, June 05), some days are entirely sunny (June 13, June 30) while most days experienced a sky characterized by different degrees of variability. Accurate nowcasting of solar irradiance is more challenging when the variability in the state-of-the-sky is higher.

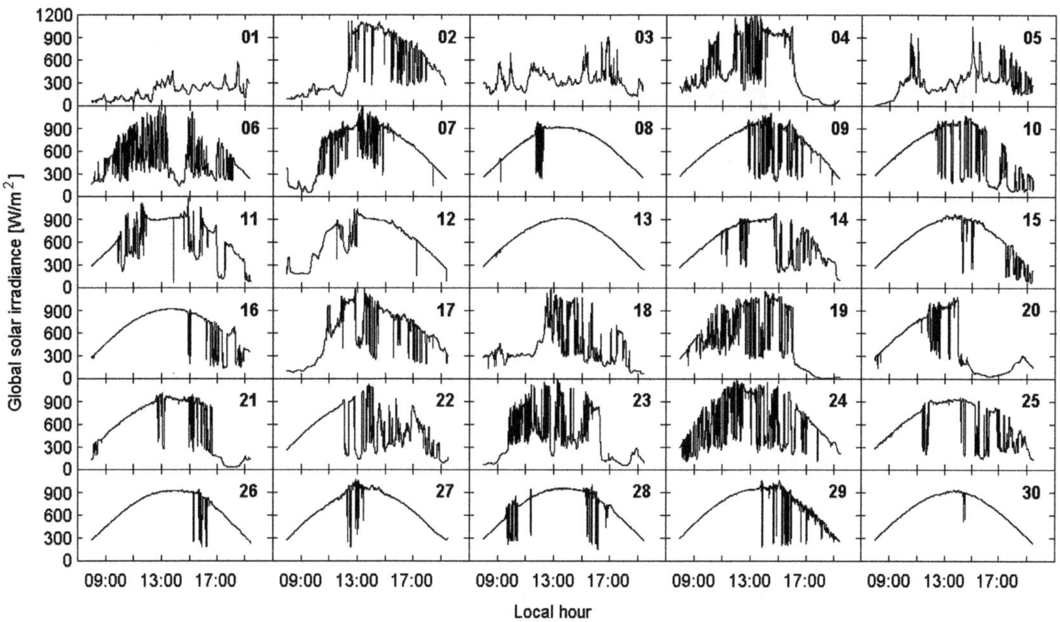

FIGURE 9.3 Global solar irradiance measured in every day of June 2019 on the Solar Platform. Only data measured for $h > 15$ degrees are plotted.

The next step in preparing the database was the postprocessing of raw data. Three parameters were inferred in this study: clearness index, sunshine number, and the sunshine stability number. These parameters are defined next.

9.2.3 Postprocessed data

Clearness index is defined as the ratio of global solar irradiance measured at ground level G and its counterpart estimated at the top of the atmosphere G_{ext} (Liu and Jordan, 1960):

$$k_t = \frac{G}{G_{ext}} \tag{9.3}$$

G_{ext} can be evaluated as a function of the solar elevation angle h, $G_{ext} = G_{SC}\,\varepsilon \sin h$, where $G_{SC} = 1361.1$ W/m (Gueymard, 2018) is the solar constant and ε is the Earth's orbit eccentricity correction factor that can be calculated with Spencer's equation (Spencer, 1971). In many practical applications, ε is computed with the simplified equation $\varepsilon = 1 + 0.0342 \cdot \cos[2\pi(j-1)/365]$, where j is the Julian day. This is a very good approximation of the Spencer's equation.

The actual level of solar irradiance at the Earth's surface results in the superposition of deterministic factors (e.g., movements of the Earth) and stochastic factors (e.g., clouds field evolution). As it is defined by Eq. (9.3), the clearness index has the ability to isolate the stochastic component in a global solar irradiance time series. Fig. 9.4

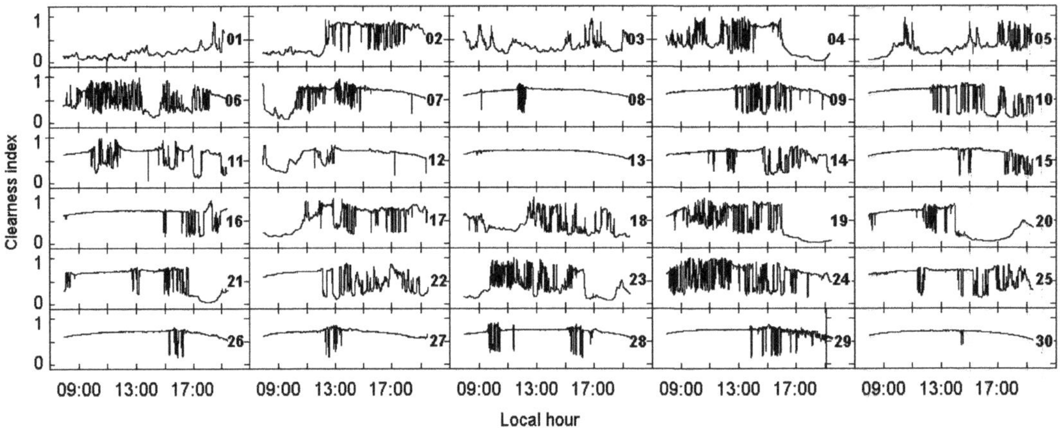

FIGURE 9.4 Clearness index in every day of June 2019 on the Solar Platform. Only data measured for $h > 15$ degrees are plotted.

illustrates the pattern of the clearness index computed with Eq. (9.3) for all the days of the database. For every day, the time series of the clearness index is random and almost stationary.

The traditional approach in nowcasting solar irradiance is based on the extrapolation of the clearness index time series. In order to develop a model, a series of historical data is required. This is obtained by staking the individual data series measured day by day. The model equation is fitted on this historical data series and, generally, the fitted coefficients are frozen and used at all times in the future for generating the forecasts.

The *Sunshine number* is defined as a time-dependent random binary variable (Badescu and Paulescu, 2011a):

$$SSN = \begin{cases} 0 & \text{if the sun is covered by clouds at time } t \\ 1 & \text{otherwise} \end{cases} \tag{9.4}$$

Let's suppose that during a given period Δt we performed N measurements of SSN at equal time intervals $\Delta\tau$. The result of the measurement consists of a SSN time series consisting of N_0 values $SSN = 0$ and N_1 values $SSN = 1$. If $\Delta\tau$ is a short enough interval (say $\Delta\tau = 15$ seconds, the interval between two measurements on Solar Platform), there is a very high probability that the state-of-the-sky does not change during $\Delta\tau$. The average value of SSN during Δt can be computed successively as follows:

$$\overline{SSN} = \frac{1}{N}\sum_{t=1}^{N} SSN_t = \frac{1}{N}(N_1 \times 1 + N_0 \times 0) = \frac{N_1}{N} = \frac{N_1 \times \Delta\tau}{N \times \Delta\tau} = \frac{\Delta t_1}{\Delta t} \tag{9.5}$$

where $\Delta t_1 = N_1 \times \Delta\tau$ is the sunshine duration period during Δt. The ratio $\Delta t_1/\Delta t$ defines a common indicator for the state-of-the-sky, namely the *relative sunshine* σ. A low value of σ is an indirect indication of a high cloud cover amount. Therefore the average value of SSN over a given period Δt equals the relative sunshine σ during Δt:

$$\overline{SSN} = \sigma \tag{9.6}$$

Series of SSN values can be derived from the series of measured global and diffuse solar irradiance values by means of the World Meteorological Organization sunshine criterion (WMO, 2008):

$$SSN = \begin{cases} 1 & \text{if} \quad (G - G_d)/\sin h > 120 \quad W/m^2 \\ 0 & \text{otherwise} \end{cases} \tag{9.7}$$

Fig. 9.5 shows the variation of the sunshine number for all the days of the database.

The Sunshine stability number, a parameter related to SSN, was introduced by Paulescu and Badescu (2011) as a quantifier for the state-of-the-sky. Let us assume an equidistant time moment series t_j $(j = 1, \ldots, n)$ during a time interval of duration Δt. One denotes $\Delta\tau \equiv t_{j+1} - t_j$ $(j = 1, \ldots, n-1)$. The fluctuations of SSN during Δt may be described by using the *sunshine stability number* $SSSN(t_j, \Delta\tau)$ $(j = 2, \ldots, n)$, which is a random Boolean variable defined by:

$$SSSN(t_{j \geq 2}, \Delta\tau) \equiv \begin{cases} 1 \text{ if} \begin{cases} SSN(t_j) < SSN(t_{j-1}) & (\text{when} \quad SSN(t_1) = 1) \text{ or} \\ SSN(t_j) > SSN(t_{j-1}) & (\text{when} \quad SSN(t_1) = 0) \end{cases} \\ 0 \text{ otherwise} \end{cases} \tag{9.8}$$

A few comments about the definition Eq. (9.8) are useful. First, one assumes that early in the morning the Sun is not covered by clouds (i.e., $SSN(t_1) = 1$). According to Eq. (9.8), $SSSN = 1$ only for those moments when the Sun is just covered by clouds. Thus counting the nonnull values of SSSN provides a measure for the frequency of the Sun's disappearance from the sky. Second, one assumes that early in the morning the Sun is covered by clouds (i.e., $SSN(t_1) = 0$). Then, according to Eq. (9.8), $SSSN = 1$ only for those moments when the Sun is just released from the clouds. This time, counting the nonnull values of SSSN provides a measure for the frequency of the Sun's apparition on the sky. To conclude, depending on the initial value of

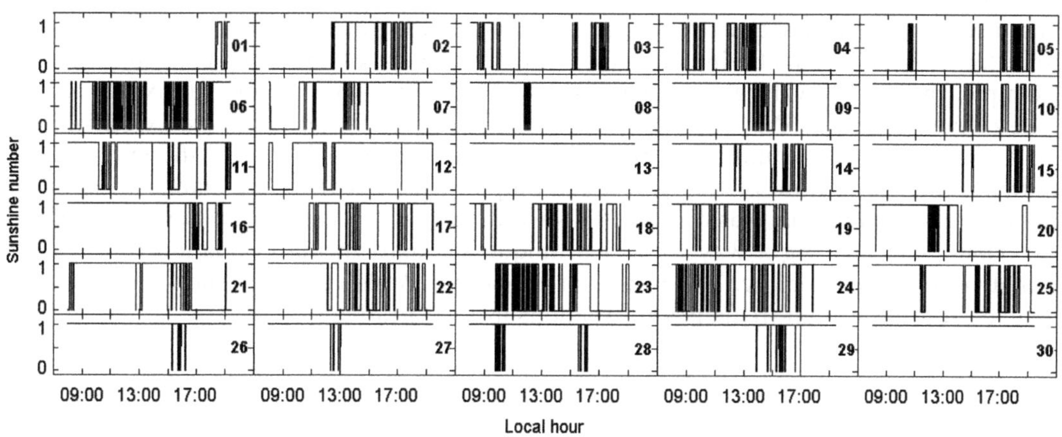

FIGURE 9.5 Sunshine number in every day of June 2019 on the Solar Platform. Only data measured for $h > 15$ degrees are plotted.

SSN, Eq. (9.8) quantifies the frequency of just one of the two different phenomena: Sun's appearance and Sun's disappearance on/from the sky, respectively.

The average value of SSSN during a time interval Δt is denoted by $\overline{SSSN}$. Note that $\overline{SSSN}$ is not a binary variable. It ranges between 0 (when the instantaneous values of SSN are all 0 or 1, respectively, for all time moments t during Δt) and 0.5 (when the SSN time series in Δt consists of alternating 0 and 1 values). The radiative regime is *fully stable* in the first case and *fully unstable* in the last case. Elementary statistical and sequential properties of both SSN and SSSN are presented in Badescu and Paulescu (2011b).

9.2.4 Solar irradiance variability

The variability of solar irradiance may cause erratic variations in the power output of a PV plant at different time scales. Compact cloud fields lead to low-frequency variation of the solar irradiance causing a significant steep increase/decrease in the PV power output. The transition occurs between two states, each of them being stable for a quite long period of time. Scattered clouds may cause high-frequency variation of the solar irradiance inducing a massive fluctuation in the power output of a PV plant. Due to a passing cloud, changes in solar irradiance measured by a pyranometer can exceed more than half of its peak in seconds. The time it takes for a moving cloud to shade the entire array of modules of a PV system depends on various factors, including the PV system size and cloud speed. Mills et al. (2011) show that in case of a 13.2-MW PV plant in Nevada, a 75% solar irradiance ramp in 10-seconds measured by a pyranometer was associated with 20% variation in power output in the same 10-second ramp. A severe event that changed the output of a pyranometer by 80% in 60 seconds led to a 50% change of the power output in the same time. Therefore understanding the variability of solar irradiance at various time scales is a key issue for solar plant deployment.

Generally, the actual level of solar irradiance at the Earth's surface results from the synergistic action of two factors: a deterministic one associated with the Earth's movement and a stochastic one associated with passing clouds. A huge effort has been invested over time for accurately quantifying the stochastic component of the solar irradiance. As an example, the study by Blaga and Paulescu (2018) analyzes the ability of six quantifiers (one of them being $SSSN$) to capture the particularities of the solar irradiance variability. Since the individual quantifiers express the same level of variability at different scales, a normalization procedure of the quantifiers output (inspired from the standardization of the normal distributions) was applied. This was a key task in comparing the output of the six quantifiers, opening a way toward a unique standard in the evaluation of the solar radiative regime variability. The results of Blaga and Paulescu (2018) show that $SSSN$ is able to capture the high-frequency fluctuations in solar irradiance.

9.2.5 Data series used in this study

All the measurements associated with $h < 15$ degrees were removed from the datasets. There are two reasons to do this: (1) the pyranometer accuracy around sunrise and sunset is

TABLE 9.1 Averages of, respectively, global solar irradiance $\overline{G}$, relative sunshine σ, sunshine stability number $\overline{SSSN}$, and clearness index k_t on the test days.

					k_t			
Day	N	$\overline{G}$ **(W/m²)**	σ	$\overline{SSSN}$	**Min**	**Mean**	**Max**	**SD**
6/21	2764	563	0.698	0.0076	0.038	0.530	0.801	0.258
6/22	2763	513	0.695	0.0105	0.105	0.517	0.871	0.214
6/23	2764	402	0.303	0.0336	0.068	0.375	0.981	0.238
6/24	2763	604	0.773	0.0261	0.142	0.600	0.993	0.228
6/25	2763	590	0.811	0.0098	0.122	0.587	0.836	0.189
6/26	2762	677	0.988	0.0033	0.156	0.668	0.785	0.075
6/27	2761	682	0.991	0.0014	0.148	0.673	0.834	0.077
6/28	2759	692	0.968	0.0098	0.134	0.684	0.892	0.102
6/29	2758	695	0.974	0.0065	0.140	0.687	0.853	0.104
6/30	2757	672	1	0	0.401	0.664	0.718	0.048
All	27614	608	0.820	0.0109	0.038	0.599	0.993	0.196

N and SD represent the number of measurements and standard deviation, respectively. Only data series measured for $h > 15$ degrees were considered.

questionable, and (2) $\sin h \rightarrow 0$ when $h \rightarrow 0$ and, consequently, Eq. (9.7) tends to provide $SSN = 1$, regardless of the values of G and G_d. From a practical point of view, this restriction limits the time interval in which forecasts are made at 12 hours, roughly from 8:00 to 20:00 local time, when most of the PV plant electricity is produced. Taking into account all the days of June 2010, the collectable solar energy for $h > 15$ degrees is 189.0 kWh/m^2 representing 95.5% of the total energy collectable between sunrise and sunset.

For fitting the ARIMA models on the clearness index time series data recorded during the first 15 days of June were considered. The data series, consisting of 41,102 records of each quantity measured and postprocessed, was used to fit the parameters of the forecasting models. The second data set contains 27,614 records from the last 10 days of June 2019. This data set was used to assess the performance of the forecasting models. Table 9.1 displays a summary of the main quantifiers of the solar radiative regime in the test days.

Visual inspection of Table 9.1 shows that the database contains a perfect sunny day (June 30), mostly sunny days (June 26, 27), days characterized by a high variability in solar irradiance (June 23, 24) and days experiencing cloudy periods (such as June 21). Testing the forecasting models against such a data set is always a challenge the results give a real picture of the model's behavior in real meteorological conditions. To illustrate the dependence of the models' performance on the variability in solar irradiance, the models were also tested on four independent days with a very different variability in the state-of-the-sky (June 24, 25, 27, and 30).

9.3 Statistical models for short-time solar irradiance

The brief introduction to nowcasting solar resources presented at the beginning of this chapter shows that a large diversity of models for intrahour forecasting of solar irradiance have been developed. The results of testing three different statistical models (persistence, ARIMA forecasting of clearness index, and the two-state model) are discussed in this chapter, aiming to emphasize the strengths and weaknesses in intrahour forecasting of solar irradiance. The three models, significantly different in nature and degree of complexity, are described next.

In the following, a forecast of the physical quantity X will be denoted $\hat{X}$.

9.3.1 Persistence

The persistence represents probably the simplest way of producing a forecast. A persistence model assumes that the future value of a time series is calculated under the assumption that nothing changes between the current time and the forecast time. In terms of solar irradiance, the persistence model estimates that the solar irradiance at the time $t+1$ equals the solar irradiance at the time t:

$$\hat{G}_{t+1} = G_t \tag{9.9}$$

The model accuracy decreases with the increasing of the forecast time horizon and, generally, it is not adequate for a time horizon greater than 1 hour. Here, the persistence model is considered as a reference in evaluating the accuracy of the other two models.

9.3.2 ARIMA modeling of clearness index

Autoregressive integrated moving average (ARIMA) designates a class of time series models, very often used in nowcasting solar irradiance. An ARIMA (p,d,q) model includes the following terms: (1) $AR(p)$ the autoregressive term of order p; (2) $MA(q)$ the moving average term of order q; and (3) $I(d)$ the nonseasonal differencing of order d. The general equation of an ARIMA (p,d,q) model is provided by the classical Box–Jenkins theory (Box and Jenkins, 1970):

$$\underbrace{\left(1-\phi_1 B-\phi_2 B^2-\ldots-\phi_p B^p\right)}_{AR(p)}\underbrace{(1-B)^d}_{I(d)} G_t = c + \underbrace{\left(1-\theta_1 B-\theta_2 B^2-\ldots-\theta_q B^q\right)}_{MA(q)}\varepsilon_t \tag{9.10}$$

where c is a constant term, B is the backshift operator $BG_t = G_{t-1}$ and ε_t is the estimated shock at time t (white noise). Even if Eq. (9.10) looks rather complicated, in practice it reduces to a simple linear equation.

In the first test the model's accuracy in forecasting solar irradiance one step ahead was evaluated at four horizons of time: 1, 5, 10, and 20 minutes. In order to perform the test, for each time horizon, a subseries of data was subtracted from the test data series. The

TABLE 9.2 The coefficients ϕ_1 and θ_1 in Eq. (9.11) fitted to subseries subtracted from the fit data series of clearness index at different lags. The fit quality is indicated by three statistical indicators: root mean square error (*RMSE*) and mean absolute percentage error (*MAPE*).

Lag	*N*	ϕ_1	θ_1	*RMSE*	*SD*	*MAPE* (%)
1 min	10275	0.64296	0.89618	0.084	0.0840	9.5
5 min	2055	0.19388	0.71985	0.127	0.1271	20.9
10 min	1027	0.20111	0.74626	0.145	0.1455	27.8
20 min	513	0.27418	0.74451	0.152	0.1525	32.0

The standard deviation SD quantifies the data dispersion. *N* is the number of data in a data set.

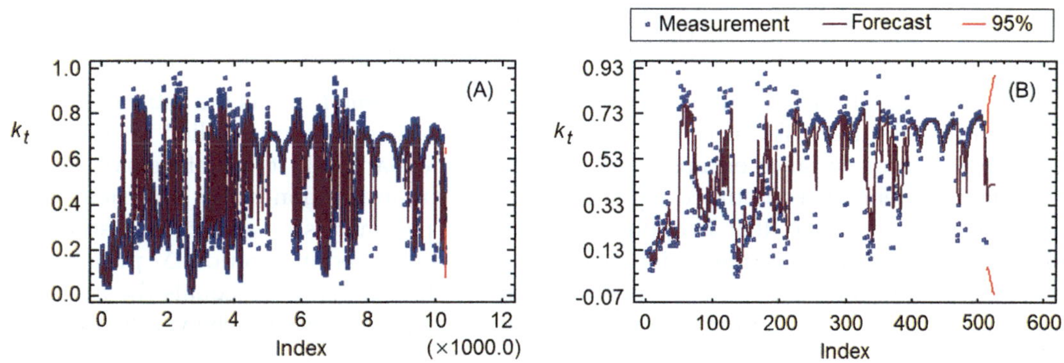

FIGURE 9.6 Comparison of forecasts with the measured clearness index for two subseries subtracted from the fit data series at two different lags: (A) 1 min and (B) 20 min.

ARIMA model was fitted on these subseries. On the basis of the parsimony principle, the ARIMA model selected for forecasting the clearness index is ARIMA(1,1,1), given by:

$$\hat{G}_t = G_{t-1} + \phi_1(G_{t-1} - G_{t-2}) + \theta_1\left(\hat{G}_{t-1} - G_{t-1}\right) + \varepsilon_t \tag{9.11}$$

The coefficients ϕ_1 and θ_1 were obtained by using the maximum likelihood method (see, e.g., Brockwell and Davis, 2016). Aiming to provide the reader with a clear perspective on the forecasting models, Table 9.2 presents the fitted coefficients and a statistic summary of the fit accuracy. Fig. 9.6 illustrates a time sequence plot of the forecasted and measured clearness index for two subseries subtracted from the fit data series at lags 1 and 20 minutes, respectively.

9.3.3 The two-state model

Basically, the two-state model connects an empirical model for estimating the clear-sky solar irradiance with a statistical model for nowcasting sunshine number (Paulescu et al., 2014). The model operates as follows: (1) if the Sun shines, the future value in the solar irradiance time series is estimated with an empirical clear-sky model (see, e.g., Badescu et al. (2013) for a comprehensive review on clear-sky solar irradiance models); (2) if the Sun is covered by clouds, the clear-sky solar irradiance estimate is adjusted to the actual value of the cloud transmittance. The difference between the two states is made by the forecasted value of SSN. Note that SSN is the sole quantity forecasted within the two-state model. However, forecasting on a random binary time series is always a big challenge. A huge effort with partial success was paid for by enhancing the accuracy of SSN forecasts (Brabec et al., 2014).

An upgraded version of the two-state model for nowcasting solar irradiance was reported by Paulescu et al. (2018). In addition to its first version, the estimated atmospheric transmittance is adjusted not only for the state SSN = 0, but also for the state SSN = 1. The model equation reads :

$$\hat{G}_t = \begin{cases} c_{cs} \cdot G_{0,t} \text{if} & S\hat{S}N_t = 1 \\ \overline{\tau}_c \cdot G_{0,t} \text{if} & S\hat{S}N_t = 0 \end{cases} \tag{9.12}$$

where $\hat{G}_t$ is the forecasted value of the solar irradiance at time t, $G_{0,t}$ is the estimated solar irradiance under clear sky at time t, and c_{cs} is a dynamic correction applied to the mean atmospheric transmittance encapsulated into the $G_{0,t}$ equation. In the first version of the two-state model c_{cs} was assumed equal to one. $\overline{\tau}_c$ is an attenuation factor consistent with the cloud transmittance, that is applied to $G_{0,t}$. Both parameters c_{cs} and $\overline{\tau}_c$ are estimated simultaneously by a linear regression applied to data measured in a period Δt prior to the forecasting moment. Regarding the period Δt, it was found that if Δt equals the forecasting time horizon, the minimum error in estimating the parameters is achieved (Costa and Mares, 2014).

For this study, the two-state model was implemented as follows. Irrespective of the time horizon, the ARIMA(0,1,0) model was applied to forecast *SSN*. This option is motivated by the results from Badescu and Paulescu (2011b) and Paulescu et al. (2014). Using different approaches, the two studies argue that ARIMA(1,0,0) and ARIMA(0,1,0) are adequate solutions for nowcasting *SSN*. The very simple clear-sky solar irradiance model of Adnot et al. (1979) was used for computing $G_{0,t}$ in Eq. (9.12):

$$G_{0,t}\left[W/m^2\right] = 951.39(\sin h_t)^{1.15} \tag{9.13}$$

Using the model by Adnot et al., is not a critical choice. The reason is that, before performing a forecast, on the basis of the most recent available measurements, the two-state model adjusts the atmospheric transmittance encapsulated into the empirical coefficients of the clear-sky model.

9.3.4 Accuracy metrics

A large diversity of metrics has been proposed and used to quantify the accuracy of solar irradiance forecasts. Currently, there is no consensus on which metrics are the most suitable for quantifying the performance of a forecasting model.

The following statistical indicators are very often used for measuring the model's performance in forecasting solar irradiance: mean bias error *MBE*, mean absolute error *MAE*, root mean square error *RMSE*, and mean absolute percentage error *MAPE*. Considering a sample of size M and denoted by c and m the computed (forecasted in this case) and measured values, respectively, the four statistical indicators are defined as follows:

$$MBE = \frac{1}{M}\sum_{i=1}^{N}(c_i - m_i) \quad (9.14)$$

$$MAE = \frac{1}{M}\sum_{i=1}^{N}|c_i - m_i| \quad (9.15)$$

$$RMSE = \left[\frac{1}{M}\sum_{i=1}^{N}(c_i - m_i)^2\right]^{1/2} \quad (9.16)$$

$$MAPE = 100 \times \frac{1}{M}\sum_{i=1}^{N}\frac{|c_i - m_i|}{m_i} \quad (9.17)$$

MBE is measure of systematic errors (or bias). It encapsulates the systematic tendency of a forecast model to under- or overestimate a future value. *MAE* quantifies the average magnitude of forecast errors and indicates the closeness of forecasts to measurements. *RMSE* is a measure of random errors giving more weight to the largest errors. *MAPE* represents a direct measure of the prediction accuracy of a forecasting model. Note that *MAPE* is not defined when the measured value is equal to zero. The units for *MBE*, *MAE*, and *RMSE* are identical to the units of c and m while *MAPE* is expressed as a percentage.

In order to compare the performance of a model applied to samples with different magnitudes of the components, the first three indicators are normalized with reference to the mean value $\mu = \frac{1}{M}\sum_{i=1}^{N} m_i$ of the measured data:

$$nMBE = \frac{MBE}{\mu} \quad (9.18)$$

$$nMAE = \frac{MAE}{\mu} \quad (9.19)$$

$$nRMSE = \frac{RMSE}{\mu} \quad (9.20)$$

The forecast accuracy is strongly dependent on location, forecast time horizon, variability in data series used for evaluation, and many other factors. Consequently, it is difficult to evaluate the quality of a forecasting model from accuracy metrics alone. Blaga et al. (2019) recommend that the presentation of forecast performances is to be done in terms of *nMBE* and *nRMSE*. Additional statistical error measures like the Kolmogorov–Smirnoff integral (see, e.g., Espinar et al., 2009), which compares the frequency distribution of forecasted and measured solar irradiance, is also recommended. More insight forecasting performance can be gained by comparing the accuracies of different forecasting models against the accuracy of a simple benchmark model. The skill score (or forecast skill) is such a comparative measure associated with a specified accuracy metric and a particular reference forecast. Commonly used as reference is the persistence model. An example of a skill score (SS) calculation which uses the accuracy metric *RMSE* (Eq. 9.20) and the persistence as a reference is

$$SS = 1 - \frac{RMSE_{model}}{RMSE_{persistence}} \tag{9.21}$$

In the case of a perfect forecast, $RMSE_{model} = 0$ and $SS = 1$. A value of SS between 0 and 1 indicates the improvement in *RMSE* over persistence. Multiplied by 100, SS expresses the improvement as a percentage. A forecast with $RMSE_{model} = RMSE_{persistence}$ results in a skill score $SS = 0$. A forecast which is less accurate than the persistence forecast results in a negative skill score.

9.4 Performance of the solar irradiance forecast

In this section the performance of forecasts performed by three models (persistence, ARIMA, and two-state) are evaluated against measured data on the Platform of West University of Timisoara, Romania. The data set used for testing the models was described in Section 9.2.5. The tests were made from three perspectives: forecast accuracy at different time horizons, forecast precision, and the influence of the variability in the state-of-the-sky on the forecast accuracy.

9.4.1 Time horizon

This test is focused on the models' accuracy in forecasting solar irradiance one step ahead for four different time horizons: 1, 5, 10, and 20 minutes. In order to perform the test, for each time horizon an appropriate subseries of data was subtracted from the test data series. The results are presented in Fig. 9.7.

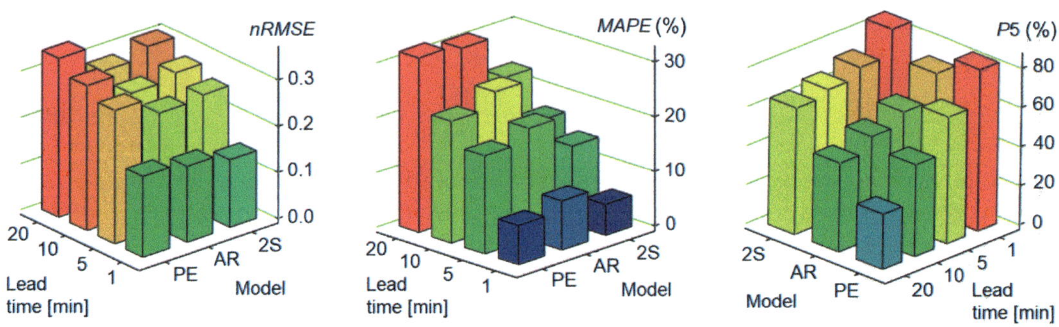

FIGURE 9.7 Performance of the persistence (PE), ARIMA (AR) and two-state (2S) models in forecasting global solar irradiance at four time horizons (1, 5, 10, and 20 min) measured by three different statistical indicators of accuracy (see Section 9.3.4 for definitions): normalized root mean square error *nRMSE*, mean absolute percentage error *MAPE*, and the percentage of forecast accurate to within 5% around measured value *P5*.

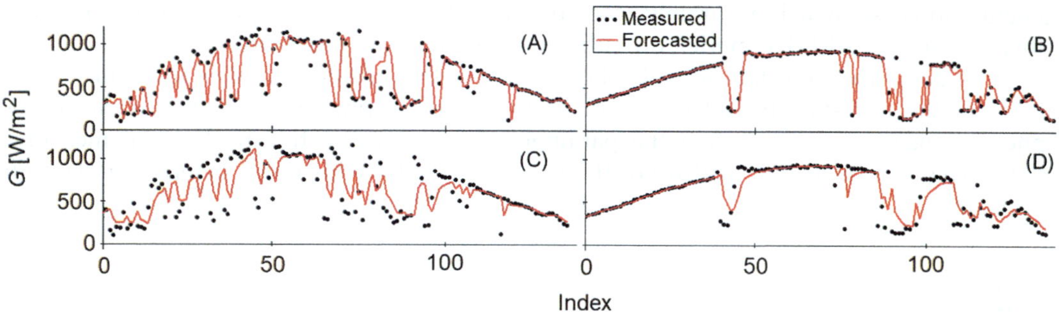

FIGURE 9.8 Measured and forecasted global solar irradiance in (A, C) June 24, 2019 and June 25, 2019 by (A, B) the two-state model and (C, D) ARIMA model at a time horizon of 5 min.

First of all, regardless of the model, no noticeable bias in forecast was observed, the maximum value of *nMBE* being recorded for the ARIMA model at 20 minutes time horizon, $nMBE = 0.003$. For this reason, *nMBE* is not plotted in Fig. 9.7. Generally, the models' performance decreases with the increasing of the forecast time horizon. At a given time horizon the performance of the three models is rather comparable. However, different metrics indicate different hierarchy. Looking at *nRMSE* (Fig. 9.8A), at very short-time horizon, the two-state model is best ranked followed by ARIMA and persistence (with *nRMSE* at 1 minutes time horizon 0.147, 0.165, and 0.175, respectively). At larger time horizon ARIMA is in first place followed closely by the two-state model and persistence (with *nRMSE* for 20 minutes time horizon 0.284, 0.305 and 0.344, respectively).

In contrast to *nRMSE*, *MAPE* makes a clearer distinction between the models' performances. The first position in the hierarchy is occupied by the two-state model ($5.7\% < MAPE < 23.0\%$). The second position is occupied by persistence ($7.2\% < \text{MAPE} < 31.9\%$). A few explanations follow. Persistence forecasts the current state forever in the future. For very short-time horizons, as long as the state-of-the-sky is stable, this is a good, even very good, approximation. Large errors in forecasting may arise if the state-of-the-sky changes between the time when a forecast was generated and the time for which the forecast was generated. Thus persistence somewhat mimics the two-state model behavior and its performance. The percentage of forecast accurate to within a given interval (5% in Fig. 9.7) centered on measurements is a measure of the model's precision, which is discussed in the next section. At the same time this percentage is a complementary indicator of the forecast accuracy, providing additional information. For example, *nRMSE* penalizes the large error in the forecast. A model which experiences a quite large *nRMSE* may be considered less accurate. A high value of the percentage of forecasts which are accurate to within a small interval centered on the measurements indicates that only few forecasts are accompanied by significant errors. This is the case of the two-state model: if *SSN* is correctly forecasted, the solar irradiance is also very accurate forecasted; if *SSN* is wrongly forecasted, a large error in forecasting solar irradiance occurs.

Fig. 9.8 displays the forecasted solar irradiance time series by the two-state and ARIMA models, superimposed on measurements recorded in two days of June 2019. This picture emphasizes a basic difference between the two models. When the Sun shines, both models generate forecasts which accurately track the measurements. During the periods of instability, while the ARIMA model provides a smoother curve (specific to the conditional mean models), the two-state model follows the abrupt variation of solar irradiance. This ability of the two-state model is due to its permanently fine-tuning to the actual atmospheric transmittance, by fitting the parameters c_{cs} and $\overline{\tau}_c$ on the most recent measurements. This is a distinctive feature of the two-state model which also determines its high precision.

9.4.2 Precision

An appropriate measure for the forecasts precision is the percentage of forecasts $P[\%]$ accurate to within a given interval $T[\%]$, centered on measurements. As Fig. 9.7 shows, the two-state model is by far the most precise model. More details on its precision are presented in Fig. 9.9, where the percentage P is plotted against T at different time horizons in four test days that are differentiated by the variability in the state-of-the-sky (see Table 9.1).

In general, the model precision decreases by increasing the time horizon of the forecast. For example, on June 25, for $T = 5\%$, at a time horizon of 1 minutes, $P = 84.0\%$ which means that 84.0% of forecasts (578 forecasts from a total of 689) are within the interval$G_{measured} \pm 0.05 \cdot G_{measured}$. For the same $T = 5\%$, this percentage decreases to $P = 71.5\%$ at a time horizon of 5 minutes and to $P = 57.5\%$ at a time horizon of 20 minutes. Another feature that emerges from Fig. 9.9 is the strong dependence of P on the variability in solar irradiance. June 24 is a day characterized by a high variability in the state-of-the-sky ($\sigma = 0.773$ and $\overline{SSSN} = 0.0261$), $P = 37.9\%$. This percentage increases to 71.5% on June 25, a day with a moderate variability ($\sigma = 0.811$ and $\overline{SSSN} = 0.0098$) and exceeds 99% on the perfectly clear and stable day, June 30.

9.4.3 Stability of the solar radiative regime

In this test, the model performance in forecasting solar irradiance for 5-minutes ahead was evaluated in four days from June 2019. These days experienced very dissimilar

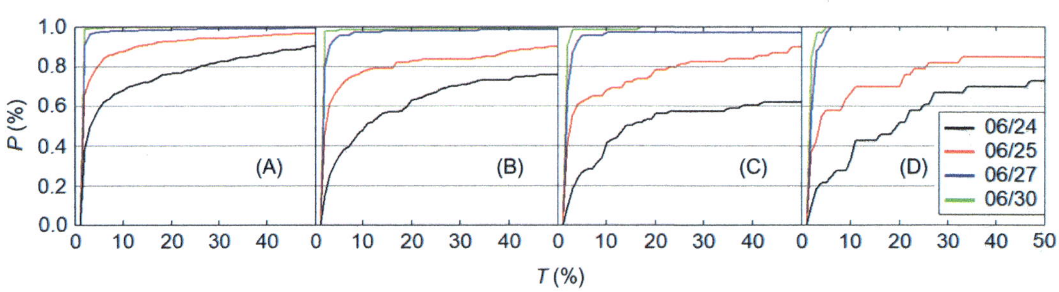

FIGURE 9.9 Percentage P of forecasts generated by the two-state model accurate to within a given interval T centered on the measurements. Four different time horizons are considered: (A) 1 min, (B) 5 min, (C) 10 min and (D) 20 min.

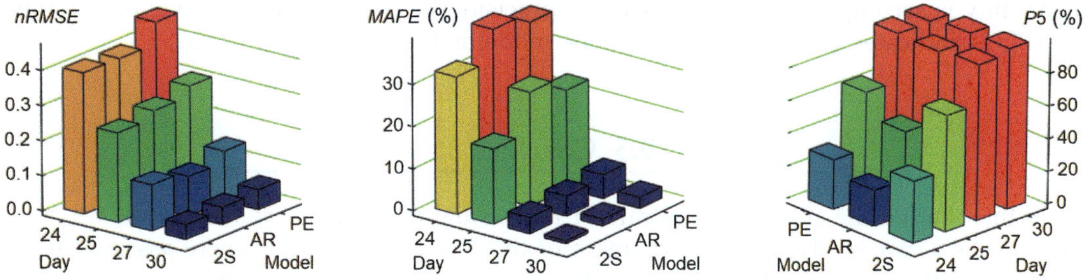

FIGURE 9.10 Performance of the persistence (PE), ARIMA, and the two-state models in forecasting global solar irradiance at a time horizon of 5 min, during four days of June 2019. The days are differentiated by the variability in the state-of-the-sky (see Table 9.1). Three different statistical indicators of accuracy (see Section 9.3.4 for definitions): normalized root mean square error *nRMSE*, mean absolute percentage error *MAPE*, and the percentage of forecast accurate to within 5% around measured value *P*5 are displayed.

patterns in solar irradiance variability (see Fig. 9.3). The days are: June 24, a day with a very unstable solar radiative regime ($\sigma = 0.773$ and $\overline{SSSN} = 0.0261$); June 25, a day with a moderate instability episode which occurred on a sunny background ($\sigma = 0.811$ and $\overline{SSSN} = 0.0098$); June 27, a day mostly sunny with an episode of instability in the state-of-the-sky at noon ($\sigma = 0.991$ and $\overline{SSSN} = 0.0014$); and June 30, a perfectly clear-sky day ($\sigma = 1$ and $\overline{SSSN} = 0$). Fig. 9.10 summarizes the results. Irrespective of model and statistical indicator, the picture highlights a strong influence of the state-of-the-sky variability in the forecast accuracy. The models' accuracy suddenly decreases by increasing the instability in the state-of-the-sky. For example, in the case of the two-state model, *nRMSE* increases about 10 times from 0.043 on the stable day June 30 to 0.403 on the unstable day June 24. The same behavior is also noticed for precision. *P*[5%] decreases about four times, from about 98.5% on June 30 to 37.9% on June 24. In terms of *MAPE*, the decrease in the models' accuracy with increasing the instability in the state-of-the-sky is steeper. For the two-state model *MAPE* increases about 40 times, from 0.37% on June 30 to 32.7% on June 24. Regardless of the level of the instability, the forecasts are not biased ($|nMBE| < 0.007$).

Concerning the response of an individual model to the change in the solar irradiance variability, some differences are observed. On June 24, in very unstable conditions, in terms of *nRMSE*, ARIMA and the two-state model achieve the same performance ($nRMSE = 0.40$) while the persistence model records a poorer performance ($nRMSE = 0.464$). This picture is preserved in all the test days. The low performance of the two-state model when the state-of-the-sky is highly variable is due to the increasing uncertainty in sunshine number forecasting with increasing variability in the state-of-the-sky. When the sunshine number is not predicted correctly, a large error in nowcasting solar irradiance occurs. Since *nRMSE* sums up the square of errors, it penalizes large errors more. In a stable solar radiative regime, there are very few changes of sunshine number, and therefore there is a low probability for an incorrect forecast of sunshine number. The variability in the state-of-the-sky has also a strong influence on the forecasting precision. For all the models, the transition of state-of-the-sky from stability to instability is accompanied by a decrease in precision. Irrespective of the level of variability, $P[5\%]$ indicates again that the two-state model is the most performand.

9.5 Conclusions

In this chapter the accuracy of intrahour forecasting of solar irradiance models was discussed. As a case study, the performances of three statistical models were assessed from different perspectives: the forecast's time horizon, the forecast's precision, and the variability in the state-of-the-sky. The three models are persistence, autoregressive integrated moving average (ARIMA), and the two-state model. The two-state model, developed by our team, works as follows: (1) if the Sun is shining, the solar irradiance is estimated by a clear-sky model whose atmospheric transmittance is adjusted in real-time; otherwise (2) the clear-sky estimation of solar irradiance is adjusted with the cloud transmittance. These two states are differentiated by the sunshine number, which is the sole quantity directly forecasted.

The results of testing the models against data measured on the Solar Platform of the West University of Timisoara (Romania) during June 2019 led to the following conclusions. Generally, the models' performances decrease with an increase in the time horizon of the forecasts. Regardless of the model, no noticeable bias in forecasts is observed. Different statistical indicators capture different particularities of the models. Looking at the relative root mean square error, a rather surprising picture was noted: to some extent the models' accuracies are the same. This is a consequence of the large errors experienced by all forecasts in days with high variability in the state-of-the-sky, which are strongly penalized by the relative root mean square error. Differently, the mean absolute percentage error clearly classifies the models. The model precision was measured by the percentage of forecasts accurate to within a given interval centered on measurements. A very good precision of the two-state model was noted. This is explained by its structure, which allows a permanent adjustment to the actual value of the atmospheric transmittance. Thus always when sunshine number is correctly predicted, the forecasts trace the measurements with accuracy. This gives a significative advantage to the two-state model in terms of precision. The variability in the state-of-the-sky has a strong influence on the accuracy of forecasts: the models' performances decrease when the instability increases. This major limitation is associated with the persistence property, that is, the general tendency of a statistical model to extrapolate the current state in the future. A substantial advance in the fight against the persistence effects may be obtained by looking deeper into the solar radiative transfer through the atmosphere and bringing it into the forecasting models.

References

Adnot, J., Bourges, B., Campana, D., Gicquel, R., 1979. Utilisation des courbes de frequence cumulees pour le calcul des installation solaires. In: Lestienne, R. (Ed.), Analise Statistique des Processus Meteorologiques Appliquee a l'Energie Solaire. Editions du CNRS, Paris, pp. 9–40.

Badescu, V., Paulescu, M., 2011a. Statistical properties of the sunshine number illustrated with measurements from Timisoara (Romania). Atmos. Res. 101, 194–204.

Badescu, V., Paulescu, M., 2011b. Autocorrelation properties of the sunshine number and sunshine stability number. Meteorol. Atmos. Phys. 112 (3-4), 139–154.

Badescu, V., Gueymard, C.A., Cheval, S., Oprea, C., Baciu, M., Dumitrescu, M., et al., 2013. Accuracy analysis for fifty-four clear-sky solar radiation models using routine hourly global irradiance measurements in Romania. Renew. Energy 55, 85–103.

Blaga, R., Paulescu, M., 2018. Quantifiers for the solar irradiance variability: a new perspective. Sol. Energy 174, 606–616.

Blaga, R., Sabadus, A., Stefu, N., Dughir, C., Paulescu, M., Badescu, V., 2019. A current perspective on the accuracy of incoming solar energy forecasting. Prog. Energy Combust. Sci. 70 (119-144), 2019.

Boland, J., David, M., Lauret, P., 2016. Short term solar radiation forecasting: island versus continental sites. Energy 113, 186–192.

Box, G.E.P., Jenkins, G.M., 1970. Time Series Analysis. Forecasting and Control. Holden-Day, San Francisco.

Brabec, M., Paulescu, M., Badescu, V., 2014. Generalized additive models for nowcasting cloud shading. Sol. Energy 101, 272–282.

Brockwell, P.J., Davis, R.A., 2016. Introduction to Time Series and Forecasting, third ed Springer International Publishing.

Costa, D., Mares, O., 2014. Estimation of the cloud transmittance from radiometric measurements at the ground level. AIP Conf. Proc. 1634, 151–155.

Dong, Z., Yang, D., Reindl, T., Walsh, W.M., 2013. Short-term solar irradiance forecasting using exponential smoothing state space model. Energy 55, 1104–1113.

Espinar, B., Ramírez, L., Drews, A., Beyer, H.G., Zarzalejo, L.F., Polo, J., et al., 2009. Analysis of different comparison parameters applied to solar radiation data from satellite and German radiometric stations. Sol. Energy 83, 118–125.

Fu, C.-L., Cheng, H.-Y., 2013. Predicting solar irradiance with all-sky image features via regression. Sol. Energy 97, 537–550.

Gueymard, C., 2018. A reevaluation of the solar constant based on a 42-year total solar irradiance time series and a reconciliation of spaceborne observations. Sol. Energy 168, 2–9.

Inman, R.H., Pedro, H.T.C., Coimbra, C.F.M., 2013. Solar forecasting methods for renewable energy integration. Prog. Energy Combust. Sci. 39, 535–576.

Kottek, M., Grieser, J., Beck, C., Rudolf, B., Rubel, F., 2006. World Map of the Köppen-Geiger climate classification updated. Meteorol. Z. 15, 259–263.

Liu, B., Jordan, R., 1960. The interrelationship and characteristic distribution of direct, diffuse and total solar radiation. Sol. Energy 4 (3), 1–19.

Mills, A., Ahlstrom, M., Brower, M., Ellis, A., George, R., Hoff, T., et al., 2011. Dark shadows. Understanding variability and uncertainty of photovoltaics for integration with the electric power system. IEEE Power Energy M. 9, 33–41.

Paulescu, M., Badescu, V., 2011. New approach to measure the stability of the solar radiative regime. Theor. Appl. Climatol. 103, 459–470.

Paulescu, M., Paulescu, E., Gravila, P., Badescu, V., 2013. Weather Modeling and Forecasting of PV Systems Operation. Springer, London.

Paulescu, M., Mares, O., Paulescu, E., Stefu, N., Pacurar, A., Calinoiu, D., et al., 2014. Nowcasting solar irradiance using the sunshine number. Energy Convers. Manag. 79, 690–697.

Paulescu, M., Mares, O., Dughir, C., Paulescu, E., 2018. Nowcasting the output power of PV systems. E3S Web Conf. 61, Article number 00010.

Paulescu, M., Paulescu, E., 2019. Short-term forecasting of solar irradiance. Renew. Energy 143, 985–994.

Pedro, H.T.C., Coimbra, C.F.M., 2015. Short-term irradiance forecastability for various solar micro-climates. Sol. Energy 122, 587–602.

Perez, R., David, M., Hoff, T.E., Jamaly, M., Kivalov, S., Kleissl, J., et al., 2016. Spatial and temporal variability of solar energy. Found. Trends Renew. Ener. 1 (1), 1–44.

Reikard, G., Haupt, S.E., Jensen, T., 2017. Forecasting ground-level irradiance over short horizons: time series, meteorological, and time-varying parameter models. Renew. Energy 112, 474–485.

Schmela, M., Beauvais, A., Chevillard, N., Paredes, M.G., Heisz, M., Rossi, R., et al. 2019. Global Market Outlook 2019–2023. Solar Power Europe <http://www.solarpowereurope.org/>.

Sepasi, S., Reihani, R., Howlader, A.M., Roose, L.R., Matsuura, M.M., 2017. Very short-term load forecasting of a distribution system with high PV penetration. Renew. Energy 106, 142–148.

Solar Platform, 2019. Solar Platform of the West University of Timisoara, Romania. <http://solar.physics.uvt.ro/srms>.

Spencer, W., 1971. Fourier series representation of the position of the sun. Search 2, 172.

Vignola, F., Michalsky, J., Stoffel, T., 2012. Solar and Infrared Radiation Measurements. CRC Press, Boca Raton.

WMO 2008. World Meteorological Organization. Guide to Meteorological Instruments and Methods of Observation, WMO-No.8/2008 <http://www.wmo.int/>. (accessed May 2020).

Yan, J., Liu, Y., Han, S., Wang, Y., Feng, S., 2015. Reviews on uncertainty analysis of wind power forecasting. Renew. Sustain. Energy Rev. 52, 1322–1330.

CHAPTER

10

Short-term electrical energy demand prediction under heat island effects using emotional neural network integrated with genetic algorithm

Sagthitharan Karalasingham[1], *Ravinesh Deo*[1] *and Ramendra Prasad*[1]

[1]School of Sciences, University of Southern Queensland, Springfield, QLD, Australia
[2]Department of Science, School of Science and Technology, The University of Fiji, Saweni, Lautoka, Fiji

10.1 Introduction

Part of our modern life, the seamless availability of electricity is taken for granted by most. Nevertheless, the effective delivery of electricity to consumers is critical to the social and economic well-being of nations. The transformative nature of electricity in a modern economy is such that electricity grids are considered part of the critical energy infrastructure (Amin, 2003; Räikkönen et al., 2016). However, the stability and resilience of this critical infrastructure are frequently exposed by blackouts and brownouts leaving millions without power, and thus exposing deep vulnerabilities in the production, transmission, and consumption of electricity (Panteli and Mancarella, 2015; Amin, 2003).

In an urbanized and digitalized world, where more than 50% of the world's population already lives in cities, the load on the electrical grid is increasingly shaped by the consumption patterns of its inhabitants in response to the climate experienced. In recent years, new uncertainties in weather and climate patterns have emerged compounding the uncertainties already impacting on short, medium, and long-term demand forecasting. The long-term trend of anthropogenic climate change, its potential impact on our cities, and

Predictive Modelling for Energy Management and Power Systems Engineering
DOI: https://doi.org/10.1016/B978-0-12-817772-3.00010-0

the unrelenting urbanization has sharpened focus on changes occurring in our local climate and their potential impact on energy consumption.

The accelerating shift from rural to urban living has contributed to the emergence of urban-scale climatic conditions (Esch et al., 2017). The urban transition has resulted in both increased urban footprint, and at the same time, higher density as efforts shift toward reducing the per-person footprint with medium and high-density developments (Jackson et al., 2016). As Oke et al. (2017) state, the reclaiming of rural landmass for an expansive urban footprint, changes to the built layout, the addition of public infrastructure, moves toward higher urban densities, and changes to economic structure have become "main drivers of urban climatic effects." Mestayer (1998) also found that overall changes to tree canopy and land usage patterns which are predominantly anthropogenic in nature fundamentally changed the urban canopy exerting control over wind, temperature, and humidity, giving rise to local climate zones (Fig. 10.1).

Further, the thermal characteristics of the built environment (Oke et al., 2017) and the changes to air circulation patterns (Mestayer, 1998) have given rise to local climate zones which are capable of altering surface climates within urban areas. The temperature differences between the near-surface air within the urban canopy and an equivalent rural setting give rise to urban heat islands (UHI) where the average temperature within urban settings tend to be higher. The adverse effects of UHI on urban communities range from the impact on human health to higher running costs of buildings.

The urban climatic effects are further exacerbated by macroscale climate change which intensifies urban climate phenomena. With the increasing prevalence of heatwaves, cold snaps, and other weather patterns the dynamics between urban-scale climate phenomena and macroscale climate change are set to have a significant impact on our energy consumption patterns. Studies in Athens, Greece by Founda and Santamouris (2017) and in the United States by Zhao et al. (2018) found that the interactions between the two phenomena intensified temperatures by up to 3.5 K in Athens and between 0.4 K–2.8 K in regions affected by UHI in the United States. Similarly, a survey of studies conducted by Santamouris (2014) on urban warming due to both UHI and climate change found that the average energy penalty of UHI that is proportional to the building density in a city stood at 2.4 ($\pm$ 1.5) kWh/m^2 per unit city surface (GEPS). Other studies have found that the number of heating and cooling degree days is comparatively different within an UHI than rural settings (Kolokotroni et al., 2010).

One of the contributors to the instability remains extreme weather events, such as heatwaves, ice storms, and storms. As Panteli and Mancarella (2015) state the long-term trend of climate change and the associated possibility of extreme weather events require the adaptation of electrical grids to "high-impact low-probability events" (Panteli and Mancarella, 2015). However, the probability of extreme events based on climate change is further exacerbated by urban climatic phenomena such as UHI and urban heat continent (UHC). In particular, the intensity and duration of extreme heat events such as heatwaves are expected to be magnified by UHI and UHC (Zhao et al., 2018). Ramamurthy and Bou-Zeid (2017) found the intensity of heatwaves increased when combined with UHI in the densest and populous cities in the United States resulting in temperature increases of 1.5 K–2 K while cities with lower density did not experience noticeable change, thus highlighting the diverse nature of the effects of urban environments.

Further, Zhao et al. (2018) also found that the use of climate-conditioning devices during heat waves contributed to an increase in anthropogenic heat intensifying the heat impact. Mapping the linkages between UHI and energy demand, a survey of studies conducted by Santamouris (2014) concluded that UHI imposed an average energy penalty per unit of city surface (GEPS) of 2.4 ($\pm$1.5) kWh/m^2. GEPS as a measure is directly proportional to the building density in a city.

Other than magnifying electricity demand, the effect of UHIs in amplifying the effects of macroscale weather events injects a higher degree of uncertainty in the functioning of electrical grids, thereby exacerbating grid instability, especially during summer seasons. Chapman et al. (2013) considered the interplay between urban heat and networks (energy, transport, broadband) and concluded that the electricity networks were most at risk of temperature-related events due to the operating limits of electricity generation and transmission equipment and potentially shortening their service life. The heatwave of February 2017, which resulted in the loss of more than 14% of the power generation capacity affecting the states of South Australia, New South Wales (NSW), and Queensland in Australia resulted due to a mix of equipment failure and exceeding of capacity limits (Australia Institute, 2017).

In the context of electricity demand and the overall stability and resilience of the electricity grid, the modeling of the relationship between UHI-affected regions and electricity in this research is critical to electricity demand planning and safe distribution during the summer seasons. Furthermore, the "high-resolution monitoring and modeling" (Chapman et al., 2013) taking into account the spatial and temporal differences of the effects of UHI is critical for identifying the disparate patterns in electricity demand requests from our urban environments.

More broadly, the data-driven modeling of the spatial and temporal differences in electricity demand would enable the consideration of energy efficiency on an urban scale. Since the structure of the urban environment imposes an energy penalty on their potential residents, initiatives toward energy efficiency at the building level could be enhanced through urban-scale initiatives aimed at mitigating heat intensity, resulting in an overall decrease in energy demand and reduction in energy peaks.

In this context, the aim of this project is to develop a data-driven model to predict short-term electricity demand, day-ahead, incorporating granular climate data capturing the effects of urban-scale climatic phenomena such as UHI. In particular the development of energy demand prediction models which take into account climatic variability in space and time, while being computationally efficient will be of practical use for the players in generating near-real-time predictions for the electricity market and policy planners.

The models evaluated against UHI-affected sites provide an important tool in capturing the shifts in electricity consumption, thereby influencing the application of electricity demand modeling toward the study of energy efficiency at the urban scale contributing to the design of energy-efficient cities.

10.2 Theoretical overview

10.2.1 Hybrid winner-take-all emotional neural network

Emotional neural networks (ENNs), the most recent evolution of the artificial neural networks (ANNs), are modeled on the human emotional brain or the limbic system, which regulates emotional behavior and responses and is known for its quick reaction to stimuli (Zamirpour and Mosleh, 2018). Unlike the traditional ANNs, which suffer from dimensionality and hence increasing computational complexity as the numbers of inputs and outputs increase, ENNs models the emotional states in learning algorithms with low computational complexity, quick reaction with the ability to inhibit incorrect response, thereby accelerating learning speed (Lotfi et al., 2014). However, generations of ENN architectures have remained application-specific due to their limited ability to store patterns, leading to lower information capacity (IC) and model uncertainty (Zamirpour and Mosleh, 2018; Lofti and Akbarzadeh, 2014).

Within the brain emotional learning (BEL) framework, ENNs replicate the biological equivalent of the emotional brain, by mapping the pathways of stimuli from the thalamus to the orbito-frontal cortex (OFC) and amygdala via shorter and long pathways and capturing the excitatory and inhibitory interactions between the amygdala and OFC to produce an output or response (Lotfi et al., 2014). However, with the more expansive view of brain response to stimuli, one needs to take into account the role of neural competition, which models the competition between excitatory and inhibitory connections resulting in the selection of the most favored inputs (Wu et al., 2014; Lotfi and Akbarzadeh-T, 2016).

Representing the "winner-take-all behavior," this study considers an ENN which incorporates neural competition in the form of a winner-take-all competition layer within the WTAENN, as proposed by Lotfi and Akbarzadeh-T (2016), for the prediction of short-term electricity demand forecasting for UHI-affected areas within ACT and NSW.

The winner-take-all architecture proposed by Lotfi and Akbarzadeh-T (2016) expands the previous models by allowing the input received by the thalamus to be distributed across multiple sensory cortexes, where only a single sensory cortex is ultimately selected as the winner resulting in the stimuli being fed to the OFC and amygdala resulting in an output. The revised winner-take-all architecture has expanded the IC of the single layer, addressing one of the limitations of ENNs highlighted earlier, while limiting computational cost and computational complexity (Lotfi and Akbarzadeh-T, 2016).

The mathematical equivalent of the basic ENN model by Lotfi et al. (2014) is as follows:

Consider the input vector:

$$\boldsymbol{P} = [p_1, p_2, p_3 \ldots p_n]$$

where $p_1 \ldots p_n$ are inputs to thalamus which are then passed on to sensory cortex.

The imprecise output of thalamus, P_{n+1}, represented by the mean operator is as follows (Lofti et al., 2014):

$$P_{n+1} = \text{mean}_{j=1\times n}(p_j) \tag{10.1}$$

The output P_{n+1} and outputs from sensory cortex become the inputs of amygdala where the weighted output of amygdala is (Lofti et al., 2014):

$$E_a = \sum_{j=1}^{n+1} (v_{j*}p_j) \tag{10.2}$$

where v_j is the amygdala weights.

The outputs from sensory cortex also become the inputs of OFC where the weighted output of OFC is (Lofti et al., 2014):

$$E_o = \sum_{j=1}^{n+1} \left(w_{j*}p_j\right) + b \tag{10.3}$$

where w_j is the OFC weights, B is the OFC bias, and p_j is the inputs

The outputs from OFC and amygdala result in the final output (Lofti et al., 2014):

$$E = E_a - E_o \tag{10.4}$$

The revised, winner-take-all model incorporating neural competition can be stated mathematically as (Lotfi and Akbarzadeh-T, 2016):

Consider the original input vector

$$\boldsymbol{P} = [p_1, p_2, p_3 \ldots p_n]$$

where $p_1 \ldots p_n$ are inputs to thalamus which are then passed on to sensory cortex modules.

The number of sensory cortex modules denoted by m and learning weights denoted as $c_1, c_2, \ldots c_n$ the single winning input i^* to sensory cortex, OFC, and amygdale are defined as (Lotfi and Akbarzadeh-T, 2016)

$$\forall_i \| [p_1, p_2, \ldots, p_n] - [c_{1,i*}, c_{2,i*}, \ldots, c_{n,i*}] \| \leq \| [p_1, p_2, \ldots, p_n] - [c_{1,i*}, c_{2,i*}, \ldots, c_{n,i*}] \|, \tag{10.5}$$

where

$1 \leq i \leq m$

The winning input from (10.5) produces the final output (E^*) as defined below (Lotfi and Akbarzadeh-T, 2016):

$$E(\vec{p}) = \rho_1 E_1(\vec{p}) + \rho_2 E_2(\vec{p}) + \ldots + \rho_i E_i(\vec{p}) + \ldots + \rho_m E_m(\vec{p}) \tag{10.6}$$

where

$$\rho i = \begin{cases} 1 \ for \ i = i* \\ 0 \ for \ i \neq i*, i = 1 \ldots m, \end{cases} \tag{10.7}$$

The weights of the OFC $[w_{1,i}, w_{2,i}, \ldots, w_{n,i}]$ and weights of amygdala $[v_{1,i}, v_{2,i}, \ldots, v_{n,i}]$ of each module output as per (Lotfi and Akbarzadeh-T, 2016):

$$\mathrm{E}(\vec{p}) = f(E_{ai} - E_{oi}) = f\left(\sum_{j=1}^{n+1}(v_{j,i}p_j) - \sum_{j=1}^{n}(w_{j,i}p_j) - b_i\right) \tag{10.8}$$

where $f \in \leftarrow C(I^n)$ is a nonconstant, bounded monotonically-increasing continuous activation function $C(I^n)$, and I^n is a unit hypercube

The increase in sensory cortex modules is fed by an expanded thalamus as below (Lotfi and Akbarzadeh-T, 2016):

$$pn + 1 = \max_{j=1.n}(p_j) \tag{10.9}$$

where p_j is the input pattern and $pn + 1$ is the expanded feature

Under the winner-take-all model, the output of the OFC is defined by Lotfi and Akbarzadeh-T (2016) as:

$$E_{oi}(\vec{p}) = \sum_{j=1}^{n}(w_{j,i}p_j) + b_i \tag{10.10}$$

where weights of the OFC $[w_{1,i}, w_{2,i}, \ldots, w_{n,i}]$ and b_i are the related bias.

The favored winning input and the final output are derived through the learning process of adjusting of competitive weights, amygdale weights, and OFC weights until an optimum is achieved (Lotfi and Akbarzadeh-T, 2016). The learning process in the proposed model is enhanced through the use of the genetic algorithm (GA), which is used to adjust weights of the ENN to minimize the cost function in search of the optimal solution.

An initial population of $2n+2$ genes is formed for each chromosome where (Lotfi et al., 2014):

$$Chrom_i = [v_1, v_2, \ldots V_n, V_{n+1}, w_1, w_2 \ldots W_n, b] \tag{10.11}$$

where n is equal to the number of input features fed to the ENN model

The output of the model for the k-th input pattern with given weights for $Chrom_i$ (Lotfi et al., 2014):

$$Y^k = E\left(P^k; Chrom_i\right) \tag{10.12}$$

Thus the fitness function for each individual is (Lotfi et al., 2014):

$$\text{fitness}_{\text{train}}(Chrom_i) = \frac{1}{m}\left[\sum_{k=1}^{m}\left(Y^k - T^k\right)^2\right]^{0.5} \tag{10.13}$$

where Y_k is response to k-th input pattern, T^k is related target, and m is the number of pattern targets

Until stop criteria $W(i)$ = Best Chrom,

- Select individuals
- Mating
- Mutating
- Replace weaker individuals with stronger offspring and create the next generation
- Calculate fitness as per Eq. (10.7)

10.2.2 Random forest model

The Random Forest (RF), an ensemble-based model, is frequently used in modeling power consumption and has demonstrated performance comparable to ANNs (Ahmad et al., 2017). Considered to be superior at handling categorical variables, complex data, data with missing values, and large datasets, the RF incorporating an ensemble-based learning algorithm generates a large number of decision trees based on the samples from the training set, hence it creates a forest (Ahmad et al., 2017, Nallathambi and Ramasamy, 2017). Independently trained, the predictions are averaged to obtain a forecast (Breiman, 2001, Denil et al., 2014). The application of the model requires some specification of parameters for the construction of the decision tree, including the number of trees, leaves, the predictor of choice for the leaf, and the randomization method (Denil et al., 2014). Being a black-box model with a robust technique for modeling nonlinear relationships between input and output variables, the interpretability of RF has spurred research into defining the theoretical basis of the model (Breiman, 2001). Breiman (2001) provided proofs based on Strong's law of large numbers to explain the ability of RF in avoiding overfitting, an issue which has plagued other models.

Theoretically, the convergence of the RF model is defined by Breiman (2001) as follows:

For classifiers $h_1(x)$, $h_2(x)$,..., $h_K(x)$ and random selection of inputs from vector Y, Breiman (2001) defines the margin function X which measures the magnitude of the margin between the votes for the winning class and the average vote of other classes.

$$mg(X, Y) = av_k I\left(\left(h_k(X) = Y\right) - \max_{j \neq Y} av_k I\left(h_k(X) = j\right)\right) \tag{10.14}$$

where $I(.)$ is the indicator function

The magnitude of the margin indicates the degree of confidence in the classification process (Breiman, 2001). The generalization error is then defined by Breiman (2001) as:

$$PE^* = P_{X,Y}\left(mg(X, Y) < 0\right) \tag{10.15}$$

where $P_{X,\,Y}$ is the probability over X, Y space

In defining the convergence process of the model Breiman (2001) uses Strong's theorem of large numbers and states, since in random forests $h_k(X) = h(X, \Theta_k)$

It converges to:

$$P_{X,Y}\left(P_\Theta(hX, \Theta) = Y - \max_{j \neq Y} P_\Theta\left(h(X, \Theta) = j\right) < 0\right) \tag{10.16}$$

In defining the convergence process of the model, Breiman (2001) uses Strong's theorem of large numbers and states:

10.2.3 Multiple linear regression

Regression modeling is a technique that is used in analyzing the relationship between a dependent variable and one or more independent variables. In the context of modeling the short-term electricity demand, a regression model which best accounts for the relations between the dependent variable, the load, and the explanatory variables can be mapped using the following general form of the multiple linear regression (MLR) equation (Amral et al., 2007; Bianco et al., 2009).

$$y = \beta_0 + \beta_1 x_1 + \beta_2 x_2 + \ldots + \beta_k x_k + E \tag{10.17}$$

where y is the dependent variable, x_n is the independent variables, and β are the regression parameters for each x_n and E is the error term. The MLR attempts to explain the variations in the response variable through the determination of regression coefficients for each predictor such that the final model minimizes the amount of unexplained variation in the response variables (Deo and Şahin, 2017)

10.3 Study area and data

10.3.1 Study area

The study focuses on urban centers in Australian Capital Territory (ACT) and NSW in Australia. A number of sites were identified from prior studies conducted by the

Commonwealth Scientific and Industrial Research Organisation (CSIRO). In their study, Caccetta et al. (2017) of CSIRO utilized Landsat 8 thermal infrared satellite imagery to identify urban centers in Australia impacted by UHI by estimating the land surface temperature (LST) and land surface emissivity (LSE), a factor influenced by vegetation, moisture, and land surface characteristics. Similarly, Meyers et al. (2017) used Landsat 8 and moderate resolution imaging spectroradiometer on MODIS satellites to study LSTs in another CSIRO study and mapped surface urban heat in Canberra.

From the sites identified by CSIRO, a shortlist of eight sites were chosen (Table 10.1) based on the availability of 30-minutes air temperature, 30-minutes electricity demand data at zone substation level, and their heat map rating.

10.3.2 Data

10.3.2.1 Input variable selection

In the context of the UHIs and electricity demand, the literature on essential demand drivers was analyzed to identify a suitable input variable subject to data availability. The temperature was identified as the principal contributor to electricity demand within a 24-hours band (Clements et al., 2016). Clements et al. (2016), in a study of the Queensland region in Australia, also noted that temperatures between 22°C and 30°C had an effect on electricity load, and beyond 30°C, the effect on demand was limited by the capabilities of climate control appliances. Further studies by Shahmohamadi et al. (2010) and Santamouris et al. (2018) found that the presence of UHI increased building energy consumption. Accordingly, air temperature was chosen as the primary climate variable.

10.3.2.2 Temporal and spatial resolution

The study of the UHI phenomena, in general, is hampered by the lack of temporal and spatial resolution of the air temperature datasets (Azevedo et al., 2016). Furthermore, studies on the effects of climate variables have found lagged effects of such variables on

TABLE 10.1 Selection of urban heat island/continent impacted sites, their physical and thermal characteristics.

State	Study sites	Latitude (°S)	Longitude (°E)	Hottest 2.5%	Hottest 8%	Hottest 16%
ACT	Gunghalin	35.1831	149.1330	✓	✓	✓
ACT	Belconnen	35.2370	149.0650	✓	✓	✓
ACT	Latham	35.2172	149.0314	✓	✓	✓
ACT	Gilmore	35.4210	149.1330		✓	✓
ACT	Theodore	35.4460	149.1220	✓	✓	✓
NSW	Westmead	33.8086	150.9870		✓	✓
NSW	Marrickville	33.9087	151.1524			✓
NSW	Campsie	33.9130	151.1041	✓	✓	✓

electricity demand. Clements et al. (2016) found the effect of temperature in the 30 minutes window preceding the electricity load request in a study in Queensland, Australia. Despite the availability of remote sensing sources, the temperature sensing capabilities are restricted to LST of high spatial resolution instead of the air temperature or canopy air temperature (Kaplan et al., 2018; Azevedo et al., 2016).

Given the above insights, the first part of the study utilizes air temperature data from fixed weather stations with high temporal coverage of air temperature and relative proximity to the study sites. The 30-minutes air temperature data for two weather stations, Sydney International Airport, and Canberra Airport, for the period of years 2008–2018, were obtained from the National Oceanic and Atmospheric Administration (NOAA). Table 10.2 provides the physical characteristics of the weather stations used in this study and Table 10.3 provides descriptive statistics associated with the air temperature datasets from the fixed weather stations.

However, due to the sparse distribution of fixed weather stations (Table 10.4), for the second part of the study, on the comparative predictive performance of demand prediction models using spatially and temporally relevant weather data, air temperature data for the period from 2008–2018, was obtained from the ERA5 reanalysis, representing the fifth generation of the European Centre for Medium-Range Weather Forecasts (ECMWF). Table 10.5 provides descriptive statistics associated with the air temperature datasets from ERA5 reanalysis.

The dataset with the spatial resolution of 0.1 degree and hourly temporal resolution was extracted for seven sites based on their longitude and latitude using the Climate Data Store toolbox of the Copernicus Earth Observation Programme of the European Union.

10.3.2.3 Electricity demand data

Half hourly electricity demand data from electricity substations which provide electricity coverage to the eight study sites for the period of 2008–2018 were obtained from Evo Energy (ACT) and Ausgrid (NSW).

TABLE 10.2 Physical characteristics of the weather stations used in this study.

Weather station	Latitude (°S)	Longitude (°E)	Elevation (m)
Canberra Airport	35.307	149.195	574.9
Sydney International Airport	33.946	151.177	6.4

TABLE 10.3 Descriptive statistics associated with the air temperature datasets for the summer months.

Weather station	Minimum (°C)	Maximum (°C)	Mean (°C)	Skewness	Kurtosis
Canberra Airport	0.9	41.0	20.12	0.4965	−0.05604
Sydney International Airport	0.9	45.5	23.09	0.8893	2.24950

TABLE 10.4 Distribution of weather stations, distance from study sites and other physical characteristics.

Study sites	Weather station	Direct distance from nearest weather station (km)
Gunghalin	Canberra Airport	14.35
Belconnen	Canberra Airport	13.65
Latham	Canberra Airport	17.10
Gilmore	Canberra Airport	13.20
Theodore	Canberra Airport	17.00
Westmead	Sydney International Airport	23.60
Marrickville	Sydney International Airport	4.10
Campsie	Sydney International Airport	7.90

TABLE 10.5 Descriptive statistics associated with the ERA5 reanalysis, air temperature datasets for the summer months.

Study sites	Minimum (°C)	Maximum (°C)	Mean (°C)	Skewness	Kurtosis
Gunghalin	4.83	37.30	19.50	0.4373	−0.2225
Belconnen	10.22	44.1	22.41	0.9648	1.2700
Theodore	3.18	36.93	19.00	0.4489	−0.2181
Gilmore	3.18	36.93	19.16	0.4401	−0.2224
Marrickville	12.32	41.60	22.53	0.9661	1.6564
Campsie	27.82	41.60	12.77	−5.173	24.91
Westmead	10.96	43.76	22.41	1.0194	1.529

10.4 Predictive model development

The models, WTAENN, RF, and MLR were developed in MATLAB on a cloud-based multiprocessor virtual machine environment. WTAENN model was built using the ENN with GA learning Version 3.0 toolbox developed by Lotfi et al. (2014) with additional codes added to make use of parallel GA optimization features of MATLAB (Fig. 10.1).

10.4.1 Feature engineering

UHI intensity and hence the potential influence of temperature on the thermal comfort of the inhabitants is likely to experience intraday variations. Studies by Solatani et al. (2017), Chow and Roth (2006), and Meng et al. (2018) have identified such spatial–temporal trends in UHI intensity including the finding that temperatures peak in the late afternoon and the degree of UHI intensity is influenced by factors including anthropogenic heat and the green canopy. Following on from the above studies,

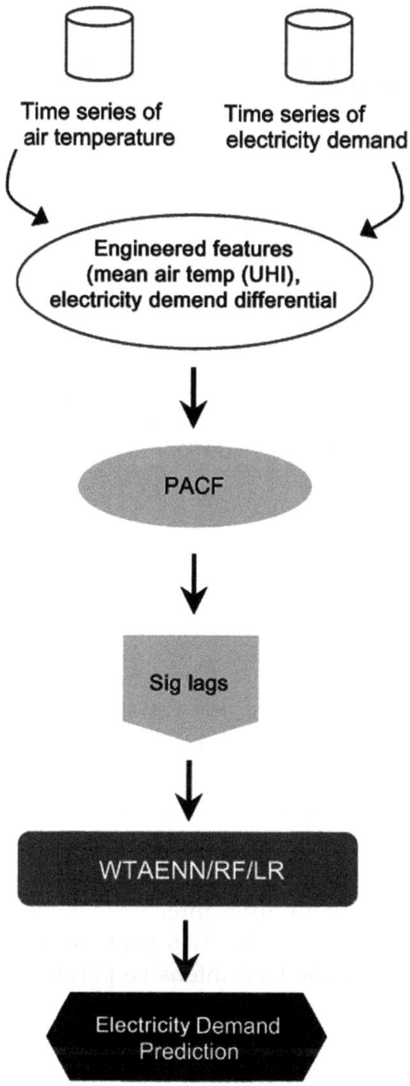

FIGURE 10.1 Model development stages.

exploratory data analysis, and the finding from the study conducted by Clements et al. (2016), which found the effect of temperature in the 30 minutes window preceding the electricity load, we constructed a new feature to model a stronger relationship between the independent and dependent variables. Notably, feature engineering is one of the essential methods employed by researchers in the study of building energy consumption and energy demand forecasting (Bansal et al., 2019; Zhang et al., 2018).

The feature, daily mean air temperature for the UHI intensive period from 1:00 pm to 12:00 am (T_{uhi}), was developed as the input variable with the inclusion of air temperature

TABLE 10.6 Descriptive statistics associated with mean air temperature for the UHI intensive period including lag time (12:30–23:30).

Weather station	Minimum (°C)	Maximum (°C)	Mean (°C)	Skewness (°C)	Kurtosis (°C)
Mean air temperature for the UHI intensive period incl lag time (12:30–23:30)					
Canberra Airport	11.27	35.8	22.81	0.2875	−0.2872
Sydney International Airport	15.2	39.05	23.70	0.4516	1.2798

TABLE 10.7 Descriptive statistics associated with electricity demand differential for the summer months from 2008 to 2018.

Substation site	Minimum (MW)	Maximum (MW)	Mean (MW)	Skewness (MW)	Kurtosis (MW)	Hottest 2.5%	Hottest 8%	Hottest 16%
Belconnen	−1.64	17.00	4.07	1.09	3.67	✓	✓	✓
Gilmore	−0.9	4.65	1.52	0.95	0.71		✓	✓
Latham	0.35	11.37	4.42	1.10	0.68	✓	✓	✓
Theodore	−0.29	10.67	4.02	1.08	0.63	✓	✓	✓
Gunghalin	0.31	13.35	3.89	1.52	2.15	✓	✓	✓
Campsie	−2.11	11.7	4.02	1.44	4.65	✓	✓	✓
Marrickville	−0.79	7.65	2.61	0.79	2.81			✓
Westmead	−1.44	5.86	0.74	2.16	6.02		✓	✓

from the $t-1$ time period. Therefore the calculated mean of the air temperature includes the temperature of the preceding 30 minutes interval by taking the mean of the air temperature readings from the 12:30–23:30. Table 10.6 provides descriptive statistics associated with the mean air temperature for the UHI intensive period.

$$T_{\text{uhi}} = \frac{\sum(T_{12:30} - T_{23:30})}{n} \quad n = \text{number of } 30 - \text{minutes observations} \tag{10.18}$$

Similarly, a response variable, electricity demand differential (D_{uhi}), was defined to represent the electricity demand for the UHI period between 13:00 and 24:00. The variable was calculated as the electricity demand difference between hourly mean electricity demand for the UHI period between 13:00 and 24:00 hours and the full 24-hours period capturing the variations in electricity consumption during the UHI intensive period. Table 10.7 provides descriptive statistics associated with the electricity demand datasets.

$$D_{\text{uhi}} = \frac{\sum_{t=1}^{n}\left(D_t - \overline{D_{Daily}}\right)}{n} \quad n = \text{number of } 30 - \text{minutes observations} \tag{10.19}$$

10.4.2 Normalization

The normalization of the dataset is an essential requirement which standardizes the values of variables to be within a comparable range (Deo et al., 2016). The input patterns and input targets were normalized between 0 and 1. Within WTAENN the myNorm function was used for normalization. The normalization could be mathematically defined as:

$$X_{norm} = \frac{(X - X_{min})}{(X_{max} - X_{min})} \tag{10.20}$$

where X_{norm} is standardized values of the input or output values, X is input or output values, X_{min} is minimum of the input or output range, and X_{max} is maximum of the input or output range

10.4.3 Significant lags

Environmental inputs such a temperature, humidity, LST, etc. frequently consist of an underlying physical structure, including seasonality and trends. Therefore using the classical statistical approach, we utilized autocorrelation functions within the Box–Jenkins algorithm of graphical tools to identify temporary autocorrelation, thereby minimizing one of the causes of model estimation errors (Friedrich and Afshari, 2015). The partial autocorrelation function (PACF) maps the relationship between the observation at a time (t) with preceding observations at various time lags ($t-1$ ... $t-n$) in the absence of other time lags, thus providing an illustration of significant lags (Taneja et al., 2016). PACF was performed on the air temperature datasets and time lags of significance (Fig. 10.2) were included as lagged inputs for the models.

10.4.4 Testing and training sets

The datasets for Mean Air Temperature for UHI time periods and Electricity Demand Differential were sequentially divided into training and testing subsets. Since there is no set method for partition datasets (Deo et al., 2016), we arrive at the best ratio between the training and testing subsets through the testing of different combinations (70%:30%,

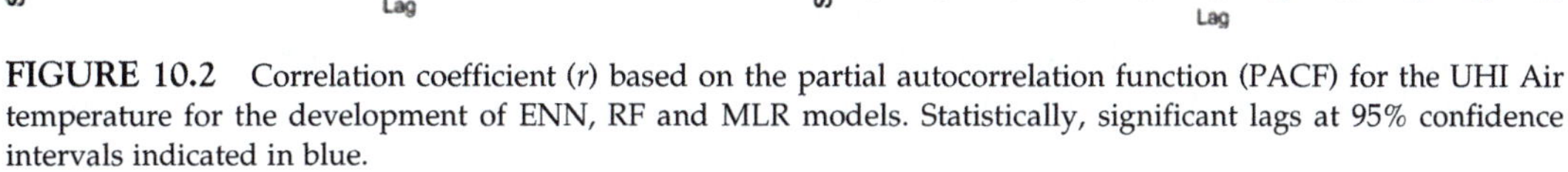

FIGURE 10.2 Correlation coefficient (r) based on the partial autocorrelation function (PACF) for the UHI Air temperature for the development of ENN, RF and MLR models. Statistically, significant lags at 95% confidence intervals indicated in blue.

TABLE 10.8 Details on data partitions used in this study.

States	Period	Datum points	Training	Testing
ACT	Summer days from 2007 December to 2018 February	953	905 (95%)	48 (5%)
NSW		844–939	802–892 (95%)	42–48 (5%)

80%:20%, 90%:10%, 95%:5%) with best performing subset, 95%:5%, selected on the basis on root mean squared values. Table 10.8 provides descriptive statistics associated with the training and testing partitions.

10.4.5 Winner-take-all emotional neural network model mevelopment

A WTAENN model was developed utilizing ENN with GA learning algorithm toolbox (version 3.0) provided by Lotfi and Akbarzadeh-T (2016). The model requires the specification of the number of competitive parts m, representing the number of sensory cortex modules (Lotfi et al.). Normalized targets were used with a hyperbolic tangent activation function.

As proposed by Lotfi and Akbarzadeh-T (2016) GA with g generations and p population was used for the optimization of competitive weights of sensory cortex, OFC and Amydala modules. The GA is applied to train the model and find the optimum weights. The choice of optimum model parameters, p and g for GA is crucial to avoid premature convergence and improve crossover combination (Pandey, 2016, Vrajitoru, 1998). In order to strike a balance between optimum performance and acceptable computational time, several independent runs of GA with population sizes between 20 and 300 and generations between 100 and 500 were performed. The best performing combination of G and P were selected in terms of their root mean square error (*RMSE*).

Similarly, several runs were performed with m between 1 and 50 with the best performing combination along with g and p were selected based on *RMSE* to form an ensemble of 10 parameter combinations to be tested across the study sites. Though the literature suggested a higher m for increased approximation accuracy, it also cautioned the possibility of overfitting in prediction or classification problems mainly if the problem space is restricted with a smaller dataset of patterns (Lotfi and Akbarzadeh-T, 2016). Considering this possibility, the selected ensemble of parameter combinations tests sets the parameter m between 5 and 50 with varying combinations of G and P parameters for the GA algorithm.

10.4.6 Random forest model development

The RF model was configured to optimise the number of ensemble-based decision trees through multiple runs with various combinations of *leaf* (1, 3, 5, 10, 20), *ntrees* (50, 200, 800, 1600) and *fboot* (0.4, 0.8, 1). The best performing model was selected based on root mean square performance.

10.4.7 Multiple linear regression model

A Multiple Linear Regression (MLR) model was developed by estimating the coefficient for the predictor. The model utilized predictor data in the training set, which consisted of one predictor variables and a check dataset for testing the model for their predictive performance.

10.4.8 Model performance assessment

Assessing the quality of the models developed and hence their predictive power is critical in model performance assessment. The developed models are assessed for predictive accuracy using a combination of model evaluation methods as no single method is sufficient to assess performance on its own. Deo and Şahin (2017), based on the recommendations of the American Society for Civil Engineers (ASCE), applied an evaluation framework with the combination of visual, descriptive (i.e., mean, variance), and standardized metrics methods such as *RMSE* and mean absolute error (MAE) in the forecasting of long-term global solar radiation.

In this study, a series of standardized metrics were considered to assess the performance of the WTAENN, RF, and MLR models. Each metric utilizes variances, means, and residuals in their underlying methodology to produce metrics indicating model performance. However, due to the differences in the underlying methodology employed, each assessment metric has its drawbacks. In this study, a set of model assessment metrics—*RMSE*, coefficient of determination (r^2), Nash–Sutcliffe coefficient of efficiency (E), Willmott's index of agreement (W), MAE, and Legates–McCabe's index (*L*)—provides us with a statistical means to evaluate the goodness of fit between the observed and predicted values.

10.5 Results and discussion

10.5.1 Demand predictions utilizing air temperature data from fixed weather stations

The models—WTAENN, MLR, and RF—were developed and evaluated in the prediction of short-term electricity demand in UHI-affected study sites in NSW and ACT. utilizing air temperature data from fixed weather stations with high temporal coverage. Assessment of the models with the help of *RMSE*, MAE, and r values indicates that overall the RF model presents the strongest relationship between the Mmean air temperature for UHI intensive periods and the electricity demand differential. However, the comparatively higher *RMSE* indicates the model accuracy and hence predictive accuracy is compromised. In contrast, the MLR model has relatively lower *RMSE* indicating a higher model accuracy and higher r values indicating a stronger linear relationship for Gungahlin ($r = 0.9071$, $RMSE = 0.1299$), Belconnen ($r = 0.918$, $RMSE = 0.1904$), Campsie ($r = 0.9112$, $RMSE = 0.5131$), and Westmead ($r = 0.7922$, $RMSE = 0.1481$) (Table 10.9).

We now asses each model on a site by site basis with Legates–McCabe's index (*L*), Willmot's index, and Nash–Sutcliffe efficiency (E). Legates–McCabe's index produces interpretable metrics which indicate the degree of explanatory power of the model

TABLE 10.9 Model performance assessment of WTAENN versus benchmark models (RF and MLR) in short-term electricity forecasting during the testing period for each study site based on Pearson's correlation coefficient (r), mean absolute error (MAE), and root mean square error (*RMSE*). Models with the lowest *RMSE* are marked in red and higher r are marked in blue.

	WTAENN			RF			MLR		
Study sites	r	*MAE (MW)*	*RMSE (MW)*	r	*MAE (MW)*	*RMSE (MW)*	r	*MAE (MW)*	*RMSE (MW)*
Gunghalin	0.890	1.182	1.443	0.988	0.0380	5.866	0.9071	0.1033	0.1299
Belconnen	0.901	0.768	0.996	0.986	0.0362	3.956	0.918	0.1429	0.1904
Theodore	0.045	1.881	2.227	0.974	0.0695	4.645	0.4389	0.198	0.2445
Gilmore	0.742	0.626	0.744	0.981	0.0439	1.351	0.791	0.1335	0.1616
Latham	0.012	1.950	2.375	0.9732	0.0892	5.0914	0.0497	0.2333	0.2845
Marrickville	0.053	0.798	1.049	0.957	0.072	13.155	−0.1962	0.2399	0.2921
Campsie	0.869	1.339	1.496	0.994	0.0270	13.488	0.9112	0.5041	0.5132
Westmead	0.867	0.649	0.756	0.984	0.0349	12.621	0.7922	0.1145	0.1481
Auburn	−0.057	0.400	0.482	0.950	0.0782	10.438	−0.0771	0.1987	0.2483

(Legates and McCabe, 2012). Overall the RF model has L values significantly higher than 0, indicating the model has more explanatory power than the observed mean. The WTAENN and MLR models, on the contrary, have values lower than zero or weaker positive values indicating the developed model has less explanatory power than the observed mean. On the basis of Legates–McCabe's index the RF model is best suited to predicting short-term electricity demand.

Assessment of the short-term forecasting using the combination of Willmot index (W), Nash–Sutcliffe efficiency (E), and the Legates–McCabe's index (L) indicates the RF model is outperforming WTAENN and MLR models (Table 10.10).

A visual assessment, of the models on a site-by-site basis, is performed using scatter plots, as in Fig. 10.3. As evidenced by the performance metrics earlier, the scatter plots for the RF model indicate a better fit for all sites, while the MLR model displays a strong positive relationship for Gungahlin, Belconnen, and Latham, as indicated by r^2 values.

The results demonstrate that RF models are superior to the hybrid WTAENN based on the model performance across the 10 sites tested.

10.5.2 Demand predictions utilizing air temperature data from reanalysis

Changes in the demand patterns are not uniform due to the spatial differences in relationships between climate, land cover, and urban land usage (Xisheng Hu, 2019; Eliasson, 2003). Microclimatic studies by Eliasson et al. (2003) and Santamouris et al. (2001) have highlighted the spatial variability of urban climate factors through in situ observation of climate variables, including air temperature. Further, studies also found statically significant intraurban variations in air temperature, due to variations in land usage and land cover (Eliasson, 2003).

TABLE 10.10 Model performance assessment of WTAENN versus benchmark models (RF and MLR) in short-term electricity forecasting during the testing period for each study site based on Legates–McCabe's index (*L*), Willmott's index of agreement (*W*), and Nash–Sutcliffe coefficient of efficiency (*E*). Optimal models with the lowest *L* are marked in red.

	WTAENN			RF			MLR		
Study sites	*L*	*W*	*E*	*L*	*W*	*E*	*L*	*W*	*E*
Gunghalin	0.380	0.877	0.777	0.812	0.985	0.967	0.370	0.814	0.735
Belconnen	−0.034	0.737	0.669	0.786	0.980	0.961	−0.160	0.087	0.372
Theodore	−4.636	−0.029	−0.018	0.560	0.887	0.934	−1.214	0.323	0.193
Gilmore	−0.171	0.715	0.539	0.730	0.970	0.943	−0.189	0.511	0.485
Latham	−3.156	0.054	−0.055	0.414	0.912	0.854	−1.694	−0.047	−0.124
Marrickville	−1.009	0.351	−0.175	0.519	0.928	0.869	−0.668	0.395	−0.504
Campsie	−0.206	0.739	0.298	0.735	0.978	0.946	0	0.528	−4.309
Westmead	0.087	0.418	0.372	0.747	0.97	0.948	−0.442	0.257	0.426
Auburn	−3.200	0.239	−0.042	0.314	0.967	0.9421	−0.919	0.358	−0.230

Incorporating the spatial heterogeneity in the urban climate patterns, to improve the accuracy of short-term electricity demand forecasting, requires spatially relevant air temperature observations. Further, to best capture the temperature effects during UHI intensive periods requires temporal relevance. In the section ahead, the short-term prediction models, hybrid WTAENN, RF, and MLR, are employed with spatially and temporally relevant air temperature data for the seven sites in NSW and ACT. The results indicate that the RF model utilizing the air temperature from ERA5 reanalysis produced the best model with higher model accuracy and a stronger model relationship, although the model is less interpretable than the ones using observed air temperatures.

The models based on daily mean air temperature from ERA5 reanalysis were first assessed for model performance with *RMSE*, MAE, and r values (Table 10.11). Although the MLR models show relatively lower *RMSE* indicating higher model accuracy, overall RF models presented the strongest relationship between the mean air temperature for the UHI intensive period (ERA5) and the electricity demand differential for all study sites. The MLR model for Gungahlin ($r = 0.9071$, $RMSE = 0.1299$), Theodore ($r = 0.827$, $RMSE = 0.162$), and Gilmore ($r = 0.745$, $RMSE = 0.171$) has the best combination of lower *RMSE*, indicating higher model accuracy, and higher *r* values, indicating a stronger relationship between the variables. The WTAENN model comparatively underperformed across all sites producing higher *RMSE*, indicating higher model errors and lower *r* values than the MLR model, indicating a weaker relationship between mean air temperature for the UHI intensive period (ERA5) and the electricity demand differential.

The models were assessed on a site-by-site basis based on Legates–McCabe's index (*L*), Willmot's index (W), and Nash–Sutcliffe efficiency (E) (Table 10.12). The WTAENN produced *L* values lower than zero or weak positive values for most sites, with the exception

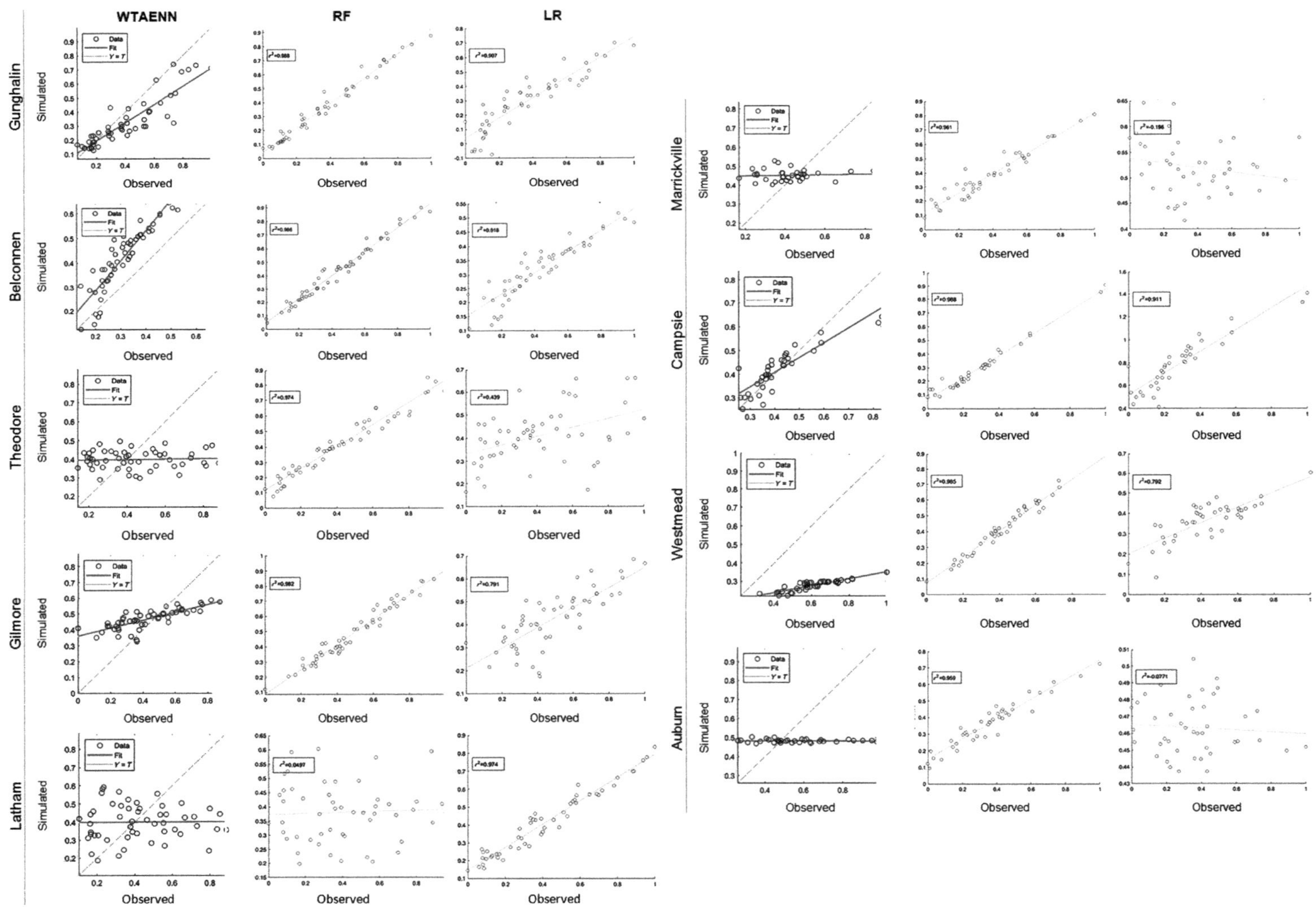

FIGURE 10.3 Side by side scatter plots of simulated versus observed for each of the models for the study sites in ACT and NSW.

of Theodore ($L = 0.270$), indicating the developed models either have less explanatory power than the observed mean or marginally better explanatory power than the observed mean. The MLR model produced weak L values for all sites except Gungahlin ($L = 0.207$) and Theodore ($L = 0.404$).

Overall the RF model produced L values significantly higher than 0, ranging from $L = 0.534$ to $L = 0.721$, indicating that the model has more explanatory power than the observed mean.

TABLE 10.11 Model performance assessment of WTAENN versus benchmark models (RF and MLR) in short-term electricity forecasting during the testing period for each study site based on Pearson's correlation coefficient (r), mean absolute error (MAE), and root mean square error (*RMSE*). Optimal models with the lowest *RMSE* are marked in red and models with high r are marked in blue.

	WTAENN			RF			MLR		
Study sites	*r*	*MAE (MW)*	*RMSE (MW)*	*r*	*MAE (MW)*	*RMSE (MW)*	*r*	*MAE (MW)*	*RMSE (MW)*
Gunghalin	0.786	1.713	2.195	0.9645	0.7201	0.945	0.795	0.164	0.210
Belconnen	0.477	1.189	1.622	0.930	0.589	0.717	0.478	0.150	0.208
Theodore	0.782	1.259	1.696	0.962	0.492	0.677	0.827	0.123	0.162
Gilmore	0.475	0.795	1.039	0.951	0.291	0.372	0.745	0.138	0.171
Marrickville	0.519	0.661	0.841	0.932	0.058	2.743	0.610	0.126	0.157
Campsie	0.589	1.432	1.786	0.950	0.060	5.235	0.659	0.122	0.178
Westmead	0.624	0.746	0.982	0.950	0.050	3.272	0.621	0.143	0.188

TABLE 10.12 Model performance assessment of WTAENN versus benchmark models (RF and MLR) in short-term electricity forecasting during the testing period for each study site based on Legates—McCabe's index (L), Willmott's index of agreement (W), and Nash—Sutcliffe coefficient of efficiency (E). Optimal models with the lowest L are marked in red.

	WTAENN			RF			MLR		
	L	*W*	*E*	*L*	*W*	*E*	*L*	*W*	*E*
Gunghalin	0.015	0.666	0.559	0.708	0.959	0.918	0.207	0.483	0.408
Belconnen	−0.716	0.144	0.213	0.521	0.906	0.846	−0.174	0.373	0.203
Theodore	0.270	0.821	0.471	0.721	0.958	0.916	0.404	0.775	0.633
Gilmore	−0.328	0.588	0.084	0.617	0.936	0.882	0.041	0.678	0.538
Marrickville	−1.981	0.282	0.224	0.534	0.908	0.849	−0.166	0.606	0.333
Campsie	−1.032	0.354	0.301	0.574	0.925	0.874	−0.198	0.189	0.349
Westmead	−0.295	−0.714	0.064	0.605	0.943	0.898	−0.562	−0.826	0.048

The models were assessed with the help of scatter plots, as in Fig. 10.4. The plots for WTAENN models indicate a weaker linear relationship between the simulated and observed electricity demand differential. Similarly, the MLR models exhibit a weaker relationship for most sites except Gunghalin, Theodore, and Gilmore, which exhibit a stronger linear positive relationship between the variables. Overall, as confirmed by model performance metrics, the plots for the RF model indicate the best fit between the simulated and observed demand differential.

The performance of the models based on daily mean air temperature from ERA5 reanalysis, T_{ERA}, were contrasted with the performance of models using observed air temperature, as shown in Figs. 10.8–10.10. Sites which underperformed under the WTAENN model using observed air temperature, Theodore and Marrickville, based on r values, showed improved performance under the same model using air temperature from ERA5 reanalysis. However, all other sites showed lower performance under the WTAENN model using the ERA5 dataset. The underperformance of the ERA5 dataset is notable given the underperforming sites consist of areas that are among the hottest 2.5%–8% of the areas in that state (Fig. 10.5).

Similar performance improvement was found under the MLR model using the air temperature from ERA5 reanalysis for Theodore and Marrickville, as shown in Fig. 10.6. The RF model using the observed air temperature values outperformed the models using air temperature from ERA5 reanalysis across all sites, as shown in Fig. 10.5. However, the performance of the RF models using observed air temperature values was between 1.2%–5.7% higher with Belconnen showing the highest performance (Fig 10.7).

A comparative assessment of the WTAENN, RF, and MLR models using observed air temperature against models using air temperature from ERA5 reanalysis based on *RMSE* was performed as shown in Figs. 10.8–10.10. Under the WTAENN model, the Theodore ($RMSE = 1.696$) and Marrickville ($RMSE = 0.841$) indicated lower *RMSE* and hence higher model accuracy under the ERA5 reanalysis datasets. The sites modeled under MLR using ERA5 reanalysis datasets showed lower *RMSE* across all sites with Campsie showing much lower *RMSE*. Similarly, the RF models using ERA5 reanalysis datasets showed lower *RMSE* and therefore higher model accuracy. The *RMSE* values under ERA5 reanalysis datasets were between 61%–85% lower than under observed air temperature datasets.

Assessment of the model performance of RF using observed air temperature against models using ERA5 reanalysis air temperature with Legates–McCabe's index (L), as shown in Fig. 10.11, indicates higher performance for Theodore (28% improvement) and Marrickville (2.9% improvement) while others experienced between 12%–34% drop in performance, indicating the models are less interpretable under ERA5 reanalysis datasets (Table 10.13).

10.6 Conclusions and remarks

Though the WTAENN model was considered to be more adept at modeling complex forecasting problems, such as wind power forecasting, the results from modeling short-term electricity demand in the context of UHI demonstrates that the model is sensitive to the inputs chosen. WTAENN's ability to produce better fitting models for study areas with an intense

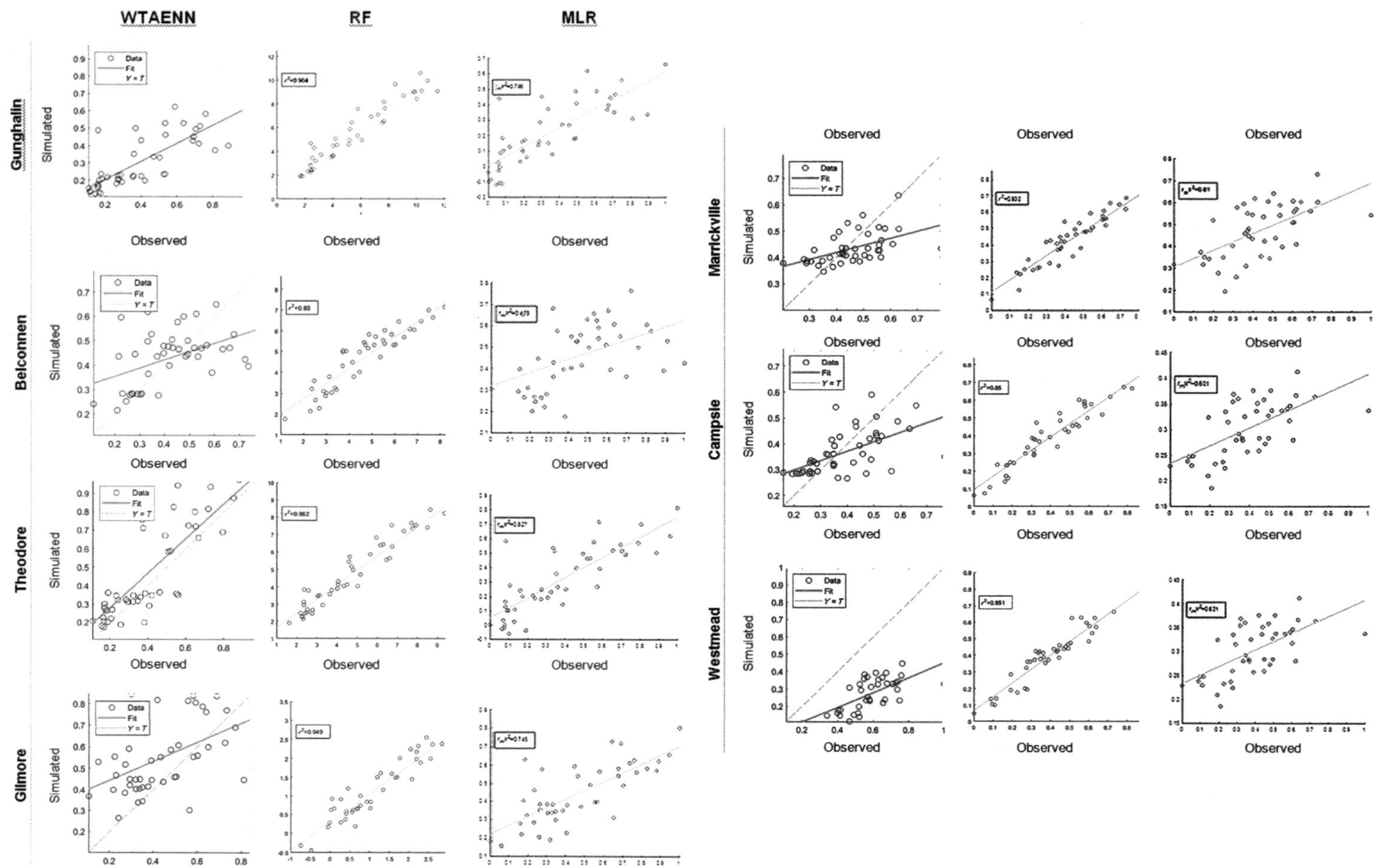

FIGURE 10.4 Side by side scatter plots of simulated versus observed for each of the models for the study sites in ACT and NSW.

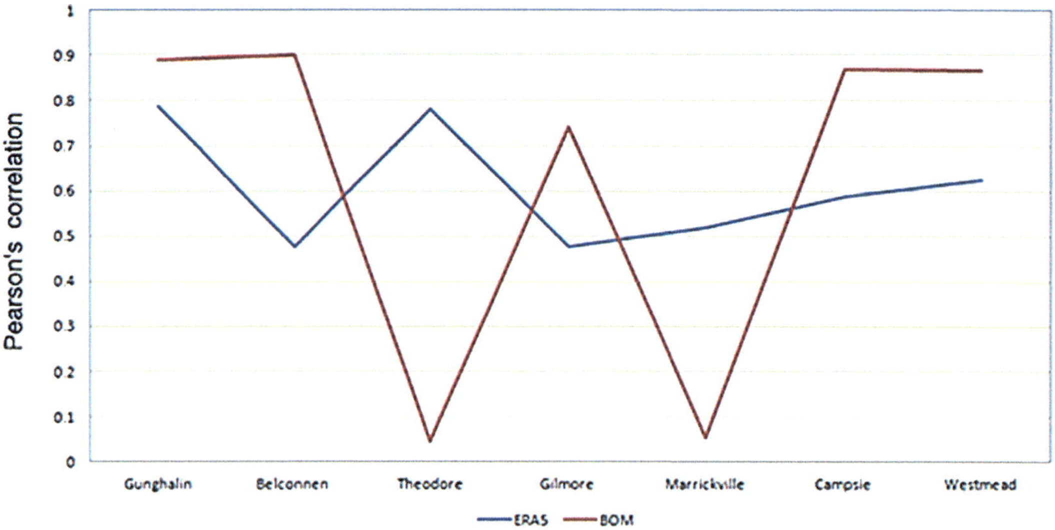

FIGURE 10.5 Plot of correlations between WTAENN model using observed air temperature versus air temperatures from ERA5 reanalysis datasets.

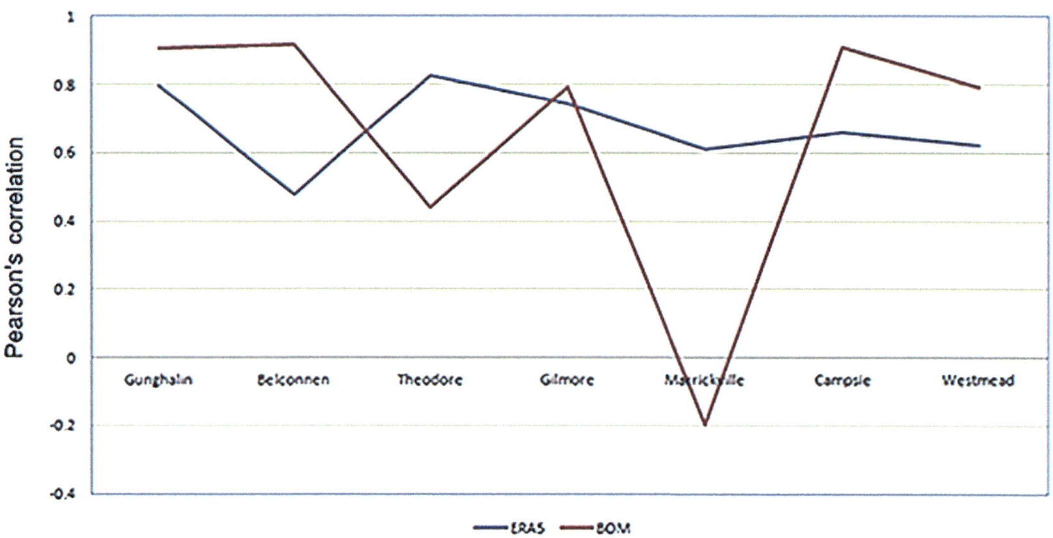

FIGURE 10.6 Plot of correlations between MLR model using observed air temperature versus air temperatures from ERA5 reanalysis datasets.

thermal profile and in close proximity to weather stations suggests the model performance could be enhanced with spatially and temporally relevant climate variables.

The WTAENN model when compared with the best performing benchmark model, the RF model, produced models with r values between 10% and 98% worse than RF and *RMSE* values between 44%–95% better than the RF model. Overall the WTAENN model produced

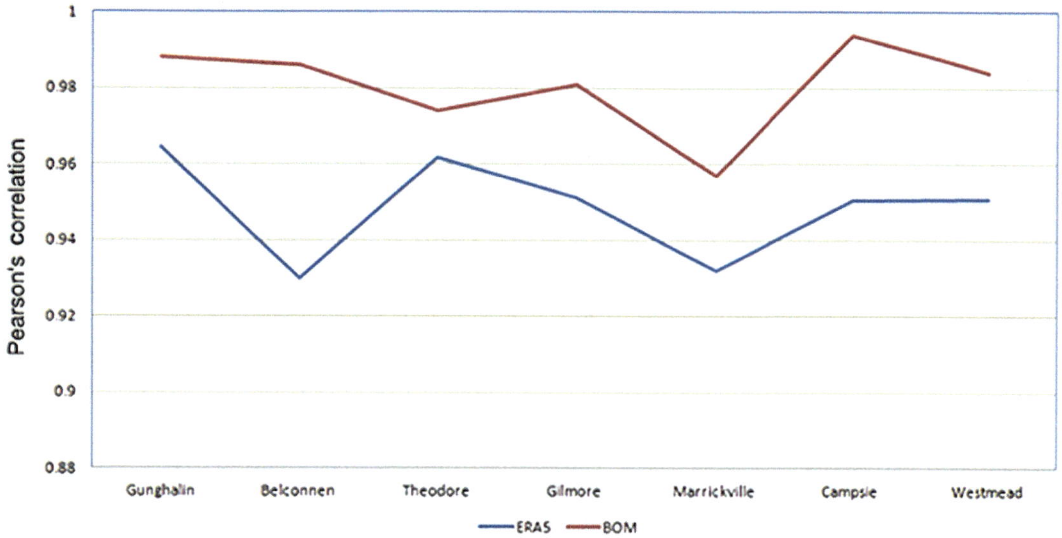

FIGURE 10.7 Plot of correlations between RF model using observed air temperature versus air temperatures from ERA5 reanalysis datasets.

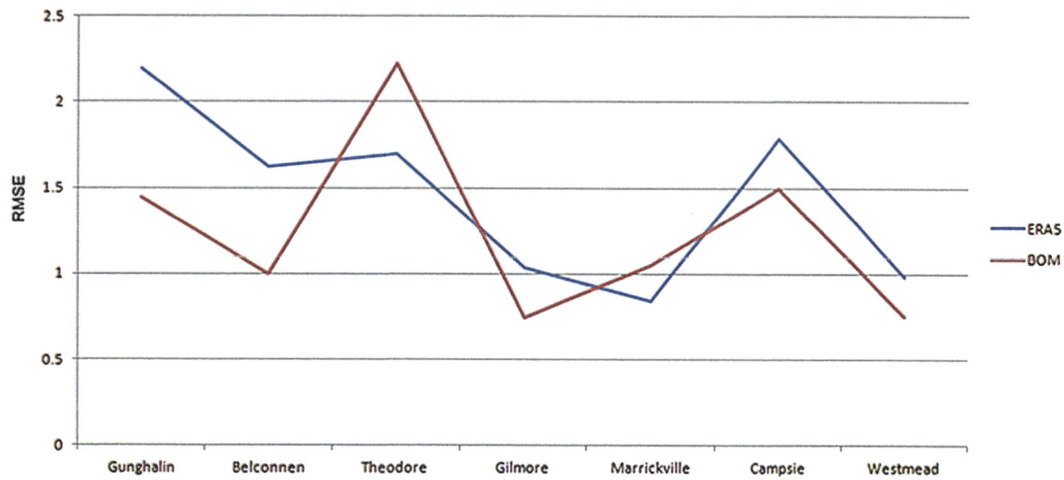

FIGURE 10.8 Plot of *RMSE* values between WTAENN model using observed air temperature versus air temperatures from ERA5 reanalysis datasets.

models which indicate a weaker association between the variables of mean air temperature for the UHI period and the demand differential. This finding is reinforced by the lower *L* values indicating the WTAENN models for most sites are no better than the observed mean.

On the contrary, the RF model outperformed across all sites, indicating a stronger association between the variables, and *L* values in a stronger positive territory, suggesting an adequate model for short-term demand prediction.

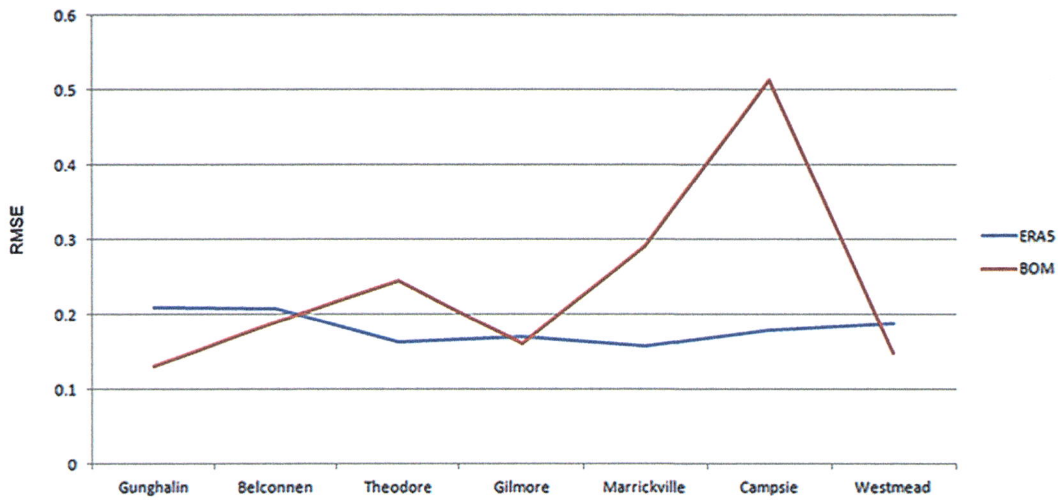

FIGURE 10.9 Plot of *RMSE* values between MLR model using observed air temperature versus air temperatures from ERA5 reanalysis datasets.

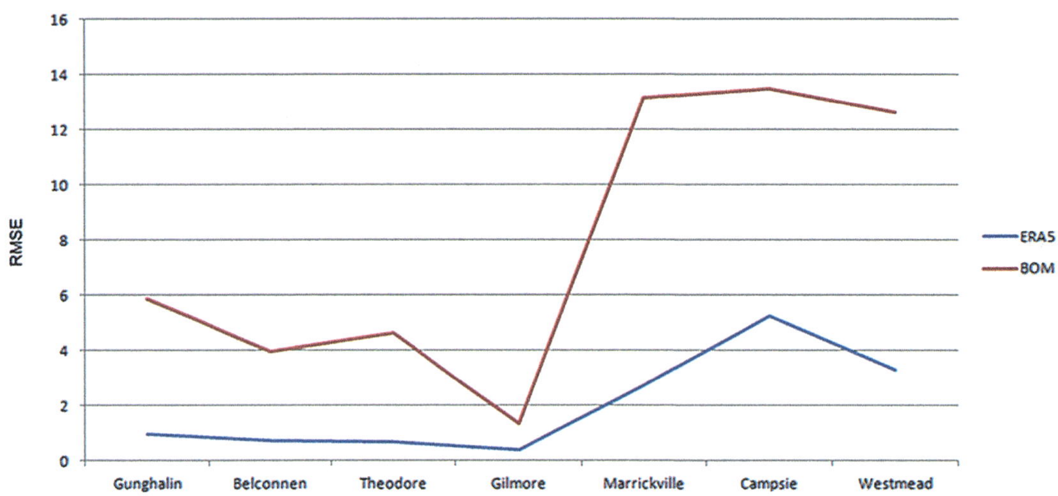

FIGURE 10.10 Plot of *RMSE* values between RF model using observed air temperature versus air temperatures from ERA5 reanalysis datasets.

With the spatially and temporally relevant ERA5 reanalysis datasets the WTAENN model continued to underperform in comparison to the best performing benchmark model RF. However, it is notable that the performance in terms of r values improved for sites which underperformed earlier using the standard observed air temperature from fixed weather stations. Theodore and Marrickville registered an improvement in r values between 89% and 94% when compared with the earlier models. At the

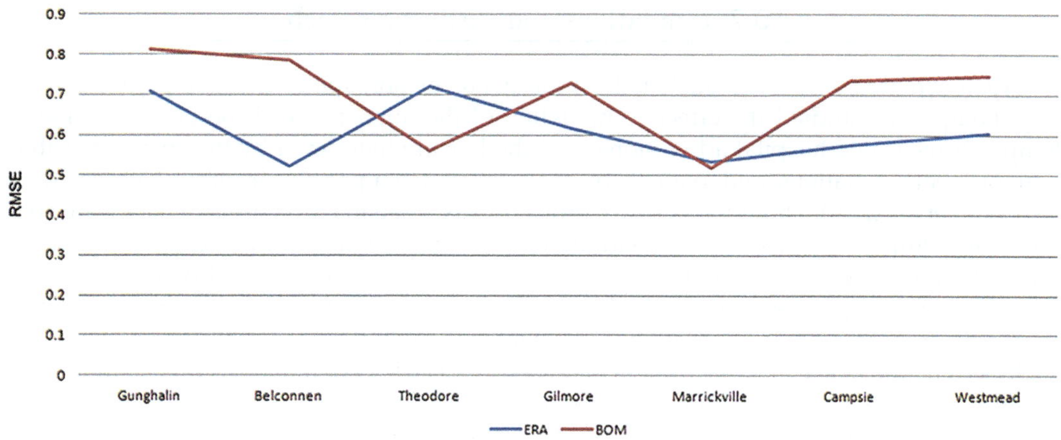

FIGURE 10.11 Plot of *L* values between RF model using observed air temperature versus air temperatures from ERA5 reanalysis datasets.

TABLE 10.13 Model parameters for the best performing WTAENN and RF models by site at each of the selected sites.

	WTAENN			RF		
Study sites	*m*	*g*	*p*	*leaf*	*trees*	*boot*
Gunghalin	5	100	30	1	200	1
Belconnen	5	100	20	1	50	1
Theodore	5	100	20	1	200	1
Gilmore	10	200	30	1	1600	1
Marrickville	5	30	100	1	50	1
Campsie	5	30	100	1	50	1
Westmead	1	100	20	1	200	1

same time, the sites which produced a better performance with the standard air temperature datasets suffered a performance decline between 13% and 56% with the ERA5 reanalysis datasets. Overall the WTAENN models continued to be inferior to RF model with lower *L* values indicating the WTAENN models are no better than the observed mean for most sites.

Taking into consideration the significant improvement in *RMSE* values and a marginally lower reduction in *r* values, the RF model utilizing the air temperature from ERA5 reanalysis produced the best model with higher model accuracy and stronger model relationship, though the lower *L* values indicate the models are less interpretable than the ones using observed air temperatures.

10.7 Limitations and further research

Urban climate phenomena such as UHI and UHC exhibit spatial and temporal sensitivity. Though well studied the effective modeling of the impact of UHI on energy demand, temperature-related hospital admissions, and the like are hampered by the sparse distribution of weather stations and paucity of spatially and temporally accurate weather data (Azevedo et al., 2016). Though this study utilized ERA5 reanalysis datasets, which were spatially and temporally sensitive, Seneviratne (2017) highlighted that the weak or loose coupling of land processes in reanalysis datasets such as ERA impact its ability to capture the real effect of land processes on land climate. Therefore without the full integration of land processes, reanalysis datasets would be inadequate to represent urban-scale land phenomena such as UHI, and hence their usefulness in modeling electricity demand would be limited.

Further, this study considered the effect of a single weather variable on short-term energy demand modeling. However, studies have shown that factors such as relative humidity and wind speed could also act as indicators of UHI intensity (Wang, 2015). Therefore the inclusion of a range of weather variables to better model UHI intensity could contribute to better modeling of the relationship between UHI and short-term electricity demand modeling.

References

Ahmad, M., Mourshed, M., Rezgui, Y., 2017. Trees vs neurons: comparison between random forest and ANN for high-resolution prediction of building energy consumption. Energy Build. 147, 77–89.

Amin, M., 2003. North America's electricity infrastructure: are we ready for more perfect storms? IEEE J. Magaz. Ieeexplore.ieee.org. Available from: <https://ieeexplore.ieee.org/stamp/stamp.jsp?tp = & arnumber = 1236231>.

Amral, N., Ozveren, C., King, D., 2007. Short term load forecasting using multiple linear regression. In: 42nd International Universities Power Engineering Conference, Brighton, 2007, pp. 1192–1198. doi:10.1109/UPEC.2007.4469121.

The Australia Institute Can't stand the heat the energy security risk of Australia's reliance on coal and gas generators in an era of increasing heatwaves, Aust. Inst. 2017, 2019. apo.org.au. [online]. Available from: <https://apo.org.au/node/121471>.

Azevedo, J., Chapman, L., Muller, C., 2016. Quantifying the daytime and night-time urban heat Island in Birmingham, UK: a comparison of satellite derived land surface temperature and high resolution air temperature observations. Remote. Sens. 8 (2), 153.

Bansal, A., Rompikuntla, S., Gopinadhan, J., Kaur, A., Kazi, Z., 2019. Energy consumption forecasting for smart meters. arXiv.org. Available from: <https://arxiv.org/abs/1512.05979>

Bianco, V., Manca, O., Nardini, S., 2009. Electricity consumption forecasting in Italy using linear regression models. Energy 34 (9), 1413–1421.

Breiman, L., 2001. Random forests. Mach. Learn. 45 (1), 5–32. pp.

Caccetta, P., Devereux, D., Amati, M., Boruff, B., Kaspar, J., Phelan, K. et al., 2017. Land surface temperature and urban heat island estimates for Australian urban centres. vol. 2. CSIRO. Data Collection.

Chow, W., Roth, M., 2006. Temporal dynamics of the urban heat island of Singapore. Int. J. Climatol. 26 (15), 2243–2260.

Clements, A., Hurn, A., Li, Z., 2016. Forecasting day-ahead electricity load using a multiple equation time series approach. Eur. J. Oper. Res. 251 (2), 522–530.

Denil, M., Matheson, D., de Freitas, N., 2014. Narrowing the gap: random forests in theory and in practice. In: Proceedings of the 31st International Conference on Machine Learning, Beijing, China, 2014, p. 32.

Deo, R., Şahin, M., 2017. Forecasting long-term global solar radiation with an ANN algorithm coupled with satellite-derived (MODIS) land surface temperature (LST) for regional locations in Queensland. Renew. Sustain. Energy Rev. 72, 828–848.

Esch, T., Heldens, W., Hirner, A., Keil, M., Marconcini, M., Roth, A., et al., 2017. Breaking new ground in mapping human settlements from space—the global urban footprint. ISPRS J. Photogramm. Remote. Sens. 134, 30–42.

Founda, D., Santamouris, M., 2017. Synergies between urban heat island and heat waves in Athens (Greece), during an extremely hot summer (2012). Sci. Rep. 7, 1.

Friedrich, L., Afshari, A., 2015. Short-term forecasting of the Abu Dhabi electricity load using multiple weather variables. Energy Procedia 75, 3014–3026.

Jackson, W.J., Argent, R.M., Bax, N.J., Bui, E., Clark, G.F., Coleman S. et al., 2016. Overview. In: Australia State of the Environment 2016, Australian Government Department of the Environment and Energy, Canberra. Available from: <https://soe.environment.gov.au/theme/overview>. doi:10.4226/94/58b65510c633b.

Kaplan, G., Avdan, U., Avdan, Z., 2018. Urban heat island analysis using the landsat 8 satellite data: a case study in Skopje, Macedonia. Proceedings 2 (7), 358.

Kolokotroni, M., Davies, M., Croxford, B., Bhuiyan, S., Mavrogianni, A., 2010. A validated methodology for the prediction of heating and cooling energy demand for buildings within the urban heat island: case-study of London. Sol. Energy 84 (12), 2246–2255.

Legates, D., McCabe, G., 2012. A refined index of model performance: a rejoinder. Int. J. Climatol. 33 (4), 1053–1056.

Lofti, E., Akbarzadeh, M.R., 2014. Practical emotional neural networks. Neural Netw. 59 (November 2014), 61–72.

Lotfi, E., Akbarzadeh-T, M., 2016. A winner-take-all approach to emotional neural networks with universal approximation property. Inf. Sci. 346-347, 369–388.

Lotfi, E., Khosravi, A., Akbarzadeh-T, M.R., Nahavandi, S., 2014. Wind power forecasting using emotional neural networks. In: 2014 IEEE International Conference on Systems, Man, and Cybernetics (SMC), San Diego, CA, 2014.

Meng, Q., Zhang, L., Sun, Z., Meng, F., Wang, L., Sun, Y., 2018. Characterizing spatial and temporal trends of surface urban heat island effect in an urban main built-up area: a 12-year case study in Beijing, China. Remote. Sens. Environ. 204, 826–837.

Mestayer, P.G., 1998. Urban Scale Models. Urban Air Pollution—European Aspects. Environmental Pollution, vol 1. Springer, Dordrecht..

Meyers, J., Devereux, D., Van Niel, T., Barnett, G., 2017. Mapping Surface Urban Heat in Canberra. CSIRO, Australia.

Nallathambi, S., Ramasamy, K., 2017. Prediction of electricity consumption based on DT and RF: an application on USA country power consumption. In: 2017 IEEE International Conference on Electrical, Instrumentation and Communication Engineering (ICEICE), Karur, 2017, pp. 1–7. doi:10.1109/ICEICE.2017.8191939.

Oke, T., Mills, G., Christen, A., 2017. Urban Climates. Cambridge University Press, Cambridge.

Pandey, H., 2016. Performance evaluation of selection methods of genetic algorithm and network security concerns. Procedia Comput. Sci. 78, 13–18.

Panteli, M., Mancarella, P. 2015. Influence of extreme weather and climate change on the resilience of power systems: impacts and possible mitigation strategies. Electr. Pow. Syst. Res. 127, 259–270.

Räikkönen, M., Tagg, A., Maki, K., Roca, M. 2016. Impact of extreme weather on critical infrastructure: the EU-INTACT risk framework. Available from: <https://www.researchgate.net/publication/309345137_Impact_of_extreme_weather_on_critical_infrastructure_the_EU-INTACT_risk_framework>.

Ramamurthy, P., Bou-Zeid, E., 2017. Heatwaves and urban heat islands: a comparative analysis of multiple cities. J. Geophys. Res.-Atmos. 122 (1), 168–178.

Santamouris, M., 2014. On the energy impact of urban heat island and global warming on buildings. Energy Build. 82, 100–113.

Santamouris, M., Papanikolaou, N., Livada, I., Koronakis, I., Georgakis, C., Argiriou, A., et al., 2001. On the impact of urban climate on the energy consumption of buildings. Sol. Energy 70 (3), 201–216.

Santamouris, M., Haddad, S., Saliari, M., Vasilakopoulou, K., Synnefa, A., Paolini, R., et al., 2018. On the energy impact of urban heat island in Sydney: climate and energy potential of mitigation technologies. Energy Build. 2018 (166), 154–164.

Seneviratne, S., 2017. Future earth system reanalyses: a land perspective. In: 5th International Conference on Reanalyses (ICR5), Rome, Italy, November 14, 2017. Institute for Atmospheric and Climate Science, ETH Zurich, Switzerland.

Shahmohamadi, P., Ani, A., Ramly, A., Maulud, K., Nor, M., 2010. Reducing urban heat island effects: a systematic review to achieve energy consumption balance. The National University of Malaysia. Available from: <https://ukm.pure.elsevier.com/en/publications/reducing-urban-heat-island-effects-a-systematic-review-to-achieve>.

Taneja, K., Ahmad, S., Ahmad, K., Attri, S., 2016. Time series analysis of aerosol optical depth over New Delhi using box–Jenkins ARIMA modeling approach. Atmos. Pollut. Res. 7 (4), 585–596.

Vrajitoru, D., 1998. Crossover improvement for the genetic algorithm in information retrieval. Inf. Process. Manag. 34 (4), 405–415.

Wu, A., Zeng, Z., Chen, J., 2014. Analysis and design of winner-take-all behavior based on a novel memristive neural network. Neural Comput. Appl. 24 (7-8), 1595–1600.

Zamirpour, E., Mosleh, M., 2018. A biological brain-inspired fuzzy neural network: Fuzzy emotional neural network. Biol. Inspir. Cogn. Arc. 26, 80–90.

Zhang, C., Cao, L., Romagnoli, A., 2018. On the feature engineering of building energy data mining. Sustain. Cities Soc. 39, 508–518.

Zhao, L., Oppenheimer, M., Zhu, Q., Baldwin, J., Ebi, K., Bou-Zeid, E., et al. 2018. Interactions between urban heat islands and heat waves. Environ. Res. Lett. 13(3), 034003.

CHAPTER

11

Artificial neural networks and adaptive neuro-fuzzy inference system in energy modeling of agricultural products

Ashkan Nabavi-Pelesaraei[1,2], *Shahin Rafiee*[1], *Fatemeh Hosseini-Fashami*[1] *and Kwok-wing Chau*[3]

[1]Department of Agricultural Machinery Engineering, Faculty of Agricultural Engineering and Technology, University of Tehran, Karaj, Iran [2]Management of Fruit and Vegetables Organizations, Tehran Municipality, Tehran, Iran [3]Department of Civil and Environmental Engineering, Hong Kong Polytechnic University, Kowloon, Hong Kong

11.1 Introduction

One of the important factors for value-adding in the agricultural sector is energy for production and cultivation of crops and agricultural processing. For cultivation and production of crops, from the beginning to the end, extensive factors including agricultural machinery, human labor, and animal are used. When comparisons are made with other areas within the global economy, energy consumption in agricultural production has been growing rapidly, becoming more mechanized, with enhancements in commercial fertilizers, and thus superannuating traditional farming. Auditing of energy can be employed as a significant tool to evaluate agricultural products' life cycle assessments, which becomes the initial level for recognizing the crop production that leads to increasing efficiency. For rising productivity, climbing food security, and helping to develop the rural economy, energy is an important input, which is used in different sectors, among agriculture (Ghorbani et al., 2011). The process of building computer models of energy systems to analyze them is energy system modeling. These models often study diverse assumptions about the technical and economic conditions at play using scenario analysis. Results may

Predictive Modelling for Energy Management and Power Systems Engineering
DOI: https://doi.org/10.1016/B978-0-12-817772-3.00011-2

contain greenhouse emissions, the system's feasibility, energy efficiency, and accumulative fiscal costs of the system under investigation. A vast range of techniques are used, ranging from broadly economic to broadly engineering. Basically, mathematical optimization is used to estimate the least-cost in some criteria. Models have different types in their field, for instance, international, national, municipal, and regional or self-governing. By maintaining national energy models, governments develop national policy. In nature, intelligence seen by humans or other animals is called natural intelligence, while intelligence demonstrated by a machine is called artificial intelligence (AI) or machine intelligence. AI understands its environment and it increases the probability of reaching its goals to the maximum, by taking actions. Obviously the term "artificial intelligence" is applied when a machine uses "cognitive" functions for which human beings are dependent on other human minds, such as "learning" and "problem solving" (Kakani et al., 2020).

An artificial neural network (ANN) is known as an immense distributed and parallel data processing system having specific functional properties similar to those of the human brain's biological nervous system (Momenzadeh et al., 2011). It is able to replace analytical approaches by providing the benefits such as no required knowledge of the parameters of the internal system and terse solutions to problems with different variables. This algorithm can be a reliable alternative method to predictions for improving energy use (Shabanzadeh-Khoshrody et al., 2016; Nabavi-Pelesaraei et al., 2018). For modeling energy consumption in the last decade, different techniques have been used from AI. An arithmetic method for solving problems in different aspects, including pattern recognition, simulation, optimization, estimation of nonlinear functions, and clustering, is ANN. On issues where methods other than ANN may fail or cannot deliver an acceptable result, ANN can be useful for analyzing the predicted data in these issues (Safa and Samarasinghe, 2011).

The first fuzzy-rule is presented by Zadeh (1965) and is one of the favorite and most powerful fuzzy logic and fuzzy set modeling methods. Sugeno and Yasukawa (1993) defined fuzzy-rule modeling as the qualitative modeling plan to employ a natural language in defining the system behavior. Nowadays, the merger of fuzzy logic and neural networks has led to a novel study called the adaptive neuro-fuzzy inference system (ANFIS). The computational procedure relies upon the self-learning capability of neural network and linguistic transparency of fuzzy inference system, and thus contains the benefits of both neural networks and fuzzy systems (Kaab et al., 2019). The ability to manage great data in nonlinear and dynamic systems, particularly for cases where full understanding of the physical relationships amongst them are not attained, is one of the great advantages of ANFIS over convectional modeling (Nayak et al., 2004). ANFIS are expanded for data classification, computer vision, database management information retrieval, and automatic control of signal processing (Jang, 1993).

The relation between input and output energies in agricultural systems is very complicated. In the most items, determination of a direct relation is not practical for management production in this sector. So, the use of a comprehensive model is pretty essential. In this chapter we surveyed the application of the ANN and ANFIS methods to a model of an energy system. The mentioned methods are expanded and different elements are explained. Moreover, several examples are given for a better understanding of AI utilization in energy modeling. Finally, the results were investigated and interpreted.

11.2 Data collection and energy calculation

11.2.1 Data collection

The studies should be focused on a special region in a country such as a city. The situation of farms, including weather, soil type, operation, implementation, etc., is variable in the different case studies. Of course it may be that in a study the energy use pattern of a country is surveyed, but the data collection should be done in different regions of that country, separately. The expression of information about the cropping system of the desired products including statistics of areas under cultivation, annual production, etc. can be effective to better recognize the importance of desired product in the case study. Moreover, referring to the geographical positions including latitude, longitude, type of soil, and the elevation can make it possible to compare or generalize the results.

11.2.1.1 Sample size method

In most cases, due to the number of farmers, a census is an activity that will take too much time and cost for the research team. For this reason, collecting all data from farmers in a case study is often a route to failure. One of the most suitable solutions to this problem is the sample method. In general, one of the most important statistical steps is calculating the proper sample size, despite observations indicating that this specific stage is ignored in most studies. The calculation of the sample size for a special study should be done correctly, so that the results of the sample are valid and do not lead to conclusions being mistaken.

Three basic formulas of sampling methods are used in the determination of sample size:

1. Cochran method: Cochran noted in statistics that sample size can be diminished a little bit for a limited population. The reason is that a huge population offers relatively more information than a small one. He presented an adjustment formula to compute the entailed sample size under this condition, as follows (Cochran, 1977):

$$n = \frac{\frac{z^2pq}{d^2}}{1 + \frac{1}{N}\left(\frac{z^2pq}{d^2} - 1\right)} \tag{11.1}$$

   where d denotes the allowable error ratio deviation from the mean population (=0.05), p is the computed proportion of an attribute within the population (=0.5), q is $1-p$ (=0.5), n is the entailed sample size, z is the reliability coefficient (=1.96 at 95% confidence level), and N is the ratio of the number of milling factories to the target population.

2. Yamane (1967) proposed another simple equation to replace the Cochran equation. He recommended that the sample size for $P = .5$ and 95% confidence level is:

$$n = \frac{N}{1 + N(e^2)} \tag{11.2}$$

   where e is the precision level (95%) and N is the population size.

3. Neyman allocation: this method of sample allocation can be employed with categorized samples. The key objective is for maximization of the accuracy of the examination for a specific sample size. The "best" sample size for stratum h by employing this allocation method is shown as follows (Mohammadi and Omid, 2010):

$$n = \frac{(\sum N_h S_h)}{N^2 D^2 + \sum N_h S_h^2} \tag{11.3}$$

where $D^2 = d^2/z^2$; z represents the reliability coefficient (1.96 denoting 95% reliability); d denotes the precision where $(x^- - X^-)$; ${S_h}^2$ represents the variance of h stratification; S_h denotes the standard deviation in the h stratification, N_h represents the number of the population in the h stratification; N denotes the number of holdings in target population, and n depicts the entailed sample size.

Other formulas are available in different literature, but the three abovementioned formulas are much more widely used than others. Table 11.1 illustrates the sample size determination in several studies.

TABLE 11.1 A summary of the previous studies conducted on energy in agricultural systems from sample size prospective.

Surveyed study	Sample size method	Calculated rate of sample size
Erdal et al. (2007)	Neyman	146
Esengun et al. (2007)	Neyman	97
Mohammadi et al. (2008)	Neyman	100
Cetin and Vardar (2008)	Neyman	22
Bayramoglu and Gundogmus (2009)	Neyman	33
Kizilaslan (2009)	Cochran	87
Mobtaker et al. (2010)	Cochran	67
Ghorbani et al. (2011)	Neyman	100
Ozkan et al. (2011)	Neyman	50
Ren et al. (2012)	Yamane	110
Bojacá et al. (2012)	Yamane	14
Neira et al. (2013)	Yamane	250
Nabavi-Pelesaraei et al. (2014a)	Cochran	52
Soheili-Fard and Salvatian (2015)	Cochran	30
Sefeedpari et al. (2016)	Cochran	40
Mostashari-Rad et al. (2019)	Cochran	120
Saber et al. (2020)	Cochran	213

11.2.1.2 The design of the questionnaire

A questionnaire is a research instrument, which is performed after having determined the sample size for data collecting. The questionnaire includes a set of questions (or other kinds of prompts) in order to collect information from farmers for an agricultural study. According to Mellenbergh (2008), there are several kinds of questionnaire, such as fully computerized, adaptive computerized questionnaire with administration, paper and pencil, face-to-face, etc. The first one is a type of questionnaire where a set of questions are presented on the computer screen and the next set of questions are designed by the computers, which depend on the answers of the previous questions. In the second one, interviewers have to answer the questions through computers. In the third one, as the name speaks for itself, the items are written on paper. In the last one, there is a set of oral questions asked by an interviewer. There are two factors, namely, reliability and validity, which should be realized in designing a questionnaire.

11.2.1.2.1 Reliability of the questionnaire

Reliability refers to the consistency of a major to collect data. To have a valid questionnaire, the reliability of the questionnaire must be obtained. In order to have a valid questionnaire, the questionnaire should be reliable. A measure should be chosen while the construct of a study is being examined. Construct refers to measuring an assumptive variable and one of the media is a questionnaire, which are among the measurement procedures which lead to accurate representation in terms of stability or consistency. The main assumption of reliability is to measure the stability and consistency of the questionnaire. A measurement procedure is reliable when the same (or nearly the same) result is obtained when used in the same conditions and by the same individuals. However, there may be tweaks in reliability which are categorized in the systematic and unsystematic part shown in Fig. 11.1.

Two forms of reliability include: "Test–retest reliability" and "Reliability within a scale." In "Test–retest reliability" the same result is obtained while the same conditions are used. Reliability within a scale is where one particular trait is measured by every individual question.

11.2.1.2.2 Validity of questionnaire

Validity is the extent to which the accuracy and consistency of research is measured. Criterion validity, content validity, construct validity, and face validity are several kinds of validity which are divided into categories, internal and external validities. Internal validity is a type of validity measuring the research accuracy in what it is supposed to measure. External validity is a type of validity in which the results of a study can be generalized to other situations and to other people where the sample was shown (Bolarinwa, 2015). Graphical representations of the subtypes of various forms of validity tests are demonstrated in Fig. 11.2.

The agricultural energy research questionnaire has various questions regarding the consumption of various inputs such as electricity, fertilizer, fuel, biocides, the amount of cultivated land, and the production of the resulting product each year, as well as the total

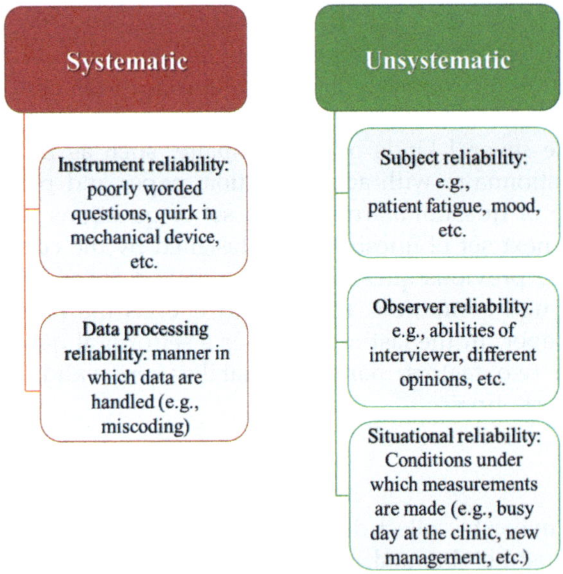

FIGURE 11.1 Systematic and unsystematic categories of reliability of questionnaire.

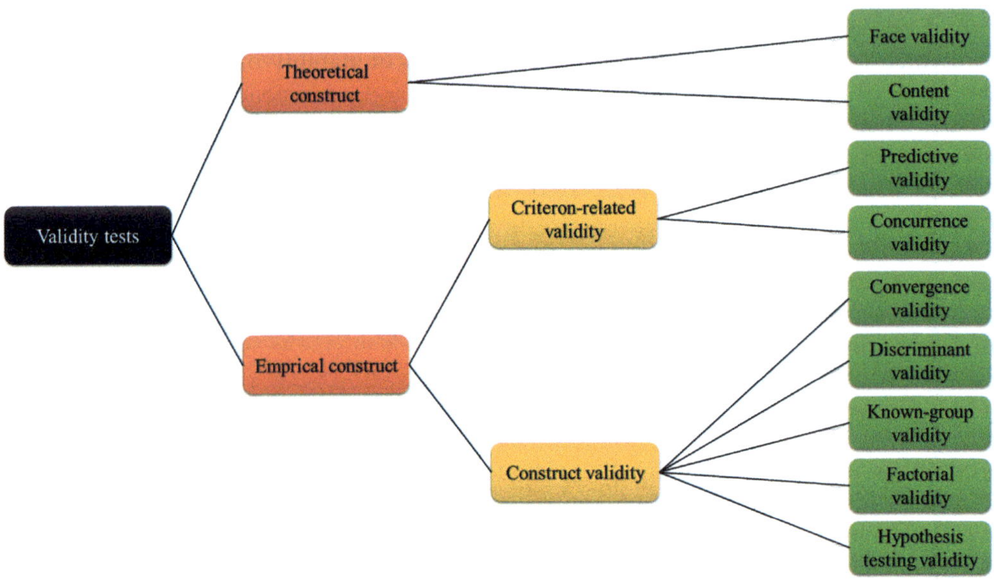

FIGURE 11.2 The subtypes of various forms of validity tests.

hours of work from the preparation of land to harvest and total hours of equipment and machinery, and so on. Table 11.2 summarizes a sample questionnaire.

A complete flowchart of the information collecting process for energy studies is shown in Fig. 11.3.

TABLE 11.2 A brief summary of the sample questionnaire.

Questionnaire No: . . ..
Data: 20. . ./. . ./. . .
The total area under cultivation or the number of calf, fattening, broiler, etc.: . . ..
Duration of the production: . . ..
Total weight of seed use or average weight of calf, fattening, broiler, etc. beginning of the period: . . ..
Yield per functional unit (ha, 1000 calf, 1000 broiler, etc.)[a]: . . ..
Number of fixed labors: . . ..
Number of daily labors: . . ..
Daily working hours: . . ..
Machinery operation used: . . ..
Types of machinery used: . . ..
Total weight of machinery per year: . . ..
Types of fuel used: . . ..
Total fuel consumption: . . ..
Types of chemical fertilizers: . . ..
Total weight of chemical fertilizers from each type: . . ..
Types of chemical biocides: . . ..
Total weight of biocides from each type: . . ..
Total weight of FYM use: . . ..
Rate of electricity consumption for each implement: . . ..
Total electricity consumption: . . ..

[a]*The unit of measurement depended on the scale of energy use and production.*

11.2.1.3 Datasets

A dataset (or data set) is an assembly of data. Often the dataset is related to the contents of a monolith database table or an integrated data matrix, with each column of the table showing a particular variable, and each row belonging to a member given from the question set. Generally, there do not exist unique datasets for recording input–output energies in the world, but governmental organizations (GO$_S$) of countries usually have the main datasets in agricultural products. They have data about the range of input consumption and yields in several categories including product base (furnishing data based on production target such as oilseeds, industrial crops, horticultural crops, etc.), annual report (furnishing data based on crop year), and geographical base (furnishing data based on production in different regions of a regions). These data can be used by researchers in energy analysis or AI models. It should be noted that the access of these datasets is limited to local producers. On a global scale, FAOSTAT provides free access to food and agriculture data for over 245 countries and

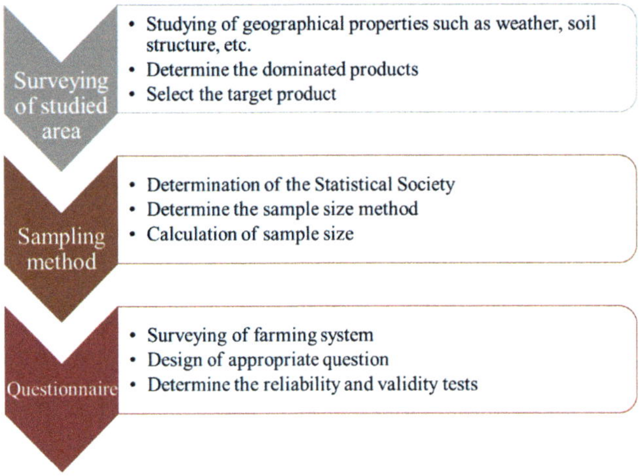

FIGURE 11.3 Information collecting process for energy studies in agricultural products.

territories and covers all FAO regional groupings from 1961 to the most recent year available (FAO, 2016). Of course these data are not applicable for energy computation because inputs are not specified. GOs and nongovernmental organizations (NGOs) that are active in agricultural activates can collect data based on the above sections and then furnish them to producers and, in the next target, they can make a dataset for agricultural production.

11.2.2 Energy analysis

11.2.2.1 Input–output energy

Input and output energy are two categories in the agricultural production from an energy point of view. Input energy sources consist of all of the inputs of the consumed energy in the production process. The main inputs include seed, farmyard manure, electricity, fuel (natural gas), machinery, biocides (fungicides, herbicides, pesticides, etc.), chemical fertilizers (potassium, phosphate, nitrogen, etc.), and human labor. Output energy often includes only harvested product, but in several systems there are some other sources such as waste (Fig. 11.4).

Covering the production and supply phases is an important issue in examining consumption of energy for agricultural production. Therefore, a specific range should be determined for the production system and the total energy consumption should be obtained within the specified range of production systems.

These activities include planting, spraying, and fertilization, land preparation and tillage, and so on, which leads to increased product performance and activities that bring energy use to the final product.

Due to the limitation of inputs and outputs in the agricultural sector, all have standard energy standard coefficients. As a result, energy values of every input and output are computed by multiplying the energy coefficient with the corresponding physical value. Table 11.3 lists the input energy coefficients, which are obtained from previous studies.

They are not related to the crop type and have fixed values. For instance, the same kind of diesel fuel and human labor have constant coefficients in converting their equivalent

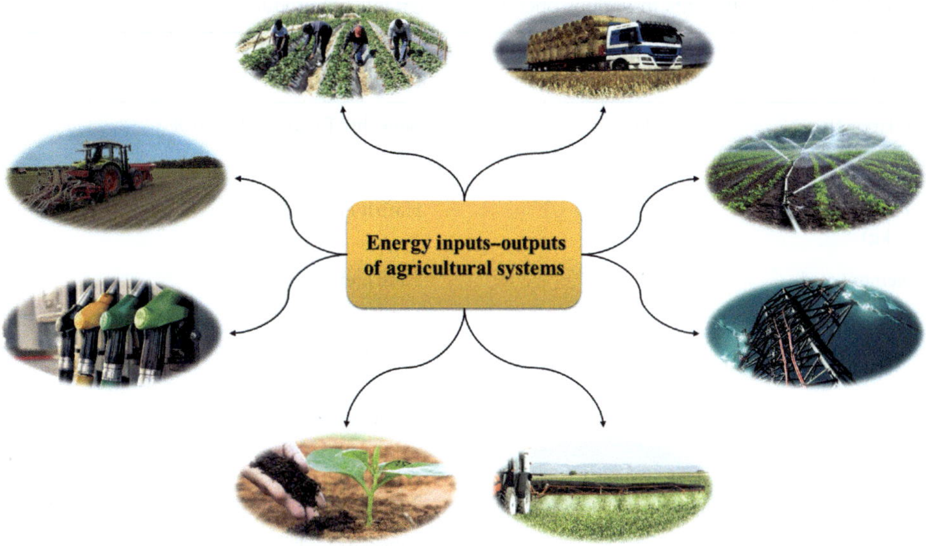

FIGURE 11.4 A summary of input–output energy of agricultural systems.

energy consumption in different types of output. Hence, the input value becomes the major difference for agricultural production in energy usage. In addition, the results are compared with other studies applying the standard energy coefficient to calculate energy consumption. The input and output energies will be computed by multiplying each item by their energy equation from Table 11.3.

The energy tantamount for machinery is declared in the following equation (Mousavi-Avval et al., 2017; Hatirli et al., 2005; Mandal et al., 2015):

$$ME = \frac{ELG}{TC_a} \tag{11.4}$$

where C denotes the effective field capacity (ha/h) computed following Hatirli et al. Hatirli et al. (2005), T denotes the machinery's economic life (h), G denotes the machine's mass (kg), L denotes the machine's useful life (years), E denotes the machine's production energy (GJ/kg/y) from Table 11.3, and ME is the machine energy (GJ/ha).

$$C_a = \frac{SWE_f}{10} \tag{11.5}$$

where E_f denotes the field efficiency, W denotes the working width (m), and S denotes the working speed (km/h) for an economic life of the machine of y years.

11.2.3 Energy indices

The energy ratio (Eq. 11.6) between the input and output can be used to evaluate the energetic efficiency of the agricultural system. One of the main methods to examine the productivity of different agricultural products and gardening is to study various energy

TABLE 11.3 Energy coefficients in various operations of agricultural production.

Inputs (unit)	Energy equivalent (MJ/unit)	References
1. Human labor (h)	1.96	Beheshti Tabar et al. (2010)
2. Machinery		
(a) Tractor and self-propelled (kg)	9–10	Hatirli et al. (2005)
(b) Implement and machinery (kg)	6–8	Hatirli et al. (2005)
(c) General machinery (kg)	142.70	Kaltsas et al. (2007)
(d) Machinery used (h)	62.70	Mobtaker et al. (2010)
3. Fuel		
(a) Diesel fuel (L)	56.31	Nabavi-Pelesaraei et al. (2018), Kizilaslan (2009)
(b) Gasoline (L)	46.3	Hosseinzadeh-Bandbafha et al. (2018)
(c) Oil (L)	36.7	Hosseinzadeh-Bandbafha et al. (2017)
(d) Natural gas (m^3)	49.5	Hosseinzadeh-Bandbafha et al. (2017)
(e) Lubricant (L)	43.80	Zentner et al. (2004)
4. Chemical fertilizers (kg)		
(a) Nitrogen	66.14	Erdal et al. (2007), Bakhtiari et al. (2015)
(b) Phosphate (P_2O_5)	12.44	Nabavi-Pelesaraei et al. (2013), Unakitan et al. (2010)
(c) Potassium (K_2O)	11.15	Nabavi-Pelesaraei et al. (2019a), Erdal et al. (2007)
(d) Zinc (Zn)	8.40	Kazemi et al. (2015)
(e) Sulfur (K_2O)	1.1	Mousavi-Avval et al. (2011)
(f) Ferrum (Fe^{2+})	6.3	Nabavi-Pelesaraei et al. (2014a)
5. Farmyard manure (kg)	0.3	Kizilaslan (2009)
6. Biocides (kg)		
(a) Herbicides	85	Pishgar-Komleh et al. (2012)
(b) Insecticides	199	Nabavi-Pelesaraei et al. (2018), Ozkan et al. (2004)
(c) Fungicides	92	Nabavi-Pelesaraei et al. (2018), Ozkan et al. (2004)
7. Electricity (kWh)	11.93	Jekayinfa et al. (2013)
8. Feed		
(a) Concentrate (kg)	13.6	Hosseinzadeh-Bandbafha et al. (2017)
(b) Maize silage (kg)	10.41	Hosseinzadeh-Bandbafha et al. (2017)
(c) Dry alfalfa (kg)	10.92	Hosseinzadeh-Bandbafha et al. (2017)

(Continued)

TABLE 11.3 (Continued)

Inputs (unit)	Energy equivalent (MJ/unit)	References
(d) Barley (kg)	15.28	Hosseinzadeh-Bandbafha et al. (2017)
(e) Soybean meal (kg)	12.06	Atilgan and Hayati (2006)
(f) Dicalcium phosphate (kg)	10	Atilgan and Hayati (2006)
(g) Fatty acid (kg)	9	Heidari et al. (2011)
(h) Minerals and vitamins (m^3)	1.59	Heidari et al. (2011)
9. Water for irrigation (m^3)	0.63	Hatirli et al. (2006)
10. Nylon (kg)	17.91	Kitani (1999)
11. Transportation (t km)	4.5	Tabatabaeefar et al. (2009)
12. Seed (kg)	As output of production system	

indices including net energy, energy productivity, specific energy, and energy use efficiency. These indicators were computed according to energy tantamount to inputs offered in Table 11.3.

They are computed in five different ways:

$$\text{Energy use efficiency} = \frac{\text{Output energy(MJ /ha)}}{\text{Input energy(MJ /ha)}} \tag{11.6}$$

$$\text{Energy productivity} = \frac{\text{Yield(kg/ ha)}}{\text{Input energy(MJ /ha)}} \tag{11.7}$$

$$\text{Specific energy} = \frac{\text{Output energy(MJ/ ha)}}{\text{Yield(kg/ha)}} \tag{11.8}$$

$$\text{Net energy} = \text{Output energy(MJ /ha)} - \text{Input energy(MJ/ha)} \tag{11.9}$$

$$\text{Energy intensiveness} = \frac{\text{Input energy(MJ/ ha)}}{\text{Total production cost(\$/ ha)}} \tag{11.10}$$

The ratio of input–output energy, or energy use efficiency, is an indicator that determines the agricultural system's energy (Pishgar-Komleh et al., 2012). Energy efficiency, defined as the ratio between output and energy consumption, is one of the most useful and beneficial indicators for understanding the economy or energy efficiency of agricultural system. Energy per unit mass and energy density are specific energy, but more accurately it means "energy density" in the sense of energy per unit volume. The concept of pure energy in the energy economy expresses the difference between the energy utilized to harvest an energy source and the energy acquired from that withdrawal. Energy intensiveness is a measure of the potential amount of economic output produced by a standardized unit or that is needed to generate a dollar value for the economic outflow (Sefeedpari et al., 2016).

11.3 Artificial neural network

The development of ANNs has been inspired by the specific function of the human marrow, although they are only remotely interconnected. There are two major likeness between them, despite the fact that ANNs do not approach the complexity of the marrow. They are namely, their building blocks are extremely interrelated and are easy types of computational devices and the function of the network is specified by the linkage between neurons. The 10^{10} computational neurons in the human brain are linked through one network with 10^4 connections or so for each element. ANNs acts as parallel and distributed computing networks, which have some of the same basic characteristics as the biological nervous systems (Fig. 11.5). Fig. 11.5 provides the information that $[X = (x_1, x_2, \ldots, x_n)]$ denotes input signal to neurons. The relative weight $[W = (w_1, w_2, \ldots, w_n)]$ of each input carries its impact, which is analogous to biological neurons' synaptic strengths. It can be observed that some inputs are more imperative than others for combining to produce a work. The input signal's intensity is determined by the adaptive weight values within the network. In producing the output signal of a neuron, a block in the body associated with biological cells play a key role in which all the weighted inputs are added algebraically (Zurada, 1992; Park and Lek, 2016):

By using appropriate elementary and enough information, ANNs are able to furnish optimal solutions via tuning their internal structures. By applying proper inputs to the ANN, knowledge will be obtained from the environment, in an imitation of brain function, and users will later be able to recall this knowledge. Although ANN has a lot of diversity and variety, with regard to the learning process the two main categories are easily recognizable (Park and Lek, 2016):

Supervised learning and unsupervised learning. In the first category, using a teacher in the learning step, ANN is said to be how well it is done or what correct behavior should be. In the second category, analysis of the properties of the dataset are undertaken automatically and its output is able to reflect these lectures.

Both groups of ANNs have been employed in energy in agriculture, that is, multilayer perception (MLP) with backpropagation algorithm for supervised learning, and self-organizing map (SOM) for unsupervised learning.

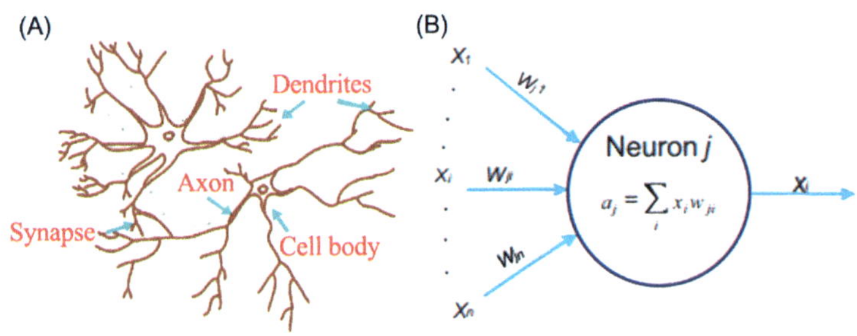

FIGURE 11.5 Schematic diagrams of (A) a biological neuron and (B) an artificial neuron as a basic processing element in a network.

11.3.1 Multilayer perception structure

A MLP is a species of feedforward ANN including at least three layers of knots: an output layer, a hidden layer, and an input layer. Except for the input nodes, a nonlinear activation function is employed to each node or neuron, the task of detecting MLP from a linear perception lies with multiple layers and nonlinear activation. It is able to discern information that is not linearly detachable. MLP is the most popular neural model for prediction. As shown in Fig. 11.6, in feedforward networks the product of input elements (a_i) and weights (w_{ij}), together with the bias (b_j), are summed at nodes (Kalogirou and Bojic, 2000; Ghritlahre and Prasad, 2018):

$$x = \left(\sum_{i=1}^{n} w_{ij}a_i\right) + b_j \tag{11.11}$$

An output is produced after a transfer function F has been imposed on X.

$$F(X) = F\left[\left(\sum_{i=1}^{n} w_{ij}a_i\right) + b_j\right] \tag{11.12}$$

In the above equations, there are two roots, which comprise weight and bias. The notion of weight is a principal concept in ANNs. A set of weighted inputs allows each neuron or artificial knot in the production system to produce related outputs. Professional working

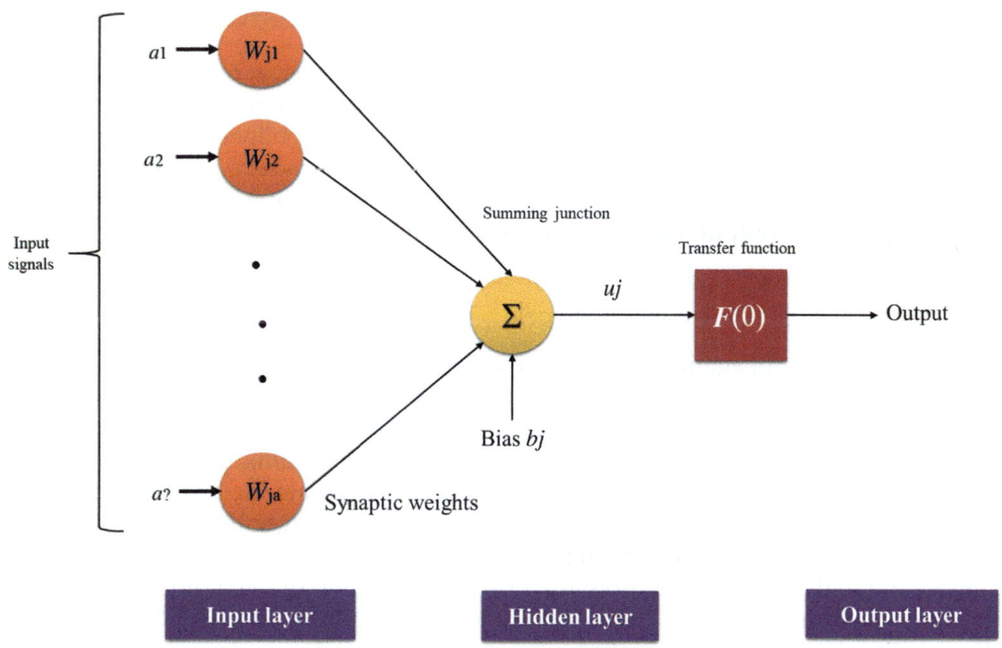

FIGURE 11.6 General structure of MLP feedforward neural network.

in the field of machine learning and AI projects, in which ANNs are used often, speak of weight as a function of biological and technological systems. The bias node in a neural network is a node that is always there. In other words, its value is set to 1, regardless of the data in a specific pattern. This is similar to interception in a regression model and has a similar function. If a neural network does not have a bias node in a data layer, it will not be able to generate an output in the next layer of 0 (on a linear scale or the value that converts to 0 when it passes through the activation function).

MLP-based models will be able to accommodate larger dimensional input spaces better than polynomial expansions, which is an interesting feature of many real-life problems, where one must model dependencies of several variables. In agricultural studies, variables are multidimensional. Therefore the MLP structure can connect the energy model to real conditions in this field.

11.3.2 Feedforward neural network

A feedforward neural network is different from frequent neural networks in that connections between nodes do not form a cycle. It was the initial and easiest type of ANN. Its information only moves in the forward direction, namely from input nodes, via hidden nodes, to output nodes (Schmidhuber, 2015). No computations are performed when inputs arrive at the first layer. Computations begin in the hidden layer, on which a linear or nonlinear activation function is imposed as follows (Schmidhuber, 2015):

$$Z_j = V_j\left(\sum_{i=1}^{p} w_{ij}^{h} x_i + p_j\right), j = 1, 2, \ldots, m. \tag{11.13}$$

where q denotes bias term, h denotes the symbol of hidden layer, p represents input number, and v_j denotes the activation function in the hidden layer. Activation functions are so significant for an ANN to learn and realize that complex and nonlinear complex operating values exist between inputs and response variables. They introduce nonlinear properties to the network. Their main purpose is to turn the input signal of a node into an ANN to an output signal. This output signal is then used as an input in the next layer on the stack.

11.3.3 Backpropagation neural network

In ANN, the backpropagation method is used to calculate the gradient needed to calculate the weight used in the network. This method is good way to educate profound neural networks having one or more hidden layers (Schmidhuber, 2015). One of the automatic differentiation techniques is backpropagation. During the learning process, the gradient descent optimization algorithm uses backpropagation to calculate the loss function's gradient for adjusting weights of neurons. The backpropagation algorithm is the most commonly used algorithm in ANNs. During feedforwarding, knowledge is generated through the processing to the output layer from the input layer. Earlier defined difference tolerance is compared with the difference between computed and desired output values and during feedforwarding in the backpropagation phase. This error value is then propagated backward for updating the connections in the input layer (Oğuz et al., 2010).

11.3.4 Levenberg–Marquardt learning algorithm

To provide a numerical solution for reducing a problem, the Levenberg–Marquardt algorithm is used. It possesses stable convergence, is fast, and is suitable to train both small- and medium-sized problems in the field of ANN (Wilamowski and Yu, 2010). In mathematics and computing, the LM algorithm is used for solving nonlinear least-squares problem, which is also known as the damped least-squares (DLS) method. These problems are minimized, especially in the least squared fit curve. In energy farming studies, for inputs of the extended model, inputs of energy with area (in multiple studies) must be selected, while the output is the product's yield. In recent years, environmental impacts and economic indicators have been added to model outcomes. By applying principles of ANN, usually, 60%, 25%, and 15% information were employed to educate, examine, and validate the ANN model, respectively. Random modules were selected amongst all samples. By employing the experimental data from the network, several structures were appraised so as to select the best predicted model. Relying on the number of inputs and outputs for agricultural production, the numbers of neurons for input and output layers were then determined. Besides, both one and two hidden layers were attempted for the ANN model. One of them is then proposed for modeling after having considered the best results amongst them. The Levenberg–Marquardt algorithm is the most popular optimization and learning algorithm for training purposes.

It performs better than simple gradient fall and fusion gradient methods in a diversity of problems (Ranganathan, 2004). A typical ANN structure often comprises a layer of input neurons, one or more hidden layers to connect between input and output layers, and a layer of output neurons (Khoshnevisan et al., 2014c). The structure of ANN models in the agricultural products are displayed in Fig. 11.7. The weight of the input and output matrix is defined as follows: all connections between hidden and input layers composed the input weight matrix, and all connections between output and hidden layers composed the output weight matrix. The propagation (x) and the output (o) of each neuron are controlled by weight (w) which is modified using the output of the previous layer based on Eq. (11.14) (Zhao et al., 2009):

$$O = f\left(T + \sum w_i x_i\right) \tag{11.14}$$

where f' is a monotonically increasing nonlinear sigmoid function and T is a threshold bias value for each neuron.

At the end of training and testing periods, the error was calculated according to the difference between computed and targeted outputs. The following error function is used here (Kiani et al., 2010):

$$E = \frac{1}{p}\sum_{p}\sum_{k}\left(t_{pk} - z_{pk}\right)^2 \tag{11.15}$$

where t_{pk} is the kth element of the pth desired pattern vector, zpk is the kth element of the output vector for pth pattern input, k is the element index in the output vector, and p is the index of the p training pairs of vectors.

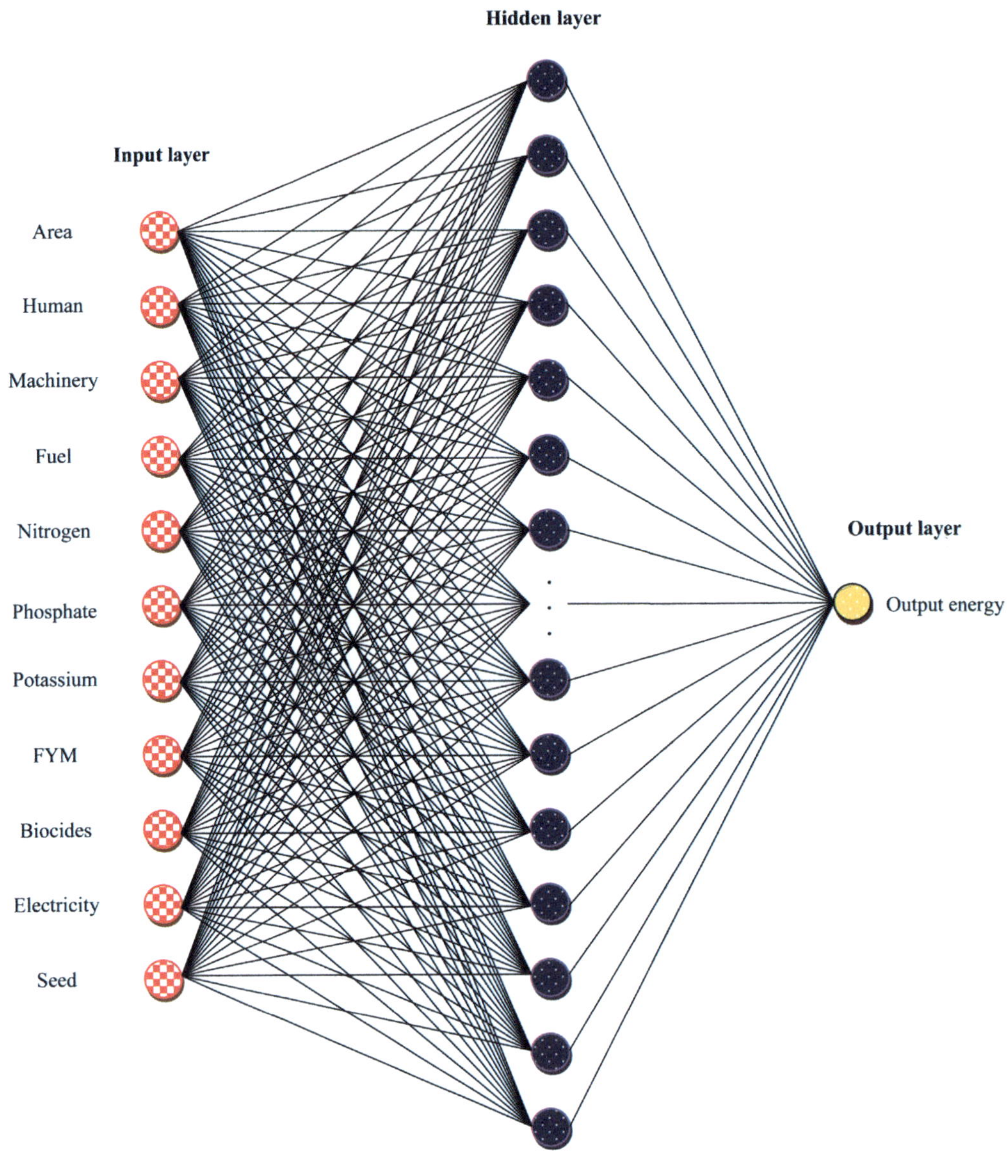

FIGURE 11.7 Schematic diagram of an example of ANN model in agricultural crops.

11.3.5 Overfitting

Loss of generalization by the model and the inability to use it for other datasets that were not used for network training are among the reasons for the importance of avoiding overfitting (or overtraining) in model development. An overfitted network possesses a good memory for particular data, and thus can learn perfectly for them, but is not able to figure out the general features from the training process. Researchers have developed different rules to determine the appropriate network parameters for the avoidance of overfitting.

Lek and Guegan (2000) reported that the number of neurons for each hidden layer, the number of hidden layers and the number of epochs are accountable for this phenomenon.

It is noted that, while the error in the training set is continuously reduced in the learning process, the error in the testing set, after having reached a minimum value, can be increased again. When the error in the testing dataset is the lowermost, the training procedure must be terminated. Geman et al. (1992) furnished a perfect review of the overfitting issue, which affected the extension of neural networks.

11.3.6 Sensitivity analysis

The ranking and selection of the main input variables of ANN can be attained via sensitivity analysis. Sensitivity analysis with sectional differentiation depends upon a computation of the output, weight, and input variables from ANN modeling results. The sensitivity is computed as follows (Sung, 1998):

$$S = \frac{\partial O}{\partial I} = O'\left(\sum_{J=1}^{J} w_{ij}^{1} H' w_{ij}^{2}\right) \tag{11.16}$$

$$S = \frac{\partial f(O)}{\partial X} \sum_{J=1}^{J} \left(w_{ij}^{1} \frac{\partial f(H)}{\partial X} w_{ij}^{2} \right) \tag{11.17}$$

where w_{ij}^1 and w_{ij}^2 denote weights for connections of the hidden layers as outputs from hidden layer nodes, H denotes the hidden node to be differentiated, I denotes the input, and O denotes the output. The order of the connections is that between the hidden and input layers and that between the output and hidden layers.

11.4 Adaptive neuro-fuzzy inference system

Although ANN is very powerful for modeling real-world problems, it has some defects. If the input data is less precise or obscure, ANN may not be able to handle them properly and a fuzzy system like ANFIS may be a better option (Moghaddamnia et al., 2009). One of the features of a neuro-fuzzy system is the combination of the learning capability of ANN and the fuzzification technique of fuzzy logic. Hence, it contains the advantages of the two techniques and is able to suit more literally the training data. Neural network techniques perform two important tasks, first of all, they help the fuzzy modeling procedure in learning information from the dataset. Secondly, they calculate the membership function parameters for the specific input–output data in the pertinent fuzzy inference system (FIS) (Suparta and Alhasa, 2016).

11.4.1 Fuzzy inference system

The membership function of the fuzzy set; the choice of (if–then) fuzzy logic rules; and the argumentation of the fuzzy inference techniques of the basic rules for output, are three

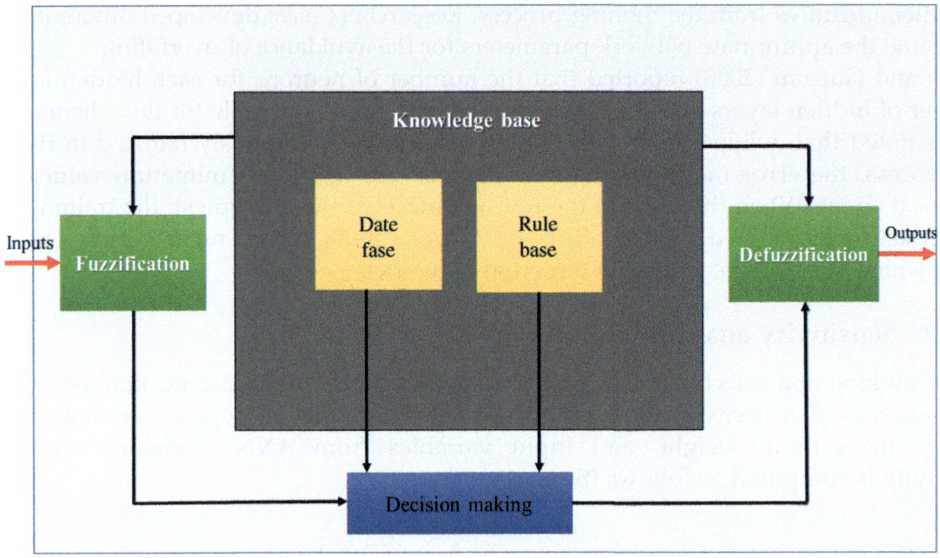

FIGURE 11.8 Schematic diagram of FIS.

main components that a FIS built on then. Fig. 11.8 displayed the detailed structure of the FIS. The condition for working out the FIS is that, in the fuzzification process, the raw value input is converted to a fuzzy value by a membership function with fuzzy value varying from 0 to 1. Knowledge base includes two major elements in the decision-making, namely, basic rules and databases. Typically, the database comprises description-like data in the fuzzy set parameter determined for an available linguistic variable. Generally, the database is developed as follows: for each linguistic variable, the number of linguistic values to be employed and the associated membership function are established and determined. It includes a conditional "If–Then" statement and fuzzy logic operators, which depend upon the rules. Basic rules are made by automatic generation or a human whilst numeric input–output data are employed in search rules. Several types of FIS exist, namely, Takagi–Sugeno, Mamdani, and Tsukamoto (Cheng et al., 2005). Amongst them, the Takagi–Sugeno model is a favorite for ANFIS.

Fig. 11.8 provides information in a FIS: four basic processing sections exist. By using specific fuzzy rules, a knowledge-based part and the dataset are defined by the corresponding MFs in the first section. An inference engine, for applying inference variation and modifications to the rules, is the second section.

A fuzzification inference is performed in the third section, in which conversion is made from the crisp input data to corresponding matching levels of linguistic terms. On the contrary, defuzzification inference is implemented in the fourth section for converting the fuzzy result back to an output with a crisp value (Barati-Harooni et al., 2016).

11.4.2 Adaptive network

Feedforward's neural network with multiple layers is an example of an adaptive network (Fig. 11.9). A supervised learning algorithm is usually employed by these networks

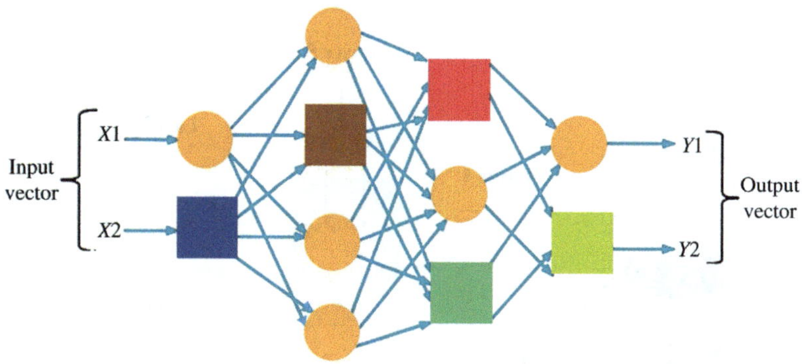

FIGURE 11.9 Schematic diagram of adaptive network.

in the learning process. Additionally, the characteristics of adaptive network architecture include a number of compatible nodes that connect directly between them without any weight value. There are different functions and tasks for each node in this network. The input signals and parameters in the node are the factors that the output depends on. The adopted learning rule is able to affect the node parameters and is able to decrease the occurrence of errors at the output of the network (Jang, 1993). Backpropagation or gradients, together with the chain rule are typically employed in learning a base adaptive network. Nowadays, backpropagation or gradient is employed for learning in a typical adaptive network. However, the weaknesses in the backpropagation algorithm can lead to a reduction in the accuracy and capacity of adaptive networks in decision-making. Two difficulties in the backpropagation algorithm are the tendency for getting stuck in a local minima and the slow convergence rate. Hence, an alternative learning algorithm, with better capabilities such as accelerating convergence and avoiding inference in local minima, was proposed by Jang (Jang, 1993).

11.4.3 Adaptive neuro-fuzzy inference system architecture

ANFIS is a category of adaptive network tantamount to FIS. In ANFIS, the membership function parameters are adjusted or tuned by employing either a hybrid learning algorithm (a hybrid combination of the least-squares method and backpropagation) or a backpropagation algorithm with the specific input–output data. Fig. 11.10 demonstrates the architecture of a five layer ANFIS with two inputs (x and y) and an output (z). The following displays a rule set with four fuzzy if–then rules (Suparta and Alhasa, 2016):

$$\text{Rule 1: if } x \text{ is } A_1 \text{ and } y \text{ is } B_1 \text{ then } f_1 = p_1x + q_1y + r_1 \tag{11.18}$$

$$\text{Rule 2: if } x \text{ is } A_2 \text{ and } y \text{ is } B_1 \text{ then } f_2 = p_2x + q_2y + r_2 \tag{11.19}$$

$$\text{Rule 3: if } x \text{ is } A_1 \text{ and } y \text{ is } B_1 \text{ then } f_1 = p_1x + q_1y + r_1 \tag{11.20}$$

$$\text{Rule 4: if } x \text{ is } A_2 \text{ and } y \text{ is } B_1 \text{ then } f_2 = p_2x + q_2y + r_2 \tag{11.21}$$

where B_1 and B_2 are fuzzy sets for variable y and A_1 and A_2 are fuzzy sets for variable x.

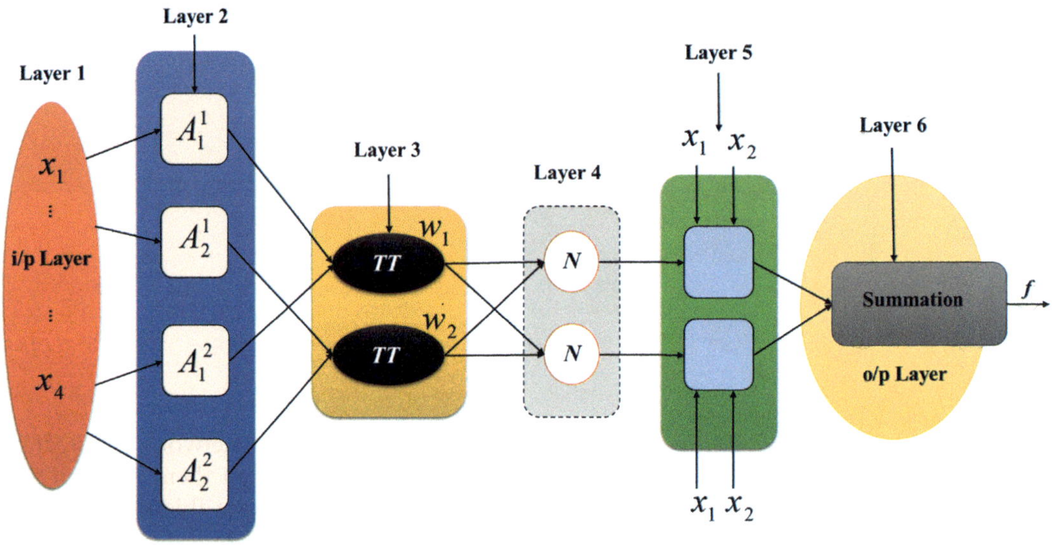

FIGURE 11.10 The architecture of ANFIS model.

The ANFIS layers shown in Fig. 11.10 are expressed as follows:

Layer 1: Each adaptive node in this layer possesses the following node function:

$$O_{1,i} = \mu_{Ai}(x) \text{for } i = 1,\ 2 \text{ or } O_{1,i} = \mu_{Bi-2}(y) \text{for } i = 3,\ 4 \tag{11.22}$$

where O_{1i} denotes the membership grade of fuzzy set A $(=A_1,\ A_2)$, A_i and B_i denote linguistic level pertinent to this node, and x and y denote the input to node i.

Layer 2: All nodes in this layer are fixed nodes with label $\prod$. The product of all inputting signals generates a node output here:

$$O_{2,i} = \mu_A(x)\mu_B(y) = w_i \text{ for } i = 1,\ 2,\ 3,\ 4 \tag{11.23}$$

All node outputs denote the firing strength (w_i) of the rule. It should be noted that the node function can be any other T-norm operators, which are able to perform fuzzy.

Layer 3: All nodes in this layer are fixed nodes with label N. The ith neuron then computes the ratio of the firing strength of the ith rule to the summation of firing strengths of all rules. The normalized firing strength represents its output here:

$$o_{3,i} = \frac{w_i}{w_1 + w_2 + w_3 + w_4} = \overline{w}_i \tag{11.24}$$

Layer 4: All nodes in this layer are adaptive nodes with the following node function:

$$o_{4,i} = \overline{w}_i f_i = \overline{w}_i(p_i x + q_i x + r_i) \tag{11.25}$$

where $\{p_i,\ q_i,\ r_i\}$ denotes this node's parameter set and w_i denotes the normalized firing strength from layer 3. Consequent parameters represent parameters in this layer.

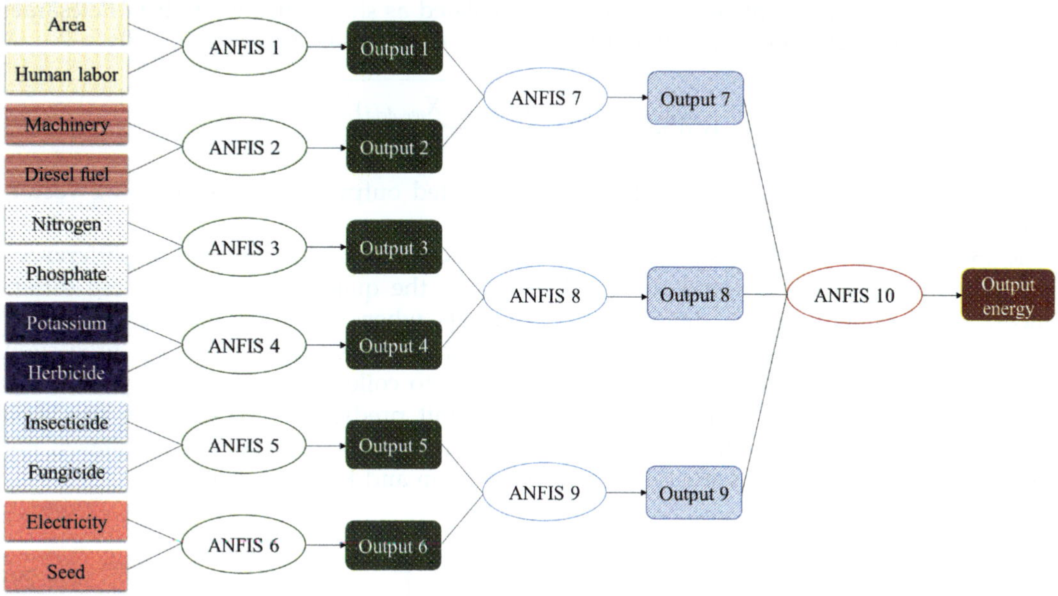

FIGURE 11.11 An example of three level ANFIS structure to predict output energy.

Layer 5: The fixed node with label $\sum$, which computes the overall output as the sum of all inputting signals from layer 4, is the only node in this layer:

$$\text{Overall output} = o_{5,i} = \sum \overline{w}_i f_i = \frac{\sum w_i f_i}{\sum w_i} \tag{11.26}$$

An important issue is that by increasing the number of ANFIs entries to more than five, the number of rules and computing time increases, which limits the use of ANFIS to solve this problem in energy studies, methods such as data clustering are used where large data are divided into natural groups to reduce the downside of the system's behavior (Mousavi-Avval et al., 2017).

Two or three level ANFIS structures are usually used for prediction of output energy in agricultural systems. Fig. 11.11 displays an example of a three level ANFIS structure in the output energy of agricultural production.

11.5 Validation of artificial neural network and adaptive neuro-fuzzy inference system model

The precision of the models was tested using some performance criteria. Root mean square error (RMSE), coefficient of determination (R^2), and mean absolute percentage error (MAPE) were selected to help us to elect the best model with the highest accuracy and minimal error. The performance criteria were calculated as shown below (Nabavi-Pelesaraei et al., 2018; Khoshnevisan et al., 2014c; Khoshnevisan et al., 2014b; Roy et al., 2018):

$$\mathrm{RMSE} = \sqrt{\frac{\sum_{i=1}^{n} (X_{obs,i} - X_{mo\ del,i})^2}{n}} \tag{11.27}$$

where X_{obs} and X_{model} are the actual and the predicted output for the ith training vector, and n is the total number of training vectors.

RMSE demonstrates the square root of the second sample moment of differences between anticipated values and observed values or the quadratic mean of these differences. These deflections are denominated residuals when computations are performed over the data model that is used for approximation and are called errors (or prediction errors) for out-of-sample computation. RMSE serves to collect error values in predictions for different times to the same degree of power unit prediction. RMSE is the precision measurement to assess prediction errors of different models for a particular dataset, not between datasets since it is scale dependent (Hyndman and Koehler, 2006).

$$R^2 = 1 - \left(\frac{\sum_{i=1}^{n} (t_i - z_i)^2}{\sum_{i=1}^{n} t_i^2} \right) \tag{11.28}$$

$$\mathrm{MAPE} = \frac{100}{n} \sum_{t=1}^{n} \left| (t_i - z_i) \right| \tag{11.29}$$

where n is the number of the points in the dataset, and t and z are actual and predicted output sets, respectively.

The original goal of computing R^2, or predicting future results, or testing hypotheses, is based on other relevant data. This method measures how well results are attained by the repeated model, based on the ratio of the total variation of results as described by the model (Nabavi-Pelesaraei et al., 2019b).

MAPE is a statistical measure, which predicts the accuracy of the system. This coefficient measures this accuracy and can be depicted as the mean absolute error percentage for each real-time equivalent period of values (de Myttenaere et al., 2016).

11.6 Other models of machine learning

Further to the above models, there are many ways of machine learning for energy modeling studies. Several of them are briefly defined as follows:

11.6.1 Support vector machine

Support vector machines (SVMs) are monitored learning models associated with learning algorithms, which analyze data applied for assortment analysis and regression analysis. Given a set of educating samples, each marked as attached to one or the other of two classes, a SVM teaching algorithm builds an example to allocate new specimens to one category or the other, making it a nonprobabilistic binary linear classifier (although methods such as Platt scaling exist to use SVM in a probabilistic classification setting). A SVM sample is an envoy of examples as spots in space, recorded so that examples of discrete classes are ordered by a comprehensible gap as vast as possible. The latest specimens are then represented into that identical space and forecasted to belong to a category stemming from which side of the gap they fall (Cortes and Vapnik, 1995). The foremost scheme of this method is to convert a nonlinear input area into a large-scale area and to find a mapper with nonlinear mappings. SVMs are predominantly performed for categorization, sample identification, and analysis of regression and generally out-accomplish other methodologies such as long-established statistical models (Huang et al., 2002).

Generally, attributes of the SVM procedure can be briefly expressed as (Zendehboudi et al., 2018):

- Appreciably exact and strong.
- Capable to model complex nonlinear determination borders.
- Less prone to overfitting compared to other patterns.
- Display a well-set explanation of the learned model.
- Inherent of effectuation in pattern identification, regression and categorization.

11.6.2 Bayesian network

Bayesian network (BNs) are randomized graphs, which associate variables with prospected probabilities, while outputs of the probability model are computed by using the Bayesian case. BN modeling is beneficial for data mining, determining and presenting crystal clear relationships surrounded by elements, illustrating expert knowledge and incorporating expert knowledge and experiential information, and identifying key hesitancy (Marcot and Penman, 2018).

BNs are accomplished in puzzling out troubles with (Borunda et al., 2016):

- Hesitancy leading to incomplete opposed, complexity, inconsistent knowledge between specialist or incomplete knowledge.
- Randomness due to random phenomena, irregularities, lack of patterns and predictability.
- BNs are helpful in resolving many difficulties of anticipation, data analysis and upgrade, detection, optimization, divergence detection, and decision-making based on the attained facts (Weber et al., 2012).
- In particular, principal applicable fields of BNs are:
- To provide global reliability estimates—BNs allow different types of knowledge in a model to include information from the feedback experience, expert knowledge such as logical rules, equations or probabilities, system behavior through functional or inactive analyzes, and observations.
- To review and analyze intricate systems—BNs create cause–effect relationships between various types of factors involved in the model, and thus modeling interactions (Borunda et al., 2016).

- Reliability—BNs are able to provide parameter determination, which becomes an input data in the decision stage. Some features, such as multivariate elements, failure dependencies, system deviations, dynamic evolution, and operating conditions, should be considered. Reliability analysis is used to deal with reliability, availability, and maintenance capability (Borunda et al., 2016).
- Risk Analysis—BNs are capable of identifying, distinguishing, measuring, and apprising the occurrence of a critical event or hazard, including estimating its probability and its consequences due to its ability to measure low probability events (Cornalba and Giudici, 2004).
- Maintenance—BNs are fine for considering principal doers involved in maintenance: (1) factors that technically keep each system in need; (2) factors that affect the relationship between dissimilar systems; and (3) to explain and describe factors of public organizational structure (Waeyenbergh and Pintelon, 2004).
- Direct applications are related to maintenance decisions and appraisal display. Defect Detection—BNs can be performed in methods where all process variables and data are encoded in the mechanical apparatus during the lattice process because they are suitable for dealing with intricate difficulty.

The strength of BNs lies in their potential, which lets them address various issues. Their graphical representation enables them to understand the intricacy of the model in one outlook. On the other hand, their weakness is that there is no particular semantic to guide the model development and to guarantee its correlation. Hence, validation of the model with the actual system must be performed (Borunda et al., 2016).

11.6.3 Genetic algorithm

Genetic algorithm (GA) is similar to the natural development process where a crowd of special species modifies the natural environment under deliberation; a population of designs is created and then allowed to evolve in order to adapt to the design environment. These algorithms were reported by Goldberg (1989) to resolve optimizing problems (Azadeh et al., 2006).

A GA possesses three basic operators: (1) selection, (2) crossover, and (3) mutation, as explained in the following (Azadeh et al., 2006):

- Selection operator searches according to a part of the members. The fitness value of the *i*th member in the population can participate in this operation on the basis of probabilities. In the selection process, members with compatibility can take part more than once, while members with inferior fitness can be omitted in order to get a bigger fit.
- Crossover action allows an exchange of the design attributes between two mating parents. This action is performed by selecting two mating parents on which two random places are elected on each chromosome string and the strings between these two places amid the mates are exchanged.
- Mutation operator is another vital operator in the process of a GA and operates on each chromosome after crossover operator, so that a random number is generated for each bite of a chromosome. This operation prevents the loss of unforeseen genetic data in the population in option and intercourse.

11.7 Interpretation of results

11.7.1 Estimation of energy consumption

As already mentioned, there were several studies in last decade regarding energy usage of agricultural products, that is, the quantity of inputs of energy and output, the percentage distribution of the associated energy, and energy indices reported in all of them. For better understanding, the percentages of each input were demonstrated in a pie chart.

Generally, the results are classified in two main categories:

1. Energy use efficiency equivalent or more than 1: when the quantity of input and output energy calculated are equal or output energy is greater than input energy. There are positive amounts of net energy and the production of studied products is acceptable from the perspective of energy. Of course, the researchers should advise the improver solutions to achieve more efficiency in the production process.
2. Energy use efficiency less than 1: actually, in these studies, the amount of output energy is less than input energy and the net energy amount is negative. Obviously, the production of these products is not reasonable from an energy point of view. In these studies, the importance of surveying energy use of each input is increased. Because, if the researchers want to introduce an appropriate discussion, they will need to share each input in the total energy consumption. For example, in most of studies that are given in Fig. 11.12, the highest percentages of energy use belonged to fuels and chemical fertilizers (nitrogen). So, among the issues that lead to lower fuel consumption are:
   a. Couple the tractor to the load or operation.
   b. Avoid unnecessary trips back to the service region from the field.
   c. Fill fuel tanks in the morning.
   d. Remove all unurgent machinery operations.
   e. Adjust the tires according to the recommended pressure and always hold at the same pressure.
   f. Check tractor owners handy for tire distribution and inflation pressure.
   g. Reduce the number of passes by combining operations where possible.
   h. Careful handling and appropriate maintenance to farm equipment such as tractors, harvesters, loaders, and other farm vehicles.
   i. Use oil of recommended viscosity to maximize engine efficiency.
   j. Alternative messy air cleaners which limit airflow needed for the combustion process.

Several results of energy analysis are presented in Table 11.4 and Fig. 11.12.

11.7.2 Artificial neural network model results

In order to represent the best model (with the maximum R and minimum error), various networks have been tried. In terms of the number of hidden layers the number of neurons and learning parameters, the type of activation functions, and network architecture, the implemented ANNs vary. With the help of neurosolution or Matlab software packages, data are trained and developed. In most studies, these models were feedforward backpropagation neural networks relying on the Levenberg–Marquardt training algorithm

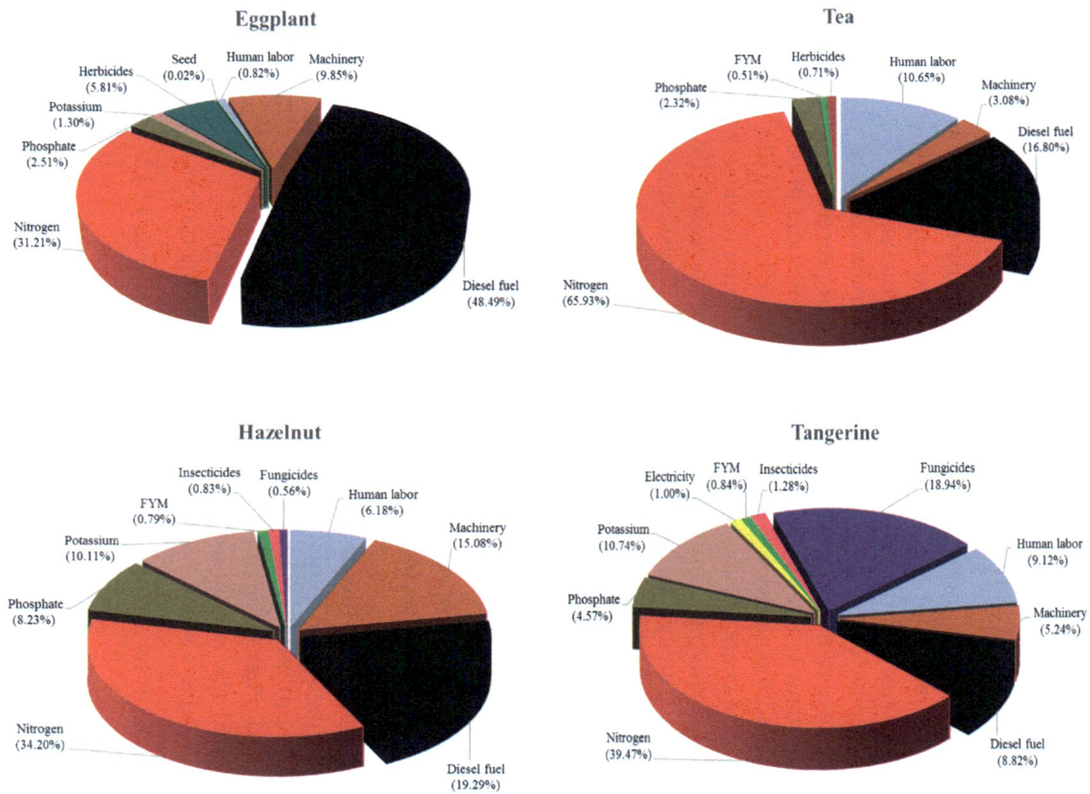

FIGURE 11.12 The share of each input in total energy use for several examples of agricultural products.

TABLE 11.4 A summary of the previous studies conducted on energy analysis in agricultural systems.

Surveyed study	Geographical scale	Product	Product functional unit	Total energy use (MJ product functional per unit)	Total output energy (MJ product functional per unit)	Energy use efficiency	Net energy
Erdal et al. (2007)	Turkey-Tokat	Sugar beet	ha	39,685.51	1,021,776	25.75	982,090.49
Esengun et al. (2007)	Turkey	Apricot	ha	28,647.03	35,462.84	1.24	6815.81
Mohammadi et al. (2008)	Iran-Ardabil	Potato	ha	81,624.96	102,432.99	1.25	20,808.03
Cetin and Vardar (2008)	Turkey-Marmara	Tomato	ha	45,538.60	36,287	0.80	− 9251.60
Bayramoglu and Gundogmus (2009)	Turkey-Antalya	Tomato	ha	90,288.50	71,536.10	0.79	− 18,752.40
Bayramoglu and Gundogmus (2009)	Turkey-Tokat	Cherry	ha	48,667	46,801	0.96	−1866
Mobtaker et al. (2010)	Iran-Hamedan	Barley	ha	25,027.47	71,525.37	2.86	46,497.90
Ghorbani et al. (2011)	Iran-Khorasan	Wheat	ha	45,367.63	65,336.32	1.44	19,968.69
Ozkan et al. (2011)	Turkey-Antalya	Tomato	ha	63,023.20	17,914.30	0.28	− 45,108.90
Alluvione et al. (2011)	Italy-Po River plain	Maize	ha	20,300	201,600	9.93	181,300
Ren et al. (2012)	China-Wudi	Sweet sorghum	ha	27,089	335,608	12.39	308,519
Bojacá et al. (2012)	Colombia	Agricultural crops	ha	141,800	96,400	0.68	−45,400
Neira et al. (2013)	Spain-Andalusia	Organic citrus	ha	57,730	22,514.70	0.39	− 35,215.30
Jekayinfa et al. (2013)	Nigeria	Pineapple-Ilbadan	ha	6117.81	21,760	3.56	15,642.19
Nabavi-Pelesaraei et al. (2014a)	Iran-Guilan	Orange	ha	25,582	47,025	1.84	21,443
Astier et al. (2014)	Mexico-Cupatitzio	Organic avocado	ha	54,809	70,599	1.29	15,790
Tian et al. (2015)	China	Grape	ha	57,697.84	299,333.22	5.19	241,635.38
Sefeedpari et al. (2016)	Iran-Alborz	Egg	1000 birds	712,464.83	151,021.80	0.21	− 561,443.03
Amid and Mesri Gundoshmian (2017)	Iran-Ardabil	Broiler	1000 birds	153,338.84	27,447.26	0.18	− 125,891.58
Hosseinzadeh-Bandbafha et al. (2017)	Iran-Qazvin	Fattening	Calf	24,003	3236	0.13	−20,767
Nabavi-Pelesaraei et al. (2018)	Iran-Guilan	Paddy	ha	51,585.61	66,112.94	1.28	14,527.33

TABLE 11.5 A summary of the previous studies conducted on ANN modeling of energy–environmental management in agricultural systems.

		ANN properties			
Surveyed study	**Product**	**Best structure**	R^2	**RMSE**	**MAPE (%)**
Taki et al. (2012b)	Corn silage	8-5-5-1	0.980	0.046	1.020
Taki et al. (2012a)	Wheat	6-8-8-1	0.870	0.164	6.070
Pahlavan et al. (2012)	Basil	7-20-20-1	0.976	0.046	3.500
Khoshnevisan et al. (2013a)	Open field strawberry	11-6-10-2	0.930	0.127	10.300
Khoshnevisan et al. (2013b)	Wheat	11-5-5-2	0.999	0.105	0.730
Nabavi-Pelesaraei et al. (2014b)	Tangerine	10-8-5	0.971	0.043	0.131
Khoshnevisan et al. (2014c)	Potato	12-8-2	0.988	0.102	7.910
Taghavifar and Mardani (2015)	Apple	8-10-2	0.988	0.110	0.680
Ahmadvand (2016)	Alfalfa	7-12-12-1	0.960	0.005	0.060
Nabavi-Pelesaraei et al. (2016)	Kiwifruit	12-9-9-2	0.987	0.054	3.200
Kalhor et al. (2016)	Broiler	8-8-13-1	0.974	0.042	3.370
Mardani and Taghavifar (2016)	Grape	8-10-2	0.993	0.217	0.880
Amid and Mesri Gundoshmian (2017)	Broiler	5-13-1	0.882	0.077	8.722
Hosseinzadeh-Bandbafha et al. (2017)	Fattening	6-16-2	0.721	0.055	0.900
Taheri-Rad et al. (2017)	Paddy	8-25-1	0.990	1.932	2.211
Nabavi-Pelesaraei et al. (2018)	Paddy	12-6-8-1	0.810	7629.791	14.200

with activation functions of sigmoid and linear for the hidden layer and output layer, respectively. Trial and error are two factors for determining the number of neurons and the number of hidden layers. ANN with an input layer with neurons in the number of inputs, one or two hidden layers with several neurons, and an output layer (output energy) or more output layers (output energy, environmental impacts) is elected as the best model to anticipate yield–environment. Finally the best of several structures are shown in Table 11.5. In all of them, the highest rank of R, RMSE, and MAEP are used to designate the best structure.

The display of sporadic plots of predicted output energy or emission against actual values for the training and examining datasets is a tool for showing ANN model performance. The predicted and actual values should be in good agreement. An example of scatter plots of ANN modeling is shown in Fig. 11.13.

11.7.2.1 Sensitivity analysis results

In energy studies of agricultural crops, in order to investigate the effects of input parameters on selected outputs, sensitivity analysis was performed. In this analysis, first, the

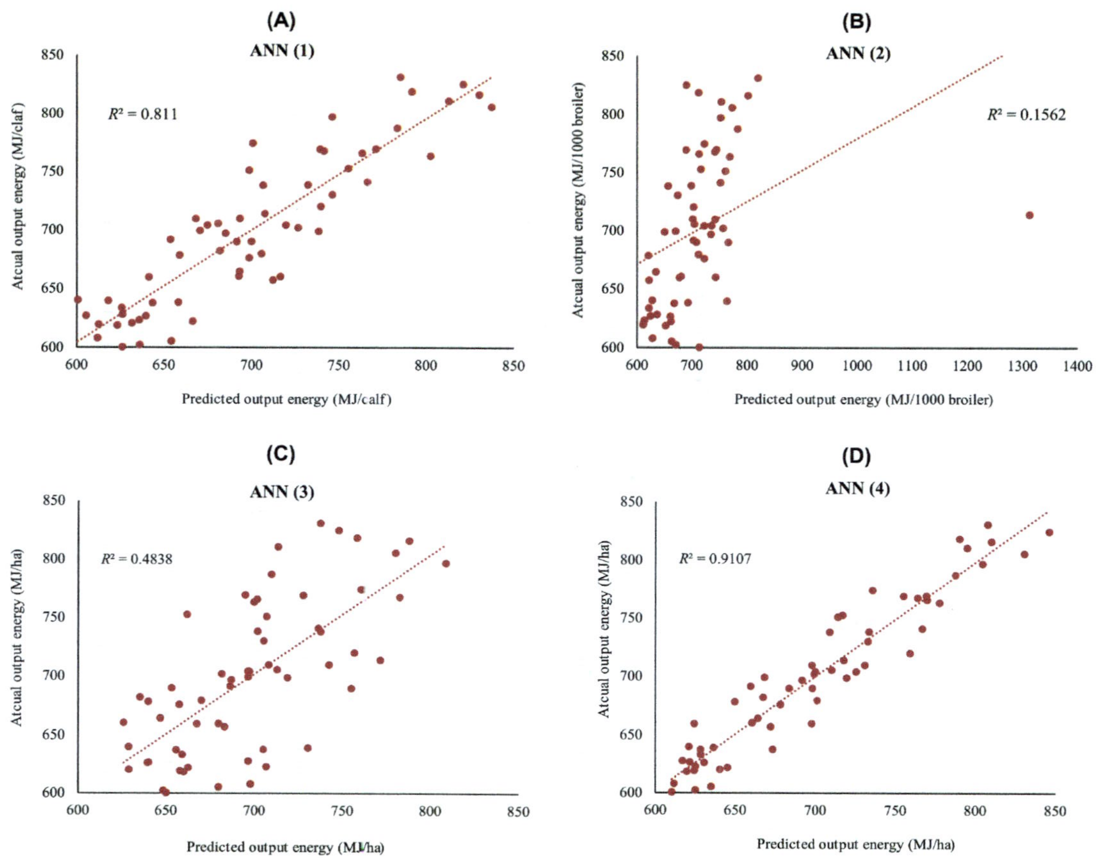

FIGURE 11.13 An example of cross-correlation of predicted and observed values for energy output in ANN modeling.

TABLE 11.6 An example of sensitivity coefficients for input energies of agricultural products (Pahlavan et al., 2012).

Inputs	Basil yield
Human labor	0.0232
Diesel fuel	0.0599
Chemical fertilizers	0.0806
FYM	0.0636
Chemicals	0.0536
Electricity	0.0363
Transportation	0.0022

TABLE 11.7 A summary of the previous studies conducted on characteristics of the best structure of first ANFIS architecture for energy–environmental management in agricultural systems.

		ANFIS properties							
		Type of MF		Number of MF					
Surveyed study	Product	Input	Output	Input	Epoch	Learning method	R^2	RMSE	MAPE (%)
Naderloo et al. (2012)	Wheat	Gauss	Linear	5	20	Hybrid	0.996	0.013	0.005
Khoshnevisan et al. (2014a)	Strawberry	Gbell	Linear	5,6	32	Hybrid	0.963	0.017	0.003
Sefeedpari et al. (2014)	Dairy	Trimf	Linear	2	60	Hybrid	0.790	0.110	0.007
Khoshnevisan et al. (2014d)	Potato	Gbell	Linear	7,7	40	Hybrid	0.974	0.029	0.200
Hosseinzadeh-Bandbafha et al. (2016)	Cow	Gbell	Linear	2,3	30	Hybrid	0.995	0.019	2.838
Elhami et al. (2016)	Lentil	Gbell	Linear	7,7	100	Hybrid	0.999	0.046	0.020
Sefeedpari et al. (2016)	Egg	Trimf	Linear	2	60	Hybrid	0.920	0.014	448.126
Elhami et al. (2016)	Chickpea	Gbell	Linear	7,7	100	Hybrid	0.986	0.053	0.090
Hosseinzadeh-Bandbafha et al. (2017)	Fattening	Gbell	Linear	2,3	50	Hybrid	0.996	0.005	0.362
Mousavi-Avval et al. (2017)	Canola	Gbell	Linear	5,6	32	Hybrid	0.900	5.350	3.633
Nabavi-Pelesaraei et al. (2018)	Paddy	Gbell	Linear	8,9	32	Hybrid	0.740	8910.59	18.67

network was trained and the weights of the connections were fixed. Then, one at a time each input parameter was varied around its mean value while the other inputs remained stable at their mean values, and the change in the output parameter was computed. The results of sensitivity analysis of an example are shown in Table 11.6. As shown in

Table 11.6, the greatest susceptibility factor is found in chemical fertilizers and farm fertilizers. (0.081 and 0.064). This means that extra consumption of 1 MW chemical fertilizer and the fertilizer energy of the farm causes an increase in the basil yield by 0.081 and 0.064 kg, respectively.

11.7.3 Adaptive neuro-fuzzy inference system simulation results

In the investigation of energy modeling in agricultural systems, ANFIS output energy modeling is carried out based on input energy consumption during the agricultural production process. The performance of the model is shown by the rate of statistical parameters. Furthermore, the kind and number of membership functions and learning methods are introduced as ANFIS parameters in the energy studies. The criteria of the best structure on ANFIS model in several researches are shown in Table 11.7. It should be noted the Gbell is more widely used in comparsion with different membership functions including Triangular, Trapezoidal, Generalized bell, Gaussian curve, Gaussian combination, Q-shaped and selected Generalized bell. Moreover, the best combination to each ANFIS is linear MFs for input and output layers. There is also a reliance on the hybrid learning method (to specify the relationships between input and output to determine the optimized distribution of MFs); high precision of this learning method was confirmed in various studies, such as Mousavi-Avval et al. (2017) for modeling canola production and Sefeedpari et al. (2016) in modeling egg production.

The interpretation of statistical parameters results is similar to ANN modeling. Results of different studies revealed that whichever uncertainty increased, the ANFIS model can offer better performance in comparison with ANN modeling.

11.7.4 Adaptive neuro-fuzzy inference system simulation results

Research into the agricultural extension method revealed several methods are very appropriate to educate farmers regarding new technology, such as energy modeling, including demonstrations (with efficacy of 80%), farm visits (with efficacy of 94%), field days (with efficacy of 88%), discussions (with efficacy of 87%), and one-on-ones (with efficacy of 87%) (Franz et al., 2009). All the above methods can be applied to transfer the concepts of energy models to farmers with online education in different sites that are the main references for farmers such as FAO. For example, different NGOs can offer the ANN or ANFIS models of cultivation of determined areas on their sites. So, farmers can enter the different amount of inputs into the models and observe the results of the output energy.

11.8 Conclusion

The final objective is a higher yield with less energy consumption with sustainable development. For this purpose, the optimization of energy consumption as an applicable method for attaining sustainable development is necessary in future research. Determination of the input consumption process and its effect on agricultural yields can be obtained by AI energy models. Researchers can insert results of their optimized patterns and compute the yield without plant farming. Actually, these models can save cost and time in future studies.

Acknowledgment

We gratefully acknowledge the financial support provided by the Management of Fruit and Vegetables Organizations of Tehran Municipality, Iran and also would like to express our gratitude to Ms. Fatemeh Mostashari-Rad for his constructive comments and suggestions in the revision process.

References

Ahmadvand, M., 2016. Determine of energy indicators and modeling of output energy for alfalfa production using artificial neural network in Lorestan province of Iran. Int. J. Farm. Allied Sci. 5, 253–259.

Alluvione, F., Moretti, B., Sacco, D., Grignani, C., 2011. EUE (energy use efficiency) of cropping systems for a sustainable agriculture. Energy 36, 4468–4481.

Amid, S., Mesri Gundoshmian, T., 2017. Prediction of output energies for broiler production using linear regression, ANN (MLP, RBF), and ANFIS models. Env. Prog. Sustain. Energy 36, 577–585.

Astier, M., Merlín-Uribe, Y., Villamil-Echeverri, L., Garciarreal, A., Gavito, M.E., Masera, O.R., 2014. Energy balance and greenhouse gas emissions in organic and conventional avocado orchards in Mexico. Ecol. Indic. 43, 281–287.

Atilgan, A., Hayati, K., 2006. Cultural energy analysis on broilers reared in different capacity poultry houses. Ital. J. Anim. Sci. 5, 393–400.

Azadeh, A., Ghaderi, S.F., Tarverdian, S., Saberi, M., 2006. Integration of artificial neural networks and genetic algorithm to predict electrical energy consumption. IECON Proc. Ind. Electron. Conf. 186, 2552–2557.

Bakhtiari, A.A., Hematian, A., Sharifi, A., 2015. Energy analyses and greenhouse gas emissions assessment for saffron production cycle. Env. Sci. Pollut. Res. 22, 16184–16201.

Barati-Harooni, A., Najafi-Marghmaleki, A., Tatar, A., Mohammadi, A.H., 2016. Experimental and modeling studies on adsorption of a nonionic surfactant on sandstone minerals in enhanced oil recovery process with surfactant flooding. J. Mol. Liq. 220, 1022–1032.

Bayramoglu, Z., Gundogmus, E., 2009. The effect of EurepGAP standards on energy input use: a comparative analysis between certified and uncertified greenhouse tomato producers in Turkey. Energy Convers. Manag. 50, 52–56.

Beheshti Tabar, I., Keyhani, A., Rafiee, S., 2010. Energy balance in Iran's agronomy (1990–2006). Renew. Sustain. Energy Rev. 14, 849–855.

Bojacá, C.R., Casilimas, H.A., Gil, R., Schrevens, E., 2012. Extending the input–output energy balance methodology in agriculture through cluster analysis. Energy 47, 465–470.

Bolarinwa, O., 2015. Principles and methods of validity and reliability testing of questionnaires used in social and health science researches. Niger. Postgrad. Med. J. 22, 195–201.

Borunda, M., Jaramillo, O.A., Reyes, A., Ibargüengoytia, P.H., 2016. Bayesian networks in renewable energy systems: a bibliographical survey. Renew. Sustain. Energy Rev. 62, 32–45.

Cetin, B., Vardar, A., 2008. An economic analysis of energy requirements and input costs for tomato production in Turkey. Renew. Energy 33, 428–433.

Cheng, C.-T., Lin, J.-Y., Sun, Y.-G., Chau, K.W., 2005. Long-term prediction of discharges in Manwan Hydropower using adaptive-network-based fuzzy inference systems models. Int. Conf. Nat. Comput. Springer, Berlin, Heidelberg, pp. 1152–1161.

Cochran, W.G., 1977. The estimation of sample size. Sampl. Tech. 3, 72–90.

Cornalba, C., Giudici, P., 2004. Statistical models for operational risk management. Phys. A Stat. Mech. Appl. 338, 166–172.

Cortes, C., Vapnik, V., 1995. Support-vector networks. Mach. Learn. 20, 273–297.

de Myttenaere, A., Golden, B., Le Grand, B., Rossi, F., 2016. Mean absolute percentage error for regression models. Neurocomputing 192, 38–48.

Elhami, B., Akram, A., Khanali, M., Mousavi-Avval, S.H., 2016. Application of ANFIS and linear regression models to analyze the energy and economics of lentil and chickpea production in Iran. Energy Equip. Syst. 4, 255–270.

Erdal, G., Esengün, K., Erdal, H., Gündüz, O., 2007. Energy use and economical analysis of sugar beet production in Tokat province of Turkey. Energy 32, 35–41.

Esengun, K., Gündüz, O., Erdal, G., 2007. Input–output energy analysis in dry apricot production of Turkey. Energy Convers. Manag. 48, 592–598.

FAO, 2016. Food and Agricultural Organization Statistical Yearbook <http://www.fao.org>.

Franz, N.K., Piercy, F.P., Donaldson, J., Westbrook, J., Richard, R., 2009. How farmers learn: improving sustainable agricultural education executive summary. Research Brief.

Geman, S., Bienenstock, F., Doursat, R., 1992. Neural networks and the bias/valance dilemma. Neural Comput. 4, 1–58.

Ghorbani, R., Mondani, F., Amirmoradi, S., Feizi, H., Khorramdel, S., Teimouri, M., et al., 2011. A case study of energy use and economical analysis of irrigated and dryland wheat production systems. Appl. Energy 88, 283–288.

Ghritlahre, H.K., Prasad, R.K., 2018. Exergetic performance prediction of solar air heater using MLP, GRNN and RBF models of artificial neural network technique. J. Env. Manage. 223, 566–575.

Goldberg, D., 1989. Genetic Algorithm in Search, Optimization and Machine Learning. Addison-Wesley, Harlow.

Hatirli, S.A., Ozkan, B., Fert, C., 2005. An econometric analysis of energy input–output in Turkish agriculture. Renew. Sustain. Energy Rev. 9, 608–623.

Hatirli, S.A., Ozkan, B., Fert, C., 2006. Energy inputs and crop yield relationship in greenhouse tomato production. Renew. Energy 31, 427–438.

Heidari, M.D., Omid, M., Akram, A., 2011. Energy efficiency and econometric analysis of broiler production farms. Energy 36, 6536–6541.

Hosseinzadeh-Bandbafha, H., Safarzadeh, D., Ahmadi, E., Nabavi-Pelesaraei, A., 2016. Modelling output energy and greenhouse gas emissions of dairy farms using adaptive neural fuzzy inference system. Agric Commun 4, 14–23.

Hosseinzadeh-Bandbafha, H., Safarzadeh, D., Ahmadi, E., Nabavi-Pelesaraei, A., 2018. Optimization of energy consumption of dairy farms using data envelopment analysis–A case study: Qazvin city of Iran. J. Saud. Soc. Agric. Sci 17, 217–228.

Hosseinzadeh-Bandbafha, H., Nabavi-Pelesaraei, A., Shamshirband, S., 2017. Investigations of energy consumption and greenhouse gas emissions of fattening farms using artificial intelligence methods. Env. Prog. Sustain. Energy 36, 1546–1559.

Huang, C., Davis, L.S., Townshend, J.R.G., 2002. An assessment of support vector machines for land cover classification. Int. J. Remote. Sens. 23, 725–749.

Hyndman, R.J., Koehler, A.B., 2006. Another look at measures of forecast accuracy. Int. J. Forecast. 22, 679–688.

Jang, J.S.R., 1993. ANFIS: adaptive-network-based fuzzy inference system. IEEE Trans. Syst. Man. Cybern. 23, 665–685.

Jekayinfa, S.O., Adebayo, A.O., Afolayan, S.O., Daramola, E., 2013. On-farm energetics of mango production in Nigeria. Renew. Energy 51, 60–63.

Kaab, A., Sharifi, M., Mobli, H., Nabavi-Pelesaraei, A., Chau, K.W., 2019. Combined life cycle assessment and artificial intelligence for prediction of output energy and environmental impacts of sugarcane production. Sci. Total. Env. 664, 1005–1019.

Kakani, V., Nguyen, V.H., Kumar, B.P., Kim, H., Pasupuleti, V.R., 2020. A critical review on computer vision and artificial intelligence in food industry. J. Agric. Food Res. 2, 100033.

Kalhor, T., Rajabipour, A., Akram, A., Sharifi, M., 2016. Modeling of energy ratio index in broiler production units using artificial neural networks. Sustain. Energy Technol. Assess. 17, 50–55.

Kalogirou, S., Bojic, M., 2000. Artificial neural networks for the prediction of the energy consumption of a passive solar building. Energy 25, 479–491.

Kaltsas, A.M., Mamolos, A.P., Tsatsarelis, C.A., Nanos, G.D., Kalburtji, K.L., 2007. Energy budget in organic and conventional olive groves. Agric. Ecosyst. Env. 122, 243–251.

Kazemi, H., Kamkar, B., Lakzaei, S., Badsar, M., Shahbyki, M., 2015. Energy flow analysis for rice production in different geographical regions of Iran. Energy 84, 390–396.

Khoshnevisan, B., Rafiee, S., Omid, M., Mousazadeh, H., 2013a. Developing an artificial neural networks model for predicting output energy and GHG emission of strawberry production. Int. J. Appl. 3, 43–54.
Khoshnevisan, B., Rafiee, S., Omid, M., Yousefi, M., Movahedi, M., 2013b. Modeling of energy consumption and GHG (greenhouse gas) emissions in wheat production in Esfahan province of Iran using artificial neural networks. Energy 52, 333–338.
Khoshnevisan, B., Rafiee, S., Mousazadeh, H., 2014a. Application of multi-layer adaptive neuro-fuzzy inference system for estimation of greenhouse strawberry yield. Measurement 47, 903–910.
Khoshnevisan, B., Rafiee, S., Omid, M., Mousazadeh, H., Clark, S., 2014b. Environmental impact assessment of tomato and cucumber cultivation in greenhouses using life cycle assessment and adaptive neuro-fuzzy inference system. J. Clean. Prod. 73, 183–192.
Khoshnevisan, B., Rafiee, S., Omid, M., Mousazadeh, H., Rajaeifar, M.A., 2014c. Application of artificial neural networks for prediction of output energy and GHG emissions in potato production in Iran. Agric. Syst. 123, 120–127.
Khoshnevisan, B., Rafiee, S., Omid, M., Mousazadeh, H., 2014d. Prediction of potato yield based on energy inputs using multi-layer adaptive neuro-fuzzy inference system. Measurement 47, 521–530.
Kitani, O., 1999. CIGR Handbook of Agricultural Engineering, American Society of Agricultural Engineers, USA, 1–330.
Kiani, M.K.D., Ghobadian, B., Tavakoli, T., Nikbakht, A.M., Najafi, G., 2010. Application of artificial neural networks for the prediction of performance and exhaust emissions in SI engine using ethanol-gasoline blends. Energy 35, 65–69.
Kizilaslan, H., 2009. Input–output energy analysis of cherries production in Tokat Province of Turkey. Appl. Energy 86, 1354–1358.
Lek, S., Guegan, J., 2000. Artificial Neuronal Networks: Application to Ecology and Evolution. Springer, Berlin.
Mandal, S., Roy, S., Das, A., Ramkrushna, G.I., Lal, R., Verma, B.C., et al., 2015. Energy efficiency and economics of rice cultivation systems under subtropical Eastern Himalaya. Energy Sustain. Dev. 28, 115–121.
Marcot, B.G., Penman, T.D., 2018. Advances in Bayesian network modelling: integration of modelling technologies. Env. Model. Softw. 111, 386–393.
Mardani, A., Taghavifar, H., 2016. An overview on energy inputs and environmental emissions of grape production in West Azerbayjan of Iran. Renew. Sustain. Energy Rev. 54, 918–924.
Mellenbergh, G., 2008. Tests and Questionnaires: Construction and Administration. Advising on research methods: A consultant's companion, 235–268.
Mobtaker, H.G., Keyhani, A., Mohammadi, A., Rafiee, S., Akram, A., 2010. Sensitivity analysis of energy inputs for barley production in Hamedan Province of Iran. Agric. Ecosyst. Env. 137, 367–372.
Moghaddamnia, A., Ghafari Gousheh, M., Piri, J., Amin, S., Han, D., 2009. Evaporation estimation using artificial neural networks and adaptive neuro-fuzzy inference system techniques. Adv. Water Resour. 32, 88–97.
Mohammadi, A., Omid, M., 2010. Economical analysis and relation between energy inputs and yield of greenhouse cucumber production in Iran. Appl. Energy 87, 191–196.
Mohammadi, A., Tabatabaeefar, A., Shahin, S., Rafiee, S., Keyhani, A., 2008. Energy use and economical analysis of potato production in Iran a case study: Ardabil province. Energy Convers. Manag. 49, 3566–3570.
Momenzadeh, L., Zomorodian, A., Mowla, D., 2011. Experimental and theoretical investigation of shelled corn drying in a microwave-assisted fluidized bed dryer using artificial neural network. Food Bioprod. Process. 89, 15–21.
Mostashari-Rad, F., Nabavi-Pelesaraei, A., Soheilifard, F., Hosseini-Fashami, F., Chau, K.W., 2019. Energy optimization and greenhouse gas emissions mitigation for agricultural and horticultural systems in Northern Iran. Energy 186, 115845.
Mousavi-Avval, S.H., Rafiee, S., Jafari, A., Mohammadi, A., 2011. Energy flow modeling and sensitivity analysis of inputs for canola production in Iran. J. Clean. Prod. 19, 1464–1470.
Mousavi-Avval, S.H., Rafiee, S., Sharifi, M., Hosseinpour, S., Shah, A., 2017. Combined application of life cycle assessment and adaptive neuro-fuzzy inference system for modeling energy and environmental emissions of oilseed production. Renew. Sustain. Energy Rev. 78, 807–820.
Nabavi-Pelesaraei, A., Abdi, R., Rafiee, S., 2013. Energy use pattern and sensitivity analysis of energy inputs and economical models for peanut production in Iran. Int. J. Agric. Crop. Sci. 5, 2193–2202.

Nabavi-Pelesaraei, A., Abdi, R., Rafiee, S., Mobtaker, H.G., 2014a. Optimization of energy required and greenhouse gas emissions analysis for orange producers using data envelopment analysis approach. J. Clean. Prod. 65, 311–317.

Nabavi-Pelesaraei, A., Fatehi, F., Mahmoudi, A., 2014b. Prediction of yield and economic indices for tangerine production using artificial neural networks based on energy consumption. Int. J. Agro. Agri. Res. 4, 57–64.

Nabavi-Pelesaraei, A., Rafiee, S., Hosseinzadeh-Bandbafha, H., Shamshirband, S., 2016. Modeling energy consumption and greenhouse gas emissions for kiwifruit production using artificial neural networks. J. Clean. Prod. 133, 924–931.

Nabavi-Pelesaraei, A., Rafiee, S., Mohtasebi, S.S., Hosseinzadeh-Bandbafha, H., Chau, K.W., 2018. Integration of artificial intelligence methods and life cycle assessment to predict energy output and environmental impacts of paddy production. Sci. Total. Env. 631–632, 1279–1294.

Nabavi-Pelesaraei, A., Rafiee, S., Mohtasebi, S.S., Hosseinzadeh-Bandbafha, H., Chau, K.W., 2019a. Assessment of optimized pattern in milling factories of rice production based on energy, environmental and economic objectives. Energy 169, 1259–1273.

Nabavi-Pelesaraei, A., Rafiee, S., Mohtasebi, S.S., Hosseinzadeh-Bandbafha, H., Chau, K.W., 2019b. Comprehensive model of energy, environmental impacts and economic in rice milling factories by coupling adaptive neuro-fuzzy inference system and life cycle assessment. J. Clean. Prod. 217, 742–756.

Naderloo, L., Alimardani, R., Omid, M., Sarmadian, F., Javadikia, P., Torabi, M.Y., et al., 2012. Application of ANFIS to predict crop yield based on different energy inputs. Measurement 45, 1406–1413.

Nayak, P.C., Sudheer, K.P., Rangan, D.M., Ramasastri, K.S., 2004. A neuro-fuzzy computing technique for modeling hydrological time series. J. Hydrol. 291, 52–66.

Neira, D.P., Montiel, M.S., Fernández, X.S., 2013. Energy analysis of organic farming in Andalusia (Spain). Agroecol. Sustain. Food Syst. 37, 231–256.

Oğuz, H., Sarıtas, I., Baydan, H.E., 2010. Prediction of diesel engine performance using biofuels with artificial neural network. Expert. Syst. Appl. 37, 6579–6586.

Ozkan, B., Akcaoz, H., Karadeniz, F., 2004. Energy requirement and economic analysis of citrus production in Turkey. Energy Convers. Manag. 45, 1821–1830.

Ozkan, B., Ceylan, R.F., Kizilay, H., 2011. Energy inputs and crop yield relationships in greenhouse winter crop tomato production. Renew. Energy 36, 3217–3221.

Pahlavan, R., Omid, M., Akram, A., 2012. Energy input–output analysis and application of artificial neural networks for predicting greenhouse basil production. Energy 37, 171–176.

Park, Y.S., Lek, S., 2016. Artificial neural networks: multilayer perceptron for ecological modeling. Develop. Env. Model 28, 123–140.

Pishgar-Komleh, S.H., Ghahderijani, M., Sefeedpari, P., 2012. Energy consumption and CO2 emissions analysis of potato production based on different farm size levels in Iran. J. Clean. Prod. 33, 183–191.

Ranganathan, A., 2004. The Levenberg-Marquardt algorithm. Tutoral LM Algorithm 11, 101–110.

Ren, L.T., Liu, Z.X., Wei, T.Y., Xie, G.H., 2012. Evaluation of energy input and output of sweet sorghum grown as a bioenergy crop on coastal saline-alkali land. Energy 47, 166–173.

Roy, S.S., Roy, R., Balas, V.E., 2018. Estimating heating load in buildings using multivariate adaptive regression splines, extreme learning machine, a hybrid model of MARS and ELM. Renew. Sustain. Energy Rev. 82, 4256–4268.

Saber, Z., Esmaeili, M.A., Pirdashti, H., Motevali, A., Nabavi-Pelesaraei, A., 2020. Exergoenvironmental-Life cycle cost analysis for conventional, low external input and organic systems of rice paddy production. J. Clean. Prod. 263, 121529.

Safa, M., Samarasinghe, S., 2011. Determination and modelling of energy consumption in wheat production using neural networks: "A case study in Canterbury province, New Zealand". Energy 36, 5140–5147.

Schmidhuber, J., 2015. Deep learning in neural networks: an overview. Neural Netw. 61, 85–117.

Sefeedpari, P., Rafiee, S., Akram, A., Komleh, S.H.P., 2014. Modeling output energy based on fossil fuels and electricity energy consumption on dairy farms of Iran: application of adaptive neural-fuzzy inference system technique. Comput. Electron. Agric. 109, 80–85.

Sefeedpari, P., Rafiee, S., Akram, A., Chau, K.W., Pishgar-Komleh, S.H., 2016. Prophesying egg production based on energy consumption using multi-layered adaptive neural fuzzy inference system approach. Comput. Electron. Agric. 131, 10–19.

Shabanzadeh-Khoshrody, M., Azadi, H., Khajooeipour, A., Nabavi-Pelesaraei, A., 2016. Analytical investigation of the effects of dam construction on the productivity and efficiency of farmers. J. Clean. Prod. 135, 549–557.
Soheili-Fard, F., Salvatian, S.B., 2015. Forecasting of tea yield based on energy inputs using artificial neural networks (a case study: Guilan province of Iran). Biological Forum 7, 1432–1438.
Sugeno, M., Yasukawa, T., 1993. A fuzzy-logic-based approach to qualitative modeling. IEEE Trans. Fuzzy 1, 7–31.
Sung, A.H., 1998. Ranking importance of input parameters of neural networks. Expert. Syst. Appl. 15, 405–411.
Suparta, W., Alhasa, K.M., 2016. Adaptive neuro-fuzzy interference system model. Tropospheric Delays Using ANFIS. Springer, Cham, pp. 5–18.
Tabatabaeefar, A., Emamzadeh, H., Ghasemi Varnamkhasti, M., Rahimizadeh, R., Karimi, M., 2009. Comparison of energy of tillage systems in wheat production. Energy 34, 41–45.
Taghavifar, H., Mardani, A., 2015. Prognostication of energy consumption and greenhouse gas (GHG) emissions analysis of apple production in West Azarbayjan of Iran using Artificial Neural Network. J. Clean. Prod. 87, 159–167.
Taheri-Rad, A., Khojastehpour, M., Rohani, A., Khoramdel, S., Nikkhah, A., 2017. Energy flow modeling and predicting the yield of Iranian paddy cultivars using artificial neural networks. Energy 135, 405–412.
Taki, M., Ajabshirchi, Y., Mahmoudi, A., 2012a. Prediction of output energy for wheat production using artificial neural networks in Esfahan province of Iran. J. Agric. Technol. 8, 1229–1242.
Taki, M., Mahmoudi, A., Mobtaker, H.G., Rahbari, H., 2012b. Energy consumption and modeling of output energy with multilayer feed-forward neural network for corn silage in Iran. Agric. Eng. Int. CIGR J. 14, 93–101.
Tian, D., Wang, J., Feng, J., 2015. Analysis of energy consumption for protected grape production system in China. Adv. Mater. Res. 1073–1076, 2468–2472.
Unakitan, G., Hurma, H., Yilmaz, F., 2010. An analysis of energy use efficiency of canola production in Turkey. Energy 35, 3623–3627.
Waeyenbergh, G., Pintelon, L., 2004. Maintenance concept development: a case study. Int. J. Prod. Econ. 89, 395–405.
Weber, P., Medina-Oliva, G., Simon, C., Iung, B., 2012. Overview on Bayesian networks applications for dependability, risk analysis and maintenance areas. Eng. Appl. Artif. Intell. 25, 671–682.
Wilamowski, B., Yu, H., 2010. Improved computation for Levenberg–Marquardt training. IEEE Trans. Neural Netw. 21, 930–937.
Yamane, T., 1967. Elementary Sampling Theory. New Jersey, USA Prentice-Hall Englewood Cliffs.
Zadeh, L.A., 1965. Fuzzy sets. Inf. Control. 8, 338–353.
Zendehboudi, A., Baseer, M.A., Saidur, R., 2018. Application of support vector machine models for forecasting solar and wind energy resources: a review. J. Clean. Prod. 199, 272–285.
Zentner, R., Lafond, G., Derksen, D., Nagy, C., Wall, D., May, W., 2004. Effects of tillage method and crop rotation on non-renewable energy use efficiency for a thin Black Chernozem in the Canadian Prairies. Soil. Tillage Res. 77, 125–136.
Zhao, Z., Chow, T.L., Rees, H.W., Yang, Q., Xing, Z., Meng, F.-R., 2009. Predict soil texture distributions using an artificial neural network model. Comput. Electron. Agric. 65, 36–48.
Zurada, J., 1992. Introduction to Artificial Neural Systems. West Publishing Company, New York.

CHAPTER

12

Support vector machine model for multistep wind speed forecasting

Shobna Mohini Mala Prasad[1], *Thong Nguyen-Huy*[2,3] *and Ravinesh Deo*[1]

[1]School of Sciences, University of Southern Queensland, Springfield, QLD, Australia [2]Centre for Applied Climate Sciences, University of Southern Queensland, Toowoomba, QLD, Australia [3]Vietnam National Space Center, Vietnam Academy of Science and Technology, Hanoi, Vietnam

12.1 Introduction

Wind energy is emerging to be a promising renewable energy source due to its advantages of independence from fossil fuels, reduction in greenhouse emissions, and environmental preservation. Wind energy is clean, inexhaustible, inexpensive, and widely distributed (Chitsazan et al., 2019). The fast deployment time and relatively low establishment and maintenance costs have made it stand out from other renewable electricity generation technologies, even in a poor economic climate (Ahmed and Khalid, 2018; Wang et al., 2018). Due to the above reasons, wind generation will constitute a large portion of the future renewable electricity market. For this growth to become increasingly expressive and inserted into the energy mix, it is necessary to explore a new mechanism that can provide the future potential for increasing wind energy as a respective renewable energy investment (Zhang et al., 2017). This will provide increased sustainable energy and concurrently help in the reduction of atmospheric pollutants associated with fossil fuels.

According to the Intergovernmental Panel on Climate Change (IPCC, 2007), wind energy offers a noteworthy potential for greenhouse gas emission reductions, both for a short-term period (Jun et al., 2018; Vapnik, 1998) and the long-term period (2050). The Global Wind Energy Council (GWEC) projections suggest that wind power is expected to supply as much as 20% of world's electricity by 2030, reducing CO2 emissions by more than 3.3 billion tonnes per year. On average each kilowatt-hour of wind power generated avoids 600 g of CO_2 by displacing the need for the generation of the same unit of electricity from

Predictive Modelling for Energy Management and Power Systems Engineering
DOI: https://doi.org/10.1016/B978-0-12-817772-3.00012-4

conventional energy sources such as coal, oil, or gas (Letcher, 2017). Several countries are beginning to recognize that wind power provides a significant opportunity for future power generation and according to the wind energy and Greenpeace organization plan, 12% of all electricity generation should be achieved through wind power by 2020.

According to the GWEC, wind energy utilization continues to increase, which is led by Denmark pushing 40%, followed by Uruguay, Portugal, and Ireland with well over 20%, Spain and Cyprus around 20%, Germany at 16%; and the big markets of China, the United States, and Canada get 4%, 5.5%, and 6% of their power from wind, respectively. In India the wind energy sector dominates all other renewable sources of energy, which is about 65.2% (i.e., 23.76 GW) of total renewable installed capacity (Agrawal and Sandhu, 2016).

Although the widespread use of wind energy elicits many benefits, nevertheless, it can be a challenging task to perform reliable and seasonable wind power management due to the stochastic and intermittent characteristics of wind speed (Hu and Chen, 2018). Wind energy resources produce power that is dependable on the external irregular source (Du et al., 2017) resulting in the variability, unpredictability, and uncertainty of wind resources (Agrawal and Sandhu, 2016). The nonstationary wind speed will influence heavily the security and stability of the power system (Jiang and Li, 2018), which will further impact power quality and peak regulation. Wind farms also have difficulties with system scheduling and energy dispatching because the availability of wind power is not known in advance. The integration of wind facilities to the electrical grid presents a major challenge to the power system operators. Such integration has a significant impact on the optimum power flow, transmission congestion, power quality issues, system stability, load dispatch, and economic analysis (Tian et al., 2018). In addition, the stochastic and intermittent characteristics of wind speed pose new challenges for managing transmission and distribution networks, which will further impact wind farms (Yu et al., 2017).

Accurate forecasting of wind farm power is an important part of wind power operations management, and wind speed forecasting is the premise behind power forecasting (Yang and Wang, 2018). For the operation of a wind power system, the wind speed must be forecast on a real-time basis to ensure enough of a stability margin exists (Tian et al., 2018). The power system operators need real-time information, not only the system's current state but also possible changes in its state in the next few minutes (Liu et al., 2018a). Therefore, developing a robust and innovative wind speed forecasting model is of great importance in the energy conversion and management domain that can concurrently provide the solution to the energy supply and demand balance problem and enhance the viability of wind by dealing with its intermittent nature (Liu et al., 2018a).

Most of the wind speed forecasting techniques can be clustered into two main groups, namely physical methods, and data-driven methods. The physical forecasting models use detailed physical properties and historical wind speed data to construct forecasting models. These models are used to simulate changes to wind speed trends at a geographical position in three dimensions by considering wind speed drivers, such as temperature, air pressure, terrain roughness, and obstructions (Wang et al., 2018). Physical models do not need to be trained using historical data, and when more detailed background information, such as time and geography, is considered, it is possible to obtain more accurate long-term forecasting values. However, physical models are very restrictive, they require considerable amounts of observed data at limited simulation scales, and consume excessive computing resources, which are expensive and difficult to obtain (He et al., 2018), and have a high computation complexity in solving the model that is not effective for short-term predictions (Wang et al., 2015).

With enhanced computational systems, many wind power estimation researches are directed utilizing data-driven models (Samadianfard et al., 2018). The essence of the data-driven model approach is to find the relationship between historical wind speed through different artificial intelligence (AI) models such as time series models and support vector machines (SVMs). Many methods of data-driven models are better than physical methods and have less forecasting error than other forecasting models.

Australia's wind energy market is only approximately 4.5% (Howard et al., 2018), despite Australia having excellent wind resources, particularly in the coastal regions, and being unconstrained by land or resource availability or material supply and security issues (Blakers et al., 2017). Thus research into wind energy feasibility studies are important for Australia's renewable energy sector as wind power is expected to play a major role in helping Australia's transition to a low-carbon economy. Furthermore, while data-driven models have been developed and have been widely used by many countries, such as India, Iran, China, and the United States (Ahmad et al., 2018), the prediction of wind speed using data-driven models has rarely been performed in Australia, although some studies have explored statistical models.

Considering that the forecasting of wind speed is of great importance to the performance and reliability of a wind power conversion system (Ahmad et al., 2018) and modern wind farm management and Australia having limited research in the wind modeling arena, this chapter has investigated how a data-driven hybrid SVM model can be used for wind speed forecasting in Australia. For this purpose, the present study is going to investigate the research question "Given the highly stochastic, rapidly changing nature of wind, is it possible to develop a data-smart model for wind speed forecasting, especially on multiple forecast horizons?" and to address this research question, the supremacy of data-driven methods (Howard et al., 2018) will be used extensively for modeling short-, medium-, and long-term wind speed for Australia.

Short-term wind speed forecasting provides information about the wind speed and power that can be expected in the next few minutes, hours, or days which is an important factor in decision-making required by wind farmers (Hong et al., 2019; Li et al., 2018a). The short-term forecasting requires the rolling forecasting of the future 0–6 hours of wind power output, which is generally used for real-time scheduling of power systems (Hong et al., 2019; Cassola and Burlando, 2012). According to Schicker (Schicker et al., 2017), medium-term wind speed forecasts can be used for real-time energy market operations, ancillary service management, transmission congestion management, and regulatory actions. Load imbalances frequently result in wind energy curtailments, the frequency and length of curtailments are often due to underutilization of the information contained within a wind energy forecast (Tian et al., 2018; Moreno and dos Santos Coelho, 2018). Accurate forecasting allows operators to achieve favorable trading performances on the electricity markets (Payero and Irmak, 2013). Network operators require reliable and accurate data to absorb the growing share of wind power and anticipate shortages due to rapid changes in wind speed and direction (Letcher, 2017; Wang et al., 2016). Long-term wind speed forecasts can help to achieve a low spinning reserve and optimal operating costs by maintaining the scheduling of the wind turbines (Wang et al., 2011a). Wind operators may be exposed to the uncertainty of the imbalance markets, therefore, a more detailed and accurate modeling of monthly wind speed is required, capturing long-term and seasonal cycle variations that wind speed may follow over different time periods which can be very useful for in maintenance planning, financial estimates, and predictions of electricity generation for network management (Alonzo et al., 2017; Hu et al., 2013).

12.2 Literature review

12.2.1 Physical methods

The physical forecasting methods use historical wind speed data, terrain feature data, and many meteorological data (atmospheric pressure, temperature, and humidity) to forecast the wind speed of the considered site and several models such as Numeric Weather Prediction (NWP) have been developed based on atmospheric dataset for wind speed forecasting (Pearre and Swan, 2018; Mughal et al., 2018). The performance of the physical methods is often satisfactory for long-term horizons (more than 24 hours ahead) and as the atmospheric situation is stable, one can expect more accurate predictions (Wu and Hong, 2007).

Cassola and Burlando (2012) presented a novel wind speed forecasting approach by applying a Kalman filter to produce the generations of an NWP model. The Kalman filter method utilized recursive observations and model generations to minimize forecasting errors. Hoolohan et al. (2018) studied the performance of different models for wind speed forecasting, including the NWP and Gaussian process regression. However, these models are inappropriate for short-term prediction (several minutes to 1 hour) due to the difficulty of information acquisition and complicated computation (Liu et al., 2018a). An unstable atmospheric situation can lead to very poor numerical weather predictions (Ahmed and Khalid, 2018) and thus to inaccurate wind power forecasting.

The mathematical complexity of such models is also a concern, as the physical interactions between these atmospheric datasets are incorporated in predictions by means of differential equations (Deo et al., 2018). The physical methods of mainframe computers require a large amount of computing time. Additionally, the forecasting strategy of the physical methods is primarily based on grid technology (Niu et al., 2018), which significantly restricts its practical applicability to local forecasting and different site characteristics can affect how predictions are adjusted to hub height wind speed predictions (Liu et al., 2018b; Li et al., 2018a). Considering the drawbacks of the physical methods, a data-driven method is employed for the present study.

12.2.2 Statistical methods

Statistical forecasting methods aim at modeling the relationship of the historical data (Liu et al., 2010). These models include the autoregressive (AR), autoregressive moving average (Barman et al., 2018), and AR integrated moving average (ARIMA) which has a higher prediction accuracy when the wind speed data shows linearity and stationarity (Yuan et al., 2017; Xiao et al., 2018).

Statistical models have been broadly applied for forecasting wind by several researchers (Pearre and Swan, 2018; Erdem and Shi, 2011; Cassola and Burlando, 2012). The implementation of an ARFIMA process has the capability to extract valid information from the past more effectively than a single AR–moving-average model (Barman et al., 2018). Taylor et al. (2009) applied an ARFIMA-generalized AR conditional heteroscedasticity (GARCH) model to forecast hourly wind speed series and presented a comparison between the model results and meteorological forecasts.

Although the time series model establishes a mapped relationship between wind speed and wind power by extracting information in historical wind speed signals, it has the disadvantages of nonlinear fitting capability weakness (Song et al., 2018; Liu et al., 2018a), the requirement of large historical records and difficulty in modeling nonlinear problems. To overcome the disadvantages, data-driven models are being put forward for wind speed forecasting.

12.2.3 Data-driven methods

To ameliorate wind speed forecasting issues, the forecasting ability of data-driven methods offer feasible alternatives. With the emergence of big data analytics and the renaissance of AI, many researchers are utilizing data-driven methods to forecast wind speed (Khosravi et al., 2018a). The data-driven method offers different processes exhibiting the behavior of phenomena being modeled using historical data like the statistical approach. It has gained remarkable attention in the circle of analysts and forecasters efficiently reflecting nonlinear behavior in wind speed forecasting (Wang et al., 2017; Samadianfard et al., 2018; Laqrichi et al., 2015; Jiang and Li, 2018). This model presents a distinct advantage over other wind speed forecasting models as the methodologies do not require any predefined mathematical models and it does not require a statistical assessment of links between the inputs (i.e., predictors) and the output (objective) variable (Deo et al., 2018).

Among the data-driven models, SVM, time series models, artificial neural network (ANN), neuro-fuzzy inference system (ANFIS), fuzzy logic methods, neuro-fuzzy network, and evolutionary optimization algorithms are among the most extensively used approaches for the prediction of wind speed. To assemble the pertinent features of the wind speed, the SVM model has been used in this study and to benchmark SVM model, a second-order Volterra model has been designed which yields good performance with reasonable prediction accuracy in the forecasting of streamflow (Rhein et al., 2013).

It is proposed and later tested in this chapter, that the SVM model can effectively enhance the prediction accuracy of wind speed. The application of the SVM model for wind speed prediction is driven, in part, by the merits of SVM models utilized in other renewable energy studies such as forecasting of electricity load (Yang et al., 2016), hydropower consumption (Wang et al., 2011b), solar radiance (Jiang and Dong, 2017), heating demand (Izadyar et al., 2015), electricity consumption (Barman et al., 2018), and electricity demand (Al-Musaylh et al., 2018a).

The SVM has been very popular among researchers, as this approach ranks very high in the context of accuracy and can solve the nonlinear problems (Zendehboudi et al., 2018; Dahhani et al., 2018). SVM is related to the supervised learning methods applied for categorization and regression and was introduced by Vapnik (1998). While other statistical models are estimated following the empirical risk minimization principle, that is the minimization of loss function over the learning set and checking of the generalization ability with some criteria, the SVM theory is based on the structural risk minimization principle with the idea to minimize the upper bounds of error of the objective function (Lazos et al., 2014).

Zendehboudi et al. (2018) investigated 35 articles on wind energy and it was concluded that the majority of researchers implemented SVM approaches to forecast wind speed (69%), followed by wind power (29%) and wind direction (2%), as illustrated in Fig. 12.1. The increasing number of studies can be explained due to the fast and accurate outputs by SVM models.

The SVM is the predominant algorithm developed in the field of wind speed forecasting for use with statistical methods. It has excellent generalization ability for complex models (Yuan et al., 2017) and is used effectively for nonlinear regression problems, it has the ability to model complex nonlinear decision boundaries, is less prone to overfitting than other forecasting models, and can be used for prediction as well as classification (Li et al., 2018b; Kong et al., 2015).

In the United States, the study of Song et al. (2018) explored two models for short-term wind direction forecasting based on ANN and SVR models. Both models used the information acquired from the previous recordings of the wind direction to forecast the near-future values. Experiment results demonstrated that the ANN forecast based on the ensemble average of 10 networks showed a larger mean absolute error and a similar mean effectiveness index than the SVM forecast; SVM outperforms ANN, both in terms of mean error and computational time. Similar investigation of the wind energy potential in Iran was implemented by Khosravi et al. (2018a) and they presented three models of machine learning algorithms: multilayer feedforward neural network (MLFFNN), support vector regression with a radial basis function (SVR-RBF), and adaptive neuro-fuzzy inference system (ANFIS), to predict wind speed, wind direction, and output power of a wind turbine to show that the SVR-RBF model outperforms the MLFFNN and ANFIS-PSO models.

Similar work in relation to the current study was carried out by Zhang et al. (2015), who investigated the performance of the SVM algorithm against multilayer perceptron

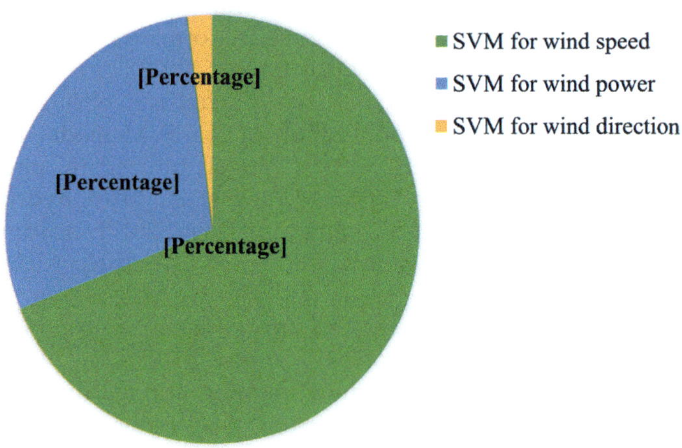

FIGURE 12.1 Distribution of publications across wind energy (Zendehboudi et al., 2018).

(MLP) and proved the effectiveness of SVM for wind forecasting. Also in China, Kong et al. (2015) employed a wind speed prediction concept with SVM for data regression (RSVR) and concluded that RSVM effectively enhanced the prediction accuracy of wind speed while Yuan et al. (2017) employed a hybrid AR fractionally integrated moving average and least square SVM model to forecast short-term wind power and ultimately achieved a higher forecasting precision with the hybrid model by choosing wind speed and direction as the input variables of the model. Thus, with the capability to track complex nonlinearity systems, SVM techniques have received considerable attention in the wind energy field.

12.2.4 Hybrid methods

Data-driven methods can better handle nonlinear relationships and thus are more flexible (Yu et al., 2017). However, it is imperative to note that the temporal behavior of wind and its related atmospheric input variables is complex and is composed of nonstationarity behaviors and large or small-scale periodic/random fluctuations over the various temporal scales (Han et al., 2017; Tian et al., 2018). This behavior can compound the ability of standalone data-driven models in accurately simulating the wind speed (Song et al., 2018) and researchers also argue that forecasting of wind speed must explore hybrid (rather than standalone) models building on the strengths of individual data-driven models (Yuan et al., 2017; Yang and Wang, 2018; Xiao et al., 2018; He et al., 2018). With the insight to address this issue, complete ensemble empirical mode decomposition with adaptive noise (CEEMDAN) coupled with a SVM model is explored in this study.

CEEMDAN was proposed by (Torres et al., 2011). It aims to segregate a higher frequency input series into lower frequency-resolved parts to extract and isolate prominent features representing the physical structure of the data. CEEMDAN is a powerful tool as it can be used to decompose the original data into high and low-frequency subseries to address the issues of nonstationary, repeats/periods and jump-type perturbations before such data are utilized for prediction purposes (Zhang et al., 2017; Jun et al., 2018; Wu and Huang, 2009).

CEEMDAN technique have merits over conventional approaches such as the wavelet transform (WT) (Deo et al., 2016; Nourani et al., 2014; Liu et al., 2018a), maximum overlap discrete WT (MODWT) (Prasad et al., 2017), empirical mode decomposition (EMD) (Santhosh et al., 2018), ensemble EMD (EEMD) (Wang et al., 2016), principal component analysis (PCA) (Skittides and Früh, 2014), and singular spectrum analysis (SSA) (Moreno and dos Santos Coelho, 2018). Recent studies show major weaknesses in WT-based models, particularly in their forecasting ability, which is limited by the adoption of noncausal filters constructed with discrete WT (DWT) algorithms (Al-Musaylh et al., 2018b) and the choice of the mother wavelet with MODWT is a major concern. In addition, there is no explicit rule to select an optimal wavelet other than by an iterative trial and error process (Prasad et al., 2018).

Moreover, CEEMDAN solves the "mode mixing" issue of EMD, achieved by the addition of Gaussian white noise to the undecomposed series (Yu et al., 2017; Jun et al., 2018). During CEEMDAN-based decomposition, Gaussian white noise with unit variance and noise coefficient is added sequentially at each decomposition stage. Although this does have limitations on parallel computing, the reconstruction of CEEMDAN decomposed data is complete and noise-free (Prasad et al., 2018; Jun et al., 2018; Zhang et al., 2017). Despite the advantages and self-adaptability making it suited for practical applications, CEEMDAN has not been broadly applied in wind speed forecasting applications in Australia.

CEEMDAN–based data-driven models have been applied extensively for analyzing nonlinear stochastic signals. Yu et al. (2017) performed a comparative test on EMD and three developed versions (EEMD, CEEMD, and CEEMDAN) for wind speed forecasting; the results showed that CEEMDAN outperforms the other three methods. Zhang et al. (2017) carried out wind speed forecasting combining (CEEMDAN) with a flower pollination algorithm with a chaotic local search (CLSFPA).

Likewise, CEEMDAN–based model was explored (Jiao et al., 2016) in precipitation forecasting. Prasad et al. (2018) systematically and comprehensively investigated the applicability of CEEMDAN in forecasting soil moisture and the results showed that the CEEMDAN approach is a viable option for forecasting and demonstrated its efficiency. CEEMDAN was employed (Al-Musaylh et al., 2018b) for electricity demand forecasting and the study showed the advantages of the CEEMDAN over other models. In terms of estimation error, CEEMDAN was comparable with wavelet-decomposition (Afanasyev and Fedorova, 2016).

Although these studies found that the models generated improved forecasts, very limited application of CEEMDAN–SVM in wind speed forecasting has been carried out in Australia. Thus the purpose of this research study is to design and evaluate a hybrid data-driven CEEMDAN–SVM model, overcoming nonstationarity issues in forecasting wind speed with potential for practical applications.

12.3 Materials and method

12.3.1 Theoretical background

12.3.1.1 Support vector machine model

SVM was proposed by Vapnik (1998) and it has been an increasingly attractive technique in multiple areas. Compared with other statistical models, support vector models have an edge on solving nonlinear problems due to the application of kernel functions (Yang et al., 2016; Wang et al., 2015; Li et al., 2018b; Kong et al., 2015). The idea of the kernel function is to map the data vectors from a low-dimension space to a high-dimension space where it is not necessary to represent the mapping function explicitly. This allows support vector models to transform a nonlinear problem into a very high-dimensional linear problem that gives more accurate predictions (Yao et al., 2017; Deo et al., 2016). A simplified schematic architecture of the SVM model build for this study is presented in Fig. 12.2.

Assume that we have a set of predictor (input) variables that are related to wind speed. The data points can be written as $G = \{(xi, yi)\}_{ni}$. The meteorological data has n variables and the objective variable is comprised of the wind speed time series, *wind speed* $= y_t$, $(t = 1, \ldots, N)$ and the meteorological input x_t is $N \times (m-1)$ matrix. In this study the $x_t\ (i)$ is the ith $(i = 1, \ldots, m-1)$ column of the input matrix x_t to represent the time series of the ith meteorological variable. The $x^{(i)}(t)$ represents the value of x_t at the time t and the

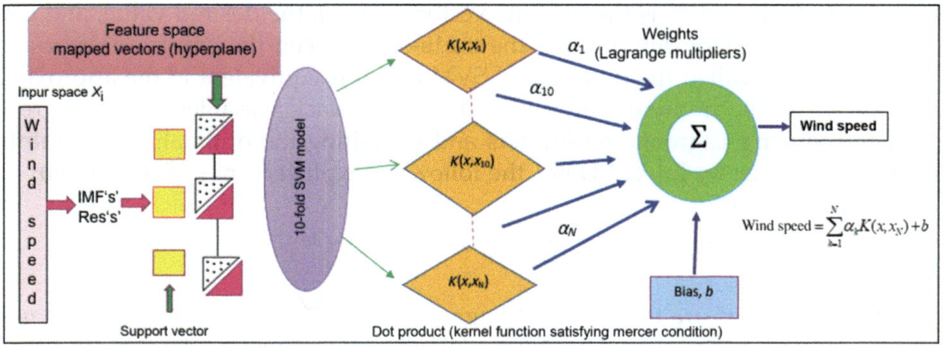

FIGURE 12.2 The architecture of the support vector machine model. Details of the input variables intrinsic mode functions (IMFs) and residual (res) are provided in the modeling framework.

predictor vector, $x = [wind\ speed, IMF, Res]$. The purpose of the SVM model is to extract the optimum set of patterns (predictive features) contained in the x time series to forecast the objective variable, wind speed (m/s).

An SVM model can provide solutions to regression problems with multiple predictors $X = \{xi\}_{i=1}^{i=n}$, where n is the number of predictor variables and each x_i has N variables. These are linked to an objective variable $Y = \{yi\}_{i=1}^{i=N}$. The matrix X is converted to a higher-dimensional feature space, in accordance with the original, but constitutes a lower-dimensional input space (Mohammadi et al., 2015; Al-Musaylh et al., 2018a). The nonlinear regression problem is defined as (Vapnik, 1998):

$$y(x) = w^T \varphi(x) + b \tag{12.1}$$

where, $\varphi(x)$ represents the high-dimensional feature spaces, which is nonlinearly mapped from the input space x. The coefficients w and b are estimated by minimizing the regularized function:

$$R(C) = \frac{1}{2} \| w \|^2 + \frac{C}{n} \sum_{i=1}^{n} L_e(d_i, y_i) \tag{12.2}$$

$$L_e(d_i, y_i) = \begin{cases} |d_i - y_i| - \varepsilon, & |d_i - y_i| \geq \varepsilon \\ 0 & \text{otherwise} \end{cases} \tag{12.3}$$

To obtain the estimation of w and b, Eq. (12.2) is transformed into the primal function given by Eq. (12.4) by introducing the positive slack variables ξ and ξ^* as follows (Vapnik, 1998):

$$\text{Minimize } R(w, \xi^*) = \frac{1}{2} \left\| w \right\|^2 + C \sum_{i=1}^{n} (\xi_i + \xi_i^*) \tag{12.4}$$

$$\text{subject to} \begin{cases} d_i - w\phi(x_i) - b_i \leq \varepsilon + \xi_i \\ w\phi(x_i) + b_i - y_i \leq \varepsilon + \xi_i^* \\ \xi_i, \xi_i^* \geq 0 \end{cases}$$

The first term $(1/2)|*w|*^2$ is the weights vector norm, d_i the desired value, and C refers to the regularized constant determining the trade-off between the empirical error and the regularized term. ε is called the tube size of SVM and it is equivalent to the approximation accuracy placed on the training data points. Here, the slack variables ξ and ξ^* are introduced. By introducing Lagrange multipliers and exploiting the optimality constraints, the decision function given by Eq. (12.1) has the following explicit form (Vapnik, 1998):

$$y(x) = \sum_{i=1}^{n}(a_i - a_i^*)K(x, x_i) + b \tag{12.5}$$

In Eq. (12.5), a_i and a_i^* are the so-called Lagrange multipliers. They satisfy the equalities $a_i \times a_i^* = 0$, $a_i \geq 0$ and $a^*_i \geq 0$ where $i = 1,2,\ldots, n$ and are obtained by maximizing the dual function of Eq. (12.4) which has the following form (Vapnik, 1998):

$$R(a_i, a_i^*) = \sum_{i=1}^{n} d_i(a_i - a_i^*) - \varepsilon\sum_{i=1}^{n} d_i(a_i - a_i^*) - \frac{1}{2}\sum_{i=1}^{n}\sum_{j=1}^{m}(a_i - a_i^*)(a_j - a_j^*)K(x_i, x_j) \tag{12.6}$$

with the constraints

$$\sum_{i=1}^{n} a_i = \sum_{i=1}^{n} a_i^*$$
$$0 \leq a_i \leq C, \; i = 1,2, \ldots, n$$
$$0 \leq a_i^* \leq C, \; i = 1,2, \ldots, n$$

$K(x_i x_j)$ is defined as the kernel function. The value of the kernel is equal to the inner product of two vectors X_i and X_j in the feature space $\varphi(x_i)$ and $\varphi(x_i)$, that is, $K(x_i x_j) = \Phi(x_i)x\Phi(x_i)$ the typical examples of the kernel function are as follows (Vapnik, 1998):

$$\text{Linear: } K(x_i, x_j) = x_i^T x_j \tag{12.7}$$

$$\text{Sigmoid: } K(x_i, x_j) = \tanh(\gamma x_i^T x_j + r) \tag{12.8}$$

$$\text{Polynomial: } K(x_i, x_j) = (\gamma x_i^T x_j + r)^4, \; \gamma > 0 \tag{12.9}$$

$$\text{Radial basis function (RBF): } K(x_i, x_j) = \exp(-\gamma\|x_i - x_j\|^2), \; \gamma > 0 \tag{12.10}$$

The selection of three characteristics of SVM have a strong affect on the performance of SVM, namely kernel function, optimization of hyperparameter, and loss function parameter ε (Jing et al., 2018), which are selected as follows:

The radial basis function (RBF) (Suykens et al., 2002) is selected as the kernel function after training all other kernel functions and observing the performance of minimization of error (Deo et al., 2016; Al-Musaylh et al., 2018a). The RBF was employed in developing the SVM model as it gave the optimum results.

Optimization of hyperparameters that minimize 10-fold cross-validation loss was found by using automatic hyperparameter optimization (Jing et al., 2018).

Ten-fold cross-validation is conducted for the optimal penalty parameters and loss function parameter ε. The parameters that minimize the cross-validation loss are selected, and shown in Table 12.3D.

It is important to note that the grid search (Hsu et al., 2003) was initially employed to select the optimal parameters but due to the large dataset (56,980 data points for 6-hourly), the process was very time-consuming. Thus 10-fold cross-validation was used which gave relatively good results as a grid search. Appendix shows comparative results of the grid search and 10-fold SVM model for monthly forecasting horizons.

12.3.1.2 Second-order Volterra model

The Volterra series is an important representation of a nonlinear system (Kashani et al., 2016; Prasad et al., 2017). A second-order representation has been adopted, as substantiated by previous studies (Gruber et al., 2011; Rathinasamy et al., 2013; Kashani et al., 2016; Prasad et al., 2017). With $x(t)$ as the model output and t as the tth instances, the second-order Volterra expansion could be expressed as (Schetzen, 1980):

$$x(t) = \int_{\tau_1=0}^{\tau_1=t} k_1(\tau_1)X(\tau-\tau_1)d\tau_1 + \int_{\tau_2=0}^{\tau_2=t}\int_{\tau_1=0}^{\tau_1=t} k_2(\tau_1,\ \tau_2)X(\tau-\tau_1)X(\tau-\tau_2)d\tau_1 d\tau_2 \tag{12.11}$$

where $k_1(\tau_1)$ and $k_2(\tau_1,\ \tau_2)$ are the Volterra kernels. In a condensed notation Eq. (12.7) gives:

$$x(t) = K_1[y(t)] + K_2[y(t)] \tag{12.12}$$

where $K_1[y(t)]$ and $K_2[y(t)]$ are the first- and second-order Volterra operators, respectively. For models requiring multiple forecaster inputs, the Volterra series expansion for a multiple input single output system is expressed as (Koukoulas and Kalouptsidis, 2000):

$$x(t) = \sum_{n=1}^{N}\sum_{\beta=1}^{M} k_1^{(n)}(\beta)x_n(t-\beta) + \sum_{n=1}^{N}\sum_{\alpha=1}^{M}\sum_{\beta=1}^{M} k_{2s}^{(n)}(\alpha,\ \beta)x_n(t-\beta)x_n(t-\beta) + \sum_{n1}^{N}\sum_{n2=1}^{n1-1}\sum_{i=1}^{M}\sum_{\beta=1}^{M} k_{2x}^{(n1.n2)}(\alpha,\ \beta)x_{n1}(t-\alpha)x_{n2}(t-\beta) \tag{12.13}$$

where N_i is the number of inputs, M represents the memory length of each significant lagged input variable; the $k_1^{(n)}$ is the first-order kernels, $k_{2s}^{(n)}$ is the second-order self-kernels, and $k_{2x}^{(n1,\ n2)}$ is the second-order cross-kernels. As suggested by Billings et al. (1988), the estimation of Volterra kernels was achieved through the principle of orthogonal least squares (OLS) since OLS suitably handles collinearity amongst predictor inputs.

12.3.1.3 Autoregressive integrated moving average model

An ARIMA model was used to forecast wind speed for 6-hourly, daily, and monthly forecasting horizons. The modeling strategy followed Box–Jenkins (Box and Jenkins, 1990; Box et al., 2015) methodology through the identification, estimation, diagnostic checking, and forecasting stages. To develop the ARIMA model, two types of linear-regression are integrated: The AR and the moving average (MA). The AR model equation is (Box and Jenkins, 1990):

$$y_t = c + a_1 y_{t-1} + \cdots + a_p y_{t-p} + u_t \tag{12.14}$$

$a_1, \ldots, a_p$ is the AR parameters, c is a constant, p is the order of the AR, and u_t is the white noise.

Similarly, the MA model is represented as (Box and Jenkins, 1990):

$$y_t = \mu + u_t + m_1 u_{t-1} + \cdots + m_q u_{t-q} \tag{12.15}$$

where $m, \ldots, m_q$ are the MA parameters, q is the order of MA, $u_t, u_{t-1}, \ldots, u_{t-q}$ are the (error) terms, and μ is the expectation of y_t. By integrating these two models with the same training data, the ARIMA model becomes (Box and Jenkins, 1990):

$$y_t = c + a_1 y_{t-1} + \cdots + a_p y_t - p + u_t + m_1 u_{t-1} + \cdots + m_q u_{t-q} \tag{12.16}$$

where p and q are the autoregressive and moving average terms, respectively.

The ARIMA model needs the statistical properties satisfying the stationarity of mean and variance in the model (Shukur and Lee, 2015). As the wind speed data is nonstationary, the ARIMA model requires differenced data to transform it to stationarity. This is denoted as ARIMA (p, q, d), where d is the degree of differencing (Yuan et al., 2016; Al-Musaylh et al., 2018a).

12.3.1.4 Complete ensemble empirical mode decomposition with adaptive noise

The CEEMDAN technique, proposed by Torres et al. (2011), is an adaptive data analysis method and has been applied in analyzing nonlinear and nonstationary data and has advantages as being self-adaptive and one that avoids mode mixing problems (Prasad et al., 2018; Zhang et al., 2017).

It decomposes complex signals into intrinsic mode functions (IMFs), which satisfy the following conditions (Wu and Huang, 2009; Huang et al., 1971): (1) in the entire data sequence, the number of extrema and the number of zero crossings in an entire sampled dataset must either be equal or differ at most by one; and (2) the mean value at any point of the envelope defined by the local maxima and the envelope defined by the local minima is zero.

During the CEEMDAN decomposition process, Gaussian white noise with unit variance and noise coefficient is added at each of the decomposition stages (Huang et al., 1971). This noise-added signal is decomposed via EMD to obtain the first IMF and the subsequent residual component. This noise is eliminated through averaging of corresponding IMFs and the residual component (Wu and Huang, 2009).

A brief realization of the CEEMDAN algorithm is as follows (Torres et al., 2011; Prasad et al., 2018; Zhang et al., 2017): consider an unresolved signal, $x(t)$, and added white noise, $k_n(t)$. To obtain the first IMF by CEEMDAN (IMF_1) for every $n = 1, \ldots N$ decompose each $x_n'(t) = x(t) + \varepsilon k_n(t)$ via EMD where N = ensemble number and ε = amplitude of the added noise. Note that at each subsequent stage, the coefficient ε_p allows an appropriate selection of the signal to noise ratio of the white noise. Then collate the first IMF produced by EMD (d_1) and compute the ensemble average as follows:

$$\text{IMF}_1(t) = \frac{1}{N}\sum_{n=1}^{N} d_1 \tag{12.17}$$

Once IMF_1 is obtained, the remaining component is described as:

$$r_1(t) = x(t) - IMF_1(t) \tag{12.18}$$

Next IMF_2 is computed from the remaining component, where $F_j(.)$ is the operator which produces the jth IMF obtained via EMD, is stated as:

$$IMF_2(t) = \frac{1}{N}\sum_{n=1}^{N} F_1[r_1(t) + \varepsilon_1 F_1(k_n(t))] \tag{12.19}$$

Hereafter, the successive IMFs ($p = 2, 3, \ldots, P$) are found using a similar step as above. The remaining component, $r_p(t)$, is repeatedly decomposed using EMD to obtain subsequent IMFs until the residue does not satisfy the conditions of IMFs. The final residual (R) component is expressed as:

$$R(t) = x(t) - \sum_{p=1}^{P} IMF_p(t) \tag{12.20}$$

Two important parameters, the ensemble number and the amplitude of added white noise, must be appropriately defined to attain optimum results and to cancel out the added white noise series from the unresolved signal (Prasad et al., 2018; Jun et al., 2018; Zhang et al., 2017). The statistical rule (Wu and Huang, 2009):

$$e_n = \frac{\varepsilon}{\sqrt{N}} \tag{12.21}$$

was used to control the effect of the added white noise where the N = ensemble number, ε = amplitude of the added noise, and e_n = final standard deviation (SD), which is the difference between the input signal and corresponding IMFs. The recommended amplitude of the added white noise is 20% of SD (Wu and Huang, 2009). Therefore based on similar studies (Prasad et al., 2018; Jun et al., 2018; Zhang et al., 2017) $e_n = 0.2$ and $N = 20$ (Torres et al., 2011) were used in this study. Fig. 12.3 presents the original wind speed series decomposed by CEEMDAN.

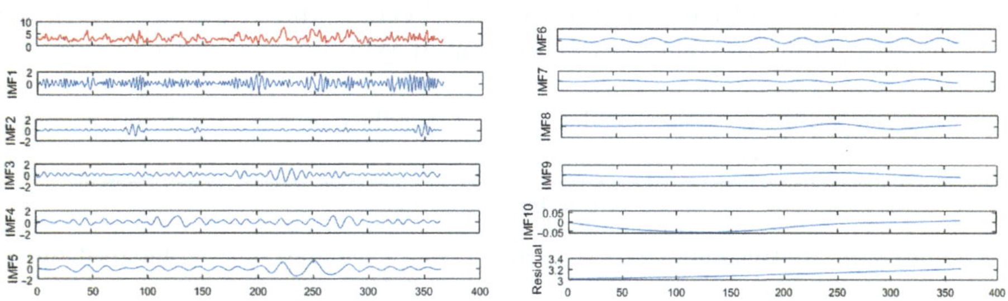

FIGURE 12.3 Temporal waveforms of IMFs and the residual from CEEMDAN transformation of unresolved time series (lag 0) of wind speed (m/s) at best performing site Capital during the training period (January 1–December 31, 1979) for daily forecasting horizon (The definitions of acronyms used here are as follows: *CEEMDAN*, complete ensemble empirical mode decomposition with adaptive noise; *IMF*, intrinsic mode functions).

12.3.2 Study area and data

12.3.2.1 Study area

Australia has excellent wind resources particularly in coastal regions (Blakers et al., 2017) owing to its geographical characteristics. Because there is a tremendous wind power potential, it is very important to investigate its actual wind power based on the precise forecasting of wind speed. The study sites Albany, Capital, Macarthur, and Woolnorth are located on wind farms, which are located around the coastal areas in Australia. The four selected study sites are summarized in Table 12.1 with distinct geophysical characteristics acquired from *Australian Geological Provinces* (Gallagher et al., 2013). Fig. 12.4 shows the physical locations of the study sites.

12.3.2.2 Data

In this study, data-driven models were developed for 6-hourly, daily, and monthly wind speed (m/s) forecasting for four study sites in Australia. The net wind speed data comprised of 6-hourly (four times per day) records for a period between January 1, 1979 and December 31, 2017 were acquired from the European Centre for Medium-Range Weather Forecasts (ECMWF) (Dee et al., 2011). To expand the forecasting horizon to daily periods, arithmetic averaging of the 6-hourly data was performed and, likewise, the monthly wind speed data was compiled by averaging the daily data.

Table 12.2 summarizes statistics of the wind speed datasets. The stochastic components present in wind speed data for all the timescales are confirmed by the high degree of skewness observed. In Table 12.2, variations in climatological patterns in wind speed (m/s) geographically-diverse sites are apparent. The skewness of 6-hourly wind speed at study site Albany and Woolnorth were closer to zero confirming near-normal distributions. Study site Capital had a skewness of 1.24 and Woolnorth had skewness of 0.538. Similarly, kurtosis (kurt $\leq$ 2) for 6-hourly wind speed data illustrated that the distributions had fewer and less extreme outliers for all study sites (Celikoglu and Tirnakli, 2018). Similar data patterns and descriptives were observed for the daily and monthly timescale across all the four study sites. None of the sites had extreme outliers for all three timescales.

Fig. 12.5 illustrates wind rose showing wind direction and wind speed frequency for monthly forecasting horizon. The wind direction at Woolnorth site was mostly from the

TABLE 12.1 Geographic locations, physical characteristics, and the capacities of wind energy conversion systems at the selected study sites.

		Physical characteristics			
Site no.	**Study sites**	**Longitude (°E)**	**Latitude (°S)**	**Elevation (m)**	**Installed capacity (MW)**
1	Albany	117.79	35.07	80	89
2	Capital	149.52	35.17	80	140.7
3	Macarthur	142.19	38.05	85	420
4	Woolnorth	144.72	40.69	60	140

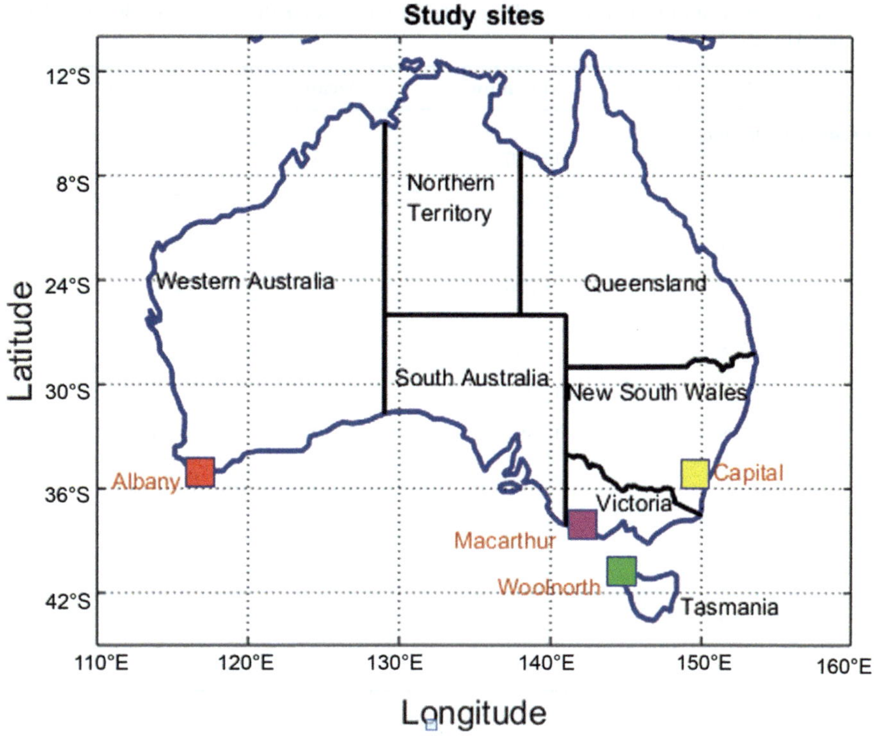

FIGURE 12.4 Map of the study region showing the selected stations and their geographical locations.

northeast with almost 50% of the time having a wind speed above 9 m/s, Woolnorth recorded the maximum wind speed of 13 m/s. Wind speed was lowest at Capital site 4.5 m/s mostly blowing from the northwest direction. Albany had wind mostly from the southeast while Macarthur had it from southwest, the wind speed for these two sites were less than 7 m/s. Wind speed at each site exhibits distinct patterns, and so it would be worth investigating through the respective models (Chitsazan et al., 2019; Dadkhah et al., 2018).

12.3.3 Predictive model development

Four predictive models were developed in MATLAB environment running on Intel *i7*, 3.40 GHz processor, namely, SVM, Volterra, CEEMDAN–SVM, and CEEMDAN–Volterra while the ARIMA model was developed using the *R* package (Adamowski et al., 2012). These models were constructed to forecast 6-hourly, daily, and monthly wind speed.

A major task was to determine the training data to construct the predictive model and testing data to evaluate its performance. There is no set rule for data partitioning (Deo et al., 2016; Prasad et al., 2018). The work of Qingqing (He et al., 2018) used 80% of inputs for training and the remaining 20% for testing the models in their hybrid model while

TABLE 12.2 Descriptive statistics of the wind speed (m/s) datasets for the present study sites (January 1, 1979–December 31, 2017).

Study sites	Minimum	Maximum	Mean	Skewness	Kurtosis
6-Hourly forecasting horizons					
Albany	0.072	14.780	5.098	0.242	−0.301
Capital	0.014	13.069	2.954	1.239	1.722
Macarthur	0.011	12.987	3.607	0.538	0.834
Woolnorth	0.019	21.736	7.081	0.297	−0.155
Daily forecasting horizons					
Albany	1.228	11.940	5.098	0.272	−0.389
Capital	0.283	11.242	2.954	1.198	1.589
Macarthur	0.598	10.887	3.607	0.838	0.554
Woolnorth	0.597	18.324	7.305	0.364	−0.209
Monthly forecasting horizons					
Albany	3.359	7.017	5.101	0.160	−0.905
Capital	1.832	5.239	2.952	1.014	1.477
Macarthur	2.497	5.600	3.605	0.975	1.352
Woolnorth	6.029	9.947	7.304	0.780	0.299

forecasting short-term wind speed, so the subsets in our study had first 80% for training and the next 20% testing (Table 12.3) for all timescales.

The sequential division method was adopted to avoid distortion of the natural embedded frequencies within the wind speed time series data (Zhang et al., 2017). This is to allow the multiresolution analysis utilities CEEMDAN to appropriately unveil and extract these entrenched features for respective data-smart models that otherwise would not have been possible (Prasad et al., 2018; Jing et al., 2018).

For the development of forecasting models, the scaling of all input variables is a mandatory requirement (Deo et al., 2016). This is because of the chaotic nature of the input, where changes in wind speed seem to occur at a higher frequency, the data required appropriate scaling to avoid predictor values (and associated patterns/attributes) with large numeric ranges from dominating attributes with narrower ones (Lin, 2003). Data were therefore normalized and bounded by zero and one through the following expression:

$$X_{norm} = \frac{(X - X_{min})}{(X_{max} - X_{min})} \tag{12.22}$$

where X = any data (input or output), X_{min} = minimum value of the entire dataset, X_{max} = maximum value of the entire dataset, and X_{norm} = normalized value of the data.

Data-driven models incorporate historical wind speed data to forecast future wind speed. The initial selection of (lagged) input variables to determine the predictors is critical for developing a robust SVM model (Deo et al., 2016; Wang et al., 2015; Li et al., 2018b). The literature outlines partial autocorrelation function (PACF) as one of the selection

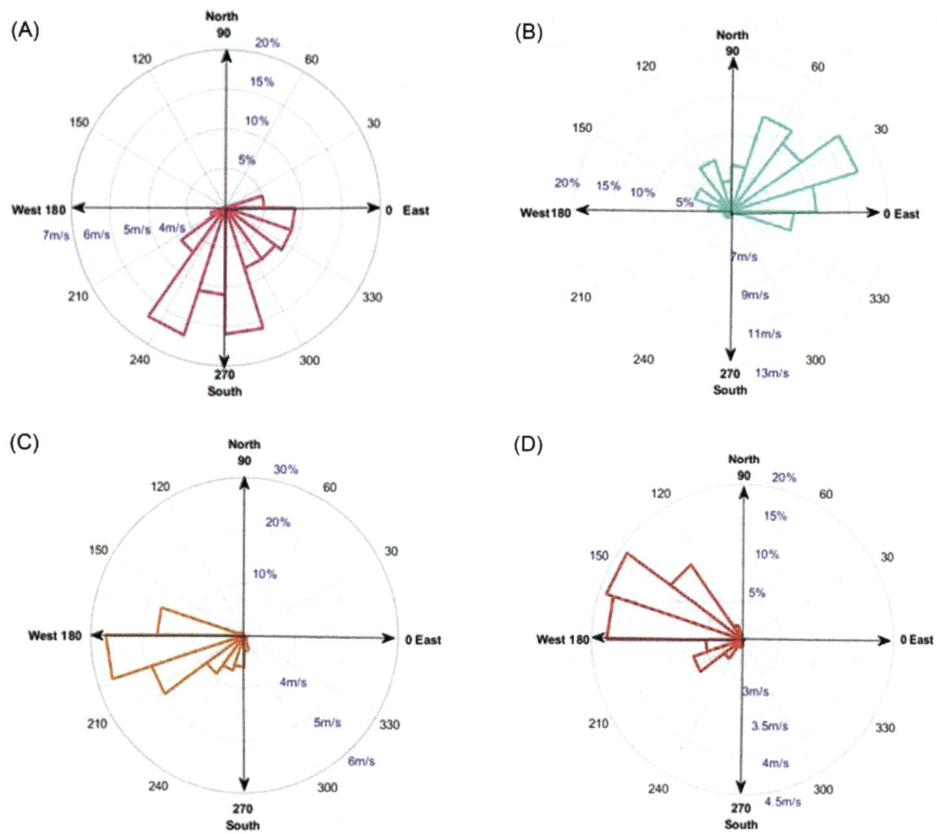

FIGURE 12.5 Wind rose plots showing wind direction and wind speed frequency for monthly forecasting horizon in the testing period (January 1, 2011–December 31, 2017) for (A) Albany, (B) Woolnorth, (C) Macarthur, and (D) Capital study site.

methods for determining the sequential time series of lagged wind speed values that provide an optimal performance (Tiwari and Chatterjee, 2010; Wang et al., 2015; Al-Musaylh et al., 2018a). Hence PACF has been adopted and lagged series with a statistically significant relationship (i.e., at 95% confidence interval) were screened as prominent inputs.

The approach employed time-lagged information to analyze the period between current and antecedent wind speed values at specific points in the past (i.e., applying a time lag) and assessed any temporal dependencies existing in the time series.

Two different modeling techniques have been adopted for this study. Firstly, the conventional modeling approach was adopted for the standalone SVM and Volterra models. For standalone models, PACF was applied to the 6-hourly, daily, and monthly intact wind speed time series (i.e., the time series without CEEMDAN analysis) and subsequently, inputs for each time lag (6-hourly, daily, monthly) were identified and analyzed using correlation analysis where the coefficient (r) between the measured wind speed and the predictors were computed one by one.

TABLE 12.3 Data partitions used in this study.

Study sites	Period	Number of datum points	Data partition	
			Training	Testing
All sites 6-hourly wind speed	January 1979–December 2017	56,980	80%	20%
			45,584	11,396
			1979–March 2010	April 2010–2017
All sites daily wind speed		14,245	80%	20%
			11,396	2849
			1979–March 2010	April 2010–2017
All sites monthly wind speed		468	80%	20%
			374	94
			1979–March 2010	April 2010–2017

Fig. 12.6 plots the magnitude of the r at the 95% confidence interval outside which the r is deemed as statistically significant for 6-hourly forecasting horizons, other timescales' statistically significant sets of predictor variables are summarized in Table 12.4. These were channeled into the SVM and Volterra models for forecasting of 6-hourly, daily, and monthly wind speed as illustrated in the schematic view of the model development in Fig. 12.7.

Alternatively, the CEEMDAN modeling process as illustrated in Fig. 12.7 was used, which can be summarized as follows:

CEEMDAN decompositions: The 6-hourly, daily, and monthly wind speed time series data (without CEEMDAN analysis) were decomposed into respective 6-hourly, daily, and monthly IMFs and a residual component using CEEMDAN procedures as explained in Section 12.4 for training and testing dataset. An equal number of IMFs were selected for both training and testing datasets by controlling the number of IMFs. An example of the IMFs and the residual component from CEEMDAN at Capital site has been illustrated in Fig. 12.3.

Model input data: *PACF* was applied to each of the 6-hourly, daily, and monthly IMFs and residual component time series generated in the above phase. Prominent lagged inputs of each IMF and residual component were determined. The individual input matrix was created for each IMF and the residual component containing its respective significant lags is summarized in Table 12.4. An example of the r based on the PACF for the IMF and residual (Res) for 6-hourly wind speed for the developing CEEMDAN model for the best performing site, Macarthur, is shown in Fig. 12.8.

Forecasting: These individual input conditions were used to forecast the respective future IMFs and the residual components using the SVM and Volterra models for 6-hourly, daily, and monthly forecasting horizons. Then the forecasted IMFs and residual components

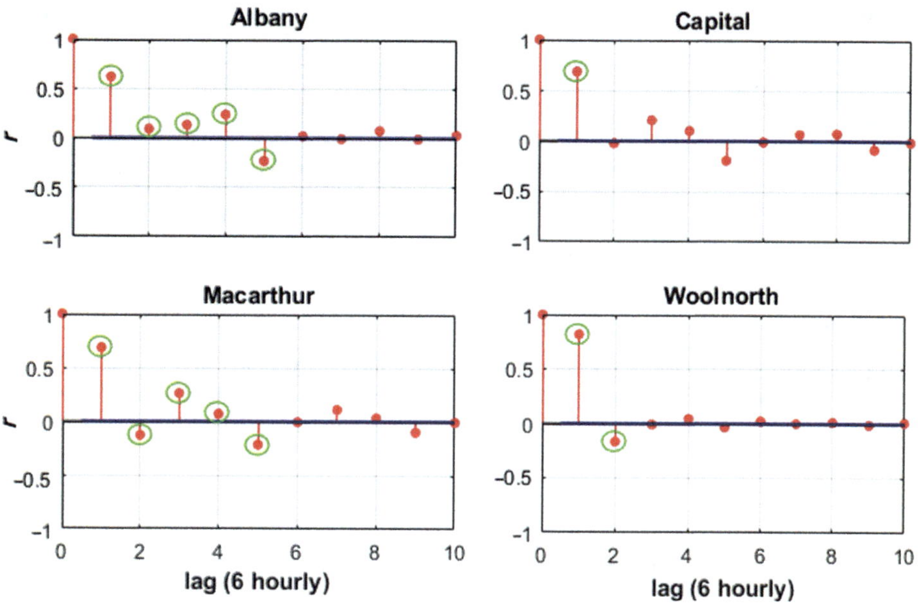

FIGURE 12.6 Correlation coefficient (r) based on the partial autocorrelation function for the 6-hourly wind speed with its predictor (i.e., wind speed) used for developing the support vector machine model and Volterra model for all sites. Statistically, significant lags at the 95% confidence interval are marked (*green*).

TABLE 12.4 The number of statistically significant lags used for developing standalone and CEEMDAN hybrid SVM and Volterra models.

Study sites	Number of significant lags used at respective sites			
	Albany	**Capital**	**Macarthur**	**Woolnorth**
6-Hourly forecasting horizons				
Standalone SVM/Volterra models (i.e., without CEEMDAN analysis)	1–6	1–5	1–5	1–2
CEEMDAN (SVM/Volterra models) IMF-1 to IMF-12	1–5	1–5	1–5	1–5
Residual	1–5	1–5	1–5	1–5
Daily forecasting horizons				
Standalone SVM/Volterra models (i.e., without CEEMDAN analysis)	1–3	1–2	1–2	1–2
CEEMDAN (SVM/Volterra models) IMF-1 to IMF10	1–5	1–6	1–5	1–5
Residual	1–5	1–6	1–5	1–5
Monthly forecasting horizons				
Standalone SVM/Volterra models (i.e., without CEEMDAN analysis)	1–9	1–9	1–8	1–5
CEEMDAN (hybrid SVM/Volterra models) IMF-1 to IMF-5	1–3	1–3	1–3	1–3
Residual	1–3	1–3	1–3	1–3

were integrated at the end to generate the 6-hourly, daily, and monthly forecasts of wind speed values.

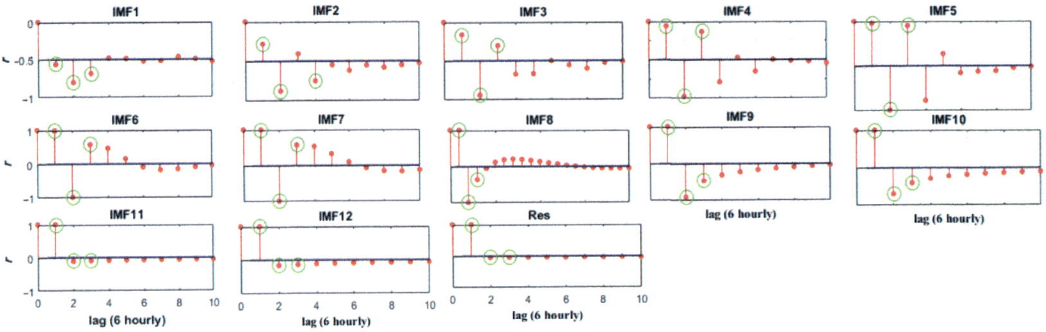

FIGURE 12.7 r based on the PACF for the Intrinsic Mode Function (IMF) and residual (res) for hourly wind speed for developing complete ensemble empirical mode decomposition with adaptive noise models for the best performing site, Macarthur. Statistically significant lags at the 95% confidence interval are marked (*green*).

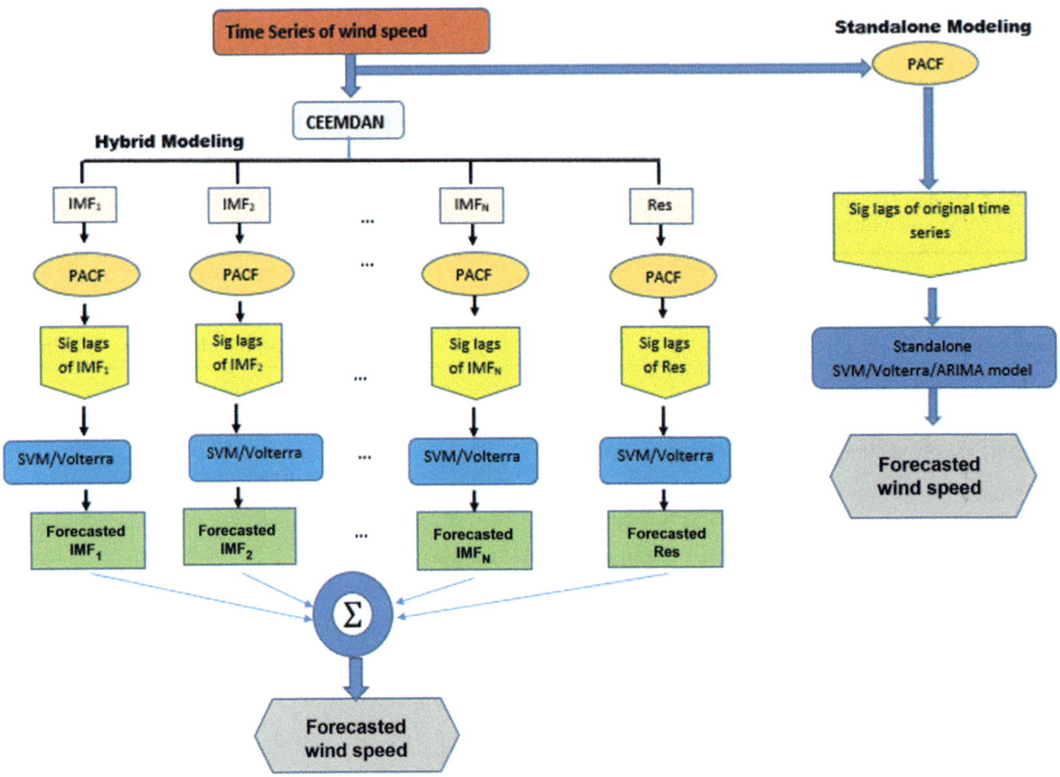

FIGURE 12.8 A schematic view of the model development stages (The definitions of acronyms used here are as follows: *CEEMDAN*, complete ensemble empirical mode decomposition with adaptive noise; *IMF*, intrinsic mode functions; *PACF*, partial autocorrelation function; *Res.*, residual; *Sig.*, significant; subscript N represents the IMF number(s); *SVM*, support vector machine model).

For 6-hourly, daily, and monthly forecasting horizons, the models employed data from January 1, 1979–December 31, 2017. For 6-hourly forecasting horizons, the standalone and hybrid SVM and Volterra models were built with 1–5 statistically significant lagged input combinations (5 representing the maximum significant number of lags of wind speed data) for study sites Capital, Macarthur, and Woolnorth, while for Albany site, 1–6 statistically significant lags were used. It is important to note that the number of IMFs and the residual component generated for hybrid models are contingent upon the nature of data which in turn determines the ensemble numbers (Prasad et al., 2018).

Similarly, for the daily forecasting horizon, the standalone SVM and Volterra were built with 1–2 statistically significant lagged input combinations for Macarthur, Capital, and Woolnorth, and 1–3 statistically significant lagged input combinations for Albany. The CEEMDAN–SVM and Volterra models were built with 1–5 statistically significant lagged input combinations for Albany, Macarthur, and Woolnorth, while Capital had 1–6 statistically significant lagged input combinations. For monthly forecasting horizons, standalone models had 1–9 statistically significant lagged input combinations for Albany and Capital, 1–8 statistically significant lagged input combinations for Macarthur, and 1–5 statistically significant lagged input combinations for Woolnorth. The hybrid models had 1–3 statistically significant lagged input combinations for all sites in the monthly forecasting horizons.

TABLE 12.5 Parameters of the ARIMA model presented in the training period for 6-hourly, daily, and monthly forecasting horizons.

Study sites	*p*	*d*	*q*	σ^2	*L*	*AIC*
6-Hourly forecasting horizons						
Albany	45	1	3	2.130	−81,950	163,954
Capital	32	1	7	1.280	−70,284	140,601
Macarthur	23	1	4	2.831	−83,182	163,179
Woolnorth	41	1	8	3.261	−91,582	183,182
Daily forecasting horizons						
Albany	27	1	6	1.650	−19,021	38,131
Capital	34	1	4	1.300	−17,648	35,329
Macarthur	26	1	3	1.460	−18,325	36,690
Woolnorth	31	0	5	2.150	−17,226	38,812
Monthly forecasting horizons						
Albany	10	1	4	0.610	−430	864
Capital	15	1	6	0.168	−195	418
Macarthur	16	1	3	0.160	−191	412
Woolnorth	17	1	5	0.778	−476	973

AIC, Akaike information criterion; *d*, degree of differencing; *L*, log-likelihood; *p*, autoregressive term; *q*, moving an average term; R^2, coefficient of determination; σ^2, variance.

On the other hand, the univariate ARIMA model's mechanism differs as it creates its own lagged data through the p and q parameters developed in its identification phase (Shukur and Lee, 2015), as seen in Table 12.5. Therefore all historical wind speed data were used as a single input (with no lags) to identify the ARIMA model for all forecasting horizons.

The selection of the optimal value of p, q, and d to build the ARIMA model was based on the model giving the largest value of log-likelihood and the smallest value of Akaike information criterion (*AIC*) (Erdem and Shi, 2011; Yuan et al., 2016). Firstly, the optimal value of d was selected while keeping p and q as 0. Once the optimal value of d was achieved, p values were trialed, and the best value of p was selected based on the lowest *AIC* and highest log-likelihood. The same procedure was repeated to achieve the optimal value of q and the results are tabulated in Table 12.5, illustrating the best values of p, d, and q with respective values of *AIC* and log-likelihood for all forecasting horizon and study sites.

For this study, the SVM model was developed using MATLAB LibSVM Toolbox (Chang and Lin, 2011). The RBF was employed in developing the SVM model after trialing other functions and it gave the optimum results in terms of mean absolute error and root mean square error in comparison with other functions. The RBF (Eq. 12.10) was used to map nonlinear input samples onto a high-dimensional feature space; it examines the nonlinearities between target and input data (Lin, 2003) and outperforms linear-kernel-based models in terms of accuracy and is also faster in the training phase (Gill et al., 2006; Wang et al., 2015; Al-Musaylh et al., 2018a).

It is important to note that the standalone SVM models developed in this research have input as the significant lags of the intact wind speed, that is, the series without (CEEMDAN analysis). On the other hand, in CEEMDAN–SVM models, the inputs represent the significant lagged series of the intrinsic mode functions (IMFs) and the residual component, which result after CEEMDAN transformations of the original input data series. Ten-fold cross-validation was conducted for the optimal parameters of the SVM model (Jing et al., 2018). The 10-fold cross-validation generated 10 models in the training period; the model with the smallest mean square error (*MSE*), was used to establish the optimal parameters. Table 12.6 lists the optimal values of each SVM model for 6-hourly, daily, and monthly wind speed forecasting horizons.

12.3.4 Model performance evaluation criteria

A wide range of statistical metrics are used to assess the performance of the developed models in this study. A comprehensive and robust model assessment requires both objective and subjective evaluations (Deo et al., 2018; Willmott, 1984; Nash and Sutcliffe, 1970; Chai and Draxler, 2014) as no single statistical measure is purely definitive (Dawson et al., 2007). The definition of the mathematical formulas are as follows (Willmott, 1984; Nash and Sutcliffe, 1970; Legates and McCabe, 1999; Wallach et al., 2019):

1. Correlation coefficient (r):

$$r = \frac{\sum_{i=1}^{N}\left(Data\_obs_i - \overline{Data\_obs}\right)\left(Data\_sim_i - \overline{Data\_sim}\right)}{\sqrt{\sum_{i=1}^{N}\left(Data\_obs_i - \overline{Data\_obs}\right)^2}\sqrt{\sum_{i=1}^{N}\left(Data\_sim_i - \overline{Data\_sim}\right)^2}}; \quad (-1 \le r \le 1) \tag{12.23}$$

2. Root mean square error (*RMSE*):

$$RSME = \sqrt{\frac{1}{N}\sum_{i=1}^{N}(Data\_sim_i - Data\_obs_i)^2}; \quad (0 \le RMSE \le 1) \tag{12.24}$$

3. Mean absolute error (Samadianfard et al., 2018):

$$MAE = \frac{1}{N}\sum_{i=1}^{N}\left|(Data\_sim_{i-}Data\_obs_i)\right|; \quad (0 \le MAE \le 1) \tag{12.25}$$

4. Willmott's index (*WI*):

$$WI = 1 - \left[\frac{\sum_{i=1}^{N}(Data\_sim_i - Data\_obs_i)^2}{\sum_{i=1}^{N}\left(\left|Data\_sim_i - \overline{Data\_sim}_i\right| + \left|Data\_obs_i - \overline{Data\_obs}_i\right|\right)^2}\right]; \quad (0 \le WI \le 1) \tag{12.26}$$

TABLE 12.6 The parameters of the support vector machine.

	SVM		
Study sites	**mse (m/s)**	**σ**	***E***
6-Hourly forecasting horizons			
Albany	0.009	2.54	0.02
Capital	0.007	3.06	0.01
Macarthur	0.008	2.32	0.01
Woolnorth	0.001	0.21	0.02
Daily hourly forecasting horizons			
Albany	0.01	4.48	0.02
Capital	0.01	3.31	0.01
Macarthur	0.01	4.05	0.01
Woolnorth	0.02	2.03	0.02
Monthly hourly forecasting horizons			
Albany	0.02	3.09	0.03
Capital	0.18	3.28	0.04
Macarthur	0.01	3,64	0.01
Woolnorth	0.02	3.21	0.02

mse, mean square error of the best model, σ, kernel scale; ε, loss function parameter for the present study sites.

5. Nash–Sutcliffe efficiency (E_{NS}):

$$E_{NS} = 1 - \left(\frac{\sum_{i=1}^{N} (Data\_obs_i - Data\_sim_i)^2}{\sum_{i=1}^{N} (Data\_obs_i - \overline{Data\_obs})^2} \right); \quad (0 \leq E_{NS} \leq 1) \tag{12.27}$$

6. Legates–McCabe's index (L):

$$L = 1 - \left[\frac{\sum_{i=1}^{N} |Data\_sim_i - Data\_obs_i|}{\sum_{i=1}^{N} \left|Data\_sim_i - \overline{Data\_obs_i}\right|} \right]; \quad (-\infty < L \leq 1) \tag{12.28}$$

In these equations, $Data\_obs$ = observed wind speed (m/s), $Data\_sim_i$ = forecasted wind speed (m/s), $\overline{Data\_obs_i}$ = mean of observed wind speed, $\overline{Data\_sim_i}$ = mean of forecasted wind speed, and N = total number of data points. The first evaluation metric, Pearson's correlation coefficient (r) provides information on the strength and direction of the agreement between $Data\_obs_i$ and $Data\_sim_i$. In general, the higher the r, the better the model fits our data. For the best model, r is expected to be close to unity. The equation, however, is based on a consideration of linear relationships and therefore is limited as it standardizes to the observed and modeled means and variances.

The absolute error measures, *RMSE* and *MAE*, assemble information on the average discrepancies between forecasted and observed values (Legates and McCabe, 1999). RMSE measures the goodness-of-fit relevant to high values. It follows an assumption that error is unbiased and follows a normal distribution and can assess the model with a higher level of skill compared with the correlation coefficient. The *RMSE*, from 0 to ∞, is a measure of the difference between observed and forecasted values, with a small *RMSE* value indicating a more efficient model to estimate the parameter.

The *MAE* is the average over the test sample of the absolute differences between prediction and actual observation where all individual differences have equal weight (Legates and McCabe, 1999). It is not weighted toward higher magnitude or lower magnitude events, but instead evaluates all deviations of the forecasted values from the observed values, in an equal manner and regardless of sign. It is a linear scoring rule that describes only the average errors, ignoring their direction of variation from the measured values and it is used for non-Gaussian cases. *MAE* takes values from 0 (perfect fit) to ∞ (worst fit). However, *MAE* does not provide information about under/overpredictions, while *RMSE* is oversensitive to peak wind speed values and insensitive to low levels (Hora and Campos, 2015; Willems, 2009).

The E_{NS} is a normalized metric that determines the relative magnitude of the residual variance of forecasted data in comparison to the measured variance. It provides a measure of the ability of a model to forecast values that are different from the mean (Willems, 2009) and it indicates how well the plot of observed versus simulated data fits the 1:1 line: 1 = perfect model; 0 = no predictive advantage; and negative values when forecasted values diverge (Legates and McCabe, 1999). The index of agreement WI was proposed by Willmott (1984) to overcome the disadvantage of E_{NS} and r. WI is important because it is used to assess the accuracy of the predicted values with respect to observed values of the data by considering the ratio of mean squared error instead of the square of the error differences for sites whose climatology are different.

Owing to the squared values of residual terms, both WI and E_{NS} are oversensitive to the peak residual values (Legates and McCabe, 1999; Dawson et al., 2007; Willmott, 1984). In comparison, the L is not overestimated since it takes absolute values into account and gives errors and differences the appropriate weights (Legates and McCabe, 1999). L is also simple, easy to interpret, and is acclaimed to yield a relative assessment of model performances.

Considering geographic differences between study sites, which lead to different distributions of wind speed, we also adopted normalized metrics: *RRMSE* and *MAPE*, as alternatives to assess the model's performance (Zhang et al., 2017). We note the absolute error measures (*RMSE* and *MAE*) are in real units which limit their ability to assess the model performances across various case study sites. Hence, the percentage error measures, viz., *RRMSE* and *MAPE* are used, as shown below:

7. Relative root mean square error (*RRMSE*, %):

$$RRMSE = \frac{\sqrt{\frac{1}{N}\sum_{i=1}^{N}(Data\_sim_i - Data\_obs_i)^2}}{\frac{1}{N}\sum_{i=1}^{N}(Data\_sim_i)} \times 100 \quad (12.29)$$

8. Mean absolute percentage error (*MAPE*, %):

$$MAPE = \frac{1}{N}\sum_{i=1}^{N}\left|\frac{(Data\_sim_i - Data\_obs_i}{Data\_obs_i}\right| \times 100 \quad (12.30)$$

RMSE and *MAE* are used to determine model performance with simultaneous monitoring of correlation r whereas WI, E_{NS}, and L provides further goodness-of-fit assessments and eventually, *RRMSE* and *MAPE* compared models at different study sites. However, the limitation of the abovementioned objective metrics is the quantification of assessment in a few numbers (Willems, 2009). Thus to get a better insight, individual model performance assessments through various diagnostic plots, for example, box plots, forecasting error histograms, time series graphs, and polar plots are also performed.

12.4 Results and discussion

12.4.1 Short-term wind speed forecasting

Fig. 12.9 depicts fluctuation components in the wind speed data at the 6-hourly forecasting horizons which reflect the intermittent nature of wind (Zhang et al., 2015). This is

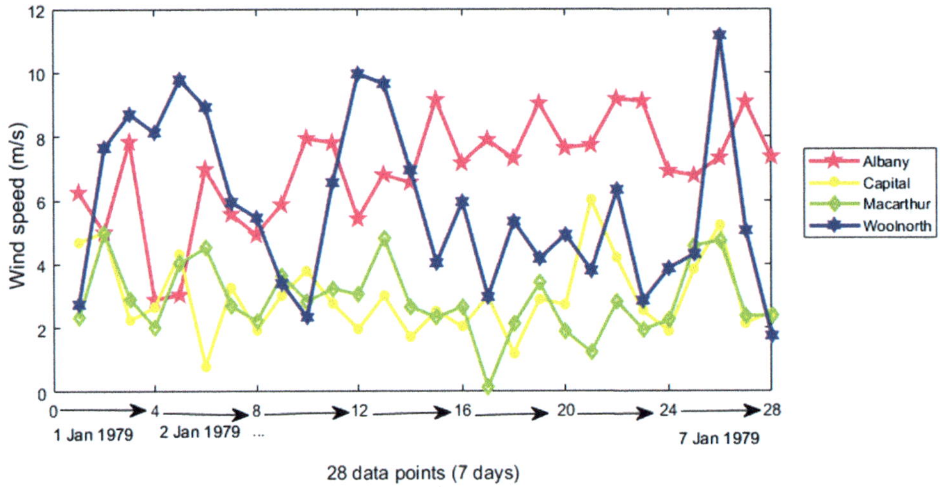

FIGURE 12.9 6-Hourly variations of wind speed (m/s) for the study sites Capital, Albany, Macarthur, Woolnorth in the training period for the case of January 1–January 7, 1979.

confirmed by a high degree of skewness observed for the 6-hourly wind speed for each site as shown in Table 12.2 for all study sites. Table 12.7 with the MSE endorses the best model highlighted in green.

The proposed model CEEMDAN–SVM has outstanding performance in comparison with other comparative models (SVM, Volterra, ARIMA, and CEEMDAN–Volterra) as shown in Table 12.8. In forecasting the 6-hourly wind speed, for all four sites (Albany, Capital, Macarthur, and Woolnorth), CEEMDAN–SVM models enumerated the maximum r values, Macarthur site recorded the maximum $r = 0.984$, while Albany attained the lowest $r = 0.936$. Consequently, the lowest *RMSE* and *MAE* values were attained by the hybrid CEEMDAN–SVM model at all four sites. Site Macarthur showed the lowest errors from the hybrid CEEMDAN–SVM model with $RMSE = 0.309$ and $MAE = 0.183$ and the highest percentage increase in r in comparison with the SVM model (30.85%), indicating it is the best performing site for 6-hourly forecasting horizons. It is evident so far that the hybridized SVM models are better in comparison to their standalone counterparts and other comparative models.

The evaluation of the hybrid CEEMDAN was undertaken with a time series plot of wind speed (Fig. 12.10). These results confirm that CEEMDAN–based models attain better accuracy for all four sites. The agreement between observed data and forecasted wind speed values by the optimal hybrid CEEMDAN–SVM model was very good, while the standalone models (SVM, Volterra, and ARIMA) divert from the observed values.

The assessment of 6-hourly wind speed forecasting demonstrated intense improvements in the performance of the hybrid CEEMDAN–SVM model (Table 12.9). Forecasting results at site Albany, Capital, and Woolnorth in terms of all three measures showed that CEEMDAN–SVM is outperforming all other models. However, at site Macarthur, the CEEMDAN–Volterra model attained a greater value of $L = 0.982$.

While comparing the hybrid CEEMDAN–SVM model with the standalone SVM from the perspective of L, which takes the primacy based on benefits discussed earlier, at all

TABLE 12.7 The performances of CEEMDAN–SVM versus the comparative models in the model development, that is, training phase based on the Pearson's correlation coefficient (*r*), root mean square error *RMSE*, and the mean absolute error (*MAE*) for 6-hourly wind speed forecasting.

	CEEMDAN–SVM			SVM			CEEMDAN–Volterra			Volterra			ARIMA		
Study sites	*r*	*RMSE*	*MAE*	*r*	*RMSE*	*MAE*	**R**	*RMSE*	*MAE*	*r*	*RMSE*	*MAE*	*r*	*RMSE*	*MAE*
6-Hourly forecasting horizons															
Albany	**0.961**	**0.689**	**0.498**	0.734	1.380	1.030	0.932	0.739	0.554	0.682	1.486	1.143	0.367	2.167	1.741
Capital	**0.953**	**0.543**	**0.412**	0.773	1.080	0.798	0.934	0.651	0.486	0.728	1.164	0.885	0.342	1.542	1.120
Macarthur	**0.981**	**0.287**	**0.170**	0.767	1.114	0.835	0.951	0.544	0.421	0.730	1.183	0.912	0.452	1.728	1.245
Woolnorth	**0.987**	**0.788**	**0.584**	0.833	1.772	1.367	0.912	1.423	1.135	0.820	1.844	1.421	0.410	2.987	2.437

The models with the largest value of the *r* and the smallest value of *MAE* and *RMSE* at each site are in boldface in green.

TABLE 12.8 The performance of CEEMDAN–SCM versus the comparative models in forecasting 6-hourly wind speed in the testing period, based on Pearson's correlation coefficient (*r*), root mean square error (*RMSE*), and mean absolute error (*MAE*).

	CEEMDAN–SVM			SVM			CEEMDAN–Volterra			Volterra			ARIMA		
Study sites	*r*	*RMSE*	*MAE*	*r*	*RMSE*	*MAE*	**R**	*RMSE*	*MAE*	*r*	*RMSE*	*MAE*	*r*	*RMSE*	*MAE*
Albany	0.936	**0.706**	0.507	0.718	1.387	1.056	0.928	0.739	0.561	0.672	1.478	1.141	0.018	2.142	1.763
Capital	0.941	**0.560**	0.423	0.736	1.119	0.838	0.924	0.667	0.501	0.705	1.177	0.898	0.066	1.645	1.259
Macarthur	0.984	**0.309**	0.183	0.752	1.156	0.868	0.946	0.568	0.431	0.721	1.126	0.938	0.005	1.778	1.368
Woolnorth	0.969	**0.798**	0.607	0.816	1.889	1.457	0.900	1.456	1.145	0.806	1.949	1.502	0.012	3.264	2.618

The optimal models yielding the lowest *RMSE* at each site are shown in boldface in red.

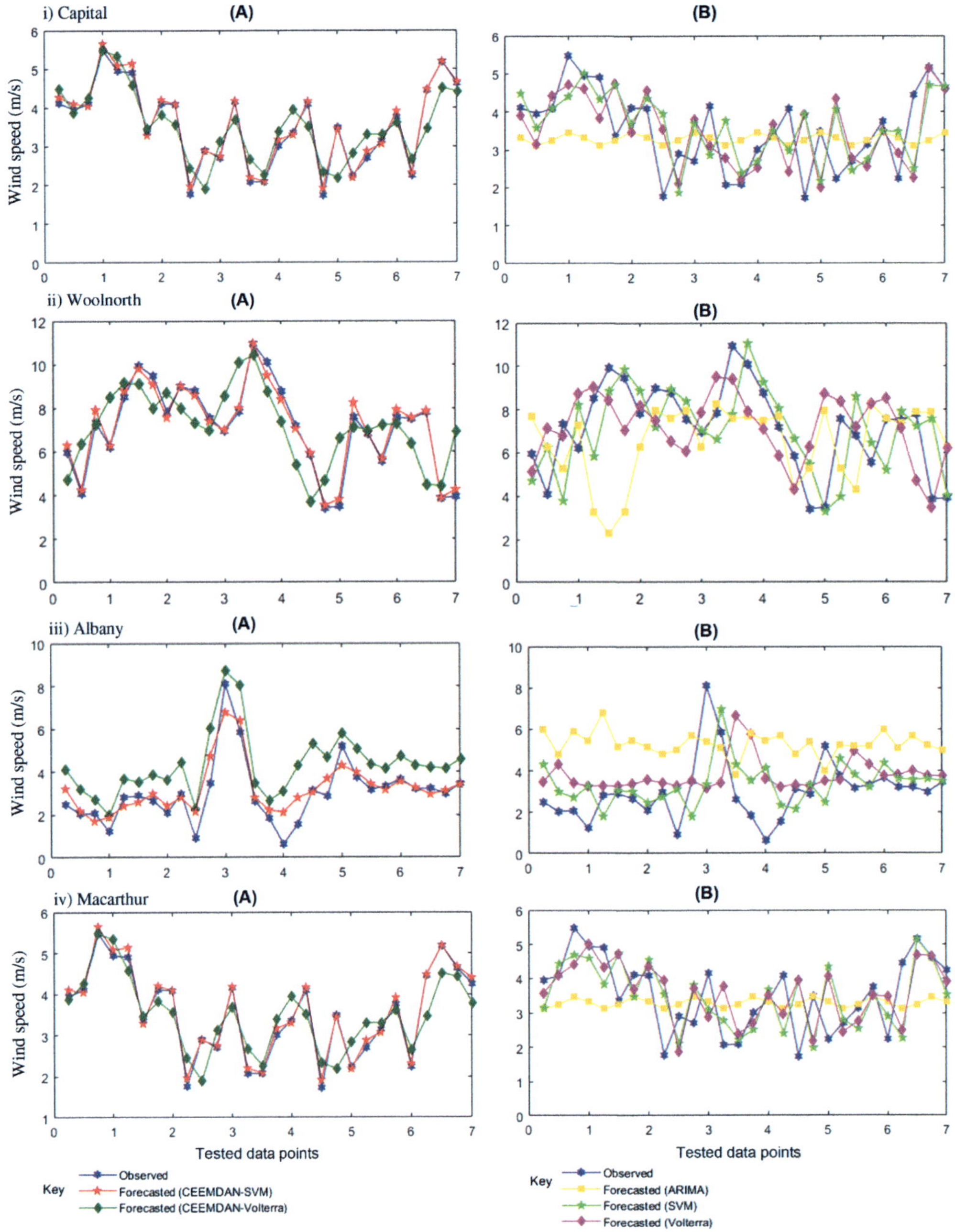

FIGURE 12.10 Observed and forecasted wind speed (m/s) during the test period (January 1–January 7, 2017), from (A) hybrid models (CEEMDAN–SVM, CEEMDAN–Volterra), (B) standalone models (SVM, Volterra, and ARIMA) for 6-hourly forecasting horizons for four study sites (Best study site: Macarthur, worst study site: Albany for all the models).

TABLE 12.9 The performance of CEEMDAN–SVM versus the comparative models in forecasting 6-hourly wind speed (m/s) in the testing period based on Willmott's Index (*WI*), Nash–Sutcliffe Efficiency (E_{NS}), and Legates–McCabe's index (*L*).

	CEEMDAN–SVM			SVM			CEEMDAN–Volterra			Volterra				ARIMA	
Study sites	*WI*	E_{NS}	*L*	*WI*	E_{NS}	*L*	*WI*	E_{NS}	*L*	*WI*	E_{NS}	*L*	*WI*	E_{NS}	*L*
Albany	0.939	0.874	**0.689**	0.713	0.514	0.353	0.929	0.862	0.565	0.659	0.447	0.302	0.361	0.159	− 0.08
Capital	0.953	0.884	**0.661**	0.772	0.538	0.328	0.941	0.836	0.599	0.747	0.489	0.280	0.083	0.002	− 0.01
Macarthur	0.987	0.969	0.869	0.768	0.565	0.379	0.954	0.895	**0.982**	0.738	0.512	0.332	0.208	0.029	0.021
Woolnorth	0.971	0.940	**0.768**	0.815	0.665	0.444	0.909	0.801	0.563	0.805	0.643	0.427	0.056	− 0.01	0.001

The models with the largest value of the Legates–McCabe's index at each site are in boldface in red.

four sites CEEMDAN–SVM performs better. The percentage increase in L at all four sites in comparison to the best standalone model, that is, SVM was 95.18% (site Albany), 101.50% (site Capital), 129.29% (site Macarthur), and 72.97% (site Woolnorth). Similarly, the hybrid CEEMDAN–SVM model did register higher values of WI and E_{NS} than all other models. The highest percentage increase of WI was 31.69% (site Albany) and the lowest increase of 19.14% (site Woolnorth) compared to the best standalone model (SVM). E_{NS} showed a similar increment, the maximum percentage increase was 71.51% (site Macarthur) and the lowest percentage increase of 41.35% (site Woolnorth).

With the value of L taking precedence based on the benefits discussed earlier, it can be noted that the hybrid CEEMDAN–SVM model has had a better performance in forecasting wind speed at three sites (sites Albany, Capital, and Macarthur) than the hybrid CEEMDAN–Volterra. Although at the site Macarthur, the CEEMDAN–Volterra had a slightly better performance, the superiority of the hybrid CEEMDAN–SVM in comparison with the hybrid CEEMDAN–Volterra was demonstrated by increments in L value, of about 21.95%, 10.35%, and 36.41% at sites Albany, Capital, and Woolnorth, respectively. The values of WI and E_{NS} demonstrated that CEEMDAN–SVM is outperforming CEEMDAN–Volterra for all sites. Hence, it is evident that the hybrid models based on CEEMDAN had enhanced the performance of the standalone SVM and Volterra models for short-term wind speed forecasting.

While the statistical metrics construed over the test period exemplify good forecasting capabilities of the CEEMDAN–SVM model compared to the comparative models, it is also of interest to this study to further explore the suitability of CEEMDAN–SVM in short-term wind speed forecasting using diagnostic plots. Fig. 12.11 shows scatterplots of

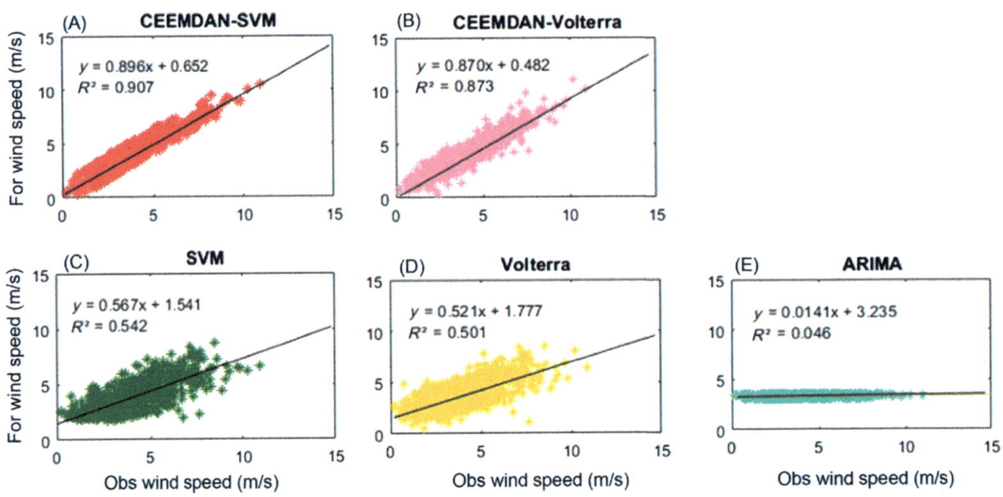

FIGURE 12.11 Scatterplots of the forecasted (for) 6-hourly wind speed versus the observed (*obs*) in m/s in the testing period (January 1–December 31, 2017) using (A) CEEMDAN–SVM, (B) CEEMDAN–Volterra, (C) SVM, (D) Volterra, and (E) ARIMA models for the best study site. Least square regression lines, its equation, and the square of the regression coefficient are shown in each panel (NB: Best models for hourly wind speed forecasting were for site Macarthur for all the models).

the observed and forecasted wind speed during the test period (January 1–December 31, 2017) from all five models at the best performing site, Macarthur. The linear fit equation and the R^2 are also provided. R^2 ranges from 0 to 1, where R^2 values close to 1 indicate a better fit (Hora and Campos, 2015).

The plots clearly show that standalone models SVM, Volterra, and the ARIMA models underperformed as the scatter points diverted from the $y = x$ linear form. On the other hand, the hybrid CEEMDAN–SVM had a very similar performance to the hybrid CEEMDAN–Volterra model with comparable R^2 values. In congruence with the outcomes of *WI*, E_{NS}, and *L*, the scatterplots of site Macarthur confirmed the superior performance of CEEMDAN–SVM. The R^2 of CEEMDAN–SVM is the maximum $R^2 = 0.91$ revealing that even at a worst case scenario, an overall 91% of the observed wind speed values could be forecasted using the CEEMDAN–SVM model while CEEMDAN–Volterra had a slightly lower value of $R^2 = 0.87$. ARIMA was the worst performing model with $R^2 = 0.05$. The gradient of the linear fit [Ideal value = +1] and the *y*-intercept [Ideal value = 0] (Wallach et al., 2019) for CEEMDAN–SVM model was close to idyllic magnitudes, further supporting the outcomes of the scatterplots and the predictor metrics (Tables 12.8 and 12.9).

Moreover, to further verify the CEEMDAN–SVM model, the frequency of the error encounter in various error brackets was also assessed as this information can assist in a better understanding of the model accuracy for practical applications. To analyze the forecasting error (*FE*) when different models are used to forecast wind speed, a histogram of the frequency of *FE* for the five models for the best performing site, Macarthur, has been prepared Fig. 12.12. Consistent with the earlier result, the most accurate forecasting is

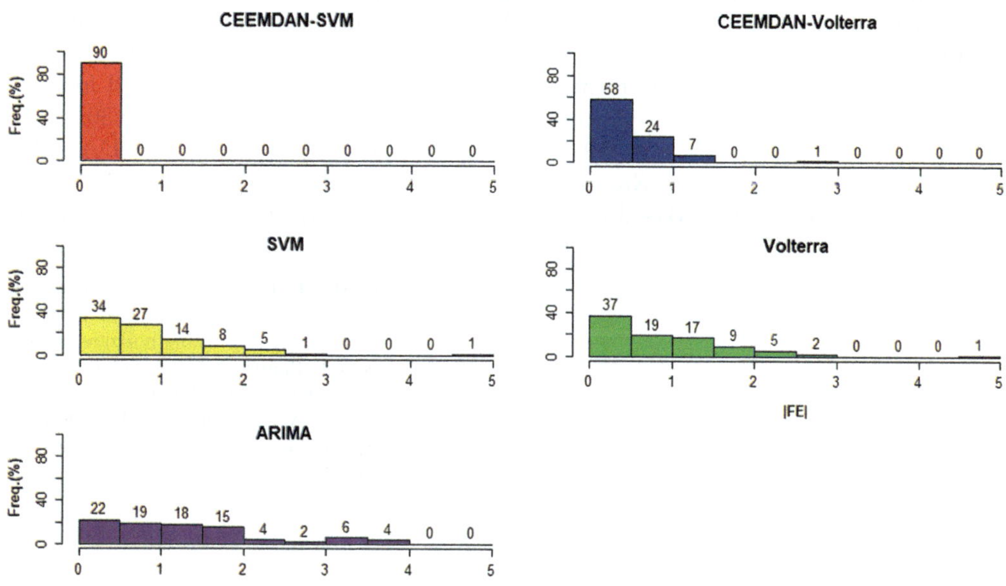

FIGURE 12.12 Histogram of the frequency distribution (Freq.) of the forecasted errors (FE) for 6-hourly forecasting horizons in the test period (January 2017) using (A) CEEMDAN–SVM model, (B) CEEMDAN–Volterra model, (C) SVM model, (D) Volterra model, and (E) ARIMA model.

obtained by the CEEMDAN–SVM model. Outstandingly, about 90% of forecasting errors are seen to lie within the lowest error margin for the CEEMDAN–SVM model. By contrast, the frequency of forecasting error for the CEEMDAN–Volterra model is only 58% within the lowest error margin, while standalone model SVM had 34%, Volterra 37%, and ARIMA 22%. Again, this confirms that the forecasting of short-term wind speed was more accurate with the CEEMDAN–SVM model compared to other models.

Furthermore, the model evaluations were carried out by the box plots that compare the distribution of model estimation error for short-term wind speed forecasting (Fig. 12.13) for the best performing site, Macarthur, where the lower quartile, upper quartile, median, and data outliers (i.e., the extreme error values) have been enumerated. It is evident that the CEEMDAN–SVM model outperforms all the other comparative models since the spread of the model errors is more concentrated toward the zero lines and the data outliers (i.e., overestimated errors) occupy a smaller magnitude, this is an evidence of a better ability of CEEMDAN–SVM models to generate smaller errors for the 6-hourly wind speed data. It is important to note that the standalone models generate significant outliers, indicating a larger inaccuracy for the forecasting of wind speed. Consequently, this finding accedes those of earlier results, providing confirmatory evidence that the present CEEMDAN–SVM model remains the ideal choice for short-term wind speed forecasting.

The forecasting ability of hybrid CEEMDAN–SVM is further explored based on relative measures. Since the sites are having different geographical, physical, and climatic characteristics (Fig. 12.4 and Table 12.2), *RRMSE* and *MAPE* were alternatively used. Table 12.10 compares model performances at these sites.

Evidently, the results in Table 12.10 showed that the lowest value of *RRMSE* and *MAPE* for all four sites was from the CEEMDAN–SVM model indicating it is the optimal model. A comparison among the standalone models (SVM, Volterra, and ARIMA) was also made, the standalone SVM model outperformed Volterra and ARIMA models for all four sites, having the best performance for Woolnorth ($RRMSE = 25.16\%$ and $MAPE = 30.76\%$). In comparison with the hybrid CEEMDAN–Volterra models, the hybrid CEEMDAN–SVM showed the largest reduction in *RRMSE* value for Macarthur (83.41%) while Albany had the lowest reduction (4.64%). Based on the least relative error, the forecasted wind speed

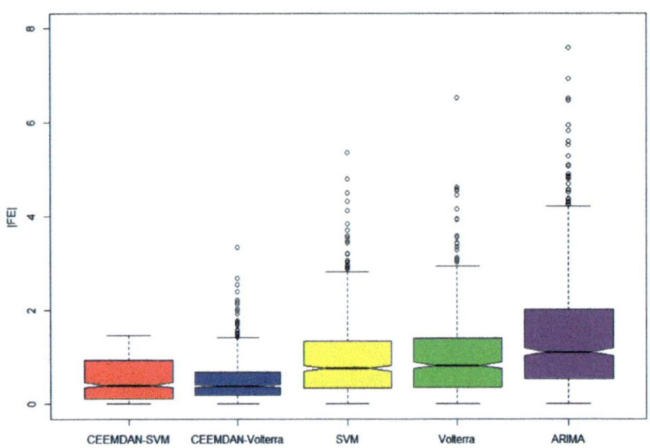

FIGURE 12.13 A boxplot of the absolute value of |FE| forecasted error (m/s) for 6-hourly forecasting horizons in the test period (January 1–December 31, 2017) encountered by the CEEMDAN–SVM, CEEMDAN–Volterra, SVM, Volterra, and ARIMA models for the best performing site Macarthur.

TABLE 12.10 A comparison at the different sites' performances using the relative error measures, relative root mean square error (*RRMSE*), and mean absolute percentage error (*MAPE*) for the forecasting of 6-hourly wind speed (m/s).

		Wind speed (m/s)			
Model	**Performance metrics (%)**	**Albany**	**Capital**	**Macarthur**	**Woolnorth**
CEEMDAN–SVM	*RRMSE*	**14.116**	**19.297**	**8.645**	**10.779**
	MAPE	**13.517**	**20.648**	**7.039**	**11.775**
SVM	*RRMSE*	27.721	38.537	32.320	25.515
	MAPE	31.442	41.547	38.974	30.764
CEEMDAN–Volterra	*RRMSE*	14.771	22.962	15.856	19.669
	MAPE	14.893	24.232	16.734	22.087
Volterra	*RRMSE*	29.537	40.535	33.960	26.326
	MAPE	33.803	49.335	41.479	30.509
ARIMA	*RRMSE*	42.799	56.668	49.684	44.108
	MAPE	60.324	78.092	59.895	65.597

The optimal model with the lowest relative (%) error is shown in boldface in red.

for Macarthur was most accurate ($RRMSE = 8.65\%$ and $MAPE = 7.04\%$), achieved via CEEMDAN–SVM model, whereas the standalone SVM model yielded an RRMSE error of 32.32% and MAPE = 38.97% for the same site confirming unarguably that hybridization is improving the performance of short-term wind speed forecasting. In accordance with the outcomes from absolute measures, the relative measures concur on the suitability of the hybrid CEEMDAN–SVM model for short-term wind speed forecasting. This finding accedes with the outcomes of other CEEMDAN-based studies (Zhang et al., 2017; Yu et al., 2017; Wang et al., 2016; Santhosh et al., 2018; Prasad et al., 2018).

12.4.2 Medium-term wind speed forecasting

Fig. 12.14 shows the *PACF* analysis for the daily wind speed time series for all the sites. For example, lag 1, lag 2, and lag 3 were used as input variables to build the standalone models for Albany. The model with the lowest value of *MAE* and *RMSE* and the maximum value of r was endorsed as the best model, as highlighted in green in Table 12.11, in the training period.

The CEEMDAN–SVM model attained the largest magnitudes of r for all sites and lowest *RMSE* for all four study sites in the training period. *MAE* of the CEEMDAN–SVM model was lowest for site Albany ($MAE = 0.543$) and Capital ($MAE = 0.401$), while CEEMDAN–Volterra achieved the lowest value of MSE for site Macarthur ($MAE = 0.231$) and Woolnorth ($MAE = 1.117$). Overall, the training performance of the CEEMDAN–SVM

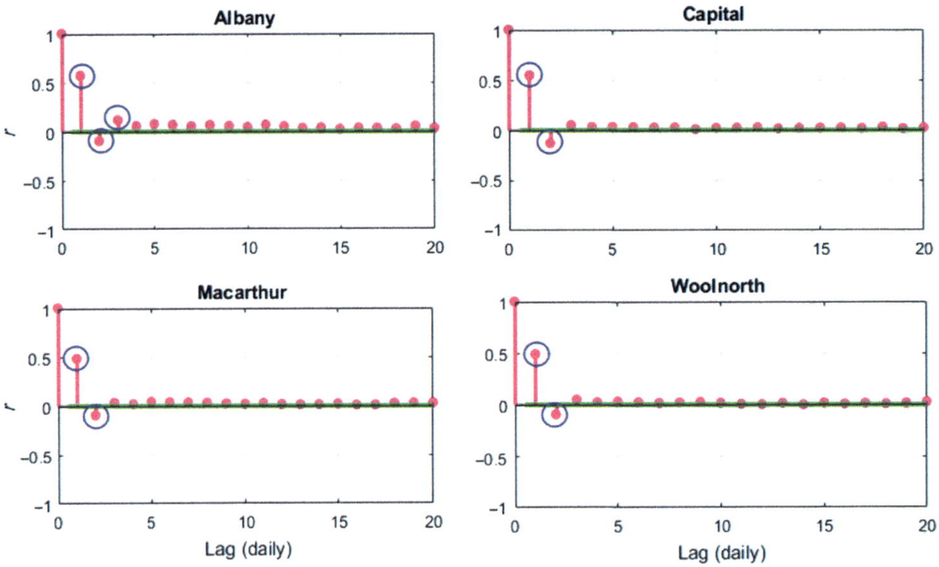

FIGURE 12.14 The correlation coefficient (r) based on the partial autocorrelation function for the daily wind speed with its predictor (i.e., wind speed) used for developing the support vector machine model and Volterra model for all sites.

model was considerably high for all the study sites. It is thus envisaged that the CEEMDAN–SVM model testing performance, as seen later, will be relatively accurate for forecasting the daily wind speed at these sites.

An evaluation of the preciseness of the hybrid CEEMDAN–SVM was applied to the study sites with the predictor data as wind speed is undertaken with respect to the standalone SVM, Volterra and ARIMA models, and CEEMDAN–Volterra model. With an aim to improve the accuracy of the objective model, the CEEMDAN–SVM technique was designed to derive different components from the given wind data series and make constructive contributions to the final forecasting (Jun et al., 2018). In the model development phase, the results showed an improved training performance, as stated in Table 12.11 where the *RMSE* and *MAE* of the CEEMDAN-based models were lower than the standalone models in the training period.

Table 12.12 compares the testing performance of the hybrid CEEMDAN–SVM versus the comparative models for four study sites. To evaluate the extent of agreement between observed and predicted wind speed, a summary is presented of the testing performance with r, *RMSE*, *MAE* coefficients. Consistent with relatively superior training performance attained by the CEEMDAN–SVM hybrid model for study sites in consideration (Table 12.11), the testing performance was outstanding compared to the SVM, Volterra, ARIMA, and CEEMDAN–Volterra models. CEEMDAN–SVM models computed the maximum r values for three sites: Capital (0.90), Macarthur site recorded the maximum $r = 0.89$, while Woolnorth attained the $r = 0.85$. For Albany, $r = 0.82$ was recorded by both the hybrid models CEEMDAN–SVM and CEEMDAN–Volterra.

TABLE 12.11 The performances of CEEMDAN–SVM versus the comparative models in the model development, that is, training phase based on the Pearson's correlation coefficient (*r*), root mean square error (*RMSE*), and the mean absolute error (*MAE*).

	CEEMDAN–SVM			SVM			CEEMDAN–Volterra			Volterra			ARIMA		
Study sites	*r*	*RMSE*	*MAE*	*r*	*RMSE*	*MAE*	*r*	*RMSE*	*MAE*	*r*	*RMSE*	*MAE*	*r*	*RMSE*	*MAE*
Daily forecasting horizon															
Albany	**0.841**	**0.671**	**0.543**	0.646	1.230	0.929	0.831	0.872	0.631	0.596	1.113	0.912	0.401	1.901	1.421
Capital	**0.923**	**0.563**	**0.401**	0.556	0.109	0.812	0.912	0.563	0.431	0.544	1.153	0.879	0.436	1.310	0.998
Macarthur	**0.891**	**0.614**	**0.469**	0.528	1.195	0.898	0.881	0.675	0.231	0.478	1.128	0.957	0.389	1.445	0.076
Woolnorth	**0.850**	**1.498**	**1.132**	0.504	2.429	1.961	0.839	1.561	1.117	0.450	2.553	2.069	0.452	2.761	2.296

The models with the largest value of the *r* and the smallest value of *MAE* and *RMSE* at each site are in boldface in green.

TABLE 12.12 The performance of CEEMDAN–SVM versus the comparative models in forecasting daily wind speed (m/s) in the testing period, based on Pearson's correlation coefficient (*r*), root mean square error (*RMSE*), and mean absolute error (*MAE*).

	CEEMDAN–SVM			SVM			CEEMDAN–Volterra			Volterra			ARIMA		
Study sites	*r*	*RMSE*	*MAE*	*r*	*RMSE*	*MAE*	*r*	*RMSE*	*MAE*	*r*	*RMSE*	*MAE*	*r*	*RMSE*	*MAE*
Albany	0.823	**0.719**	0.567	0.603	1.252	0.967	0.823	0.888	0.697	0.586	1.265	0.990	0.028	1.793	1.484
Capital	0.902	**0.573**	0.422	0.557	1.108	0.822	0.899	0.583	0.456	0.549	1.108	0.843	0.027	1.328	0.999
Macarthur	0.889	**0.645**	0.479	0.496	1.231	0.941	0.874	0.685	0.536	0.482	1.130	0.969	0.035	1.447	1.088
Woolnorth	0.847	**1.519**	1.174	0.507	2.467	1.981	0.831	1.592	1.277	0.441	2.604	2.609	0.038	2.856	2.297

The optimal models yielding the lowest *RMSE* at each site are shown in boldface in red.

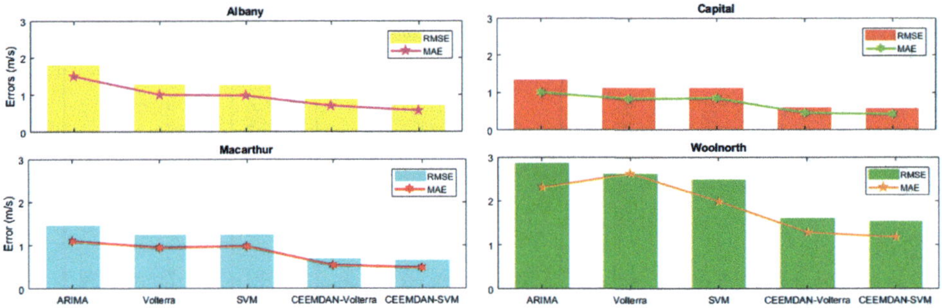

FIGURE 12.15 The root mean square error *RMSE* and mean absolute error *MAE* for daily forecasting horizon over the test period encountered by the CEEMDAN–SVM, CEEMDAN–Volterra, SVM, Volterra, and ARIMA models for Capital, Macarthur, Woolnorth, and Albany study sites.

The standalone models had *RMSE* in the range of 1.10–2.86 but this value was reduced to occupy the range of 0.57–1.52 for the case of the hybridized models (CEEMDAN–SVM) as depicted in Fig. 12.15. Similarly, the *MAE* was lowest for the CEEMDAN–SVM model for all four sites: Albany (0.57), Capital (0.42), Macarthur (0.48) and Woolnorth (1.17). In fact, a significant decrease in *MAE* was found for all study sites when CEEMDAN–SVM was compared with the standalone SVM model: Albany (≈41%), Capital (≈48.60%), Macarthur (≈49.09%) and Woolnorth (≈40.73%). Correspondingly, the magnitude of testing *RMSE* was reduced by almost 42.57% (Albany), 48.29% (Capital), 47.60% (Macarthur), and 38.43% (Woolnorth). Closer inspection of Table 12.12 shows Capital is the best performing site as it attained the lowest *RMSE* and *MAE* and the largest value of *r* while Woolnorth is the worst performing site with the largest *RMSE* and *MAE* errors across all models.

Table 12.13 shows the accuracy of CEEMDAN–SVM versus comparative models in the testing period in terms of forecasting accuracy. Again, the CEEMDAN–SVM model was superior for all study sites in terms of *WI* and *L*, while in terms of E_{NS}, CEEMDAN–SVM performed well for three sites, Capital, Macarthur, and Woolnorth, and CEEMDAN–Volterra performed well achieving better results for Albany.

According to the results (Table 12.13), the CEEMDAN–SVM model returns the highest values of *WI* for all sites as Albany (0.82), Capital (0.92), Macarthur (0.91), and Woolnorth (0.86). CEEMDAN–Volterra also performed equally well giving good results but not superior to CEEMDAN–SVM. The maximum value of $E_{NS=}0.81$ was attained at the site Capital by CEEMDAN–SVM model while the lowest value of $E_{NS=}0.064$ was attained at the site Macarthur by ARIMA model. The percentage increased in *L* at all four sites in comparison to the best standalone model, that is, SVM was 84.74% (Albany), 67.41% (Capital), 73.67% (Macarthur) and 71.31% (Woolnorth). It must be noted that at all sites, the CEEMDAN–SVM model registered very high-performance indicator values in comparison with other comparative models. Hence, the hybrid CEEMDAN–SVM model has enhanced the performance of the daily wind speed forecasting.

Moreover, the empirical cumulative distribution function (ECDF) was plotted at each study site for different forecasting abilities in Fig. 12.16. This figure illustrates the percentage of the absolute value of *FE* encountered through the ECDF for optimal models at daily forecasting horizons. The figure clearly confirms that the CEEMDAN–SVM model was better

TABLE 12.13 The performance of CEEMDAN–SVM versus the comparative models in forecasting daily wind speed (m/s) in the testing period based on Willmott's Index (*WI*), Nash–Sutcliffe Efficiency (E_{NS}), and Legates–McCabe's index (L).

	CEEMDAN–SVM			**SVM**			**CEEMDAN–Volterra**			**Volterra**			**ARIMA**		
Study sites	*WI*	E_{NS}	L	*WI*	E_{NS}	L	*WI*	E_{NS}	L	*WI*	E_{NS}	L	*WI*	E_{NS}	L
Albany	0.824	0.673	**0.460**	0.567	0.357	0.249	0.818	0.676	0.443	0.543	0.342	0.232	0.415	0.321	0.149
Capital	0.919	0.813	**0.586**	0.495	0.298	0.191	0.914	0.806	0.552	0.560	0.299	0.171	0.122	− 0.077	0.017
Macarthur	0.905	0.789	**0.566**	0.369	0.231	0.149	0.889	0.762	0.515	0.435	0.231	0.122	0.273	0.064	0.015
Woolnorth	0.857	0.716	**0.489**	0.423	0.253	0.140	0.833	0.688	0.445	0.376	0.167	0.088	0.067	− 0.001	0.002

The models with the largest value of the Legates–McCabe's index at each site are in boldface.

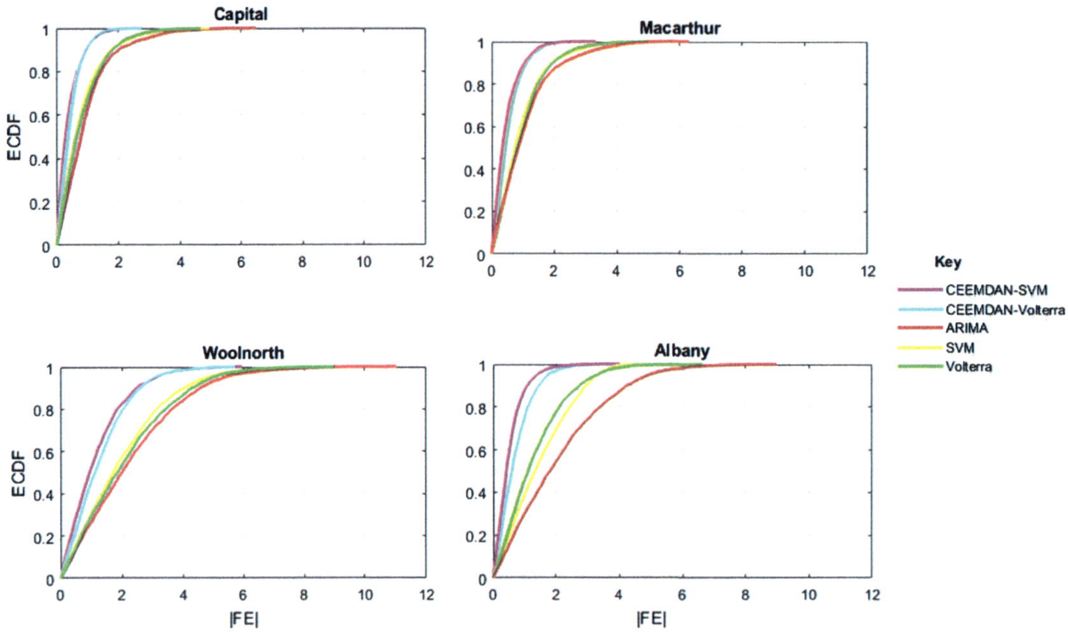

FIGURE 12.16 Empirical cumulative distribution function (ECDF) of the forecasted error |FE| (m/s) for daily forecasting horizon generated by the proposed hybrid CEEMDAN–SVM model versus its counterpart models in the testing period (January 1, 2010–December 31, 2017) for Capital, Macarthur, Woolnorth, and Albany study sites.

than CEEMDAN–Volterra and other standalones for all study sites based on the percentage of errors in the error bracket (0 to $\pm$ 12 m/s). Thus, the performance of CEEMDAN–SVM is more accurate for daily wind speed forecasting for the selected study sites.

While the ECDF plot exemplifies good forecasting skills of the CEEMDAN–SVM compared to the classical model, it is also of interest to this study that the time series of observed and forecasted values of wind speed are checked with the model *FE* to analyze the daily forecasting of wind speed. Fig. 12.17 plots the time series of the comparative and CEEMDAN–SVM model observed and forecasted values of wind speed and *FE*, |*FE*| = |*Observed-forecasted*| deduced for the period of January 2017 in the testing period for the best performing site Capital. It is clear from Fig. 12.17 that the standalone model and CEEMDAN–Volterra exhibit higher amplitude in the fluctuation of the *FE* compared to the CEEMDAN–SVM model whose *FE* is in the range of [0:1] confirming that CEEMDAN–SVM based models attain better accuracy for this site. In addition, the agreement between observed data and forecasted wind speed values by the optimal hybrid CEEMDAN–SVM model was very good, while the standalone models (SVM, Volterra, and ARIMA) divert from the observed values. Again, this confirms that the CEEMDAN–SVM model is outperforming and more accurate compared to the other comparative models used in this study.

For more perspicacity, the Taylor diagrams were used (Fig. 12.18), providing a more concrete and conclusive argument about the statistical summary of how well the forecasted daily wind speed matched with the observed in terms of their correlation and the

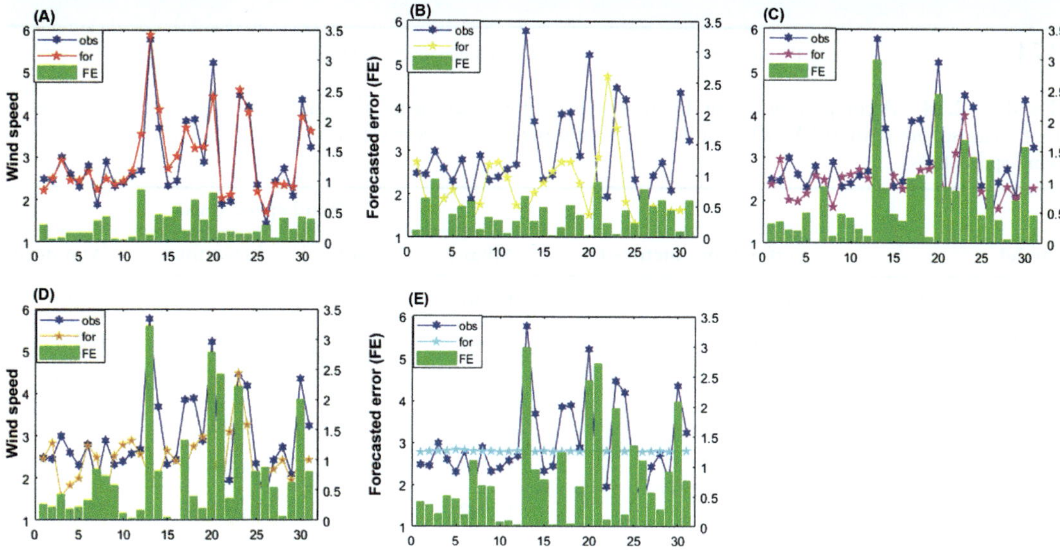

FIGURE 12.17 The observed (obs) and the forecasted (for) daily wind speed (m/s) with the corresponding forecasted error |FE| over the test period (January 2017) encountered by the (A) CEEMDAN–SVM, (B) CEEMDAN–Volterra, (C) SVM, (D) Volterra, and (E) ARIMA models for the best performing site, Capital.

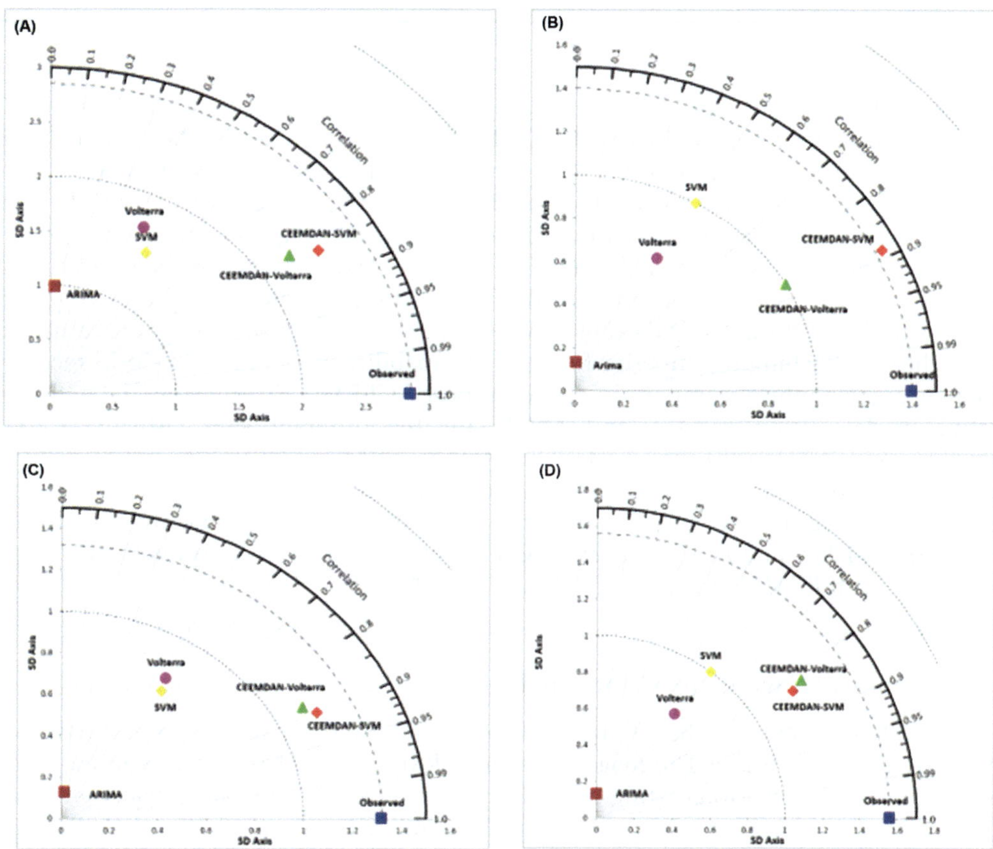

FIGURE 12.18 Taylor diagram showing the correlation coefficient between observed and forecasted and standard deviation for each model utilizing (A) Woolnorth daily forecasting horizons, (B) Capital daily forecasting horizons, (C) Macarthur Daily forecasting horizons, and (D) Albany daily forecasting horizons.

TABLE 12.14 A comparison at the different sites' performances using the relative error measures, relative root mean square error (*RRMSE*), and mean absolute percentage error (*MAPE*) for the forecasting of daily wind speed (m/s).

		Wind speed (m/s)			
Model	**Performance metrics (%)**	**Albany**	**Capital**	**Macarthur**	**Woolnorth**
CEEMDAN–SVM	*RRMSE*	**19.704**	**14.247**	**18.012**	**20.508**
	MAPE	**16.446**	**12.077**	**15.267**	**19.151**
SVM	*RRMSE*	25.023	38.178	34.390	33.331
	MAPE	21.971	31.991	28.481	33.275
CEEMDAN–Volterra	*RRMSE*	17.753	20.073	19.123	21.482
	MAPE	15.942	17.916	16.965	21.101
Volterra	*RRMSE*	25.291	38.149	34.361	35.168
	MAPE	22.845	34.510	30.780	35.095
ARIMA	*RRMSE*	35.853	45.745	40.415	38.572
	MAPE	39.549	40.295	32.294	39.883

The optimal model with lowest relative (%) error is shown in boldface in red.

SD on a single figure (Taylor, 2001). For Capital, the correlation of the CEEMDAN–SVM model was about 0.90 followed by CEEMDAN–Volterra≈0.89, SVM≈0.56, Volterra≈0.55, and ARIMA≈0.03. The CEEMDAN–SVM model was closer to the observed wind speed as its correlation is about 0.90 to other models. Similarly, the CEEMDAN–SVM again appeared to be the best model for Albany, Macarthur, and Woolnorth sites because its correlation lies within the close neighborhood of the observed wind speed data. Overall, the Taylor plot vividly asserted that in both the best performing site Capital and worst performing site Albany, the CEEMDAN–SVM was better than all the comparative models, thus further establishing the superior forecasting capability of CEEMDAN–SVM model.

Finally, a geographical comparison of the proposed CEEMDAN–SVM model with comparative models using *RRMSE* and *MAPE* for the different sites (Albany, Capital, Macarthur, and Woolnorth) was made. CEEMDAN–SVM model was found to yield the lowest relative percentage errors (*RRMSE* and *MAPE*) across all study site as shown in Table 12.14 and Fig. 12.19 confirming that it is the optimal model. Capital attained the lowest $RRMSE \approx 14.28\%$ and $MAPE \approx 12.08\%$. On the other hand, Woolnorth is the worst performing site with $RRMSE \approx 20.57\%$ and $MAPE \approx 12.08\%$.

12.4.3 Long-term wind speed forecasting

The results generated by the CEEMDAN–SVM and the comparative models have been summarized in Table 12.15. The magnitudes of *r*, *RMSE*, and *MAE* attained in training of the monthly wind speed forecasting by CEEMDAN–SVM model at site Albany were seen to be $r = 0.931$, $RMSE = 0.291$, $MAE = 0.223$; for Capital $r = 0.745$, $RMSE = 0.341$,

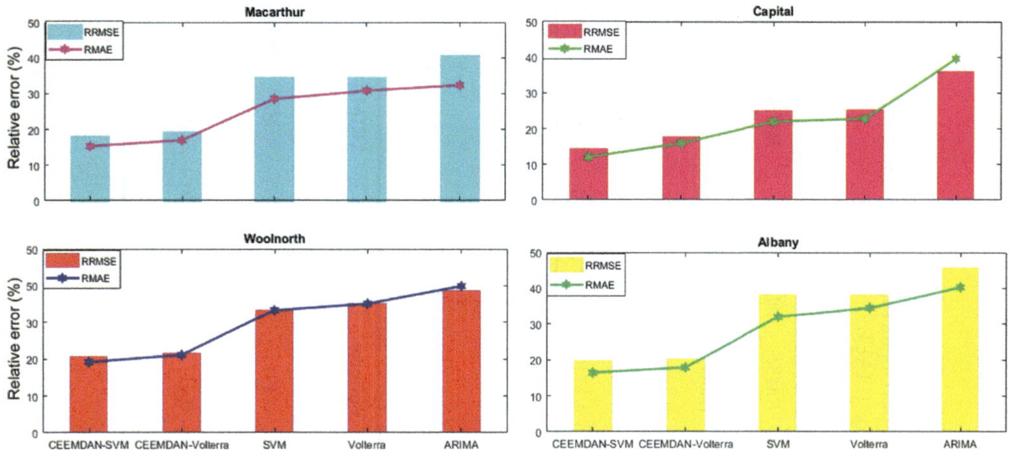

FIGURE 12.19 The relative root mean square error (RRMSE) and relative mean square error (RMSE) for daily forecasting horizon over the test period encountered by the CEEMDAN–SVM, CEEMDAN–Volterra, SVM, Volterra and ARIMA models for Capital, Macarthur, Woolnorth, and Albany study sites.

$MAE = 0.253$; for Macarthur $r = 0.732$, $RMSE = 0.371$, $MAE = 0.286$; and for Woolnorth $r = 0.788$, $RMSE = 0.631$, $MAE = 0.467$. Overall, the training performance of the CEEMDAN–SVM model was noticeably good for all four study sites. It is thus envisioned that the proposed hybrid CEEMDAN–SVM model is a promising model for wind speed forecasting.

The results of CEEMDAN–SVM and the comparative models for forecasting monthly wind speed for the testing period for all study sites shown in Table 12.16. CEEMDAN–SVM model registered the maximum value of r for the study site Albany and Capital, while Macarthur and Woolnorth had the highest value reported by CEEMDAN–Volterra. Albany recorded the maximum value of $r = 0.922$ compared to the other three sites. Accordingly, the lowest *RMSE* values were attained by the hybrid CEEMDAN–SVM model at two sites (Capital and Woolnorth) and the other two sites (Albany and Macarthur) showed the lowest errors from the hybrid CEEMDAN–Volterra model. Interestingly, the CEEMDAN–SVM model recorded the lowest *MAE* for all study sites as Albany (0.248), Capital (0.264), Macarthur (0.268) and Woolnorth (0.481). It is evident so far that the hybridized ensemble SVM and Volterra models indeed are better in comparison to their standalone counterparts. However, due to the rather unclear outcomes from these metrics, the decision to determine the optimal hybrid model can be obscured. The accuracy of the other data-intelligent model CEEMDAN–Volterra was similar and confirmed that the hybrid models had a better potential to generate accurate wind speed forecasts.

Table 12.17 shows that the hybrid CEEMDAN models demonstrated a dramatic improvement in comparison to the standalone models for wind speed forecasting. More closely with the value of L taking the superiority based on benefits discussed earlier, it can be noted that the hybrid CEEMDAN–SVM model had a better performance in forecasting monthly wind speed at all four sites, achieving the maximum value of L in Albany ($L = 0.62$), Macarthur ($L = 0.37$), Woolnorth ($L = 0.41$), and the lowest value was attained

TABLE 12.15 The performances of CEEMDAN–SVM versus the comparative models in the model development, that is, training phase based on the Pearson's correlation coefficient (r), root mean square error (*RMSE*), and the mean absolute error (*MAE*).

	CEEMDAN–SVM			SVM			CEEMDAN–Volterra			Volterra			ARIMA		
Study sites	*r*	*RMSE*	*MAE*	*r*	*RMSE*	*MAE*	*r*	*RMSE*	*MAE*	*r*	*RMSE*	*MAE*	*r*	*RMSE*	*MAE*
Monthly forecasting horizon															
Albany	0.931	0.291	0.223	0.871	0.384	0.282	0.839	0.479	0.363	0.801	0.467	0.372	0.745	0.532	0.461
Capital	0.745	0.341	0.253	0.691	0.376	0.264	0.588	0.459	0.723	0.579	0.414	0.317	0.497	0.465	0.393
Macarthur	0.732	0.371	0.286	0.614	1.088	0.801	0.751	0.334	0.245	0.452	1.201	0.942	0.456	0.491	0.368
Woolnorth	0.788	0.631	0.467	0.332	0.892	0.662	0.817	0.598	0.461	0.214	1.171	0.917	0.335	0.976	0.743

TABLE 12.16 The performance of CEEMDAN–SVM versus the comparative models in forecasting monthly wind speed (m/s) in the testing period, based on Pearson's correlation coefficient (r); root mean square error (*RMSE*), and mean absolute error (*MAE*).

	CEEMDAN–SVM			SVM			CEEMDAN–Volterra			Volterra			ARIMA		
Study sites	*r*	*RMSE*	*MAE*	*r*	*RMSE*	*MAE*	*r*	*RMSE*	*MAE*	*r*	*RMSE*	*MAE*	*r*	*RMSE*	*MAE*
Albany	0.922	0.311	0.249	0.824	0.448	0.357	0.919	**0.307**	0.248	0.802	0.491	0.387	0.727	0.582	0.474
Capital	0.706	**0.360**	0.264	0.506	0.441	0.329	0.569	0.462	0.287	0.483	0.454	0.347	0.476	0.475	0.394
Macarthur	0.709	0.381	0.268	0.417	1.278	0.982	0.740	**0.364**	0.285	0.338	0.517	0.394	0.426	0.495	0.378
Woolnorth	0.772	**0.601**	0.481	0.203	1.021	0.766	0.803	0.612	0.496	0.224	1.267	1.015	0.315	0.985	0.757

The models yielding the lowest *RMSE* at each site are shown in boldface in red.

TABLE 12.17 The performance of CEEMDAN–SVM versus the comparative models in forecasting monthly wind speed (m/s) in the testing period based on Willmott's Index (*WI*), Nash–Sutcliffe efficiency (E_{NS}), and Legates–McCabe's index (*L*).

	CEEMDAN–SVM			SVM			CEEMDAN–Volterra			Volterra			ARIMA		
Study sites	*WI*	E_{NS}	*L*	*WI*	E_{NS}	*L*	*WI*	E_{NS}	*L*	*WI*	E_{NS}	*L*	*WI*	E_{NS}	*L*
Albany	0.917	0.835	**0.617**	0.839	0.657	0.450	0.910	0.837	0.615	0.817	0.589	0.405	0.692	0.423	0.272
Capital	0.732	0.493	**0.327**	0.529	0.248	0.163	0.622	0.166	0.278	0.547	0.201	0.117	0.542	0.122	−0.001
Macarthur	0.742	0.501	**0.369**	0.547	0.154	0.112	0.779	0.544	0.331	0.335	0.092	0.077	0.421	0.163	0.112
Woolnorth	0.771	0.592	**0.408**	0.331	0.011	0.149	0.806	0.638	0.391	0.208	−0.549	−0.259	0.381	0.059	0.063

The models with the largest value of the Legates–McCabe's index at each site are in boldface.

by Capital ($L = 0.33$). Comparing the hybrid CEEMDAN–SVM model with the standalone SVM noted the significant increment in the value of L by about 37.11% at site Albany, a substantial increment by about 100.61% at site Capital, a large increment by about 229.48% at site Macarthur, and 173.83% at site Woolnorth.

Similarly, the value of WI increased for all sites with the CEEMDAN–SVM model in comparison with the SVM model with Albany having $WI = 0.92$ (i.e., incremented by 9.29%), Capital attained $WI = 0.732$ (increased by 38.37%), Macarthur achieved $WI = 0.74$ (growth by 35.65%), and Woolnorth noted the maximum rise with $WI = 0.77$ (significant growth by 132.93%). The performance of CEEMDAN–Volterra models was superior in terms of E_{NS} compared with CEEMDAN–SVM as it achieved higher values of E_{NS} for three sites, Albany (0.837), Macarthur (0.544), and Woolnorth (0.638), while site Capital had $E_{NS} = 0.493$ achieved by the CEEMDAN–SVM model which had a significant growth of 196.98% in comparison with the $E_{NS} = 0.166$ achieved by the CEEMDAN–Volterra model.

Although at three sites (Albany, Macarthur, and Woolnorth), the CEEMDAN–Volterra had a slightly better performance in terms of performance assessment of Nash–Sutcliffe efficiency (E_{NS}), the superiority of the hybrid CEEMDAN–SVM model in comparison with the hybrid CEEMDAN–Volterra model was demonstrated by significant increments in L value and WI, respectively. Hence, it is evident that the hybridization based on CEEMDAN did enhance the performance of the standalone SVM models for monthly wind speed forecasting.

To further verify the CEEMDAN–SVM model, the forecasting error $|FE|$ was assessed for each model for the best study site Albany using a diagnostic violin plot (Fig. 12.20), as

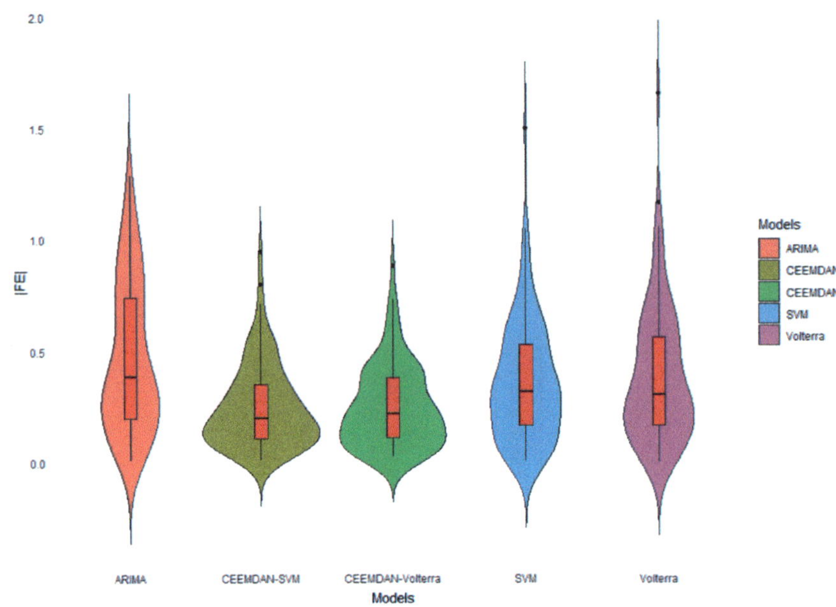

FIGURE 12.20 The absolute value of $|FE|$ forecasted error (m/s) for monthly forecasting horizons in the testing period using the CEEMDAN–SVM model, CEEMDAN–Volterra model, SVM model, Volterra model, and ARIMA model Albany study sites.

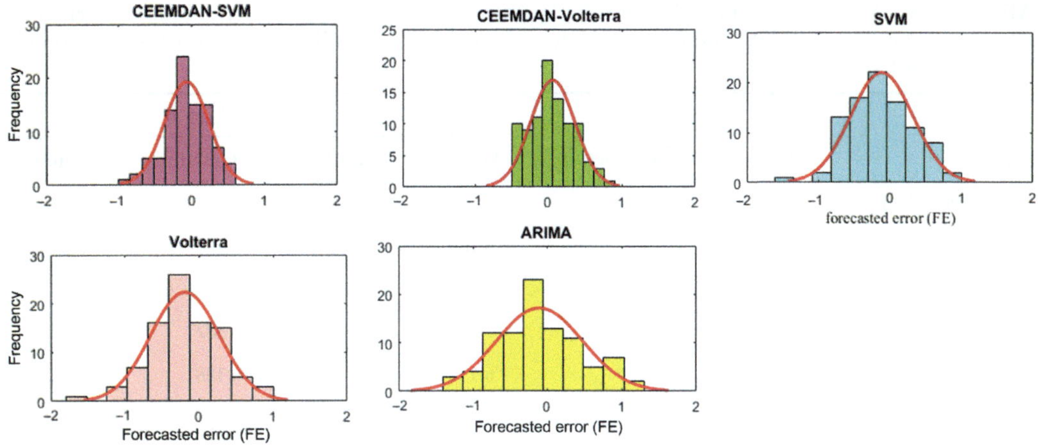

FIGURE 12.21 Histogram of the frequency distribution of the forecasting errors (FE) for monthly forecasting horizons in the testing period using the CEEMDAN–SVM model, CEEMDAN–Volterra model, SVM model, Volterra model and ARIMA model for the best performing site Albany.

this information can assist in a better understanding of the model accuracy for practical applications. In line with the earlier results, the most accurate forecasting is obtained by the CEEMDAN–SVM model. It is clear from the shape of the distribution (narrow in the middle) illustrated by standalone models (SVM, Volterra, and ARIMA) that the weights of forecasting errors are widely spread out, ranging from 0 to 2 m/s, while the hybrid models CEEMDAN–SVM and CEEMDAN–Volterra had a shape (extremely skinny on each end and wide in the middle) indicating that the forecasting errors are highly concentrated around the median, and thus had smaller error brackets of approximately 0–1 m/s. The median error for hybrid CEEMDAN models was less than 0.25 m/s as indicated by the black marker in the boxplot element. The 95% confidence interval of |FE| by CEEMDAN models was approximately 0–0.75 m/s, which is far better than the standalone models. Again, this confirms that the monthly wind speed forecasting by hybrid CEEMDAN models was more accurate compared with the standalone models.

Moreover, Fig. 12.21 shows a histogram of *FE* where the *FE* is accumulated in error brackets of step-size 0.2. *FE* is the difference between forecasted and observed wind speed during the test period and is computed as *FE* = *forecasted–observed*, Idyllic value = 0. Hence, a better model is bound to have higher occurrences of *FE* closer to zero. For succinctness, the diagram for the best performing site is shown in Fig. 12.21. The standalone models (SVM, Volterra, and ARIMA) showed more spread of forecasting errors in between $-2.0 \leq FE \leq 1.5$. Conversely, the hybrid models (CEEMDAN–SVM and CEEMDAN–Volterra) registered small spreads in forecasting error between $-1.0 \leq FE \leq 1.0$. Accordingly, the forecasting error histogram confirmed the suitability of hybrid CEEMDAN models as lower forecasting errors and improved accuracies are apparent.

Furthermore, Table 12.18 shows a geographical comparison of the proposed CEEMDAN–SVM model with its comparative models SVM, Volterra, ARIMA, and CEEMDAN–Volterra for all four study sites (Albany, Capital, Macarthur, and Woolnorth).

TABLE 12.18 A comparison at the different sites' performances using the relative error measures, relative root mean square error (RRMSE) and mean absolute percentage error (MAPE), for the forecasting of monthly wind speed (m/s).

		Wind speed (m/s)			
Model	**Performance metrics (%)**	**Albany**	**Capital**	**Macarthur**	**Woolnorth**
CEEMDAN–SVM	*RRMSE*	6.211	**12.403**	10.447	**8.176**
	MAPE	5.100	**9.144**	**7.434**	**6.310**
SVM	*RRMSE*	8.962	15.071	15.436	13.825
	MAPE	7.394	11.358	19.587	9.841
CEEMDAN–Volterra	*RRMSE*	**6.152**	15.834	**10.186**	8.267
	MAPE	**4.965**	9.793	7.965	6.511
Volterra	*RRMSE*	9.902	15.543	14.363	17.110
	MAPE	8.165	12.161	10.841	13.350
ARIMA	*RRMSE*	11.629	16.320	13.865	13.296
	MAPE	9.838	14.120	10.352	9.803

The optimal model with lowest relative (%) error is shown in boldface in red.

Study site Albany appears to be the most accurate station in forecasting monthly wind speed with the CEEMDAN–Volterra model attaining $RRMSE \approx 6.15\%$ and $MAPE \approx 4.97\%$. In terms of individual model performance, CEEMDAN–SVM model outperformed other models at sites Capital ($RRMSE \approx 12.40\%$, $MAPE \approx 9.14\%$) and Woolnorth ($RRMSE \approx 8.18\%$, $MAPE \approx 6.31\%$), while CEEMDAN–Volterra performed well for Albany ($RRMSE \approx 6.15\%$, $MAPE \approx 4.96\%$) and achieved the smallest value of $RRMSE \approx 10.19\%$ at Macarthur.

Convincing evidence of the supremacy of the hybrid CEEMDAN–SVM model, in terms of monthly wind speed forecasting skills, has been noted so far. However, for better integration of wind power into the electric grid, a seasonal analysis of wind speed forecasting accuracy is imperative (Xiao et al., 2018). Fig. 12.22 shows the magnitude of the average values of absolute forecasting errors $|FE|$ accumulated over the monthly timescale for the proposed CEEMDAN–SVM and the comparative models (CEEMDAN–Volterra, SVM, Volterra, and ARIMA) in the testing period for the year 2017. For succinctness, the worst performing site Capital was selected, the polar plots revealed that the CEEMDAN–SVM model apparently has the lowest *FE* for most of the months except for February and November whereby CEEMDAN–Volterra achieved minimum errors for these months. The hybrid CEEMDAN–SVM proved to be the best with FE less than 2.5 m/s for 10 months. The CEEMDAN attained minimum forecasting errors in January (FE = 0.15 m/s), which was consistent with other models at site Capital for this month, while the maximum magnitude was recorded in November (FE = 3.27 m/s). Interestingly, CEEMDAN–Volterra achieved much lower errors (FE = 1.89 m/s) during November. Focusing on standalone models, SVM is outperforming the other standalone models Volterra and ARIMA in terms of monthly FE, however, the FE spread is in a much wider

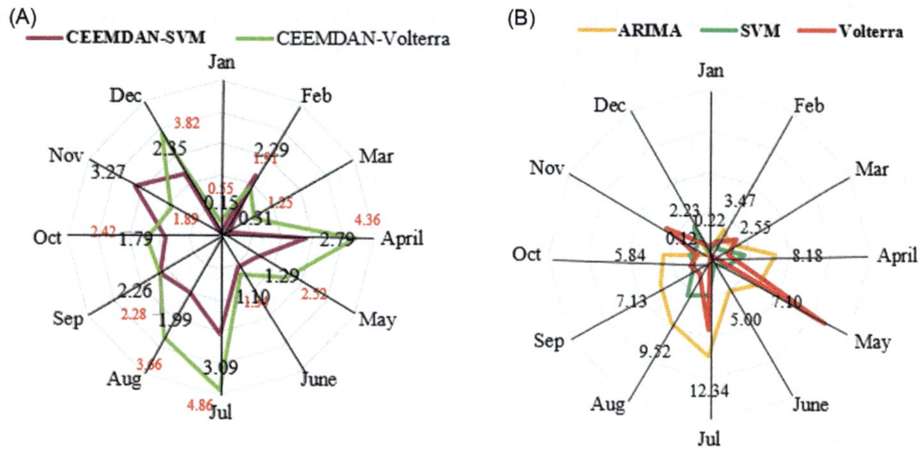

FIGURE 12.22 Polar plots showing monthly wind speed relative error analysis from (A) hybrid models (CEEMDAN–SVM, CEEMDAN–Volterra), (B) standalone models (SVM, Volterra, and ARIMA) in the testing period from January 1, 2011 to December 31, 2017 for Capital.

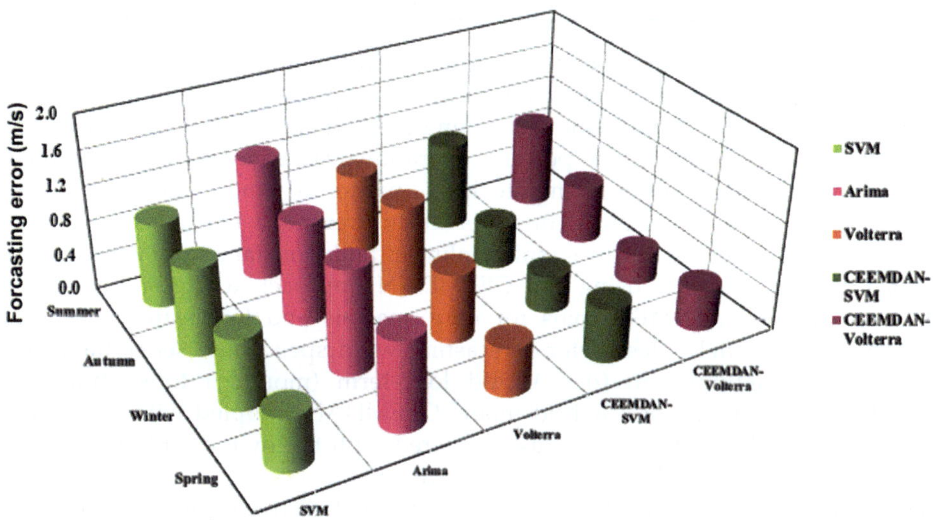

FIGURE 12.23 Bar graphs of average seasonal forecasting errors (Summer, Autumn, Winter, Spring) in forecasting monthly wind speed using the best: SVM, Volterra, Arima, CEEMDAN–SVM, and CEEMDAN–Volterra models (NB: Best models for monthly *wind speed* forecasting were for Albany site for all the models).

range (0.22, 12.34 m/s) for standalone models. This evidence suggests that the forecasts generated by the hybrid CEEMDAN models are more stable, specifically, CEEMDAN–SVM provided better performance for most of the months.

Fig. 12.23 shows a three-dimensional bar graph of the FE generated by hybrid CEEMDAN–SVM and the comparative models for major seasons (i.e.,

Summer = December–January–February; Autumn = March–April–May; Winter = June–July–August; and Spring = September–October–November) analyzed using the monthly forecasted data. For concision, the results for the most accurately forecasted site (Albany) are presented. Congruent with earlier results analyzed for monthly forecast assessment, the seasonal analysis of the forecasted wind speed revealed a greater efficiency of the hybrid CEEMDAN–SVM model compared with the comparative models, FE generated by the CEEMDAN–SVM model were significantly lower for two seasons Autumn and Summer with minimum forecasted error less than 0.4 m/s in Autumn and maximum forecasted error (FE = 0.8 m/s) in Summer. An interesting feature was recorded during summer, the *FE* values were at maximum for all models. It is important to note that CEEMDAN–Volterra performed well for seasonal (spring and winter) wind speed forecasting, while standalone models (SVM, Volterra, and ARIMA) generated large errors, with ARIMA being the worst performing model for all four seasons. This indicates that the newly constructed hybrid model can attain a much more accurate performance compared to the standalone models, which is the benefit of integrating the CEEMDAN tool for wind speed forecasting of the study sites.

12.5 Conclusion

Accurate forecasting of wind speed is essential for the planning, scheduling, and control of wind energy generation and conversion in the wind power industry (Kong et al., 2015; He et al., 2018). However, standalone forecasting methods may not be sufficiently reliable and accurate for decision-makers to perform operational strategies purely when the uncertainty level increases due to the strong stochastic nature of wind (Yang and Wang, 2018; Xiao et al., 2018; Liu et al., 2018c). Thus this study presents a hybrid CEEMDAN–SVM model framework whereby the original nonlinear and nonstationary wind speed time series data are decomposed using the CEEMDAN approach to decompose raw wind speed series into several IMF signals for easy analysis and forecasting. The proposed CEEMDAN–SVM model is tested for forecasting wind speed time series data for short-term (6-hourly), medium-term (daily), and long-term (monthly) forecasting horizons, respectively. The suitability of a hybrid model CEEMDAN, coupled with the SVM model was examined and the performance was compared with the hybrid CEEMDAN–Volterra model and standalone models (SVM, Volterra, ARIMA).

Overall, the CEEMDAN–SVM models, outperformed all the approaches under investigation (i.e., standalone models and CEEMDAN–Volterra model) and showed its better capability for modeling and simulating the 6-hourly, daily, and monthly wind speed. It is noteworthy that if climatological forcing and wind parameters are not available in real-time, the data-intelligent models based on the CEEMDAN ensemble approach with historically forecasted wind speed can amicably be incorporated into the decision-making process by wind farmers, wind farm managers, and relevant energy decision-makers to promote the utilization of wind energy in grid-connected and isolated power systems.

The results showed that for a short-term forecasting horizon (6-hourly), the standalone SVM model outperformed Volterra and ARIMA models at all study sites with the largest value of r, WI, E_{NS}, and L and smallest value of $RMSE$, MAE, $RRMSE$, and $MAPE$. While for a medium-term and long-term forecasting horizon, SVM and Volterra equally performed well. ARIMA was the worst performing model at all sites for all timescales. The incorporation of CEEMDAN algorithms led to enhanced accuracy of the standalone models resolving the issues related to "nonstationarity" (varying mean) and "seasonality" (changes in variance) within the model input series.

The precise selection of SVM's parameters has a huge effect on their accuracies. Gradient descent and grid search (Hsu et al., 2003) algorithms, which can be considered as the conventional methods, could not be selected as the best choice due to their drawbacks, such as computational complexity and their very time-consuming nature. A combination of CEEMDAN–SVM with a genetic algorithm could be considered as one of the best approaches, which has been successfully applied in the field of wind speed forecasting (Khosravi et al., 2018b). Also, the performance of particle swarm optimization (PSO) to search optimal parameters could not be neglected, since it has shown excellent results in many studies (Eseye et al., 2018; Al-Musaylh et al., 2018b; Bamakan et al., 2016). Moreover, a combination of CEEMDAN–SVM with PSO should be tested and further researched due to the merits of PSO in solving optimization problems.

In this study, the forecasting horizon was restricted to the minimum of 6-hourly it is needed to emulate the wind speed over a much shorter and a practically realistic timescale (e.g., hourly or subhourly). To address this, the application of the hourly wind speed forecasting model within a CEEMDAN–SVM based approach can lead to a new paradigm for real-time wind speed forecasting.

Additionally, in this study, only net wind speed was used as the input variable to build all the forecasting models and the forecasting capability of the proposed hybrid CEEMDAN–SVM was powerful. However, feeding other factors such as local site topography, pressure gradient, local weather, and other terrain conditions into the hybrid model can help improve the forecasting accuracy. In accordance with the present findings, the hybrid CEEMDAN–SVM model could be explored further in wind speed forecasting problems, where other climate or atmospheric input data, especially from space-borne measurements primarily from satellites, can provide an alternative for the development of observations of wind speed over larger spatial extents. Thus, satellite datasets from Giovanni NASA MODIS can be utilized to build the assessment and forecasting models which can further enhance the forecasting performance.

References

Adamowski, J., et al., 2012. Comparison of multiple linear and nonlinear regression, autoregressive integrated moving average, artificial neural network, and wavelet artificial neural network methods for urban water demand forecasting in Montreal, Canada. Water Resour. Res. 48 (1).

Afanasyev, D.O., Fedorova, E.A., 2016. The long-term trends on the electricity markets: comparison of empirical mode and wavelet decompositions. Energy Econ. 56, 432–442.

Agrawal, A., Sandhu, K.S., 2016. Most influential parametrical and data needs for realistic wind speed prediction. Renew. Energy 94, 452–465.

Ahmed, A., Khalid, M., 2018. An intelligent framework for short-term multi-step wind speed forecasting based on functional networks. Appl. Energy 225, 902–911.

Ahmad, T., et al., 2018. A comprehensive overview on the data driven and large scale based approaches for forecasting of building energy demand: a review. Energy Build. 165, 301–320.

Al-Musaylh, M.S., et al., 2018a. Short-term electricity demand forecasting with MARS, SVR and ARIMA models using aggregated demand data in Queensland, Australia. Adv. Eng. Inform. 35, 1–16.
Al-Musaylh, M.S., et al., 2018b. Two-phase particle swarm optimized-support vector regression hybrid model integrated with improved empirical mode decomposition with adaptive noise for multiple-horizon electricity demand forecasting. Appl. Energy 217, 422–439.
Alonzo, B., et al., 2017. Modelling the variability of the wind energy resource on monthly and seasonal timescales. Renew. Energy 113, 1434–1446.
Bamakan, S.M.H., Wang, H., Ravasan, A.Z., 2016. Parameters optimization for nonparallel support vector machine by particle swarm optimization. Procedia Comput. Sci. 91, 482–491.
Barman, M., Dev Choudhury, N.B., Sutradhar, S., 2018. A regional hybrid GOA-SVM model based on similar day approach for short-term load forecasting in Assam, India. Energy 145, 710–720.
Billings, S.A., Korenberg, M.J., Chen, S., 1988. Identification of non-linear output-affine systems using an orthogonal least-squares algorithm. Int. J. Syst. Sci. 19 (8), 1559–1568.
Blakers, A., Lu, B., Stocks, M., 2017. 100% renewable electricity in Australia. Energy 133, 471–482.
Box, G.E.P., Jenkins, G., 1990. Time Series Analysis, Forecasting and Control. Holden-Day, Inc, p. 500.
Box, G.E., et al., 2015. Time Series Analysis: Forecasting and Control. John Wiley & Sons.
Cassola, F., Burlando, M., 2012. Wind speed and wind energy forecast through Kalman filtering of numerical weather prediction model output. Appl. Energy 99, 154–166.
Celikoglu, A., Tirnakli, U., 2018. Skewness and kurtosis analysis for non-Gaussian distributions. Phys. A: Stat. Mech. Appl. 499, 325–334.
Chai, T., Draxler, R.R., 2014. Root mean square error (RMSE) or mean absolute error (MAE)?—arguments against avoiding RMSE in the literature. Geosci. Model. Dev. 7 (3), 1247–1250.
Chang, C.-C., Lin, C.-J., 2011. LIBSVM: a library for support vector machines. ACM Trans. Intell. Syst. Technol. 2 (3), 27.
Chitsazan, M.A., Sami Fadali, M., Trzynadlowski, A.M., 2019. Wind speed and wind direction forecasting using echo state network with nonlinear functions. Renew. Energy 131, 879–889.
Dadkhah, M., Jahangoshai Rezaee, M., Zare Chavoshi, A., 2018. Short-term power output forecasting of hourly operation in power plant based on climate factors and effects of wind direction and wind speed. Energy 148, 775–788.
Dahhani, O., El-Jouni, A., Boumhidi, I., 2018. Assessment and control of wind turbine by support vector machines. Sustain. Energy Technol. Assess. 27, 167–179.
Dawson, C.W., Abrahart, R.J., See, L.M., 2007. HydroTest: a web-based toolbox of evaluation metrics for the standardised assessment of hydrological forecasts. Environ. Model. Softw. 22 (7), 1034–1052.
Dee, D.P., et al., 2011. The ERA-Interim reanalysis: configuration and performance of the data assimilation system. Q. J. R. Meteorol. Soc. 137 (656), 553–597.
Deo, R.C., et al., 2018. Multi-layer perceptron hybrid model integrated with the firefly optimizer algorithm for windspeed prediction of target site using a limited set of neighboring reference station data. Renew. Energy 116, 309–323.
Deo, R.C., Wen, X., Qi, F., 2016. A wavelet-coupled support vector machine model for forecasting global incident solar radiation using limited meteorological dataset. Appl. Energy 168, 568–593.
Du, P., et al., 2017. Research and application of a novel hybrid forecasting system based on multi-objective optimization for wind speed forecasting. Energy Convers. Manag. 150, 90–107.
Erdem, E., Shi, J., 2011. ARMA based approaches for forecasting the tuple of wind speed and direction. Appl. Energy 88 (4), 1405–1414.
Eseye, A.T., Zhang, J., Zheng, D., 2018. Short-term photovoltaic solar power forecasting using a hybrid Wavelet-PSO-SVM model based on SCADA and meteorological information. Renew. Energy 118, 357–367.
Gallagher, R., et al., 2013. Australian Geological Provinces, 2013, first ed. Geoscience Australia.
Gill, M.K., et al., 2006. Soil moisture prediction using support vector machines. JAWRA J. Am. Water Resour. Assoc. 42 (4), 1033–1046.
Gruber, J.K., et al., 2011. Nonlinear MPC based on a Volterra series model for greenhouse temperature control using natural ventilation. Control. Eng. Pract. 19 (4), 354–366.
Han, Q., et al., 2017. Non-parametric hybrid models for wind speed forecasting. Energy Convers. Manag. 148, 554–568.

He, Q., Wang, J., Lu, H., 2018. A hybrid system for short-term wind speed forecasting. Appl. Energy 226, 756–771.

Hong, D.Y., et al., 2019. Ultra-short-term forecast of wind speed and wind power based on morphological high frequency filter and double similarity search algorithm. Int. J. Electr. Power Energy Syst. 104, 868–879.

Hoolohan, V., Tomlin, A.S., Cockerill, T., 2018. Improved near surface wind speed predictions using Gaussian process regression combined with numerical weather predictions and observed meteorological data. Renew. Energy 126, 1043–1054.

Hora, J., Campos, P., 2015. A review of performance criteria to validate simulation models. Expert. Syst. 32 (5), 578–595.

Howard, B.S., et al., 2018. Modeling the carbon budget of the Australian electricity Sector's transition to renewable energy. Renew. Energy .

Hsu, C.W., Chang, C.C., Lin, C.J., 2003. A Practical Guide to Support Vector Classification, vol. 101, pp. 1396–1400. https://www.csie.ntu.edu.tw/~cjlin/papers/guide/guide.pdf.

Huang, N.E., et al., 1998. The empirical mode decomposition and the Hilbert spectrum for nonlinear and non-stationary time series analysis. Proc. R. Soc. London. Ser. A: Math. Phys. Eng. Sci. 454, 903–995.

Hu, Y.-L., Chen, L., 2018. A nonlinear hybrid wind speed forecasting model using LSTM network, hysteretic ELM and differential evolution algorithm. Energy Convers. Manag. 173, 123–142.

Hu, J., Wang, J., Zeng, G., 2013. A hybrid forecasting approach applied to wind speed time series. Renew. Energy 60, 185–194.

IPCC, 2007. Climate Change 2007—Mitigation of Climate Change: Working Group III Contribution to the Fourth Assessment Report of the IPCC. Cambridge University Press, Cambridge.

Izadyar, N., et al., 2015. Appraisal of the support vector machine to forecast residential heating demand for the district heating system based on the monthly overall natural gas consumption. Energy 93, 1558–1567.

Jiang, H., Dong, Y., 2017. Global horizontal radiation forecast using forward regression on a quadratic kernel support vector machine: case study of the Tibet Autonomous Region in China. Energy 133, 270–283.

Jiang, P., Li, C., 2018. Research and application of an innovative combined model based on a modified optimization algorithm for wind speed forecasting. Measurement 124, 395–412.

Jiao, G., Guo, T., Ding, Y., 2016. A new hybrid forecasting approach applied to hydrological data: a case study on precipitation in Northwestern China. Water 8 (9), 367.

Jing, G., et al., 2018. An air balancing method using support vector machine for a ventilation system. Build. Environ. 143, 487–495.

Jun, W., et al., 2018. A new weighted CEEMDAN–based prediction model: an experimental investigation of decomposition and non-decomposition approaches. Knowl. Syst. 160, 188–199.

Kashani, M.H., et al., 2016. Integration of Volterra model with artificial neural networks for rainfall-runoff simulation in forested catchment of northern Iran. J. Hydrol. 540, 340–354.

Khosravi, A., et al., 2018a. Prediction of wind speed and wind direction using artificial neural network, support vector regression and adaptive neuro-fuzzy inference system. Sustain. Energy Technol. Assess. 25, 146–160.

Khosravi, A., Machado, L., Nunes, R.O., 2018b. Time series prediction of wind speed using machine learning algorithms: a case study Osorio wind farm, Brazil. Appl. Energy 224, 550–566.

Kong, X., et al., 2015. Wind speed prediction using reduced support vector machines with feature selection. Neurocomputing 169, 449–456.

Koukoulas, P., Kalouptsidis, N., 2000. Second-order Volterra system identification. IEEE Trans. Signal. Process. 48 (12), 3574–3577.

Laqrichi, S., et al., 2015. Integrating uncertainty in software effort estimation using Bootstrap based neural networks. IFAC-PapersOnLine 48 (3), 954–959.

Lazos, D., Sproul, A.B., Kay, M., 2014. Optimisation of energy management in commercial buildings with weather forecasting inputs: a review. Renew. Sustain. Energy Rev. 39, 587–603.

Legates, D.R., McCabe, G.J., 1999. Evaluating the use of "goodness-of-fit" measures in hydrologic and hydroclimatic model validation. Water Resour. Res. 35 (1), 233–241.

Letcher, T.M., 2017. Wind Energy Engineering: A Handbook for Onshore and Offshore Wind Turbines. Elsevier Science & Technology, San Diego, CA.

Li, H., et al., 2018a. Research and application of a combined model based on variable weight for short term wind speed forecasting. Renew. Energy 116, 669–684.

Li, C., et al., 2018b. Short-term wind power prediction based on data mining technology and improved support vector machine method: a case study in Northwest China. J. Clean. Prod. 205, 909–922.
Lin, H., 2003. Hydropedology. Vadose Zone J. 2 (1), 1–11.
Liu, H., et al., 2010. A hybrid statistical method to predict wind speed and wind power. Renew. Energy 35 (8), 1857–1861.
Liu, H., et al., 2018a. A novel ensemble model of different mother wavelets for wind speed multi-step forecasting. Appl. Energy 228, 1783–1800.
Liu, H., Mi, X., Li, Y., 2018b. An experimental investigation of three new hybrid wind speed forecasting models using multi-decomposing strategy and ELM algorithm. Renew. Energy 123, 694–705.
Liu, H., Mi, X., Li, Y., 2018c. Comparison of two new intelligent wind speed forecasting approaches based on wavelet packet decomposition, complete ensemble empirical mode decomposition with adaptive noise and artificial neural networks. Energy Convers. Manag. 155, 188–200.
Mohammadi, K., et al., 2015. Support vector regression based prediction of global solar radiation on a horizontal surface. Energy Convers. Manag. 91, 433–441.
Moreno, S.R., dos Santos Coelho, L., 2018. Wind speed forecasting approach based on singular spectrum analysis and adaptive neuro fuzzy inference system. Renew. Energy 126, 736–754.
Mughal, M.O., et al., 2018. Forecasting and verification of winds in an East African complex terrain using coupled mesoscale—and micro-scale models. J. Wind. Eng. Ind. Aerodyn. 176, 13–20.
Nash, J.E., Sutcliffe, J.V., 1970. River flow forecasting through conceptual models part I—A discussion of principles. J. Hydrol. 10 (3), 282–290.
Niu, T., et al., 2018. Multi-step-ahead wind speed forecasting based on optimal feature selection and a modified bat algorithm with the cognition strategy. Renew. Energy 118, 213–229.
Nourani, V., et al., 2014. Applications of hybrid wavelet–artificial intelligence models in hydrology: a review. J. Hydrol. 514, 358–377.
Payero, J.O., Irmak, S., 2013. Daily energy fluxes, evapotranspiration and crop coefficient of soybean. Agric. Water Manag. 129, 31–43.
Pearre, N.S., Swan, L.G., 2018. Statistical approach for improved wind speed forecasting for wind power production. Sustain. Energy Technol. Assess. 27, 180–191.
Prasad, R., et al., 2017. Input selection and performance optimization of ANN-based streamflow forecasts in the drought-prone Murray Darling Basin region using IIS and MODWT algorithm. Atmos. Res. 197, 42–63.
Prasad, R., et al., 2018. Soil moisture forecasting by a hybrid machine learning technique: ELM integrated with ensemble empirical mode decomposition. Geoderma 330, 136–161.
Rathinasamy, M., Adamowski, J., Khosa, R., 2013. Multiscale streamflow forecasting using a new Bayesian model average based ensemble multi-wavelet Volterra nonlinear method. J. Hydrol. 507, 186–200.
Rhein, M., et al., 2013. Observations: ocean. In: Stocker, T.F., et al., (Eds.), Climate Change 2013: The Physical Science Basis. Contribution of Working Group I to the Fifth Assessment Report of the Intergovernmental Panel on Climate Change. Cambridge University Press, Cambridge and New York, pp. 255–316.
Samadianfard, S., et al., 2018. Wavelet neural networks and gene expression programming models to predict short-term soil temperature at different depths. Soil. Tillage Res. 175, 37–50.
Santhosh, M., Venkaiah, C., Vinod Kumar, D.M., 2018. Ensemble empirical mode decomposition based adaptive wavelet neural network method for wind speed prediction. Energy Convers. Manag. 168, 482–493.
Schetzen, M., 1980. The Volterra and Wiener Theories of Nonlinear Systems. Wiley.
Schicker, I., et al., 2017. Short-range wind speed predictions for complex terrain using an interval-artificial neural network. Energy Procedia 125, 199–206.
Shukur, O.B., Lee, M.H., 2015. Daily wind speed forecasting through hybrid KF-ANN model based on ARIMA. Renew. Energy 76, 637–647.
Skittides, C., Früh, W.-G., 2014. Wind forecasting using principal component analysis. Renew. Energy 69, 365–374.
Song, J., Wang, J., Lu, H., 2018. A novel combined model based on advanced optimization algorithm for short-term wind speed forecasting. Appl. Energy 215, 643–658.
Suykens, J.A.K., et al., 2002. Weighted least squares support vector machines: robustness and sparse approximation. Neurocomputing 48 (1), 85–105.

Taylor, K.E., 2001. Summarizing multiple aspects of model performance in a single diagram. J. Geophys. Res. Atmos. 106 (D7), 7183–7192.

Taylor, J.W., McSharry, P.E., Buizza, R., 2009. Wind power density forecasting using ensemble predictions and time series models. IEEE Trans. Energy Convers. 24 (3), 775–782.

Tian, C., Hao, Y., Hu, J., 2018. A novel wind speed forecasting system based on hybrid data preprocessing and multi-objective optimization. Appl. Energy 231, 301–319.

Tiwari, M.K., Chatterjee, C., 2010. Uncertainty assessment and ensemble flood forecasting using bootstrap based artificial neural networks (BANNs). J. Hydrol. 382 (1), 20–33.

Torres, M.E., et al., 2011. A complete ensemble empirical mode decomposition with adaptive noise. In: 2011 IEEE International Conference on Acoustics, Speech and Signal Processing (ICASSP).

Vapnik, V., 1998. The Nature of Statistical Learning Theory. Springer Science & Business Media.

Wallach, D., et al., 2019. In: Wallach, D., et al., (Eds.), Chapter 9—Model Evaluation, in Working with Dynamic Crop Models, third ed. Academic Press, pp. 311–373.

Wang, X., Guo, P., Huang, X., 2011a. A review of wind power forecasting models. Energy Procedia 12, 770–778.

Wang, S., et al., 2011b. A novel seasonal decomposition based least squares support vector regression ensemble learning approach for hydropower consumption forecasting in China. Energy 36 (11), 6542–6554.

Wang, J., et al., 2015. Short-term wind speed forecasting using support vector regression optimized by cuckoo optimization algorithm. Math. Probl. Eng. 2015, 1–13.

Wang, S., et al., 2016. Wind speed forecasting based on the hybrid ensemble empirical mode decomposition and GA-BP neural network method. Renew. Energy 94, 629–636.

Wang, J., et al., 2017. Research and application of a combined model based on multi-objective optimization for multi-step ahead wind speed forecasting. Energy 125, 591–613.

Wang, Y., et al., 2018. Correlation aware multi-step ahead wind speed forecasting with heteroscedastic multi-kernel learning. Energy Convers. Manag. 163, 384–406.

Willems, P., 2009. A time series tool to support the multi-criteria performance evaluation of rainfall-runoff models. Environ. Model. Softw. 24 (3), 311–321.

Willmott, C.J., 1984. On the evaluation of model performance in physical geography. In: Gaile, G.L., Willmott, C.J. (Eds.), Spatial Statistics and Models. Springer, The Netherlands; Dordrecht, pp. 443–460.

Wu, Y.K., Hong, J.S., 2007. A literature review of wind forecasting technology in the world. In: 2007 IEEE Lausanne Power Tech.

Wu, Z., Huang, N.E., 2009. Ensemble empirical mode decomposition: a noise-assisted data analysis method. Adv. Adapt. Data Anal. 01 (01), 1–41.

Xiao, L., Dong, Y., Dong, Y., 2018. An improved combination approach based on Adaboost algorithm for wind speed time series forecasting. Energy Convers. Manag. 160, 273–288.

Yang, Z., Wang, J., 2018. A hybrid forecasting approach applied in wind speed forecasting based on a data processing strategy and an optimized artificial intelligence algorithm. Energy 160, 87–100.

Yang, Y., et al., 2016. An incremental electric load forecasting model based on support vector regression. Energy 113, 796–808.

Yao, X., Crook, J., Andreeva, G., 2017. Enhancing two-stage modelling methodology for loss given default with support vector machines. Eur. J. Oper. Res. 263 (2), 679–689.

Yuan, C., Liu, S., Fang, Z., 2016. Comparison of China's primary energy consumption forecasting by using ARIMA (the autoregressive integrated moving average) model and GM(1,1) model. Energy 100, 384–390.

Yuan, X., et al., 2017. Wind power prediction using hybrid autoregressive fractionally integrated moving average and least square support vector machine. Energy 129, 122–137.

Yu, C., Li, Y., Zhang, M., 2017. Comparative study on three new hybrid models using Elman neural network and empirical mode decomposition based technologies improved by singular spectrum analysis for hour-ahead wind speed forecasting. Energy Convers. Manag. 147, 75–85.

Zendehboudi, A., Baseer, M.A., Saidur, R., 2018. Application of support vector machine models for forecasting solar and wind energy resources: a review. J. Clean. Prod. 199, 272–285.

Zhang, Y., et al., 2015. Wind power prediction considering nonlinear atmospheric disturbances. Energies 8 (1), 475.

Zhang, W., et al., 2017. A combined model based on CEEMDAN and modified flower pollination algorithm for wind speed forecasting. Energy Convers. Manag. 136, 439–451.

Appendix

It is important to note that, the selection of model parameters is crucial to obtain an accurate SVM model. For this reason, a grid search procedure, over a wide range of values seeking the smallest *MSE*, was also used to establish the optimal parameters. Fig. 12.A1 illustrates a surface plot of the *MSE* with respect to different regularization constants *C* and σ (kernel width) values for the SVM model used in monthly wind speed forecasting for site Albany. However, the major drawback of the grid search method is that it is very time-consuming. For this experiment, it took 6 days to find the optimal parameters of the daily SVM model which is not desirable for real-life applications. Thus to obviate this limitation, the 10-fold cross-validation SVM model was used. The results of the 10-fold cross-validation SVM model were compared with the results generated by grid search SVM models as shown in Table 12.A1.

It is clear from the results that the 10-fold cross-validation SVM model generated equally good results as grid search SVM model, but it has better generalization capability

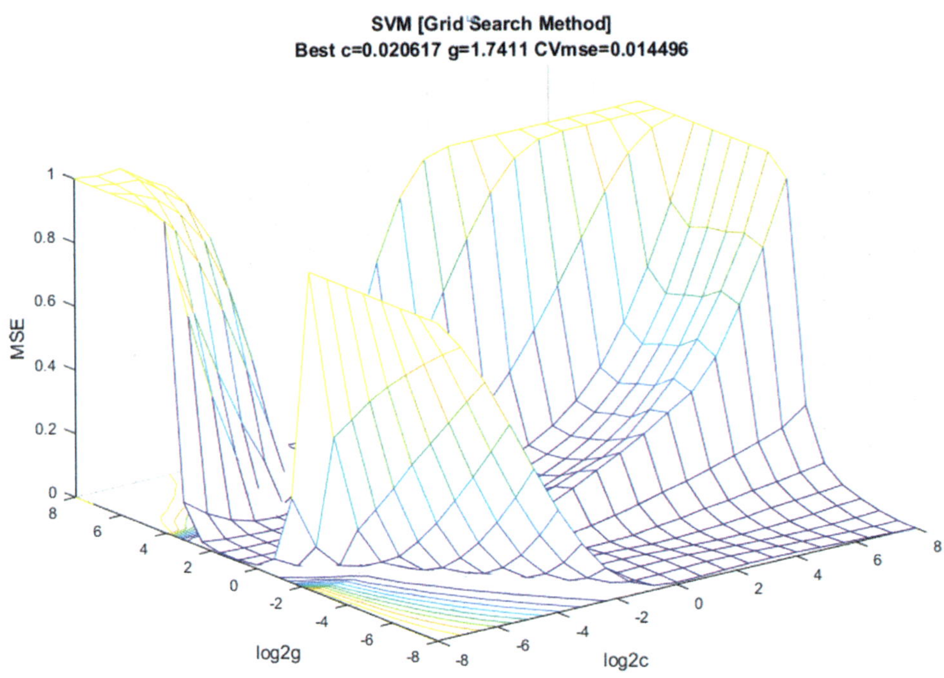

FIGURE 12.A1 SVM model parameters for daily wind speed forecasting at site Albany.

TABLE 12.A1 Comparison of 10-fold cross-validation SVM model and grid search SVM model.

	r	RMSE	MAE
Albany daily SVM model	0.6036	1.2516	0.9679
Albany daily SVM grid search model	0.6129	1.2392	0.9621
Albany monthly 10-fold SVM model	0.8245	0.4488	0.3576
Albany monthly SVM grid search model	0.8249	0.4519	0.3574
Woolnorth daily SVM model	0.5071	2.4674	1.9806
Woolnorth daily SVM grid search model	0.5077	2.4647	1.9794
Woolnorth monthly SVM model	0.2039	1.0218	0.7663
Woolnorth monthly SVM grid search model	0.2036	1.01	0.7721

and it is able to handle large-scale data with computationally fast forecasting results (in less than 1 hour) giving it an added versatility for such decision-support systems.

CHAPTER

13

MARS model for prediction of short- and long-term global solar radiation

Dilki T. Balalla[1], Thong Nguyen-Huy[2,3] and Ravinesh Deo[1]

[1]School of Sciences, University of Southern Queensland, Springfield, QLD, Australia [2]Centre for Applied Climate Sciences, University of Southern Queensland, Toowoomba, QLD, Australia [3]Vietnam National Space Center, Vietnam Academy of Science and Technology, Hanoi, Vietnam

13.1 Introduction

Forecasting solar energy is an important research area for energy producers, policymakers, and engineers since the application of renewable energy instead of fossil fuel is a capable outlook in addressing climate change (Solangi et al., 2011). Over the last decade, people have been dependent on fossil resources, which have started to decline rapidly compared to the significant rise in population. Hence, scientists have started discussing the perspective of renewable energy (wind, solar, wave, biomass, etc.) in the making of strategies for sustainable development because being dependent on fossil fuels has brought some downsides.

According to IPCC 2012, using fossil fuels has led to the emission of greenhouse gases (GHG) which significantly contributed to global warming. Lund (2007) says the strategies for sustainable development involve three major technological changes consisting of energy saving of the demand sector, high-efficiency rate of energy production, and replacement of fossil fuels by other renewable energy resources. As for his conclusion, converting the present energy systems into a 100% renewable energy system is not impossible. The exquisiteness of renewable energy is that it can be harnessed from various sources that are endless, unlike fossil fuel. Panwar et al. (2011) state there are many advantages in using renewable energy in energy systems. The first benefit would be that renewable energy sources are harmless to nature, rather than conventional energy innovations. The energy from these sources is harnessed without doing any harm to the environment. Another benefit is that renewable energy will not deplete like fossil fuels or other sources

Predictive Modelling for Energy Management and Power Systems Engineering
DOI: https://doi.org/10.1016/B978-0-12-817772-3.00013-6

of energy. Frondel et al. (2010) believe that more industrialized nations have started implementing finances for sustainable power sources, as it is an environmentally healthy, abundant, and cost-effective energy source. They further state that energy security is a side advantage of using renewable energy, as renewable energy allows you to be independent of imported energy sources. It is therefore important that researchers are focused on investigating the future use of renewable energy.

In 2001 the Australian Government presented a Mandatory Renewable Energy Target (MRET) program with the objective of expanding uptake of the sustainable power source in Australia's power supply. In 2007 the Renewable Energy Target (RET) implemented by the government focused on guaranteeing that 23.5% of Australia's power supply originates from sustainable power sources by 2020 (Hunt and Macfarlane, 2015)

Solar radiation is a global natural phenomenon consisting of electromagnetic waves including visible, ultraviolet light, and infrared radiation. According to Lewis and Nocera (2006), solar power has eventually become the most exploitable energy resource as it provides more energy in 1 hour to Earth than all of the energy consumed by humans in 1 year. Using solar energy effectively will be advantageous, as it reduces the electricity bill; solar energy extraction, conversion, and transmission is cheaper compared to nonrenewable energy resources; and solar energy is available throughout the year (Chwieduk, 2004). The advantages for industries would be solar energy can adequately supplement power supply from a power transmission matrix, for example, when power request crests in the summer. Hence, the researchers are encouraged to try applying new mechanisms for forecasting solar radiation as the information based on the forecast horizon is essential for the efficient use of solar radiation.

The Australian mainland has become a finer potential solar energy asset on the planet by having an abundance of solar radiation of approximately 3.2 TW h/(capita year $m^{2)}$. Australia gets solar radiation of roughly $58 million PJ every year, which is 10,000 times bigger than its annual cumulative energy consumption yet solar energy accounts for only 0.2% of its total energy consumption (Geoscience Australia, 2010). The areas with the highest solar radiation are the regional areas in the northwest and central mainland. The renewed focus on climate change in Australia at the national level has resulted in $44.9 billion being invested in clean energy in 2011 and in the installation of over 1.4 million solar systems (Council, 2015). Bahadori and Nwaoha (2013) states that though solar energy is mainly used for small applications such as thermal heating and accounts for only 0.1% of the essential energy utilization, solar energy use in Australia is likely to increase by 5.9% per year to 24PJ in 2029–30. Though the critical development in solar radiation forecast models is inescapable, versatile forecast models of solar radiation stay unexplored.

Despite Australia being highly abundant in solar energy, especially Queensland (Yusaf et al., 2011; Beath, 2012) which is known as "Sunshine State" and receives solar radiation 10,000 times bigger than its annual energy consumption, not even 1% of it has been used in proper energy production (Geoscience Australia, 2010). Hence, the Government has been encouraging researchers (Haidar et al., 2015) to model solar radiation in Australia's diverse locations with high accuracy and then to harness this energy where it is economically sustainable.

Since the use of solar energy in Australia is predicted to increase and critical development in solar radiation forecast models is inescapable, there are not enough versatile forecast models of solar radiation constructed in regional Queensland, which is the research problem encountered in this research.

The motivation comes from a need for a versatile solar radiation forecasting method using easily and locally available data for daily and monthly scheduling of energy resources in regional Queensland. The intention of this research was to try and address the research question "Is machine learning algorithm, Multivariate Adaptive Regression Splines model, a versatile forecasting model for solar radiation?" The objective of this chapter is to develop a machine learning (ML) algorithm to validate and assess errors for the method used to forecast solar radiation based on historical data. The specific aims are to construct (1) short-term (daily) global solar radiation model using the MARS algorithm considering the nonlinear behavior of surface-level solar radiation with its predictor variables; and (2) long-term (monthly) global solar radiation model using the MARS algorithm to enable the solar energy assessment over a long-term period and considering.

This chapter carried out short-term and long-term solar radiation forecasting model development for regional Queensland. Short-term forecasting provides predictions up to 7 days ahead. These forecasts are valuable for grid operators in order to make important decisions for grid operation. It will provide valuable information regarding the time scheduling of power systems (Wan et al., 2015). Long-term forecasting has been carried out considering 1-month ahead, 3-month, and 6-month ahead forecast. This is useful for energy companies to make decisions and negotiate contracts with energy producers (Martín et al., 2010) and also for effective operation and maintenance planning of solar power systems (Koca et al., 2011). The information gathered from the seasonal analysis can be used for studying the seasonal patterns of the solar energy and for Seasonal Thermal Energy Storage (i.e., STES) (Allen et al., 1984) where the heat acquired from solar collectors in hot months can be stored for future use when needed, including during winter months.

13.2 Literature review

Modeling strategies connected for solar radiation prediction use two kinds of models: deterministic (or a mathematical) or data-driven (or a black-box). The deterministic model is one in which results are accurately decided through known connections among states and occasions, with no space for arbitrary variety. In addition, these models are very easy to develop and have very good precision. Autoregressive integrated moving average (ARIMA), multivariate adaptive regression splines (MARS), artificial neural network (ANN), and support vector regression (SVR) are among the most used forecasting tools at the present.

13.2.1 The advantages of machine learning models

The most significant advantage of the data-driven models over physical models is that they could be used as a universal model as they do not need any initial conditions. Voyant et al. (2017) said that the ML models are superior in finding familiarities between predictor variables and the dependent variable though it seems that they are not correlated. Parmar et al. (2015) stated that, by using ML methods, precise and consistent results can be obtained in radiomic applications in clinical care. In cancer research, ML methods have the ability to extract key features or trends from complex data set (Kourou et al., 2015).

Further, it says that an appropriate level of validation is needed for these methods to be applied in everyday clinical practice. Olden et al. (2008) specified that ML techniques are capable of handling complex problems with multiple interacting elements and they outperform the classical approaches, thus making ML best for modeling ecological systems. Hindman (2015) states that ML places great prominence on model assessing and model adjusting of forecasts toward the mean to reduce overfitting. Dalaka et al. (2000) provides many advantages of ML algorithms over ordinary curve fitting methods. Their flexibility to select between accuracy and overview and their robustness against outlier and mixtures of different responses is the main advantage of ML models. He further states that the models are transparent and easily interpreted. They seem to be able to handle quite complex dependencies among attributes, not requiring prior expert knowledge. While these models are very complex to develop and have a good accuracy, their ability to remove the correlated highlights in indicators that are not joined in their unique scientific plans, is constrained (Deo and Şahin, 2017).

13.2.2 Studies on machine learning methods used as a universal model

Wang (2015) used the support vector machine (SVM) algorithm in bioinformatics for protein production. He further used SVM to develop a new predictor algorithm, which showed a higher success rate. Parmar et al. (2015) identified the Wilcoxon test-based feature selection method (WLCX) and a classification method random forest (RF) had the highest performance in the radiomic study. Kourou et al. (2015) had stated that the ANN, Bayesian network (BN), SVM, and decision trees (DT) are some of the ML methods which have been widely used in cancer research and have produced effective and accurate results. Vempala and Russo (2017) applied RF, ANN, and linear regression algorithms in ML methods to model the music emotion judgments. Malhotra (2014) used ML methods in investigating the performance of the fault-proneness predictions using static code metrics. Wu et al. (2009) used ML methods in diabetes disease prediction. The ML method used is Laplacian/SVM (LapSVM) and the results suggested that it would be a promising method in further relative studies. Kong et al. (2017) used ML methods for automatic detection of acromegaly from facial photographs. Nanayakkara et al. (2018) has used logistic regression and three ML approaches in comparison to the APACHE III score for predicting in-hospital mortality after cardiac arrest developed from demographic, physiological, and biochemical information available within 24 hours of admission to an Intensive Care Unit (ICU). Arslan et al. (2017) has used ML techniques to create a secure biometric system in securing the distinctive feature extraction, providing secure data storage in the biometric database, maintaining transmission channels used in biometric applications from vulnerabilities, and in ensuring the correctness of the results obtained from intelligent decision mechanism.

Likewise, there are a vast number of instances where ML methods are used in various industries such as cancer research, clinical research, radiomic study, bioinformatics study, weather forecasting, and software engineering. From the literature, it is evident that all these studies have provided accurate and effective results in decision-making and for further research.

13.2.3 Studies on machine learning methods used in solar radiation forecasting

Data-driven models, specifically, can precisely predict solar power, which is viewed as a testing assignment. Solar radiation forecasts are usually performed using numerical weather prediction (NWP) models. But now ML models have come into the light and produce more accurate results rather than NWP models as NWP models are only available at a few spatial points (Aggarwal and Saini, 2014).

Angstrom (1924) elaborated that as the diffused sky radiation plays a dominant role at high latitudes, the use of deterministic models for solar radiation forecasting is better than using mathematical models. He is certain that without modeling of temperature and radiation together with diffused radiation, no relation can be built up between them. As an alternative strategy for solar energy forecasting, ML has been used to extricate climatic examples implanted in air information. Al-Musaylh et al. (2018a) stated that the models' huge computational power has encouraged a rise in their selection.

In the present period, where large amounts of information are gathered from estimations and physical models and are moreover upgraded through improved information items by satellites and reanalysis schemes, calculations offer a guarantee for demonstrating solar energy from known practices of solar variability (Deo and Şahin, 2017).

Solar radiation forecasting is a field where ML methods are vastly used and taken advantage of. Here, some papers are discussed of how they have used ML methods in solar radiation forecasting and how they have helped in building up a versatile model.

The SVM model has been used by Belaid and Mellit (2016) in predicting daily and monthly global solar radiation in an arid climate in Ghardaïa (Algeria). The obtained results appeared to be accurate. Ramli et al. (2015) was successful in building up SVM and ANN models to investigate the performance in predicting solar radiation in Saudi Arabia. SVM had higher accuracy over ANN and it was faster in predicting the radiation on the tilted surfaces. Quej et al. (2017) applied ANFIS, SVM, and ANN models to predict daily global solar radiation in a warm subhumid environment. There also, SVM performed better than the other techniques. This paper suggests SVM as a promising alternative model to predict solar radiation. Here, it can be thought that the SVM model has performed well in predicting solar radiation.

A study on applying a nonlinear SVM model with a hard penalty function based on glow-worm swarm optimization for forecasting daily global solar radiation has been carried out by Jiang and Dong (Jiang and Dong, 2016). In this paper, they reviewed more research that has applied ML methods. The results from his model showed that the model was effective and showed the best forecasting capability at all four sites. Deo et al. (2016) adopted a wavelet-coupled SVM (W-SVM) model to forecast global solar radiation. He has stated that ML methods have been used widely to extract climatic features in atmospheric data. He further stated that ML methods would be beneficial in understanding the nonlinear effects of solar energy and the natural dynamicity in modeling solar radiation.

Olatomiwa et al. (2015) discusses the application of a hybrid ML technique, namely SVM with a firefly algorithm (SVM-FFA), for predicting solar radiation. The result indicated that the SVM-FFA model can be categorized as an efficient ML method for horizontal solar radiation prediction. Paoli et al. (2010) developed an ANN model, a multilayer perceptron (MLP) network, and an ad hoc time series preprocessing to forecast the daily global solar radiation on a horizontal surface. MLP and ARMA have been used (Voyant et al., 2013) to predict global solar radiation and photovoltaic power for forecast horizons of 1 day, 2 days, and 3 days, and of 1 hour and 24 hours, and for the next 5 minutes.

Aggarwal and Saini (2014) further says that among ML models that are used to predict solar radiation, most abundant models are ANN, artificial network-based fuzzy inference system (ANFIS), SVM, extreme ML (ELM), and linear least square regression (LSR).

Hence, by this, and the literature that has been reviewed, it is clear that the MARS model has not been used much in solar radiation forecasting. In addition, by reviewing the literature, it was evident that there were not enough solar forecasting models developed in regional Queensland.

13.2.4 General applications of mars model used in previous research

The MARS model, which was implemented by Friedman, describes the piecewise regression process (Friedman, 1991) of using the commitments of each input, where intuitive impacts from exploratory terms are used to model the predictions (Deo et al., 2017). The main advantage of the MARS model is that it can analyze nonlinear features in target and descriptive relationships. This analysis is performed without taking the assumptions on the relationships between the predictor (descriptive variable) and the predictand (target variable) to consideration. MARS only generates the forecasts based on the learned associations from the segregated training data into splines over an identical interim (Deo et al., 2017). Al-Musaylh et al. (2018b) says that the MARS model is a fast and quite sensible ML model, which can adopt a piecewise (linear or cubic) basis function.

A forecasting model development for electricity demand using MARS, ARIMA, and SVR models has been carried out by Al-Musaylh et al. (2018a). The results showed that MARS and SVR models performed better and are suitable for electricity demand forecasting in Queensland. Sigauke and Chikobvu (2010) have carried out a study in South Africa for daily peak electricity load forecasting using a MARS model. The performances of the model have been evaluated by comparing it to a piecewise linear regression model. The results have shown that the MARS models achieved better forecasting as a lower RMSE was found.

MARS algorithm has been applied by Sekhar Roy et al. (2018) in estimating heating load in buildings. Further, this paper emphasizes MARS over other ML models, which has more computational complexity and takes high computation time for their model building process. The results showed that MARS produced higher accuracy results than others. Deo et al. (2017) applied MARS, least squares SVM (LSSVM), and M5 model tree (M5Tree) in forecasting drought in Eastern Australia. MARS and M5Tree have produced more accurate results than LSSVM. Zhang and Goh (2013) used MARS in modeling highly nonlinear multivariate engineering problems where MARS was able to assess the relative importance of each variable through an analysis of variance (ANOVA) process.

Wang et al. (2017) used MARS, ANN, and SVM in estimating terrestrial latent heat flux over North America. Though all three models proved to be accurate, ANN slightly outperformed the other two algorithms. Deo et al. (2015) applied relevance vector machine (RVM), ELM, and MARS for the prediction of evaporative loss where the RVM model appeared to be more accurate despite slight differences in the overall prediction skill.

All the literature apart from Wang et al. (2017) and Deo et al. (2015) shows us that the MARS model outperformed the other models used in model development.

13.2.5 MARS model applications in solar radiation forecasting

Keshtegar et al. (2018) developed four regression techniques to compare and forecast solar radiation. The specific research compared the Kriging method vs. MARS, response surface method, and M5Tree in modeling solar radiation in two stations located in the Eastern Mediterranean Region of Turkey. The results showed that MARS produced the best forecasting results for one station (Adana), though the Kriging method outperformed all other techniques in the Antakya station.

Li et al. (2016) has applied MARS on forecasting the daily power output of a grid-connected photovoltaic system. They carried out comparison models to evaluate the accuracy of the MARS model. The results showed that the nonlinear models outperformed the linear models on average and MARS was able to provide reliable forecast results in this study due to its nonlinear property.

Salcedo-Sanz et al. (2018) applied coral reefs optimization (CRO) integrated with an ELM model to predict global solar radiation in Queensland, which was afterward evaluated against MARS, SVR, and a multiple linear regression (MLR) algorithm. Hybrid CRO-(ELM)-ELM system was able to yield promising results.

Although the MARS model has been used in diverse fields, its application in solar radiation forecasting is very limited. Through this chapter, I intend to find answers to whether MARS is a versatile forecast model for daily and monthly solar radiation forecasting in regional Queensland. The ARIMA model was developed to offer a comparative framework for the MARS model.

13.3 Materials and methodology

13.3.1 Study area

A set of solar radiation prediction models is to be constructed using the MARS model for regional sites in Queensland. Queensland is Australia's second-largest state covering 1.9 million km^2 and having a population of about 4.8 million citizens. Queensland is known as the Sunshine State as it is perceived to have more days of sunshine per year than the other states as it has a more consistent climate. The target sites that have been selected for this chapter are Toowoomba (27.56°S, 151.95°E) in Darling Downs South West QLD, Cairns (16.90°S, 145.71°E) in Far North QLD, Mount Isa (20.72°S, 139.49°E) in North QLD, and Longreach (23.44°S, 144.25°E) in Central QLD (Fig. 13.1).

In selecting the target sites, the following climatic variables are looked upon (Table 13.1).

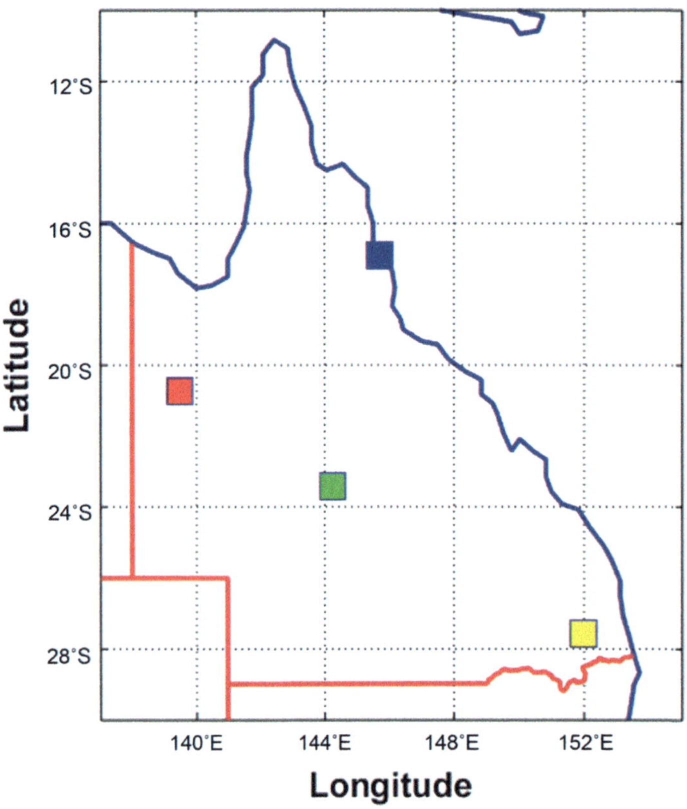

FIGURE 13.1 The study site map for daily and monthly solar radiation prediction. Toowoomba (27.56°S, 151.95°E) in Darling Downs South West QLD, Cairns (16.90°S, 145.71°E) in Far North QLD, Mount Isa (20.72°S, 139.49°E) in North QLD, and Longreach (23.44°S, 144.25°E) in Central QLD.

TABLE 13.1 The weather statistics for the four target sites.

	Cairns	Mount Isa	Toowoomba	Longreach
Temperature (°C)	27–30	30–33	24–27	30–33
Rainfall (mm)	2000–3000	400–600	600–1000	400–600
Humidity (%)	90–100	40–50	90–100	40–50
Solar exposure (MJ/m^2)	18–21	21–24	18–21	21–24

Temperature—average daily maximum temperature annual; rainfall—average rainfall annual; humidity—average daily 9 a.m. relative humidity annual; solar exposure—average daily solar exposure annual.
Bureau of Meteorology, Australian Government.

13.3.2 Data

The data for the period January 1, 1950 to December 31, 2017 is extracted from Scientific Information for Land Owners (SILO) (Jeffrey et al., 2001) operated by the Queensland Department of Environment and Resource Management. SILO data contain Australian climate data from 1889.

The input data comprise seven primary variables including maximum and minimum air temperature, evaporation, solar radiation, vapor pressure, and mean sea level pressure and nine derived variables from the primary variables. For this research, only seven input variables comprising maximum and minimum temperature, rainfall, evaporation, vapor pressure, relative humidity at maximum and minimum temperature against the target variable solar radiation have been used.

The advantages of SILO data are that it provides the data for the whole of Australia with no missing data as the missing values are interpolated as per statistical strategies. The datasets are ready to use in a variety of formats and coordinated effort is cultivated by the utilization of normal datasets. Hence using SILO data for this research is preferred. Table 13.2 describes the summary statistics of input and output variables for the present study sites.

13.3.3 Methodology

13.3.3.1 Theory of MARS model

The MARS model is a nonparametric regression technique first introduced by Jerome H. Friedman in 1991 (Friedman, 1991). It builds MLR models across the range of predictor values. MARS has the ability to explore complex and nonlinear behaviors between predictors and predictand (Deo et al., 2015) and to analyze each input's influence on modeling the objective variable. The main advantage of the MARS model is that it makes no assumptions on the relationship between the response variable and the predictor variables (Friedman, 1991; Deo et al., 2017; Butte et al., 2010).

MARS builds a model in two steps. First, it creates a collection of basis functions based on the trends from training data partitioned into splines (Friedman, 1991; Deo et al., 2017). For each spline, a separate linear regression is modeled, each with its own slope. The inputs are divided into subgroups and knots within the spline (Friedman, 1991; Sephton, 2001). Knots are the connections between the separate regression lines. MARS automatically searches for the best spots to place the knots.

Fig. 13.2 shows a diagram of the MARS model (Deo et al., 2017). In this study, the predictors for MARS, X defined by (maxT, minT, Rainfall, VP, Evap, RHmaxT, RHminT) are utilized to model the objective variable Y, solar radiation (Radn).

Let suppose that X is the predictor vector (X_1, X_2,. . ., X_N), the predictand Y is expressed as:

$$Y = g(X) + \varepsilon \tag{13.1}$$

where ε is the model error distribution and N is the number of training data points (Deo et al., 2017; Deo et al., 2015).

Then MARS applies BF(x) with piecewise linear functions max(0, $x - c$) to try to approximate $g(.)$. The knot appears at the position c (Deo et al., 2017; Zhang and Goh, 2013):

$$\max(0, x - c) = \begin{cases} x - c, & c \geq t \\ 0, & \text{otherwise} \end{cases} \tag{13.2}$$

Thus the MARS model relationship between X and Y is exhibited as:

$$Y = g(X) = a_0 + \sum_{n=1}^{N} a_m BF(X) \tag{13.3}$$

TABLE 13.2 The descriptive statistics of input and output variables for the present study sites (January 1, 1950–December 31, 2017).

Variable	Notation	Minimum	Maximum	Mean	Standard deviation	Skewness	Kurtosis
Cairns (16.90°S, 145.71°E)							
Maximum temperature	Tmax (°C)	18	39.5	28.7	2.7	− 0.141	− 0.413
Minimum temperature	Tmin (°C)	6.5	27.5	20.5	3.1	− 0.803	0.469
Rainfall	Rain (mm)	0	408.3	5.7	18.3	7.079	72.313
Evaporation	Evap (mm)	0.2	34.6	5.5	1.6	0.447	4.392
Vapor pressure	VP (hPa)	5	34	23.1	4.6	− 0.408	− 0.243
Relative humidity at maximum temperature	RHmaxT (%)	16.3	95	58.4	9.5	0.144	0.496
Relative humidity at minimum temperature	RHminT (%)	34.1	100	93.7	7.2	− 1.912	5.71
Solar radiation	Radn (MJ/m^2)	4	30	19	5.4	− 0.427	− 0.505
Longreach (23.44°S, 144.25°E)							
Maximum temperature	Tmax (°C)	8.5	47	31.4	6.1	− 0.266	− 0.704
Minimum temperature	Tmin (°C)	− 2	31	15.9	6.6	− 0.351	− 0.879
Rainfall	Rain (mm)	0	195.6	1.2	6.3	10.531	168.974
Evaporation	Evap (mm)	0.2	23.2	8	3.2	0.274	− 0.605
Vapor pressure	VP (hPa)	0	39	14	6.2	0.295	− 0.822
Relative humidity at maximum temperature	RHmaxT (%)	0	96.6	30.2	13.4	1.046	1.837
Relative humidity at minimum temperature	RHminT (%)	0	100	72.4	20.3	− 0.425	− 0.592
Solar radiation	Radn (MJ/m^2)	4	32	21.3	5.8	− 0.381	− 0.313
Mount Isa (20.72°S, 139.49°E)							
Minimum temperature	Tmax (°C)	9	45.5	31.7	5.7	− 0.354	− 0.616
Maximum temperature	Tmin (°C)	− 2	31.5	17.5	6.2	− 0.452	− 0.751
Rainfall	Rain (mm)	0	189.2	1.2	6	9.671	140.372
Evaporation	Evap (mm)	0.4	21.2	8.3	2.8	0.352	− 0.134

(Continued)

TABLE 13.2 (Continued)

Variable	Notation	Minimum	Maximum	Mean	Standard deviation	Skewness	Kurtosis
Vapor pressure	VP (hPa)	0	42	12.8	6.8	0.594	− 0.755
Relative humidity at maximum temperature	RHmaxT (%)	0	96.6	27.1	13.8	1.078	1.539
Relative humidity at minimum temperature	RHminT (%)	0	100	60.3	22	− 0.017	− 0.831
Solar radiation	Radn (MJ/m^2)	4	32	21.7	5.3	− 0.471	0.132
Toowoomba (27.56°S, 151.95°E)							
Minimum temperature	Tmax (°C)	7	41	23.1	5.1	0.041	− 0.501
Maximum temperature	Tmin (°C)	− 6.5	25.5	11.1	5.6	− 0.495	− 0.586
Rainfall	Rain (mm)	0	164.4	2.5	8.1	6.888	72.692
Evaporation	Evap (mm)	0.2	15.2	4.4	2	0.372	− 0.411
Vapor pressure	VP (hPa)	3	27	14.3	4.6	0.015	− 0.711
Relative humidity at maximum temperature	RHmaxT (%)	10.9	100	49.9	13.5	0.386	0.552
Relative humidity at minimum temperature	RHminT (%)	28.2	100	95.6	28.2	− 2.922	10.348
Solar radiation	Radn (MJ/m^2)	3	32	18.5	6.4	− 0.099	− 0.709

where a_0 is a constant, $\{a_n\}_1^N$ is a model coefficient, N is the number of subregions, and $BF(x)$ is a spline function defined as $C\ (X|s, t_1, t, t_2)$ where $t_1 < t < t_2$ and $-1 < s < +1$.

In the forward phase, MARS repeatedly adds basis functions in pairs to the model and tries to find the pairs of basis functions that give the maximum reduction in the residual error sum of squares. It continues until the change in residual error is too small to continue or until the maximum number of terms is reached. The backward phase of the MARS model prunes the overfitted model. It removes terms one by one discarding the least effective term according to the generalized cross-validation criterion (*GCV*) until it finds the best submodel (Deo et al., 2017):

$$GCV = \frac{MSE}{\left[1 - \frac{\tilde{G}(M)}{N}\right]^2} \tag{13.4}$$

where $MSE = \frac{1}{N}\sum_{i=1}^{N}[Yi - \hat{f}(Xi)]^2$ and $\left[1 - \frac{\tilde{G}(M)}{N}\right]^2$ is a penalty that accounts for an increasing variance from a complex model.

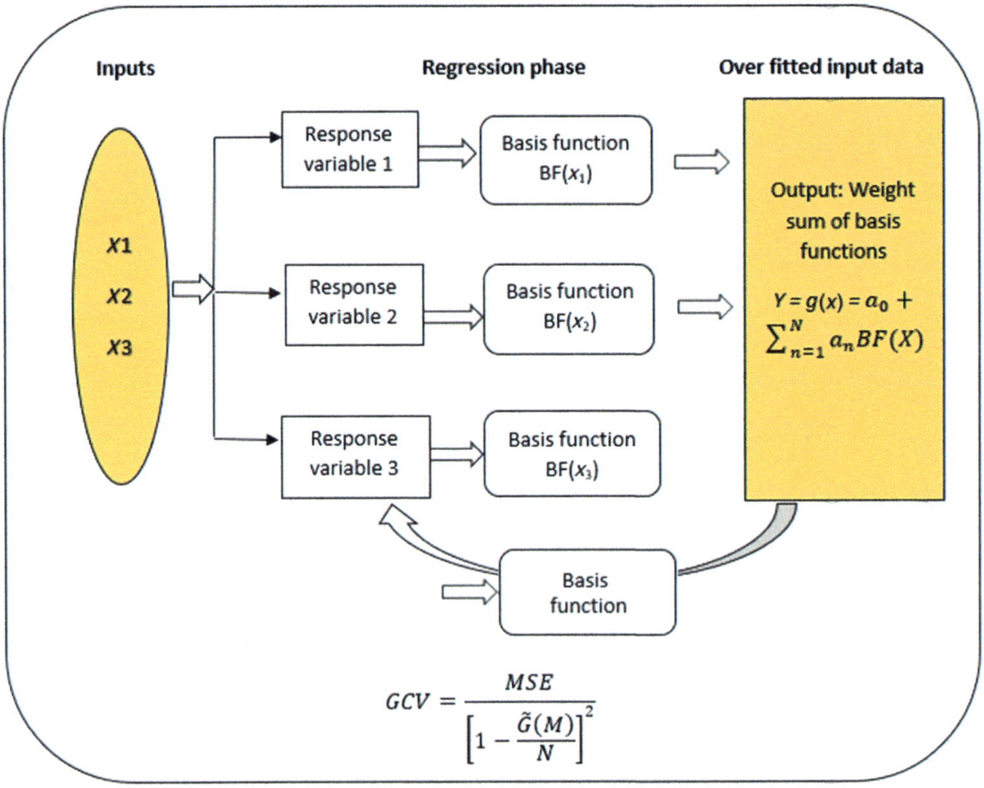

FIGURE 13.2 The schematic view of the MARS structure adopted for forecasting daily and monthly solar radiation. BF(X1), BF(X2), and BF(X3) are basis functions of input 1, 2, and 3, respectively. GCV denotes generalized cross-validation criterion, and MSE is mean square error, where N is the number of subregions, and $\tilde{G}(M)$ is a penalty factor.

$\tilde{G}(M) = C(M) + \nu.M$ where ν is a penalty factor with a characteristic value of $v = 3$ and $C(M)$ is the number of parameters being fitted.

The MARS model with the lowest value of the *GCV* for the training data set is considered the optimal model (Al-Musaylh et al., 2018a).

13.3.3.2 *Theory of the autoregressive integrated moving average model*

In time series analysis, ARIMA is a generalization of an ARMA model. ARIMA models can be discussed using the Box–Jenkins approach (Box and Jenkins, 1976).

To develop the ARIMA model, two types of linear regressions are integrated together. The autoregressive (AR) part indicates that the predictand is regressed on its prior or the lagged values and is written as (Al-Musaylh et al., 2018a; Box and Jenkins, 1976):

$$y_t = c + a_1 y_{t-1} + \ldots + a_p y_{t-p} + u_t \tag{13.5}$$

where $a_1, \ldots a_p$ are AR parameters, c is a constant, p is the order of AR, and u_t is the white noise.

The moving average (MA) part shows that the regression error is made linearly combining the error terms which occurred at various times (Al-Musaylh et al., 2018a; Box and Jenkins, 1976):

$$y_t = \mu + m_1 u_{t-1} + \ \ldots \ + m_q u_{t-q} + u_t \tag{13.6}$$

where $m_1, \ldots, m_p$ are MA parameters, μ is the expectation of y_t, q is the order of AR and $u_{t-1}, \ldots, u_{t-q}, u_t$ are the white noise.

ARIMA models are generally denoted as ARIMA (p, d, q) where parameters p, d, and q represent the order of AR model, degree of differencing, and the order of the MA model, respectively.

To select the best order of a nonseasonal ARIMA model, two criteria are used. They are Akaike information criteria (AIC) and Bayesian information criteria (BIC). The difference between AIC and BIC is that AIC tries to fit the best realistic model while BIC tries to find the optimal fit:

$$AIC = -2\log(L) + 2(p + q + k) \tag{13.7}$$

$$BIC = AIC + (\log(T) - 2)(p + q + k) \tag{13.8}$$

where p is the order of the AR part, q is the order of MA part, k is the intercept of the model, and L is the likelihood of the data. The lowest values for AIC and BIC represent the best model.

As identifying these optimal p, d, and q consumes time, in this chapter p, d, and q have been chosen using the auto ARIMA method which optimizes the parameters within the model itself, and then the modeling was done. The ARIMA model was developed using the "R" software with its architecture established by an iterative modeling process.

13.3.3.3 The model development

The predictive models were developed by the MATLAB-based ARESLab toolbox. The main purpose of this research is to explore the suitability of the MARS model in forecasting global solar radiation. One of the most difficult tasks was to decide the size of data partitions as the training set to build up the model and testing set to assess the performance. The researchers have used different data partitioning for their works, as there is no rule for choosing the size of the datasets.

To avoid numeric issues caused by data attributes or large fluctuations, the scaling of input variables and solar radiation was performed as a normalization between 0 and 1:

$$x_{\text{norm}} = \frac{(x - x_{\text{min}})}{(x_{\text{max}} - x_{\text{min}})} \tag{13.9}$$

where x is the input variable or output variable, x_{min} is the minimum of the input, x_{max} is the maximum of the input, and x_{norm} is the normalized input.

In this research, four data partitions have been used in order to check which data partition yields the best results. The data were partitioned into 40% training, 30% validation, and 30% testing; 60% training, 20% validation, and 20% testing; 70% training, 15% validation, and 15% testing; and 80% training, 10% validation, and 10% testing. Both daily and monthly forecast models were developed utilizing these data divisions.

There are two possible ways of arranging data for forecasting. Statistically significant lags of inputs are being used for forecasting. This requires creating lagged data. The second way is shifting the target variable to get different forecast horizons without any changes being done to the input variables. In this research, the second method has been used. For each daily and monthly time series, three forecast horizons have been developed. The daily model was developed using 1-day, 3-day, and 7-day ahead forecast horizons and the monthly model was developed using 1-month, 3-month, and 6-month ahead forecast horizons.

Tables 13.3 and 13.4 show different data partitions for the 1-day, 3-day, and 7-day ahead forecast horizons and 1-month, 3-month, and 6-month ahead forecast horizons in Cairns.

13.3.3.4 Model evaluation criteria

To assess the veracity of the MARS model for forecasting global solar radiation, a thorough statistical evaluation was performed using statistical measures:

I. Correlation coefficient (r) is expressed as:

$$r = \left[\frac{\sum_{i=1}^{N} (Data\_obs_i - <Data\_obs_i>)(Data\_sim_i - <Data\_sim_i>)}{\sqrt{\sum_{i=1}^{N} (Data\_obs_i - <Data\_obs>)^2} \sqrt{\sum_{i=1}^{N} (Data\_sim_i - <Data\_sim>)^2}} \right] \quad (13.10)$$

where r should fall within 0 and 1. When it is closer to +1, the fit is better between the 2 factors. $r = 0$ means that it cannot be predicted using the independent variable.

II. Nash–Sutcliffe coefficient (E_{NS}) is expressed as:

$$E_{NS} = 1 - \left[\frac{\sum_{i=1}^{N} (Data\_obs_i - Data\_sim_i)^2}{\sum_{i=1}^{N} (Data\_obs_i - <Data\_obs>)^2} \right] \quad (13.11)$$

where E_{NS} can take values from 0 to 1. $E_{NS} = 1$ indicates a perfect match of modeled discharge to the observed data.

III. Wilmott's index (WI) is expressed as:

$$WI = 1 - \left[\frac{\sum_{i=1}^{N} (Data\_obs_i - Data\_sim_i)^2}{\sum_{i=1}^{N} (|Data\_sim_i - <Data\_obs>| + |Data\_obs_i - <Data\_sim>|)^2} \right] \quad (13.12)$$

where WI can take values from 0 to 1. $WI = 1$ indicates a perfect match of modeled discharge to the observed data.

TABLE 13.3 The descriptive table of different data partitions for the daily data set of Cairns (January 1, 1950–December 31, 2017).

Daily	Shift of the target variable	Data size	Data partition		
			Training	**Validation**	**Testing**
Cairns	None	24,837	40%	30%	30%
		01-01-1950 to 31-12-2017	01-01-1950 to 14-03-1977	15-03-1977 to 07-08-1997	08-08-1997 to 31-12-2017
			60%	20%	20%
			01-01-1950 to 20-10-1990	21-10-1990 to 25-05-2004	26-05-2004 to 31-12-2017
			70%	15%	15%
			01-01-1950 to 06-08-1997	07-08-1997 to 19-10-2007	20-10-2007 to 31-12-2017
			80%	10%	10%
			01-01-1950 to 25-05-2004	26-05-2004 to 14-03-2011	15-03-2011 to 31-12-2017
	1	24,836	40%	30%	30%
		01-01-1950 to 30-12-2017	01-01-1950 to 13-03-1977	14-03-1997 to 06-08-1997	07-08-1997 to 30-12-2017
			60%	20%	20%
			01-01-1950 to 19-10-1990	20-10-1990 to 24-05-2004	25-05-2004 to 30-12-2017
			70%	15%	15%
			01-01-1950 to 07-08-1997	08-08-1997 to 19-10-2007	20-10-2007 to 30-12-2017
			80%	10%	10%
			01-01-1950 to 24-05-2004	25-05-2004 to 13-03-2011	14-03-2011 to 30-12-2017
	3	24,834	40%	30%	30%
		01-01-1950 to 28-12-2017	01-01-1950 to 13-03-1977	14-03-1997 to 05-08-1997	06-08-1997 to 28-12-2017
			60%	20%	20%
			01-01-1950 to 17-10-1990	18-10-1990 to 22-05-2004	23-05-2004 to 28-12-2017
			70%	15%	15%
			01-01-1950 to 05-08-1997	06-08-1997 to 17-10-2007	18-10-2007 to 28-12-2017

(Continued)

TABLE 13.3 (Continued)

Daily	Shift of the target variable	Data size	Training	Data partition Validation	Testing
			80%	10%	10%
			01-01-1950 to 24-05-2004	25-05-2004 to 12-03-2011	13-03-2011 to 28-12-2017
	7	24,830	40%	30%	30%
		01-01-1950 to 24-12-2017	01-01-1950 to 11-03-1977	12-03-1997 to 02-08-1997	03-08-1997 to 24-12-2017
			60%	20%	20%
			01-01-1950 to 15-10-1990	16-10-1990 to 19-05-2004	20-05-2004 to 24-12-2017
			70%	15%	15%
			01-01-1950 to 03-08-1997	04-08-1997 to 14-10-2007	15-10-2007 to 24-12-2017
			80%	10%	10%
			01-01-1950 to 20-05-2004	21-05-2004 to 08-03-2011	09-03-2011 to 24-12-2017

IV. Root mean square error (*RMSE*: MJ/m^2) is expressed as:

$$RMSE = \sqrt{\frac{1}{N}\sum_{i=1}^{N}(Data\_sim_i - Data\_obs_i)^2} \tag{13.13}$$

where the lower the *RMSE*, the better match the modeled solar radiation is to the observed data.

V. Mean absolute error (*MAE*: MJ/m^2) is expressed as:

$$MAE = \frac{1}{N}\sum_{i=1}^{N}\left|(Data\_sim_i - Data\_obs_i)\right| \tag{13.14}$$

where the lower the *MAE*, the better the match of the observed and forecasted solar radiation.

XVI. Relative root mean square error (*RRMSE*: %) is expressed as:

$$RRMSE = \frac{\sqrt{\frac{1}{N}\sum_{i=1}^{N}(Data\_sim_i - Data\_obs_i)^2}}{\frac{1}{N}\sum_{i=1}^{N}(Data\_sim_i)^2} \tag{13.15}$$

where the lower the *RRMSE*, the better the model is performing.

TABLE 13.4 The descriptive table of different data partitions for the monthly data set of Cairns (January 1, 1950—December 31, 2017).

Monthly	Shift of the target variable	Data size	Training	Data partition	
				Validation	Testing
Cairns	None	816	40%	30%	30%
		01-1950 to 12-2017	01-1950 to 02-1977	03-1977 to 07-1997	08-1997 to 12-2017
			60%	20%	20%
			01-1950 to 10-1990	11-1990 to 05-2004	06-2004 to 12-2017
			70%	15%	15%
			01-1950 to 08-1997	09-1997 to 10-2007	11-2007 to 12-2017
			80%	10%	10%
			01-1950 to 04-2004	05-2004 to 02-2011	03-2011 to 12-2017
	1	815	40%	30%	30%
		01-1950 to 11-2017	01-1950 to 02-1977	03-1977 to 07-1997	08-1997 to 11-2017
			60%	20%	20%
			01-1950 to 09-1990	10-1990 to 04-2004	05-2004 to 11-2017
			70%	15%	15%
			01-1950 to 07-1997	08-1997 to 09-2007	10-2007 to 11-2017
			80%	10%	10%
			01-1950 to 03-2004	04-2004 to 01-2011	02-2011 to 11-2017
	3	813	40%	30%	30%
		01-1950 to 09-2017	01-1950 to 01-1977	02-1977 to 05-1997	06-1997 to 09-2017
			60%	20%	20%
			01-1950 to 06-1990	07-1990 to 01-2004	02-2004 to 09-2017
			70%	15%	15%
			01-1950 to 05-1997	06-1997 to 07-2007	08-2007 to 09-2017
			80%	10%	10%
			01-1950 to 03-2004	04-2004 to 12-2010	01-2011 to 09-2017
	6	810	40%	30%	30%
		01-1950 to 06-2017	01-1950 to 12-1976	01-1977 to 03-1997	04-1997 to 06-2017
			60%	20%	20%
			01-1950 to 06-1990	07-1990 to 12-2003	01-2004 to 06-2017

(Continued)

TABLE 13.4 (Continued)

Monthly	Shift of the target variable	Data size	Training	Data partition Validation	Testing
			70%	15%	15%
			01-1950 to 04-1997	05-1997 to 05-2007	06-2007 to 06-2017
			80%	10%	10%
			01-1950 to 12-2003	01-2004 to 09-2010	10-2010 to 06-2017

XVII. Legates and Nash–Sutcliffe coefficient (L) is expressed as:

$$L = 1 - \frac{\sum_{i=1}^{N} |Data\_sim_i - Data\_obs_i|}{|Data\_obs_i - < Data\_obs > |} \quad (13.16)$$

where higher values attained for L show that the model is performing better.

$Data\_obs_i$ and $Data\_sim_i$ are measured and forecasted solar radiation, respectively. $< Data_{obsi} >$ and $< Data_{sim_i} >$ were the average observed and forecasted solar radiation in the test or train period, N was the number of data points in the respective period.

13.4 Results and discussion

13.4.1 MARS model for short-term forecasting

This section illustrates the results of the short-term (daily) forecasting model for solar radiation in regional Queensland. It presents the model development criteria as in the model performance matrix for training and validation and the estimation and forecasting techniques of the model.

13.4.1.1 Model development

To develop a robust MARS model for the daily forecasting, a vital task was to optimize the architecture of the model by deciding the best data partition and the best forecast horizon for daily forecasting. Hence, to determine the effect of training data length, four different sets of data partitions have been used involving 40% of the data for training, 60% of the data for training, 70% of the data for training, and 80% of the data for training. The target variable has been shifted by 1, 3, and 7 to develop 1-day, 3-day, and 7-day ahead forecast horizons. The MARS model has been technologically advanced to generate cubic and linear forecasting models. As the results showed that the linear model slightly outperformed the cubic model, only the linear results are shown here.

Table 13.5 shows the model performance in r, *RMSE*, and *MAE* in training and validation data sets in the four target sites. r shows the linear relationship of the predictive variables and the predictand, while *RMSE* and *MAE* show the overall error of the model.

TABLE 13.5 The model performance criteria Pearson's correlation coefficient (*r*), root mean square error (*RMSE*), and mean absolute error (*MAE*) in training and validation data sets for different target shifts and different data partitions for all the target sites.

		Training			Validation		
Shift of the target variable	**Data partition**	*r*	*RMSE* **(MJ/m²)**	*MAE* **(MJ/m²)**	*r*	*RMSE* **(MJ/m²)**	*MAE* **(MJ/m²)**
Cairns							
No shift	40/30/30	0.853	2.868	2.185	0.892	2.517	1.960
	60/20/20	0.866	2.742	2.088	0.892	2.401	1.866
	70/15/15	0.870	2.686	2.046	0.894	2.397	1.854
	80/10/10	**0.872**	**2.653**	**2.025**	**0.895**	**2.372**	**1.806**
1-day shift	**40/30/30**	**0.690**	**3.980**	**3.088**	0.638	4.233	3.272
	60/20/20	0.673	4.049	3.146	0.639	**4.084**	**3.146**
	70/15/15	0.670	4.044	3.144	**0.645**	4.088	3.149
	80/10/10	0.665	4.050	3.151	0.634	4.110	3.186
3-day shift	**40/30/30**	**0.565**	**4.537**	**3.600**	0.428	4.964	3.892
	60/20/20	0.524	4.662	3.704	0.437	**4.769**	**3.716**
	70/15/15	0.516	4.664	3.701	0.443	4.798	3.762
	80/10/10	0.503	4.683	3.715	**0.445**	4.772	3.745
7-day shift	**40/30/30**	**0.514**	**4.716**	**3.789**	0.373	5.061	4.052
	60/20/20	0.469	4.835	3.885	**0.396**	**4.848**	**3.847**
	70/15/15	0.467	4.813	3.862	0.365	4.988	3.951
	80/20/20	0.453	4.831	3.871	0.387	4.913	3.892
Longreach							
No shift	40/30/30	**0.846**	3.155	**2.291**	**0.841**	3.220	2.457
	60/20/20	0.845	3.164	2.348	0.825	3.154	2.425
	70/15/15	0.844	3.146	2.341	0.832	3.052	2.334
	80/10/10	0.841	**3.146**	2.349	0.850	**2.949**	**2.272**
1-day shift	**40/30/30**	**0.706**	**4.200**	**2.978**	0.666	4.370	3.222
	60/20/20	0.695	4.248	3.073	0.651	4.222	3.144
	70/15/15	0.692	4.238	3.074	0.667	**4.058**	2.993
	80/10/10	0.686	4.232	3.079	**0.679**	4.074	**2.982**
3-day shift	**40/30/30**	**0.629**	**4.609**	**3.307**	0.557	4.842	3.614
	60/20/20	0.609	4.687	3.425	0.553	4.619	3.465

(*Continued*)

TABLE 13.5 (Continued)

Shift of the target variable	Data partition	Training			Validation		
		r	*RMSE* (MJ/m²)	*MAE* (MJ/m²)	*r*	*RMSE* (MJ/m²)	*MAE* (MJ/m²)
	70/15/15	0.605	4.674	3.427	0.576	**4.441**	**3.321**
	80/10/10	0.598	4.660	3.426	**0.581**	4.508	3.354
7-day shift	**40/30/30**	**0.596**	**4.759**	**3.459**	0.513	5.003	3.752
	60/20/20	0.573	4.845	3.575	0.505	4.784	3.616
	70/15/15	0.569	4.828	3.578	0.521	**4.638**	**3.511**
	80/20/20	0.561	4.814	3.581	**0.541**	4.655	3.534
Mount Isa							
No shift	40/30/30	0.826	3.008	2.175	0.838	2.934	2.196
	60/20/20	**0.836**	2.943	2.160	0.818	2.970	2.247
	70/15/15	**0.836**	**2.928**	**2.155**	0.833	2.866	2.178
	80/10/10	0.832	2.934	2.168	**0.850**	**2.811**	**2.106**
1-day shift	**40/30/30**	**0.704**	**3.788**	**2.667**	0.656	4.088	2.908
	60/20/20	0.691	3.877	2.746	0.621	4.047	2.927
	70/15/15	0.687	3.873	2.755	0.627	4.002	2.852
	80/10/10	0.675	3.904	2.780	**0.669**	**3.930**	**2.817**
3-day shift	**40/30/30**	**0.587**	**4.320**	**3.093**	0.512	4.624	3.349
	60/20/20	0.570	4.407	3.169	0.483	**4.499**	3.283
	70/15/15	0.564	4.403	3.167	0.477	4.502	**3.231**
	80/10/10	0.551	4.419	3.185	**0.518**	4.509	3.265
7-day shift	**40/30/30**	**0.559**	**4.423**	**3.197**	0.400	4.979	3.674
	60/20/20	0.522	4.574	3.333	0.391	4.739	3.500
	70/15/15	0.510	4.584	3.344	0.408	**4.675**	**3.430**
	80/20/20	0.497	4.595	3.358	**0.436**	4.748	3.471
Toowoomba							
No shift	40/30/30	0.867	3.200	2.457	0.881	3.218	2.535
	60/20/20	0.871	3.167	2.452	0.881	3.126	2.465
	70/15/15	**0.874**	3.144	2.430	0.878	3.066	2.420
	80/10/10	0.873	**3.135**	**2.422**	**0.882**	**2.984**	**2.345**
1-day shift	**40/30/30**	**0.675**	**4.746**	3.644	0.660	4.951	3.742
	60/20/20	0.670	4.784	3.672	**0.670**	4.787	3.615

(*Continued*)

TABLE 13.5 (Continued)

Shift of the target variable	Data partition	Training *r*	Training *RMSE* (MJ/m^2)	Training *MAE* (MJ/m^2)	Validation *r*	Validation *RMSE* (MJ/m^2)	Validation *MAE* (MJ/m^2)
	70/15/15	0.673	4.776	3.653	0.668	**4.671**	**3.547**
	80/10/10	0.672	4.767	**3.636**	0.665	4.669	3.604
3-day shift	**40/30/30**	**0.574**	**5.265**	**4.120**	0.517	5.589	4.338
	60/20/20	0.557	5.356	4.177	**0.547**	5.364	4.144
	70/15/15	0.561	5.350	4.153	0.546	**5.228**	**4.032**
	80/10/10	0.558	5.339	4.137	0.526	5.300	4.162
7-day shift	**40/30/30**	**0.539**	**5.419**	**4.254**	0.480	5.719	4.502
	60/20/20	0.522	5.500	4.308	**0.522**	5.467	4.285
	70/15/15	0.522	5.500	4.308	**0.522**	5.467	4.285
	80/20/20	0.525	5.477	4.274	0.495	**5.414**	**4.272**

The optimal data partitions yielding the highest *r*, and lowest *RMSE* and *MAE* for each shift and the target site are in red bold-faced.

RMSE and *MAE* can express more about the model as *r* will not describe the nonlinear relationships in the model or the effect of outliers.

Training and validation parameters for estimating and forecasting the daily solar radiation are good, so it can be assumed that the model will perform well in the testing data set too. As highlighted, 80% for training data set performed best for all the target sites when there are no shifts for the target, that is when the results are estimated from the observed values. For all the forecast horizons, 40% of data for the training set has produced the best results.

According to the estimation results, Cairns performed the best in estimating the solar radiation as it has yielded an *r* value of 0.872 for training and has the lowest *RMSE* value of 2.653. By comparing *r* and *RMSE* for all the sites, Longreach yielded the second-lowest *r* and highest *RMSE*. Hence, Longreach can be mentioned as the lowest performing target site for estimating solar radiation.

For the forecasting horizons, the 1-day ahead forecast has produced the best results in high *r* and low *RMSE* and *MAE* for all the sites. Hence, for daily solar radiation forecasting, 40% data in training set and the 1-day ahead forecast have yielded the best simulated results.

The MARS model is a weighted sum of basis functions while the MARS equation is a summation of the basis functions multiplied by its coefficient. Table 13.6 shows the basis functions and the MARS equation for all the target sites with no shift of the target variable and 80% data for training set.

The MARS equation for Cairns target site is:

$$y = -1.5216 + 0.37212 \times BF1 - 1.8822 \times BF2 - 1.1543 \times BF3 + 0.67887 \times BF4 - 0.89141 \times BF5 - 0.74797 \times BF6 + 0.34617 \times BF7 + 17.591 \times BF8 + 1.5935 \times BF9 + 0.76221 \times BF10 - 0.60283 \times BF11 + 0.96039 \times BF12 - 17.659 \times BF13 + 18.169 \times BF14 + 0.50935 \times BF15 - 0.34461 \times BF16 \quad (13.17)$$

TABLE 13.6 The basis functions of the MARS model for estimating daily solar radiation in four target sites.

Cairns	Longreach
BF1 = max(0, Evap −0.2093)	BF1 = max(0, Evap −0.43478)
BF2 = max(0,0.2093 − Evap)	BF2 = max(0, RHmaxT −0.47308)
BF3 = max(0, RHmaxT −0.50064)	BF3 = max(0,0.47308 − RHmaxT)
BF4 = max(0,0.50064 − RHmaxT)	BF4 = max(0, RHminT −0.691)
BF5 = max(0, maxT −0.65116)	BF5 = max(0, maxT −0.45455)
BF6 = max(0, minT −0.61905)	BF6 = max(0, minT −0.68182)
BF7 = max(0,0.61905 − minT)	BF7 = max(0, Evap −0.17391)
BF8 = max(0, Rainfall −0.00097967)	BF8 = max(0, VP −0.64103)
BF9 = max(0,0.098837 − Evap)	BF9 = max(0,0.64103 − VP)
BF10 = max(0, VP −0.7931)	BF10 = max(0, minT −0.24242)
BF11 = max(0,0.7931 − VP)	BF11 = max(0,0.75325 − maxT)
BF12 = max(0, RHmaxT −0.7967)	BF12 = max(0,0.50649 − maxT)
BF13 = max(0, Rainfall −0.13079)	
BF14 = max(0,0.13079 − Rainfall)	
BF15 = max(0, maxT −0.39535)	
BF16 = max(0,0.39535 − maxT)	
Mount Isa	**Toowoomba**
BF1 = max(0, Evap −0.52885)	BF1 = max(0,0.38159 − RHmaxT)
BF2 = max(0,0.52885 − Evap)	BF2 = max(0, RHminT −0.88997)
BF3 = max(0, RHmaxT −0.49379)	BF3 = max(0,0.88997 − RHminT)
BF4 = max(0, maxT −0.50685)	BF4 = max(0, minT −0.625)
BF5 = max(0,0.50685 − maxT)	BF5 = max(0,0.625 − minT)
BF6 = max(0, minT −0.74627)	BF6 = max(0, maxT −0.36765)
BF7 = max(0,0.74627 − minT)	BF7 = max(0, Evap −0.4)
BF8 = max(0, VP −0.47619)	BF8 = max(0,0.4 − Evap)
BF9 = max(0,0.47619 − VP)	BF9 = max(0, Evap −0.12)
BF10 = max(0, Rainfall −0.00052854)	BF10 = max(0, RHmaxT −0.75309)
BF11 = max(0,0.00052854 − Rainfall)	BF11 = max(0,0.75309 − RHmaxT)
BF12 = max(0, RHminT −0.783)	BF12 = max(0, VP −0.70833)
BF13 = max(0,0.22115 − Evap)	BF13 = max(0,0.70833 − VP)
BF14 = max(0,0.62008 − RHmaxT)	BF14 = max(0,0.4375 − minT)
BF15 = max(0, maxT −0.80822)	

The definition of acronyms used here are as follows: *BF*, Basis function; *Evap*, evaporation; *maxT*, maximum temperature; *minT*, minimum temperature; *Rainfall*, amount of rainfall; *RHmaxT*, relative humidity at maximum temperature; *RHminT*, relative humidity at minimum temperature; *VP*, vapor pressure.

The MARS equation for Longreach target site is:

$$y = 0.52107 - 0.69237 \times BF1 - 0.83535 \times BF2 + 0.65451 \times BF3 + 0.15771 \times BF4 + 0.24392 \times BF5 - 0.35565 \times BF6 + 1.0503 \times BF7 + 0.715 \times BF8 - 0.26435 \times BF9 - 0.32339 \times BF10 - 0.31573 \times BF11 + 0.36689 \times BF12 \quad (13.18)$$

The MARS equation for Mount Isa target site is:

$$y = 0.46972 + 0.23384 \times BF1 - 0.74019 \times BF2 - 0.58955 \times BF3 + 0.58677 \times BF4 - 0.088748 \times BF5 - 0.68414 \times BF6 + 0.22179 \times BF7 + 0.43388 \times BF8 - 0.096172 \times BF9 - 0.075182 \times BF10 + 49.674 \times BF11 + 0.21767 \times BF12 + 0.37511 \times BF13 + 0.47276 \times BF14 - 0.35633 \times BF15 \quad (13.19)$$

The MARS equation for Toowoomba target site is:

$$y = 0.20342 - 0.59565 \times BF1 + 0.16462 \times BF2 + 0.066377 \times BF3 - 0.73044 \times BF4 + 0.22158 \times BF5 + 0.21853 \times BF6 - 0.40813 \times BF7 - 0.27239 \times BF8 + 0.78908 \times BF9 - 0.29743 \times BF10 + 0.88602 \times BF11 + 0.36567 \times BF12 - 0.21693 \times BF13 - 0.21682 \times BF14 \quad (13.20)$$

where y is the estimated daily solar radiation and BF_i are the basis functions.

Table 13.7 shows the basis functions and the MARS equation for all four target sites with 1-day shift of the target variable and 40% data for training set.

The respective MARS equation for Cairns target site is:

$$y = 0.27366 - 2.9618 \times BF1 - 0.13595 \times BF2 + 0.20023 \times BF3 - 0.70995 \times BF4 - 2.1703 \times BF5 + 0.38587 \times BF6 + 2.499 \times BF7 - 0.66292 \times BF8 + 66.558 \times BF9 - 0.29844 \times BF10 + 0.677 \times BF11 - 0.47033 \times BF12 + 0.59014 \times BF13 + 0.2667 \times BF14 \quad (13.21)$$

The respective MARS equation for Longreach target site is:

$$y = 0.4479 - 0.47663 \times BF1 - 0.64329 \times BF2 + 0.97447 \times BF3 + 0.15687 \times BF4 + 1.3374 \times BF5 - 0.47836 \times BF6 + 0.45015 \times BF7 + 0.17384 \times BF8 - 0.40137 \times BF9 - 0.56888 \times BF10 + 0.24025 \times BF11 - 0.09161 \times BF12 - 1.9087 \times BF13 - 0.27305 \times BF14 \quad (13.22)$$

The respective MARS equation for Mount Isa target site is:

$$y = 0.57232 - 0.69042 \times BF1 - 0.63996 \times BF2 + 0.88881 \times BF3 + 15.379 \times BF4 - 0.1714 \times BF5 + 2.9653 \times BF6 + 0.12614 \times BF7 - 0.42422 \times BF8 - 0.4827 \times BF9 + 0.88277 \times BF10 - 2.615 \times BF11 - 0.29637 \times BF12 + 0.2348 \times BF13 \quad (13.23)$$

The respective MARS equation for Toowoomba target site is:

$$y = -0.064368 - 0.64704 \times BF1 - 0.43227 \times BF2 + 0.15997 \times BF3 + 2.7213 \times BF4 - 0.58054 \times BF5 + 0.25387 * BF6 + 0.094886 \times BF7 - 0.93845 \times BF8 - 3.1144 \times BF9 + 1.9246 \times BF10 + 0.093083 \times BF11 + 0.11702 \times BF12 + 0.6918 \times BF13 + 0.17216 \times BF14 \quad (13.24)$$

TABLE 13.7 The basis functions of the MARS model for forecasting 1-day ahead solar radiation.

Cairns	Longreach
BF1 = max(0, Evap −0.2093)	BF1 = max(0, Evap −0.47826)
BF2 = max(0, RHmaxT −0.44981)	BF2 = max(0, RHmaxT −0.32919)
BF3 = max(0, VP −0.58621)	BF3 = max(0,0.32919 − RHmaxT)
BF4 = max(0,0.58621 − VP)	BF4 = max(0, RHminT −0.653)
BF5 = max(0, minT −0.90476)	BF5 = max(0, Evap −0.16522)
BF6 = max(0,0.90476 − minT)	BF6 = max(0, minT −0.74242)
BF7 = max(0, Evap −0.11047)	BF7 = max(0,0.45455 − maxT)
BF8 = max(0, Rainfall −0.00024492)	BF8 = max(0, VP −0.46154)
BF9 = max(0,0.00024492 − Rainfall)	BF9 = max(0,0.46154 − VP)
BF10 = max(0, RHmaxT −0.6582)	BF10 = max(0, Evap −0.33913)
BF11 = max(0,0.6582 -RHmaxT)	BF11 = max(0,0.33913 − Evap)
BF12 = max(0, maxT −0.62791)	BF12 = max(0, Rainfall −0.025051)
BF13 = max(0, Rainfall −0.08817)	BF13 = max(0,0.025051 − Rainfall)
BF14 = max(0, maxT −0.34884)	BF14 = max(0, maxT −0.72727)

Mount Isa	Toowoomba
BF1 = max(0, Evap −0.52885)	BF1 = max(0,0.25333 − Evap)
BF2 = max(0, RHmaxT −0.20083)	BF2 = max(0, RHmaxT −0.35354)
BF3 = max(0,0.20083 − RHmaxT)	BF3 = max(0,0.35354 − RHmaxT)
BF4 = max(0, RHminT −0.997)	BF4 = max(0, Evap −0.12)
BF5 = max(0,0.997 − RHminT)	BF5 = max(0, minT −0.59375)
BF6 = max(0,0.23077 − Evap)	BF6 = max(0, maxT −0.38235)
BF7 = max(0, maxT −0.45205)	BF7 = max(0,0.38235 − maxT)
BF8 = max(0, minT −0.77612)	BF8 = max(0,0.044404 − Rainfall)
BF9 = max(0,0.32692 − Evap)	BF9 = max(0, Evap −0.4)
BF10 = max(0, Evap −0.20192)	BF10 = max(0,0.4 − Evap)
BF11 = max(0,0.20192 − Evap)	BF11 = max(0, RHminT −0.83983)
BF12 = max(0,0.29851 − minT)	BF12 = max(0,0.83983 − RHminT)
BF13 = max(0,0.53425 − maxT)	BF13 = max(0, Evap −0.34667)
	BF14 = max(0, VP −0.66667)

The definition of acronyms used here are as follows: *BF*, Basis function; *Evap*, evaporation; *maxT*, maximum temperature; *minT*, minimum temperature; *Rainfall*, amount of rainfall; *RHmaxT*, relative humidity at maximum temperature; *RHminT*, relative humidity at minimum temperature; *VP*, vapor pressure.

TABLE 13.8 ARIMA with structure (p, d, q) with d = differencing, p and q = order of autoregressive (AR) and moving average (MA) term.

Target site	ARIMA structure (p, d, q)	AIC	Log-likelihood	BIC	Variance
Cairns	ARIMA (3,0,1)	101050.8	− 50519.4	101097.4	19.567
Longreach	ARIMA (2,0,1)	94614.59	− 47302.3	94653.41	13.51363
Mount Isa	ARIMA (2,0,2)	91898.46	− 45943.2	91945.46	11.55851
Toowoomba	ARIMA (3,0,2)	101156.7	− 51678.2	101153.8	19.478

Akaike's information criterion (AIC) is used to identify the model in conjunction with log-likelihood, Bayesian information criterion (BIC), variance and correlation coefficient.

Table 13.8 displays the ARIMA model's architecture and the respective goodness-of-fit tests performed to construct the forecasting model.

13.4.1.2 Estimation of daily solar radiation in regional Queensland

Estimation of the target variable suggests that there is no forecasting being done. Better results for estimation show that the model is versatile for forecasting the solar radiation. In the previous section, it has been proved that the data partitioning 80% for training set performed the best for estimating the daily solar radiation.

Fig. 13.3 shows the scatterplots for all the target sites in estimating solar radiation

A least squares regression line, $y = ax + b$ is also used to illustrate the relationship between observed solar radiation and estimated solar radiation where b is the intercept of the regression line and a is the coefficient. Cairns shows the best relationship between estimated and observed values as it has the highest coefficient of determination while Mount Isa shows the lowest.

Table 13.9 depicts the performance parameters for the MARS model in estimating the daily solar radiation. Although Toowoomba shows the best r, its *RMSE* and *MAE* are comparatively higher compared to Longreach, which has the second-highest r. Hence, it is difficult to imply which target site is performing best by only referring to these parameters. Longreach and Toowoomba have performed similarly considering these parameters. The geographical parameters like *RRMSE* and *RMAE* give information for selecting the best target site. As Longreach has the lowest *RRMSE* and *RMAE* values, the conclusion is made as Longreach has performed the best out of four target sites in estimating solar radiation.

13.4.1.3 Forecasting daily solar radiation in regional Queensland

In this section, results gained from MARS and ARIMA for forecasting daily solar radiation in regional Queensland are assessed to validate the adequacy of the MARS model and to resolve whether MARS is an adequate model in solar radiation forecasting.

In the model development section, it has been shown that in forecasting solar radiation, 40% for training data partition performed best.

Fig. 13.4 shows the boxplots for 1-day, 3-day, and 7-day ahead forecasted results in training data set for Cairns. The box plot of 1-day ahead forecast has a lower median, upper quartile, maximum, and minimum compared to the 3-day and 7-days ahead

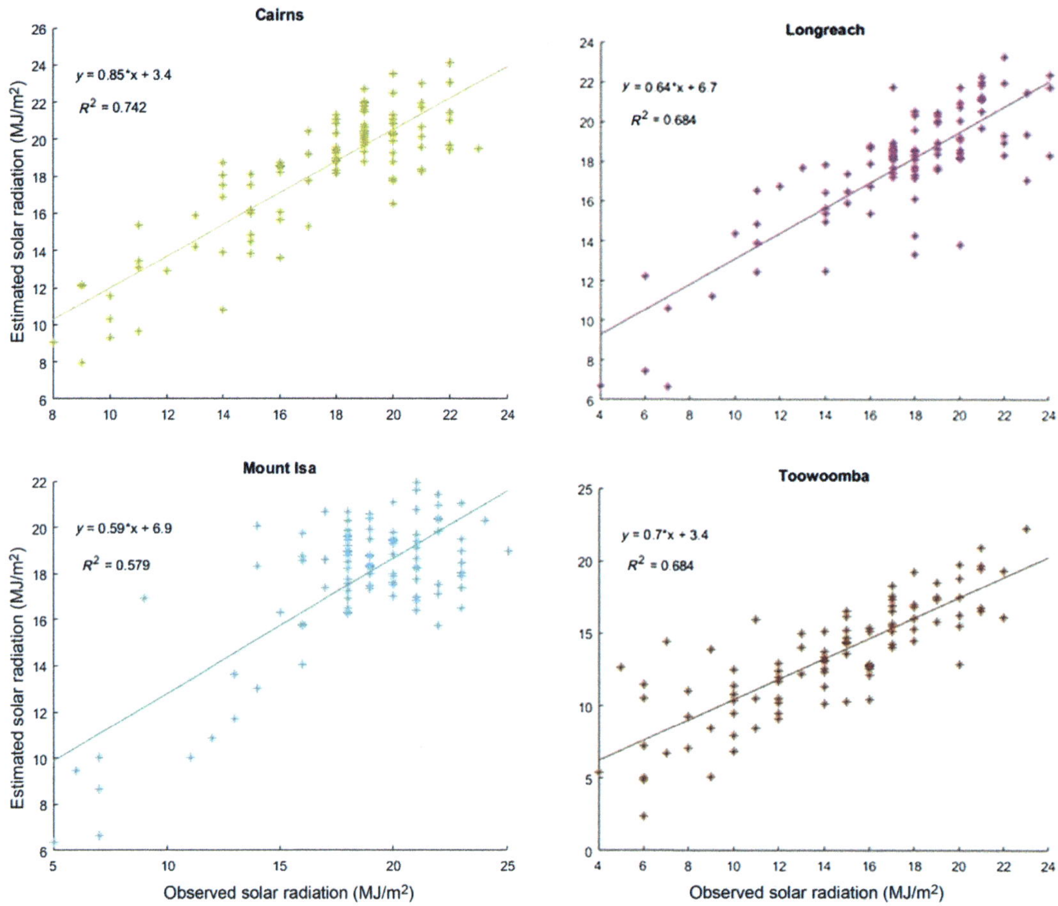

FIGURE 13.3 The scatterplots of observed and estimated daily solar radiation (MJ/m^2) in testing data set (10% data in testing, data: 100 data points from March 16, 2011 to June 23, 2011) for Cairns, Longreach, Mount Isa, and Toowoomba. For each scatterplot, least squares fitting line and respective R^2 are shown. (Note: $R^2 = 1 - \frac{\sum_{i=1}^{N}(\text{data} - \text{estimated data})^2}{\sum_{i=1}^{N}(\text{data} - \text{mean observed data})}$).

boxplots. It says that the errors of the 1-day ahead forecast are comparatively lower than the other forecast horizons. Hence, by this and from the 13.4.1 model development section, it is evident that the 1-day ahead forecast produced the best results in daily forecasting of solar radiation in regional Queensland.

Fig. 13.5 shows a graphical representation of the *RMSE* and *MAE* for all the target sites.

The bar graph shows that the error values of the ARIMA model are higher than those of the MARS model (Fig. 13.5). Hence, it can be concluded that MARS has outperformed the ARIMA model regarding errors.

TABLE 13.9 Correlation coefficient (*r*), root mean square error (*RMSE*), mean absolute error (*MAE*), Wilmott's index (*WI*), Nash–Sutcliffe coefficient (E_{NS}), Legates and McCabe's Index (*L*), geographical parameters relative root mean square error (*RRMSE*), and relative mean absolute error (*RMAE*) values in the testing period for the selected MARS model (80% training) in estimating daily solar radiation (MJ/m^2) for all target sites.

Target site	*r*	*RMSE* (MJ/m^2)	*MAE* (MJ/m^2)	*WI*	E_{NS}	*L*	*RRMSE* (%)	*RMAE* (%)
Cairns	0.831	3.169	2.381	0.795	0.634	0.438	16.676	15.588
Longreach	0.859	2.921	2.228	0.848	0.72	0.499	13.936	12.955
Mount Isa	0.796	3.405	2.552	0.723	0.545	0.362	15.995	14.571
Toowoomba	0.876	3.127	2.432	0.88	0.764	0.545	16.989	16.328

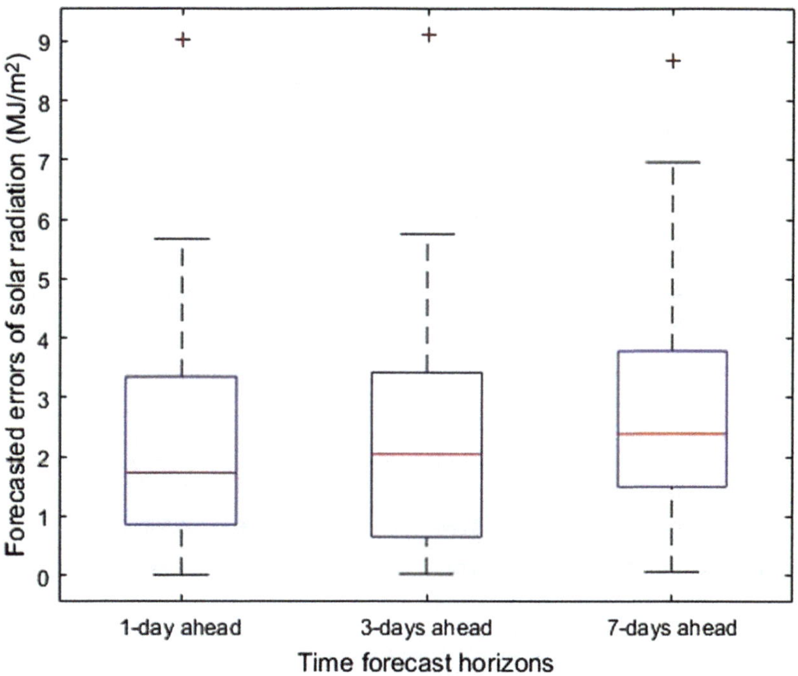

FIGURE 13.4 The box plots of daily forecasted errors in solar radiation (MJ/m^2) in testing data set (30% data in testing, data: 50 data points from August 10, 1997 to September 27, 1997) for Longreach.

Tables 13.10 and 13.11 show the model performance metrics in the testing data set for the MARS model and the comparative model, the ARIMA model. Clearly, the MARS model has performed much better than the ARIMA model, as the ARIMA model's error values are higher than the MARS error values. In addition, ARIMA has very low Wilmott's index, Nash–Sutcliffe coefficient, and Legates and McCabe's index values compared to ARIMA.

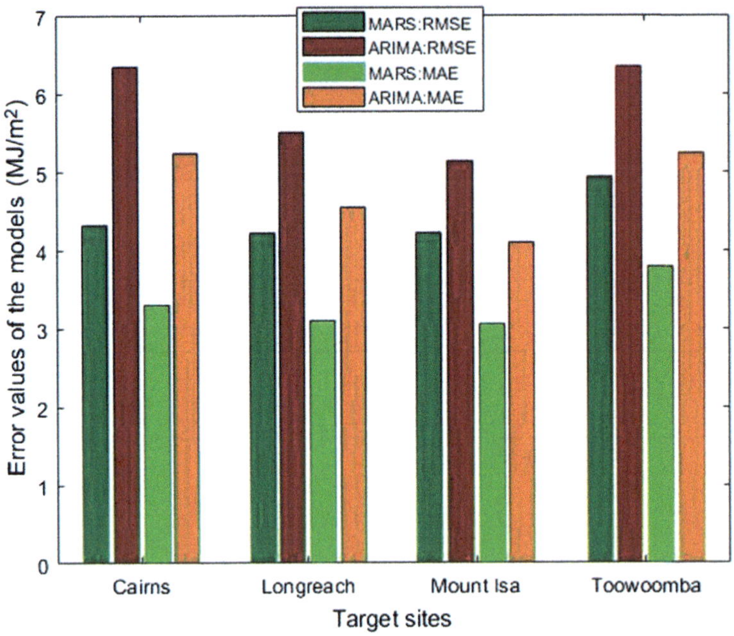

FIGURE 13.5 The bar graph of *RMSE* (MJ/m^2) and *MAE* (MJ/m^2) for daily forecasted solar radiation in 1-day ahead forecast horizon testing data set (30% data in testing, data: 7451 data points from August 8, 1997 to December 31, 2017) for Longreach.

TABLE 13.10 Comparison of correlation coefficient (r), root mean square error (*RMSE*), and mean absolute error (*MAE*) for MARS and ARIMA models in 1-day ahead forecasting of daily solar radiation (MJ/m^2) for all the target sites.

	MARS			ARIMA		
Target site	r	***RMSE* (MJ/m^2)**	***MAE* (MJ/m^2)**	r	***RMSE* (MJ/m^2)**	***MAE* (MJ/m^2)**
Cairns	0.612	4.307	3.298	− 0.053	6.330	5.240
Longreach	**0.655**	**4.215**	3.086	− 0.046	5.507	4.543
Mount Isa	0.607	4.217	**3.043**	− 0.041	5.139	4.088
Toowoomba	0.644	4.928	3.779	− 0.053	6.330	5.240

The best target sites yielding the highest r and lowest *RMSE* and *MAE* are in red bold-face.

Table 13.12 shows the geographical parameters, *RRMSE* and *RMAE*, for the testing data for the MARS model in forecasting 1-day ahead solar radiation. Toowoomba has the highest *RRMSE* and *RMAE* values while the target site Mount Isa has the lowest *RRMSE* and *RMAE* values. However, Longreach has produced the best r, *RMSE*, *WI*, E_{NSi} and L values (shown in Table 13.10 and 13.11). Hence, it can be concluded that the developed MARS model has performed best in forecasting solar radiation for Longreach and it has performed lowest in Toowoomba. Fig. 13.6 shows the time series plot for the best performing site Longreach.

TABLE 13.11 Comparison of Wilmott's Index (*WI*), Nash–Sutcliffe coefficient (E_{NS}), and Legates and McCabe's index (*L*) for MARS and ARIMA models in 1-day ahead forecasting of daily solar radiation (MJ/m^2) for all the target sites.

	MARS			ARIMA		
Target site	*WI*	E_{NS}	*L*	*WI*	E_{NS}	*L*
Cairns	0.602	0.33	0.229	0.045	− 0.009	− 0.005
Longreach	**0.648**	**0.411**	**0.318**	0.018	− 0.006	− 0.005
Mount Isa	0.582	0.323	0.254	0.052	− 0.006	− 0.002
Toowoomba	0.643	0.389	0.275	0.045	− 0.009	− 0.005

The best target sites yielding the highest *WI*, E_{NS}, and *L* are in red bold-face.

TABLE 13.12 The geographical parameters relative root mean square error (*RRMSE*) and relative mean absolute error (*RMAE*) for the MARS model in the 1-day ahead forecast of solar radiation (MJ/m^2).

Target site	*RRMSE* (%)	*RMAE* (%)
Cairns	22.785	21.935
Longreach	19.864	18.162
Mount Isa	**19.509**	**17.470**
Toowoomba	26.447	25.869

The target sites yielding the lowest *RRMSE* and *RMAE* are in red bold-face.

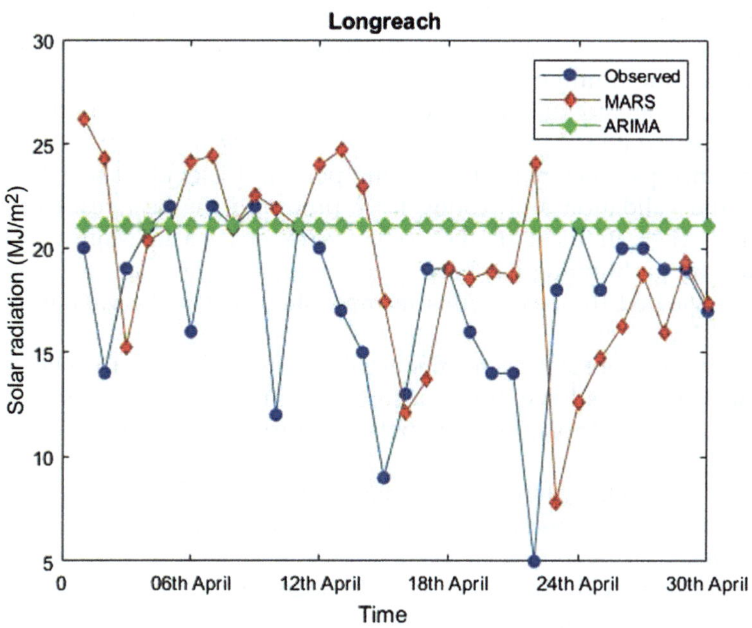

FIGURE 13.6 The time series of observed versus forecasted solar radiation (MJ/m^2) in MARS and ARIMA for the testing data set (30 data points from April 1, 1998 to April 30, 1998) for the best performing site, Toowoomba, and the worst performing site, Mount Isa.

The time series plot of the observed solar radiation versus the forecasted solar radiation in MARS and ARIMA shows that the MARS model-generated forecasted solar radiation is a better fit to the observed solar radiation than that of ARIMA.

These findings indicate that the MARS model outperforms the ARIMA model in forecasting daily solar radiation and Longreach has performed best in forecasting daily solar radiation.

13.4.2 MARS model for long-term solar radiation forecasting and seasonal analysis

This section describes the results of the long-term (monthly) solar radiation estimation and forecasting for regional Queensland using the MARS model. The model development subsection represents the information on data partitioning and the shift of the target variable and the best performance model. The results of the seasonal analysis of solar radiation for selected target sites are also illustrated.

13.4.2.1 Model development

As mentioned in Section 13.3.3.3, it is crucial to identify the best data partition and the best forecast horizon for long-term (monthly) forecasting. The same sets of data partitioning were used, consisting of 40% data for training, 60% data for training, 70% data for training, and 80% data for training. The target variable has been shifted by 1, 3, and 6 to develop 1-month, 3-month, and 6-month ahead solar radiation forecasts.

Table 13.13 shows the training and validation performance metrics for the four target sites.

However, training parameters for estimation (no shift of the target variable) are inconsistent; performance parameters in the validation period have been consistent with having the best results for 80% data in the training set. Hence, for estimating solar radiation in these four sites, 80% data in training and 10% data each in the validation and testing have been chosen as the best data partitioning.

By examining the performance parameters in the training period, it is evident that 40% in the training and 60% in the validation and testing have produced the best results in high r, and low *RMSE*, and *MAE* in most of the scenarios except for in Mount Isa. Nevertheless, the particular data partitioning has been used for forecasting solar radiation in the selected sites as it has yielded the best results in most sites and in the validation period for Mount Isa.

As the long-term solar radiation forecasting results are not consistent with the short-term forecasting results, the target sites have different forecast horizons, which has produced the best results for each site. Table 13.14 depicts a comparison of the forecast horizons with the target sites.

For Cairns and Mount Isa target sites, shifting the target variable by 3 months yielded the best results, while shifting the target variable by 6 months produced the best results for Longreach and Toowoomba.

The basis functions multiplied with their coefficients together make the MARS equation for their respective model. Tables 13.15–13.17 show the basis functions of the

TABLE 13.13 The model performance criteria Pearson's correlation coefficient (*r*), root mean square error (*RMSE*), and mean absolute error (*MAE*) in training and validation data sets for different target shifts and different data partitions for all the target sites.

		Training			Validation		
Shift of the target variable	**Data partition**	*r*	***RMSE*** **(MJ/m^2)**	***MAE*** **(MJ/m^2)**	*r*	***RMSE*** **(MJ/m^2)**	***MAE*** **(MJ/m^2)**
Cairns							
No shift	40/30/30	0.959	0.977	0.752	0.954	1.161	0.915
	60/20/20	**0.961**	**0.959**	0.748	0.952	1.081	0.869
	70/15/15	**0.961**	0.964	**0.743**	0.956	1.012	0.805
	80/10/10	0.959	0.972	0.754	**0.963**	**0.902**	**0.727**
1-Month shift	**40/30/30**	**0.822**	**1.972**	**1.552**	0.690	2.654	2.084
	60/20/20	0.791	2.128	1.684	0.664	2.538	1.992
	70/15/15	0.781	2.173	1.698	0.681	2.451	1.929
	80/10/10	0.772	2.186	1.727	**0.698**	**2.397**	**1.884**
3-Month shift	**40/30/30**	**0.845**	**1.856**	**1.495**	0.749	2.392	1.859
	60/20/20	0.831	1.936	1.508	0.719	2.395	1.951
	70/15/15	0.824	1.977	1.581	0.731	2.316	1.854
	80/10/10	0.816	1.991	1.579	**0.783**	**2.056**	**1.619**
6-Month shift	**40/30/30**	**0.794**	**2.108**	**1.681**	**0.640**	2.823	2.160
	60/20/20	0.786	2.154	1.720	0.629	2.631	**2.044**
	70/15/15	0.774	2.205	1.775	0.599	2.736	2.149
	80/20/20	0.755	2.254	1.772	0.612	**2.617**	2.205
Longreach							
No shift	40/30/30	**0.970**	**1.005**	**0.782**	0.922	1.697	1.317
	60/20/20	0.958	1.201	0.931	0.921	1.573	1.214
	70/15/15	0.956	1.220	0.964	0.939	1.364	1.102
	80/10/10	0.953	1.255	0.989	**0.948**	**1.279**	**1.038**
1-Month shift	**40/30/30**	**0.838**	**2.255**	**1.741**	0.689	3.113	2.546
	60/20/20	0.807	2.465	1.969	0.719	2.848	2.204
	70/15/15	0.803	2.483	1.989	**0.745**	**2.686**	2.086
	80/10/10	0.792	2.525	2.019	0.708	2.817	**2.046**
3-Month shift	**40/30/30**	**0.835**	**2.275**	**1.791**	0.691	**2.590**	3.068
	60/20/20	0.826	2.356	1.857	**0.728**	2.905	2.275

(*Continued*)

TABLE 13.13 (Continued)

Shift of the target variable	Data partition	Training r	Training RMSE (MJ/m^2)	Training MAE (MJ/m^2)	Validation r	Validation RMSE (MJ/m^2)	Validation MAE (MJ/m^2)
	70/15/15	0.821	2.380	1.870	0.707	2.945	**2.161**
	80/10/10	0.800	2.481	1.953	0.709	2.835	2.193
6-Month shift	**40/30/30**	**0.881**	**1.958**	**1.528**	**0.818**	2.466	1.930
	60/20/20	0.851	2.192	1.718	0.800	**2.391**	**1.859**
	70/15/15	0.858	2.137	1.680	0.753	2.638	2.050
	80/20/20	0.845	2.206	1.723	0.765	2.590	2.013
Mount Isa							
No shift	40/30 /30	0.899	1.277	0.999	0.804	2.027	1.678
	60/20/20	0.916	**1.228**	**0.936**	**0.809**	2.746	2.444
	70/15/15	0.909	1.246	0.972	0.802	1.977	1.687
	80/10/10	**0.917**	1.273	0.984	0.762	**1.505**	**1.255**
1-Month shift	**40/30/30**	0.667	2.176	1.663	**0.719**	**2.533**	**2.053**
	60/20/20	**0.769**	**1.958**	**1.459**	0.687	3.701	3.397
	70/15/15	0.752	1.973	1.509	0.168	3.400	2.955
	80/10/10	**0.769**	2.048	1.568	− 0.042	2.732	2.191
3-Month shift	40/30/30	0.680	2.152	1.687	**0.516**	**3.326**	**2.732**
	60/20/20	**0.774**	**1.942**	**1.476**	0.436	5.098	4.745
	70/15/15	0.741	2.008	1.543	− 0.131	4.194	3.736
	80/10/10	0.754	2.107	1.625	− 0.453	4.093	3.692
6-Month shift	**40/30/30**	0.591	2.372	1.844	0.318	2.996	2.280
	60/20/20	0.741	**2.054**	**1.553**	**0.460**	4.505	4.118
	70/15/15	0.715	2.087	1.591	0.111	3.210	2.722
	80/20/20	**0.744**	2.147	1.645	0.152	**2.604**	**2.216**
Toowoomba							
No shift	40/30 /30	**0.964**	**1.116**	**0.859**	0.952	1.537	1.219
	60/20/20	0.961	1.184	0.930	0.950	1.506	1.213
	70/15/15	0.962	1.180	0.928	0.952	1.478	1.170
	80/10/10	0.962	1.181	0.927	**0.956**	**1.210**	**0.971**
1-Month shift	**40/30/30**	**0.850**	**2.214**	**1.781**	0.764	3.133	2.489
	60/20/20	0.816	2.469	1.963	0.829	2.749	2.228

(Continued)

TABLE 13.13 (Continued)

Shift of the target variable	Data partition	Training r	Training RMSE (MJ/m^2)	Training MAE (MJ/m^2)	Validation r	Validation RMSE (MJ/m^2)	Validation MAE (MJ/m^2)
	70/15/15	0.832	2.398	1.896	**0.830**	2.632	**2.024**
	80/10/10	0.825	2.446	1.949	0.808	**2.591**	2.039
3-Month shift	**40/30/30**	**0.852**	**2.203**	**1.732**	0.710	3.409	2.454
	60/20/20	0.826	2.411	1.903	0.795	**2.927**	2.328
	70/15/15	0.821	2.471	1.978	**0.798**	3.001	2.346
	80/10/10	0.818	2.489	1.956	0.723	2.966	**2.233**
6-Month shift	**40/30/30**	**0.891**	**1.906**	**1.471**	0.814	2.849	2.216
	60/20/20	0.854	2.228	1.755	**0.866**	2.332	1.949
	70/15/15	0.863	2.181	1.736	0.858	**2.220**	**1.859**
	80/20/20	0.861	2.201	1.754	0.755	2.769	2.168

The optimal data partitions yielding the highest *r*, and lowest *RMSE* and *MAE* for each shift and the target site are in red bold-face.

TABLE 13.14 Performance parameters correlation coefficient (*r*), root mean square error (*RMSE*), and mean absolute error (*MAE*) in training period for selecting the best forecast horizon in long-term solar radiation prediction for the all target sites.

Target Site	Shift of the target variable	Data partition	r	RMSE (MJ/m^2)	MAE (MJ/m^2)
Cairns	1-month	40/30/30	**0.822**	1.972	1.552
	3-month	40/30/30	0.845	**1.856**	**1.495**
	6-month	40/30/30	0.794	2.108	1.681
Longreach	1-month	40/30/30	0.838	2.255	1.741
	3-month	40/30/30	0.835	2.275	1.791
	6-month	40/30/30	**0.881**	**1.958**	**1.528**
Mount Isa	1-month	40/30/30	0.667	2.176	**1.663**
	3-month	40/30/30	**0.680**	**2.152**	1.687
	6-month	40/30/30	0.591	2.372	1.844
Toowoomba	1-month	40/30/30	0.850	2.214	1.781
	3-month	40/30/30	0.852	2.203	1.732
	6-month	40/30/30	**0.891**	**1.906**	**1.471**

The optimal data partitions yielding highest *r*, and lowest *RMSE* and *MAE* are in red bold-face.

TABLE 13.15 The basis functions of the MARS model for estimating monthly solar radiation.

Cairns	Longreach
BF1 = max(0,0.69679 − Rainfall)	BF1 = max(0,0.44863 − Rainfall)
BF2 = max(0, VP −0.69237)	BF2 = max(0, VP −0.34776)
BF3 = max(0,0.69237 − VP)	BF3 = max(0,0.34776 − VP)
BF4 = max(0,0.37298 − Evap)	BF4 = max(0,0.096531 − Evap)
BF5 = max(0, RHmaxT −0.54006)	BF5 = max(0, RHmaxT −0.91252)
BF6 = max(0,0.54006 -RHmaxT)	BF6 = max(0,0.91252 -RHmaxT)
BF7 = max(0,0.37961 -minT)	BF7 = max(0,0.024546 -minT)
BF8 = max(0, Evap −0.93952)	BF8 = max(0, Rainfall −0.12443)
BF9 = max(0, maxT −0.74109)	BF9 = max(0, Rainfall −0.82078)
BF10 = max(0,0.65417 − Evap)	BF10 = max(0, Evap −0.7995)
BF11 = max(0, maxT −0.55759)	BF11 = max(0,0.7995 − Evap)
BF12 = max(0,0.55759 − maxT)	BF12 = max(0,0.40943 − maxT)
BF13 = max(0,0.19323 − minT)	
Mount Isa	**Toowoomba**
BF1 = max(0, Rainfall −0.43215)	BF1 = max(0, Rainfall −0.41012)
BF2 = max(0, VP −0.39471)	BF2 = max(0,0.41012 − Rainfall)
BF3 = max(0,0.39471 − VP)	BF3 = max(0, VP −0.18228)
BF4 = max(0, RHmaxT −0.51941)	BF4 = max(0, Evap −0.25751)
BF5 = max(0,0.51941 − RHmaxT)	BF5 = max(0,0.25751 -Evap)
BF6 = max(0,0.054943 − minT)	BF6 = max(0, RHmaxT −0.4307)
BF7 = max(0,0.30341 − Evap)	BF7 = max(0,0.11299 − minT)
BF8 = max(0, maxT −0.9153)	BF8 = max(0, Rainfall −0.10208)
BF9 = max(0,0.9153 − maxT)	BF9 = max(0,0.53478 − Rainfall)
BF10 = max(0,0.56359 − VP)	BF10 = max(0, Rainfall −0.51581)
BF11 = max(0,0.50907 − Rainfall)	BF11 = max(0,0.53659 − Rainfall)
BF12 = max(0,0.46717 − Rainfall)	

The definition of acronyms used here are as follows: *BF*, Basis function; *Evap*, evaporation; *maxT*, maximum temperature; *minT*, minimum temperature; *Rainfall*, amount of rainfall; *RHmaxT*, relative humidity at maximum temperature; *RHminT*, relative humidity at minimum temperature; *VP*, vapor pressure.

long-term estimation (80% data in training) and forecasting (40% data in training) of solar radiation for the four target sites. For Cairns and Mount Isa, 3 months target shift and for Longreach and Toowoomba 6 months target shift have been selected to develop the best model.

TABLE 13.16 The basis functions of the MARS model for forecasting monthly solar radiation.

Cairns	Longreach
BF1 = max(0, Evap −0.81042)	BF1 = max(0, Rainfall −0.53025)
BF2 = max(0, Rainfall −0.12723)	BF2 = max(0,0.57396 − Evap)
BF3 = max(0,0.12723 − Rainfall)	BF3 = max(0,0.030682 − minT)
BF4 = max(0,0.60759 − RHmaxT)	BF4 = max(0, maxT −0.57732)
BF5 = max(0, maxT −0.64936)	BF5 = max(0,0.48361 − VP)
BF6 = max(0, maxT −0.38421)	BF6 = max(0, RHmaxT −0.88758)
BF7 = max(0,0.91129 − Evap)	BF7 = max(0,0.50457 − Rainfall)
	BF8 = max(0, Rainfall −0.58333)
	BF9 = max(0,0.58333 − Rainfall)
	BF10 = max(0, maxT −0.78203)
	BF11 = max(0,0.81688 − RHmaxT)
Mount Isa	**Toowoomba**
BF1 = max(0,0.38948 − VP)	BF1 = max(0,0.53478 − Rainfall)
BF2 = max(0,0.86747 − Evap)	BF2 = max(0,0.79614 − Evap)
BF3 = max(0, maxT −0.80255)	BF3 = max(0,0.52755 − Rainfall)
BF4 = max(0,0.60593 − maxT)	BF4 = max(0,0.58266 − Rainfall)
BF5 = max(0,0.82828 − Evap)	BF5 = max(0, maxT −0.61324)
BF6 = max(0,0.88785 − Evap)	BF6 = max(0,0.17328 − minT)
	BF7 = max(0,0.97349 − RHmaxT)
	BF8 = max(0,0.50768 − Rainfall)

For Cairns and Mount Isa, the 3-month ahead forecasting model and for Longreach and Toowoomba the 6-month ahead forecasting model have been shown. The definition of acronyms used here are as follows: *BF*, basis function; *Evap*, evaporation; *maxT*, maximum temperature; *minT*, minimum temperature; *Rainfall*, amount of rainfall; *RHmaxT*, relative humidity at maximum temperature; *RHminT*, relative humidity at minimum temperature; *VP*, vapor pressure.

TABLE 13.17 ARIMA with structure (p, d, q) with d = differencing, p and q = order of autoregressive (AR) and moving average (MA) term.

Target site	ARIMA structure (p,d,q)	AIC	Log-likelihood	BIC	Variance
Cairns	ARIMA (10,0,0)	2410.679	− 1193.339	2462.848	3.876
Longreach	ARIMA (10,0,1)	2459.415	− 1216.707	2515.931	4.216
Mount Isa	ARIMA (5,1,6)	2336.566	− 1156.28	2388.714	3.442
Toowoomba	ARIMA (7,0,0)	2456.912	− 1219.46	2496.039	4.227

Akaike's Information Criterion (AIC) used to identify the model in conjunction with log-likelihood, Bayesian information criterion (BIC), variance and correlation coefficient.

The respective MARS equation for estimation in Cairns:

$$y = 0.087651 - 0.54254 \times BF1 - 0.22773 \times BF2 + 0.95109 \times BF3 - 0.43026 \times BF4 + 0.39352 \times BF5 - 0.58899 \times BF6 + 0.49405 \times BF7 + 2.6505 \times BF8 - 0.95546 \times BF9 + 0.73322 \times BF10 + 1.2784 \times BF11 - 0.84824 \times BF12 - 0.29626 \times BF13 \tag{13.25}$$

The respective MARS equation for estimation in Longreach:

$$y = 0.56086 - 0.45177 \times BF1 - 0.6517 \times BF2 + 1.3245 \times BF3 - 1.1402 \times BF4 + 1.1475 \times BF5 - 0.22362 \times BF6 - 1.865 \times BF7 + 0.49469 \times BF8 - 0.67595 \times BF9 + 0.75972 \times BF10 - 0.24882 \times BF11 + 0.13846 \times BF12 \tag{13.26}$$

The respective MARS equation for estimation in Mount Isa:

$$y = 0.48121 + 0.50353 \times BF1 - 0.89131 \times BF2 + 0.94277 \times BF3 + 0.89125 \times BF4 - 0.49022 \times BF5 - 1.0176 \times BF6 + 0.44877 \times BF7 + 0.98446 \times BF8 - 0.42246 \times BF9 + 0.36402 \times BF10 + 1.6699 \times BF11 - 2.3615 \times BF12 \tag{13.27}$$

The respective MARS equation for estimation in Toowoomba:

$$y = -0.28997 + 5.6439 \times BF1 - 5.6824 \times BF2 - 0.59385 \times BF3 + 0.084613 \times BF4 - 0.24856 \times BF5 + 0.11114 \times BF6 - 0.37937 \times BF7 + 1.2294 \times BF8 + 65.511 \times BF9 - 6.5297 \times BF10 - 59.557 \times BF11 \tag{13.28}$$

The respective MARS equation for forecasting in Cairns:

$$y = 0.2663 + 1.1624 \times BF1 + 0.21975 \times BF2 - 2.5789 \times BF3 - 0.34069 \times BF4 + 0.85176 \times BF5 - 0.67715 \times BF6 + 0.72259 \times BF7 \tag{13.29}$$

The respective MARS equation for forecasting in Longreach:

$$y = 1.0772 - 11.429 \times BF1 - 0.42506 \times BF2 + 2.0892 \times BF3 - 1.1758 \times BF4 - 0.69171 \times BF5 - 1.4738 \times BF6 + 6.7178 \times BF7 + 10.917 \times BF8 - 5.9893 \times BF9 + 0.84909 \times BF10 + 0.55508 \times BF11 \tag{13.30}$$

The respective MARS equation for forecasting in Mount Isa:

$$y = 0.55613 + 0.51684 \times BF1 + 12.713 \times BF2 + 0.74946 \times BF3 - 0.3258 \times BF4 - 5.9547 \times BF5 - 7.2962 \times BF6 \tag{13.31}$$

The respective MARS equation for forecasting in Toowoomba:

$$y = 0.41299 - 26.481 \times BF1 - 0.45065 \times BF2 + 15.867 \times BF3 + 7.5265 \times BF4 - 0.87019 \times BF5 + 0.45523 \times BF6 + 0.097793 \times BF7 + 4.1034 \times BF8 \tag{13.32}$$

Table 13.17 displays the ARIMA model's architecture and the respective goodness-of-fit tests performed to construct the forecasting model.

13.4.2.2 Estimation of monthly solar radiation in regional Queensland

Estimation of long-term solar radiation is obliging as it will provide resilience in that the developed model will perform better or not in forecasting solar radiation. In the model development section, the data partitioning 80% for the training set yielded the best results out of the other data partition sets.

Table 13.18 shows the performance parameters for the best long-term estimation models for solar radiation for all the target sites.

The highest r has been obtained for the estimation model for the Toowoomba site while the lowest is for Mount Isa. *RMSE* and *MAE* which show the errors of the model should be lower for the optimal model. Toowoomba has the highest *WI*, E_{NS}, and L compared to all the other sites while Mount Isa has the lowest of them all. In contrast, Mount Isa has the best *RRMSE* (5.8%) and *RMAE* (4.6%). However, as the estimation model for Longreach had yielded the best performance parameters in all the other metrics, Longreach site can be considered as the best site for estimating long-term solar radiation from these four sites.

Surprisingly, the estimation models of solar radiation for the short term and long term have performed similarly when comparing their best performing site. For short-term and long-term estimation models, 80% data in training was the best data partition and for both occasions, the Longreach target site performed the best while Mount Isa performed worst.

Fig. 13.7 shows the scatterplots representing the relationship between the observed and estimated values for all the target sites.

The scatterplots of Cairns, Longreach, and Toowoomba show that the estimated and observed solar radiation have a good relationship with each other as the correlation coefficient (R^2) is very close to +1. Toowoomba has the best R^2 value, so it shows that approximately 91% of the estimated solar radiation can be explained by the observed solar radiation. Mount Isa represents the weakest relationship between the observed and the estimated solar radiation. Approximately 50% of the estimated solar radiation can be explained by the observed solar radiation in Mount Isa. This shows that the MARS model can be utilized to generate solar radiation forecasts in regional Queensland.

TABLE 13.18 Correlation coefficient (*r*), root mean square error (*RMSE*), and mean absolute error (*MAE*), Wilmott's index (*WI*), Nash–Sutcliffe coefficient (E_{NS}), and Legates and McCabe's index (*L*), geographical parameters, relative root mean square error (*RRMSE*), and relative mean absolute error (*RMAE*) values in the testing period for the selected MARS model (80% training) in estimating monthly solar radiation (MJ/m^2) for all target sites.

Target sites	*r*	*RMSE* (MJ/m^2)	*MAE* (MJ/m^2)	*WI*	E_{NS}	*L*	*RRMSE* (%)	*RMAE* (%)
Cairns	0.941	1.669	1.374	0.830	0.753	0.515	8.791	7.708
Longreach	0.953	1.434	1.115	0.937	0.878	0.685	6.847	5.677
Mount Isa	0.701	1.454	1.145	0.762	0.402	0.192	5.801	4.566
Toowoomba	0.954	1.384	1.133	0.954	0.910	0.715	7.518	6.572

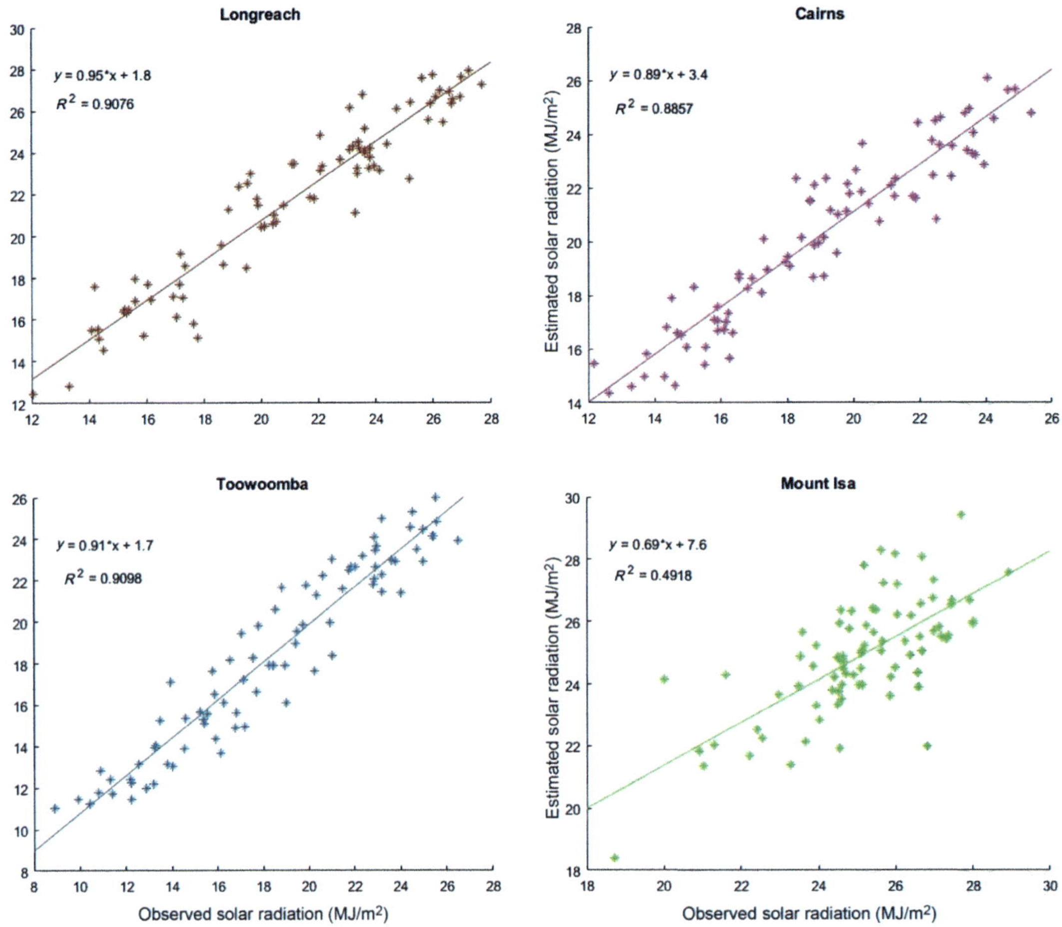

FIGURE 13.7 The scatterplots of observed and estimated monthly solar radiation (MJ/m^2) in testing data set (10% data: 82 data points from March 2011 to December 2017) for Cairns, Longreach, Mount Isa, and Toowoomba. For each scatterplot, least squares fitting line and respective R^2 are shown.

13.4.2.3 Forecasting monthly solar radiation in regional Queensland

In this section, a MARS model is developed and discussed, for forecasting long-term solar radiation in regional Queensland and is assessed by a comparative model framework.

In the model development section, it has been shown that in forecasting long-term solar radiation, 40% for training data produced the best results. Nevertheless, a 3-month ahead forecast horizon was able to harvest the optimal results for Cairns and Mount Isa, while a 6-month ahead forecast yielded the best outcomes for Longreach and Toowoomba.

Fig. 13.8 shows the forecasting errors for monthly forecasting horizons in the testing period for MARS and ARIMA models in Cairns, Longreach, Mount Isa, and Toowoomba.

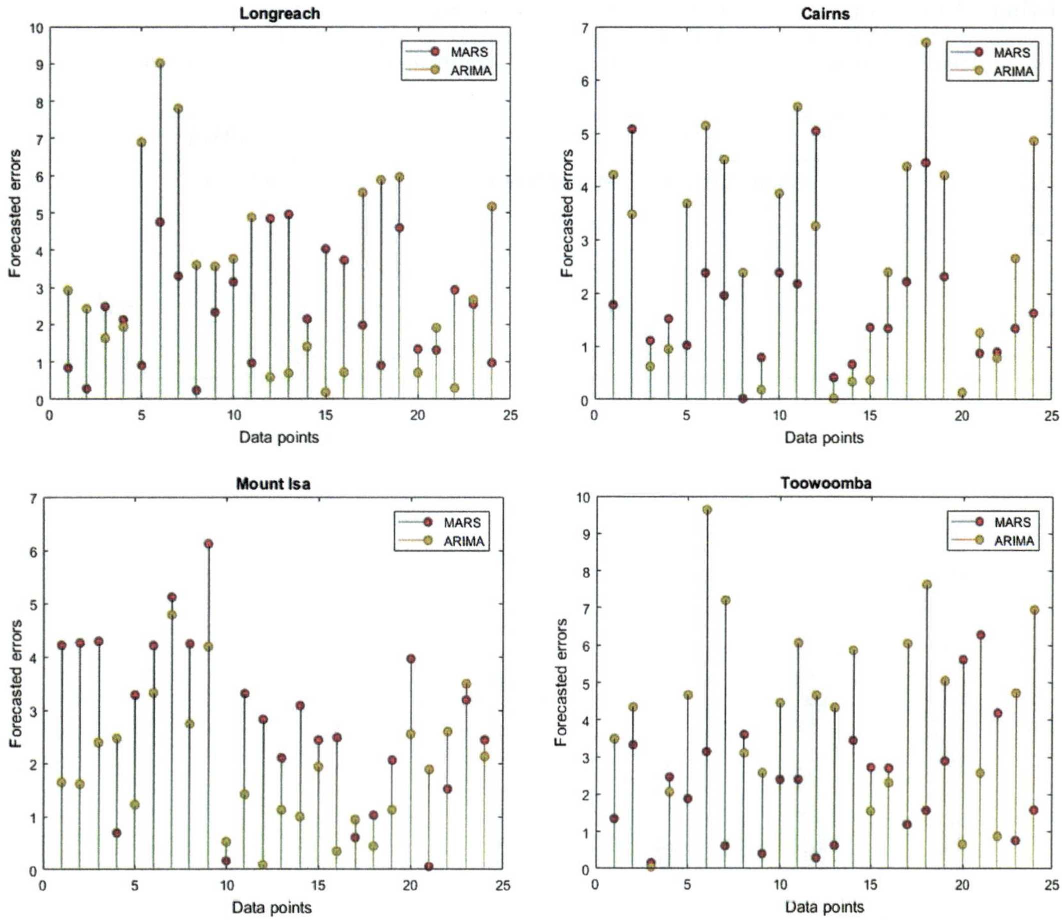

FIGURE 13.8 The stem plots showing the forecasted error of monthly solar radiation (MJ/m^2) in testing data set (24 data points from January 2016 to December 2017) for Cairns, Longreach, Mount Isa, and Toowoomba.

Except for Mount Isa, all the other target sites show higher error values in the ARIMA model compared to the MARS model, showing that the MARS model can forecast monthly solar radiation more accurately than the ARIMA model (Table 5.9) (Table 13.19).

Fig. 13.9 depicts that the most accurate MARS model was for Toowoomba as it produced 0.779 of r while ARIMA produced only 0.447 compared to that. All the sites except Mount Isa have produced r values higher than 0.7, so it is proved that the MARS model has performed better compared to the ARIMA model for forecasting long-term solar radiation (Table 13.20).

Fig. 13.10 shows the WI, E_{NS}, and L for the best performing site and worst performing site in predicting long-term solar radiation using MARS. Toowoomba was considered to produce the most accurate results as it yielded very high results. Compared to

TABLE 13.19 Comparison of correlation coefficient (*r*), root mean square error (*RMSE*), and mean absolute error (*MAE*) for MARS and ARIMA models in optimal scenarios for the target sites: Cairns: 3-month ahead forecast; Longreach: 6-month ahead forecast; Mount Isa: 3-month ahead forecast; Toowoomba: 6-month ahead forecast.

	MARS			ARIMA		
Target site	*r*	***RMSE*** **(MJ/m²)**	***MAE*** **(MJ/m²)**	*r*	***RMSE*** **(MJ/m²)**	***MAE*** **(MJ/m²)**
Cairns	0.729	**2.293**	**1.813**	0.564	2.810	2.324
Longreach	0.740	2.816	2.206	0.647	3.248	2.669
Mount Isa	0.098	5.174	4.796	− 0.074	2.515	2.092
Toowoomba	**0.779**	2.886	2.327	0.447	3.935	3.307

The best sites having the highest *r* and lowest *RMSE* and *MAE* are in red bold-face.

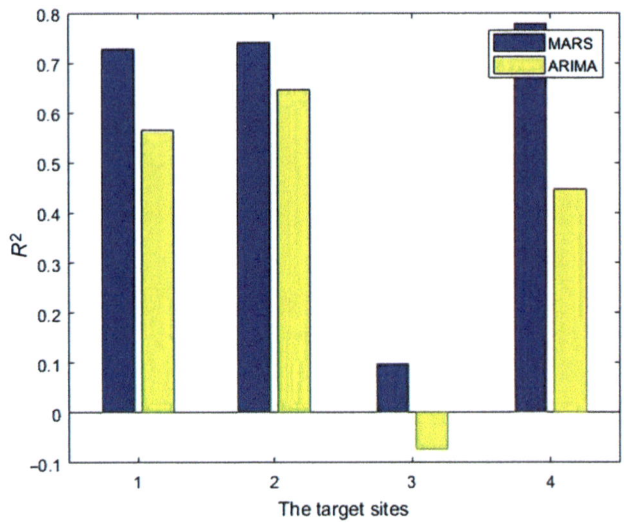

FIGURE 13.9 The bar graph depicting the *r* values in the testing data set (30% data: 243 data points from October 1997 to December 2017) of MARS and ARIMA models for all the target sites (1: Cairns; 2: Longreach; 3: Mount Isa; 4: Toowoomba).

TABLE 13.20 Comparison of Wilmott's index (*WI*), Nash–Sutcliffe coefficient (E_{NS}), and Legates and McCabe's Index (*L*) for MARS and ARIMA models in optimal scenarios for the target sites.

	MARS			ARIMA		
Target site	*WI*	E_{NS}	*L*	*WI*	E_{NS}	*L*
Cairns	0.722	0.507	0.347	0.509	0.257	0.161
Longreach	0.754	0.500	0.358	0.579	0.331	0.220
Mount Isa	0.346	− 6.845	− 2.293	0.439	− 0.854	− 0.434
Toowoomba	**0.799**	**0.561**	**0.383**	0.423	0.177	0.117

The best sites having the highest *r* and lowest *RMSE* and *MAE* are in red bold-face.

Toowoomba, Mount Isa performed very badly as it could not even get reach 0.1 for *WI*, E_{NS}, and *L*. Here also, MARS is superior compared to ARIMA as it yielded low results.

The geographical parameters show that Mount Isa has performed the worst in forecasting monthly solar radiation as it has yielded high relative errors (Table 13.21). Though Toowoomba has received the second-highest errors, when considering the other parameters, Toowoomba can be seen as the highest performing site for forecasting long-term solar radiation using the MARS model.

Fig. 13.11 depicts the time series of observed versus the forecasted solar radiation using MARS and ARIMA models.

The time series of the best performing site and the worst performing site both show that the forecasted solar radiation generated by the MARS model more closely follows the observed solar radiation than that of ARIMA.

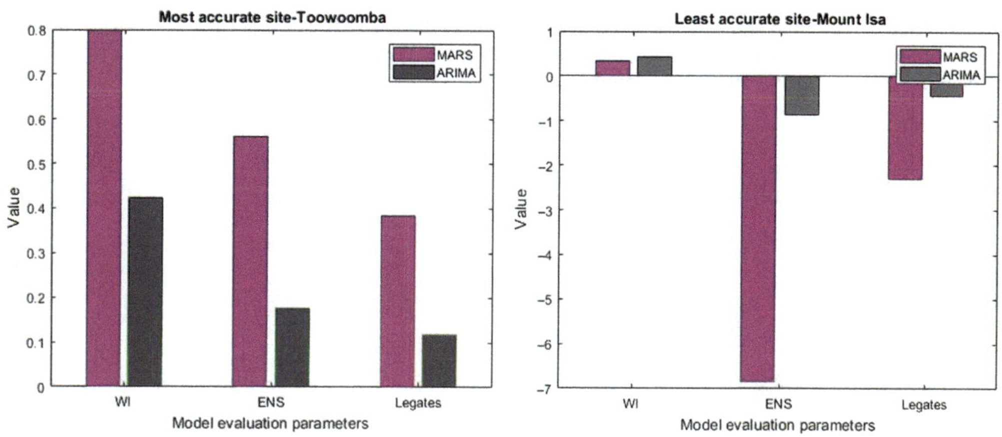

FIGURE 13.10 The bar graph depicting the Wilmott's index (*WI*), Nash–Sutcliffe coefficient (E_{NS}), and Legates and McCabe's (*L*) parameters in the testing data set (30% data: 243 data points from October 1997 to December 2017) for the best performing site, Toowoomba, and worst performing site, Mount Isa.

TABLE 13.21 The geographical parameters in the testing data set: relative root mean square error (*RRMSE*) and relative mean absolute error (*RMAE*) values for the MARS model in forecasting monthly solar radiation (MJ/m^2).

Target site	*RRMSE* (%)	*RMAE* (%)
Cairns	**12.126**	**9.785**
Longreach	13.273	11.028
Mount Isa	20.531	18.715
Toowoomba	15.486	13.283

The sites having the *RRMSE* and *RMAE* are in red bold-face.

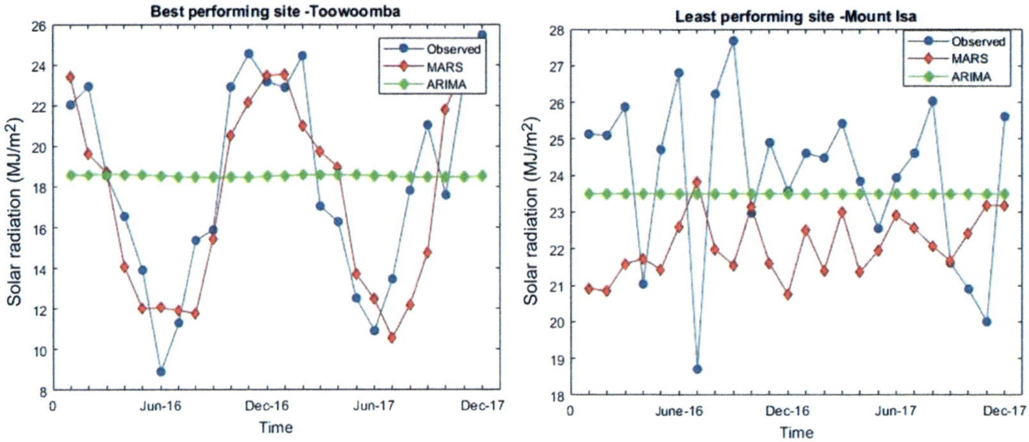

FIGURE 13.11 The time series of observed versus forecasted solar radiation (MJ/m^2) in MARS and ARIMA for the testing data set (24 data points from January 2016 to December 2017) for the best performing site, Toowoomba, and worst performing site, Mount Isa.

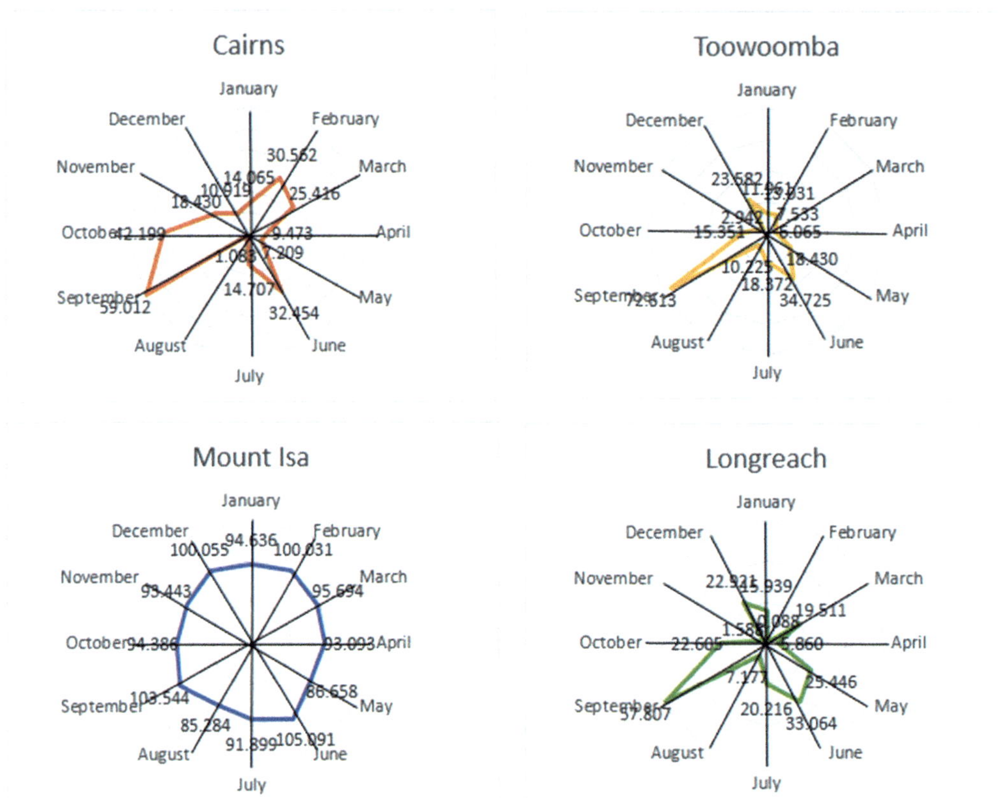

FIGURE 13.12 The polar plots showing monthly solar radiation (MJ/m^2) forecasted error in the testing data set (Cairns and Mount Isa: September 1997 to November 2017; Longreach and Toowoomba: December 1997 to November 2017) for all the sites.

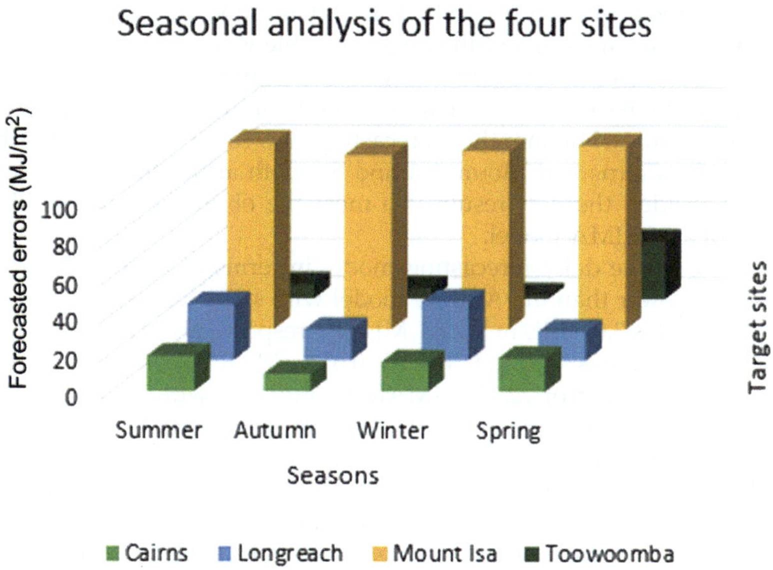

FIGURE 13.13 3D bar graph showing seasonal solar radiation forecasted errors (MJ/m^2) in the testing data set (Cairns and Mount Isa: September 1997 to November 2017; Longreach and Toowoomba: December 1997 to November 2017) for all the sites. (Note: Summer, December–February; Autumn, March–May; Winter, June–August; Spring, September–November).

13.4.2.4 Seasonal analysis for the four target sites

To draw more solid conclusive arguments about the forecasting skill of the MARS model, seasonal analysis for the results has been carried out. The forecasted errors of observed and forecasted solar radiation have been calculated for each month and season. Fig. 13.12 shows the forecasting errors calculated for the months for each site.

In all target sites, September has produced the highest forecasted errors. The forecasted errors in Mount Isa throughout the year are very high compared to the other sites. The lowest forecasted errors were yielded in August, February, and November for Cairns, Longreach, and Toowoomba, respectively. April and August's months have produced lower forecasting errors in all the target sites.

Fig. 13.13 shows the graph depicting the monthly forecasted errors in seasons for all the target sites.

Mount Isa has produced the highest forecasted errors throughout the four seasons. The MARS model will perform best in autumn in Cairns as the forecasted errors are low. Though Toowoomba has obtained high forecasting errors in the spring, it has relatively low forecasting errors for all the seasons. Hence, Toowoomba is better for forecasting monthly and seasonal solar radiation.

13.5 Conclusion

This chapter established the preciseness of a MARS model for forecasting solar radiation for regional Queensland. A data set from 1950 to 2017 for Cairns, Longreach, Mount

Isa, and Toowoomba was utilized from SILO. To enhance the veracity of the MARS model, several data partitioning and shifts of the target variable were trialed.

For daily and monthly estimation, 80% data for the training set was adopted as the best partitioning. For daily forecasting, 40% data for training and a 1-day ahead forecast horizon produced the best results. For monthly forecasting, the same data partition but 3-month ahead forecasting for Cairns and Mount Isa and 6-month ahead forecasting for Longreach and Toowoomba yielded the best results. To meet the objectives, the MARS model was benchmarked with an ARIMA model.

The performance of the daily forecasting model in terms of its *RMSE* and *MAE* in the training period was lower than the ARIMA model with site-averaged $RMSE \sim 4.18$ MJ/m^2, 3.98 MJ/m^2 and $MAE \sim 3.09$ MJ/m^2, 2.85 MJ/m^2, respectively. However, in the testing period performance metrics yielded *WI* of 0.618 (MARS) compared with only 0.04 (ARIMA) and a *relative error* of 22.15% (MARS) compared with 29.42% (ARIMA), when averaged for all the sites. Therefore, it can be ascertained that MARS performed better over ARIMA in daily forecasting.

The monthly forecasting model attained better results for MARS in terms of its *WI* in testing, which is 0.655 and 0.488 for ARIMA. MARS performed similarly to ARIMA with respect to a *relative error* of which they obtained 15.3% and 15.4%. The difference in performance between MARS and ARIMA was high in daily forecasting.

By an analysis of relative percentage error and the other performance parameters between the four sites, variability in the model's accuracy was evident with the optimal performance for daily and monthly forecasting being obtained for Longreach and Toowoomba, while the worst performances were attained for Toowoomba and Mount Isa, respectively.

Although this project has emphasized the importance of the MARS model in solar radiation forecasting, the testing of the model for shorter timescales (e.g., hourly) and a wider range of regional areas needs to be performed to enhance the veracity of the model. The accuracy and the performance of the model can be improved if the model can be hybridized with algorithms such as complete ensemble empirical mode decomposition with adaptive noise (CEEMDAN). To improve its practicality for real-life applications, a few further tests are necessary with an assessment of alternative models in extreme weather, model performance with potential uncertainties like a machine or human error, and the response of the model in terms of climatic anomalies like ENSO events.

References

Aggarwal, S.K., Saini, L.M., 2014. Solar energy prediction using linear and nonlinear regularization models: a study on AMS (American Meteorological Society) 2013–14 solar energy prediction contest. Energy 78, 247–256.

Allen, R., Kannberg, L., Raymond, J., 1984. Seasonal Thermal Energy Storage. Pacific Northwest Lab., Richland, WA.

Al-Musaylh, M.S., et al., 2018a. Short-term electricity demand forecasting with MARS, SVR and ARIMA models using aggregated demand data in Queensland, Australia. Adv. Eng. Inform. 35, 1–16.

Al-Musaylh, M.S., et al., 2018b. Two-phase particle swarm optimized-support vector regression hybrid model integrated with improved empirical mode decomposition with adaptive noise for multiple-horizon electricity demand forecasting. Appl. Energy 217, 422–439.

Angstrom, A., 1924. Solar and terrestrial radiation: report to the international commission for solar research on actinometric investigations of solar and atmospheric radiation. Q. J. R. Meteorol. Soc. 50 (210), 121–126.

Arslan, B., Ulker, M., Sagiroglu, S., 2017. Machine learning methods used in evaluations of secure biometric system components. In: 2017 16th IEEE International Conference on Machine Learning and Applications (ICMLA).

Bahadori, A., Nwaoha, C., 2013. A review on solar energy utilisation in Australia. Renew. Sustain. Energy Rev. 18, 1–5.

Beath, A.C., 2012. Industrial energy usage in Australia and the potential for implementation of solar thermal heat and power. Energy 43 (1), 261–272.

Belaid, S., Mellit, A., 2016. Prediction of daily and mean monthly global solar radiation using support vector machine in an arid climate. Energy Convers. Manag. 118, 105–118.

Box, G.E., Jenkins, G.M., 1976. Time Series Analysis: forecasting and control, rev. edn. Holder-Day, San Francisco, CA.

Butte, N.F., et al., 2010. Validation of cross-sectional time series and multivariate adaptive regression splines models for the prediction of energy expenditure in children and adolescents using doubly labeled water. J. Nutr. 140 (8), 1516–1523.

Chwieduk, D., 2004. Solar energy utilization. Opto-Electron. Rev. 12 (1), 13–20.

Council, C.E., 2015. Clean Energy Australia Report 2014.

Dalaka, A., et al., 2000. Modelling the effects of environmental conditions on apparent photosynthesis of Stipa bromoides by machine learning tools. Ecol. Model. 129 (2), 245–257.

Deo, R.C., Şahin, M., 2017. Forecasting long-term global solar radiation with an ANN algorithm coupled with satellite-derived (MODIS) land surface temperature (LST) for regional locations in Queensland. Renew. Sustain. Energy Rev. 72, 828–848.

Deo, R.C., Samui, P., Kim, D., 2015. Estimation of monthly evaporative loss using relevance vector machine, extreme learning machine and multivariate adaptive regression spline models. Stoch. Environ. Res. Risk Assess. 30(6), 1769-1784.

Deo, R.C., Kisi, O., Singh, V.P., 2017. Drought forecasting in eastern Australia using multivariate adaptive regression spline, least square support vector machine and M5Tree model. Atmos. Res. 184, 149–175.

Deo, R.C., Wen, X., Qi, F., 2016. A wavelet-coupled support vector machine model for forecasting global incident solar radiation using limited meteorological dataset. Appl. Energy 168, 568–593.

Friedman, J.H., 1991. Multivariate adaptive regression splines. Ann. Statist. 19 (1), 1–67.

Frondel, M., et al., 2010. Economic impacts from the promotion of renewable energy technologies: the German experience. Energy Policy 38 (8), 4048–4056.

Geoscience Australia, 2010. Australian Energy Resource Assessment. Australian Government, Canberra.

Haidar, A.M., Muttaqi, K., Sutanto, D., 2015. Smart grid and its future perspectives in Australia. Renew. Sustain. Energy Rev. 51, 1375–1389.

Hindman, M., 2015. Building better models: prediction, replication, and machine learning in the social sciences. Ann. Am. Acad. Pol. Soc. Sci. 659 (1), 48–62.

Hunt, G., Macfarlane, I., 2015. Certainty and Growth for Renewable Energy. Media Release, Canberra, Australia.

Jeffrey, S.J., et al., 2001. Using spatial interpolation to construct a comprehensive archive of Australian climate data. Environ. Model. Softw. 16 (4), 309–330.

Jiang, H., Dong, Y., 2016. A nonlinear support vector machine model with hard penalty function based on glowworm swarm optimization for forecasting daily global solar radiation. Energy Convers. Manag. 126, 991–1002.

Keshtegar, B., Mert, C., Kisi, O., 2018. Comparison of four heuristic regression techniques in solar radiation modeling: Kriging method vs RSM, MARS and M5 model tree. Renew. Sustain. Energy Rev. 81, 330–341.

Koca, A., et al., 2011. Estimation of solar radiation using artificial neural networks with different input parameters for Mediterranean region of Anatolia in Turkey. Expert. Syst. Appl. 38 (7), 8756–8762.

Kong, X., et al., 2017. Automatic Detection of Acromegaly From Facial Photographs Using Machine Learning Methods. EBioMedicine.

Kourou, K., et al., 2015. Machine learning applications in cancer prognosis and prediction. Comput. Struct. Biotechnol. J. 13, 8–17.

Lewis, N.S., Nocera, D.G., 2006. Powering the planet: chemical challenges in solar energy utilization. Proc. Natl Acad. Sci. 103 (43), 15729–15735.

Li, Y., et al., 2016. Forecasting the daily power output of a grid-connected photovoltaic system based on multivariate adaptive regression splines. Appl. Energy 180, 392–401.

Lund, H., 2007. Renewable energy strategies for sustainable development. Energy 32 (6), 912–919.

Malhotra, R., 2014. Comparative analysis of statistical and machine learning methods for predicting faulty modules. Appl. Soft Comput. 21, 286–297.

Martín, L., et al., 2010. Prediction of global solar irradiance based on time series analysis: application to solar thermal power plants energy production planning. Sol. Energy 84 (10), 1772–1781.

Nanayakkara, S., et al., 2018. Machine learning models significant improve outcome prediction after cardiac arrest. J. Am. Coll. Cardiol. 71 (11, Suppl.), A775.

Olatomiwa, L., et al., 2015. A support vector machine–firefly algorithm-based model for global solar radiation prediction. Sol. Energy 115, 632–644.

Olden, Julian D., Joshua, J. Lawler, Poff, N.L., 2008. Machine learning methods without tears: a primer for ecologists. Q. Rev. Biol. 83 (2), 171–193.

Panwar, N.L., Kaushik, S.C., Kothari, S., 2011. Role of renewable energy sources in environmental protection: a review. Renew. Sustain. Energy Rev. 15 (3), 1513–1524.

Paoli, C., et al., 2010. Forecasting of preprocessed daily solar radiation time series using neural networks. Sol. Energy 84 (12), 2146–2160.

Parmar, C., et al., 2015. Machine learning methods for quantitative radiomic biomarkers. Sci. Rep. 5, 13087.

Quej, V.H., et al., 2017. ANFIS, SVM and ANN soft-computing techniques to estimate daily global solar radiation in a warm sub-humid environment. J. Atmos. Sol.-Terr. Phys. 155, 62–70.

Ramli, M.A.M., Twaha, S., Al-Turki, Y.A., 2015. Investigating the performance of support vector machine and artificial neural networks in predicting solar radiation on a tilted surface: Saudi Arabia case study. Energy Convers. Manag. 105, 442–452.

Salcedo-Sanz, S., et al., 2018. An efficient neuro-evolutionary hybrid modelling mechanism for the estimation of daily global solar radiation in the sunshine state of Australia. Appl. Energy 209, 79–94.

Sekhar Roy, S., Roy, R., Balas, V.E., 2018. Estimating heating load in buildings using multivariate adaptive regression splines, extreme learning machine, a hybrid model of MARS and ELM. Renew. Sustain. Energy Rev. 82, 4256–4268.

Sephton, P., 2001. Forecasting Recessions: Can We Do Better on Mars, 83. Federal Reserve Bank of St. Louis Review (March/April 2001).

Sigauke, C., Chikobvu, D., 2010. Daily peak electricity load forecasting in South Africa using a multivariate nonparametric regression approach. ORiON 26(2).

Solangi, K., et al., 2011. A review on global solar energy policy. Renew. Sustain. Energy Rev. 15 (4), 2149–2163.

Vempala, N.N., Russo, F.A., 2017. Modeling music emotion judgments using machine learning methods. Front. Psychol. 8.

Voyant, C., et al., 2013. Multi-horizon solar radiation forecasting for Mediterranean locations using time series models. Renew. Sustain. Energy Rev. 28, 44–52.

Voyant, C., et al., 2017. Machine learning methods for solar radiation forecasting: a review. Renew. Energy 105, 569–582.

Wan, C., et al., 2015. Photovoltaic and solar power forecasting for smart grid energy management. CSEE J. Power Energy Syst. 1 (4), 38–46.

Wang, C., 2015. A modified machine learning method used in protein prediction in bioinformatics. Int. J. Bioautomation 19 (1), 25–36.

Wang, X., et al., 2017. MODIS-based estimation of terrestrial latent heat flux over north america using three machine learning algorithms. Remote. Sens. 9 (12), 1326.

Wu, J., et al., 2009. A semi-supervised learning based method: Laplacian support vector machine used in diabetes disease diagnosis. Interdiscip. Sci. Comput. Life Sci. 1 (2), 151.

Yusaf, T., Goh, S., Borserio, J., 2011. Potential of renewable energy alternatives in Australia. Renew. Sustain. Energy Rev. 15 (5), 2214–2221.

Zhang, W.G., Goh, A.T.C., 2013. Multivariate adaptive regression splines for analysis of geotechnical engineering systems. Comput. Geotech. 48, 82–95.

CHAPTER

14

Wind speed forecasting in Nepal using self-organizing map-based online sequential extreme learning machine

Neelesh Sharma and Ravinesh Deo

University of Southern Queensland, Springfield, Springfield, QLD, Australia

14.1 Introduction

Renewable energy is often depicted as a clean source of energy and could have an effect in minimizing environmental impacts by reducing global warming and mitigating the greenhouse effect (Panwar et al., 2011; Wang and Han, 2014). A transition from fossil fuels to renewable sources has been growing in the past 30 years since the price of oil seems to increase constantly and, in contrast, there is a decline in the cost of renewable sources of energy (Herzog et al., 2018). Renewable energy was first adopted by most countries as an integral aspect of national energy policy goals after the 1973 oil crisis (Nepal, 2012). It also ignited interest in wind energy, water pumps, power supply in remote areas, and production of grid electricity powered by wind (Herzog et al., 2018).

Among other renewable sources, wind energy has become a major attraction in today's world because of its low pollution emissions and high efficiency (Wu and Hong, 2007). Wind energy is accepted globally as a clean energy source and the cheapest replacement to coal (Nepal, 2012; Herzog et al., 2018). A 24% average annual growth in wind energy production across the world has been observed since 1990. Based on the trending decline in cost, analysts had forecasted that by the end of 2015 the electricity production cost of wind reached 2.5 US cents/KWh; lower than most fossils fuels (Herzog et al., 2018). According to a Greenpeace organization plan, by 2020 12% of all electricity production should be achieved from wind energy (Wang et al., 2011).

Predictive Modelling for Energy Management and Power Systems Engineering
DOI: https://doi.org/10.1016/B978-0-12-817772-3.00014-8

Nepal, the focus of this chapter, is one of the developing countries that has been in the dark ages of the electricity crisis, with persistent power cuts since 2006 (Laudari, 2016). At present, hydroelectricity is the main source of grid electricity in Nepal, and only 56% of households in Nepal have access to the grid. Fig. 14.1 shows the power supply and demand for electricity from the year 2009 to 2018. It is observed that both supply and demand is increasing constantly throughout the year, however, the supply of power has not met its demand. According to Solar and Wind Energy Resource Assessment (SWERA), Nepal has the potential to generate 2100 MW electricity from solar, 50 MW from microhydro, and 3000 MW from wind (SWERA, 2006; Gurung et al., 2013). Thus wind energy seems to be a viable solution as an alternative source to overcome the yearly drought of electricity in Nepal.

Wind speed is stochastic in nature and is affected by the topology, time, and different daily meteorological factors. The intermittency in wind speed proves to be a challenge for a steady wind supply and an uninterrupted electricity generation (Soman et al., 2010). The variability in wind speed has a negative impact on the output and efficiency of wind turbines (Laudari, 2016). These fluctuations may also mount pressure on the operating cost of wind turbines, disrupting the supply and demand for power (Foley et al., 2012; Keren and Sabitha, 2016).

Wind speed forecasting is proving to be an alluring tool to overcome the abovementioned issues associated with power generation through the wind. Wind speed forecasting is useful for trading companies to assure their commodity by the authenticity and availability of their power source. Wind speed forecasting also facilitates power system operators to manage grid operations and schedule spinning reverse capacity (Chang, 2014). Wind speed forecasting is generally classified as long term and short term. Long-term forecasting enables scheduling and planning of grid maintenance, maintenance of outages, wind farms, and energy storage. Whereas short-term forecasting is essential for real-time grid operations, electricity market clearing, and operational security in market trading (Foley et al., 2012; Chang, 2014). Hence, a reliable wind forecasting model has positive impacts on both the technical and economic aspects of wind power generation.

Nepal has a high availability of wind potential that has not been explored yet. On the other hand, the power generation through other sources like hydroelectricity and solar is unable to meet the power demands (Fig. 14.1). Although Nepal has the capacity to

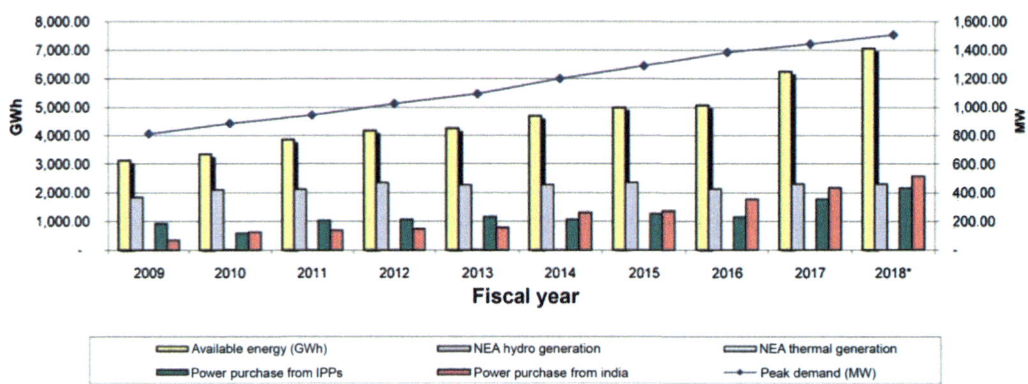

FIGURE 14.1 Power supply versus demand for the fiscal year 2009-17 in Nepal (NEA, 2018).

generate 3000 MW from wind (SWERA, 2006), it has been purchasing a high volume of power from its neighboring country India (NEA, 2018) but is still unable to meet the demand for electricity. Therefore it needs to harness wind energy as a viable alternative power source. Wind energy can help to mitigate the difference in the power demand and supply in Nepal, as seen in Fig. 14.1. A sustainable and reliable power source from wind can be established with the help of an authentic tool to study nonlinear wind speed patterns and forecast wind speed. Considering the necessity of an alternative source of power in Nepal, the availability of wind, the issues with the nonlinear nature of wind speed, and the importance of wind forecasting, this chapter is focused on building a versatile and reliable decision support system for wind farms to forecast wind speed at 6-hourly, daily, and monthly intervals for five diverse geographic locations in Nepal.

This chapter aims to forecast wind speed for three forecast horizons, short-term (6-hourly), daily, and monthly. The short-term forecasting will be useful to wind farms for real-time market trading and make the decision for the decrement and increment of loads. The daily forecasting will be applicable for the planning of daily grid maintenance and outage plans. Monthly forecasting aids the wind farms to reserve maximum power for unseen outage plans. Thus the study will facilitate wind farms with a reliable decision support system that will help wind farms to study the change in nature of wind speed in short term, daily, and monthly forecast horizons, and provide a reliable and sustainable source of energy in Nepal.

14.1.1 Research objectives

The purpose of this chapter is to determine the authenticity of a hybrid Self-Organizing Map-based Online Sequential Extreme Learning Machine (SOM–OSELM) for wind speed forecasting that can act as a reliable decision support system for wind farms. The self-organizing map (SOM) is used as a data splitting process and online sequential extreme learning machine (OSELM) is the machine-learning algorithm to forecast wind speed. The following objectives are identified to fulfill the chapter objectives.

1. To develop and evaluate the suitability of the SOM–OSELM model to forecast short-term (6-hourly), daily, and monthly wind speeds at five geographically diverse sites in Nepal.
2. To benchmark the SOM–OSELM algorithm with existing methods: OSELM, M5, SOM–M5, and ARIMA.
3. To observe if SOM is a more effective data partitioning tool in wind forecasting than traditional data splitting.

Through these objectives, this chapter aims to generate a systematic decision support system for wind turbines that will help to minimize their operating cost, solve the issue of interruption in the supply of electricity, and establish a sustainable source of wind energy.

14.2 Literature review

Wind energy has merit in both economical and climatic aspects over fossil fuels. Wind energy is a cheaper and clean source of energy compared with nonrenewable sources, but it comes with its own challenge because of its intermittent and stochastic nature (Wu and

Hong, 2007; Colak et al., 2012). Wind speed forecasting plays a crucial role to establish wind energy as a sustainable and clean source of energy (Deo et al., 2018). However, wind speed randomly changes and is affected by topology, time, and different daily meteorological factors like temperature, humidity, and wind direction. This variability in wind speed has a negative impact on the output and efficiency of wind turbines (Laudari, 2016).

The fluctuations in wind speed prove to be a challenging factors in terms of a steady wind supply and an uninterrupted electricity generation. These fluctuations may also mount pressure on the operating cost of wind turbines disrupting the supply and demand in electricity (Foley et al., 2012; Keren and Sabitha, 2016). An Institute of Electrical and Electronics Engineers (IEEE)/Power and Energy Society (PES) summary stated that system operating cost, system peak load, and per-unit cost of wind electricity would increase up to 20% due to wind uncertainty and variability (Wu and Hong, 2007).

Wind forecasting plays an important role in wind assessment as it helps facilitate proper scheduling of spinning reverse capacity, wind station planning, energy storage, and commerce (Foley et al., 2012; Deo et al., 2018). Therefore improving the prediction of wind speed and energy generation helps to minimize risks and assists the wind turbines in being operated in a cost-efficient manner.

The literature review aims to highlight and point us in a direction to achieve the chapter aims and objectives of the chapter. This chapter aims to design a decision support system for wind farms, based on a hybrid SOM coupled with OSELM, to forecast wind speed for diverse sites in Nepal.

The chapter also proposes to test the proposed models against the existing M5 and ARIMA models. This literature is focused on wind forecasting in different forecast horizons and their applications for the benefits of a wind farm. Secondly, it will review the existing models used for data portioning used in forecasting. Finally, the literature will review existing learning algorithms used for wind speed forecasting and then emphasize the extreme learning machine (ELM) and online sequential extreme learning machine (OSELM) model used in forecasting.

14.2.1 Wind speed forecasting and forecast horizon

In literature, wind forecasting is generally carried out for different time domains, that is, long term (greater than 3 days), medium term (few hours to 3 days) and short term (few minutes to hours). However, the time intervals and subintervals are not fixed and may vary depending on the necessity and the researchers. For example (Chandra et al., 2013; Keren and Sabitha, 2016; Lazarevska, 2016; Alencar et al., 2017) in their study on wind speed forecasting, have classified time scales into four categories, that is, very/ultra-short term, short term, medium forecast, and long term forecast with slightly different time intervals. These researchers applied very short-term forecasting for regulations of turbines and electricity marketing, short-term forecasting for load increment and decrement decision, medium-term forecating for generator online–offline decisions, and long-term forecasting for maintenance scheduling and reserve requirements. Whereas Azad et al. (2014) have mentioned time intervals of a few hours for short domain, a few hours to 3 days as medium term, and greater than 6 days as long term, which helps in wind turbine switching and low spinning reserve. Similarly, Wang et al. (2011) divided the wind forecasting into three-time domains: intermediate short term (8 hours ahead), short term (daily), and long term (multiple days). The authors classified intermediate short term as being applicable for real-time grid operations, short term for the planning of economic dispatch load, and long term for maintenance and operational planning.

From previous studies, we can see that short-, very short-, or medium-term forecasting does not have an explicit time difference amongst them and their roles are interchangeable for wind forecasting. Also, no evidence was found for a 6-hourly real-time prediction, which will be carried out in this chapter. Our dataset contains a consistent dataset for 6 hours, thus in our chapter wind forecasting will be basically divided into only three domains, that is, short-term 6-hourly, mid-term daily, and monthly. The monthly forecast will enable us to schedule and plan grid maintenance and maintenance of outages. They also are useful factors while surveying a new location for a wind farm. Also, monthly forecasts will help to plan proper reserve and low spinning reverse (Foley et al., 2012; Chandra et al., 2013; Azad et al., 2014). On the other hand, the daily forecast will provide information on daily maintenance and generation online and offline decisions. The short-term forecast will be essential for wind turbine control, market trading, and load increment—decrement decisions (Foley et al., 2012; Chandra et al., 2013) (Table 14.1).

14.2.2 Data partitioning in forecasting

The data partition is an important aspect, which is often considered trivial, while in the development of a feedforward neural network cross-validation is commonly used to ensure generalization. Data generalization is the ability of a machine-learning algorithm to accurately represent the unseen datasets (May et al., 2010). The use of cross-validation, however, is the use of simple random sampling (SRS) in data splitting that may lead to inappropriate data splitting (Reitermanov, 2010). Inappropriate data splitting methods may lead to high variability in model performance and inaccuracy. There are various approaches (SRS, systematic sampling, trial and error, stratified sampling) proposed by statisticians to overcome the issue of inappropriate data splitting (Reitermanov, 2010), among which stratified sampling is considered the more sophisticated approach. May et al. (2010) compared different data splitting methods (SRS, DUPLEX, stratified sampling) for ANN development. The study concluded that the stratified sampling method using SOM produced the lowest variance and bias and the lowest model error in ANN. Reitermanov (2010) performed a similar study where a comparison among SRS, DUPLEX, and stratified random sampling showed that SRS showed higher variance and bias in the model error for high-dimensional datasets and concluded that stratified random sampling was the most appropriate method for high dimension and complex datasets. Thus the data sampling method should be an important aspect for data modelers, since it is performed only once during the initial stage of data modeling and the model's performance will be dependent on it.

14.2.3 Wind forecasting models

Studies show that mathematical or physical method-based predictions for vertical extrapolation are only reliable for moderate and strong winds, whereas for turbulent winds the estimation error is very high (Laudari, 2016) and these models have high

TABLE 14.1 Summary of classification of wind speed forecasting based on forecast horizons.

Research	Time domain	Time range	Applications
Chandra et al. (2013), Lazarevska (2016)	Very short-term	Within 30 min interval	Electricity market trading
	Short term	30 min–6 h	Load increment–decrement
	Medium term	6 h–1 day	Generator online–offline decision
	Long-term	1 day to a week and more	Maintenance scheduling and reserve requirement
Keren and Sabitha (2016), Alencar et al. (2017)	Ultra-short-term	Within 60 min interval	Regulations of turbines and electricity marketing
	Short term	1 h to many hours	Load increment and decrement decision
	Medium term	Many hours to 1 week	Generator online–offline decision
	Long term	1 week to a year or more	Maintenance scheduling and reserve requirement
Azad et al. (2014)	Short term	Few hours	–
	Medium term	Few hours to 3 days	–
	Long term	Greater than 6 days	Wind turbine switching and spinning reserve
Wang et al. (2011)	Intermediate short term	8 h ahead	Real-time grid operations
	Short term	Daily ahead	Economic dispatch load planning
	Long term	Multiple days ahead	Maintenance and operational planning
Proposed work	Short term	6 h	Wind turbine control, market trading
			Load increment–decrement decisions
	Mid term	Daily	Daily maintenance generator online–offline decision
	Long term	Monthly	Schedule and plan grid maintenance and outage
			Surveying a new location of wind farms
			Plan proper reserve and low spinning reverse

computation time (Mohammadi et al., 2015). Also, different statistical and physical models are researched in papers for short- and long-term wind forecasting. Physical models are based on meteorological and atmospheric parameters like pressure, temperature, surface obstacles, and roughness and need more computational time (Keren and Sabitha, 2016). Physical models are reliable for long-term predictions, whereas they are unreliable for short-term predictions. On the other hand, statistical methods are consistent for short-term forecasting (Wang et al., 2018). Also unlike the physical model, the statistical method does not utilize meteorological processes, the operating cost is low, but the forecasting error increases with the increase in time of prediction (Keren and Sabitha, 2016). The statistical method is also depicted as a "black box" as it establishes a statistical relationship between the power output and the weather prediction and converts the input variable to output in a single step (Foley et al., 2012).

Machine-learning methods such as support vector machines (SVM) and artificial neural network (ANN) have been the key interest of researchers in recent years for wind forecasting. These methods rely on historical data to forecast wind and have good error tolerance (Mi et al., 2017; Wang et al., 2018). Machine-learning methods make a prediction from historical time series data using the relation between forecasted electricity output and predicted winds (Foley et al., 2012). ANN is also referred to as the data-driven approach since it learns from experience and has been proven to solve complex nonlinear problems (Foley et al., 2012; Mohammadi et al., 2015). A major disadvantage of popular machine-learning methods such as ANN and SVM is that the learning time is extremely high. Therefore researchers have designed a relatively new algorithm, the extreme learning machine (ELM) with good generalization performance and a fast learning rate (Mohammadi et al., 2015). ELM overcomes other challenges faced by SVM and ANN, like poor computational scalability, intervention from users, and iterative tuning of parameters (Huang et al., 2011; Ding et al., 2013). Studies have shown that ELM can be highly efficient for automated prediction systems, for example, wind prediction in real time (Foley et al., 2012; Mohammadi et al., 2015; Keren and Sabitha, 2016; Mi et al., 2017). These matters of concern have initiated new forecasting methods to be implemented in wind energy. Since we are using historical data, ELM is beneficial over other machine-learning methods and is automated.

14.2.4 Extreme learning machine

ELM was first introduced to improve the efficiency and speed of a single-hidden-layer feedforward network (SLFNs) (Huang et al., 2011). As opposed to the conventional belief of neural network generalization theory, linear theory, and control theory, the ELM algorithm does not require hidden nodes/neurons to be tuned. Unlike ANN, that periodically assigns hidden nodes, ELM randomly assigns hidden nodes, constructs biases and input weights of hidden layers, and determines the output weights using least squares methods. This justifies the low computational time of ELM and thus is preferred by researchers over ANN.

Lazarevska (2016) and Syed and Aggarwal (2016) used a standalone ELM which was tested against existing persistence models. Lazarevska (2016) used *RMSE* for wind predictions whereas Syed and Aggarwal (2016) used *MAE* for hourly to 12-hourly predictions. Lazarevska (2016) outlined that the performance of a traditional ELM depends on randomness mechanism, that is, as the ELM parameters are randomly chosen an uncertainty problem arises and it's difficult to know if the model has the best solution. To solve the issue, the paper used using an iterative model design where the model error is compared to the desired error tolerance. Syed and Aggarwal (2016) based their studies on three wind farms of Ontario, Canada. The paper used three parameters: wind speed, wind direction, and wind power, as inputs. The study showed that the ELM model outperformed the persistence models for 12-hour predictions with *MAE* ranging from 5 to 21 for the three sites. The paper showed that the *MAE* for both the methods increased successively as hours increased from 1 to 12. The paper concluded that ELM is notably fast and doesn't encounter local minima and overfitting, it also tends to reach the smallest training error and reaches the smallest norm of weights, unlike traditional gradient-based learning algorithms.

Li et al. (2016) concluded that a standalone ELM did not perform well for short-term forecasting on data from wind farms located in Northern China at 15 minutes interval. The research observed a large error around the peak of the curve; the paper used an error correction model to improve the performance of ELM. Also, for better performance, the research used min/max normalization for even better accuracy. The performance of the model was compared with existing persistence methods. The accuracy of the hybrid error correction ELM was verified using a normalized root mean square method (*NRMSE*) value of 5.76% which was much less than the standalone ELM (21.09%). The paper suggested ELM doesn't perform well due to the stochastic nature of wind which is improved in terms of *NRMSE*.

Wang et al. (2018) in his research proposed an extreme machine-learning model for short-term forecasting for three sites in China. In this research, ELM was coupled with improved complementary ensemble empirical mode decomposition (EMD) with adaptive noise (ICEEMDAN) to solve the issues of the stochastic nature of wind by performing nonstationary decomposition. Their model also included autoregressive integrated moving average (ARIMA) for feature selection. ICEEMDAN–ARIMA was used for preprocessing and postprocessing (error correction) of dataset similarly for postprocessing, that is, to reduce ELM's output fluctuations the authors used simple ensemble methods (arithmetic average). Ensemble models help to reduce the uncertainty of the predictions (Foley et al., 2012; Wang et al., 2018). The paper evaluated its model by comparing ICEEMDAN–ARIMA with other pre- and postprocessing models coupled with ELM, like EMD, EEMD, CEEMDAN, and standalone methods ARIMA and ELM using *MAE*, *RMSE*, and *MAPE*. Their studies showed that the ELM–ICEEMDAN–ARIMA hybrid model outperformed other models. The paper concluded that a hybrid model has significantly higher accuracy than standalone methods and improved the *MAPE* by around 33%.

Mi et al. (2017) used a hybrid model of an ELM model coupled with wavelet domain denoising (WDD), wavelet packet decomposition (WPD), EMD, ARIMA, and outlier correction method. WDD was used to reduce the noise present in the original wind data, WPD was used to decompose wind speed in nonstationary layers, similar to Wang et al. (2018) who used ICEEMDAN. ARIMA, and ELM to develop the multistep forecasting model. Also, an outlier correction method was used to increase the robustness of ARIMA and ELM forecasting. The outlier correction method (outlier detection and outlier correction) was applied to mitigate the overfitting problems in ELM and ARIMA. The hybrid model was evaluated using *MAPE*, *MAE*, and *RMSE* and was compared against existing standalone ARIMA and ELM model and hybrid models WPD–ELM. The paper hybrid model WDD–WPD–EMD–ARIMA–ELM proved to be an appropriate model for stochastic wind speed and outperformed other benchmarked models.

Liu et al. (2015) analyzed four different hybrid models, combining signal decomposing algorithms with ELM. The signal decomposing models used were WD, WPD, EMD and fast ensemble EMD (FEEMD). The paper used *MAE*, *MAPE*, and *RMSE* to test the performance accuracy and they were tested against single ARIMA, ELM, and MLP models. The study showed that single ELM outperforms the classical ARIMA and MLP algorithms. However, for a better generalization capacity and accuracy decomposing algorithms, WD/WPD/EMD/FEEMD-ELM can be combined with ELM since all the proposed hybrid models showed much higher accuracy than a single ELM model. Xiao et al. (2016a,b) used a variant of ELM, that is, SaDE−ELM for electricity forecasting and studies proved that the self-adaptive differential algorithm improves the performance of ELM. Mahmoud et al. (2018) applied an improved variant of ELM, the SaDE-ELM for wind forecasting in Australia. SaDE-ELM also outperforms ELM with computational time since it can self-adaptively determine the control parameters and generation techniques involved in differential evolution (Cao et al., 2012; Mahmoud et al., 2018). These papers outlined that, although ELM outperforms other existing persistence models, ELM alone is prone to the variability of wind speeds and suggests that it is feasible to adopt the decomposition of wind data before applying the ELM models.

14.2.5 Online sequential extreme learning machine

From the above literature, ELM has been identified as an effective approach for wind forecasting. However, for data with high dimensions, ELM takes time to converge to an optimal region when it just depends on stochastic search methods (Xiao et al., 2016a,b). ELM is quite sensitive to the number of nodes/neurons in the hidden layers and is quite difficult to solve (Wang et al., 2016). Also, for ELM all the training data should be used for training, which may not be the case in the real-world where the data may be acquired one by one or in chunks (Ding et al., 2013). An improved variant of ELM named OSELM is adopted by researchers instead of ELM to fulfill the requirement of online and real-time forecasting (Wang and Han, 2015). OSELM follows a sequential learning method which is favored over batch learning algorithms, since it does not require retraining of the data when new data is added (Huang et al., 2011), thus for time series data OSELM proves to be a highly efficient model.

Wang and Han (2014) used OSELM with different kernels to predict nonstationary time series data and compared the performance with conventional ELM with kernels and online support vector regression (OSLSVR). The proposed OSELM outperformed the comparative models with better accuracies and minimum learning time and they concluded OSELM can automatically tune weights when new data samples are removed or added.

There is not much literature found in the area of wind speed prediction using OSELM models. OSELM is widely used for forecasting stochastic data like air quality, financial time series forecasting, and yearly sunspots (Peng, 2013; Wang and Han, 2014). Peng (2013) used OSELM for air quality forecasting in Canada and other countries and concluded OSELM performed better than MLR, OSMLR, and MLP-NN. Also, it was observed that the use of the data splitting method is often not considered while building the model.

These studies show that the use of a self-organizing map improves the performance of feedforward neural networks and an online sequential method is an effective way to improve the performance of the traditional ELM. Although OSELM and SOM-based stratified sampling for data splitting algorithm were used for forecasting of time series data, these two models have not been implemented together to forecast wind speed, to the best of the authors' knowledge. Thus the variant of traditional ELM, that is, OSELM combined with SOM as a data splitting method will be utilized and compared with other models in this chapter (Table 14.2).

Previous studies indicate that ELM has been widely adopted because of its forecasting accuracy and low computational time over preexisting benchmark models like ARIMA, SVM, and ANN. However, to the researcher's knowledge, wind forecasting has not been explored using SOM–OSELM. Researchers have also applied different hybrid techniques mainly for wave decomposition to improve the efficiency of ELM. However, the data splitting methods are often ignored in these studies. Data splitting methods are only carried out once at the start of model building and the performance of the model highly depends on it, thus they should be taken into consideration while building a model. Also, it was observed from the literature review, that the wind energy in Nepal has not been explored

TABLE 14.2 Summary of models used for wind speed forecasting based on literature where *MAE*, mean absolute error, *RMSE*, root mean square error, *NRMSE*, normalized *RMSE*.

Reference	Proposed models	Evaluation methods	Study region	Remarks
Syed and Aggarwal (2016)	ELM	*MAE*	Ontario, Canada	ELM tends to reach the smallest training error and norm of weights ELM does not encounter local minima and overfitting
Lazarevska (2016)	ELM	*RMSE*	–	Randomly assigned ELM parameters to cause uncertainty in identifying the best model
Li et al. (2016)	ELM + Hybrid Error Correction	NRMSE	Northern China	Stand-alone ELM does not perform well for short-term forecasting
Wang et al. (2018)	ELM-ICEEMDAN–ARIMA	*MAE, RMSE,* and *MAPE*	China	• ICEEMDAN for nonstationary decomposition • ARIMA for feature selection
Mi et al. (2017)	WDD-WPD-EMD-ARIMA-ELM	*MAE, RMSE,* and *MAPE*	–	• Outlier correction method was used to mitigate overfitting in ELM • WDD-WPD-EMD for noise reduction and wave decomposition
Liu et al. (2015)	WD/WPD/EMD/FEEMD-ELM	*MAE, RMSE,* and *MAPE*	–	• Single ELM performs better than ARIMA and MLE • Signal decomposing algorithm WD/WPD/EMD/FEEMD coupled improves the performance of ELM
Mahmoud et al. (2018)	SaDE-ELM	*RMSE* and CPU computational time	Australia	SaDE-ELM was found to have higher reliability and fast speed than ANN, SVM, ELM

using the proposed model. Thus this paper considers SOM as the data splitting method to build a hybrid SOM–OSELM for the fulfillment of the objectives of the study.

14.3 Materials and methods

14.3.1 Study area and dataset

The chapter uses data from the European Centre for Medium-Range Weather Forecasts (ECMWF) (Dee et al., 2011; Hagspiel et al., 2011) for five wind hotspot locations in Nepal, that is, Biratnagar, Tarhara, Okhaldhunga, Jumla, and Khumaltar, as shown in Fig. 14.2. These locations are of wind turbines in Nepal, installed in different topology.

To evaluate the versatility of the hybrid SOM–OSELM model for wind forecasting, the study sites were chosen carefully to ensure they represent different geographic locations of Nepal. Thus based on the available monthly maximum value of windspeed in 2008 (Chhetri and Shakya, 2010), topology and altitude, the five sites were selected as shown in Table 14.3. The sites Biratnagar and Tarhara lie at low altitudes (72–200 m), that is, the Terai belt of Nepal whereas Okhaldhunga (1720 m), Khumaltar (1350 m), and Jumla (2300 m) lie in hilly regions of Nepal.

From Table 14.3, it is observed that the monthly maximum value of wind in 2008 was highest in low altitude wind turbines, Biratnagar (3.11 km/h) and Khumaltar (1.08 m/s). Mostly, the wind turbines with maximum monthly value are chosen with an exception,

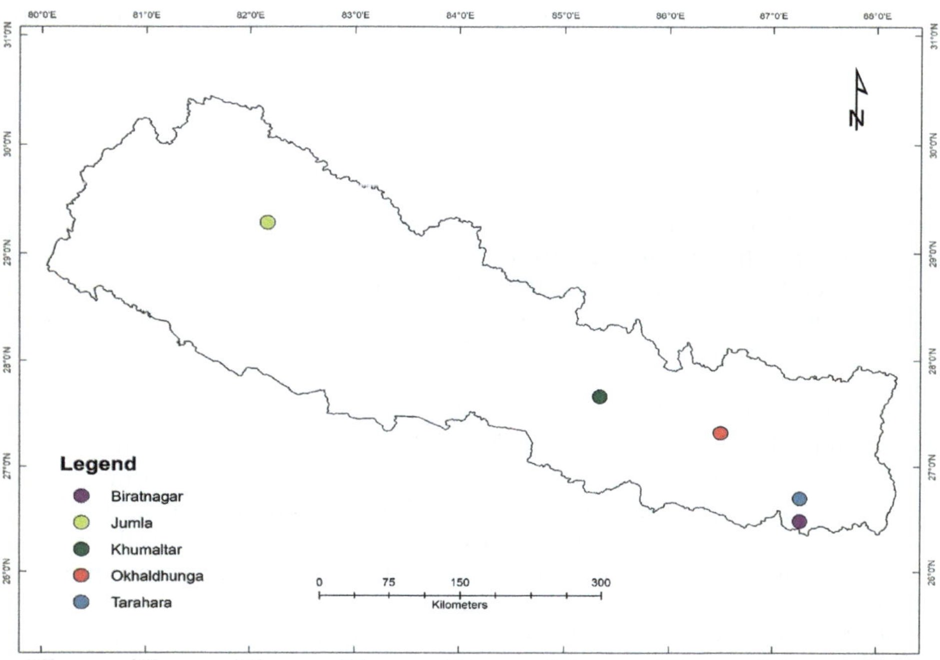

FIGURE 14.2 Study area and location of five windmill sites in Nepal.

TABLE 14.3 Monthly maximum average wind speed in m/s for wind turbines in Nepal.

Location	Altitude	Latitude	Longitude	Monthly maximum average wind speed 2008 (m/s)	Reanalysis dataset monthly maximum wind speed average 2008 (m/s)
Biratnagar	72 m	26° 29′ N	87° 16′ E	3.11	2.831
Tarahara	200 m	26° 42′ N	87° 16′ E	2.75	2.542
Okhaldhunga	1720 m	27° 19′ N	86° 30′ E	2.33	1.799
Jumla	2300 m	29° 17′N	82° 10′ E	1.77	1.988
Khumaltar	1350 m	27° 40′ N	85° 20′ E	1.08	1.715

Khumaltar, having a monthly maximum speed of 1.08 m/s, to observe if the machine-learning algorithms have similar efficiency for both high and low wind speed.

The data for the study was obtained from the ECMWF database; the ECMWF Re-Analysis (ERA) Interim Dataset from the years 1979 to 2017 was used for all five locations. One of the advantages of using the ERA-Interim Dataset is that it does not have any missing values. The ERA-Interim dataset ERA-Interim dataset provides a global view encompassing good spatial coverage consistency in climate variables with a minimum time delay (Dee et al., 2011; Hagspiel et al., 2011). The dataset parameters 10 m zonal (U) and meridional (V) component at 6-hourly intervals will be used for average wind forecasting. The net wind was calculated using Net Wind $= \sqrt{U^2 + V^2}$. The net wind speed obtained from U and V components of wind was used to forecast wind speed for all forecast horizons.

14.3.2 Theoretical details of models

14.3.2.1 Data splitting method: self-organizing map

SOM is an unsupervised ANN algorithm that represents the input space of training data into a two-dimensional discrete map or grid in an orderly fashion. The mapping without losing much information preserves the metric and topological relationship between original data (Rustam et al., 2007). The data items are mapped to one neuron in the grid and the distance between two neurons in the grid represents the similarities among the elements (Kohonen, 2014). The SOM is a clustering algorithm that clusters the data from the optimum partition of weight vectors (May et al., 2010). This feature of SOM is the base of the SOM-based stratified sampling (SOM), that is, the SOM algorithm clusters the training datasets into a [m n] dimension grid representing different clusters depending on the distribution of the weight vectors. Next, samples are drawn from each stratum and this is how data splitting is performed using SOM (May et al., 2010).

The neurons in the layer of SOM are arranged either in random, grid, or hexagonal topology, among which hexagonal provides the best visualization. Distance functions like boxdist, mandate, linkdist, and dist are used to calculate the distance between the neurons, among which linkdist is used widely (Kohonen, 2014). During the SOM training period, the weighted vector linked with the neurons moves to become the center of the cluster and the neurons with similar topology move close to each other, thus the dataset with similar variance and bias are clustered together (Kohonen, 2014).

SOM consists of a [m n] size grid, as seen in Fig. 14.3 and is equal to the total neurons associated with the dataset. Each node n has a weighted vector w with d dimension ($w = (w_{i1}, w_{i2}, \ldots w_{id})$), where i is the number of neurons. The SOM algorithm consists of the following steps (Kangas and Kohonen, 1996; Kohonen, 2013; Chaudhary et al., 2014):

1. Initially, random values are taken as the initial weight of vectors w_i.
2. A random sample of input training vector $x(t)$ is selected from the input space, this value acts as the input to all neurons.
3. Next, the winning neuron is identified that has the weight vector in the closest vicinity of the input vector. The best neuron is selected from the minimum value from $c = \arg\{\, \|w_i(f) - x(f)\| $
   where $w_i(t)$ and $x(t)$ are weight and input vector of neuron n at f iteration, respectively.
4. The weight of the neurons is updated with $w_i(f+1) = w_i(f) + h_{c,i}(f)[x(f) - w_i(f)]$, where $h_{c,i}(f)$ is the Gaussian neighborhood denoted by

$$h_{c,i}(f) = \eta(t) \times \exp\left(\frac{\|r_C - r_i\|}{2\sigma^2(t)}\right) \tag{14.1}$$

where $\eta(t)$ is the learning rate, r is the coordinate of a neuron on the grid, and $\sigma(t)$ is the width of the neighborhood radius.

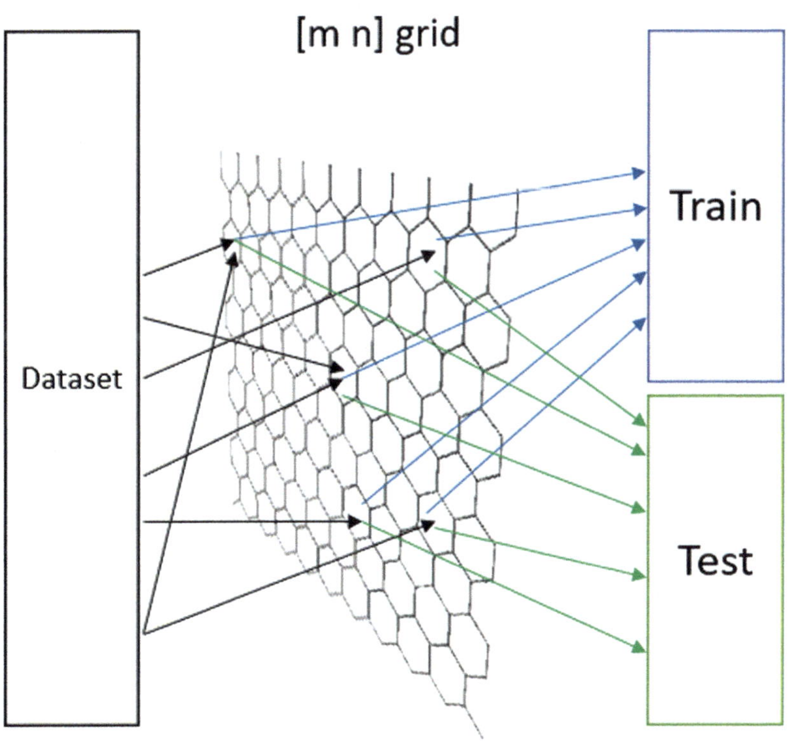

FIGURE 14.3 SOM data splitting method.

The steps from 2 to 4 are iterated until the feature map reaches the optimum value and stops changing. SOM, unlike conventional learning algorithm, updates the weights of the winning neurons along with its neurons in the vicinity. Thus the adjoining neurons incline to have analogous weight vectors and are susceptible to identical input vectors (Kohonen, 2014).

14.3.2.2 Objective model: online sequential extreme learning machine

OSELM is a variant of ELM. ELM was first introduced to increase the efficiency and speed of SLFNs (Huang et al., 2006; Huang, 2015; Potočnik and Govekar, 2015). As opposed to the conventional belief of neural network generalization theory, linear system theory, and control theory, the ELM learning algorithm does not require hidden neurons/nodes to be tuned (Huang, 2015) (Fig. 14.4).

For generalized SLFNs the output function of ELM is written as (Huang et al., 2006)

$$f_L(x) = \sum_{i=1}^{L} \beta_i h_i(x) = h(x)\beta \tag{14.2}$$

where $\beta_i = (\beta_1, \beta_{2,\ldots,}\beta_L)$ is the vector of output weights between the output neurons $(m > = 1)$ and L hidden layer neurons, $h(x) = [(h_1(x), h_2(x), \ldots\ldots\ldots h_L(x))]$ is the hidden layer output function with respect to x.

$$h_i(x) = G(a_i, b_i, x_j)a_i = R^d, b_i \epsilon R \tag{14.3}$$

where $G(a_i, b_i, x)$ are the nonlinear piecewise continuous functions, a_i is the input weight vector connecting the input neurons with ith hidden neurons and b_i is the threshold of

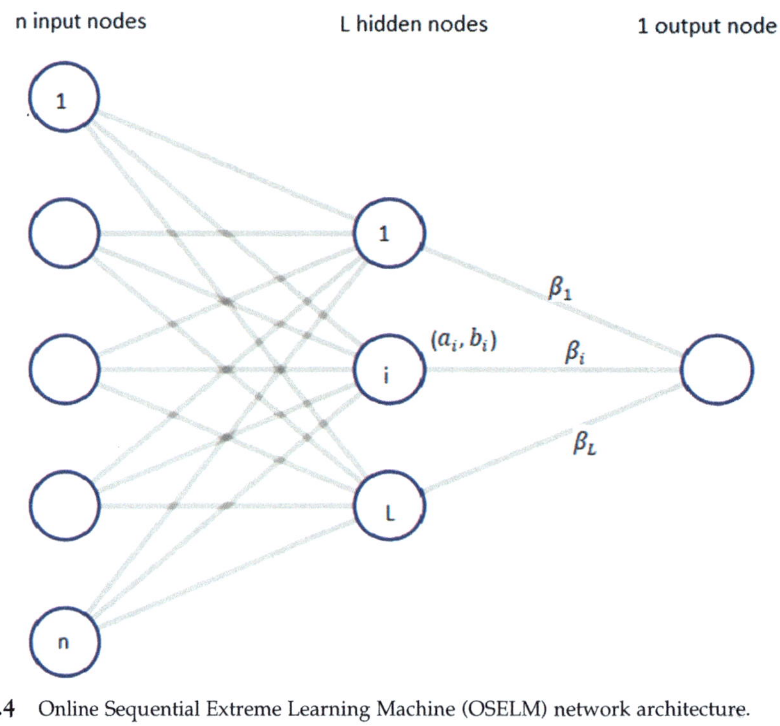

FIGURE 14.4 Online Sequential Extreme Learning Machine (OSELM) network architecture.

ith hidden node. $\sum_{i=1}^{L} \beta_i h_i(x) = t_j$ where $j = 1,2,3,\ldots N$ can also be written as (Huang et al., 2006)

$$H\beta = T \tag{14.4}$$

where,

$$H(a_1, \ldots a_L, b_1 \ldots b_L, x_1 \ldots, x_L) = \begin{bmatrix} G(a_1, b_1, x_1) & \cdots & G(a_L, b_L, x_1) \\ \vdots & \ddots & \vdots \\ G(a_1, b_1, x_N) & \cdots & G(a_L, b_L, x_N) \end{bmatrix}$$

$$\beta = \begin{bmatrix} \beta_1' \\ \vdots \\ \beta_1' \end{bmatrix}$$

$$T = \begin{bmatrix} t_1' \\ \vdots \\ t_1' \end{bmatrix}$$

H is the hidden layer output matrix, the output of ith hidden node is the ith column of H, with respect to the inputs $x_1 \ldots, x_N$ (Rong et al., 2008). From 14.3, we get the value of β

$$\hat{\beta} = H^{+}T \tag{14.5}$$

where H^{+} is Moore–Penrose pseudo-inverse.

Let us consider rank $H = L$ (number of hidden nodes), H^{+} can be denoted as (Huang et al., 2006).

$$H^{+} = \left(H'H\right)^{-1}H' \tag{14.6}$$

and

$$\hat{\beta} = \left(H'H\right)^{-1}H'T \tag{14.7}$$

ELM randomly assigns hidden nodes, and biases and weight of the hidden layers determine the output weight using the least squares method; this justifies the low computational time of ELM (Ding et al., 2013; Potočnik and Govekar, 2015), an approach that is also adopted by OSELM, an online and real-time forecasting variant of traditional ELM. OSELM adopts a sequential learning algorithm unlike other batch learning variants of ELM, and is capable of real-time nonstationary forecasting (Wang and Han, 2014). OSELM follows a sequential learning algorithm, which does not require retraining of the data when new data is added, which is very efficient when dealing with nonstationary real-time datasets (Huang et al., 2011).

OSELM is carried out in two phases: first an initial training phase and second a sequential learning phase. In the first phase of OSELM, batch learning is carried out on the training data and solves Eq. 14.6. If N_0 is the total training data and is less than the hidden neurons (L), for batch ELM $\|H_\beta - T_0\|^2$ is minimized. Thus from Eq. 14.4, we get (Lim et al., 2013; Peng, 2013)

$$\beta^0 = K_0^{-1} H_0' T_0 \tag{14.8}$$

where

$$K_0 = H_0' H_0$$

When $(k+1)^{th}$ batch of new training data is introduced, the OSELM is updated into the online learning stage by using a recursive least squares solution given by (Lim et al., 2013; Peng, 2013)

$$\beta^{k+1} = \beta^k + K_{k+1}{}^{-1} H_{k+1}{}' (T_{k+1} - H_{K+1}\beta^k) \tag{14.9}$$

where

$$K_{k+1} = K_0 + H_{k+1}{}' H_{k+1}$$

The chunk of data once used is discarded and is not used in further model updates.

14.3.2.3 Benchmark method: M5

M5 model uses the divide and conquer rule to divide the data space into subspaces and represents each subspace as a linear regression model (Taghi Sattari et al., 2013). The M5 model tree is more accurate then regression trees and also easy to interpret (Kaveh et al., 2016). The M5 model is a combination of traditional decision trees, where the leaves will possibly be represented by linear regression. M5 is a piecewise linear model, thus lies between the absolute nonlinear feedforward neural networks and truly linear ARIMA model (Solomatine and Xue, 2004).

The M5 tree model follows a similar method to the decision tree. It consists of three major steps, tree development, pruning, and smoothing. The splitting of datasets in the M5 tree depends on the standard deviation of class values that reach a node by calculating the error at the node. It then tests all attributes at the node to estimate the anticipated decline in error. The standard deviation reduction (SDR) is calculated using the following

$$\text{SDR} = sd(Z) - \Sigma \frac{|Z_i|}{|Z|} sd(Z_i) \tag{14.10}$$

where sd is the standard deviation, Z is the set of examples reaching nodes, Z_i are a subset of Z, and i represents the total outcomes of a potential set. The attribute that maximizes anticipated decline in error is selected (Kaveh et al., 2016; Khozani et al., 2017). When the M5 tree is built, it is subjected to pruning, to reduce the number of leaves and overcome overfitting. Next, the smoothing process is carried out to stop the acute discontinuousness of the leaves (Kaveh et al., 2016).

14.3.2.4 Benchmark method: autoregressive integrated moving average model

ARIMA model is a statistical model for forecasting and analyzing time series data. ARIMA model works in acquiring long-range correlation, an attribute seen in wind speeds, hence ARIMA models are widely used for wind forecasting (Kavasseri and Seetharaman, 2009). ARIMA model uses a dependent relationship between the number of lagged observations and true observations to check if the time series is nonstationary or stationary (Somvanshi et al., 2006). It then differences the observation between two different timestamps to make a stationary time series. Finally, it checks dependency between residual error and observations and applies it to lagged observations. ARIMA model standard notation is written as ARIMA (p, d, q) where p is the lag order, d is the degree of differencing, and q is the order of moving average. The next value of a variable is assumed to be a combination of a linear function of past values and random errors. ARIMA models can be generalized by the following equation.

$$y_t = \theta_0 + \varnothing_1 y_{t-1} + \varnothing_1 y_{t-1} \ldots + \varnothing_p y_{t-p} + \varepsilon_t - \theta_1 \varepsilon_{t-1} - \theta_2 \varepsilon_{t-2} - \ldots - \theta_q \varepsilon_{t-q} \quad (14.11)$$

where y_t is the actual value, ε_t is the random error at the time-period t. θ_j ($j = 0, 1, \ldots, q$) and $\varnothing_i (i = 1, 2, \ldots, p)$ are model parameters (Zhang, 2003).

14.3.2.5 Model development

The model development was carried out in three phases: data preprocessing, data splitting, and model and performance metrics and further divided into a relevant subdivision as shown in Fig. 14.5.

The net wind speed was initially subjected to the partial autocorrelation function (PACF), to calculate the significant lags from the original net wind speed data. PACF is the autocorrelation of a datapoint with itself at different intervals of time. When two variables are correlated, there lies a mutual linear dependence on other variables, which is also known as confounding. Thus for time series, the partial correlation between x_t and x_{t-s} is the correlation between x_t and x_{t-s}, after removing the linear dependence on $x_1, x_2, \ldots, x_{t-s+1}$ that comes between the time points t and $t-s$.

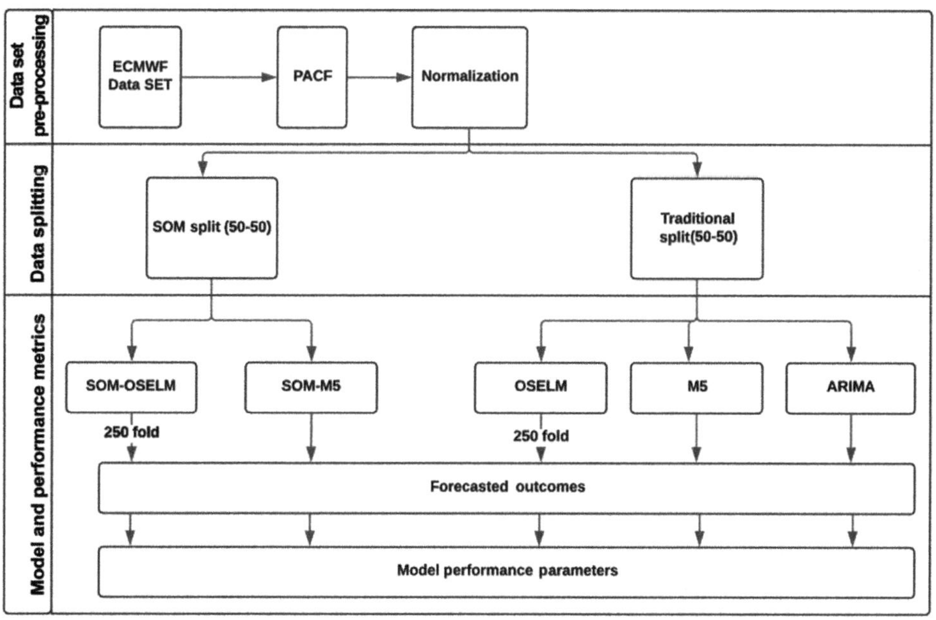

FIGURE 14.5 Model development process.

Fig. 14.6 shows the PACF for the 6-hourly wind speed of Jumla; the red dots represent the lagged values, that is, the number of time periods that separate the ordered data. If the lags are above the 95% confidence interval shown by the two blue horizontal lines, then the lag is significantly correlated with the original wind speed. These significant lags are used as the input for the learning algorithm.

After the lags are selected as inputs, the inputs are normalized using a min–max normalization method, to overcome the numerical issues caused due to data fluctuations (Hsu et al., 2003). Normalization is carried out by using Eq. 14.12.

$$D_{norm} = \frac{D - D_{min}}{D_{max} - D_{min}} \tag{14.12}$$

where D represents the data points of the variables, D_{max} is the maximum value, D_{min} denotes the minimum value, and D_{norm} stands for the normalized data points.

Once the data were normalized, the third phase of the design was data splitting. Data splitting is only carried out once during model building but is however neglected during modeling development. In this study, a SOM-based stratified sampling is applied for data partition to reduce the variance and bias of training and test datasets. The lagged inputs obtained from PACF are subjected to SOM. The SOM acts as a stratification tool, to represent the data into a [m n] grid. A hexagonal topology of SOM was used for this study. The value of [m n] represents the total number of strata, and this can have a significant influence on the mapping of the dataset to different strata.

In the previous study, to determine the optimum grid size of SOM and solve the issue of the hit and trial method, a heuristic rule was applied using the formula $K = \beta n^{0.54}$, where the value of β was 0.2 for small, 1 for medium, and 5 for large SOM size (Vesanto, 1999).

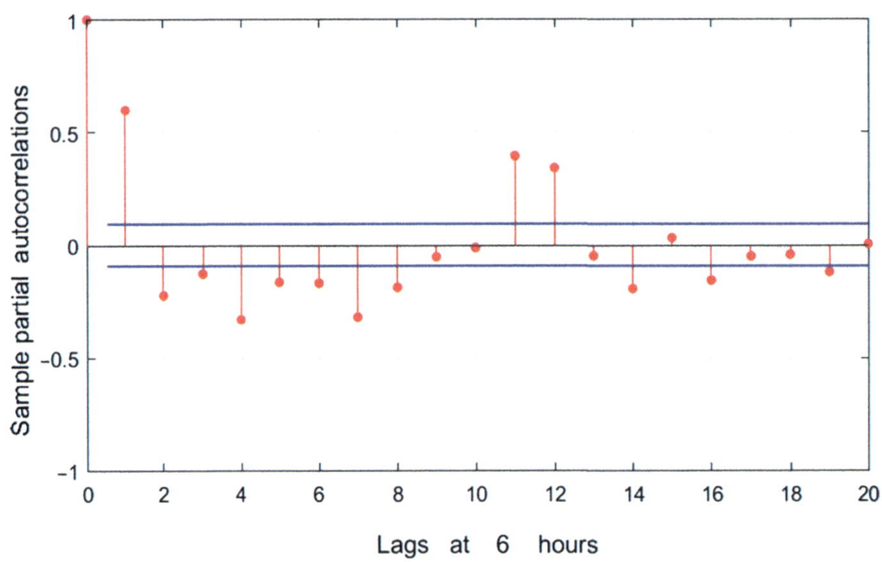

FIGURE 14.6 Partial autocorrelation of six-hourly wind speed of Jumla.

However, the author declined the suitability and theoretical basis of this rule. Thus in this study, a hit and trial method was used to determine the grid size of SOM. The best values of m and n were determined by analyzing the SOM hit plot; the hit plot shows the number of data associated with each neuron. An even distribution of neurons suggests the dimension of the grid is ideal for the dataset. Data splitting is carried out by equal allocation methods, that is, 50% data from each grid is selected as training data and the rest as the test dataset.

Next, the model is built on the training datasets, and then it is applied to the test datasets. The OSELM model takes the training datasets as input and gives the forecasted results. For, OSELM model, the number of neurons need to be specified, the algorithm creates a model on the training datasets and finally, the model is tested using the test dataset. There are no fixed methods to select the number of neurons for OSELM. Thus the OSELM model was run 250 times, increasing the number of neurons after each run and the optimum neurons and model were selected depending on the Legates (E).

To authenticate the results of the proposed model, comparative models OSELM, SOM–M5, M5, and ARIMA were constructed. The hybrid models SOM–OSELM and SOM–M5 were build using SOM-based data partition (50% test and 50% training), whereas the standalone models OSELM, M5, and ARIMA were built on a traditional data split method (50% test and 50% training).

The OSELM and M5 models were carried out in MATLAB software, whereas ARIMA was carried out in R-software. To develop the M5 algorithm, the split threshold was set to 0.0 and smoothing to 15. For ARIMA p, d, q needs to be determined, where p is the lag order, d is the degree of differencing, and q is the order of moving average. Traditionally the optimum p, d, and q are determined through an iterative process where the optimum p, d, and q are calculated iteratively. Since it is a time-consuming process, auto.Arima function in R-software solves this issue by generating the optimal p, d, and q according to either AIC or BIC values. The auto.arima function finds the possible models within the order constraints provided. The constraints for p, d, and q were set to 20 for the study.

14.3.2.6 Model evaluation methods

Model performance was accessed using the observed and simulated data. The assessment of the model in this study was carried out by the following methods.

14.3.2.6.1 Correlation coefficient (*r*)

The correlation coefficient determines the degree of a linear relationship between two variables and is denoted by r (Peng, 2013). The value of r lies between -1 and 1. A correlation 1 means the variables are perfectly positively linearly correlated and -1 denotes perfect negative correlation. If two sets have an r value of 0 then there exists no linear relationship. The correlation coefficient is denoted by (Peng, 2013)

$$r = \frac{\sum_{i=1}^{n}(O_i - |O_i|)(O_s - |O_s|)}{\sqrt{\sum_{i=1}^{n}(O_i - |O_i|)^2}\sqrt{\sum_{i=1}^{n}(O_s - |O_s|)^2}} \tag{14.13}$$

14.3.2.6.2 Root mean square error

RMSE is the square root of the mean square error of observed and simulated data. *RMSE* is the standard deviation of the prediction errors/residuals. Prediction error represents the spread of the data points from the line of best fit. For best fit, that is, for $r = 1$, the *RMSE* value is 0, since all the data points will lie on the regression line (Barnston, 1992). *RMSE* values denote the error in absolute units since they are nonnormalized matrices.

RMSE is mathematically denoted by (Chai and Draxler, 2014)

$$RMSE = \sqrt{\left(\frac{1}{n}\right)\sum_{i=1}^{n}(S_i - O_i)^2} \tag{14.14}$$

14.3.2.6.3 Mean absolute error

MAE is the mean absolute value of the absolute errors. Absolute error is the mean value between the observed and simulated values. The *MAE* value 0 represents perfect fit and the value of *MAE* is always greater than or equal to 0. Like *RMSE*, *MAE* errors are absolute units too (Chai and Draxler, 2014).

$$MAE = \frac{1}{n}\sum_{i=1}^{n}|(S_i - O_i)| \tag{14.15}$$

Due to the data squaring mechanism involved during the calculation of *MAE* and *RMSE*, these values are highly sensitive to the effects of outliers.

14.3.2.6.4 Mean absolute percentage error

MAPE is the average of the absolute percentage error of forecasted values. Smaller *MAPE* means that the forecast results are better. *MAPE* is the relative value of *MAE*, which is used for variation of a model performance at geographically diverse sites (Deo et al., 2018). *MAPE* is represented as (Chai and Draxler, 2014)

$$MAPE = \frac{1}{n}\frac{\sum_{i=1}^{n}|S_i - O_i|}{O_i} \tag{14.16}$$

14.3.2.6.5 Relative root mean squared error

RRMSE is the percentage variation in accuracy and is calculated by dividing the *RMSE* value with the average value of the observed data. *RRMSE* is the relative value of *RMSE* and is used to determine the variation of model performance at geographically diverse sites (Chai and Draxler, 2014; Deo et al., 2018).

$$RRMSE = 100 \times \frac{\sqrt{\left(\frac{1}{n}\right)\sum_{i=1}^{n}(S_i - O_i)^2}}{\left(\frac{1}{n}\right)\sum_{i=1}^{n}(O_i)} \tag{14.17}$$

14.3.2.6.6 Nash–Sutcliffe model of efficiency coefficient

NSE is used to assess the predictive power of hydrological models. Its value ranges from $-\infty$ to 1. *NSE* determines the relative magnitude of noise and quantifies how accurately the model simulation forecasts simulated variables. *NSE* is denoted by (Zhong and Dutta, 2015)

$$NSE = 1.0 - \frac{\sum_{i=1}^{n} (O_i - S_i)^2}{\sum_{i=1}^{n} (O_i - \overline{O})^2} \tag{14.18}$$

14.3.2.6.7 Willmott index

WI was introduced to overcome the insensitivity of the coefficient of determination and Nash–Sutcliffe. Its value ranges from 0 to 1. *WI* is denoted by (Willmott et al., 2011)

$$WI = 1 - \frac{\sum_{i=1}^{n} (O_i - S_i)^2}{\sum_{i=1}^{n} (|S_i - \overline{O}| + |O_i - \overline{O}|)^2} \tag{14.19}$$

14.3.2.6.8 Legates and McCabe's index

Legates and McCabe's index has a value between 0 to 1. It is a modified form of the WI, that provides integral information on overall model performance. For large wind speed prediction Legates and McCabe's index is more informative than WI and *NSE*, since for poorly fitted models the errors are not augmented by the square of predicted errors as in WI and *NSE* (Ghimire et al., 2018). *E* is denoted by (Legates and McCabe, 2013).

$$E = 1 - \frac{\sum_{i=1}^{n}(O_i - S_i)}{|O_i - \overline{O}|)} \tag{14.20}$$

where O_i is the observed wind speed and S_i is the forecasted wind speed, $|O_i|$ is the average of observed wind speed, and $|S_i|$ is the average of forecasted wind speed.

14.4 Short-term forecasting

The short-term forecasting in this chapter is defined as the prediction of wind speed in 6-hour intervals, although there can be other definitions in a different context. The short-term forecasting is important for wind turbine control, market trading, and load increment–decrement decisions (Foley et al., 2012; Azad et al., 2014). In this section, the model development process and parameters used for short-term forecasting are outlined along with the performance metrics of the models.

14.4.1 Dataset for short-term forecasting

Fig. 14.7 displays the time series of 6-hourly net wind speed for five sites in Nepal. It is observed that the 6-hourly wind speed is highly intermittent in nature for all five sites. The 6-hourly wind speed was recorded at time intervals of six hours at 06.00 a.m., 12.00 p.m., 18:00 p.m., and 00:00 are at Coordinated Universal Time (UTC) time zone, equivalent to 11:45 a.m., 17:45 a.m., 23:45 p.m., and 5:45 a.m. in Nepal time. The 6-hourly wind speed dataset consisted of 56,981 data points.

Table 14.4 shows the summary of the five sites for a 6-hourly interim dataset from 1979 to 2107. Among the five sites, the 6-hourly maximum net wind speed at 10 m is observed in Jumla at 10.81 m/s, followed by Biratnagar at 10.15 m/s. In a horizontal coordinate system with y-axis northward and x-axis eastward, the negative value of zonal (U) wind represents its direction from the east and positive means wind blows from the west. Similarly, the meridional wind is negative if it flows from the north and positive if it flows from the south.

14.4.2 Model development of short-term forecasting

Initially, the significant lags with higher correlation were selected as the input variables. Fig. 14.8 shows the partial correlation function of 6-hourly wind speed for five sites. The lags are situated at an interval of 6 hours. Seven significant lags above 95% confidence

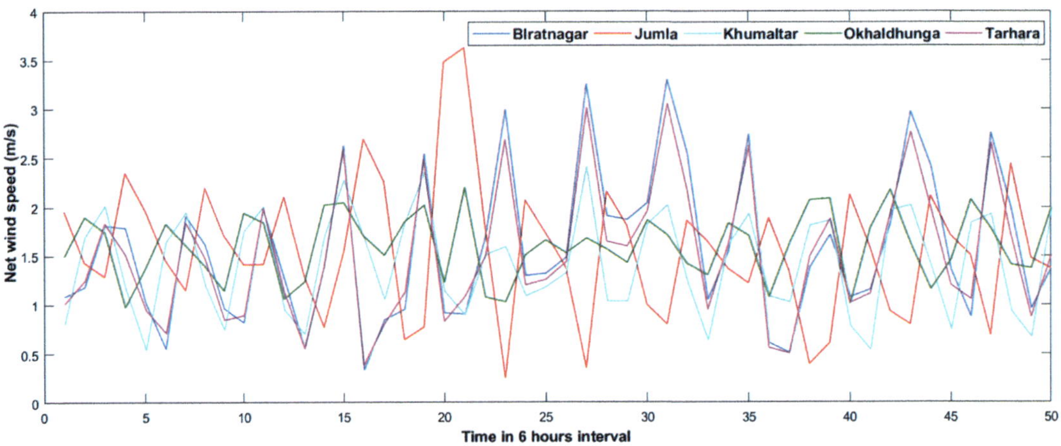

FIGURE 14.7 Time series wind speed at six-hourly intervals.

TABLE 14.4 Explanatory data analysis 6-hourly wind speed in m/s in terms of the maximum value(max), mean and standard deviation (std. dev).

Location	Zonal wind-U: Max(m/s) East–West	Zonal wind-U: Max(m/s) West–East	Zonal wind-U: Mean (m/s)	Zonal wind-U: Std. dev. (m/s)	Meridional wind-V: Max (m/s) North–South	Meridional wind-V: Max (m/s) South–North	Meridional wind-V: Mean (m/s)	Meridional wind-V: Std. dev. (m/s)	Net wind speed (m/s)
Biratnagar	−9.79	10.14	−0.39	2.19	−4.85	4.07	0.03	1.06	10.15
Tarahara	−8.72	9.18	−0.3	1.95	−4.3	3.59	0.13	1.08	9.19
Okhaldhunga	−4.82	5.49	0.16	1.11	−3.83	8.97	0.33	1.34	9.41
Jumla	−3.84	4.73	0.34	1.25	−10.21	4.12	0.26	1.27	10.81
Khumaltar	−5.17	5.13	0.26	1.1	−3.54	4.39	0.29	1.15	5.5

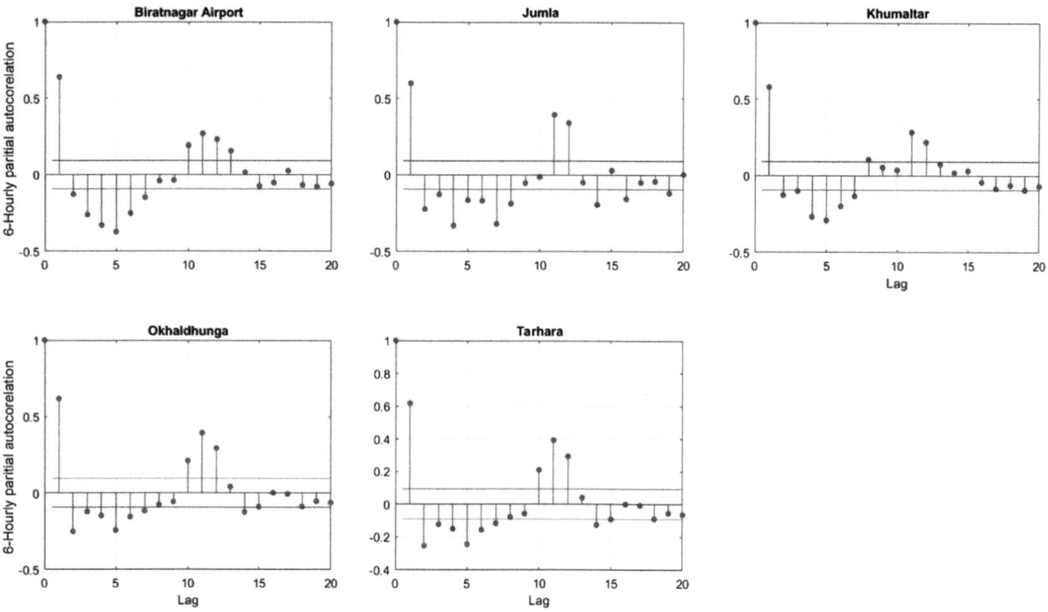

FIGURE 14.8 Partial autocorrelation function of six-hourly net wind speed (m/s).

interval were considered as inputs from all stations (Fig. 14.8), among which six act as predictor variables and the seventh as the target variable. The lag at time $t-1$ was observed to have the highest correlation (around 0.7) with the original wind speed at $t = 0$, indicating there is a high correlation of wind speed and its lag at the first 6-hour interval.

The next step, involved data splitting, two different sets of training and test data; a traditional data split and a SOM-based data splitting technique were used to build the models. The hybrid models implemented the SOM-based data split, whereas the standalone models adopted the traditional data split method. The model building parameters for SOM–OSELM are shown in Table 14.6. Biratnagar required the largest grid size and Jumla the smallest, that is, [15 8] and [8 5], respectively, for the symmetrical distribution of data points into clusters. The SOM hit plots of Jumla for a grid size [8 5] is shown in Fig. 14.9, which represents the most symmetrical distribution of data points spread across the stratum, found after a series of hit and trial method.

Table 14.5 compares the mean, standard deviation, and variance of test and training dataset for SOM data split (50–50) and traditional data split (50–50). From Table 14.5 it is observed that the difference in mean, standard deviation, and variance is less between training and test dataset for the SOM-based data split compared to the traditional data splitting method for all input elements. This ensures that the training and test dataset are more evenly distributed through the SOM-based data splitting method compared to the traditional data splitting method.

Next, the models were built on the training dataset. The hybrid models (SOM–OSELM and M5–OSELM) were built using the SOM partition and standalone models (OSELM, M5, and ARIMA) using traditional partitions. Table 14.6 shows the model development

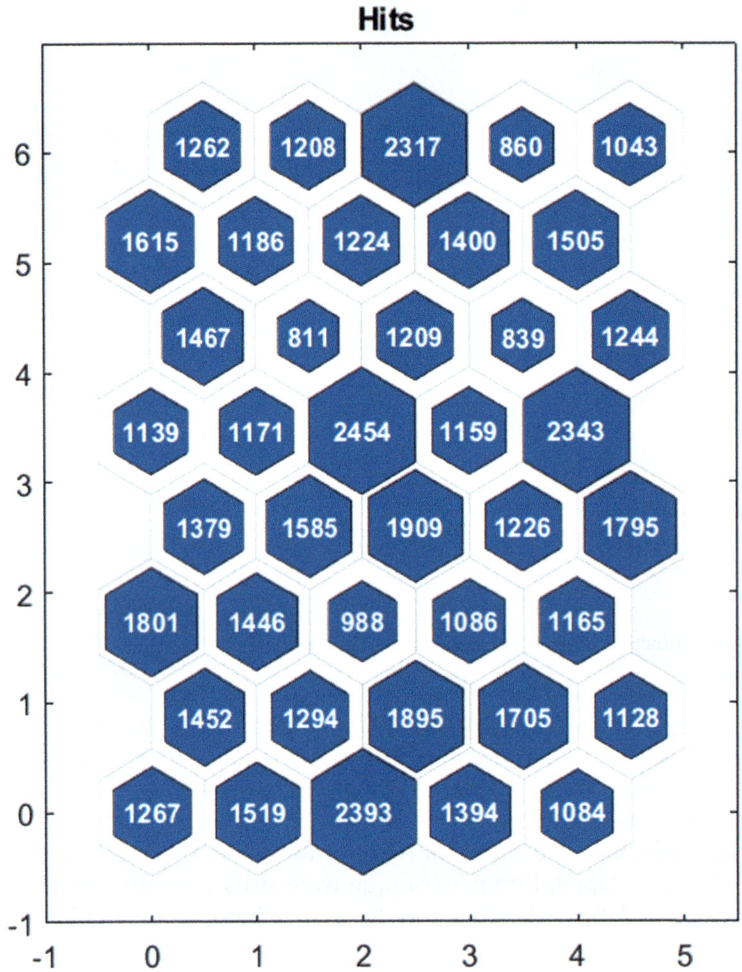

FIGURE 14.9 SOM hits plot of Jumla for six-hourly wind speed (m/s).

parameters and performance of SOM–OSELM for the training dataset. For training data, the value of the correlation coefficient lies between 0.61 to 0.73, indicating a high correlation between observed and predicted values for training data. Also, the *RMSE* ranging from 0.48 to 0.82 m/s and *MAE* 0.37 to 0.62 m/s indicate the model attained less error in the training period. Since the model attains high correlation and low mean square and absolute errors for training sets, this model is suitable for the test dataset.

14.4.3 Results for short-term forecasting

The performance of the SOM–OSELM is compared with the benchmark models, SOM–M5, OSELM, M5, and ARIMA using the performance criteria (Eqs. 14.14–14.20). Table 14.7 shows the comparative performance metrics for the models, it is observed that

TABLE 14.5 SOM data split versus traditional data split for 6-hourly wind speed (m/s) of Jumla.

	SOM data split (50-50)							Traditional 50-50 data split						
	Input 1	Input 2	Input 3	Input 4	Input 5	Input 6	Input 7	Input 1	Input 2	Input 3	Input 4	Input 5	Input 6	Input 7
Mean (m/s)														
Training	0.3498	0.3495	0.3505	0.3495	0.35	0.3499	0.3496	0.3503	0.3503	0.3503	0.3503	0.3503	0.3503	0.3503
Test	0.3495	0.3498	0.3488	0.3498	0.3493	0.3495	0.3498	0.349	0.349	0.349	0.349	0.349	0.349	0.349
Standard deviation (m/s)														
Training	0.1398	0.1407	0.14	0.1403	0.1403	0.1401	0.1402	0.1374	0.1374	0.1374	0.1374	0.1374	0.1374	0.1374
Test	0.1405	0.1395	0.1403	0.1399	0.14	0.1402	0.1401	0.1428	0.1428	0.1428	0.1428	0.1428	0.1428	0.1428
Variance (m/s)														
Train	0.0195	0.0198	0.0196	0.0197	0.0197	0.0196	0.0197	0.0189	0.0189	0.0189	0.0189	0.0189	0.0189	0.0189
Test	0.0197	0.0195	0.0197	0.0196	0.0196	0.0197	0.0196	0.0204	0.0204	0.0204	0.0204	0.0204	0.0204	0.0204

TABLE 14.6 Training parameters and results of SOM−OSELM based on correlation coefficient (r), root mean square error (*RMSE*) and mean absolute error (*MAE*).

Sites	SOM grid	Neurons arrangement (input-neurons-output)	Performance metric (training)		
			r	*RMSE* (m/s)	*MAE* (m/s)
Biratnagar	[15 10]	6-52-1	0.729	0.828	0.626
Jumla	[8 5]	6-53-1	0.619	0.533	0.412
Khumaltar	[10 8]	6-44-1	0.675	0.485	0.372
Okhaldhunga	[12 8]	6-53-1	0.647	0.506	0.366
Tarhara	[12 8]	6-44-1	0.729	0.728	0.549

SOM−OSELM model outperforms its comparative models for all the sites with 0.61−0.74 correlation between observed and forecasted wind speed. However, errors, that is, *RMSE* and *MAE* values, of SOM−OSELM model were inconsistent and showed comparatively high error for Biratnagar (*MAE*: 0.63 m/s and *RMSE*: 0.83 m/s) and Tarhara (*MAE*: 0.559 m/s and *RMSE*: 0.74 m/s) compared with OSELM, SOM−M5, and M5. ARIMA showed the lowest correlation and high mean and absolute errors for all the sites.

The proposed hybrid SOM−OSELM outperformed the comparative models in terms of the higher metrics, attaining the highest value of *NSE* (0.53) and *E* (0.3) for Biratnagar. The results from *WI*, however, were in conflict as SOM−M5's *WI* for Jumla (0.74) and Tarhara (0.827) outperformed SOM−OSELM's 0.73 and 0.825, respectively. Among the standalone models the OSELM model was found to be better than M5 and ARIMA. ARIMA was observed again as the worst performing model for the higher metrics *WI*, *NSE*, and *E*.

Further, *RRMSE* and *MAPE* were used to find the variation of the model's performance in different geographic locations. Table 14.7 shows *RRMSE* and *MAPE* values for SOM−OSELM for the five sites were low in comparison to the benchmark models. The *MAPE* (36.7%) and *RRMSE* (31.04%) for Okhaldhunga were relatively low compared to other sites for SOM−OSELM, indicating the proposed model performance for Okhaldhunga was better than for other sites. The comparative models also performed better in Okhaldhunga, compared to other sites in terms of *RRMSE* and *MAPE*. However, there was no significant difference between the *RRMSE* and *MAE* values among different sites, so it was concluded that the model's performance was independent of the topology.

Also, to test the effects of SOM-based data splitting, a comparison is made between the standalone models OSELM and M5 with their hybrid models SOM−OSELM and SOM−M5 respectively. Legates values of 0.3, 0.218, 0.281, 0.258, and 0.304 were observed for the SOM−M5 model for the five sites Biratnagar, Jumla, Khumaltar, Okhaldhunga, and Tarhara, respectively, that outperformed the 0.29, 0.22, 0.26, 0.23, and 0.29 Legates values of the standalone M5 model. Thus it was observed that both the hybrid models stood out in their performance compared with their standalone models. This implies that the SOM-based data splitting technique improves the performance of machine-learning models compared with traditional data splitting methods.

TABLE 14.7 Performance metrics for short-term forecasting in terms of correlation coefficient (*r*), root mean square error (*RMSE*), mean absolute error (*MAE*), Willmott's index (*WI*), Nash Sutcliffe model of efficiency coefficient (*NSE*), legates (*E*), relative root mean square error (*RRMSE*) and mean absolute percentage error (*MAPE*).

	SOM−OSELM			OSELM			SOM−M5			M5			ARIMA		
Study sites	*r*	*RMSE* (m/s)	*MAE* (m/s)	*r*	*RMSE* (m/s)	*MAE* (m/s)	*r*	*RMSE* (m/s)	*MAE* (m/s)	*r*	*RMSE* (m/s)	*MAE* (m/s)	*r*	*RMSE* (m/s)	*MAE* (m/s)
Biratnagar	0.72	0.833	0.63	0.72	0.729	0.55	0.71	0.758	0.57	0.7	0.758	0.57	0.038	1055	0.816
Jutnla	0.619	0.537	0.415	0.618	0.546	0.421	0.593	0.552	0.425	0.596	0.56	0.431	− 0.001	0.697	0
Khumaltar	0.666	0.49	0.375	0.649	0.501	0.382	0.647	0.503	0.383	0.63	0.514	0.393	0.006	0.659	− 0.003
Okhaldhnnga	0.641	0.512	0.382	0.63	0.516	0.387	0.632	0.518	0.387	0.609	0.529	0.399	0.007	0666	0.526
T arliara	0.723	0.74	0.559	0.722	0.729	0.548	0.707	0.758	0.571	0.696	0.758	0.57	0.038	1055	0.816

	SOM−OSELM			OSELM			SOM−M5			M5			ARIMA		
Study sites	*WI*	*NSE*	*E*	*WI*	*NSE*	*E*	*WI*	*NSE*	*E*	*WI*	*NSE*	*E*	*WI*	*NSE*	*E*
Biratnagar	0.83	0.525	0.32	0.83	0.523	0.31	0.56	0.501	0.3	0.54	0.488	0.29	0.096	0	− 0.02
Jumla	0.737	0.383	0.236	0.743	0.381	0.241	0.525	0.346	0.218	0.523	0.351	0.222	0.013	0	0
Khumaltar	0.781	0.444	0.296	0.774	0.42	0.281	0.573	0.416	0.281	0.544	0.39	0.261	0.08	− 0.003	− 0.005
Okhaldhunga	0.758	0.411	0.269	0.755	0.395	0.254	0.527	0.397	0.258	0.478	0.364	0.232	0.118	− 0.007	− 0.014
Tarhara	0.825	0.523	0.318	0.827	0.52	0.317	0.566	0.499	0.304	0.538	0.481	0.29	0.108	− 0.005	− 0.018

	SOM−OSELM		OSELM		SOM−M5		M5		ARIMA	
Study sites	*MAPE* (%)	*RRMSE* (%)	*MAPE* (%)	*RRMSE* (%)	*MAPE* (%)	*RRMSE* (%)	*MAPE* (%)	*RRMSE* (%)	*MAPE* (%)	*RRMSE* (%)
Biratnagar	45.5	38.75	46.1	38.9	45.91	39.7	46.9	40.38	56.5	71.1
Jumla	44.4	31.43	44.7	32	44.85	32.4	45.3	32.82	60.6	40.9
Khumaltar	38.5	32.67	39.1	33.8	38.83	33.5	39.9	34.64	57.8	44.4
Okhaldhunga	36.7	31.04	38.8	31.9	36.8	31.4	39.3	32.67	53.7	41.1
Tarhara	42.7	37.35	43	37.6	43.48	38.3	44.3	39.09	66.8	54.4

To further evaluate the performance of the models for Biratnagar attaining the highest Legates value, a scatter plot was visualized (Fig. 14.8) and used to compare the goodness of fit of the observed and predicted wind speed for different models. The coefficient of the determinant line (r^2) was inserted in the scatter plots to determine the covariance between observed wind speed compared to simulated wind speed. It is observed that 72.5% of the variance in the forecasted values is predictable from the observed values for SOM–OSELM compared to SOM–M5 (70%), OSELM (72.4%), M5 (70%), and ARIMA (3.8%) (Fig. 14.10).

Fig. 14.9 shows the time series plot of the observed and simulated values of Biratnagar. The red bars in the graph represent the histogram of forecasted error. The time series graphs show that for SOM–OSELM, the simulated results are in good harmony with the observed wind speed compared to the comparative models. Also, for SOM–OSELM the forecasted error histogram shows comparatively fewer errors than the benchmark models. ARIMA was found to be prone to forecasted errors (Fig. 14.11).

14.4.4 Summary for short-term forecasting

The chapter presents short-term forecasting using 6-hourly wind speeds. It was observed that the proposed hybrid model showed some inconsistency in terms of *RMSE*, *MAE*, and *WI*, however, it outperformed the benchmark models for all sites in terms of correlation coefficient, *NSE* and *E*. For SOM–OSELM, the highest value of Legates was observed for Biratnagar (0.32), followed by Tarhara (0.318), Jumla showed the lowest value of Legates (0.236). The study showed the SOM–OSELM model performance was independent of the geographic location since similar *RRMSE* (36%–45%) and *MAPE* (31%–38%) percentages were observed for all study sites.

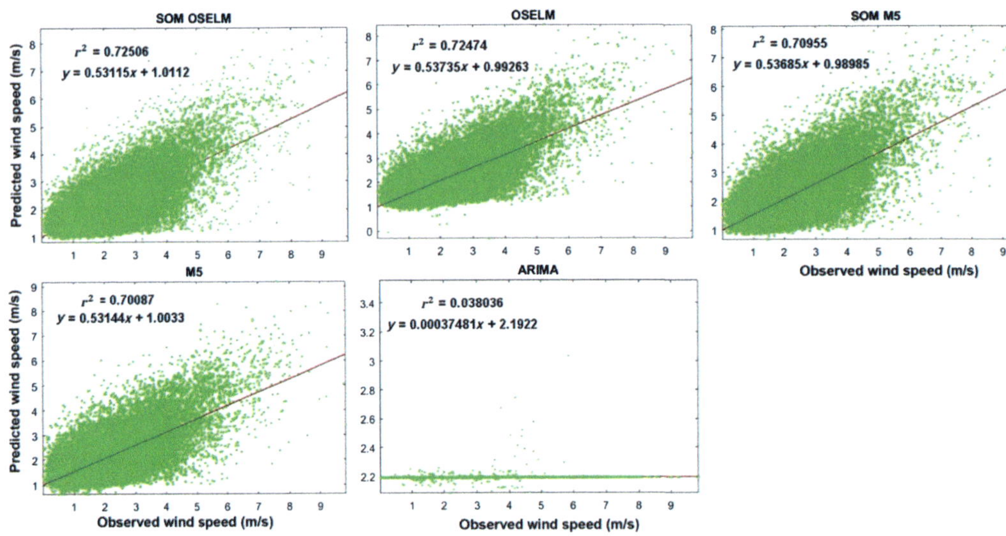

FIGURE 14.10 Scatterplots of observed vs simulated six-hourly wind speed (m/s) for Biratnagar.

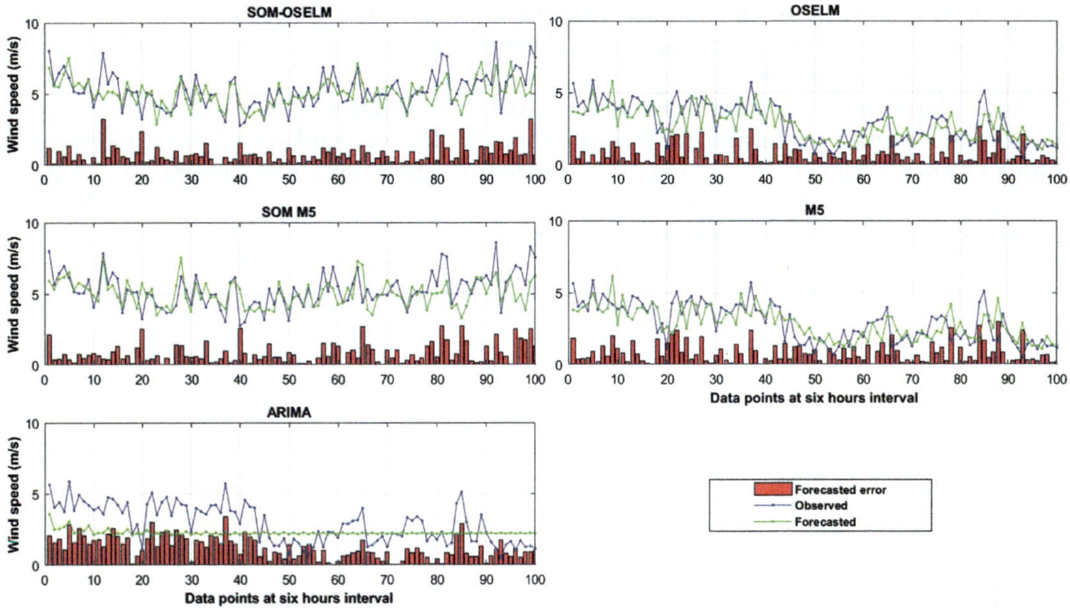

FIGURE 14.11 Time series of observed and forecasted time series with error histograms for Biratnagar six-hourly wind speed (m/s) for the first hundred data points in the testing period.

It was also observed that for the nonlinear 6-hourly wind speed ARIMA was comparatively the worst performing model. Also, it was observed that the two hybrid models built on SOM data partitioning outperformed their standalone model, validating the effectiveness of the SOM data partitioning method over traditional data partitioning. Since Legates is consistently high for the proposed model compared to the benchmark models, the SOM-coupled OSELM, proves to be an effective technique for short-term forecasting.

The short-term forecasting provides forecasting of wind in 6-hour intervals and thus will be effective for wind turbines with regard to taking decisions for grid operations and market trading in real-time.

14.5 Daily forecasting model

For daily forecasting, the daily wind speed was determined from the average value of the 6-hourly dataset. Daily forecasting is essential for wind farms for daily decisions including economic dispatch load planning, daily maintenance, and generator online/off-line decisions (Lazarevska, 2016).

14.5.1 Dataset for daily forecasting

The daily dataset consisted of 14,245 data points. Fig. 14.12 shows the time series graph for daily wind speed. The daily wind speed is less stochastic compared to the 6-hourly wind speed (Fig. 14.7). The average monthly wind speed of Khumaltar was observed to be lower than other sites. Biratnagar and Tarhara, the two sites in the vicinity, as seen in the map of Nepal (Fig. 14.2) showed similar patterns in monthly wind speeds, with Biratnagar having a higher magnitude.

An exploratory data analysis of the daily dataset is shown in Fig. 14.13. The highest value was observed for Biratnagar (8.9 m/s) and the lowest wind speed was observed for Jumla (3.5 m/s). The standard deviation of mean monthly wind was observed to be high for Biratnagar and Tarhara implying that there was more deviation from mean wind speed for Biratnagar and Tarhara than for other sites.

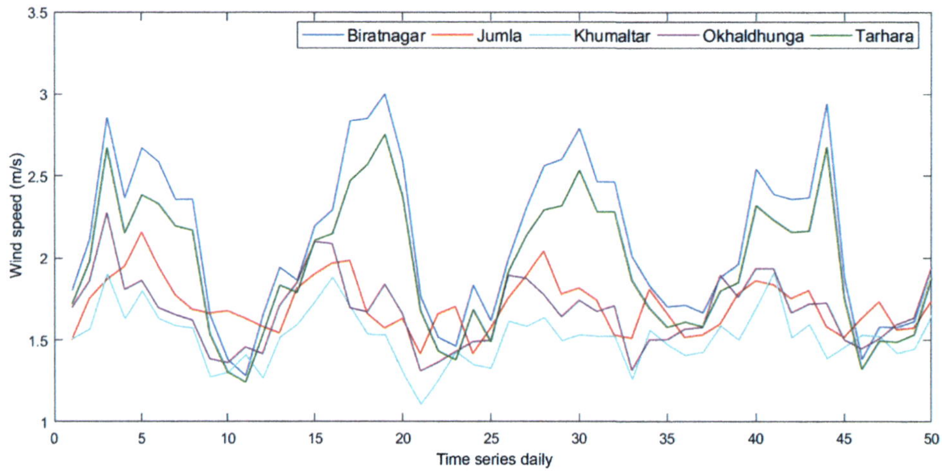

FIGURE 14.12 Time series for daily wind speed (m/s) for first fifty randomly sampled data points.

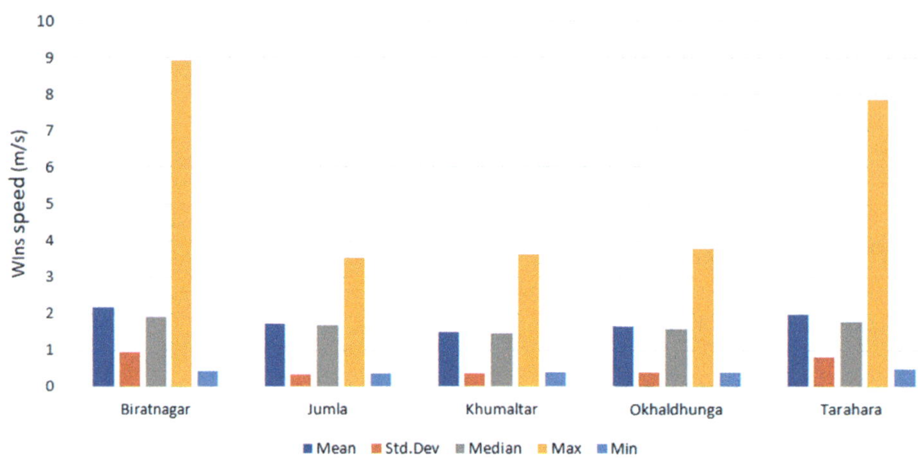

FIGURE 14.13 Explanatory data analysis for daily wind speed (m/s) in terms of mean, standard deviation (std. dev), median, minimum (min) and maximum (max) values.

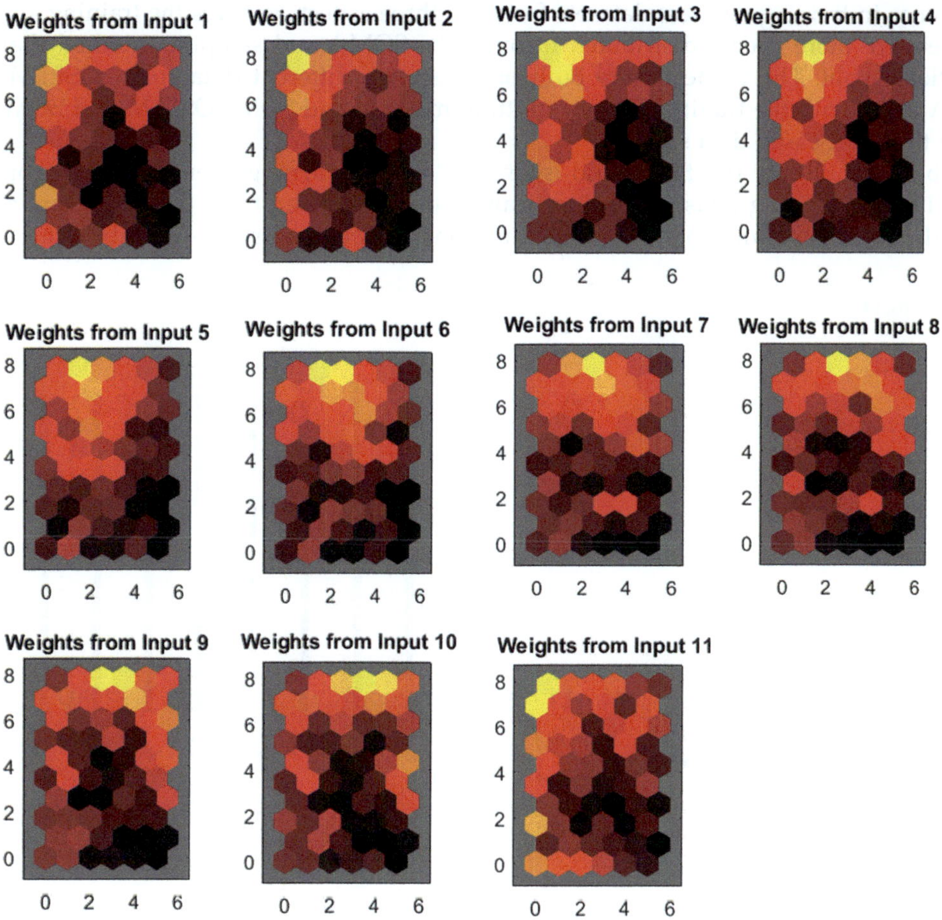

FIGURE 14.14 SOM weight planes for Khumaltar.

14.5.2 Model development for daily forecasting

The model development phase of daily forecasting was similar to that in short-term forecasting. Fig. 14.14 shows the SOM weight planes for [10 6] SOM grid of Khumaltar daily wind speed. The colors of the cells represent the weight that connects inputs with neurons. Darker and lighter colors represent larger and smaller weights, respectively. Elements having similar connection patterns in SOM weight planes show a high correlation between inputs. From Fig. 14.14, it is observed all the 11 elements have weight planes distributed in similar patterns, illustrating that the inputs have a good correlation with each other. Fifty percent of data points from each cluster were selected for training and the remaining 50% for testing.

Fig. 14.15 shows the difference in absolute mean, standard deviation, and variance through SOM partition is lower compared with the traditional partition. The low difference in training and test datasets for these three aspects ensures the training and test datasets are more evenly distributed through the SOM-based sampling method than the traditional split. The models were built on these two different data splits: OSELM, M5, and ARIMA with the traditional data splitting method; and SOM–OSELM and SOM–M5 using the SOM-based data split.

Table 14.8 shows the SOM grid size and neurons arrangement for the proposed model in the training phase. The significant features selected from the PACF, are represented in the arrangement of the neurons (input–neurons–output). The most

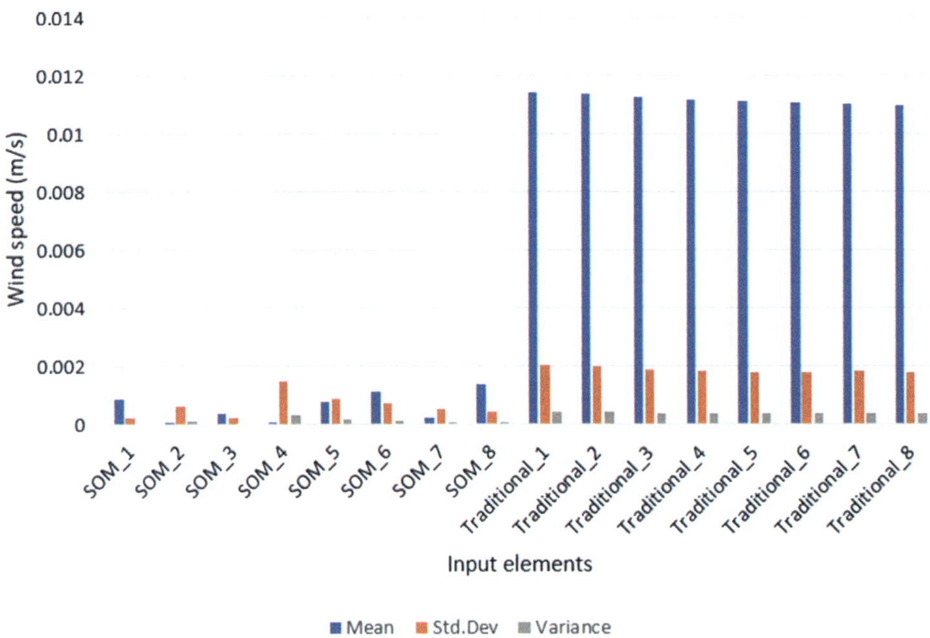

FIGURE 14.15 Absolute difference in mean, standard deviation (std. dev) and variance of SOM partition and traditional partition.

TABLE 14.8 Model evaluation and performance metric of daily forecasting for a training period in terms of correlation coefficient (r), root mean square error ($RMSE$) and mean absolute error (MAE).

Sites	SOM grid	Neurons arrangement (input-neurons-output)	Performance metric (training)		
			r	$RMSE$(m/s)	MAE(m/s)
Biratnagar	[10 8]	7–53–1	0.569	0.289	0.219
Jumla	[10 6]	7–49–1	0.547	0.267	0.202
Khumaltar	[10 6]	10–49–1	0.698	0.588	0.428
Okhaldhunga	[15 10]	7–42–1	0.568	0.290	0.220
Tarhara	[15 10]	7–44–1	0.631	0.307	0.232

symmetrical grid size for monthly wind is shown in Table 14.8, with Okhaldhunga and Tarhara having the same largest grid size, that is, [15 10]. The correlation coefficient among observed and predicted was significant, lying above 0.54, but less when compared to the 6-hourly wind speed. *RMSE* was below 0.3 m/s, except for Khumaltar (0.58 m/s), and *MAE* value below 0.23 m/s was recorded, except for Khumaltar again with *MAE* 0.43 m/s. Since the proposed model performs well in terms of *r*, *RMSE*, and *MAPE*, the model can be used to forecast wind speed for test samples.

14.5.3 Results for daily forecasting

Table 14.9 shows the results from the SOM–OSELM, OSELM, SOM–M5, M5, and ARIMA. Observing the linear relationship (*r*), the mean absolute and squared errors in the initial testing specifications of the model, it is clearly observed that the hybrid model SOM–OSELM performed better than its comparative models. The highest correlation between observed and predicted wind speed was observed using SOM–OSELM for Khumaltar (0.69). The root mean square error of the hybrid SOM–OSELM is lower than for the comparative models. The *RMSE* of SOM–OSELM was observed to be below 0.31 m/s for four sites, whereas for Khumaltar the error was increased to 0.59 m/s. Observation of *MAE* also showed similar results with a comparatively higher error of 0.43 m/s for Khumaltar. Overall, the absolute and mean square errors were low and correlations was high for SOM–OSELM, compared to the benchmark models. These initial performance metrics favor SOM–OSELM as a model with a higher linear relationship and less absolute and relative errors.

Moving to *WI*, *NSE*, and *E* to access the accuracy of the models, the values of *WI*, *NSE*, and *E* for SOM–OSELM show higher accuracy than the comparative models. The value of SOM–OSELM Legates was observed to be highest, that is, 0.3 for Khumaltar followed by Tarhara with *E* 0.23, whereas the minimum Legates (0.183) for SOM–OSELM was observed for Biratnagar and Okhaldhunga. The proposed model also showed consistent performance for *WI* and *NSE*.

It was observed that the values of *WI*, *NSE*, and *E* increased from 0.488, 0.428, and 0.252 to 0.520, 0.447, and 0.280, respectively, when the SOM-based data splitting method was introduced to the standalone M5 model for Khumaltar. Similar results were observed for other sites, where the hybrid models SOM–M5 and SOM–OSELM performed better than their standalone models in terms of *WI*, *NSE*, and *E*. Thus it is observed that the use of SOM as a data splitting method acts as a catalyst to improve the performance of standalone models using traditional data splitting techniques.

The *RRMSE* and *MAPE* values were lower for monthly wind speed compared to the 6-hourly wind speed. A lower *RRMSE* and *MAPE* indicates good model performance at different geographical sites. The *RRMSE* and *MAPE* of SOM–OSELM were low compared to benchmark models. The comparison of the performance of SOM–OSELM for different sites in terms of *RRMSE* and *MAE* showed similar results; the SOM–OSELM performance was not affected by the topology of the sites. However, the lowest values of *RRMSE*

TABLE 14.9 Daily forecasting performance metric in terms of correlation coefficient (*r*), root mean square error (*RMSE*), mean absolute error (*MAE*), Willmott's index (*WI*). Nash–Sutcliffe model of efficiency coefficient (*NSE*), legates (*E*), relative root mean square error (*RRMSE*) and mean absolute percentage error (*MAPE*).

	SOM–OSELM			OSELM			SOM–M5			M5			ARIMA		
Study sites	*r*	*RMSE* (m/s)	*MAE* (m/s)	*r*	*RMSE* (m/s)	*MAE* (m/s)	*r*	*RMSE* (m/s)	*MAE* (m/s)	*r*	*RMSE* (m/s)	*MAE* (m/s)	*r*	*RMSE* (m/s)	*MAE* (m/s)
Biratnagar	0.558	0.293	0.223	0.519	0.306	0.232	0.539	0.299	0.227	0.518	0.306	0.23	0.038	1.055	0.816
Jumla	0.536	0.267	0.202	0.44	0.304	0221	0.504	0.275	0.208	0.494	0.289	0.22	0	0.697	0.554
Khumaltar	0.692	0.592	0.431	0.662	0.611	0.443	0.67	0.609	0.443	0.658	0.613	0.45	0.006	0.659	0.534
Okhaldhunga	0.556	0292	0.223	0.519	0.306	0.232	0.524	0.301	0231	0.518	0.306	023	0.007	0.666	0.526
Tarhara	0.627	0.309	0236	0.591	0.324	0247	0.591	0.322	0.246	0.58	0.328	025	0.038	1.055	0.816

	SOM–OSELM			OSELM			SOM–M5			M5			ARIMA		
Study sites	*WI*	*NSE*	*E*	*WI*	*NSE*	*E*	*WI*	*NSE*	*E*	*WI*	*NSE*	*E*	*WI*	*NSE*	*E*
Biratnagar	0.681	0.31	0.183	0.668	0262	0.157	0.449	0286	0.167	0.363	0261	0.15	0.108	− 0.005	− 0.02
Jumla	0.665	0287	0.199	0.626	0.159	0.162	0.416	0246	0.175	0.394	0237	0.18	0.013	0	0
Khumaltar	0.799	0.478	0.3	0.786	0.433	0.26	0.52	0.447	0.28	0.488	0.428	025	0.08	− 0.003	− 0.01
Okhaldhunga	0.679	0.309	0.183	0.668	0261	0.157	0.41	0268	0.152	0.363	0261	0.15	0.118	− 0.007	− 0.01
Tarhara	0.744	0.392	023	0.723	0.346	0201	0.469	0.34	0.195	0.439	0.329	0.19	0.108	− 0.005	− 0.02

	SOM–OSELM		OSELM		SOM–M5		M5		ARIMA	
	MAPE	*RRMSE*	*MAPE*	*RRMSE*	*MAPE*	*RRMSE*	*MAPE*	*RRMSE*	*MAPE*	*RRMSE*
Study sites	(%)	(%)	(%)	(%)	(%)	(%)	(%)	(%)	(%)	(%)
Biratnagar	15.67	19.48	16.69	20.64	15.96	19.83	16.91	20.64	66.82	54.4
Jumla	12.66	15.64	13.94	17.81	13.02	16.09	13 7	16.96	60.62	40.85
Khumaltar	23.48	29.92	24.64	31.51	24.06	30.82	24.9	31.65	57.83	44.41
Okhaldhunga	15.88	19.49	16.69	20.64	16.41	16.91	20.64	31.65	53.71	41.12
Tarhara	15.32	18.82	16.37	20.03	15.96	19.61	16.63	20.28	66.82	54.4

(12.66%) and *MAE* (15.64%) were recorded for Jumla, suggesting SOM–OSELM performs better in Jumla than in other sites.

Fig. 14.16 shows the forecasted error histogram for Khumaltar, where the frequency of the error is examined using 0.5 size error bins. SOM–OSELM has 68.74% of its datum points, within the smallest error bin (0.5), compared to OSELM (66.23%), SOM–M5 (67.60%), M5

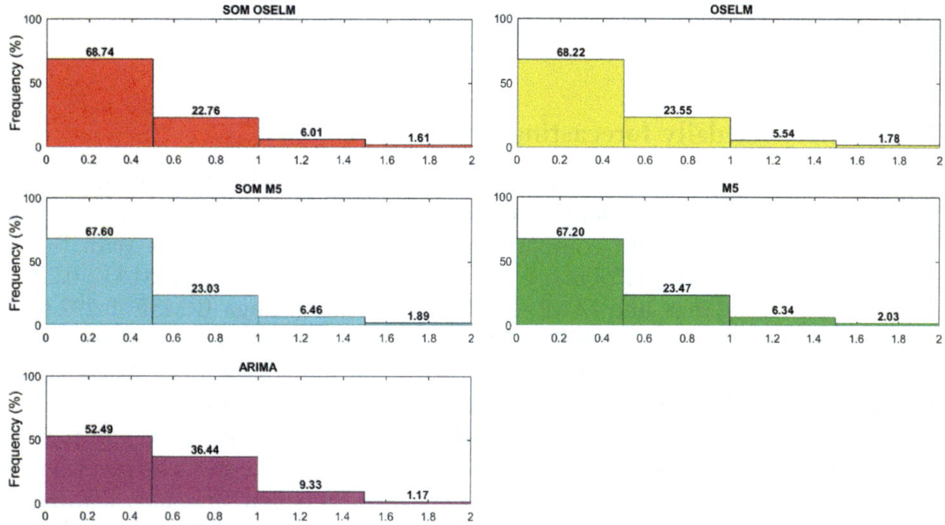

FIGURE 14.16 Error histogram for daily forecasting of Khumaltar with error bins of 0.5 m/s.

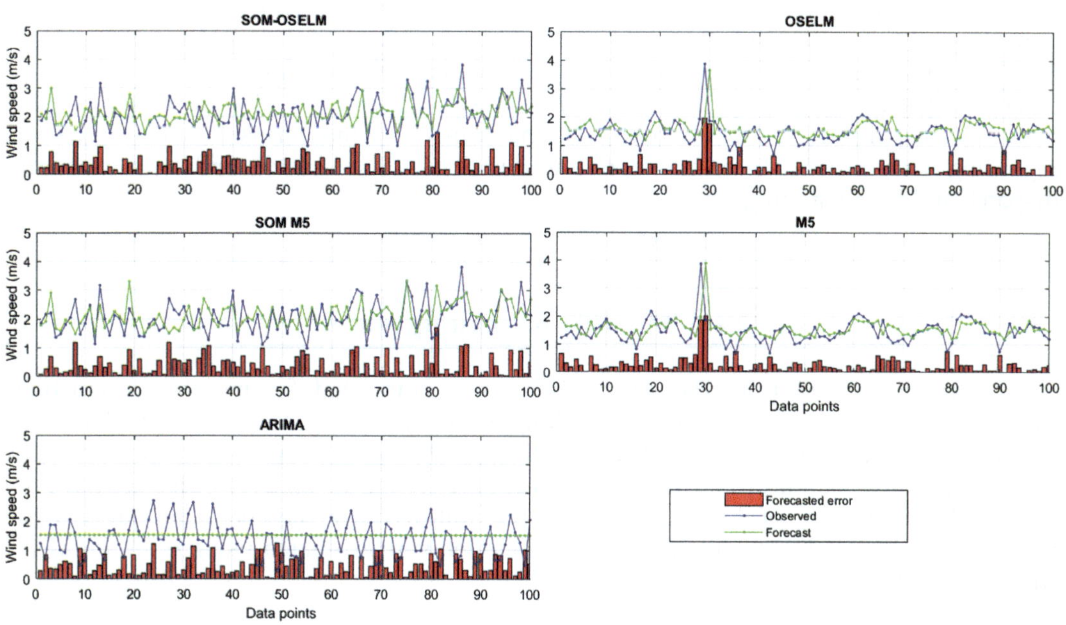

FIGURE 14.17 Observed vs forecasted time series with error histograms for Khumaltar daily wind speed (m/s) for the last hundred data points

(67.20%), and ARIMA (52.49%). Thus the majority of the wind speed was forecasted with minimum errors for the hybrid SOM–OSELM compared with its comparative models.

Fig. 14.17 shows the time series of the observed and forecasted wind speed for Khumaltar for daily forecast horizons. It was observed that error bins for the proposed SOM–OSELM-based model were relatively lower than the benchmark models. The graph shows that the predicted wind speed from the SOM–OSELM model closely follows the observed wind speed.

14.5.4 Summary for daily forecasting

This section presents the daily forecasting of wind speed. The daily average of the 6-hourly wind speed was used for daily forecasting. The *RMSE* and *MAE* value for the proposed hybrid model was low in comparison to the benchmark models with Biratnagar (RMSE: 0.293 m/s and MAE: 0.223 m/s), Jumla (RMSE: 0.267 m/s and MAE: 0.221 m/s), Khumaltar (RMSE: 0.592 m/s and MAE: 0.431 m/s), Okhaldhunga (RMSE: 0.292 m/s and MAE: 0.223 m/s), and Tarhara (RMSE: 0.309 m/s and MAE: 0.236 m/s). Also, the model outperformed the comparative models in terms of the higher metrics, *WI*, *NSE*, and *E*, attaining the highest *WI* (0.799), *NSE* (0.478), and *E* (0.3) for Khumaltar.

Jumla was seen to be the site where SOM–OSELM performed best, compared with other sites, as it had the lowest *RRMSE* (12.66%) and *MAPE* (15.64%). However, not much difference was observed in *RRMSE* and *MAPE* values between the sites, leading to the conclusion that the proposed model's performance is not affected by geographical locations. As in the short-term forecasting, the SOM-based hybrid models outperformed the standalone models, proving that the SOM-based data splitting technique is more effective than the traditional data splitting method for both short-term and daily forecasting. Thus SOM–OSELM is seen as being effective for daily wind speed forecasting.

The daily forecasting helps the wind farms to study the nature of wind speed in a 24-hour interval, making it an effective tool that can aid wind farms to make decisions regarding daily maintenance, online/offline decisions for generator, and with economic dispatch and load planning.

14.6 Monthly forecasting model

A monthly average from the daily dataset was computed for monthly forecasting. Monthly forecasting is often considered long-term forecasting and is essential for the installation of new wind stations. It also facilitates the wind farms to schedule the grid maintenance and plan a proper reserve of electricity (Wang et al., 2011). This section provides a brief analysis of monthly wind speed followed by the modeling process for monthly forecasting. Finally, the results of the model were evaluated using *r*, *RMSE*, *MAE*, *RRMSE*, *MAPE*, *WI*, *NSE*, and *E*.

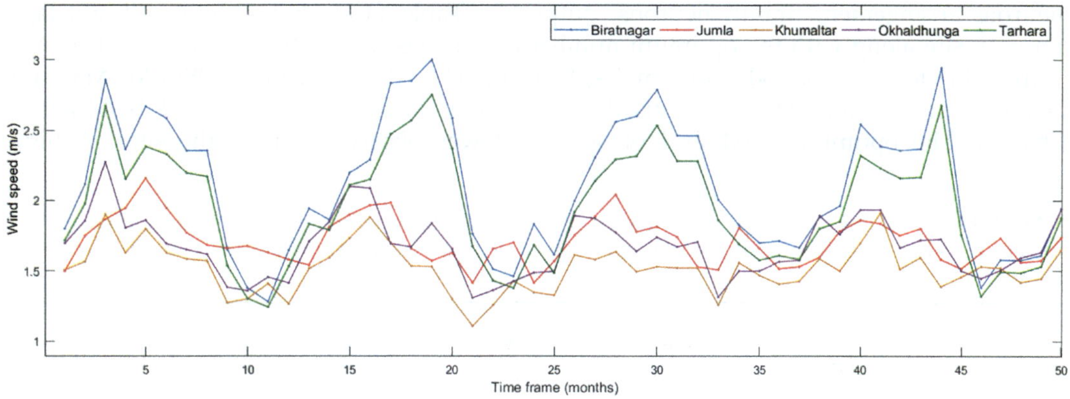

FIGURE 14.18 Monthly wind speed time series of first fifty data points.

TABLE 14.10 Exploratory data analysis for monthly wind speed in terms of mean, standard deviation, median, minimum and maximum values.

Sites	Mean (m/s)	Standard deviation (m/s)	Median (m/s)	Minimum (m/s)	Maximum (m/s)
Biratnagar	2.1507	0.4966	2.1115	1.2805	3.5574
Jumla	1.7087	0.1741	1.6908	1.3529	2.2266
Khumaltar	1.5017	0.175	1.4952	1.104	2.1248
Okhaldhunga	1.6492	0.22	1.6452	1.1784	2.4918
Tarhara	1.9798	0.4303	1.9602	1.1775	3.2363

14.6.1 Dataset for monthly forecasting

The monthly dataset consisted of 468 data points. Fig. 14.18 shows the time series graph for monthly wind speed. The monthly wind speed was seen to have a periodic curve over the months that was less stochastic than the daily and 6-hourly wind speed. In Biratnagar and Tarhara, which lie in the vicinity of each other, wind speed showed similar patterns over the months.

The value of the monthly dataset ranged from 1.10 to 3.55 m/s (Table 14.10). The standard deviation for Biratnagar and Tarhara was greater than 0.4 m/s compared to other sites, with a standard deviation of less than 0.2 m/s indicating that the wind speed for Biratnagar and Tarhara deviates more from mean wind speed than the other sites.

14.6.2 Model development for monthly forecasting

The input net monthly wind speed was subjected to PACF; the lags having correlation above 95% confidence interval were selected as the inputs for the model development. The neurons arrangement (input–neurons–output) in Table 14.11 shows the significant lags for each site along with the optimum number of neurons for the training period. The lags were subjected to a 50–50 traditional split to develop the standalone OSELM, M5, and ARIMA models, and SOM was used to split the dataset for the hybrid SOM-based OSELM. The optimum SOM grid of size [6 4] was observed for all the sites. A high

TABLE 14.11 Model parameters and performance for the monthly wind speed of SOM–OSELM for a training period in terms of correlation coefficient (*r*), root mean square error (*RMSE*) and mean absolute error (*MAE*).

Sites	SOM grid	Neurons arrangement (input-neurons-output)	Performance metrics (Training)		
			R	*RMSE* (m/s)	*MAE* (m/s)
Biratnagar	[6 4]	9–47–1	0.874	0.087	0.069
Jumla	[6 4]	9–52–1	0.851	0.269	0.206
Khumaltar	[6 4]	9–51–1	0.839	0.114	0.093
Okhaldhunga	[6 4]	9–52–1	0.856	0.233	0.176
Tarhara	[6 4]	7–43–1	0.805	0.103	0.081

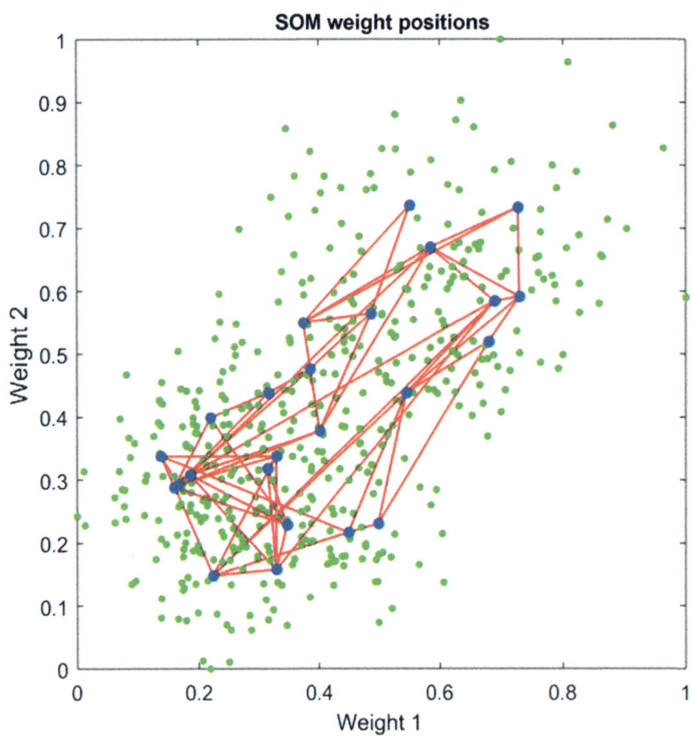

FIGURE 14.19 SOM weight positions for Jumla monthly wind speed.

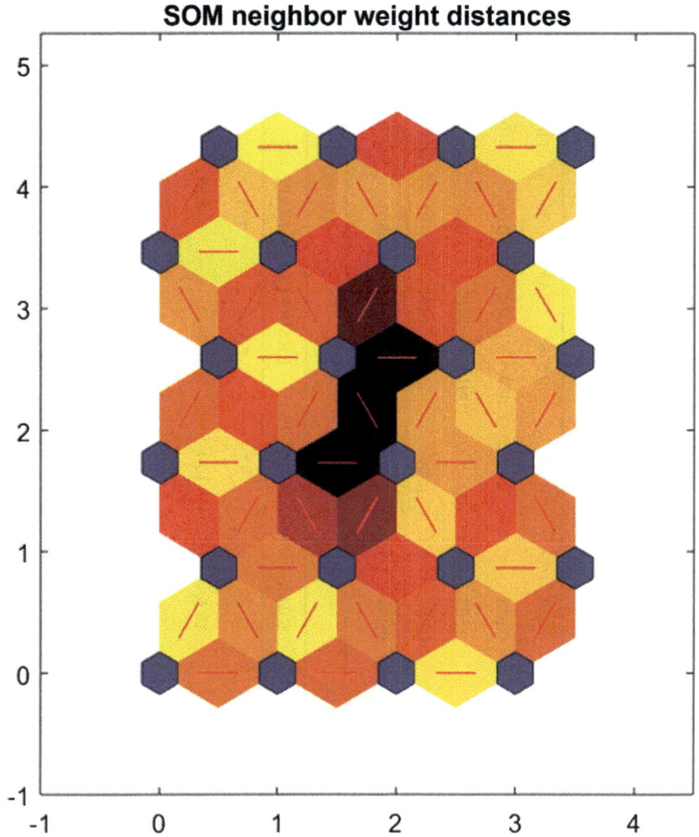

FIGURE 14.20 SOM neighbor weight distance for Jumla monthly wind speed.

correlation greater than 0.8 was observed between observed and forecasted results for all five sites. Also, the *RMSE* and *MAE* values observed were between 0.087–0.269 m/s and 0.069–0.206 m/s. Thus the proposed model showed high correlation and low errors between observed and forecasted values, so the model can be applied for the test dataset.

Fig. 14.18 shows the SOM weight positions for the Jumla dataset with the optimum grid size of [6 4]. SOM weight positions show the locations of the weight vectors and data points. The input weights are well distributed throughout the input space after 1000 iterations. The green dots in the Fig. 14.18 represent the data points, the blue points represent the neurons, and the red line represents the distance between the neighboring neurons. From Fig. 14.18, it is observed that the neurons are well distributed in the input space.

Also, another graph to visualize all the weights at the same time in 2D space is produced with SOM neighbor weight distances. Fig. 14.19 represents the distance between the neighbors, where the blue hexagons represent the neurons, and the color in the regions indicates the distance between the neurons—a light color represents a smaller distance and a dark color represents a larger distance. The light colors and even distribution throughout the plane bounded by some dark segments show that the network has clustered the data into even groups as represented by the SOM weight positions (Fig. 14.20).

TABLE 14.12 Ivalutioa metric for monthly forcca.stiq iD terms of correlatioa ooefficieat (r), root meu. squrc error (RMSE), mem absolute error (M.4E), Willmott's Iadex (Jt1), Nasii-Satcliffe model ofEflicieiiC]' C&eflicieltt (NSF), Legates (E), relative root meu squrc error (RRMSE) u.d mem absolate pm:utage error (M.4PE)

StudySitts	SOM-OSEUI			OSELM			SOM-MS			MS					
	RMSE (mls)	MAE (mls)		RMSE (mls)	MAE (mls)		RMSE (mls)	MAE (mls)		RMSE (mls)	MAE (mls)		RMSE (mls)	ARIMA	MAE (mls)
Bin apr	0.73	0.119	0.09	0.63	0.145	0.12	0.76	0.111	0.08	0.73	0.126	0.1	0.54	0.41	034
Jlllllb.	0.78	0313	014	0.67	0391	0.3	0.8	0191	0.22	0.7	0368	018	0.22	0.181	0.15
Khlllllltar	0.71	0.16	0.12	0.73	0.166	0.13	0.71	0.159	0.12	0.7	0.168	0.13	0.19	0.173	0.14
Okhaldl11mp	0.73	019	0.22	0.65	0.352	017	0.76	0.267	0.21	0.7	0324	0.25	0.25	0118	0.18
Tulwa	0.68	0.132	0.1	0.67	0.133	0.11	0.68	0.128	0.1	0.7	0.129	0.1	0.49	0311	0.3

StudySitts	SOM-OSEUI			OSELM			SOM-MS			MS			ARIMA		
	WI	NSE	E	WI	NSE	E	WI	NSE	E	WI	NSE	E	WI	NSE	E
Bin apr	0.85	0.499	032	0.78	037	011	0.72	0.569	039	0.63	0.528	0.35	0.56	0143	0.15
Jlllllb.	0.88	0.575	0.42	0.8	0312	014	0.78	0.633	0.46	0.58	039	018	0.06	0.023	0.01
Kh11111altar	0.82	0.503	031	0.84	0.40:i	014	0.65	0.507	032	037	0391	018	0.25	−0.01	−0.01
Okhaldl11mp	0.85	0.504	037	0.78	0161	011	0.76	0.58	0.4	0.56	0.375	017	−0.02	0311	−0.02
Tulwa	0.82	0.43	017	0.79	0394	013	0.62	0.459	0.29	0.46	0.435	016	0.51	0.181	0.11

StudySitts	SOM-OSEW		OSELM		SOM-YS		MS		ARWA	
	MAPE (%)	RRMSE (%)	MAPE (%)	RRMSE (%)	MAPE (%)	RRMSE (%)	MAPE (%)	RRMSE (%)	MAPE (%)	RRMSE (%)
Bin apr	5.54	6.989	6.98	8.5	4.969	6.48	5.75	7.354	173	19.4
Jllllllb.	11.6	14.65	14.5	18.5	10.83	13.6	14.4	17.44	8.68	10.6
Khlllllltar	738	9.675	831	10.3	7332	9.64	8.25	10.38	9.95	11.6
OkhaiAmlp	11.1	14.69	141	18.2	10.78	13.5	13.8	16.71	11.5	13.4
Tulwa	6.92	8.779	7.5	8.99	6.776	8.56	716	8.672	17	19.1

14.6.3 Results for monthly forecasting

Table 14.12 shows the performance metric for monthly forecasting. The performance metrics of monthly forecasting were better compared to the short-term and daily forecasting. A comparison among the data models indicated that, although SOM–OSELM proved to be the best model in terms of *WI*, the hybrid model outperforms the SOM–OSELM in terms of the correlation coefficient, *RMSE*, *MAE*, *WI*, *NSE*, *MAPE*, and *RRMSE*. The best performance was observed by SOM–M5 for Jumla: $E = 0.46$, $NSE = 0.633$, and $r = 0.80$ compared with $E = 0.42$, $NSE = 0.575$, and $r = 0.78$ for SOM–OSELM. SOM–M5 also displayed minimum errors with *RMSE* below 0.29 m/s and *MAE* below 0.27 m/s. In terms of *RRMSE* and *MAPE*, SOM–M5 had the lowest values for Biratnagar of 6.48% and 4.96%, respectively. The *RRMSE* and *MAPE* values of both models showed very little difference among the sites, indicating the performance of the models is independent of the geographical locations.

To validate the performance of SOM, a comparison between the hybrid and standalone models is observed in Fig. 14.21. It is observed that the Legates of the hybrid models outperformed their respective standalone models for all the sites. Thus it is observed that in terms of Legates the hybrid models (SOM–OSELM and SOM–M5) implementing SOM as data partition outperformed their standalone models, leading to the conclusion that the SOM-based partition is more useful in wind speed forecasting than traditional split methods.

Fig. 14.22 compares the 6-hourly, daily, and monthly wind speeds for Jumla. It is observed that the average monthly wind speed follows a periodic pattern and is less stochastic in nature than the daily and 6-hourly wind speeds. Since OSELM is a truly nonlinear model and SOM–M5 lies between a nonlinear model and linear model, the model performance of the SOM–M5 and M5 outperforming the SOM–OSELM and OSELM, respectively, can be justified by the periodic and less intermittent behavior of wind speed for the monthly period compared to daily and 6-hourly.

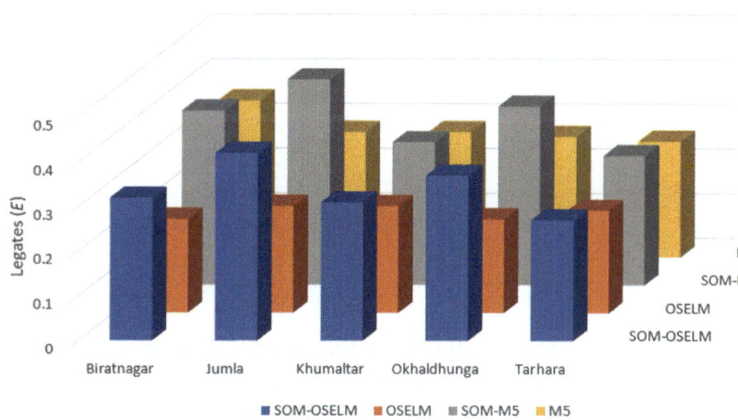

FIGURE 14.21 Comparison of performance of hybrid SOM-OSELM and SOM-M5 models with their respective standalone models OSELM and ELM in terms of Legates(*E*).

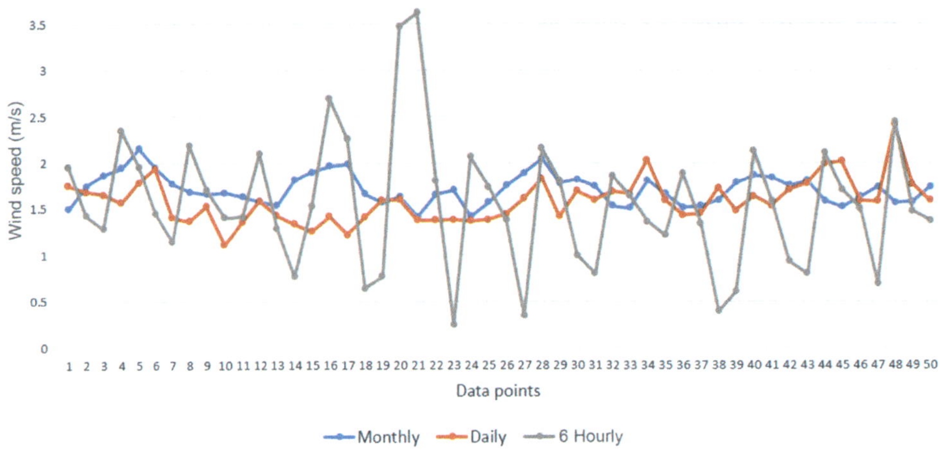

FIGURE 14.22 Comparison of wind speed of Jumla for first fifty data points.

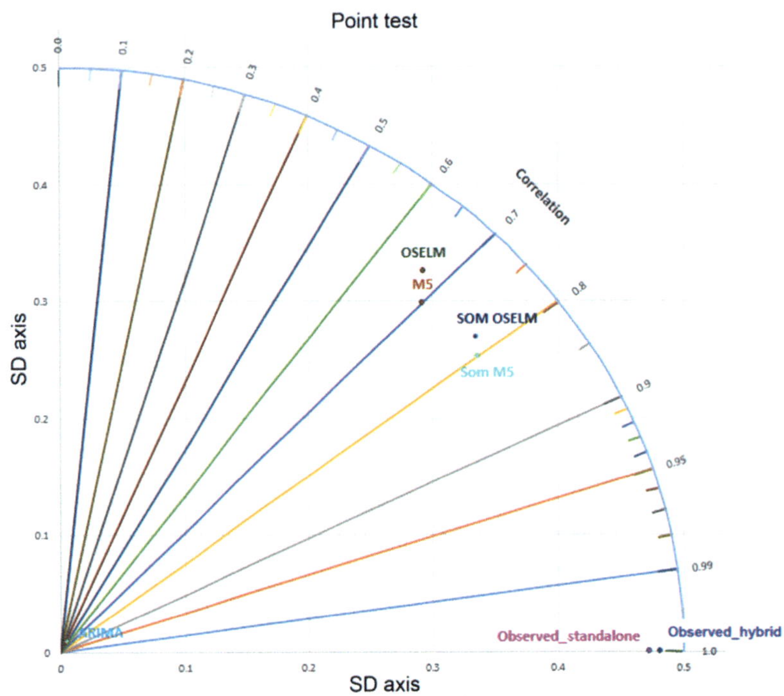

FIGURE 14.23 Taylor diagram showing the correlation coefficient (r) and standard deviation between observed and predicted wind speed for a monthly forecast horizon of Jumla.

Fig. 14.23 represents the Taylor diagram for Jumla. Taylor plots summarize how closely the forecasted wind speed matches the observed wind speed in terms of standard deviation and correlation. The polar axis is the standard deviation and the radial axis is the correlation coefficient (r), which is used to determine the closest match of the observed wind speed to the forecasted results. Since the observed dataset is different

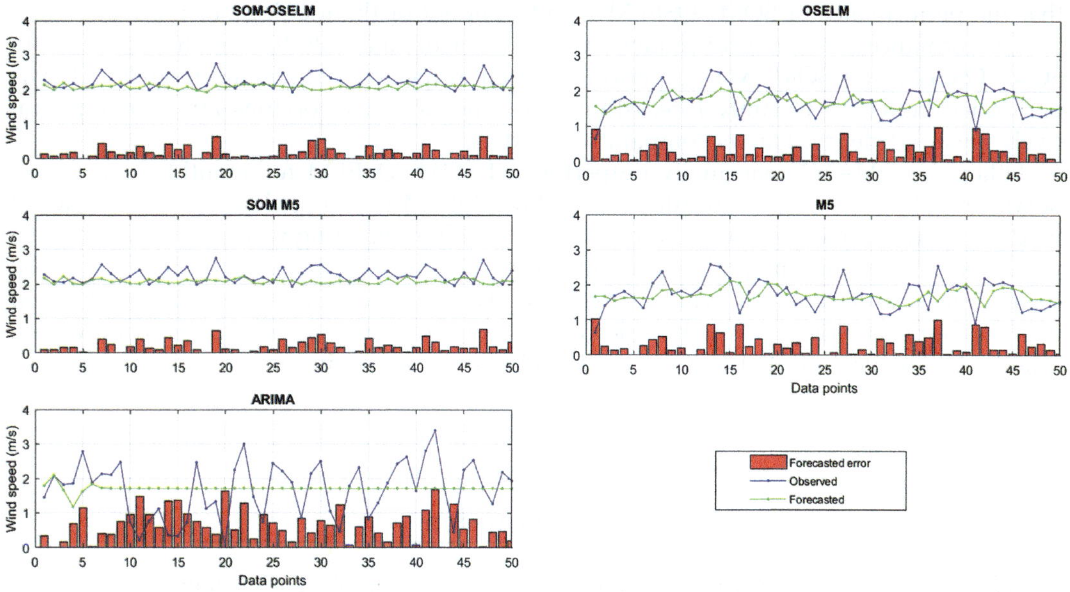

FIGURE 14.24 Time series of observed vs predicted wind speed (m/s) of the first fifty data points for Jumla.

for standalone and hybrid models, the plot consists of observed wind speed as Observed Standalone for standalone models and Observed Hybrid for hybrid models. The model values closer to the observed values represent a good model. From Fig. 14.23, it is observed that the SOM−M5 produced the highest value of *r* and generated the closest forecasted wind speed for the observed wind speed. Also, the M5 shows slightly better results than the OSELM model and the hybrid models outperform their respective standalone models.

Fig. 14.24 represents the time series plot for observed and forecasted values for Jumla, along with error bins, although the predicted line follows the observed data points closely for both SOM−OSELM and SOM−M5 model. However, the graph shows that the SOM−M5 has error bins of low magnitude compared to the SOM−OSELM.

14.6.4 Summary for monthly forecasting

In this section, monthly forecasting was carried out, which is essential for wind farms to back up reserves and for the selection of new sites to build wind farms. Unlike the daily and 6-hourly forecast horizons, the monthly wind speed was less stochastic and periodic in nature. Thus the pairwise linear model coupled with SOM, SOM−M5, outperformed the proposed model. SOM−OSELM, however, performed better than SOM−M5 in terms of *WI*, Biratnagar: 0.85; Jumla: 0.88; Khumaltar: 0.82; Okhaldhunga: 0.85; Tarhara: 0.82 of SOM−OSELM compared to Biratnagar: 0.72; Jumla: 0.78; Khumaltar: 0.65; Okhaldhunga: 0.76; Tarhara: 0.62.

The performance of the SOM−OSELM model was consistent throughout the sites in terms of RRMSE and MAPE, ranging from 5.5% to 11.6% and 6.9% to 14.6%, suggesting that the performance of SOM–OSELM is independent of the topology of the study sites. The performance of the hybrid models stood out for the monthly forecast horizon as both SOM–OSELM and SOM–M5 outperformed their respective standalone models. ARIMA performed better for monthly forecasting than other forecast horizons but had comparatively lower efficiency than its comparative models.

Although SOM–M5 marginally outperformed SOM–OSELM for monthly forecasting, SOM–OSELM showed good results, thus SOM–OSELM was considered to be an efficient model to perform monthly wind forecasting. Monthly forecasting is termed as forecasting wind speed in a time interval of a month, that is, almost 30 days. Monthly wind speed forecasting is applicable for wind farms to study the nature of variance in wind speed over different months which is beneficial for wind farms to reserve maximum power for unforeseen outages and it also assists in the selection of new wind farms at a different geographical location.

14.7 Conclusion

The accurate prediction of wind speed can help to establish a sustainable source of energy and enhance the efficiency of wind farms (Foley et al., 2012). Nepal has high potential in wind, however, more than half of the population is without a supply of electricity (Laudari, 2016). Thus Nepal was selected as the study area for the chapter. This chapter has attempted to determine an optimal data-driven machine-learning method for forecasting short-term (6-hourly), daily, and monthly wind speed for five geographically diverse sites in Nepal. The dataset used was obtained from the ERA–Interim Dataset, from the years 1979–2017. The study proposes an optimum OSELM method, coupled with an ideal data splitting technique using SOM which decreases the variance of test and training data and improves the performance of OSELM.

PACF was applied to select the significant lags as input. Two different approaches of data splitting, that is, traditional 50–50 split and SOM-based 50–50 data-splitting, were used to highlight the importance of the data splitting process in wind forecasting. The standalone OSELM, a standalone M5, a hybrid M5 coupled with SOM for data partition, and ARIMA model, were developed as benchmark models in this study. The benchmark models were compared with the proposed SOM–OSELM using different performance metrics: correlation *r*, *RMSE*, *MAE*, *RRMSE*, *MAPE*, *WI*, *NSE* and *E*.

The proposed SOM–OSELM model proved to be highly efficient for wind speed forecasting for daily and 6-hourly forecast horizons. Also, the proposed model showed consistent performance for all geographical locations, indicating SOM–OSELM to be a versatile model that is not affected by the topology. Thus the first objective of the chapter can be validated.

SOM–OSELM performance was compared with the performance of the benchmark model to fulfill the second objective of the model. SOM–OSELM outperformed the benchmark models for both short-term and daily forecast horizons. However, for monthly wind speed which was less intermittent than other forecast horizons, the piecewise linear model M5 coupled with SOM showed better results than the proposed model for all performance metrics except for *WI*. ARIMA, being the truly linear model, didn't perform well for any of the forecast horizons.

The SOM-based data splitting managed to evenly distribute the wind speed data into training and test dataset, with a similar mean, standard deviation, and variance compared to the traditional split. Also, the SOM-based hybrid models SOM–OSELM and SOM–M5 performed better than their respective standalone models, OSELM and M5, justifying the third objective of the study, that is, the SOM-based data partition is more effective in wind forecasting than the traditional split method and improves the performance of machine-learning models.

The study showed that although SOM–M5 outperformed SOM–OSELM for the monthly dataset with marginal values, SOM–OSELM displayed good performance in terms of Legates for all forecast horizons. Also, it was observed that its performance was consistent throughout the geographically diverse sites. Thus it is concluded that SOM–OSELM is a robust model for wind speed forecasting and its performance is independent of geographical locations. Thus this study advocates SOM–OSELM as a reliable decision support system for wind farms. The short-term forecasting results can assist wind farms to make the decision for load increment and decrement along with market trading. Similarly, the daily forecasting is applicable for regular maintenance of the electricity generator, and the monthly forecasting will be beneficial to schedule and plan grid maintenance and outage plans. Furthermore, monthly forecasting aids wind farms to reserve maximum power for unforeseen outages and assists in the selection of new wind farms at different geographical locations in Nepal. Thus through a reliable decision support system to forecast wind for short-term, daily, and monthly wind speed, wind farms in Nepal can establish themselves as reliable and sustainable sources of energy and help other sources (hydro and solar) to meet the electricity demand of Nepal.

References

Alencar, D., de Mattos Affonso, C., Oliveira, R., Moya Rodríguez, J., Leite, J. Reston Filho, J. 2017. Different models for forecasting wind power generation: case study . MDPI. Energies 10 (12) 1976.

Azad, H.B., Mekhilef, S., Ganapathy, V.G., 2014. Long-term wind speed forecasting and general pattern recognition using neural networks. IEEE Trans. Sustain. Energy 5 (2), 546–553.

Barnston, A.G., 1992. Correspondence among the correlation, RMSE, and Heidke forecast verification measures; refinement of the Heidke score. Weather. Forecast. 7 (4), 699–709.

Cao, J., Lin, Z., Huang, G.-B., 2012. Self-adaptive evolutionary extreme learning machine. Neural Process. Lett. 36 (3), 285–305.

Chai, T., Draxler, R.R., 2014. Root mean square error (RMSE) or mean absolute error (MAE)? – arguments against avoiding RMSE in the literature. Geosci. Model Dev. 7, 1247-1250.

Chandra, D.R., Kumari, M.S., Sydulu, M., 2013. A detailed literature review on wind forecasting. In: 2013 International Conference on Power, Energy and Control (ICPEC), pp. 630–634.

Chang, W.-Y., 2014. A literature review of wind forecasting methods. J. Power Energy Eng. 2, 161–168.

Chaudhary, V., Bhatia, R.S., Ahlawat, A.K., 2014. A novel self-organizing map (SOM) learning algorithm with nearest and farthest neurons. Alex. Eng. J. 53 (4), 827–831.

Chhetri, M.B.P., Shakya, A., 2010. Wind Intensity Measurement and Wind Loading Codes in Nepal. http://www.wind.arch.t-kougei.ac.jp/APECWW/Report/2010/Nepala.pdf.

Colak, I., Sagiroglu, S., Yesilbudak, M., 2012. Data mining and wind power prediction: a literature review. Renew. Energy 46, 241–247.

Dee, D.P., Uppala, S., Simmons, A., Berrisford, P., Poli, P., Kobayashi, S., et al., 2011. The ERA-Interim Reanalysis: Configuration and Performance of the Data Assimilation System – This Report Is Superseded by the Version Published in Q J Roy Meteor Soc. ECMWF, Shinfield Park, Reading.

Deo, R.C., Ghorbani, M.A., Samadianfard, S., Maraseni, T., Bilgili, M., Biazar, M., 2018. Multi-layer perceptron hybrid model integrated with the firefly optimizer algorithm for wind speed prediction of target site using a limited set of neighboring reference station data. Renew. Energy 116, 309–323.

Ding, S., Zhao, H., Zhang, Y., Xu, X. & Ru, N., 2013. Extreme learning machine: algorithm, theory and applications. Artif. Intell. Rev., 44, 103–115.

Foley, A.M., Leahy, P.G., Marvuglia, A., McKeogh, E.J., 2012. Current methods and advances in forecasting of wind power generation. Renew. Energy 37 (1), 1–8.

Ghimire, S., Deo, R.C., Downs, N.J., Raj, N., 2018. Self-adaptive differential evolutionary extreme learning machines for long-term solar radiation prediction with remotely-sensed MODIS satellite and reanalysis atmospheric products in solar-rich cities. Remote. Sens. Environ. 212, 176–198.

Gurung, A., Karki, R., Cho, J.S., Park, K.W., Oh, S.-E., 2013. Roles of renewable energy technologies in improving the rural energy situation in Nepal: gaps and opportunities. Energy Policy 62, 1104–1109.

Hagspiel, S., Papaemannouil, A., Schmid, M., Andersson, G., 2011. Copula-Based Modeling of Stochastic Wind Power in Europe and Implications for the Swiss Power Grid. Elsevier.

Herzog, V., Lipman, T.A., Kammen, D., 2018. Renewable Energy Sources. University of California, Berkeley, CA.

Hsu, C.-W., Chang, C.-C., Lin, C.-J., 2003. A Practical Guide to Support Vector Classification. National Taiwan University, Taiwan.

Huang, G-B., 2015. What are extreme learning machines? Filling the gap between Frank Rosenblatts dream and John von Neumanns puzzle. Cognitive Computation, 7, 263-278.

Huang, G.-B., Hui Wang, D., Lan, Y., 2011. Extreme learning machines: a survey. Int. J. Mach. Learn Cybern. 2.

Huang, G.-B., Zhu, Q.-Y., Siew, C.-K., 2006. Extreme learning machine: theory and applications. Neurocomputing 70 (1), 489–501.

Kangas, J., Kohonen, T., 1996. Developments and applications of the self-organizing map and related algorithms. Math. Comput. Simul. 41 (1), 3–12.

Kavasseri, R.G., Seetharaman, K., 2009. Day-ahead wind speed forecasting using f-ARIMA models. Renew. Energy 34 (5), 1388–1393.

Kaveh, A., Bakhshpoori, T., Hamze-Ziabari, S.M., 2016. Derivation of new equations for prediction of principal ground-motion parameters using M5′ algorithm. J. Earthq. Eng. 20 (6), 910–930.

Keren, B.S., Sabitha, K., 2016. Probabilistic forecasting of wind power generation using extreme learning machine involving bootstrap method. IJITECH 04 (04), 0729–0736.

Khozani, Z.S., Bonakdari, H., Zaji, A.H., 2017. Efficient shear stress distribution detection in circular channels using extreme learning machines and the M5 model tree algorithm. Urban. Water J. 14 (10), 999–1006.

Kohonen, T., 2013. Essentials of the self-organizing map. Neural Netw. 37, 52–65.

Kohonen, T., MATLAB Implementations and Applications of the Self-Organizing Map Unigrafia Oy, Helsinki, Finland, 2014.

Laudari, Raju. (2016). Assessment of Economic Viability of Wind Energy in Nepal: A Case Study of Ten Sites. IOE Graduate Conference, 2015 pp. 169–179, 10.13140/RG.2.1.1982.1682.

Lazarevska, E., 2016. Wind speed prediction with extreme learning machine. In: 2016 IEEE 8th International Conference on Intelligent Systems (IS), pp. 154–159.

Legates, D.R., McCabe, G., 2013. A refined index of model performance: a rejoinder, Inter. J. Clim. 33.

Li, Z., Ye, L., Zhao, Y., Song, X., Teng, J., Jin, J., 2016. Short-term wind power prediction based on extreme learning machine with error correction. Prot. Control. Mod. Power Syst. 1 (1), 1.

Lim, J.-S., Lee, S., Pang, H.-S., 2013. Low complexity adaptive forgetting factor for online sequential extreme learning machine (OS-ELM) for application to nonstationary system estimations. Neural Comput. Appl. 22 (3), 569–576.

Liu, H., Tian, H.-Q., Li, Y.-F., 2015. Four wind speed multi-step forecasting models using extreme learning machines and signal decomposing algorithms. Energy Convers. Manag. 100, 16–22.

Mahmoud, T., Dong, Z.Y., Ma, J., 2018. An advanced approach for optimal wind power generation prediction intervals by using self-adaptive evolutionary extreme learning machine. Renew. Energy 126, 254–269.

May, R.J., Maier, H.R., Dandy, G.C., 2010. Data splitting for artificial neural networks using SOM-based stratified sampling. Neural Netw. 23 (2), 283–294.

Mi, X.-w, Liu, H., Li, Y.-F., 2017. Wind speed forecasting method using wavelet, extreme learning machine and outlier correction algorithm. Energy Convers. Manag. 151, 709–722.

Mohammadi, K., Shamshirband, S., Yee, P.L., Petković, D., Zamani, M., Ch, S., 2015. Predicting the wind power density based upon extreme learning machine. Energy 86, 232–239.

NEA, 2018. A review in review: fiscal year 2017/2018. Kathmandu.

Nepal, R., 2012. Roles and potentials of renewable energy in less-developed economies: the case of Nepal. Renew. Sustain. Energy Rev. 16 (4), 2200–2206.

Panwar, N.L., Kaushik, S.C., Kothari, S., 2011. Role of renewable energy sources in environmental protection: a review. Renew. Sustain. Energy Rev. 15 (3), 1513–1524.

Peng, H., 2013. Air Quality Prediction by Machine Learning Methods. The University of British Columbia, Vanouver.

Potočnik, P., & Govekar, E. 2015. Practical Considerations in Training Extreme Learning Machines. *Proceedings of the 16th International Conference on Engineering Applications of Neural Networks (INNS).* Rhodes, Island, Greece: Association for Computing Machinery.

Reitermanova Z. Data splitting. WDS's 10 proceedings of contributed papers, Part. 2010;1:31–36. [Google Scholar].

Rong, H.-J., Huang, G.-B., Ong, Y.-S., 2008. Extreme learning machine for multicategories classification applications. 2008 IEEE International Joint Conference on Neural Networks (IEEE World Congress on Computational Intelligence). pp. 1709–1713.

Rustum, R, Adeloye, A., Simala, A., 2007. Kohonen self-organising map (KSOM) extracted features for enhancing MLP-ANN prediction models of BOD_5. In: IAHS-AISH Publication - Water Quality and Sediment Behaviour of the Future: Predictions for the 21st Century, three hundred and fourteen ed., pp. 181–187, International Symposium: Quantification and Reduction of Predictive Uncertainty for Sustainable Water Resources Management - 24th General Assembly of the International Union of Geodesy and Geophysics (IUGG), Perugia, Italy.

Solomatine, D.P., Xue, Y., 2004. M5 model trees and neural networks: application to flood forecasting in the upper reach of the Huai River in China. J. Hydrol. Eng. 9 (6).

Somvanshi, V.K., Pandey, O.P., Agrawal, P.K., Kalanker, N.V., Prakash, M.R., Chand, R., 2006. Modeling and prediction of rainfall using artificial neural network and ARIMA techniques. J. Ind. Geophys. Union. 10(2), 141–151.

Soman, S.S., Zareipour, H., Malik, O., Mandal, P., 2010. A review of wind power and wind speed forecasting methods with different time horizons. In: North American Power Symposium 2010, pp. 1–8.

SWERA, 2006, Solar and wind energy resource assessment in Nepal (SWERA). Final Report (GIS Apart).

Syed, A., Aggarwal, S.K., 2016. Wind power forecasting using extreme learning machine. Int. J. Res. Sci. Innov. 3 (3), 5.

Taghi Sattari, M., Pal, M., Apaydin, H., Ozturk, F., 2013. M5 model tree application in daily river flow forecasting in Sohu Stream, Turkey. Water Resour. 40 (3), 233–242.

Vesanto, J., 1999. SOM-based data visualization method. Intell. Data Anal. 3 (2), 111–126.

Wang, X., Han, M., 2014. Online sequential extreme learning machine with kernels for nonstationary time series prediction. Neurocomputing 145, 90–97.

Wang, X., Han, M., 2015. Improved extreme learning machine for multivariate time series online sequential prediction. Eng. Appl. Artif. Intell. 40, 28–36.

Wang, X., Guo, P., Huang, X., 2011. A review of wind power forecasting models. Energy Procedia 12, 770–778.

Wang, G.-G., Lu, M., Dong, Y.-Q., Zhao, X.-J., 2016. Self-adaptive extreme learning machine. Neural Comput. Appl. 27 (2), 291–303.

Wang, L., Li, X., Bai, Y., 2018. Short-term wind speed prediction using an extreme learning machine model with error correction. Energy Convers. Manag. 162, 239–250.

Willmott, Cort J., Robeson, S.M., Matsuuraa, K., 2011. Short communication a refined index of model performance. Int. J. Climatol 32, 2088–2094.

Wu, Y.K., Hong, J.S., 2007. A literature review of wind forecasting technology in the world. 2007 IEEE Lausanne Power Tech. 504–509.

Xiao, C., Dong, Z.Y., Xu, Y., Meng, K., Zhou, X., 2016a. Rational and self-adaptive evolutionary extreme learning machine for electricity price forecast. Res. Online Memetic Comput. 8 (3), 223–233.

Xiao, C., Dong, Z., Xu, Y., Meng, K., Zhou, X. & Zhang, X. 2016b. Rational and self-adaptive evolutionary extreme learning machine for electricity price forecast. *Memetic Computing*, 8, 223-233.

Zhang, G.P., 2003. Time series forecasting using a hybrid ARIMA and neural network model. Neurocomputing 50, 159–175.

Zhong, X., Dutta, U., 2015. Engaging Nash–Sutcliffe efficiency and model efficiency factor indicators in selecting and validating effective light rail system operation and maintenance cost models. J. Traffic Transp. Eng. 3, 255–265.

CHAPTER

15

Potential growth in small-scale distributed generation systems in Brazilian capitals

Carmen B. Rosa, Paula D. Rigo and Julio Cezar M. Siluk

Post-Graduate Program in Production Engineering, Federal University of Santa Maria (UFSM), Santa Maria, Brazil

15.1 Introduction

The development of countries is increasingly bringing the scenario of the reduction of natural resources for electricity generation. Related to this development is the consumption of electricity and the need to find sustainable solutions capable of stopping the harmful effects of greenhouse gas emissions in large urban centers, that is, actors of climate impact mitigation (Khoodaruth et al., 2017). The transition to a more significant renewable energy mixed energy system is underway, supported by technological advances and demand projections (Brummer, 2018).

According to REN21 (2018) the year 2017 was another record-breaking one for renewable energy, characterized by the largest ever increase in renewable power capacity, falling costs, increases in investment, and advances in enabling technologies. Many developments during the year impacted the deployment of renewable energy. These impacts include the lowest ever bids for renewable power in tenders throughout the world, a significant increase in attention to electrification of transport, increasing digitalization, jurisdictions pledging to become coal-free, new policies and partnerships on carbon pricing, and new initiatives and goals set by groups of governments at all levels (REN21, 2018).

Given this, there is photovoltaic (PV) solar generation; this energy represents over 88% of the distributed generation in the country. PV provides many advantages, since it can meet the energy demand of the urban environment and projects in locations with difficult access. Besides, it does not contribute to greenhouse gas emissions. This indicates that the combination of cost reduction with technological innovation and government programs allow the increasing economic viability for the installation of this system (Flowers et al., 2016). However, while the International Energy Agency predicts that PV's share in global

Predictive Modelling for Energy Management and Power Systems Engineering
DOI: https://doi.org/10.1016/B978-0-12-817772-3.00015-X

electricity will reach 16% by 2050, in Brazil the generation of electricity through the PV source was present in the electricity matrix of 2018 with only a 1.2% stake (ANEEL, 2019a). The scenario of PV solar energy in Brazil constitutes promising expectations, due to the advantageous geographic location—with almost all its territory located within the tropical range—to capture solar energy, since the degree of incidence of solar rays in this region is almost perpendicular, favoring high levels of solar irradiation (Pereira et al., 2017b). The average annual irradiation in Brazil varies between 1200 and 2400 kWh/m^2/year, being above the average of Europe (Pereira et al., 2017b). With all this potential for the generation of energy from the solar source, according to the records of the National Electric Energy Agency (ANEEL, 2019a), the country presented a total of installations of small-scale power generation solar systems in Brazil in November 2019 of 144,209, which corresponds to 1.67 GW of power.

In Brazil, electricity generation systems through renewable sources linked to the public grid—classified as Distributed Generation (DG)—grow through the offsetting of credits, in which the consumer generates their electricity through renewable sources, injects the generated energy in the public electricity grid, and subtracts the total energy consumed by the total generated. This model became possible only from ANEEL Normative Resolution 482/2012, updated by Normative Resolution ANEEL 687/2015 (ANEEL, 2012).

In the international scenario, the diffusion of PV systems to date has been unevenly distributed geographically, and until recently, some countries with ambitious subsidy schemes, most notably Germany, accounted for the bulk of the world market. Thus it is known that there are differences in the diffusion of PV between countries, although, the patterns cannot be explained solely by economic profitability (Palm, 2016). Tanaka et al. (2017) corroborate this thinking by stating that the decision-making process of a firm to invest in renewable energy is simple. However, the decision-making process of the average person who purchases durable goods related to energy use is difficult to understand. Fadly and Fontes (2019) claim in their research that given the importance of economic growth, analyses of patterns of green energy diffusion has attracted much attention. As such, understanding the drivers of diffusion of renewables appears to be an essential and timely question.

This same observation is reflected in Brazil, a country with 5570 municipalities that has significant variations in the rates of competitiveness index diffusion of the PV systems installation (Rosa et al., 2020). Brazil's DG based on PV systems presents only 7.95 W/inhabitant, there are states with a diffusion rate of less than 1, 2.60 W/inhabitant (Acre state), and states with a rate higher than 9, 19.68 W/inhabitant (Rio Grande do Sul state) (ANEEL, 2019a). Despite the need for diversification of the electric generation matrix and good solar radiation in the country, the main challenges to its widespread use can be classified as technological, legislative (political), and financial (Faria, Trigoso, Cavalcanti, 2017). Besides, an improved understanding of different perspectives of these variations can, for example, direct legislative decision-makers and regulators and provide promising market knowledge for the design of measures to promote the growth of PV solar energy installations.

With this scenario, it is possible to identify that the diffusion of energy generation through PV technology has a growth potential associated with local and regional incentives, that is, the role of the municipalities in the democratization of PV energy is decisive for the higher insertion of solar energy in the electrical matrix. Discussing municipalities, then, becomes a critical topic of great importance, particularly given the future scenario of population growth in urban centers. Given this, with the aim of identifying difficulties and opportunities to increase the participation of PV energy in this country, this work brings as a scientific contribution a ranking of potential urban sectors for the installation of PV systems based on performance indicators inserted in the mathematical model.

These results refer to the research problem considered in this work: how to measure the growth potential index of the Brazilian capitals in the installation of small-scale DG PV systems?

It is intended with the mathematical model to rank the main urban centers for the installation of PV systems, boosting the market analysis and the prospects of clients of companies installing PV systems, and scientific support to boost political and economic incentives for the growth of the participation of PV solar energy in the Brazilian electric power matrix. This study diagnosed the performance of the 26 capitals of Brazil and the Federal district. In the Brazilian scenario, this approach was discussed in the article Rosa (2020) by applying modeling with the factors selected for this analysis. However, this study broadens the research scenario in order to discuss Brazil as a whole.

To measure the growth potential of Brazilian capitals in relation to PV energy diffusion, the multiple-criteria decision method (MCDM) approach was used by combining the analysis hierarchical process and the technique for order of preference by similarity ideal solution (AHP–TOPSIS).

15.2 Distributed generation in Brazil

Historically, the supply of electricity to Brazilian consumers has been studied conventionally, under the logic of generation and distribution of electricity by centralized plants, predominantly hydroelectric. Centralized models require high-voltage transmission to carry electricity from the generation center to consumers far from the power plant (Lopes et al., 2015). The performance of the public electricity distribution service is by concessionaires, permissionaires, and authorities. In 2019 there were 109 agents, including public, private, and mixed public–private, operating in the distribution market responsible for connecting millions of consumer units to the generating plants through a set of facilities (ANEEL, 2019b).

The traditional concept of a centralized grid system is continually changing due to increasing levels of decentralized generation (Richter, 2013). The use of small electricity generators for consumer supply, mainly from renewable sources, is increasingly encouraged by government programs, in order to reduce social and environmental impacts, delay investments in expanding transmission and distribution systems, and promote the diversity of the country's energy matrix (Paris et al., 2018).

In developing countries, planning and expanding the energy system requires significant challenges for regulators, researchers, and all stakeholders in order to provide good quality, reliable, sustainable, and affordable energy to their population. As a result, the growing demand for electricity is accompanied by technological development. Also, in the last few years, society is increasingly aware of the need to achieve sustainable development, realizing the importance of avoiding significant impacts on the environment, such as greenhouse gas emissions (Camilo et al., 2017). So, efforts for this scenario are perceived in countries around the world, seeking alternatives to meet their demand for electricity, but avoiding causing more significant impacts on the environment.

A few decades ago, Europe and the United States added a new element in power supply design: DG of electricity. The definition of DG provided by article 14 of Decree-Law No. 5.163/2004 defines it as *"the production of electricity from ventures of concessionary, permissionary or authorized agents, directly connected to the buyer's electrical distribution system, except those from hydroelectric plants with installed capacity exceeding 30 MW; and thermoelectric, including cogeneration, with energy efficiency below 75%"* (Brasil, 2004). Already the Institute of Electrical and Electronics Engineers (IEEE) defines DG in the international context as *"the generation of electricity in facilities that are sufficiently smaller than power stations and, therefore, allow interconnection at almost any point in the power system"* (Richter, 2013). The National Institute of Energy Efficiency (INEE) indicates that the basic principle of DG is *"add small or medium-sized generation from the use of alternative energy sources based on different technologies in distribution and transmission systems"* (Soccol et al., 2016).

Recognizing the lack of consensus on a definition of DG, Pepermans et al. (2005) presented a review of the available definitions and categorizations of DG. They concluded that, although vague, the best definition was provided by Ackermann et al. (2001) *"a source of electricity generation that is directly connected to the distribution network."* This definition implies that DG is always connected to the measured or networked infrastructure. The concept also involves measuring, control, and command equipment that articulates the operation of the generators and the eventual control of loads so that they adapt to the energy supply. This is not always the case, particularly for developing countries, where DG can provide an alternative to grid-connected electricity supply (Ackermann et al., 2001).

From all definitions and principles in Brazil, since April 2012—when ANEEL Normative Resolution 482/2012 was created—the consumer has known the term DG and can generate his own electricity from renewable sources. This grid-connected, close-to-load generation system can be called microgeneration and minigeneration distributed power (ANEEL, 2012). DG has an important impact on the entire power transmission and distribution system as it changes the design of the current system and becomes a key issue of high criticality (Picciariello et al., 2015).

With the advent of new distributed power generation technologies, power grids have now interconnected several renewable sources of energy (Keshav and Rosenberg, 2010). Nonreservoir hydroelectric power plants, wind power generation, and solar power generation are examples of power generation that is exposed to stochastic meteorological variations. This means intermittent power generation operation (Richter, 2013). Hence, the resolution proposed the creation of an energy compensation system, known internationally as *Net Metering*. The owner of a small renewable source generator does not need to consume all the energy produced at the time of generation, because they will be able to inject into the power grid to receive kWh credits on the bill (ANEEL, 2012).

In order to reduce the costs and time for the connection of microgeneration and minigeneration, the electric power compensation system can be made compatible with the General Supply Conditions (Normative Resolution No. 414/2010), increasing the target audience, and improving the information in the invoice, as published by the regulatory body ANEEL in Normative Resolution 687/2015 revising Normative Resolution 482/2012. In this update, microgeneration is characterized by an installed power of less than or equal to 75 kW, and minigeneration is characterized by an installed power of greater than 75 kW and less than or equal to 5 MW (ANEEL, 2015). The two DG frameworks, micro- and minigeneration, are based on the sources of electricity generation: wind, PV systems, biomass, and small hydroelectric plants.

The advance of DG, as shown in Fig. 15.1, implies transformations in the electricity sector and especially in the distribution segment. This large insertion of DG and prosumers (producers + consumers) in distribution systems has led to a major change in the traditional electricity business model, as the consumer becomes part of its traditional value chain, offering new services, such as voltage control services, system information, rapid demand response, storage capacity, and more (Daza, 2018). Note that the source responsible for this growth is solar PV.

In general, the advance of the decentralization of electric power generation can promote several benefits to the consumer and the electric system, since it effectively meets the growth of energy demand (Camilo et al., 2017). A point of great attention are the water crises faced by Brazil periodically. It is known that electricity in Brazil is supplied primarily by large hydroelectric plants, so other forms of generation can be triggered to reduce this dependence (Silva et al., 2018).

Another effort concerns Brazil's federalist character, which provides state governments with financial and administrative authority to guide energy-related policies and investments (Carstens et al., 2019). In the extension of state authority, in 2015 the National Council for Finance Policy (CONFAZ) approved the agreement on Tax on Circulation of

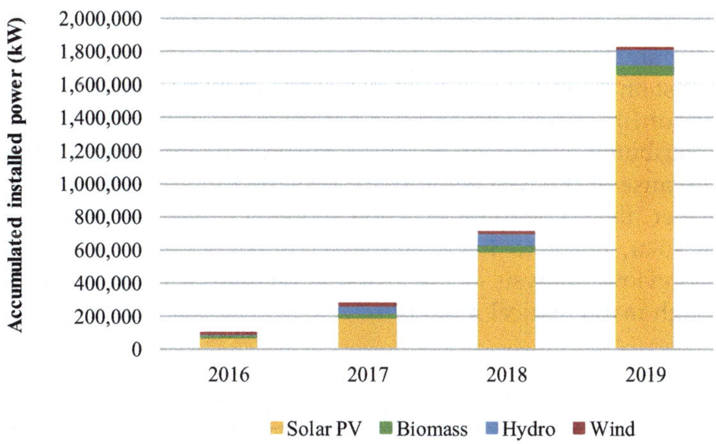

Fonte: (ANEEL, 2019c)

FIGURE 15.1 Growth of DG in Brazil Source: *Fonte: Agência Nacional de Energia Elétrica—ANEEL, 2019c. Geração distribuida [WWW Document]. Agência Nac. Energ. Elétrica. <https://app.powerbi.com/view?r = eyJrIjoiZjM4NjM0OWYtN2IwZS00YjViLTllMjItN2E5MzBkN2ZlMzVkIiwidCI6IjQwZDZmOWI4LWVjYTctNDZhMi05MmQ0LWVhNGU5YzA-xNzBlMSIsImMiOjR9> (accessed 16.11.19.).*

Goods and Services (ICMS 16/2015). This agreement grants the exemption from the tax on electricity supplied by the distributor to the consumer unit in the amount corresponding to the sum of the electricity injected into the distribution network by the same consumer unit. The idea was to make policies conditional on regional approaches to renewable energy development (Carstens et al., 2019). This autonomy given to the states resulted in a geographically unequal diffusion of the DG in the national scenario, although the patterns cannot be explained solely by the economic profitability (Fadly and Fontes, 2019).

The study of Gucciardi Garcez (2017) explained that the penetration of DG in Brazil, in terms of the number of PV systems and installed capacity, is quite incipient. Some of the identified barriers, according to the same author, are the lack of direct incentives, current energy planning guidelines (which focus on large-scale low-carbon hydropower generation), and lack of viable financing. Even with the low results that marked the 3 years after Resolution 482/2012, the DG received attention and interest in Brazil for several reasons, such as the international movement to increase small-scale and localized generation, especially in the case of PV solar energy, its application in a Smart Grid architecture, and environmental considerations such as climate change (dos Santos et al., 2018).

Gucciardi Garcez (2017) declared that future studies on the determinants of DG penetration should be carried out in order to verify the expansion potential of DG from the adhesion of micro- and minigeneration projects in Brazil. The adoption of DG systems is made by different people and companies, and their decisions are not always economically rational, and there are other factors that influence their actions. Thus designing the distributed micro- and minigeneration market is not a trivial task (Carstens et al., 2019; Rigo et al., 2019).

Considering that micro- and minigeneration in Brazil have reached a consistent degree of regulation, electricity tariffs in the country are high and tax barriers are partially equated, investments in PV systems by consumers connected to the low-voltage grid tend to become more attractive (Gucciardi Garcez, 2017). In this sense, the Energy Research Company (EPE) estimates that DG PV solar energy could account for 1.3% of the National Interconnected System (SIN) load in 2030 and 5.7% of the load in 2050. So, EPE expects an installed capacity of 10 GWp by 2030 and 78 GWp by 2050 (EPE, 2017).

Micro- and distributed minigeneration tend to become increasingly attractive to electricity consumers because of the reduction in the costs of adopting incentive policies in many countries. However, the generation diffusion dynamics, focused on the most representative source, PV solar, tends to have significant impacts on the electricity sector, being essential to the previous analysis of the growth potential of PV solar generation diffusion distributed through mathematical models.

15.3 Measurement method

The growth potential of DG in the 26 state capitals of Brazil and the Federal District was established by the multicriteria decision support analysis (MCDM) approach through the combination of hierarchical process analysis and the technique for order of preference for similarity ideal methods. solution (AHP-TOPSIS). First, we selected the relevant factors for measuring the potential for expansion of PV energy in a city, based on the study of

Rosa et al. (2020). From this reference, we considered two points of view: (1) demographic and socioeconomic development; and (2) energy sustainability, which included 11 critical success factors for analysis, presented in Table 15.1.

Based on the 11 factors, PV experts were selected to weigh the relative importance levels of each factor in assessing the potential of PV system installations in cities. Experts include professors from universities that are members of PV research groups and members of PV solar energy associations. In total, eight experts weighted the factors, an appropriate number for the AHP weighting methodology (Saaty, 1987). The method AHP is based on the pairwise comparison matrix $W = \|w_{j,k}\| \left(j, k = 1, 2, 3, \ldots 11\right)$. The experts compared all the evaluation criteria i relative to j for all 11 factors ($n = 11$). With the judgment matrix, the weight of each factor was calculated, forming the weight vector w_i presented in Table 15.2.

The three most important factors are related to the economic capacity of the city, especially the energy tariff cost. In Brazil, there is a variation in the cost of the energy tariff per city by the cost of the concessionaire, power transmission, and distribution company; and the tax rate on energy trade by state. The higher the energy tariff, the more attractive is the DG installation, as the payback time becomes smaller. But for investment to be possible, experts point to the factors "GDP per capita" and "Class of income per household" as important.

Sequentially, the factors "Solar irradiation" and "Energy demand" have similar importance, as it is the capacity of energy production added to the need to increase energy

TABLE 15.1 Factors' reference.

Factor		References
Demographic and socioeconomic development	Population	Aliyu et al. (2018), Mah et al. (2018), Rosa et al. (2020)
	Gross domestic product per capita	Gava Gastaldo et al. (2019), Carli et al. (2018), Rosa et al. (2020)
	Instruction level	Mah et al. (2018), Gava Gastaldo et al. (2019), Rosa et al. (2020)
	Territorial area	Mah et al. (2018), Rosa et al. (2020)
	Demographic density	Crowe and Li (2020), Rosa et al. (2020)
	Human development index	Padmanathan et al. (2019), Rosa et al. (2020), Araújo et al. (2019)
	Class of income per household	Padmanathan et al. (2019), Gava Gastaldo et al. (2019), Rosa et al. (2020)
Energy sustainability	Solar irradiation	Li et al. (2019), Gava Gastaldo et al. (2019), Rosa et al. (2020), Rigo et al. (2019), Araújo et al. (2019), George et al. (2019)
	Energy tariff	
	Number of photovoltaic systems	
	Energy demand	

TABLE 15.2 Weighting of experts by AHP method.

No.	Factor	w_j (%)
1	Energy tariff	11.34
2	Gross domestic product per capita	10.58
3	Class of income per household	10.33
4	Solar irradiation	9.82
5	Energy demand	9.57
6	Instruction level	9.07
7	Human development index	9.07
8	Number of photovoltaic systems	8.06
9	Population	7.81
10	Demographic density	7.56
11	Territorial area	6.80

production and diversify the Brazilian energy matrix. The "Instruction level" and "DHI" are social factors and have the same impotence. They are followed by the "number of PV systems" already existing in the capitals, which can promote future installations in the vicinity or expansions. Finally, the factors related to the physical capacities of the capitals, such as population, demographic density, and territorial area.

The database for factor measurement for each capital were collected and presented in the Table 15.3. Brazil has 26 capitals plus the Federal District, divided into five regions. The seven first columns were collected from IBGE (2019). The first column presents the estimated population of 2019, where the smallest population is from Palmas, with 299,000 inhabitants, and the largest is São Paulo, with more than 12 million inhabitants. The total sum of the population of the Brazilian capitals is 50 million, corresponding to approximately 25% of the Brazilian population. The second column presents the GDP per capita, where Brasilia, capital of Brazil, has the highest value. The third column corresponds to the population with complete higher education, totaling approximately 10 million people, 20% of the population of the capitals. The fifth column presents demographic density, which reflects the level of verticalization of the capital. In the model, it is considered a negative point, because the higher the verticalization, the more difficult is the growth in the number of PV modules per inhabitant. The North and Midwest regions have low demographic densities and the Southeast region has São Paulo with the highest population density of the country, of 8000 inhabitants/km^2. The sixth column presents the HDI of the capitals, which vary little within each region, and in all vary from 0.72 (Maceió) to 0.85 (Florianópolis and Vitória). And the seventh column presents the Class of Income per household, where São Paulo has a value much higher than all the other capitals.

The eighth column presents the Solar Irradiation data from the capitals, which range from 4.2 to 5.8 Wh/m^2.day and were collected from the Brazilian Solar Energy Atlas (Pereira et al., 2017a). The ninth column is the energy tariff per capital distributor plus

TABLE 15.3 Database of Brazilian capitals.

Region	Capital	Population (×1000)	Gross domestic product per capita (R$/hab)	Instruction level (×1000)	Territorial area (km^2)	Demographic density (hab/km^2)	HDI	Class of income per household (×1000)	Solar irradiation (Wh/m^2. dia)	Energy tariff (R $/kWh)	Number of PV systems	Annual energy demand (GWh)
North	Rio Branco	407	21,544	55	8835	46	0.73	21	4560	0.73	112	503
	Macapá	503	19,935	71	6564	77	0.73	27	4944	0.66	105	713
	Manaus	2183	35,564	295	11,401	191	0.74	107	4323	0.77	351	3244
	Belém	1493	20,350	224	1059	1409	0.75	94	4864	0.83	400	3574
	Porto Velho	494	28,836	65	34,091	14	0.74	35	4480	0.68	114	807
	Boa Vista	399	24,853	53	5687	70	0.75	20	4886	0.68	44	705
	Palmas	299	28,974	48	2219	135	0.79	22	5218	0.73	814	423
Northeast	Maceió	1019	20,853	137	509	2001	0.72	57	5523	0.64	372	1477
	Salvador	2872	20,797	535	694	4140	0.76	191	5365	0.67	504	4682
	Fortaleza	2669	23,045	391	312	8544	0.75	155	5776	0.65	1220	3537
	São Luis	1102	26,154	131	583	1890	0.77	62	5208	0.75	567	1078
	João Pessoa	809	23,346	160	211	3829	0.76	57	5530	0.68	411	1046
	Recife	1646	30,478	296	219	7520	0.77	125	5462	0.67	314	2376
	Teresina	865	22,598	103	1391	622	0.75	48	5572	0.76	1029	910
	Natal	884	24,891	164	167	5281	0.76	61	5674	0.61	598	1415
	Aracaju	657	25,718	115	182	3607	0.77	51	5496	0.64	697	1092
Midwest	Goiânia	1516	32,209	288	729	2080	0.80	149	5247	0.67	1060	3328
	Cuiabá	613	37,930	121	3267	188	0.79	55	5106	0.77	1222	1480
	Campo Grande	896	29,443	154	8093	111	0.78	71	5031	0.72	858	1743
	Brasília	3015	79,100	689	5761	523	0.82	342	5252	0.62	1330	6525

(*Continued*)

TABLE 15.3 (Continued)

Region	Capital	Population (×1000)	Gross domestic product per capita (R$/hab)	Instruction level (×1000)	Territorial area (km^2)	Demographic density (hab/km^2)	HDI	Class of income per household (×1000)	Solar irradiation (Wh/m^2. dia)	Energy tariff (R $/kWh)	Number of PV systems	Annual energy demand (GWh)
Southeast	Vitória	362	60,428	106	97	3728	0.85	51	4957	0.65	248	891
	Belo Horizonte	2512	35,122	534	331	7580	0.81	300	5129	0.78	1470	6330
	São Paulo	12,252	57,071	2693	1521	8055	0.81	1324	4450	0.62	838	34,685
	Rio de Janeiro	6719	50,691	1563	1200	5598	0.80	761	4732	0.80	1961	16,072
South	Curitiba	1933	44,239	485	435	4444	0.82	259	4194	0.66	628	5026
	Porto Alegre	1484	49,578	370	495	2995	0.81	217	4430	0.69	617	3862
	Florianópolis	501	39,048	150	675	742	0.85	75	4251	0.57	525	1679

state taxes, and ranges from 0.61 to 0.80 R$/kWh (ANEEL, 2019d). The tenth column shows the number of PV installations in each capital (ANEEL, 2019c). It is noted that the Midwest and Southeast regions are the most developed. Finally, it presents the annual electricity demand of the capital, with São Paulo having the largest demand, of approximately 34,000 GWh (EPE, 2018).

With the capital database of Table 15.3 and the weighting of the factors in Table 15.2, it was possible to use the TOPSIS method for capital evaluation. TOPSIS is based on the concept that the ranking of alternatives must have the shortest distance from the Positive Ideal Solution (PIS) and the longest distance from the Negative Ideal Solution (NIS) (Şengül et al., 2015). Thus, TOPSIS was applied with the 11 factors ($n = 11$) in the 27 capitals ($m = 27$) calculating the Euclidean distance of alternative i ($i = 1, 2, \ldots.27$) to the worst solution (d_{iw}) and the best solution (d_{ib}) and finally applying the Eq. 15.1 in alternatives.

$$S_i = \frac{d_{iw}}{(d_{iw} + d_{ib})}; 0 \leq S_i \leq 1 \qquad (15.1)$$

Then, the capital ordering is performed in relation to the value of S_i:

- $S_i = 1$ if and only if the alternative solution has the best condition; and
- $S_i = 0$ if and only if the alternative solution has the worst condition.

The growth potential of each Brazilian capital will be the value calculated by the TOPSIS method, which we named Potential Growth Index (PG_i)($i = 1, 2, \ldots.27$).

15.4 Results

The mathematical modeling based on the AHP-TOPSIS combination allowed the generation of a ranking of capitals by growth potential, as presented in Table 15.4. The PG_i is considered the growth potential of DG of each Brazilian capital. Considering that the index can vary from 0 to 1, the highest value is 0.85 for the capital São Paulo (SP), and the smallest is 0.03 for the capital Macapá (AP), reflecting a significant difference between the values.

The city of Sao Paulo has the highest potential in Brazil. In all sectors of the economy, São Paulo will always have a high potential to be explored, as it concentrates on the most significant number of people and companies. Even having a high population density—which is considered a negative point for DG—all other factors improve exploitation potential because the economic power of the city promotes a high number of people and companies with purchasing power to acquire PV systems. Another negative factor for PV energy exploration in São Paulo is the solar radiation index. Sao Paulo is known as the "drizzle land." However, its irradiation rates are significantly higher than those of high-development European countries in PV generation. This means that even if not getting the maximum efficiency of PV modules, it is still a positive investment for the residents of that city. Finally, the city's electricity demand is considered very high and is usually supplied by the largest hydroelectric plant in Brazil, Itaipu Binacional. This means that the

TABLE 15.4 Growth potential index of DG ranking.

Ranking	Region	Capital	PG_i
1	Southeast	São Paulo	0.853424567
2	Southeast	Rio de Janeiro	0.557293659
3	Midwest	Brasília	0.307647314
4	Southeast	Belo Horizonte	0.288822440
5	Northeast	Fortaleza	0.246301609
6	South	Curitiba	0.206902765
7	Northeast	Recife	0.185696290
8	Northeast	Salvador	0.183840706
9	South	Porto Alegre	0.178993161
10	Midwest	Goiânia	0.159572534
11	Southeast	Vitória	0.154778116
12	Midwest	Cuiabá	0.148483397
13	Northeast	Natal	0.139292311
14	Northeast	Teresina	0.121555792
15	North	Manaus	0.116679342
16	Northeast	Aracaju	0.116425745
17	Midwest	Campo Grande	0.110358223
18	Northeast	João Pessoa	0.106178741
19	North	Belém	0.103491506
20	North	Palmas	0.098113574
21	South	Florianópolis	0.094417091
22	Northeast	São Luís	0.093777660
23	Northeast	Maceió	0.076157108
24	North	Porto Velho	0.040593566
25	North	Rio Branco	0.033192218
26	North	Boa Vista	0.032506676
27	North	Macapá	0.027139516

diversification of the matrix through DG is positive for meeting future demand, with the city's natural economic growth.

The second PG_i is the city of Rio de Janeiro (RJ), known for its tourist potential, beautiful natural landscape and sunny beach days. The sunny days of Rio de Janeiro all year round enhance its potential for PV power generation. Also, Rio de Janeiro has a large number of people with purchasing power to invest in PV energy. A critical issue of this city is the slums, which can be a negative and/or positive factor for the diffusion of PV energy. The negative factor is the slums still have illegal electrical installations, increasing the cost of commercial losses by the city's electric utility. The positive factor is the incentives for slum development, and the need to bring electrification to all residents of these regions, who often lack the proper power transmission and distribution infrastructure. Finally, the city has the highest energy tariff in Brazil, which encourages residents to invest in PV energy as it makes the payback time lower, which is one of the reasons why Rio de Janeiro is the city with the largest number of PV systems in Brazil.

The third position is the capital of Brazil, Brasilia (DF). Brasilia has the highest GDP per capita and a low demographic density compared to the first two in the ranking. Both factors promote the diffusion of PV energy because it concentrates a large number of people with high purchasing power and household area available for the installation of PV systems. Another factor that drives distributed energy is the city's annual energy demand, which is among the highest in the country. Besides, there is Brasilia's potential for the number of highly educated people, reflecting the common sense of city residents in terms of environmental awareness and general knowledge of PV systems.

The fourth position is Belo Horizonte (MG). It is the state capital of Minas Gerais, which was the first Brazilian state to launch subsidies for distributed power generation, giving state tax exemption to Prosumers. After the publication of DG's regulations, between 2012 and 2016, the diffusion of the systems occurred with higher speed in the state of MG, and consequently in its capital. Belo Horizonte was for many years the city with the largest number of PV systems installed, being recently surpassed by Rio de Janeiro. Interestingly, it does not have as high a GDP per capita as the top three in the ranking. However, like Brasilia, it has a high annual energy demand, and like Rio de Janeiro, it has a high energy tariff. Both factors, plus tax exemption, made the capital a favorable scenario for investment in PV.

In contrast to the four capitals of higher PG_i mentioned above are Porto Velho (RO), Rio Branco (AC), Boa Vista (RR), and Macapá (AP), with the worst performance, located in the North region. These capitals have the lowest values of the factors analyzed. They are the capital cities with the lowest population, the lowest number of people with purchasing power, and lowest HDI rates. The northern region of Brazil has always suffered from energy shortages. However, with the technological advancement of PV energy, 15 indigenous villages had access to electricity by storing energy in batteries. That is, solar power generation in this region is important for electricity supply in rural communities isolated from the electricity distribution system. It is known that the DG of electric energy in Brazil is characterized by the injection of the energy generated in the distribution grid of the interconnected system. For this reason, the region has the lowest PG_i.

To summarize PG_i, we divide them into six categories. Fig. 15.2 shows these six categories arranged in Brazilian territory and signaled by colors, and the symbol marks the location of the capital in each state.

According to the map, it is possible to observe that there is a synergy in the evaluation by region of the country. Of the five Brazilian regions, Southeast concentrates the highest indexes, with capitals in the first, second, and third categories of Fig. 15.2. This region is historically the most economically developed in the country, as well as concentrates most of the Brazilian population. Many Brazilians immigrated from other regions to this region with the advent of

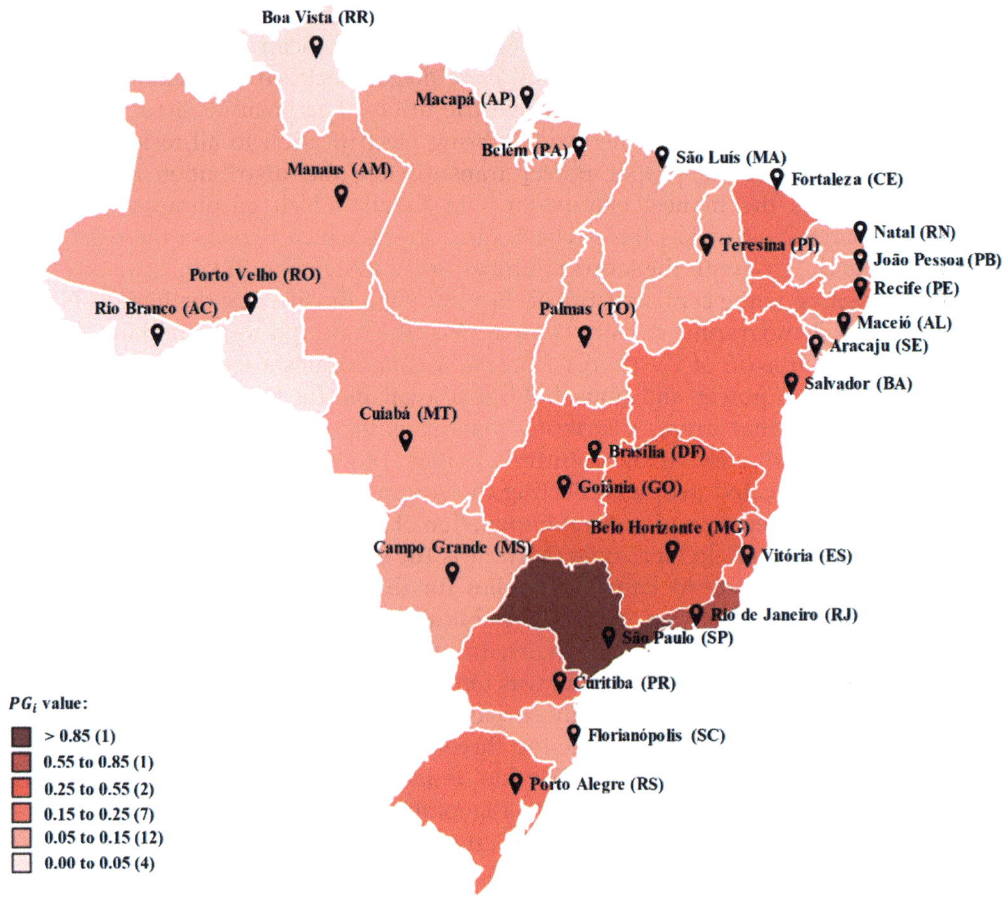

FIGURE 15.2 Map of growth potential index of DG in Brazil.

industrialization, due to its industrial development. In the second line of performance is the South region that concentrates an average potential compared to the Southeast. In the South region, two of the three capitals have potential in the fourth category of Fig. 15.2. When studying the regional climate of southern Brazil compared to the climate of other geographic regions of Brazil, it is not difficult to verify that it is considerably different: while the other regions are characterized by a warm (tropical) climate, the southern region has a temperate Mesothermal climate. Thus, this region is characterized by cold air masses from the south of the continent, which cause a sharp decline in average air temperatures, with values below 22ºC. Electricity consumption data show a significant portion of increased consumption in the cold period of this region, especially in the residential sector. This analysis associated with the high cost of electricity and GDP per capita drives the dispersion potential of distributed PV energy.

However, all capitals in the Northeast region are known for their tourism potential, but only three have potential in the fourth category of the Northeast (Fig. 15.2): Salvador (BA),

Recife (PE), and Fortaleza (CE). This is because these capitals not only developed with tourism but also have good indices of industrialization and higher education, which increased the number of people with graduates and with a class of income per household greater than five minimum wages. Then there is the Midwest region that has two of the four capitals in categories 3 and 4 of Fig. 15.2, Goiania (GO) and Brasília (DF). Finally, according to the above analysis, the region with the lowest potential indices is the North.

The investigation sequence is based on the evaluation of the factor with the highest correlation with PG_i. For this, the correlation between the capitals database and the vector of PG_i was calculated using Pearson's correlation coefficient. The Table 15.5 presents the correlation matrix between the 11 factors plus the PG_i. Values considered significant

TABLE 15.5 Correlation matrix among factors and $\mathbf{PG}_i$.

	GDP per capita	Instruction level	Territorial area	Demographic density	HDI	Class of Income per household	Solar Irradiation	Energy tariff	Number of PV Systems	Annual Energy Demand	P G
Population	0.458 p=.016	0.993 p=0.00	-0.144 p=.475	0.556 p=.003	0.236 p=.237	0.985 p=0.00	-0.223 p=.264	-0.056 p=.780	0.424 p=.027	0.987 p=0.00	0.968 p=.000
GDP per capita		0.516 p=.006	-0.060 p=.765	0.144 p=.474	0.755 p=.000	0.544 p=.003	-0.361 p=.064	-0.226 p=.257	0.396 p=.041	0.485 p=.010	0.578 p=.002
Instruction level			-0.153 p=.445	0.543 p=.003	0.310 p=.115	0.997 p=0.00	-0.263 p=.185	-0.082 p=.683	0.441 p=.021	0.992 p=0.00	0.978 p=.000
Territorial area				-0.442 p=.021	-0.358 p=.067	-0.151 p=.453	-0.365 p=.061	0.076 p=.708	-0.315 p=.110	-0.132 p=.512	-0.236 p=.237
Demographic density					0.249 p=.210	0.535 p=.004	0.246 p=.216	-0.190 p=.341	0.350 p=.074	0.511 p=.006	0.627 p=.000
HDI						0.346 p=.077	-0.276 p=.163	-0.278 p=.160	0.422 p=.028	0.290 p=.143	0.411 p=.033
Class of Income per household							-0.289 p=.143	-0.079 p=.695	0.454 p=.017	0.991 p=0.00	0.979 p=.000
Solar Irradiation								-0.062 p=.757	0.203 p=.310	-0.292 p=.140	-0.163 p=.418
Energy tariff									0.260 p=.190	-0.083 p=.679	-0.073 p=.719
Number of PV Systems										0.383 p=.048	0.565 p=.002
Annual Energy Demand											0.962 p=.000

(P-value>.05) are bold, and red bold are those considered to have strong correlation (Pearson's coefficient > 0.8).

Among the 11 factors that make up PG_i, four have a strong and significant correlation. The larger the population, the higher is the number of people with higher education levels and the number of people with a class of income greater than five minimum wages, because these factors are correlated with each other. Thus these three factors, plus the annual demand for electricity, are the factors that most impact the PG_i. Four other factors were considered significant but with moderate to weak correlation, such as number of PV systems in the capital, GDP per capita, HDI, and demographic density.

The calculation of the PG_i of the Brazilian capitals allows us to discuss if the potential is being exploited. Thus capitals were ranked by number of PV systems to compare with the ranking of ranking PG_i. Considering that the ranking positions are ideally equivalent, the difference between the rankings was calculated. In this case, it was possible to make the judgment of the current situation of the capitals. If the position in the PG_i ranking is better than the position of the Number PV systems ranking, the potential is not being explored (judgment ↓); if the position in the PG_i ranking is equivalent to the position of the Number PV systems ranking in up to three positions, the potential is being sufficiently exploited (judgment →); and if the position in the PG_i ranking is worse than the position of the number PV systems ranking, the potential is being explored (judgment ↑). Table 15.6 presented the results.

The data of Table 15.6 shows that the modeling results do not necessarily indicate the ranking of capitals with highest installed power, but the ranking of capitals that have the best ecosystem for development and investment in PV energy, when having the factors used in the construction of modeling. Of the 27 capitals, 12 are tapping their potential in a balanced way (→), eight are spreading more than their potential ↑, and only five are not tapping their potential (↓).

In general, the result of mathematical modeling to calculate the PG_i of the Brazilian capitals enables the creation of actions to expand the number of PV systems installations in cities with high market potential. An example of this is the motivation for the use of PV integration technology and architecture in buildings, an active solar strategy aligned with current international sustainability and renewable energy criteria for buildings, which is essential for the diffusion of PV energy in cities with the profile of São Paulo. Also, the ranking of capitals presents a list of cities with a significant tourism profile, which enables energy efficiency decisions and effective, sustainable policies to collaborate with the growing number of PV solar systems.

However, this research has proposed an identification of the potential for PV energy expansion through solar systems distributed among low-voltage consumers. Besides generating scientific support to boost political and economic incentives for the growth of PV energy in Brazil, the result also suggests that companies in the solar system installation area may benefit from the knowledge of potentially underrated locations for their business.

15.5 Conclusion

The results of the mathematical modeling and analysis presented in the study showed that although Brazil stands out in the international scenario of DG, the installation of solar

TABLE 15.6 PG_i versus aproveitamento da energia.

Capital	PG_i ranking oosition	Number photovoltaic systems ranking position	Difference between rankings positions	Judgment
São Paulo	1	9	−8	↓
Rio de Janeiro	2	1	1	→
Brasília	3	3	0	→
Belo Horizonte	4	2	2	→
Fortaleza	5	5	0	→
Curitiba	6	12	−6	↓
Recife	7	22	−15	↓
Salvador	8	17	−9	↓
Porto Alegre	9	13	−4	↓
Goiânia	10	6	4	↑
Vitória	11	23	−12	↓
Cuiabá	12	4	8	↑
Natal	13	14	−1	→
Teresina	14	7	7	↑
Manaus	15	21	−6	↓
Aracaju	16	11	5	↑
Campo Grande	17	8	9	↑
João Pessoa	18	18	0	→
Belém	19	19	0	→
Palmas	20	10	10	↑
Florianópolis	21	16	5	→
São Luis	22	15	7	↑
Maceió	23	20	3	→
Porto Velho	24	24	0	→
Rio Branco	25	25	0	→
Boa Vista	26	27	−1	→
Macapá	27	26	1	→

PV systems in low-voltage consumers has not been addressed the same way in all places with high potential. In general, the incentive for power generation through the solar source is believed to be predominantly given in regions with high solar incidence, which in fact does not happen. Most global demand continues to be primarily driven by government policy, energy tariff prices, and social aspects. Thus the mathematical modeling to measure the expansion potential of distributed PV solar installations identified, from 11 factors, a ranking of the capitals of Brazil with the greatest potential for investment in distributed PV generation. According to the proposal, a research instrument was applied to eight specialists in order to assess the degree of relative importance. Energy tariff, GDP per capita, and class of income per household factors were, respectively, the most important from the evaluators' point of view.

The result of the modeling generated a decreasing ranking of competitiveness levels, the discussion of the results stimulated scientific support to boost political and economic incentives for the growth of PV solar energy participation in the Brazilian electric matrix and, mainly, the need for public policy actions for the development of cities indicated as competitive for the installation of PV systems. The legal concession of Brazilian states to the option to exempt solar energy equipment was an intention to condition public policies on regional approaches to the development of renewable energy. This autonomy given to the states resulted in a geographically unequal diffusion of the DG in the national scenario, although the patterns cannot be explained solely by the economic profitability. Thus the modeling result contributed to the perception of the importance of energy efficiency campaigns in public buildings, dissemination of the benefits of DG to consumers by the political influences and municipal means of communication, proactive approach of electricity utilities, and regulatory and legal clarity by providing companies with facilities for solar generation systems under the shared generation modality.

The incentive for the installation of solar systems by public policies based on subsidy actions benefits society as a whole, through the impact on the percentage of renewable sources in the country's electric matrix and consequent decrease in the dependence on the water source, increased job offers in the cities, increased the purchasing power of consumers, reduction of environmental pollution rates and a consequent advancement of research in the field of solar energy. The discussion of the results stimulated the market analysis and the prospecting of clients by the companies installing PV systems contributed with new actions to expand the number of PV solar systems in southern Brazil in locations underestimated by the installation companies.

As for the recommendations, therefore, the identified capitals with high potential need to develop market policies and strategies that promote the growth of the number of decentralized energy consumers. The mathematical modeling developed allows nontrivial conclusions by data analysis and may be useful for similar evaluations in other countries aimed at understanding the dynamics of PV installations to increase the diffusion.

Acknowledgments

The authors thank INCTGD, CAPES, CNPq and FAPERGS for the financial support received for the development of this work.

References

Ackermann, T., Andersson, G., Söder, L., 2001. Distributed generation: a definition. Electr. Power Syst. Res. 57, 195–204. Available from: https://doi.org/10.1016/S0378-7796(01)00101-8.

Agência Nacional de Energia Elétrica—ANEEL, 2012. Resolução normativa 482, de Abril 17, 2012.

Agência Nacional de Energia Elétrica—ANEEL, 2015. Resolução Normativa 687, de 24 de Novembro de 2015.

Agência Nacional de Energia Elétrica—ANEEL, 2019a. BIG—Banco de informações de geração [WWW Document].

Agência Nacional de Energia Elétrica—ANEEL, 2019b. Serviço público de distribuição de energia elétrica—ANEEL [WWW Document].

Agência Nacional de Energia Elétrica—ANEEL, 2019c. Geração distribuida [WWW Document]. Agência Nac. Energ. Elétrica. <https://app.powerbi.com/view?r = eyJrIjoiZjM4NjM0OWYtN2IwZS00YjViLTllMjItN2E5MzBkN2ZlMzVkIiwidCI6IjQwZDZmOWI4LWVjYTctNDZhMi05MmQ0LWVhNGU5YzAxNzBlMSIsImMiOjR9> (accessed 16.11.19.).

Agência Nacional de Energia Elétrica—ANEEL, 2019d. Ranking das tarifas [WWW Document]. <https://www.aneel.gov.br/ranking-das-tarifas> (accessed 16.11.19.).

Aliyu, A.K., Modu, B., Tan, C.W., 2018. A review of renewable energy development in Africa: a focus in South Africa, Egypt and Nigeria. Renew. Sustain. Energy Rev. 81, 2502–2518. Available from: https://doi.org/10.1016/J.RSER.2017.06.055.

Araújo, K., Boucher, J.L., Aphale, O., 2019. A clean energy assessment of early adopters in electric vehicle and solar photovoltaic technology: geospatial, political and socio-demographic trends in New York. J. Clean. Prod. 216, 99–116. Available from: https://doi.org/10.1016/J.JCLEPRO.2018.12.208.

Brasil, 2004. 2004—Decretos—Portal da legislação [WWW Document].

Brummer, V., 2018. Of expertise, social capital, and democracy: assessing the organizational governance and decision-making in German renewable energy cooperatives. Energy Res. Soc. Sci. 37, 111–121. Available from: https://doi.org/10.1016/J.ERSS.2017.09.039.

Camilo, H.F., Udaeta, M.E.M., Veiga Gimenes, A.L., Grimoni, J.A.B., 2017. Assessment of photovoltaic distributed generation—issues of grid connected systems through the consumer side applied to a case study of Brazil. Renew. Sustain. Energy Rev. 71, 712–719. Available from: https://doi.org/10.1016/J.RSER.2016.12.099.

Carli, R., Dotoli, M., Pellegrino, R., 2018. Multi-criteria decision-making for sustainable metropolitan cities assessment. J. Environ. Manage 226, 46–61. Available from: https://doi.org/10.1016/J.JENVMAN.2018.07.075.

Carstens, D.D., dos, S., Cunha, S.K. da, 2019. Challenges and opportunities for the growth of solar photovoltaic energy in Brazil. Energy Policy 125, 396–404. Available from: https://doi.org/10.1016/J.ENPOL.2018.10.063.

Crowe, J.A., Li, R., 2020. Is the just transition socially accepted? Energy history, place, and support for coal and solar in Illinois, Texas, and Vermont. Energy Res. Soc. Sci. 59, 101309. Available from: https://doi.org/10.1016/J.ERSS.2019.101309.

Daza, E.F.B., 2018. Determinação de Arranjos Regulatórios e Econômicos Para Viabilizar Investimentos em Sistemas de Armazenamento de Energia em Redes de Distribuição de Energia Elétrica. Universidade Federal de Santa Maria.

dos Santos, L.L.C., Canha, L.N., Bernardon, D.P., 2018. Projection of the diffusion of photovoltaic systems in residential low voltage consumers. Renew. Energy 116, 384–401. Available from: https://doi.org/10.1016/J.RENENE.2017.09.088.

EPE, 2017. Projeção da demanda de energia elétrica.

EPE, 2018. Anuário Estatístico de Energia Elétrica [WWW Document]. <http://epe.gov.br/sites-pt/publicacoes-dados-abertos/publicacoes/PublicacoesArquivos/publicacao-160/topico-168/Anuario2018vf.pdf> (accessed 20.11.19.).

Fadly, D., Fontes, F., 2019. Geographical proximity and renewable energy diffusion: an empirical approach. Energy Policy 129, 422–435. Available from: https://doi.org/10.1016/J.ENPOL.2019.02.034.

Faria, H., Trigoso, F., Cavalcanti, J., 2017. Review of distributed generation with photovoltaic grid connected systems in Brazil: challenges and prospects. Renew. Sustain. Energy Rev. 75, 469–475. Available from: https://doi.org/10.1016/j.rser.2016.10.076.

Flowers, M.E., Smith, M.K., Parsekian, A.W., Boyuk, D.S., Mcgrath, J.K., Yates, L., 2016. Climate impacts on the cost of solar energy. Energy Policy 94, 264–273. Available from: https://doi.org/10.1016/j.enpol.2016.04.018.

Gava Gastaldo, N., Rediske, G., Donaduzzi Rigo, P., Brum Rosa, C., Michels, L., Mairesse Siluk, J.C., 2019. What is the profile of the investor in household solar photovoltaic energy systems? Energies 12. Available from: https://doi.org/10.3390/en12234451.

George, A., Boxiong, S., Arowo, M., Ndolo, P., Chepsaigutt-Chebet, Shimmon, J., 2019. Review of solar energy development in Kenya: opportunities and challenges. Renew. Energy Focus. 29, 123–140. Available from: https://doi.org/10.1016/J.REF.2019.03.007.

Gucciardi Garcez, C., 2017. Distributed electricity generation in Brazil: an analysis of policy context, design and impact. Util. Policy 49, 104–115. Available from: https://doi.org/10.1016/J.JUP.2017.06.005.

IBGE, 2019. IBGE CIDADES [WWW Document]. Indicadores Munic. <https://cidades.ibge.gov.br/> (accessed 26.11.19.).

Keshav, S., Rosenberg, C., 2010. How internet concepts and technologies can help green and smarten the electrical grid. In: Proceedings of the First ACM SIGCOMM Workshop on Green Networking, Green Networking'10. ACM, New York, NY, pp. 35–40 <https://doi.org/10.1145/1851290.1851298>.

Khoodaruth, A., Oree, V., Elahee, M.K., Clark, W.W., 2017. Exploring options for a 100% renewable energy system in Mauritius by 2050. Util. Policy 44, 38–49. Available from: https://doi.org/10.1016/J.JUP.2016.12.001.

Li, X., Lin, A., Young, C.-H., Dai, Y., Wang, C.-H., 2019. Energetic and economic evaluation of hybrid solar energy systems in a residential net-zero energy building. Appl. Energy 254, 113709. Available from: https://doi.org/10.1016/J.APENERGY.2019.113709.

Lopes, Y., Fernandes, N., Muchaluat-Saade, D., 2015. Geração Distribuída de Energia: Desafios e Perspectivas em Redes de Comunicação. In: XXXIII Simpósio Brasileiro de Redes de Computadores e Sistemas Distribuídos. Vitória.

Mah, D.N., Wang, G., Lo, K., Leung, M.K.H., Hills, P., Lo, A.Y., 2018. Barriers and policy enablers for solar photovoltaics (PV) in cities: perspectives of potential adopters in Hong Kong. Renew. Sustain. Energy Rev. 92, 921–936. Available from: https://doi.org/10.1016/J.RSER.2018.04.041.

Padmanathan, K., Govindarajan, U., Ramachandaramurthy, V.K., Rajagopalan, A., Pachaivannan, N., Sowmmiya, U., et al., 2019. A sociocultural study on solar photovoltaic energy system in India: stratification and policy implication. J. Clean. Prod. 216, 461–481. Available from: https://doi.org/10.1016/J.JCLEPRO.2018.12.225.

Palm, A., 2016. Local factors driving the diffusion of solar photovoltaics in Sweden: a case study of five municipalities in an early market. Energy Res. Soc. Sci. 14, 1–12. Available from: https://doi.org/10.1016/J.ERSS.2015.12.027.

Paris, W., Takigawa, F., Aranha Neto, E.A.C., Fernandes, R.C., 2018. Levantamento da Geração Distribuída Compartilhada no Brasil. In: VII Congresso Brasileiro de Energia Solar. Gramado.

Pepermans, G., Driesen, J., Haeseldonckx, D., Belmans, R., D'haeseleer, W., 2005. Distributed generation: definition, benefits and issues. Energy Policy 33, 787–798. Available from: https://doi.org/10.1016/J.ENPOL.2003.10.004.

Pereira, E.B., Martins, F.R., Gonçalves, A.R., Costa, R.S., Rutter, R., Abreu, S.L., et al., 2017a. Atlas Brasileiro de Energia Solar, second ed. INPE, São José dos Campos.

Pereira, E.B., Martins, F.R., Rüther, R., Abreu, S.L. de, 2017b. Atlas Brasileiro de Energia Solar, second ed. São José dos Campos.

Picciariello, A., Vergara, C., Reneses, J., Frías, P., Söder, L., 2015. Electricity distribution tariffs and distributed generation: quantifying cross-subsidies from consumers to prosumers. Util. Policy 37, 23–33. Available from: https://doi.org/10.1016/J.JUP.2015.09.007.

REN21, 2018. Renewables 2018—Global Satus Report. França.

Richter, M., 2013. German utilities and distributed PV: how to overcome barriers to business model innovation. Renew. Energy 55, 456–466. Available from: https://doi.org/10.1016/J.RENENE.2012.12.052.

Rigo, P.D., Siluk, J.C.M., Lacerda, D.P., Rosa, C.B., Rediske, G., 2019. Is the success of small-scale photovoltaic solar energy generation achievable in Brazil? J. Clean. Prod. 240, 118–243. Available from: https://doi.org/10.1016/j.jclepro.2019.118243.

Rosa, C.B., Wendt, J.F.M., Chaves, D.M.S., Thomasi, V., Michels, L., Siluk, J.C.M., 2020. Mathematical modeling for the measurement of the competitiveness index of Brazil south urban sectors for installation of photovoltaic systems. Energy Policy 136, 111048. Available from: https://doi.org/10.1016/j.enpol.2019.111048.

Saaty, R.W., 1987. The analytic hierarchy process—what it is and how it is used. Math. Model. 9, 161–176. Available from: https://doi.org/10.1016/0270-0255(87)90473-8.

Şengül, Ü., Eren, M., Eslamian Shiraz, S., Gezder, V., Sengül, A.B., 2015. Fuzzy TOPSIS method for ranking renewable energy supply systems in Turkey. Renew. Energy 75, 617–625. Available from: https://doi.org/10.1016/j.renene.2014.10.045.

Silva, J.L.S., Cavalcante, M.M., Machado, R., Silva, M.R., Delgado, D.B.M., 2018. Análise do avanço da geração distribuída no Brasil. In: VII Congresso Brasileiro de Energia Solar. Gramado.

Soccol, F., Pereira, L.A., Celeste, W.C., Coura, D.J.C., Chaves, G.L.D., 2016. Desafios para implementação da geração distribuída de energia no Brasil: Uma revisão integrativa da literatura. Braz. J. Prod. Eng. 2, 31–43

Tanaka, K., Sekito, M., Managi, S., Kaneko, S., Rai , V., 2017. Decision-making governance for purchases of solar photovoltaic systems in Japan. Ener. Pol. 111, 75–84. Available from https://doi.org/10.1016/j.enpol.2017.09.012.

CHAPTER

16

Trend of energy consumption in developing nations in the last two decades: a case study from a statistical perspective

Anshuman Dey Kirty

Vellore Institute of Technology, Vellore, India

16.1 Introduction

There are many sources of energy across the globe, but they can mainly be categorized into two: renewable and nonrenewable. The renewable sources of energy are also known as "clean" or "sustainable" energy, which is the energy that is generated by renewable sources that are naturally replenished on a human timescale. These include energy mainly received from sun (solar energy), wind, hydropower, geothermal, and bioenergy. The environmental advantages of using clean energy is immense, but the solutions using renewable sources are sometimes inefficient, expensive to harness, or have a few other disadvantages. Nonrenewable energy on the other hand cannot be replenished again for many hundreds of thousands of years. The sources for nonrenewable sources primarily are fossil fuels such as coal, petroleum, and natural gas. Carbon is one of the main elements in the fossil fuel. At this point fossil fuels are readily available and easily processed to be used as an efficient source of energy but pose a serious threat to the environment by the emission of various harmful gases. Energy consumption can be divided into four categories on a high level, as given by the Energy Information Administration of the US Department of Energy.

- Residential
- Commercial
- Industrial users
- Transportation

Predictive Modelling for Energy Management and Power Systems Engineering
DOI: https://doi.org/10.1016/B978-0-12-817772-3.00016-1

The residential consumptions of energy include running lighting, heating, and other appliances mainly in the form of electricity. Energy is commercially used for the same purpose as its residential uses but mostly in higher volumes and to indirectly gain profits. Industrial users use the energy in their daily process or business, such as in agriculture, mining, construction, and manufacturing. The last sector is the energy utilized for transportation of people, products, and goods. The United States consumes energy mostly in the industrial sector, whereas transport is the main area of energy consumption in the European Union.

The growth of all countries has been powered by energy produced by coal, petroleum, natural gas, and hydroelectric power plants. The consuption of energy has increased in the developing countries (India, China, Brazil, Egypt, and South Africa), whereas in the developed countries like the United States the consumption rates have remained constant for a few years. The primary total consumption in the United States can be seen in Table 16.1 to be almost constant from the years 2015–2019.

This case study starts with study of the behavior of the world by concentrating on the major sources of data energy. The trends of energy sources from the 1900s to 2000s have changed in terms of volume and the type. The major focus of this study will be on the following developing nations (Fig. 16.1):

- Brazil
- Chile
- China
- Egypt
- India
- South Africa

These nations were chosen as they represent several continents and cover various geographies. We will discuss the responsible nations that are pushing for the renewable resources by increasing the percentage of use of renewable sources of energy.

The prediction of the energy consumption for the developing countries is based on time series analysis.

$$y(t) = g(t) + s(t) + h(t) + \varepsilon_t$$

TABLE 16.1 Energy consumption by sector.

	Approximation of energy consumption (trillion Btu)				
January to July	**2019**	**2018**	**2017**	**2016**	**2015**
Residential	14,000	14,000	13,000	14,000	14,000
Commercial	12,000	12,000	12,000	12,00	12,000
Industrial	21,000	22,000	21,000	21,000	21,000
Transportation	18,000	19,000	19,000	19,000	18,000
Primary total	67,000	68,000	65,000	65,000	66,000

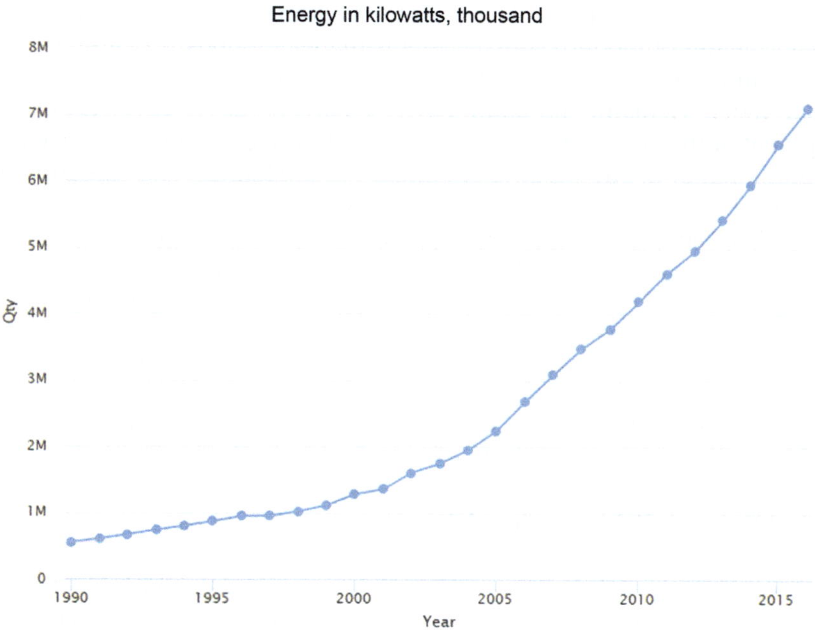

FIGURE 16.1 Energy consumption trend in China.

The above time series is decomposable with three main model components: trend, seasonality, holidays.

$g(t)$ is the trend function for the nonperiodic changes; $s(t)$ is the seasonality component that changes periodically (weekly, yearly); $h(t)$ accounts for the irregularities or the "holidays"; and ε_t is the error term representing idiosyncratic changes that are not accounted for in the model (assuming that the ε_t is normally distributed). Prophet uses time as the regressor and fits several linear and nonlinear functions of time as components.

16.2 Related work

Many countries have introduced regulations that target the residential sector to reduce energy consumption. This paper analyses how stricter regulations and standards lead to lowering of residential energy consumption. Studies have been made on electricity and nonelectricity energy consumption, which is mainly in household appliances (Aydin and Brounen, 2019; Chen et al., 2013). Economic growth has come at the cost of the degradation of the environment with the rapid generation of energy to meet the needs. The sustainable growth can only be achieved when more focus is given toward the renewable sources of energy and more investments are directed to research (Kahia et al., 2019). The increase in efficiency of energy usage is one of the reasons for China's decoupling effort but India has concentrated more on the advancement of technology. The results show that

China has performed better in the decoupling efforts than India (Wang et al., 2019). Fuel switching and the use of modern cooking fuel is the way to look forward. The use of hard fuel causes pollution but is still highly used in the rural areas in countries like Ghana, Brazil, Nepal, India, Vietnam, and South Africa (Heltberg, 2004). Foreign investments and the development of the economy have a direct effect on the greenhouse gas emissions and energy consumption in the developing countries. Discussions on Kuznets curve hypothesis show that it is effective for China and Indonesia. China, India, Iran, Indonesia, and South Africa show that the pollution haven hypothesis is valid. Modern and clean energy sources are the only way forward for sustainable growth with the environment in mind (Sarkodie and Strezov, 2019). In many countries it has been noticed that the energy preservation strategies did not have an impact on the trade and industrial growth of the emerging countries. The energy conservation policies just made it more welcoming for the renewable sources of energy, which has immense positive monetary and environmental impact (Ozcan and Ozturk, 2019). In South Africa findings suggest that at a high pace of trade and industry growth there is a lower increase in energy depletion. This suggeste a weakening in energy power while certifying energy productivity in South Africa (Bekun et al., 2019). ANN and SVM were used for the prediction of electric energy usage using the weather data (Torabi et al., 2019). Reduction in energy intensity is one of the most important reasons for less emissions of CO_2 in energy intense countries. In the industry sector phasing out superfluous size helps reduce emissions in China. Economic policies also are a significant means used to encourage the development of China's low-carbon economy in the upcoming years (Ma et al., 2019). Many policies and regulations are made around the predicted values for energy consumption in the coming years. A forecast of energy consumption has been done and is compared to the estimations made by the EIA (US Energy Information Administration) for the countries United States, India, Brazil, South Korea, Japan, Mexico, and Japan. The estimations made in the research are found to be more than the EIA estimations (Chang et al., 2012). The research is based on projections of the quantity of fossil fuels by country. The projections indicate that the fossil fuels have high growth rate till 2025 and then remain constant, after which it would start declining. World coal production in particular will peak in 2025 due to high Chinese production and only natural gas would in future show high growth (Mohr et al., 2015). The research discusses causalities between economic growth (carbon emission), economic growth (energy consumption), and energy consumption (carbon emissions). In developed and developing countries higher energy consumption has led to higher carbon emissions. In developed countries, unlike in the developing countries, carbon emission is not linked with economic development (Waheed et al., 2019). The research takes SAARC countries as the sample and finds that the rising population along with economic growth has caused growth in the CO_2 emissions. Green renewable energy should be considered as a long-term alternative to overcome the issues of emissions caused by fossil fuels like coal and natural gas (Rehman and Rashid, 2017). Trend analysis and projections are used to model the CSE (consumption of energy sources). The research and the study shows that well-developed models can be used for projections of primary energy consumption by the sources for future planning (Aydin, 2015). These research papers show the methodology that can be applied for big data engineering and how machine learning can be implemented and modeled for various use cases (Roy et al., 2018c; Samui et al., 2019; Roy et al.,

2016a,b). The implementation of deep learning and optimization algorithms can be studied to apply to the energy data so that optimization can be used on a global level to reduce and efficiently manage the energy that is being generated using the nonrenewable resources and is therefore harmful for the environment (Balas et al., 2019; Kim et al., 2018). The recent advancement in the neural networks, KNN, and Random forest algorithms can be leveraged in the energy sector to study the impacts on the environment. The growth of renewable and nonrenewable energy is disproportionate as renewable resources are erratic and not much focused research has been completed (Jayabalan et al., 2018; Roy et al., 2018a,b, 2020; Biswas et al., 2020). The energy consumption at a provincial level was broken down and studied. There has been a gradual decrease in the interrelationship between China's trade and industry growth and energy consumption progression. The western province has become the area of China with the maximum energy consumption (Liu et al., 2019). Cleaner sources of energy can be very well utilized and are being studied highly and implemented across the globe (Noel et al., 2018; Peng and Wild, 2017; Rahmani and Al-Sallal, 2020; Böttiger et al., 2016). Energy optimization and predictions based on clustering can be a great use case in the future at a large scale. The algorithms that are being studied can also be used in the field of energy (Dudin et al., 2019; Strielkowski et al., 2019; Roy et al., 2013, 2014, 2012, 2017; Kaul et al., 2016; Tiwari et al., 2012). The efficiency of energy received and then utilized is of importance; the better the optimization and utilization of energy, then less energy needs to be produced. Efficiency of energy can be improved by changes in the distribution grid, efficiency certification of machines in manufacturing, and improving household appliances for better output (Baysan et al., 2019; Zhao et al., 2019; Yang and Wei 2019; Matraeva et al., 2019; Khalil et al., 2019; Cai et al., 2019; Carlton et al., 2019; Yao et al., 2013).

16.3 Implementation

16.3.1 Data sources

The energy data collected is from the UN and IEA datastats website. The data tables were also extracted from the public opendatasoft site.

The data features are:
- Country or Area
- Commodity or Transaction
- Year
- Unit Quantity

World Energy
- Year
- Coal (unit—Kilo Tonne of Oil Equivalent (ktoe))
- Natural Gas (unit—Kilo Tonne of Oil Equivalent (ktoe))
- Nuclear (unit—Kilo Tonne of Oil Equivalent (ktoe))
- Hydro (unit—Kilo Tonne of Oil Equivalent (ktoe))
- Wind Solar etc. [unit—Kilo Tonne of Oil Equivalent (ktoe)]

- Biofuel and Waste [unit—Kilo Tonne of Oil Equivalent (ktoe)]
- Oil [unit—Kilo Tonne of Oil Equivalent (ktoe)]

Added/Derived Feature

- Renewable or nonrenewable feature based on type of energy source

16.3.2 Data exploration

The use of coal for energy production saw a rapid growth in the years 2010–15 around the world. After which there is a slight dip and it might reduce even further. The natural gas has a steady growth rate from the 1990s and is still continuing. The graph in Fig. 16.2 clearly shows that oil has been the major source of energy in the world from the start of the 1990s and is still the case even today. The wind and solar (i.e., renewable) sources of energy were almost negligible in the 1990s but have shown some growth from 2010. This growth can be attributed to the countries wanting to adopt a more sustainable and environment-friendly growth in the economy by utilizing renewable resources.

The exact quantities of the sources of energy are shown in Tables 16.2 and 16.3, along with the percentage contribution. The two tables show a comparison of data that include energy source quantities for the years 1990–2016 and just for the year 2016.

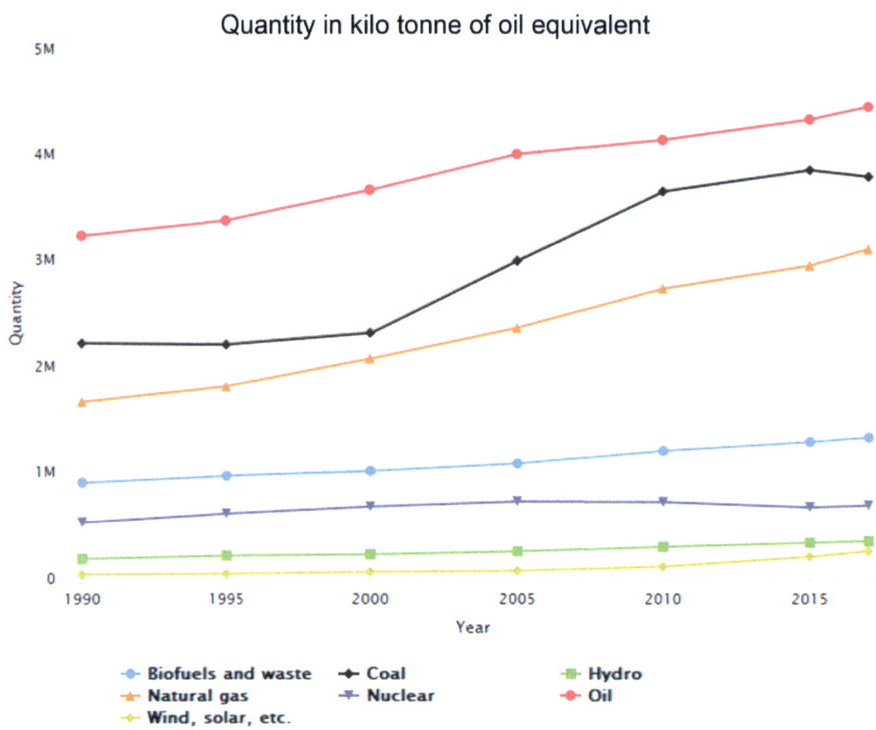

FIGURE 16.2 Trend of world energy sources.

TABLE 16.2 Percentage quantity of type of energy source 1990–2016.

Sr. no.	Type	Quantity (KTOE)	Percentage quantity (%)
1	Oil	27,181,884	34.01
2	Coal	21,034,412	26.32
3	Natural gas	16,694,006	20.89
4	Biofuels and waste	7,778,236	9.73
5	Nuclear	4,607,400	5.76
6	Hydro	1,856,576	2.32
7	Wind, solar	780,106	0.98

TABLE 16.3 Percentage quantity of type of energy source 2016.

Sr. no.	Type	Quantity (KTOE)	Percentage quantity (%)
1	Oil	4,449,499	31.85
2	Coal	3,789,934	27.13
3	Natural gas	3,106,799	22.24
4	Biofuels and waste	1,329,064	9.51
5	Nuclear	687,481	4.92
6	Hydro	351,029	2.51
7	Wind, solar	256,830	1.84

The data shows that the renewable energy sources have a higher quantity as a contribution to the whole energy sector in the year 2016. The percentage increase of the renewable resources is not the rapid level that is required to reduce depletion of the nonrenewable resources. It can be said that the renewable resources like the wind and solar energies have not yet been able to replace the energy requirements met by the oil, coal, and natural gas. There is a long way to go in terms of productionizing the renewable sources of energy at a large scale.

16.3.3 Analysis on Developing countries

The nations were carefully chosen based on their varied size, geographic location, and most importantly they were all developing countries. These are the countries that in the future will be the "developed nations," setting standards for the next set of developing countries. That is one of the main reasons why developing countries were used for this analysis (Fig. 16.3).

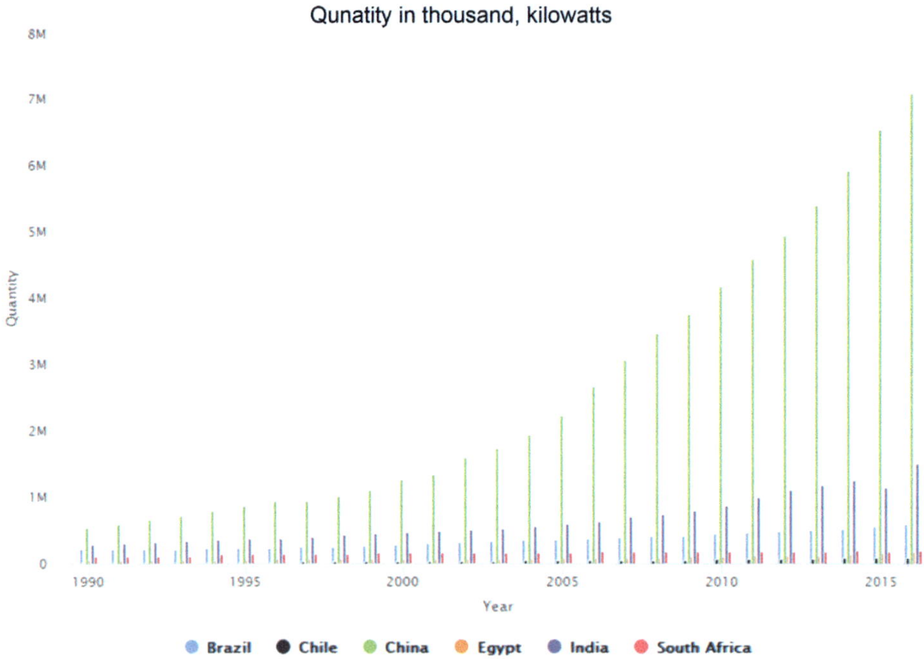

FIGURE 16.3 Trend of developing countries energy production.

Nations considered in this study:

- India
- China
- Brazil
- Chile
- South Africa
- Egypt

1. China produces more energy than India, South Africa, Chile, Egypt, and Brazil combined.
2. Energy production by South Africa has not increased at a notable pace.
3. India and China show an exponential increase in the energy that is being produced by the two countries.
4. The energy produced by China has a huge impact on the global numbers as the volumes are very high.

Energy produced overall increases year by year but the increase in the renewable sources of energy is quite slow when compared to the nonrenewable sources of energy, as shown by Fig. 16.4. This is not a good sign for the environment regarding more sustainable growth. Countries ought to make policies that place the use of renewable sources of energy and the transformation from the nonrenewable to renewable sources of energy at the core of decision-making (Tables 16.4 and 16.5).

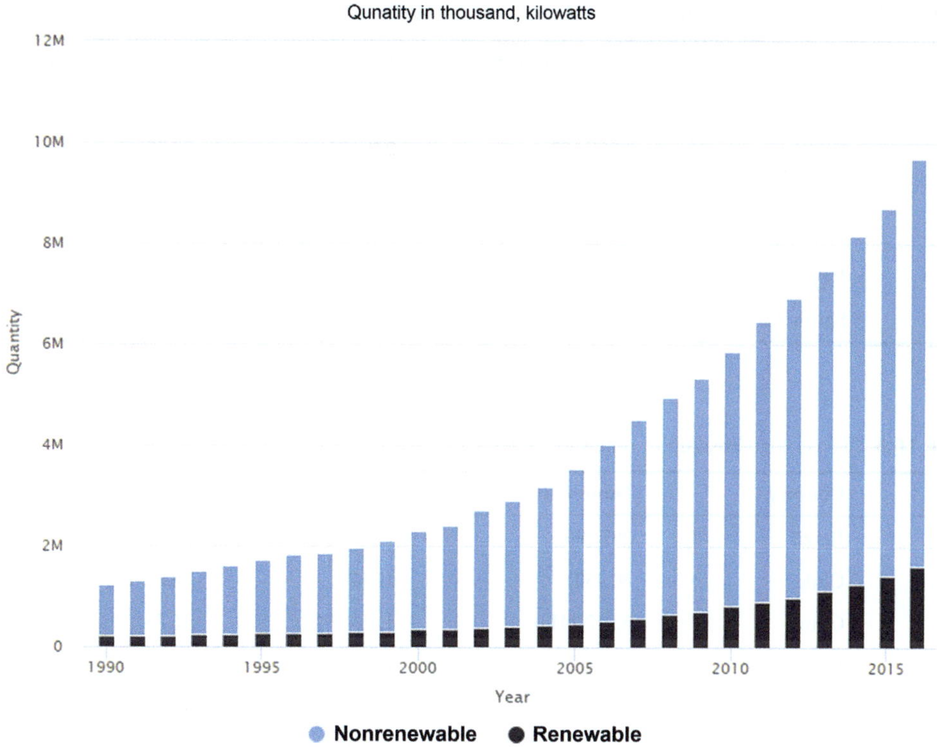

FIGURE 16.4 Trend of renewable and nonrenewable energy.

- Brazil has the highest percentage of renewable sources of energy production from 1990–2016.
- Though China has huge volumes of energy production, its usage of renewable source of energy is 13.30% which is quite significant compared to many other countries. A push for higher cleaner sources of energy in China could have much more impact than in a country like Chile. The total volume of energy production is significantly less in Chile when compared to that of India or China.
- South Africa showed the least use of clean sources of energy like solar and wind in the last two decades.
- The recent trends show that South Africa is currently increasing its renewable sources of energy at a high pace.
- Egypt on the other hand has shown signs of slowing the increase in renewable sources of energy production.

The countries could impact the overall numbers significantly if governments make policy and regulations against overuse of nonrenewable source of energy. The contribution of Brazil to the hydroenergy and renewable energy output is more than that of China, even though China has a huge volume of energy productions, as shown in Table 16.6. This

TABLE 16.4 Percentage breakup of renewable energy for 1990–2016.

Sr. no.	Country	Nonrenewable	Renewable	Percentage renewable (%)
1	Brazil	6,306,874	3,945,740	38.49
2	Chile	1,027,152	271,132	20.88
3	China	60,849,219.6	9,331,094	13.30
4	India	15,810,656	2,082,478	11.64
5	Egypt	2,237,098	164,502	6.85
6	South Africa	4,212,308	74,552	1.74

TABLE 16.5 Percentage breakup of renewable energy for 2016.

Sr. no.	Country	Nonrenewable	Renewable	Percentage renewable (%)
1	Brazil	387,396	214,292	35.62
2	Chile	80,450	17,670	18.01
3	China	5,912,241.6	1,189,692	16.75
4	Egypt	175,406	7374	4.03
5	India	1,323,328	185,160	12.27
6	South Africa	181,070	12,496	6.46

shows that Brazil's government and the industries are very aware of long-term economic and environmental development. India too has a high renewable energy consumption. Total primary energy supply and consumption is very high, at almost 83% of all the developing nations taken into consideration during our study.

16.3.4 Prediction using proposed models

For the data engineering and wrangling, libraries like Tidyverse and Dplyr were extensively used in the R programming language. The visualizations were made using the Highcharter package.

To predict the energy consumption for the coming years in the developing countries we have used time series analysis. The programing language of R was used for the analysis along with the time series prediction library Prophet.

Prophet library was developed by Facebook and made open source for the world, to enhance time series analysis at scale. Prophet uses trends, seasonality, and irregularities and models them into an equation.

Yearly seasonality was used for the model as the data points used are on a yearly basis.

TABLE 16.6 Energy consumption and output indicator.

Sr. no.	Indicator name	China quantity	Brazil quantity	India quantity	South Africa quantity	Egypt quantity	Chile quantity
1	Total primary energy supply (TJ)	62.99	8.36	20.53	4.99	2.11	1.03
2	Total final consumption (TJ)	62.29	9.73	20.92	3.70	2.16	1.19
3	Total final energy consumption (TFEC)	62.84	9.59	20.71	3.64	2.05	1.16
4	Renewable energy consumption (TJ)	49.02	13.97	33.35	2.00	0.49	1.17
5	Total electricity output (GWh)	61.35	10.60	17.69	6.56	2.55	1.24
6	Hydroenergy consumption (TJ)	40.50	44.61	9.99	0.19	1.78	2.94
7	Renewable energy electricity output (GWh)	42.08	41.02	12.11	0.21	1.71	2.88

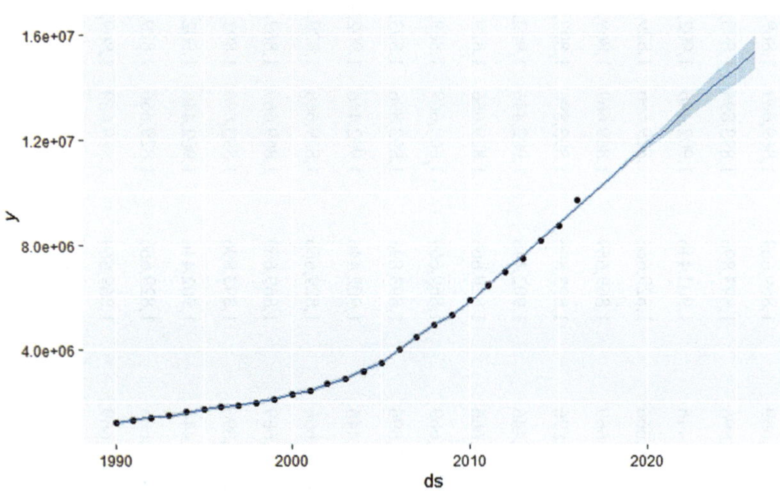

FIGURE 16.5 Projection of energy consumption to 2026.

16.3.5 Findings and output

The model was built using the Prophet library *prophet* function and the *predict* function to predict the quantities of energy production from renewable and nonrenewable sources of energy.

In the projection diagram below the quantity of energy in kilo tonne of oil equivalent (ktoe) is shown on the y axis and the *ds* represents the year on the x axis (Fig. 16.5).

The line is very well fit and the model has projected the values very accurately. The projections of the energy quantity can be done based on the line projection (Table 16.7).

TABLE 16.7 Projection output of prophet until 2026.

DS	Additive terms	Additive terms Lower	Additive Terms upper	Yearly	Yearly Lower	Yearly Upper	Yha Lower	Yhat Upper	Trend Lower	Trend Upper	Yhat
01–01–1990	1,869,669	1,869,669	1,869,669	1,869,669	1,869,669	1,869,669	1,118,761	1,285,332	−665,628	−665,628	**1,204,041**
01–01–1991	1,893,896	1,893,896	1,893,896	1,893,896	1,893,896	1,893,896	1,234,471	1,408,948	−567456	−567,456	**1,326,440**
01–01–1992	1,902,446	1,902,446	1,902,446	1,902,446	1,902,446	1,902,446	1,352,294	1,523,514	−469,283	−469,283	**1,433,163**
01–01–1993	1,829,666	1,829,666	1,829,666	1,829,666	1,829,666	1,829,666	1,374,348	1,548,860	−370,842	−370,842	**1,458,824**
01–01–1994	1,869,669	1,869,669	1,869,669	1,869,669	1,869,669	1869669	1514637	1682620	−272,669	−272,669	**1,597,000**
01–01–1995	1,893,896	1,893,896	1,893,896	1,893,896	1,893,896	1893896	1641051	1807011	−174,497	−174,497	**1,719,400**
01–01–1996	1,902,446	1,902,446	1,902,446	1,902,446	1,902,446	1902446	1743049	1906733	−76,1783	−76,1783	**1,826,268**
01–01–1997	1,829,666	1,829,666	1,829,666	1,829,666	1,829,666	1,829,666	1,762,578	1,939,090	22,496.95	22,496.95	**1,852,163**
01–01–1998	1,869,669	1,869,669	1,869,669	1,869,669	1,869,669	1,869,669	1,910,724	2,077,621	121,145.5	121,145.5	**1,990,815**
01–01–1999	1,893,896	1,893,896	1,893,896	1,893,896	1,893,896	1,893,896	2,028,282	2,195,894	219,847.3	219,847.3	**2,113,744**
01–01–2000	1,902,446	1,902,446	1,902,446	1,902,446	1,902,446	1,902,446	2,231,617	2,399,564	413,307.5	413,307.5	**2,315,754**
01–01–2001	1,829,666	1,829,666	1,829,666	1,829,666	1,829,666	1,829,666	2,349,228	2522,301	607,617.9	607,617.9	**2,437,284**
01–01–2002	1,869,669	1,869,669	1,869,669	1,869,669	1,869,669	1,869,669	2,582,931	2,756,231	801,710.1	801,710.1	**2,671,380**
01–01–2003	1,893,896	1,893,896	1,893,896	1,893,896	1,893,896	1,893,896	2,810,223	2,973,876	995,827.4	995,827.4	**2,889,724**
01–01–2004	1,902,446	1,902,446	1,902,446	1,902,446	1,902,446	1,902,446	3,164,844	3,334,199	1,341,185	1,341,185	**3,243,631**
01–01–2005	1,829,666	1,829,666	1,829,666	1,829,666	1,829,666	1,829,666	3,431,547	3,600,567	1,687,735	1,687,735	**3,517,401**
01–01–2006	1,869,669	1,869,669	1,869,669	1,869,669	1,869,669	1,869,669	3,925,801	4,086,703	2,138,338	2,138,338	**4,008,008**
01–01–2007	1,893,896	1,893,896	1,893,896	1,893,896	1,893,896	1,893,896	4,398,996	4,569,522	2,589,155	2,589,155	**4,483,051**
01–01–2008	1,902,446	1,902,446	1,902,446	1,902,446	1,902,446	1902446	4,859,516	5,023,617	3,040,038	3,040,038	**4,942,484**
01–01–2009	1,829,666	1,829,666	1,829,666	1,829,666	1,829,666	1829666	5,243,975	5,406,528	3,492,156	3,492,156	**5,321,821**
01–01–2010	1,869,669	1,869,669	1,869,669	1,869,669	1,869,669	1869669	5,729,931	5,897,821	3,943,263	3,943,263	**5,812,932**
01–01–2011	1,893,896	1,893,896	1,893,896	1,893,896	1,893,896	1893896	6,346,557	6,516,857	4,541,311	4,541,311	**6,435,207**

01–01–2012	1,902,446	1,902,446	1,902,446	1,902,446	1,902,446	1,902,446	6,952,872	7,127,988	5,139,359	5,139,359	**7,041,805**
01–01–2013	1,829,666	1,829,666	1,829,666	1,829,666	1,829,666	1,829,666	7,479,235	7,648,590	5,739,046	5,739,046	**7,568,712**
01–01–2014	1,869,669	1,869,669	1,869,669	1,869,669	1,869,669	1,869,669	8,122,041	8,294,516	6,337,094	6,337,094	**8,206,763**
01–01–2015	1,893,896	1,893,896	1,893,896	1,893,896	1,893,896	1,893,896	8,737,701	8,910,715	6,935,142	6,935,142	**8,829,039**
01–01–2016	1,902,446	1,902,446	1,902,446	1,902,446	1,902,446	1,902,446	9,352,803	9,523,906	7,533,191	7,533,191	**9,435,637**
01–01–2017	1,829,666	1,829,666	1,829,666	1,829,666	1,829,666	1,829,666	9,873,731	10,052,767	8,118,831	8,147,397	**9,962,543**
01–01–2018	1,869,669	1,869,669	1,869,669	1,869,669	1,869,669	1,869,669	10,494,452	10,704,184	8,679,604	8,783,242	**10,600,595**
01–01–2019	1,893,896	1,893,896	1,893,896	1,893,896	1,893,896	1,893,896	11,092,581	11,357,165	9,230,604	9,426,410	**11,222,870**
01–01–2020	1,902,446	1,902,446	1,902,446	1,902,446	1,902,446	1,902,446	11,657,395	12,012,141	9,773,963	10,088,086	**11,829,468**
01–01–2021	1,829,666	1,829,666	1,829,666	1,829,666	1,829,666	1,829,666	12,118,397	12,602,305	10,306,248	10,759,599	**12,356,375**
01–01–2022	1,869,669	1,869,669	1,869,669	1,869,669	1,869,669	1,869,669	12,690,130	13,311,343	10,832,913	11,429,997	**12,994,426**
01–01–2023	1,893,896	1,893,896	1,893,896	1,893,896	1,893,896	1,893,896	13,224,158	14,030,278	11,353,822	12,123,774	**13,616,701**
01–01–2024	1,902,446	1,902,446	1,902,446	1,902,446	1,902,446	1902446	13756737	14,741,172	11,856,746	12,825,735	**14,223,300**
01–01–2025	1,829,666	1,829,666	1,829,666	1,829,666	1,829,666	1,829,666	14,211,123	15,370,249	12,375,221	13,526,962	**14,750,206**
01–01–2026	1,869,669	1,869,669	1,869,669	1,869,669	1,869,669	1,869,669	14,757,243	16,101,105	12,876,200	14,239,905	**15,388,258**

The last column *yhat* in the table above shows the projections of the quantities of energy in kilo tonne of oil equivalent (ktoe) until the year 2026.

16.4 Conclusion

This chapter has provided a glimpse of the energy production and consumption quantities across the globe with a focus on developing countries. It also projected the quantities of energy that would be used in the coming years. The study clearly shows a huge imbalance in the class of energy production sources, i.e., the world relies highly on the nonrenewable resources for its economic and infrastructure prospects. The analysis shows that energy consumption in the developing countries is continuously increasing. The economic and infrastructural growths have direct correlations with the amount of energy being produced and consumed by the countries. This is an alarming sign if the countries don't plan to improve and grow on the backbone of renewable resources. The increase in the rate of adoption of the renewable resources of energy needs to be supported by the government. Brazil has shown great signs of implementation in terms of renewable sources of energy. The CO_2 emissions have caused long-term climate change and the use of oil and natural gases has depleted the natural resources. Further studies can bring many innovative ways in which the renewable sources of energy can be utilized and the consumption of the nonrenewable resources can be optimized. The future studies could include how the nonrenewable resources can be replaced by the renewable resources. The impact of the nonrenewable sources of energy on the climate and the environment is dangerous. Further research can also be done related to why cleaner energy is not much used? The projections in this chapter can further be validated with actual data.

References

Aydin, G., 2015. The modeling and projection of primary energy consumption by the sources. Energ. Source. Part B 10 (1), 67–74.

Aydin, E., Brounen, D., 2019. The impact of policy on residential energy consumption. Energy 169, 115–129.

Balas, V.E., Roy, S.S., Sharma, D., Samui, P. (Eds.), 2019. Handbook of Deep Learning Applications, vol. 136. Springer.

Baysan, S., Kabadurmus, O., Cevikcan, E., Satoglu, S.I., Durmusoglu, M.B., 2019. A simulation-based methodology for the analysis of the effect of lean tools on energy efficiency: an application in power distribution industry. J. Clean. Prod. 211, 895–908.

Bekun, F.V., Emir, F., Sarkodie, S.A., 2019. Another look at the relationship between energy consumption, carbon dioxide emissions, and economic growth in South Africa. Sci. Total Environ. 655, 759–765.

Biswas, R., Vasan, A., Roy, S.S., 2020. Dilated deep neural network for segmentation of retinal blood vessels in fundus images. Iran J. Sci. Technol. Trans. Electr. Eng. 44, 505–518.

Böttiger, M., Paulitschke, M., Bocklisch, T., 2016. Innovative reactive energy management for a photovoltaic battery system. Energy Procedia 99, 341–349.

Cai, W., Liu, C., Lai, K.H., Li, L., Cunha, J., Hu, L., 2019. Energy performance certification in mechanical manufacturing industry: a review and analysis. Energ. Convers. Manage. 186, 415–432.

Carlton, E.J., Barton, K., Shrestha, P.M., Humphrey, J., Newman, L.S., Adgate, J.L., et al., 2019. Relationships between home ventilation rates and respiratory health in the Colorado Home Energy Efficiency and Respiratory Health (CHEER) study. Environ. Res. 169, 297–307.

Chang, Y., Lee, J., Yoon, H., 2012. Alternative projection of the world energy consumption-in comparison with the 2010 international energy outlook. Energy Policy 50, 154–160.

Chen, J., Wang, X., Steemers, K., 2013. A statistical analysis of a residential energy consumption survey study in Hangzhou, China. Energ. Buildings 66, 193–202.

Dudin, M.N., Frolova, E.E., Protopopova, O.V., Mamedov, A.A., Odintsov, S.V., 2019. Study of innovative technologies in the energy industry: nontraditional and renewable energy sources. Entrepreneur. Sustain. Issues 6 (4), 1704–1713.

Heltberg, R., 2004. Fuel switching: evidence from eight developing countries. Energy Econ. 26 (5), 869–887.

Jayabalan, J., Roy, S.S., Samui, P., Kurup, P., 2018. Intelligent models applied to elastic modulus of jointed rock mass. Handbook of Research on Trends and Digital Advances in Engineering Geology. IGI Global, pp. 1–30.

Kahia, M., Jebli, M.B., Belloumi, M., 2019. Analysis of the impact of renewable energy consumption and economic growth on carbon dioxide emissions in 12 MENA countries. Clean Technol. Environ. Policy 21 (4), 871–885.

Kaul, D., Roy, S.S., Roy, R., Barna, C., Mehta, S., Misra, A., 2016. Prediction of customer satisfaction using naive bayes, multiclass classifier, K-Star and IBK. International Workshop Soft Computing Applications. Springer, Cham, pp. 153–161.

Khalil, A.J., Barhoom, A.M., Abu-Nasser, B.S., Musleh, M.M., Abu-Naser, S.S., Energy Efficiency Predicting Using Artificial Neural Network, 2019, alazhar.edu.

Kim, D., Sekhar Roy, S., Länsivaara, T., Deo, R., Samui, P. (Eds.), 2018. Handbook of Research on Predictive Modeling and Optimization Methods in Science and Engineering. IGI Global.

Liu, H., Wang, C., Tian, M., Wen, F., 2019. Analysis of regional difference decomposition of changes in energy consumption in China during 1995–2015. Energy 171, 1139–1149.

Ma, X., Wang, C., Dong, B., Gu, G., Chen, R., Li, Y., et al., 2019. Carbon emissions from energy consumption in China: its measurement and driving factors. Sci. Total Environ. 648, 1411–1420.

Matraeva, L., Solodukha, P., Erokhin, S., Babenko, M., 2019. Improvement of Russian energy efficiency strategy within the framework of "green economy" concept (based on the analysis of experience of foreign countries). Energy Policy 125, 478–486.

Mohr, S.H., Wang, J., Ellem, G., Ward, J., Giurco, D., 2015. Fuel 141, 120–135.

Noel, L., de Rubens, G.Z., Sovacool, B.K., 2018. Optimizing innovation, carbon and health in transport: assessing socially optimal electric mobility and vehicle-to-grid pathways in Denmark. Energy 153, 628–637.

Ozcan, B., Ozturk, I., 2019. Renewable energy consumption-economic growth nexus in emerging countries: a bootstrap panel causality test. Renew. Sustainable Energy Reviews 104, 30–37.

Peng, X., Wild, J., 2017. Innovative microgrid solution for renewable energy integration within the REIDS initiative. Energy Procedia 143, 599–604.

Rahmani, M., Al-Sallal, K.A., 2020. Abu Dhabi schools: optimization of building form for energy efficiency. Renewable Energy and Sustainable Buildings. Springer, Cham, pp. 623–631.

Rehman, M.U., Rashid, M., 2017. Energy consumption to environmental degradation, the growth appetite in SAARC nations. Renew. Energy 111, 284–294.

Roy, S.S., Gupta, A., Sinha, A., Ramesh, R., 2012. Cancer data investigation using variable precision rough set with flexible classification. Proceedings of the Second International Conference on Computational Science, Engineering and Information Technology. ACM, pp. 472–475.

Roy, S.S., Viswanatham, V.M., Rawat, S.S., Shah, H., 2013. Multicriteria decision examination for electrical power grid monitoring system. 2013 7th International Conference on Intelligent Systems and Control (ISCO). IEEE, pp. 26–30.

Roy, S.S., Krishna, P.V., Yenduri, S., 2014. Analyzing intrusion detection system: an ensemble based stacking approach. 2014 IEEE International Symposium on Signal Processing and Information Technology (ISSPIT). IEEE, pp. 000307–000309.

Roy, S.S., Pratyush, C., Barna, C., 2016a. Predicting ozone layer concentration using multivariate adaptive regression splines, random forest and classification and regression tree. International Workshop Soft Computing Applications. Springer, Cham, pp. 140–152.

Roy, S.S., Sinha, A., Roy, R., Barna, C., Samui, P., 2016b. Spam email detection using deep support vector machine, support vector machine and artificial neural network. International Workshop Soft Computing Applications. Springer, Cham, pp. 162–174.

Roy, S.S., Biba, M., Kumar, R., Kumar, R., Samui, P., 2017. A new SVM method for recognizing polarity of sentiments in Twitter. Handbook of Research on Soft Computing and Nature-Inspired Algorithms. IGI Global, pp. 281–291.

Roy, S.S., Roy, R., Balas, V.E., 2018a. Estimating heating load in buildings using multivariate adaptive regression splines, extreme learning machine, a hybrid model of MARS and ELM. Renew. Sustain. Energy Rev. 82, 4256–4268.

Roy, S.S., Kulshrestha, P., Samui, P., 2018b. Classifying images of drought-affected area using deep belief network, kNN, and random forest learning techniques. Deep Learning Innovations and Their Convergence With Big Data. IGI Global, pp. 102–119.

Roy, S.S., Samui, P., Deo, R., Ntalampiras, S., 2018c. Big Data in Engineering Applications, vol. 44. Springer.

Roy, S.S., Samui, P., Nagtode, I., Jain, H., Shivaramakrishnan, V., Mohammadi-ivatloo, B., 2020. Forecasting heating and cooling loads of buildings: a comparative performance analysis. J. Amb. Intel. Hum. Comp. 11 (3), 1253–1264.

Samui, P., Kim, D., Jagan, J., Roy, S.S., 2019. Determination of uplift capacity of suction caisson using gaussian process regression, minimax probability machine regression and extreme learning machine. Iran. J. Sci. Technol. Trans. Civil Eng. 43 (1), 651–657.

Sarkodie, S.A., Strezov, V., 2019. Effect of foreign direct investments, economic development and energy consumption on greenhouse gas emissions in developing countries. Sci. Total Environ. 646, 862–871.

Strielkowski, W., Volkova, E., Pushkareva, L., Streimikiene, D., 2019. Innovative policies for energy efficiency and the use of renewables in households. Energies 12 (7), 1392.

Tiwari, S., Roy, S.S., Viswanatham, V.M., 2012. Using rough sets for optimal cost evaluation in supply chain management. 2012 International Conference on Computing, Communication and Applications. IEEE, pp. 1–5.

Torabi, M., Hashemi, S., Saybani, M.R., Shamshirband, S., Mosavi, A., 2019. A hybrid clustering and classification technique for forecasting short-term energy consumption. Environ. Prog. Sustain. Energy 38 (1), 66–76.

Waheed, R., Sarwar, S., Wei, C., 2019. The survey of economic growth, energy consumption and carbon emission. Energy Rep. 5, 1103–1115.

Wang, Q., Jiang, R., Zhan, L., 2019. Is decoupling economic growth from fuel consumption possible in developing countries? – a comparison of China and India. J. Clean. Prod. 229, 806–817.

Yang, Z., Wei, X., 2019. The measurement and influences of China's urban total factor energy efficiency under environmental pollution: based on the game cross-efficiency DEA. J. Clean. Prod. 209, 439–450.

Yao, Y., Cao, Q., Vasilakos, A.V., 2013. EDAL: an energy-efficient, delay-aware, and lifetime-balancing data collection protocol for wireless sensor networks. 2013 IEEE 10th International Conference on Mobile Ad-hoc and Sensor Systems. IEEE, pp. 182–190.

Zhao, H., Guo, S., Zhao, H., 2019. Provincial energy efficiency of China quantified by three-stage data envelopment analysis. Energy 166, 96–107.

Index

Note: Page numbers followed by "*f*" and "*t*" refer to figures and tables, respectively.

E

F

G

N

Q

R

T

Y

9780128177723